Achieve Positive Learning Outcomes with WileyPLUS...

Every one of your students has the potential to make a difference. And realizing that potential starts right here, in your course.

When students succeed in your course—when they stay on-task and make the breakthrough that turns confusion into confidence—they are empowered to realize the possibilities for greatness that lie within each of them. We know your goal is to create an environment where students reach their full potential and experience the exhilaration of academic success that will last them a lifetime. *WileyPLUS* can help you reach that goal.

Wiley**PLUS** is an online suite of resources—including the complete text—that will help your students:

- come to class better prepared for your lectures
- get immediate feedback and context-sensitive help on assignments and quizzes
- track their progress throughout the course

"I just wanted to say how much this program helped me in studying... I was able to actually see my mistakes and correct them. ... I really think that other students should have the chance to use *WileyPLUS*."

Ashlee Krisko, *Oakland University*

www.wileyplus.com

88% of students surveyed said it improved their understanding of the material.*

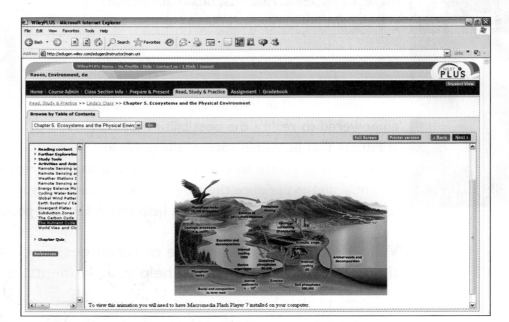

Environment 6th Edition

Environment

6th Edition

Peter H. Raven
Missouri Botanical Garden

Linda R. Berg
Former affiliations:
University of Maryland, College Park
St. Petersburg College

David M. Hassenzahl
University of Nevada, Las Vegas

WILEY

John Wiley & Sons, Inc.

To our family, friends, and colleagues who gave freely of their support and knowledge as we prepared the sixth edition of *Environment*.

Especially to:
Pat, Alice, Elizabeth, Francis, and Kate
Alan, Jennifer, and Pat
Hillary, Mikaela, and Kobe

SENIOR EDITOR   Rachel M. Falk
ASSOCIATE EDITOR   Merillat Staat
EXECUTIVE MARKETING MANAGER   Clay Stone
SENIOR PRODUCTION EDITOR   Sandra Dumas
PRODUCTION MANAGER   Dorothy Sinclair
SENIOR DESIGNER   Madelyn Lesure
PHOTO DEPARTMENT MANAGER   Hilary Newman
PHOTO EDITOR   Kathy Bendo
SENIOR MEDIA EDITOR   Linda Muriello
EDITORIAL ASSISTANT   Alissa Rufino
PRODUCTION MANAGEMENT SERVICES   Ingrao Associates
COVER PHOTO   Chris Sembrot/chrissembrot.com

This book was set in 10/12 Janson by Prepare and printed and bound by R. R. Donnelley/Jefferson City. The cover was printed by R. R. Donnelley/Jefferson City.

This book is printed on acid free paper. ∞

To order books or for customer service, please call 1-800-CALL WILEY (225-5945).

ISBN-13 978-0470-11926-6

Printed in the United States of America

10 9 8 7 6 5 4 3 2 1

Preface

This book is intended as an introductory text for undergraduate students, both science and nonscience majors. Although relevant to all students, *Environment* 6/e is particularly appropriate for those majoring in education, journalism, government and politics, and business, as well as the traditional sciences. We assume our students have very little prior knowledge of environmental science. Important ecological concepts and processes are presented in a straightforward, unambiguous manner.

The overarching concept of **environmental sustainability** has never been more important to the field of environmental science than it is today. Sustainability, a central theme of *Environment*, is integrated throughout the text. Yet the more we learn about the environment, the more we realize that interactions among different components of the environment are many and complex. Therefore, a central theme of the sixth edition of *Environment* is **environmental systems**. Understanding how change to one component affects other processes, places, and organisms is essential to managing existing problems, avoiding future problems, and improving the world we live in.

From the opening pages, we acquaint students with current environmental issues—issues that have many dimensions and that defy easy solutions. We begin by examining the scientific, historical, ethical, governmental, and economic underpinnings of environmental science. This provides a conceptual foundation for students that they can then bring to bear on the rest of the material in the book. We next explore the basic ecological principles that govern the natural world and consider the many ways in which humans affect the environment. Later chapters examine in detail the effects of human activities, including overpopulation, energy production and consumption, depletion of natural resources, and pollution.

Although we avoid unwarranted optimism when presenting these problems, we try to avoid the gloomy predictions of disaster so commonly presented by the media. Instead, students are encouraged to take active, positive roles to meet the environmental challenges of today and tomorrow.

All of the chapters have been painstakingly researched, and extraordinary efforts have been made to obtain the most recent data available. Both instructors and students will benefit from the book's currency because environmental issues and trends are continually changing.

Environment 6/e integrates important information from many different fields, such as biology, geography, chemistry, geology, physics, economics, sociology, natural resources management, law, and politics. Because environmental science is an interdisciplinary field, this book is appropriate for use in environmental science courses offered by a variety of departments, including (but not limited to) biology, geology, geography, and agriculture.

New member of author team for *Environment* 6/e

In this edition the author team has been expanded to include David M. Hassenzahl, who Chairs the Department of Environmental Studies at the University of Nevada, Las Vegas. Dr. Hassenzahl's research focuses on the role of scientific information in environmental decision making, a topic he approaches from a systems perspective. He has won awards for both his teaching and research and is well known for his work on coping with uncertainty in risk analysis. He earned his BA in Environmental Science and Paleontology from the University of California at Berkeley, and has an interdisciplinary PhD in Science, Technology, and Environmental Policy from Princeton University's Woodrow Wilson School.

Refinement of the learning system in *Environment* 6/e

Learning environmental science is a challenging endeavor. A well-developed pedagogical plan that facilitates student mastery of the material has always been a hallmark of *Environment*, and the sixth edition has refined the **learning system** to a higher, more effective level. Pedagogical features in the sixth edition include:

Chapter Introductions illustrate certain concepts in the chapter with stories about some of today's most pressing environmental issues.

World View and Closer to You reminders at the end of each chapter's introduction direct students to visit *www.wiley.com/college/raven* where they can watch, read, and listen to **local and global news stories** relating to a chapter. **Suggested Reading** lists for each chapter are also available online to provide current references for further learning.

Learning Objectives at the beginning of each section head indicate in behavioral terms what the student must be able to do to demonstrate mastery of the material in the chapter.

Review Questions at the end of each section give students the opportunity to test their comprehension of the learning objectives.

EnviroNews provide additional topical material about relevant environmental issues.

Meeting the Challenge boxes profile environmental success stories.

You Can Make a Difference boxes suggest specific courses of action or lifestyle changes students can make to improve the environment.

Tables and Graphs, with complete data sources cited at the end of the text, summarize and organize important information.

Marginal Glossaries, located within every chapter, provide handy definitions of the most important terms.

Case In Point features in the body of the chapter offer a wide variety of in-depth case studies that address important issues in the field of environmental science.

Summaries with Selected Key Terms restate the chapter learning objectives and provide a review of the material presented. Boldfaced selected key terms, including marginal glossary terms, are integrated within each summary, enabling students to study vocabulary words in the context of related concepts.

Thinking About the Environment questions, many new to this edition, encourage critical thinking and highlight important concepts and applications. At least one question in each chapter provides a systems perspective.

Take a Stand features appear at the end of every chapter and ask students to enter into a debate about an issue or controversy from the chapter. Students then visit our **Web site** to find links for researching the situation and tools for organizing their arguments.

The **Appendices** at the end of the book include a new Appendix III on Modeling.

The *2006 World Population Data Sheet*, provided by the Population Reference Bureau, is folded into the text (inside the back cover) and is intended to be pulled out for classroom use. Chapter 8 provides a student assignment of population questions based on the data sheet.

Updated and expanded art program of Environment 6/e

The sixth edition has been enhanced visually with an art program that reinforces and expands concepts discussed in the text. Numerous photographs and cartoons elaborate relevant issues and add visual detail. Because certain cartoons reflect a point of view that may be controversial, a critical thinking question has been added to the legend of each cartoon. Area maps feature insets of hemispheric locator maps to help students visualize where a particular area is on a continent. Examples of new line art include Figures 1.9 (fuel efficient hybrids); 4.2 (three domains and six kingdoms of life); 5.11 (atmospheric convection); 6.18 (zonation in the ocean); 7.15 (possible dose-response curves for a single data set); 8-15 (fertility changes in selected developing countries; 10.7 (city as a system); 11.7 (mountaintop removal coal mining); 15-3 (soil profile); 17.8 (direct and indirect causes of declining biological diversity); 19.8 (evolution of antibiotic resistance); 19-13a (collapse of the cod fishery on Georges Bank); 20.12 (Phase I vapor recovery at an underground gasoline tank); 21.5 (simulated effect of atmospheric CO_2 on coral reef calcification); 23.10 (components of an integrated pest management approach); and 25-8 (global climate system diagram).

Major changes in the sixth edition

A complete list of all changes and updates to the sixth edition of *Environment* is too long to fit in the Preface, but one or two of the more important changes to each chapter follow:

Chapter 1, Introducing Environmental Science and Sustainability, has a new chapter opener on Las Vegas that introduces systems, a new EnviroNews on green roofs, and a new section on Earth systems and environmental science.

In **Chapter 2, Environmental Laws, Economics, and Ethics,** the environmental economics section is reorganized and rewritten.

Chapter 3, Ecosystems and Energy, presents a new EnviroNews on the hippopotamus-tilapia connection.

The section on diversity in **Chapter 4, Ecosystems and Living Organisms,** now includes both domains and kingdoms of life, along with a new paragraph on the modern synthesis of evolution.

Chapter 5, Ecosystems and the Physical Environment, contains a new chapter opener on Hurricane Katrina and New Orleans, additional information on hurricanes, and a new EnviroNews on mangroves and the Asian tsunami disaster.

In **Chapter 6, Major Ecosystems of the World,** portions of marine ecosystem sections are rewritten and expanded. Chapter 6 also has a new EnviroNews on cold-water coral reefs.

A new Case In Point on Pandemic Influenza and a new table comparing epidemiology and toxicology are in **Chapter 7, Human Health and Environmental Toxicology.**

Throughout **Chapter 8, Population Change,** human population data, including all graphs and tables, are updated.

Chapter 9, Addressing Population Issues, has a new case in point on Millennium Development Goals and a new EnviroNews on the Millennium Villages Project.

Chapter 10, The Urban World, contains new information on brownfields, Portland's public transportation system, smart growth, and the importance of green space.

In **Chapter 11, Fossil Fuels,** the section on "Energy Consumption" now includes historical energy use and an energy source-consumption flow graphic. Chapter 11 also has a new EnviroNews about campus approaches to reduce fossil fuel use

Chapter 12, Nuclear Energy, contains new discussion of nuclear energy as part of our energy system and a new EnviroNews on Nuclear Power, Climate Change, and the Systems Perspective.

Chapter 13, Renewable Energy and Conservation, has a new introduction that focuses on a systems perspective and appropriate energy use. Chapter 13 also has a new table on the pros and cons of dams and a new Case in Point on the Three Gorges Dam.

A new chapter introduction in **Chapter 14, Water: A Limited Resource,** compares access to water from a systems perspective in a Canadian town and Nairobi, Kenya. Chapter 14 also has a new EnviroNews on water and sustainability at the University of British Columbia and a new section on climate change and freshwater supplies.

Chapter 15, Soils Resources, has new sections on soil salinization and agroforestry.

Chapter 16, Minerals: A Nonrenewable Resource, has a new paragraph on China's increasing production and consumption of minerals.

Chapter 17, Preserving Earth's Biological Diversity, contains new material on the Millennium Ecosystem Assessment and on declining biological diversity using a systems perspective.

New material in **Chapter 18, Land Resources** includes a chapter introduction on Congaree and Great Sand Dunes National Parks, a Meeting the Challenge box on preserving forests in the eastern United States, and an EnviroNews about the University of Washington student capstone course.

Chapter 19, Food Resources: A Challenge for Agriculture has a new chapter introduction on world food problems, a new EnviroNews on Food Safety, and new material on organic agriculture.

Chapter 20, Air Pollution, contains a new EnviroNews on the South Coast Air Quality Management District efforts to reduce ozone, a new campus sustainability EnviroNews: "Flexcar for Undergraduates," and a Case in Point on air pollution in Beijing and Mexico City.

Coverage of global climate change is updated to the most recent IPCC reports, and expanded in **Chapter 21, Global Climate Change** to including a new introduction on opportunities to reduce the rate of atmospheric carbon buildup.

Chapter 22, Water Pollution, has a new EnviroNews on contamination from Katrina and a new Case in Point on green chemistry that includes a systems perspective.

Chapter 23, The Pesticide Dilemma, has a new EnviroNews on campuses reducing pesticide use and a new Case in Point on pesticides in Central America.

Chapter 24, Solid and Hazardous Wastes, is updated throughout, including new material on deep well injection.

Chapter 25, Tomorrow's World, is completely reorganized and revised. The five parts of the new action plan should resonate more with students than the original nine-point plan. New material includes expanded coverage on agriculture, a new section on stabilizing climate, and a new section on designing sustainable cities.

Ancillaries for *Environment* 6/e

The package accompanying *Environment* 6/e includes several items developed specifically to augment students' understanding of environmental issues and concerns. Together, these ancillaries provide instructors and students with interesting and helpful teaching and learning tools and take full advantage of both electronic and print media.

For the Student

The expanded and revised *Environment* 6/e **Book Companion Site** is located at *www.wiley.com/college/raven*. In addition to access to *World View and Closer to You* news stories, the Student Resources at this site also include *Quizzes* for student self-testing authored by Blase Maffia, University of Miami; the *Biology NewsFinder; Flash Cards; and Animations* for select text concepts. Much of the site is co-authored by Chris Migliaccio and Mitch Fishkind, both of Miami-Dade Community College-Wolfson Campus. Their contributions include *Quantitative and Essay Questions, Take a Stand* (described above) and *Useful Website Links*.

For the Instructor

Book Companion Site (*www.wiley.com/college/raven*). Instructor Resources on the book companion site include a Biology Visual Library containing all of the line illustrations in the textbook in jpeg format, as well as access to numerous other life science illustrations; the Instructor's Manual; Test Bank; PowerPoint Presentations tailored to each chapter of *Environment* 6/e; and select flash animations for use in classroom presentations. Instructor Resources are password protected.

Test Bank by Jennifer Rivers Cole, Northeastern University. Containing approximately 60 multiple choice and essay test items per chapter, this test bank offers assessment of both basic understanding and conceptual applications. The *Environment, 6/e* Test Bank is offered in two formats: MS Word files and a Computerized Test Bank. The easy-to-use test-generation program fully supports graphics, print tests, student answer sheets, and answer keys. The software's advanced features allow you to create an exam to your exact specifications.

Instructor's Manual originally by our new co-author, David Hassenzahl, University of Nevada-Las Vegas and revised by Jody Terrell, Texas Woman's University. The Instructor's Manual now provides over 90 creative ideas for in-class activities. Also included are lecture outlines prepared by Joy Sales Colquitt and answers to all End-of-Chapter and Review Questions prepared by Jessica O'Leary.

All **Line Illustrations and Photos** from *Environment* 6/e, in jpeg files and PowerPoint format.

PowerPoint Presentations by Elizabeth Johnson, Post University are tailored to *Environment* 6/e's topical coverage and learning objectives. These presentations are designed to convey key text concepts, illustrated by embedded text art. An effort has been made to reduce the amount of words on each slide and increase the use of visuals to illustrate concepts.

Personal Response System questions by Peter van Walsum, Baylor University are specifically designed to foster student discussion and debate in class.

Animations. Select text concepts are illustrated using flash animation, designed for use in classroom presentations.

WileyPLUS

WileyPLUS provides an integrated suite of teaching and learning resources, including an online version of the text, in one easy-to-use website. Organized around the essential activities you perform in class, *WileyPLUS* helps you:

- **Prepare and Present.** Create class presentations using a wealth of Wiley-provided resources, including an online version of the textbook, PowerPoint slides, animations, and more—making your preparation time more efficient. You may easily adapt, customize, and add to this content to meet the needs of your course.

- **Create Assignments.** Automate the assigning and grading of homework or quizzes by using Wiley-provided question banks or by writing your own. Student results will be automatically graded and recorded in your gradebook. *WileyPLUS* can link homework problems to the relevant section of the online text, providing students with context-sensitive help.

- **Track Student Progress.** Keep track of your students' progress via an instructor's gradebook, which allows you to analyze individual and overall class results to determine student progress and level of understanding.

- **Administer Your Course.** *WileyPLUS* can easily be integrated with other course management systems, gradebooks, or other resources you are using in your class, providing you with the flexibility to build your course in your own way.

John Wiley & Sons may provide complementary instructional aids and supplements or supplement packages to those adopters qualified under our adoption policy. Please contact your sales representative for more information. If as an adopter or potential user you receive supplements you do not need, please return them to your sales representative or send them to: Attn: Wiley Returns Department, Heller Park Center, 360 Mill Road, Edison, NJ 08817.

Acknowledgments

The development and production of *Environment* 6/e was a process involving interaction and cooperation among the author team and between the authors and many individuals in our home and professional environments. We are keenly aware of the valuable input and support from editors, colleagues, and students. We also owe our families a debt of gratitude for their understanding, support, and encouragement as we struggled through many revisions and deadlines.

The Editorial Environment

Preparing this book has been an enormous undertaking, but working with the outstanding editorial and production staff at John Wiley and Sons has made it an enjoyable task. We thank our Publisher Kaye Pace and Acquisitions Editors Rachel Falk and Rebecca Hope for their support, enthusiasm, and ideas. Associate Editor, Merillat Staat expertly guided us through the revision process, coordinated the final stages of development, and provided us with valuable suggestions before the project went into production. We also thank Merillat and Senior Media Editor, Linda Muriello for overseeing and coordinating the development of the supplements and media components. Alissa Rufino and Stephen Reiss are much appreciated for their editorial assistance.

We are grateful to Executive Marketing Manager, Clay Stone for his superb marketing and sales efforts.

We thank Kathy Bendo for her contribution as photo researcher, helping us find the wonderful photographs that enhance the text. We appreciate the artistic expertise of Senior Designer, Madelyn Lesure for the striking cover and interior designs; Senior Production Editor, Sandra Dumas and Suzanne Ingrao, Ingrao Associates for their efforts in maintaining a smooth production process; and Senior Illustration Editor, Anna Melhorn, who coordinated the art development and ensured that a consistent standard of quality was maintained throughout the illustration program.

Our colleagues and students have provided us with valuable input and have played an important role in shaping *Environment* 6/e. We thank them and ask for additional comments and suggestions from instructors and students who use this text. You can reach us through our editors at John Wiley and Sons; they will see that we get your comments. Any errors can be corrected in subsequent printings of the book, and more general suggestions can be incorporated into future editions.

The Professional Environment

The success of *Environment* 6/e is due largely to the quality of the many professors and specialists who have read the manuscript during various stages of its preparation and provided us with valuable suggestions for improving it. We appreciate the efforts of Alan R. Berg, who was instrumental in researching and analyzing the data used in the sixth edition. Thanks go to Lester Brown of the Earth Policy Institute for the overall outline for reorganizing Chapter 25 and to Robert Clegern, University of Maryland University College for making the excellent suggestion to reorganize the chapter in this way. In addition, the reviewers of the first five editions made important contributions that are still part of this book.

Reviewers of the Sixth Edition

Abbed Babaei, Cleveland State University
Dora Barlaz, The Horace Mann School
Rosina Bierbaum, University of Michigan
Stefan Cairns, Central Missouri State University
Robert Clegern, University of Maryland University College
Jennifer Rivers Cole, Northeastern University
Ellen J. Crivella , Ohio State University
John H. Crow, Rutgers State University, New Jersey
Ruth Darling, Sierra College
Timothy Farnham, University of Nevada, Las Vegas
Noah Greenwould, Center for Biological Diversity
Elizabeth Johnson, Post University
Nancy Knowlton, Scripps Institution of Oceanography, University of California, San Diego
Tim Kusky, St. Louis University
Michael Larsen, Campbell University
Ernesto Lasso de la Vega, Edison College
James K. Lein, Ohio University
Michael MacCracken, Climate Change Programs Climate Institute
Blase Maffia, University Of Miami
Eric Maurer, University Of Cincinnati
William Russ Mclain, Davis and Elkins College
Helen Neill, University of Nevada, Las Vegas
Diane O'Connell, Schoolcraft College
Barry Perlmutter, College of Southern Nevada
Thomas E. Pliske, Florida International University
Sandra Postel, Global Water Policy Project
Barbara C. Reynolds, University of North Carolina, Asheville
Marvin Sigal, Gaston College
Robert Socolow, Princeton University
William J. Smith, Jr., University of Nevada, Las Vegas
Val H. Smith, University of Kansas
Krystyna Stave, University of Nevada, Las Vegas
Richard T. Stevens, Monroe Community College
Jody A. C. Terrell, Texas Woman's University
Kip Thompson, Ozarks Technical Community College
G. Peter van Walsum, Baylor University

Reviewers of Earlier Editions

David Aborn, University of Tennessee, Chattanooga
John Aliff, Georgia Perimeter College
Sylvester Allred, Northern Arizona University
Diana Anderson, Northern Arizona University
Lynne Bankert, Erie Community College
David Bass, University of Central Oklahoma
Richard Bates, Rancho Santiago Community College
George Bean, University of Maryland
Mark Belk, Brigham Young University
David Belt, Penn Valley Community College
Bruce Bennett, Community College of Rhode Island, Knight Campus
Eliezer Bermúdez, Indiana State University
Patricia Beyer, University of Pennsylvania, Bloomsburg
Rodger Bland, Central Michigan University
David Boose, Gonzaga University
Frederic Brenner, Grove City College
Christopher Brown, University of Kansas
John Dryden Burton, McLennan Community College
Kelly Cain, University of Arizona
Ann Causey, Prescott College
Jeff Chaumba, University of Georgia
Robert Chen, University of Massachusetts, Boston
Gary Clambey, North Dakota State University
George Cline, Jacksonville State University
Rebecca Cook, Lambuth University
Harold Cones, Christopher Newport College
Bruce Congdon, Seattle Pacific University
Nate Currit, Pennsylvania State University
Karen DeFries, Erie Community College
Armando de la Cruz, Mississippi State University
Michael L. Denniston, Georgia Perimeter College
Michael L. Draney, University of Wisconsin, Green Bay
Jean Dupon, Menlo College
Laurie Eberhardt, Valparaiso University
Thomas Emmel, University of Florida
Donald Emmeluth, Montgomery Community College, Fulton
Margaret "Tim" Erskine, Lansing Community College
Kate Fish, EarthWays, St. Louis, Missouri
Allen Lee Farrand, Bellevue Community College
Huan Feng, Montclair State University
Steve Fleckenstein, Sullivan County Community College
Andy Friedland, Dartmouth College
Carl Friese, University of Dayton
Bob Galbraith, Crafton Hills College
Jeffrey Gordon, Bowling Green State University
Joseph Goy, Harding University
Stan Guffey, University of Tennessee
Mark Gustafon, Texas Lutheran University
John Harley, Eastern Kentucky University
Neil Harriman, University of Wisconsin, Oshkosh
Denny Harris, University of Kentucky
Syed E. Hasan, University of Missouri
David Hassenzahl, University of Nevada, Las Vegas
Leland Holland, Pasco-Hernando Community College
James Horwitz, Palm Beach Community College
Dan Ippolito, Anderson University
Solomon Isiorho, Indiana University/Purdue University, Fort Wayne
Jef Jaeger, University of Nevada, Las Vegas
John Jahoda, Bridgewater State College
Jan Jenner, Talladega College
David Johnson, Michigan State University
Linda Johnston, University of Kansas
Patricia Johnson, Palm Beach Community College
Thomas Justice, McLennan Community College
Karyn Kakiba-Russell, Mount San Antonio College
Colleen Kelley, Northern Arizona University
Penny Koines, University of Maryland
Allen Koop, Grand Valley State University
Katrina Smith Korfmacher, Denison University

Karl F. Korfmacher, Denison University
Ericka Lawson, Baylor College of Medicine (formerly Columbia College)
Norm Leeling, Grand Valley State University
Joe Lefevre, State University of New York, Oswego
Tammy Liles, Lexington Community College
James Luken, Northern Kentucky University
Timothy Lyon, Ball State University
Andy Madison, Union University
Jane M. Magill, Texas A&M University
Alberto Mancinelli, Columbia University
Michael McCarthy, Eastern Arizona College
Linda Matson McMurry, San Pedro Academy/Texas Department of Agriculture
George Middendorf, Howard University
Chris Migliaccio, Miami-Dade Community College
Gary Miller, University of North Carolina, Asheville
Mark Morgan, Rutgers University
Michael Morgan, University of Wisconsin, Green Bay
James Morris, University of South Carolina
Andrew Neill, Joliet Junior College
Lisa Newton, Fairfield University
James T. Oris, Miami University of Ohio
Robert Paoletti, Kings College

Richard Pemble, Moorhead State University
Ervin Poduska, Kirkwood Community College
Lowell Pritchard, Emory University
James Ramsey, University of Wisconsin, Stevens Point
Elizabeth Reeder, Loyola College
James J. Reisa, Environmental Sciences and Toxicology, National Research Council
Maralyn Renner, College of the Redwoods
Howard Richardson, Central Michigan University
W. Guy Rivers, Lynchburg College
Sandy Rock, Bellevue Community College
Barbra Roller, Miami Dade Community College, South Campus
Richard Rosenberg, Pennsylvania State University
Wendy Ryan, Kutztown University
Neil Sabine, Indiana University East
Frank Schiavo, San Jose State University
Jeffery Schneider, State University of New York, Oswego
Judith Meyer Schultz, University of Cincinnati
Sullivan Sealey, University of Miami
Mark Smith, Fullerton College
Ashley Steinhart, Fort Lewis College
Lynda Swander, Johnson County Community College

Thomas Schweizer, Princeton Energy Resources International
Mark Smith, Victor Valley Community College
Nicholas Smith-Sebasto, University of Illinois
Joyce Solochek, Marquette University
Morris Sotonoff, Chicago State University
Steven Steel, Bowling Green State University
Allan Stevens, Snow College
Karen Swanson, William Paterson University
Norm Thompson, University of Georgia
Jerry Towle, California State University, Fresno
Jack Turner, University of South Carolina, Spartanburg
Eileen Van Tassel, Michigan State University
Bruce Van Zee, Sierra College
Laine Vignona, University of Wisconsin, River Falls
Matthew Wagner, Texas A&M University
Phillip Watson, Ferris State University
Alicia Whatley, Troy State University
Jeffrey White, Indiana University
Susan Whitehead, Becker College
John Wielichowski, Milwaukee Area Technical College
James Willard, Cleveland State University
Ray E. Williams, Rio Hondo College

About the Authors

PETER H. RAVEN, one of the world's leading botanists, has dedicated more than three decades to conservation and biodiversity as President of the Missouri Botanical Garden in St. Louis, where he has cultivated a world-class institution of horticultural display, education, and research. Described by Time magazine as "Hero for the Planet," Dr. Raven champions research around the world to preserve endangered species and is a leading advocate for conservation and a sustainable environment.

Dr. Raven is Chair of the National Geographic Society's Committee for Research and Exploration, and he is a Past President of the American Association for the Advancement of Science. He is the recipient of numerous prizes and awards, including the prestigious National Medal of Science in 2001, the highest award for scientific accomplishments in this country; Japan's International Prize for Biology; the Environmental Prize of the Institut de la Vie; the Volvo Environment Prize; the Tyler Prize for Environmental Achievement; and the Sasakawa Environment Prize. He also has held Guggenheim and MacArthur fellowships.

Dr. Raven received his Ph.D. from the University of California, Los Angeles, after completing his undergraduate work at the University of California, Berkeley.

LINDA R. BERG is an award-winning teacher and textbook author. She received a B.S. in science education, M.S. in botany, and a Ph.D. in plant physiology from the University of Maryland. Her research focused on the evolutionary implications of steroid biosynthetic pathways in various organisms. Her recent interests involve the Florida Everglades.

Dr. Berg formerly taught at the University of Maryland, College Park, for 17 years, and has been at St. Petersburg College in Florida for 10 years. She has taught introductory courses in environmental science, biology, and botany to thousands of students and received numerous teaching and service awards. Dr. Berg is also the recipient of many national and regional awards, including the National Science Teachers Association Award for Innovations in College Teaching, the Nation's Capital Area Disabled Student Services Award, and the Washington Academy of Sciences Award in University Science Teaching.

During her career as a professional science writer, Dr. Berg has authored or coauthored several leading college science textbooks. Her writing reflects her teaching style and love of science.

©2007 Hassenzahl

DAVID M. HASSENZAHL is an internationally recognized scholar in the area of risk analysis. His research focuses on incorporating scientific information and expertise into public decisions, with particular emphasis on the management, interpretation, and communication of uncertainty. Dr. Hassenzahl has spent the past two decades addressing environmental management from a systems perspective on subjects as diverse as nuclear waste, climate change, toxic chemicals, and public health. He holds a B.A. in Environmental Science and Paleontology from the University of California at Berkeley, and a Ph.D. from Princeton University's Program in Science, Technology, and Environmental Policy.

Dr. Hassenzahl chairs the Department of Environmental Studies at the University of Nevada, Las Vegas, where he has received the UNLV Foundation Distinguished Teaching Award, as well as the Outstanding Researcher, Outstanding Teacher, and Outstanding Service Contribution Awards from the Greenspun College of Urban Affairs. Among his numerous academic publications is a widely used risk analysis textbook. He serves on the Council of the Society for Risk Analysis and UNLV's representative to the Council of Environmental Deans and Directors. Prior to his academic career, Dr. Hassenzahl worked in the private sector as an environmental manager at a pulp and paper mill, and in the public sector as an inspector for the (San Francisco) Bay Area Air Quality Management District.

Brief Contents

Contents

10 The Urban World 214

PART FOUR

THE SEARCH FOR ENERGY

11 Fossil Fuels 233

12 Nuclear Energy 259

13 Renewable Energy and Conservation 282

Introducing Environmental Science and Sustainability

Las Vegas, Nevada. Las Vegas functions as a complex system composed of smaller interrelated systems, including the human population, water, transportation, and climate. (© Jon Hicks/Corbis)

People are often surprised to learn that Las Vegas originally attracted people for its water. Las Vegas (Spanish for "the Meadows") was the site of an Artesian spring that attracted Southern Paiutes on a seasonal basis for centuries. In the 1600s, travelers on the Santa Fe Trail began stopping at Las Vegas. The first permanent settlement began at the start of the 20th century when the railroad identified Las Vegas as a water stop. A deep aquifer was discovered, and a small town began to grow.

Las Vegas grew slowly until Hoover Dam was built across the Colorado River 48 km (30 mi) to the southeast to produce electricity and provide water to California and Arizona. The dam provided Las Vegas with an ample water supply. Around 1935, the city (and the surrounding area that makes up the Las Vegas Valley) began to grow rapidly. With over 1.5 million inhabitants, the valley faces many environmental challenges, and represents an example in which solving environmental problems requires a *systems perspective* (see photo).

Supplying water to a growing city requires making tradeoffs among the parts in a web (or system) of sources, wants, needs, and priorities. Most water used in Las Vegas originates as snowfall, coming both from nearby mountains and from the Rocky Mountains, hundreds of kilometers away. Nevada competes with other states for Colorado River water. Within Nevada, urban areas compete with rural areas. Environmental groups are concerned about organisms that rely on water that would be pumped. This issue is nothing new: groundwater pumping once threatened a subspecies of desert pupfish found only in Devil's Hole, a small pool about 145 km (90 mi) northwest of Las Vegas. In 1976, the U.S. Supreme Court ruled that the fish had a right to the water—an early case under the Endangered Species Act.

Only about 10 cm (4 in) of precipitation falls in Las Vegas each year, but it often falls so quickly that flooding is a problem. Flooding, which causes deaths and property damage, is exacerbated by paved surfaces. Since paved ground is impervious to water, water flows along streets and sidewalks rather than soaking into the ground. Runoff carries pollutants—oil from roads, animal wastes from yards, and fertilizers from golf courses—into Lake Mead. Since Las Vegas gets most of its drinking water from Lake Mead, polluted runoff increases the need for water treatment.

Transportation is a similarly complex system, and since roads are paved for transportation, changes in this system impact the water system. Las Vegas's air quality is worsening as more cars and trucks are driven increasing distances. Construction contributes air pollution as dust blows from cleared surfaces. Inversion layers trap these pollutants, which can accumulate in the atmosphere for days.

Climate change also threatens Las Vegas. Climate models predict higher temperatures and less snowfall in the southwestern United States over the next few decades. This will increase demand for air conditioning—a major energy use—and for water. Effective planning will require understanding the system interactions among population growth, regional climate, transportation, water use, and energy demand.

On the other hand, Las Vegas is an excellent place to develop innovative solutions. A large commercial photovoltaic system is being built in nearby Boulder City. Newer casinos incorporate green design elements such as energy efficiency and lower water use. Some housing developments are trying to avoid the urban sprawl that has dominated the area, and several "zero energy" homes have been built. Las Vegas represents a sort of urban sustainability laboratory: solutions that work there will have applications around the world.

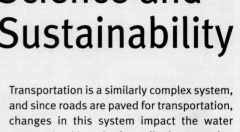

World View and Closer to You...
Real news footage relating to environmental science and sustainability around the world and in your own geographic region is available at www.wiley.com/college/raven by clicking on "World View and Closer to You."

Human Impacts on the Environment

- Define *poverty.*
- Distinguish among highly developed countries, moderately developed countries, and less developed countries.

Figure 1.1 **Planet Earth.** Earth's life-support system and the sun supply us with six essential requirements: air, water, food, energy, shelter, and clothing. However, Earth's resources are finite. (The Image Bank/Getty Images)

Of the millions of species inhabiting Earth, only one—*Homo sapiens*—has a far superior intellectual capacity. This special ability has made it possible for humans to venture into space. One of the benefits that the space program has given us is a view of Earth from space that shows the uniqueness of our planet in the solar system (**Figure 1.1**).

Earth is remarkably suited for life. Water, important both in the internal composition of organisms and as an external environmental factor affecting life, covers three-fourths of the planet. Our temperature is habitable—neither too hot, as on Mercury and Venus, nor too cold, as on Mars and the outer planets. We receive a moderate amount of sunlight—enough to power photosynthesis, which supports almost all the life forms that inhabit Earth. Our atmosphere bathes the planet in gases and provides essential oxygen and carbon dioxide that organisms require. On land, soil develops from rock and provides support and essential minerals for plants. Mountains that arise from geologic processes and then erode over vast spans of time affect weather patterns, provide minerals, and store reservoirs of fresh water as ice and snow that melt and flow to lowlands during the warmer months. Lakes and ponds, rivers and streams, wetlands, and groundwater reservoirs provide terrestrial organisms with fresh water.

Earth's abundant natural resources have provided the backdrop for a parade of living things to evolve. Life has existed on Earth for about 3.8 billion years. Although early Earth was inhospitable by modern standards, it provided the raw materials and energy needed for early life forms to arise and adapt. Some of these early cells evolved over time into simple multicellular organisms—early plants, animals, and fungi. Today, several million species inhabit the planet. A representative sample of Earth's biological diversity includes intestinal bacteria, paramecia, poisonous mushrooms, leafhoppers, prickly pear cacti, sea horses, dogwoods, angelfish, daisies, mosquitoes, pitch pines, polar bears, spider monkeys, and kingfishers (**Figure 1.2**).

About 100,000 years ago—a mere blip in Earth's 4.5-billion-year history—an evolutionary milestone began with the appearance of modern humans in Africa. Large brains and the ability to communicate made our species successful. Over time our population grew; we expanded our range throughout the planet and increasingly impacted the environment with our presence and our technologies. In many ways these technologies have made life better, at least for those of us who live in the United States, Canada, and other highly developed nations.

At the same time, there are many indications that we are headed for environmental catastrophe. Today the human species is the most significant agent of environmental change on our planet. We are overpowering the planet with our burgeoning population; transforming forests, prairies, and deserts to meet our needs and desires; and consuming ever-increasing amounts of Earth's abundant but finite resources—rich topsoil, clean water, and breathable air. We are eradicating thousands upon thousands of unique species as we destroy or alter their habitats. Evidence continues to accumulate that human-induced climate change is putting the natural environment at risk. Thus, human activities are disrupting global systems.

This book introduces the major environmental problems that humans have created. It considers ways to address these issues, yet it emphasizes that each "solution" has the potential to cause additional problems. Most important, it explains why we must

minimize human impact on our planet. We cannot afford to ignore the environment—our lives, as well as those of future generations, depend on it.

Increasing Human Numbers

Figure 1.3 is a portrait of about 440 million people. It is a satellite photograph of North America, including the United States, Mexico, and Canada, at night. The tiny specks of light represent cities, whereas the great metropolitan areas, such as New York along the northeastern seacoast, are ablaze with light.

The most important environmental problem, the one that links all others, is the many people in this picture. According to the United Nations (U.N.), in 1950 only 8 cities in the world had populations larger than 5 million, the largest being New York, with 12.3 million. By 2005 the largest city, Tokyo, Japan, had 35.2 million inhabitants, and the combined population of the world's 10 largest cities was 179.4 million (see Table 10.1).

Figure 1.3 Satellite view of North America at night. This image shows most major cities and metropolitan areas in the United States, Mexico, and Canada. (Earth Imaging/Stone/Getty Images)

Figure 1.4 **Human population numbers, 1800 to present.** It took thousands of years for the human population to reach 1 billion (in 1800) but only 130 years to reach 2 billion (1930). It only took 30 years to reach 3 billion (1960), 15 years to reach 4 billion (1975), 12 years to reach 5 billion (1987), and 12 years to reach 6 billion (1999). (Population Reference Bureau)

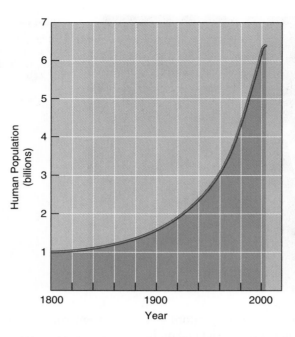

poverty: A condition in which people cannot meet their basic needs for adequate food, clothing, shelter, education, or health.

highly developed countries: Countries with complex industrialized bases, low rates of population growth, and high per capita incomes.

In 1999 the human population as a whole passed a significant milestone: 6 billion individuals. Not only is this figure incomprehensibly large, but also our population grew this large in a brief span of time. In 1960 the human population was only 3 billion (**Figure 1.4**). By 1975 there were 4 billion people, and by 1987 there were 5 billion. The more than 6 billion people who currently inhabit our planet consume great quantities of food and water, use a great deal of energy and raw materials, and produce much waste.

Despite the vigorous involvement of most countries with family planning, population growth rates are not expected to change overnight. Several billion people will be added to the world in the 21st century, so even if we remain concerned about the overpopulation problem and even if our solutions are effective, the coming decades may be clouded with tragedies. The conditions of life for many people may worsen considerably.

On a global level, nearly one in four people lives in extreme **poverty** (**Figure 1.5**). One measure of poverty is having a per capita income of less than $2 per day, expressed in U.S. dollars adjusted for purchasing power. Using this as a measure, we see that nearly 3.5 billion people—more than half the world's population—currently live at this level of poverty. Poverty is associated with low life expectancy, illiteracy, and inadequate access to health services, safe water, and balanced nutrition. According to the U.N. Food and Agricultural Organization, more than 800 million people lack access to the food needed for healthy, productive lives. This estimate includes a high percentage of children.

The world population may stabilize by the end of the 21st century, given the family planning efforts currently underway. U.N. demographers have noticed a decrease in worldwide fertility rate to a current average of about three children per family, and this overage is projected to continue to decline in coming decades. Population experts have made various projections for the world population at the end of the 21st century, from about 7.7 billion to 10.6 billion, based on how fast the fertility rate decreases.

No one knows whether Earth can support so many people indefinitely. Among the tasks we must accomplish is feeding a world population considerably larger than the present one without destroying the planet's natural resources that support us. The quality of life for our children and grandchildren will depend to a large extent on our ability to achieve this goal.

The Gap Between Rich and Poor Countries

Generally speaking, countries are divided into rich (the "haves") and poor (the "have-nots"). Rich countries are known as **highly developed countries**. The United States, Canada, Japan, and most of Europe, which represent 19% of the world's population, are highly developed countries.

Figure 1.5 **Slum in Mumbai (Bombay), India.** Many of the world's people live in extreme poverty. One trend associated with poverty is the increasing movement of poor people from rural to urban areas. As a result, the number of poor people living in or around the fringes of cities is mushrooming. (Jerry Cooke/Photo Researchers, Inc.)

Poor countries, in which 81% of the world's population live, fall into two subcategories—moderately developed and less developed. Mexico, Turkey, South Africa, and Thailand are examples of **moderately developed countries**. **Less developed countries (LDCs)** include Bangladesh, Mali, Ethiopia, and Laos. Cheap, unskilled labor is abundant in LDCs, but capital for investment is scarce. Most economies of LDCs are agriculturally based, often for only one or a few crops. As a result, crop failure or a lower world market value for that crop is catastrophic to the economy. Hunger, disease, and illiteracy are common in LDCs.

REVIEW

1. What is poverty?

2. What is a highly developed country? a moderately developed country? a less developed country?

moderately developed countries: Developing countries with a medium level of industrialization and average per capita incomes that are lower than those of highly developed countries.

less developed countries: Developing countries with a low level of industrialization, a high fertility rate, a high infant mortality rate, and a low per capita income (relative to highly developed countries).

Population, Resources, and the Environment

LEARNING OBJECTIVES

- Distinguish between people overpopulation and consumption overpopulation.
- Describe the three most important factors that determine human impact on the environment.

The relationships among population growth, use of natural resources, and environmental degradation are complex. We address the details of resource management and environmental problems in this and later chapters, but for now, let us consider two useful generalizations: (1) The resources essential to an individual's survival are small, but a rapidly increasing population (as in developing countries) tends to overwhelm and deplete a country's soils, forests, and other natural resources (**Figure 1.6a**). (2) In highly developed nations, individual resource demands are large, far above requirements for survival. To satisfy their desires rather than their basic needs, many people in affluent nations exhaust resources and degrade the global environment because they are extravagant consumers (**Figure 1.6b**).

Types of Resources

When examining the effects of humans on the environment, it is important to distinguish between the two types of natural resources: nonrenewable and renewable

Figure 1.6 Consumption of natural resources.

(a) A typical Indian family, from Ahraura Village, India, with their possessions. The rapidly increasing number of people in developing countries overwhelms their natural resources, even though individual resource requirements may be low. (Peter Ginter/Material World)

(b) A typical U.S. family, from Pearland, Texas, with all their possessions. People in highly developed countries consume a disproportionate share of natural resources. (Peter Ginter & Peter Menzel/Material World)

Figure 1.7 Natural resources. Nonrenewable resources are replaced on a geologic time scale, and their supply diminishes with use. Renewable resources are replaced on a fairly rapid time scale and, as will be explained in later chapters, are derived from the sun's energy.

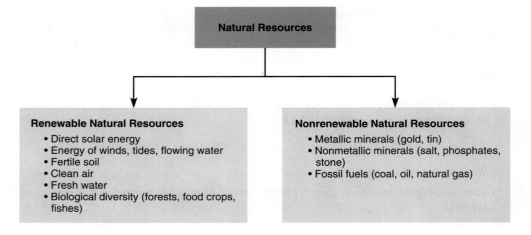

(**Figure 1.7**). **Nonrenewable resources,** which include minerals (such as aluminum, tin, and copper) and fossil fuels (coal, oil, and natural gas), are present in limited supplies and are depleted by use. Natural processes do not replenish nonrenewable resources within a reasonable period on the human time scale. Fossil fuels, for example, take millions of years to form.

In addition to a nation's population, several other factors affect how nonrenewable resources are used—including how efficiently the resource is extracted and processed as well as how much of it is required or consumed by different groups. People in the United States, Canada, and other highly developed nations tend to consume most of the world's nonrenewable resources. Nonetheless, the inescapable fact is that Earth has a finite supply of nonrenewable resources that sooner or later will be exhausted. In time, technological advances may find substitutes for nonrenewable resources. Slowing the rate of population growth and consumption will buy time to develop such alternatives.

Some examples of **renewable resources** are trees, fishes, fertile agricultural soil, and fresh water. Nature replaces these resources fairly rapidly (on a scale of days to decades), and they can be used forever as long as they are not overexploited in the short term. In developing countries, forests, fisheries, and agricultural land are particularly important renewable resources because they provide food. Indeed, many people in developing countries are subsistence farmers who harvest just enough food so they and their families can survive.

These resources are only truly renewable when the systems that support them are in balance. Rapid population growth can disrupt the systems and cause the overexploitation of renewable resources. For example, large numbers of poor people must grow crops on land inappropriate for farming—such as on mountain slopes or in tropical rain forests. Although this practice may provide a short-term solution to the need for food, it does not work in the long run: When these lands are cleared for farming, their agricultural productivity declines rapidly, and severe environmental deterioration occurs. Renewable resources, usually are only potentially renewable. They must be used in a *sustainable* way—in a manner that gives them time to replace or replenish themselves.

The effects of population growth on natural resources are particularly critical in developing countries. The economic growth of developing countries is often tied to the exploitation of their natural resources, usually for export to highly developed countries. Developing countries are faced with the difficult choice of exploiting natural resources to provide for their expanding populations in the short term (to pay for food or to cover debts) or conserving those resources for future generations.

It is instructive to note that the economic growth and development of the United States, Canada, and other highly developed nations came about through the exploitation and, in some cases the destruction, of their resources. Continued economic growth in highly developed countries now relies significantly on the importation of these resources from less developed countries.

Resource Consumption

Consumption is the human use of materials and energy. Consumption, which is both an economic and a social act, provides the consumer with a sense of identity as well as status among peers. The media, including the advertising industry, promote consumption as a way to achieve happiness. We are encouraged to spend, to consume.

People in highly developed countries are extravagant and wasteful consumers; their use of resources is greatly out of proportion to their numbers. A single child born in a highly developed country causes a greater impact on the environment and on resource depletion than 12 or more children born in a developing country. Many natural resources are used to provide the automobiles, air conditioners, disposable diapers, cell phones, DVD players, computers, clothes, newspapers, athletic shoes, furniture, boats, and other "comforts" of life in highly developed nations. Yet such consumer goods represent a small fraction of the total materials and energy required to produce and distribute these goods. According to the Worldwatch Institute, a private research institution in Washington, D.C., Americans collectively consume almost 10 billion tons of materials every year. The disproportionately large consumption of resources by highly developed countries affects natural resources and the environment as much as or more than the population explosion in the developing world.

People Overpopulation and Consumption Overpopulation

A country is *overpopulated* if the level of demand on its resource base results in damage to the environment. In comparing human impact on the environment in developing and highly developed countries, we see that a country can be overpopulated in two ways. **People overpopulation** occurs when the environment is worsening because there are too many people, even if those people consume few resources per person. People overpopulation is the current problem in many developing nations.

In contrast, **consumption overpopulation** results from the consumption-oriented lifestyles in highly developed countries. Consumption overpopulation has the same effect on the environment as people overpopulation—pollution and environmental degradation. Many affluent, highly developed nations, including the United States, Canada, Japan, and most of Europe, suffer from consumption overpopulation: *Highly developed nations represent less than 20% of the world's population, yet they consume significantly more than half of its resources.*

According to the Worldwatch Institute, highly developed nations account for the lion's share of total resources consumed:

- 86% of aluminum used
- 76% of timber harvested
- 68% of energy produced
- 61% of meat eaten
- 42% of the fresh water consumed

These nations also generate 75% of the world's pollution and waste.

Ecological Footprint   Environmental scientist **Mathis Wackernagel** developed the concept of ecological footprints to help people visualize what they use from the environment. According to Wackernagel, each person has an **ecological footprint,** an amount of productive land, fresh water, and ocean required on a continuous basis to supply that person food, wood, energy, water, housing, clothing, transportation, and waste disposal. The *2002 Living Planet* report produced by scientists at the World Wildlife Fund, the U.N. Environment Program, and other organizations, calculated that Earth has about 11.4 billion hectares (28.2 billion acres) of productive land and water. If we divide this area by the global human population, it indicates that each person is allotted about 1.9 hectares (4.7 acres). However, the average global ecological footprint of each person is about 2.3 hectares (5.7 acres), which means we humans have an *ecological deficit*—we have overshot our allotment. We can see the short-term results around us—forest destruction, degradation of croplands, loss of biological diversity,

people overpopulation: A situation in which there are too many people in a given geographic area.

consumption overpopulation: A situation that occurs when each individual in a population consumes too large a share of resources.

Figure 1.8   Ecological footprints.

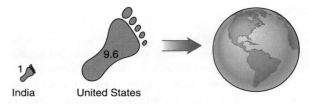

(a) In developing nations such as India, about 1 hectare (2.5 acres) is needed to meet the resource requirements of an average person, whereas the ecological footprint of each individual in a highly developed country such as the United States is 9.6 hectares (23.7 acres).

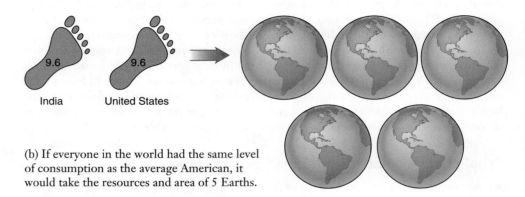

(b) If everyone in the world had the same level of consumption as the average American, it would take the resources and area of 5 Earths.

declining ocean fisheries, and local water shortages. The long-term outlook, if we do not seriously address our consumption of natural resources, is potentially disastrous.

In developing nations such as India and Nigeria, the average ecological footprint is less than 1 hectare (2.5 acres). In the United States, the average ecological footprint is about 9.6 hectares (23.7 acres). If all people in the world had the same lifestyle and level of consumption as the average North American, and assuming no changes in technology, we would need four additional planets the size of Earth (**Figure 1.8**).

As developing nations increase their economic growth and improve their standard of living, more and more people in those countries purchase consumer goods. More new cars are now sold annually in Asia than in North America and Western Europe combined. These new consumers may not consume at the high level of the average consumer in a highly developed nation, but their consumption has increasingly adverse effects on the environment. For example, air pollution caused by automotive traffic in urban centers in developing countries is terrible and getting worse every year. Millions of dollars are lost because of air pollution–related health problems in these cities. One of society's challenges is to provide new consumers in developing countries (as well as us) with less polluting, less consuming forms of transportation.

The *IPAT* Model

When you turn on the tap to brush your teeth in the morning, you probably do not think about where the water comes from or about the environmental consequences of removing it from a river or the ground. Similarly, most North Americans do not think about where the energy comes from when they flip on a light switch or start a car, van, or truck. We do not realize that all the materials in the products we use every day come from Earth, nor do we grasp that these materials eventually are returned to Earth, much of them in sanitary landfills.

Such human impacts on the environment are difficult to assess. They are estimated using the three factors most important in determining environmental impact (*I*):

- The number of people (*P*).
- The affluence per person, which is a measure of the consumption or amount of resources used per person (*A*).
- The environmental effects (resources needed and wastes produced) of the technologies used to obtain and consume the resources (*T*).

These factors are related in this way:

$$I = P \times A \times T.$$

In science, a **model** is a formal statement that describes the behavior of a system. The *IPAT* model, which biologist Paul Ehrlich and physicist John Holdren first proposed in the 1970s, shows the mathematical relationship between environmental impacts and the forces driving them. For example, to determine the environmental impact of carbon dioxide (CO_2) emissions from motor vehicles, multiply the population times the number of cars per person (affluence/consumption per person) times the average annual CO_2 emissions per year (technological impact). This model demonstrates that although increasing motor vehicle efficiency and developing cleaner technologies will reduce pollution and environmental degradation, a larger reduction will result if population and per capita consumption are also controlled.

The *IPAT* equation, though useful, must be interpreted with care, in part because we often do not understand all of the environmental impacts of a particular technology on complex environmental sysems. Motor vehicles are linked not only to global warming from CO_2 emissions but to local air pollution (tailpipe exhaust), water pollution (improper disposal of motor oil and antifreeze), stratospheric ozone depletion (from leakage of air conditioner coolants), and solid waste (disposal of automobiles in sanitary landfills). There are currently more than 600 million motor vehicles on the planet, and the number is rising rapidly.

The three factors in the *IPAT* equation are always changing in relation to each other. Consumption of a particular resource may increase, but technological advances may decrease the environmental impact of the increased consumption. Consumer trends and choices may affect environmental impact. The average fuel economy of new cars and light trucks (sport utility vehicles, vans, and pickup trucks) in the United States declined from 22.1 miles per gallon in 1988 to 20.4 miles per gallon in the early 2000s, in part because of the popularity of sport utility vehicles (SUVs). In addition to being less fuel-efficient than cars, SUVs emit more emissions per vehicle mile. More recently, the introduction of hybrids promises to increase average fuel economy, provided these cars become popular (**Figure 1.9**). Such trends and uncertainties make the *IPAT* equation of limited usefulness for long-term predictions.

The *IPAT* equation is valuable because it helps identify what we do not know or understand about consumption and its environmental impact. The National Research Council of the U.S. National Academy of Sciences has identified research areas we must address, including these: Which kinds of consumption have the greatest destructive impact on the environment? Which groups in society are responsible for the greatest environmental disruption? How can we alter the activities of these environmentally disruptive groups? It will take years to address such questions, but the answers should help decision makers in government and business formulate policies to alter consumption patterns in an environmentally responsible way. Our ultimate goal should be to reduce consumption so that our current practices do not compromise the ability of future generations to use and enjoy the riches of our planet.

model: A representation of a system; describes the system as it exists and predicts how changes in one part of the system will affect the rest of the system.

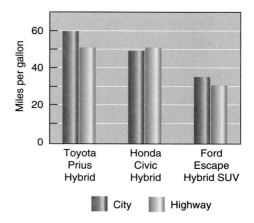

Figure 1.9   Fuel-efficient hybrids. Shown are city and highway miles per gallon for three hybrid models available in 2007.

REVIEW

1. How is human population growth related to natural resource depletion?
2. What is the difference between people overpopulation and consumption overpopulation?
3. What does the *IPAT* model demonstrate?

Environmental Sustainability

LEARNING OBJECTIVES

- Define *environmental sustainability*.
- Relate Garrett Hardin's description of the tragedy of the commons in medieval Europe to the global commons today.
- Briefly describe *sustainable development*.

environmental sustainability: The ability to meet the current human need for natural resources without compromising the ability of future generations to meet their needs.

One of the most important concepts in this text is **environmental sustainability**. *Sustainability* implies that humans can manage natural resources indefinitely without the environment going into a decline from the stresses imposed by human society on the natural systems (such as fertile soil, water, and air) that maintain life. When the environment is used sustainably, humanity's present needs are met without endangering the welfare of future generations. Environmental sustainability applies at many levels, including individual, community, regional, national, and global levels. (We will revisit environmental sustainability in Chapter 25.)

Environmental sustainability is based in part on the following ideas:

- We must consider the effects of our actions on the health and well-being of natural *ecosystems*, including all living things.
- Earth's resources are not present in infinite supply. We must live within ecological limits that let renewable resources such as fresh water regenerate for future needs.
- We must understand all the costs to the environment and to society of the products we consume.
- We must each share in the responsibility for environmental sustainability.

Many experts in environmental problems think human society is not operating sustainably because of the following human behaviors:

- We are using nonrenewable resources such as fossil fuels as if they were present in unlimited supplies.
- We are using renewable resources such as fresh water and forests faster than natural systems can replenish them (**Figure 1.10**).
- We are polluting the environment with toxins as if the capacity of the environment to absorb them is limitless.
- Our numbers continue to grow despite Earth's finite ability to feed us, sustain us, and absorb our wastes.

If left unchecked, these activities may reach the point of environmental catastrophe, threatening the life-support systems of Earth to such a degree that recovery is impossible.

At first glance, the issues may seem simple. Why do we not just stop the overconsumption, population growth, and pollution? The answer is that various interacting ecological, societal, and economic factors complicate the solutions. Our inadequate un-

Figure 1.10   A logger cuts down the last standing tree on a clear-cut forest slope. Logging has destroyed the habitat for forest organisms and increased the rate of soil erosion on steep slopes. Photographed in Canada. (Topham/The Image Works)

derstanding of how the environment works and how human choices affect the environment is a major reason that problems of environmental sustainability are difficult to resolve. The effects of many interactions between the environment and humans are unknown or difficult to predict, and we generally do not know if we should take corrective actions before our understanding is more complete.

Sustainability and the Tragedy of the Commons

Garrett Hardin (1915–2003) was a professor of human ecology at the University of California–Santa Barbara who wrote about human environmental dilemmas (**Figure 1.11**). In 1968 he published his classic essay, "The Tragedy of the Commons," in the journal *Science* in which he contended that our inability to solve many environmental problems is the result of a struggle between short-term individual welfare and long-term environmental sustainability and societal welfare.

Hardin used the commons to illustrate this struggle. In medieval Europe, the inhabitants of a village shared pastureland, called the commons, and each herder could bring animals onto the commons to graze. The more animals a herder brought onto the commons, the greater the advantage to that individual. When every herder in the village brought as many animals onto the commons as possible, the plants were killed from overgrazing, and the entire village suffered. Thus, the users inevitably destroyed the commons they depended on.

Hardin said that one of the outcomes of the eventual destruction of the commons was private ownership of land, because when each individual owned a parcel of land, it was in that individual's best interest to protect the land from overgrazing. A second outcome Hardin considered was government ownership and management of such resources, because the government's authority could impose rules on users of the resource and thereby protect it.

Hardin's paper has stimulated a great deal of research in the decades since it was published. In general, scholars agree that degradation of the self-governing commons has sometimes occurred in the past and is sometimes occurring today. However, many scholars think that such destruction is not inevitable—that is, it is possible to sustainably manage common resources without privatization (individual ownership) or government management.

As one goes from local to regional to global commons, the challenges of sustainably managing resources become more complex. In today's world, Hardin's parable has particular relevance at the global level. These modern-day commons, sometimes collectively called the **global commons**, are experiencing increasing environmental stress (see, for example, the discussion of declining fisheries in Chapter 19). No individual, jurisdiction, or country owns the global commons, and they are susceptible to overuse. Although exploitation may benefit only a few, everyone on Earth must pay for the environmental cost of exploitation.

The world needs effective legal and economic policies to prevent the short-term degradation of our global commons and ensure the long-term well-being of our natural resources. There are no quick fixes because solutions to global environmental problems are not as simple or short term as are solutions to some local problems. Most environmental ills are inextricably linked to other persistent problems such as poverty, overpopulation, and social injustice—problems beyond the ability of a single nation to resolve. The large number of participants that must organize, agree on limits, and enforce rules complicates the creation of global treaties to manage global commons. Cultural and economic differences among participants make finding solutions even more challenging.

Clearly, all people, businesses, and governments must foster a strong sense of **stewardship**—shared responsibility for the sustainable care of our planet. Cooperation and commitment at the international level are essential if we are to alleviate poverty, stabilize the human population, and preserve our environment and its resources for future generations.

Figure 1.11   Garrett Hardin. (© Vic Cox, photographer)

global commons: Those parts of our environment available to everyone but for which no single individual has responsibility—the atmosphere, fresh water, forests, wildlife, and ocean fisheries.

E N V I R O N E W S

Green Roofs

A roof that is completely or partially covered with vegetation and soil is known as a *green roof.* Also called *eco-roofs,* green roofs can provide several environmental benefits. For one thing, the plants and soil are effective insulators, reducing heating costs in winter and cooling costs in summer. The rooftop mini-ecosystem filters pollutants out of rainwater and reduces the amount of storm water draining into sewers. In urban areas, green roofs provide wildlife habitat, even on the tops of tall buildings. A city with multiple green roofs provides "stepping stones" of habitat that enable migrating birds and insects to pass unharmed through the city. Green roofs can also be used to grow vegetable and fruit crops or to provide a refuge for people living or working in the building. Green roofs allow urban systems to more closely resemble the natural systems they have replaced.

Green roofs may be added to existing buildings, but it is often easier and less expensive to install them in new buildings. Modern green roofs, which are designed to support the additional weight of soil and plants, consist of several layers that hold the soil in place, stop plant roots from growing through the rooftop, and drain excess water, thereby preventing leaks. Currently, Chicago, Illinois, is the U.S. city with the largest total area of green roofs. One of the largest individual green roofs in the United States is the Ford Motor Company's Plant in Dearborn, Michigan.

Global Plans for Sustainable Development

In 1992 representatives from most of the world's countries met in Rio de Janeiro, Brazil, for a groundbreaking summit, the *U.N. Conference on Environment and Development*. Countries attending the conference examined environmental problems that are international in scope: Pollution and deterioration of the planet's atmosphere and ocean, a decline in the number and kinds of organisms, and destruction of forests.

In addition, the participants adopted *Agenda 21*, an action plan in which future economic development, particularly in developing countries, will be reconciled with environmental protection. The goals of sustainable development are achieving improved living conditions for all people while maintaining a healthy environment in which natural resources are not overused and excessive pollution is not generated. Three factors—environmentally sound decisions, economically viable decisions, and socially equitable decisions—interact to promote sustainable development (**Figure 1.12**).

A serious application of the principles of environmental sustainability to economic development will require many changes in such fields as population policy, agriculture, industry, economics, and energy use. Agenda 21 recommended more than 2,500 actions to deal with our most urgent environmental, health, and social problems.

With few exceptions, we have made little progress in improving the quality of life for poor people or in solving the world's most serious environmental problems through implementation of the Rio conventions. The world is operating much as it always has. Many governments have shifted their attention from serious environmental priorities to other challenges such as terrorism and worsening international tensions. Meanwhile, scientific warnings on important environmental problems such as global climate change have increased.

Despite the lack of significant change at the international level since the 1992 Earth Summit, national, state, and local levels have made important environmental progress. Many countries have enacted more stringent air pollution laws, including the phasing out of leaded gasoline. More than 100 countries have created sustainable development commissions. Corporations that promote environmentally responsible business practices have joined to form the World Business Council for Sustainable Development. The World Bank, which makes loans to developing countries, has invested $8.5 billion in sustainable development projects around the world.

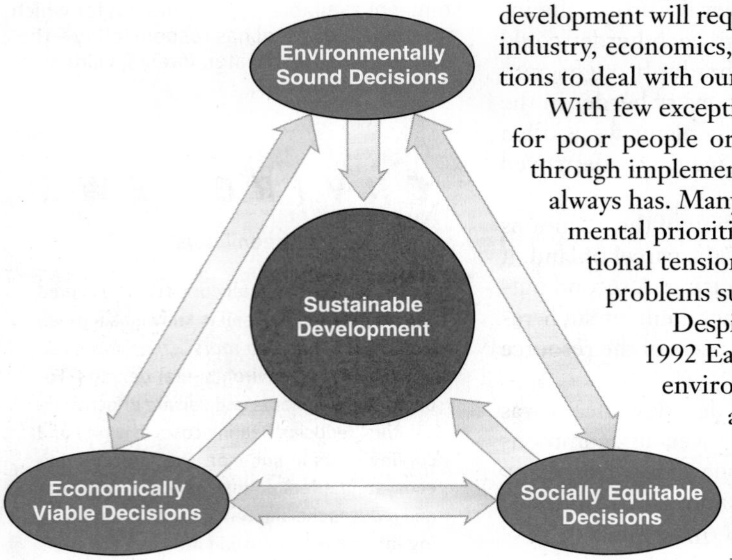

sustainable development: Economic development that meets the needs of the present generation without compromising the ability of future generations to meet their own needs.

Figure 1.12   Sustainable development. Environmentally sound decisions do not harm the environment or deplete natural resources. Economically viable decisions consider all costs, including long-term environmental and societal costs. Socially equitable decisions reflect the needs of society and ensure that all groups share costs and benefits equally.

REVIEW

1. What is environmental sustainability?
2. What is the "tragedy of the commons"?
3. How does sustainable development incorporate the principles of environmental sustainability?

Environmental Science

LEARNING OBJECTIVES

- Define *environmental science* and briefly describe the role of Earth systems in environmental science.
- Outline the steps of the scientific method.
- Distinguish between deductive and inductive reasoning.

environmental science: The interdisciplinary study of humanity's relationship with other organisms and the nonliving physical environment.

Environmental science encompasses the many interconnected issues involving human population, Earth's natural resources, and environmental pollution. Environmental science combines information from many disciplines, such as biology, geography, chemistry, geology, physics, economics, sociology, demography (the study of populations), cultural anthropology, natural resources management, agriculture, engineering, law, politics, and ethics. **Ecology,** the branch of biology that studies the interrelationships between organisms and their environment, is a basic tool of environmental science (**Figure 1.13**).

Environmental scientists try to establish general principles about how the natural world functions. They use these principles to develop viable solutions to environmental problems—solutions based as much as possible on scientific knowledge. Environmental problems are generally complex, and so our understanding of them is often less complete than we would like it to be. Environmental scientists are often asked to reach a consensus before they fully understand the systems that they study. As a result, they often make recommendations based on probabilities rather than precise answers.

Many of the environmental problems discussed in this book are serious, but environmental science is not simply a "doom and gloom" listing of problems coupled with predictions of a bleak future. To the contrary, its focus, and our focus as individuals and as world citizens, is on identifying, understanding, and solving problems that we as a society have generated. A great deal is being done, and more must be done to address the problems of today's world.

Earth Systems and Environmental Science

One of the most exciting aspects of environmental science and many other fields of science is working out how systems that consist of many interacting parts function as a whole. Las Vegas, discussed in the chapter introduction, is an urban system that in turn is composed of smaller systems, such as the transportation and water systems; these smaller systems are linked and interact with one another in the overall urban system.

The systems approach provides a broad look at overall processes, as opposed to the details of individual parts or steps. A school crossing guard in Las Vegas may be quite familiar with the detailed workings of the traffic lights at a given intersection, but that knowledge does not automatically translate into an understanding of the transportation system of the entire city. Thus, using a systems perspective helps scientists gain valuable insights that are not obvious when looking at system components.

Also, problems arise from *not* thinking about systems. For example, if a company decides to burn waste oil to avoid it leaking into groundwater, we shift pollution from groundwater to the air. A systems perspective would require company executives to think about the tradeoffs between the two disposal methods and, more importantly, about alternatives that might avoid generating waste oil in the first place.

Environmental scientists often use *models* to describe the interactions within and among environmental systems. Many of these models are computer simulations that represent the overall effect of competing factors to describe an environmental system in numerical terms. Models help us understand how the present situation developed from the past or how to predict the future course of events. Models also generate additional questions about environmental issues. (Appendix III contains an introduction to modeling.)

A natural system consisting of a community of organisms and its physical environment is known as an **ecosystem.** In ecosystems, biological processes (such as photosynthesis) interact with physical and chemical processes to modify the composition of gases in the atmosphere, transfer energy from the sun through living organisms, recycle waste products, and respond to environmental changes with resilience. Natural ecosystems are the foundation for our concept of environmental sustainability.

Ecosystems are organized into larger and larger systems that interact with one another (discussed in Chapter 3). At a global level are Earth systems, which include Earth's climate, atmosphere, land, coastal zones, and the ocean. Environmental scientists use a systems approach to try to understand how human activities are altering global environmental parameters such as temperature, carbon dioxide concentration in the atmosphere, land cover, changes in nitrogen levels in coastal waters, and declining fisheries in the ocean.

Many aspects of Earth systems are in a steady state or more accurately, a **dynamic equilibrium,** in which the rate of change in one direction is the same as the rate of change in the opposite direction. *Feedback* occurs when a change in one part of a system leads to a change in another part. Feedback can be negative or positive. In a

Figure 1.13   Ecology in action. Here a plant ecologist makes observations critical to understanding rainforest plants. Photographed in Costa Rica. (Stephen Ferry)

system: A set of components that interact and function as a whole.

negative feedback mechanism, a change in some condition triggers a response that counteracts, or reverses, the changed condition; a negative feedback mechanism works to keep an undisturbed system in dynamic equilibrium. For example, consider fish in a pond. As the number of fish increases, available food decreases and fewer fish survive; thus, the fish population declines.

In a **positive feedback mechanism,** a change in some condition triggers a response that intensifies the changing condition; a positive feedback mechanism leads to greater change from the original condition. A positive feedback mechanism can be very disruptive to an already disturbed system. For example, melting of polar and glacial ice can lead to greater absorption of solar heat by the exposed land area, which in turn leads to more rapid melting. As you will see throughout this text, many feedback mechanisms operate in the natural environment.

Science as a Process

The key to the successful solution of any environmental problem is rigorous scientific evaluation. It is important to understand clearly just what science is, as well as what it is not. Most people think of **science** as a body of knowledge—a collection of facts about the natural world and a search for relationships among these facts. However, science is also a dynamic *process*, a systematic way to investigate the natural world. Scientists seek to reduce the apparent complexity of our world to general **scientific laws** (also called *natural laws*). Scientific laws are then used to make predictions, solve problems, or provide new insights.

Scientists collect objective **data** (singular, *datum*), the information with which science works. Data are collected by observation and experimentation and then analyzed or interpreted. Conclusions are inferred from the available data and are not based on faith, emotion, or intuition.

Scientists publish their findings in scientific journals, and other scientists examine and critique their work. Confirming the validity of new results by *repeatability* is a requirement of science—observations and experiments must produce consistent results when other scientists repeat them. The scrutiny by other scientists reveals any inconsistencies in results or interpretation, and these errors are discussed openly. Thus, science is *self-correcting* over time.

There is no absolute certainty or universal agreement about anything in science (**Figure 1.14**). Science is an ongoing enterprise, and generally accepted ideas must be reevaluated in light of newly discovered data. Scientists never claim to know the "final answer" about anything because scientific understanding changes. However, this must not prevent us from using current knowledge in environmental science to make envi-

Figure 1.14   **Agreement in science.** Why do scientists never agree with absolute certainty about anything they study? (Mischa Richter/Cartoonbank.com)

ronmental decisions. Far too often, governments and businesses end up doing nothing because there is no final answer.

Uncertainty does not mean that scientific conclusions are invalid. There is overwhelming evidence linking exposure to tobacco smoke and incidence of lung cancer. We cannot state with absolute certainty which smokers will get lung cancer, but this uncertainty does not mean there is no correlation between smoking and lung cancer. On the basis of the available evidence, we say people who smoke have an increased risk of developing lung cancer.

Several areas of human endeavor are not scientific. Ethical principles often have a religious foundation, and political principles reflect social systems. However, scientific laws derive not from religion or politics but from the physical world around us. If you drop an apple, it will fall, whether or not you wish it to and despite any legislation you may pass forbidding it. Science aims to discover and better understand the general laws that govern the natural world.

The Scientific Method   The established processes scientists use to answer questions or solve problems are collectively called the **scientific method** (**Figure 1.15**). Although there are many variations of the scientific method, it basically involves five steps:

scientific method: The way a scientist approaches a problem by formulating a hypothesis and then testing it by means of an experiment.

1. **Recognize a question or unexplained occurrence in the natural world.** After a problem is recognized, one investigates relevant scientific literature to determine what is already known about it.
2. **Develop a *hypothesis*, or educated guess, to explain the problem.** A good hypothesis makes a prediction that can be tested and possibly disproved. The same factual evidence is often used to formulate several alternative hypotheses; each must be tested.

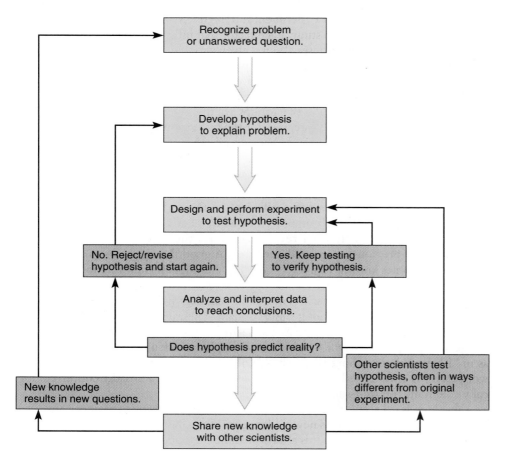

Figure 1.15   Scientific method. The basic steps of the scientific method are shown in yellow. Scientific work rarely proceeds in so straightforward a manner; examples of these paths are shown in orange.

3. **Design and perform an experiment to test the hypothesis.** An experiment involves collecting data by making careful observations and measurements. Much of the creativity in science involves designing experiments that sort out the confusion caused by competing hypotheses. The process never "proves" anything; instead, it disproves or falsifies alternative hypotheses until only the most plausible hypothesis is left.

4. **Analyze and interpret the data to reach a conclusion.** Does the evidence match the prediction stated in the hypothesis—that is, do the data support or refute the hypothesis? Should the hypothesis be modified or rejected based on the observed data?

5. **Share new knowledge.** Publishing articles in scientific journals or books and presenting the information at meetings permits others to repeat the experiment or design new experiments that either verify or refute the work.

Although the scientific method is usually described as a linear sequence of events, science is rarely as straightforward or tidy as the scientific method implies. Good science involves creativity in recognizing questions, developing hypotheses, and designing experiments. Scientific knowledge progresses by trial and error. Many creative ideas end up as dead ends, and there are often temporary setbacks or reversals of direction as knowledge progresses. Scientific knowledge often expands haphazardly, with the "big picture" emerging slowly from confusing and sometimes contradictory details.

Scientific work is often incorrectly portrayed in the media as "new facts" that have just come to light. At a later time, additional "new facts" that question the validity of the original study are reported. If one were to read the papers on which such media reports are based, one would find that the scientists made tentative conclusions based on their data. Science progresses from uncertainty to less uncertainty, not from certainty to greater certainty. Science is self-correcting over time, despite the fact that it never "proves" anything.

Controls and Variables in Experimental Design   Most often, many factors influence the processes we want to study. Each factor that influences a process is a **variable.** To evaluate alternative hypotheses about a given variable, we must hold all other variables constant so that they do not confuse or mislead us. To test a hypothesis about a variable, two forms of the experiment are done in parallel. In the experimental group, we alter the chosen variable in a known way. In the **control** group, we do not alter that variable. In all other respects the two groups are the same. We then ask, "What is the difference, if any, between the outcomes for the two groups?" Any difference is the result of the influence of that variable because all other variables remained the same. Much of the challenge of environmental science lies in designing control groups and in successfully isolating a single variable from all other variables.

Theories   Theories explain scientific laws. A **theory** is an integrated explanation of numerous hypotheses, each supported by a large body of observations and experiments and evaluated by the peer review process. A theory condenses and simplifies many data that previously appeared unrelated. A good theory grows as additional information becomes known. It predicts new data and suggests new relationships among a range of natural phenomena.

A theory simplifies and clarifies our understanding of the natural world because it demonstrates relationships among classes of data. Theories are the solid ground of science, the explanations of which we are most sure. This definition contrasts sharply with the general public's use of the word "theory," implying lack of knowledge, or a guess—as in "I have a theory about the existence of life on other planets." In this book, the word *theory* is always used in its scientific sense, to refer to a broadly conceived, logically coherent, and well-supported explanation.

Yet there is no absolute truth in science, only varying degrees of uncertainty. Science is continually evolving as new evidence comes to light, and its conclusions are always provisional or uncertain. It is always possible that the results of future experiments will contradict a prevailing theory, which will then be replaced by a new or modified theory that better explains the scientific laws of the natural world.

Inductive and Deductive Reasoning   Scientists use inductive and deductive reasoning. Discovering general principles by carefully examining specific cases is **inductive reasoning.** The scientist organizes data into manageable categories and asks the question, "What does this information have in common?" He or she then seeks a unifying explanation for the data. Inductive reasoning is the basis of modern experimental science.

As an example of inductive reasoning, consider the following:

Fact: Gold is a metal heavier than water.

Fact: Iron is a metal heavier than water.

Fact: Silver is a metal heavier than water.

Conclusion based on inductive reasoning: All metals are heavier than water.

Even if inductive reasoning makes use of correct data, the conclusion may be either true or false. As new data come to light, they may show that the generalization arrived at through inductive reasoning is false. Science has shown that the density of lithium, the lightest of all metals, is about half that of water. When one adds this information to the preceding list, a different conclusion is reached: Most metals are heavier than water. Inductive reasoning, then, produces new knowledge but is prone to error.

Science also makes use of **deductive reasoning,** which proceeds from generalities to specifics. Deductive reasoning adds nothing new to knowledge, but it makes relationships among data more apparent. Here is an example.

General rule: All birds have wings.

A specific example: Robins are birds.

Conclusion based on deductive reasoning: All robins have wings.

This is a valid argument. The conclusion that robins have wings follows inevitably from the information given. Scientists use deductive reasoning to determine the type of experiment or observations necessary to test a hypothesis.

REVIEW

1. What is environmental science? What are some of the disciplines involved in environmental science?

2. Why is a systems perspective so important in environmental science?

3. What are the steps of the scientific method? Why is each important?

4. What is the difference between inductive and deductive reasoning?

 ## Addressing Environmental Problems

LEARNING OBJECTIVES

- List and briefly describe the five stages in addressing environmental problems.
- Briefly describe the history of the Lake Washington pollution problem of the 1950s and how it was resolved.

You now know the strengths and limitations of science—what science can and cannot do. Before examining the environmental problems in the remaining chapters of this text, let us consider the elements that contribute to addressing those problems. What is the role of science? Given that we can never achieve complete certainty in science, at what point are scientific conclusions considered certain enough to warrant action? Who makes the decisions, and what is the tradeoff?

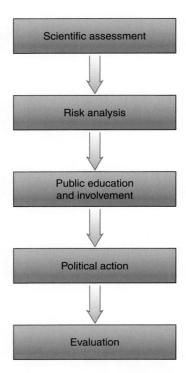

Figure 1.16 **Addressing environmental problems.** These five steps provide a framework for addressing environmental problems. Solving environmental problems rarely proceeds in such a straightforward manner.

Viewed simply, there are five stages in addressing an environmental problem (**Figure 1.16**).

1. **Scientific assessment.** The first stage in addressing any environmental problem is scientific assessment, the gathering of information. The problem is defined. Data are then collected, and experiments or simulations are performed.

2. **Risk analysis.** Using the results of a scientific investigation, we can analyze the potential effects of doing nothing or of intervening—what is expected to happen if a particular course of action is followed, including any adverse effects the action might generate. In other words, the risks of one or more remediation options are considered.

3. **Public education and involvement.** In a democracy, public awareness and endorsement is an essential part of addressing environmental problems. When alternative courses of action exist, the public must be informed. This involves explaining the problem, presenting all the available alternatives for action, and revealing the probable costs and results of each choice.

4. **Political action.** The affected parties, through their elected officials, select a course of action and implement it. Ideally, such an action is based on the best available evidence. During the political process, however, there are always differences of opinion about how this evidence should be interpreted when selecting a specific course of action. Some of these disagreements are based on economic, social, or political considerations rather than on scientific evidence.

5. **Evaluation.** The results of any action taken should be carefully monitored, both to see if the environmental problem is being addressed and to improve the initial assessment and modeling of the problem.

These five stages represent an ideal approach to systematically addressing environmental problems. In real life, addressing environmental problems is rarely so neat and tidy, particularly when the problem is of regional or global scale, or with higher costs and less obvious benefits for the money invested. Quite often, the public becomes aware of a problem, which triggers discussion regarding remediation before the problem has been clearly identified. Also, we often do not know what scientific information is needed until stages 2, 3, or even 4.

To demonstrate the five steps as they operate in an ideal situation, let us consider a relatively simple environmental problem recognized and addressed in the 1950s: Pollution in Lake Washington. This problem, unlike many environmental issues we face today, was relatively easy to diagnose and solve.

CASE IN POINT

Lake Washington

Lake Washington is a large, deep freshwater lake on the eastern boundary of the city of Seattle (**Figure 1.17**). During the first part of the 20th century, the Seattle metropolitan area expanded eastward toward the lake from the shores of Puget Sound, an inlet of the Pacific Ocean. As this expansion occurred, Lake Washington came under increasingly intense environmental pressures. Between 1941 and 1954, 10 suburban sewage treatment plants began operating at points around the lake. Each plant treated the raw sewage to break down the organic material within it and released the effluent (treated sewage) into the lake. By the mid-1950s, a great deal of treated sewage had been poured into the lake.

Scientists at the University of Washington in Seattle were the first to note the effects of this discharge on the lake. Their studies of the lake's

Figure 1.17 **Lake Washington.** This large freshwater lake forms the eastern boundary of Seattle, Washington.

organisms indicated that large masses of cyanobacteria (photosynthetic bacteria) were growing in the lake (**Figure 1.18**). Growth of such large numbers of cyanobacteria requires a plentiful supply of nutrients such as nitrogen and phosphorus, and deepwater lakes such as Lake Washington do not usually have many dissolved nutrients. The amount of filamentous cyanobacteria in Lake Washington's waters hinted that the lake was becoming richer in dissolved nutrients.

In 1955 the Washington Pollution Control Commission, citing the scientists' work, concluded that the treated sewage effluent was raising the levels of dissolved nutrients to the point of serious pollution. The sewage treatment was not eliminating many chemicals, particularly phosphorus, a major component of detergents. Mats of cyanobacteria formed a green scum over the surface of the water, and the water stank from the odor of rotting organic matter. The bacteria that decompose the dead cyanobacteria multiplied explosively, consuming vast quantities of oxygen in the process, until the lake's deeper waters were so depleted that they could no longer support many oxygen-requiring organisms such as fishes and small invertebrates.

Scientific Assessment   Scientific assessment of an environmental problem verifies that a problem exists and builds a sound set of observations on which to base a solution. Lake Washington's microscopic life was studied in detail in 1933. When the telltale signs of pollution first appeared in 1950, scientists examined and compared data from the earlier study with the present. They hypothesized that treated sewage was introducing so many nutrients into the lake that its waters were supporting the growth of cyanobacteria. They also predicted that the decline could be reversed: If the pollution were stopped, the lake would slowly recover. They outlined three steps necessary to save the lake:

1. Comprehensive regional planning by the many suburbs surrounding the lake.
2. Complete elimination of sewage discharge into the lake.
3. Research to identify the key nutrients causing the cyanobacteria to grow.

Risk Analysis   It is one thing to suggest that treated sewage no longer be added to Lake Washington and quite another to devise an acceptable remediation option. Further treatment of sewage could remove some nutrients, but it might not be practical to remove all of them. The alternative was to dump the sewage somewhere else— but where? In this case, officials weighed their options, analyzed the risks involved, and decided to discharge the treated sewage into Puget Sound. In their plan, a ring of sewers built around the lake would collect the treated sewage and treat it further before discharging it into Puget Sound.

It is important that the solution to one problem not produce another. The plans to further treat the sewage were formulated to minimize the environmental impact of diverting Lake Washington's discharge into Puget Sound. It was assumed that the treated effluent would have less of an impact on the greater quantity of water in Puget Sound. Also, nutrient chemistry in marine water is different from that in fresh water. Phosphate does not control cyanobacterial growth in Puget Sound as it does in Lake Washington. The growth of photosynthetic bacteria and algae in Puget Sound is largely limited by tides, which mix the water and transport the tiny organisms into deeper water, where they cannot get enough light to grow rapidly.

Public Education and Involvement   Despite the Washington Pollution Control Commission's conclusions, local sanitation authorities were not convinced that urgent action was necessary. Public action required further education, and scientists played a key role. They wrote articles for the general public that explained what nutrient enrichment is and what problems it causes. The general public's awareness increased as local newspapers published these articles.

Political Action   Cleaning up the lake presented serious political problems because there was no regional mechanism in place to permit the many local suburbs to act

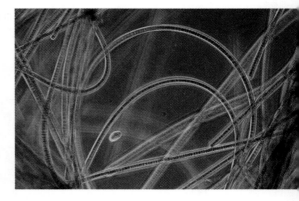

Figure 1.18   Light micrograph of *Oscillatoria*. This photosynthetic cyanobacterium grew in huge quantities in Lake Washington while it was polluted. (Sinclair Stammers//Photo Researchers, Inc.)

Figure 1.19   **Nutrients in Lake Washington compared with cyanobacterial growth.**

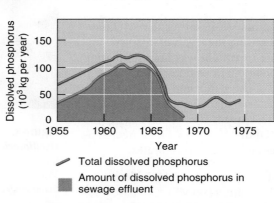

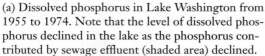

Total dissolved phosphorus

Amount of dissolved phosphorus in sewage effluent

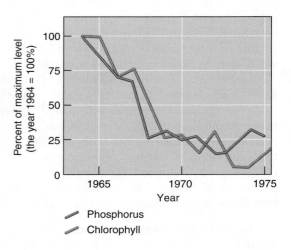

Phosphorus

Chlorophyll

(a) Dissolved phosphorus in Lake Washington from 1955 to 1974. Note that the level of dissolved phosphorus declined in the lake as the phosphorus contributed by sewage effluent (shaded area) declined.

(b) Cyanobacterial growth from 1964 to 1975, during Lake Washington's recovery, as measured indirectly by the amount of chlorophyll, the pigment involved in photosynthesis. Note that as the level of phosphorus dropped in the lake, the number of cyanobacteria (that is, the chlorophyll content) declined. (From Edmondson, W.T.)

together on matters such as sewage disposal. In late 1957, the state legislature passed a bill permitting a public referendum in the Seattle area regarding the formation of a regional government with six functions: water supply, sewage disposal, garbage disposal, transportation, parks, and planning. The referendum was defeated, apparently because suburban voters felt the plan was an attempt to tax them for the city's expenses. An advisory committee immediately submitted a revised bill limited to sewage disposal to the voters. Over the summer there was widespread discussion of the lake's future, and when the votes were counted, the revised bill passed by a wide margin.

At the time it was passed, the Lake Washington plan was the most ambitious and most expensive pollution control project in the United States. Every household in the area had to pay additional taxes for construction of a massive trunk sewer to ring the lake, collect all the effluent, treat it, and discharge it into Puget Sound. Meanwhile, the lake had deteriorated further. Visibility had declined from 4 m (12.3 ft) in 1950 to less than 1 m (3.1 ft) in 1962 because the water was clouded with cyanobacteria. In 1963 the first of the waste treatment plants around the lake began to divert its effluent into the new trunk sewer. One by one, the others diverted theirs, until the last effluent was diverted in 1968. The lake's condition began to improve (**Figure 1.19**).

Evaluation   Water transparency returned to normal within a few years. Cyanobacteria persisted until 1970 but eventually disappeared. By 1975 the lake was back to normal, and today the lake remains clear. The population in Seattle and around Lake Washington continues to grow (Seattle's population reached 3 million in 2005). The population around Lake Washington is outstripping the existing sewage collection and treatment system, so plans are underway to expand the service.

Every environmental intervention is an experiment, and continued monitoring is necessary because environmental scientists work with imperfect tools. There is a great deal we do not know, and every added bit of information increases our ability to deal with future problems. The unanticipated always lurks just beneath the surface of any experiment carried out in nature. ■

Working Together

The reversal of the pollution of Lake Washington is a particularly clear example of how environmental science identifies and addresses environmental problems. Many environmental problems facing us today are far more complex than that of Lake Washington, including those involving the global commons, and public attitudes are widely variable. Lake Washington's pollution problem was solved only because the many small towns involved in the problem cooperated in seeking a solution. Today, confrontation over an environmental problem frequently makes it difficult to reach an agreement. Even scientists disagree among themselves and call for additional research to help them arrive at a consensus. In such an atmosphere, politicians often adopt a "wait and see" approach.

Such delays are really a form of negative action because the consequences of many environmental problems are so serious they must be addressed before a consensus is reached. The need for additional studies should not prevent us from taking action on such serious regional and global issues as stratospheric ozone depletion, global climate change, pollution in the Great Lakes, and acid rain. We must recognize the uncertainty inherent in environmental problems; consider a variety of possible approaches; weigh the costs, benefits, and probable outcomes of each; and set in motion policies flexible enough to allow us to modify them as additional information becomes available.

REVIEW

1. What are the steps used to solve an environmental problem?
2. What was the Lake Washington pollution problem of the 1950s? How was it addressed?

 # REVIEW OF LEARNING OBJECTIVES WITH KEY TERMS

• Define *poverty*.

Poverty is a condition in which people cannot meet their basic needs for adequate food, clothing, shelter, education, or health.

• Distinguish among highly developed countries, moderately developed countries, and less developed countries.

Highly developed countries, such as the United States, Canada, Japan, and most of Europe, have complex industrialized bases, low rates of population growth, and high per capita incomes. Mexico, Turkey, South Africa, and Thailand are examples of **moderately developed countries**—developing countries with a medium level of industrialization and average per capita incomes lower than those of highly developed countries. **Less developed countries (LDCs),** such as Bangladesh, Mali, Ethiopia, and Laos, are developing countries with low levels of industrialization, high fertility rates, high infant mortality rates, and very low per capita incomes (relative to highly developed countries).

• Distinguish between people overpopulation and consumption overpopulation.

People overpopulation, the current problem in many developing nations, is a situation in which there are too many people in a given geographic area. **Consumption overpopulation,** which results from the consumption-oriented lifestyles in highly developed countries, occurs when each individual in a population consumes too large a share of resources.

• Describe the three most important factors that determine human impact on the environment.

A **model** is a representation of a system. A model describes the system as it exists and predicts how changes in one part of the system will affect the rest of the system. One model of environmental impact (*I*) has three factors: the number of people (*P*); the affluence per person (*A*), which is a measure of the consumption or amount of resources used per person; and the environmental effect of the technologies used to obtain and consume those resources (*T*). This model shows the mathematical relationship between environmental impacts and the forces that drive them:

$$I = P \times A \times T.$$

• Define *environmental sustainability*.

Environmental sustainability is the ability to meet the current human need for natural resources without compromising the ability of future generations to meet their needs; in other words, that humans can manage natural resources indefinitely without the environment going into a decline from the stresses imposed by human society on natural systems that maintain life.

• Relate Garrett Hardin's description of the tragedy of the commons in medieval Europe to the global commons today.

In 1968 **Garrett Hardin** published his classic essay, "The Tragedy of the Commons." He contended that our inability to solve many environmental problems is the result of a struggle between short-term individual welfare and long-term environmental sustainability and societal welfare. In today's world Hardin's parable has particular relevance at the global level. The **global commons** are those parts of our environment that are available to everyone but for which no single individual has responsibility—shared resources such as the atmosphere, fresh water, forests, wildlife, and ocean fisheries.

• **Briefly describe _sustainable development_.**

Sustainable development is economic development that meets the needs of the present generation with compromising the ability of future generations to meet their own needs. Three factors—environmentally sound decisions, economically viable decisions, and socially equitable decisions—interact to promote sustainable development.

• **Define _environmental science_ and briefly describe the role of Earth systems in environmental science.**

Environmental science is the interdisciplinary study of humanity's relationship with other organisms and the nonliving physical environment. Environmental scientists study systems; each **system** is a set of components that interact and function as a whole. A natural system consisting of a community of organisms and its physical environment is known as an **ecosystem.** Ecosystems are organized into larger and larger systems that interact with one another. At a global level are Earth systems, which include Earth's climate, atmosphere, land, coastal zones, and the ocean.

• **Outline the steps of the scientific method.**

The **scientific method** is the way a scientist approaches a problem by formulating a hypothesis and then testing it by means of an experiment. There are many variations of the scientific method, which basically involves five steps:

1. State the problem or unanswered question.
2. Develop a **hypothesis.**

3. Design and perform an experiment to test the hypothesis.
4. Analyze and interpret the **data.**
5. Share the conclusion with others.

• **Distinguish between deductive and inductive reasoning.**

Inductive reasoning begins with specific examples and seeks to draw a conclusion or discover a unifying rule on the basis of those examples. Inductive reasoning provides new knowledge but is error prone. **Deductive reasoning** operates from generalities to specifics; although it adds nothing new to knowledge, it makes relationships among data more apparent.

• **List and briefly describe the five stages in addressing environmental problems.**

Addressing environmental problems requires the application of approaches from several fields.

1. Scientific assessment involves identifying a potential environmental problem and collecting data to construct a model.
2. Risk analysis evaluates the potential effects of intervention.
3. Public education and involvement occur when the results of scientific assessment and risk analysis are placed in the public arena.
4. Political action is the implementation of a particular risk-management strategy by elected or appointed officials.
5. Evaluation monitors the effects of the action taken.

• **Briefly describe the history of the Lake Washington pollution problem of the 1950s and how it was resolved.**

Lake Washington exemplifies a successful approach to addressing a relatively simple environmental problem. The pouring of treated sewage into Lake Washington had raised its level of nutrients to the point where the lake supported the growth of filamentous cyanobacteria. Disposal of the sewage in another way solved the lake's pollution problem.

Thinking About the Environment

1. Why does a single child born in the United States have a greater effect on the environment than 12 or more children born in a developing country?

2. Do you think it is possible for the world to sustain its present population of more than 6 billion indefinitely? Why or why not?

3. In this chapter, we said the current global ecological footprint is 1.9 hectares. Do you think it will be higher, lower, or the same in 15 years? Explain your answer.

4. How are the concepts of ecological footprint and the _IPAT_ model similar? Which concept do you think is easier for people to grasp?

5. Explain the following ancient proverb as it relates to the concept of environmental sustainability: we have not inherited the world from our ancestors; we have borrowed it from our children.

6. Name an additional example of a global commons other than those mentioned in this chapter.

7. The currents at a beach community remove and bring in sand so that, over an extended period, the beach does not appear to change. How is this an example of a dynamic equilibrium? Is it an example of a negative feedback mechanism? Why or why not?

8. Thomas Henry Huxley once wrote, "The great tragedy of science—the slaying of a beautiful hypothesis by an ugly fact." Explain

what he meant, based on what you have learned about the nature of science.

9. In the chapter, the term _model is_ defined as a formal statement that describes a situation and that can be used to predict the future course of events. On the basis of this definition, is a model the same thing as a hypothesis? Explain your answer.

10. People want scientists to give them precise, definitive answers to environmental problems. Explain why this is not possible.

11. When Sherlock Holmes amazed his friend Watson by determining the general habits of a stranger on the basis of isolated observations, what kind of reasoning was he using? Explain.

12. Place the following stages in addressing environmental problems in order and briefly explain each: evaluation, public education and involvement, risk analysis, scientific assessment, political action.

13. Although the Lake Washington case demonstrates the five components of addressing an environmental problem, the final outcome—dumping treated sewage into Puget Sound—is not an ideal, long-term solution. Explain why.

14. What does the term _system_ mean in environmental science?

15. What is an Earth system?

Quantitative questions relating to this chapter are on our Web site.

Take a Stand

Visit our Web site at http://www.wiley.com/college/raven (select Chapter 1 from the Table of Contents) for links to more information about current environmental problems in both Lake Washington and Puget Sound. Find out about current political actions in progress to deal with these problems, and debate the issues with your classmates. You will find tools to help you organize your research, analyze the data, think critically about the issues, and construct a well-considered argument.

Take a Stand activities can be done individually or as a team, and as oral presentations, written exercises, or Web-based (e-mail) assignments.

Additional online materials relating to this chapter, including a Student Testing Section with study aids and self-tests, Environmental News, Activity Links, Environmental Investigations, and more, are also on our Web site.

Environmental Laws, Economics, and Ethics

Northern spotted owl. This rare species *(Strix occidentalis caurina)* is found predominantly in old-growth forests in the Pacific Northwest, from northern California to southern British Columbia. Protecting the owl's habitat benefits many other species that reside in the same environment. (Jack Wilburn/Animals Animals)

During the late 1980s and 1990s, an environmental controversy began in western Oregon, Washington, and northern California that continues today. At stake were thousands of jobs and the future of large tracts of *old-growth coniferous forest*, along with the existence of organisms that depend on the forest. The northern spotted owl came to symbolize the confrontation because of its dependence on mature forests for habitat (see photograph).

Old-growth forests have never been logged. Because most of the forests in the United States were logged at one time or another, less than 10% of old-growth forests remain, and this fraction is decreasing. Most old-growth forests in the United States are found in the Pacific Northwest and Alaska.

Old-growth forests provide biological habitats for many species, including the northern spotted owl and 40 other endangered or threatened species. Provisions of the *Endangered Species Act* require the government to protect the habitat of endangered species so that their numbers increase. To enforce this law, in 1991 a court ordered the suspension of logging in about 1.2 million hectares (3 million acres) of federal forest where the owl lives. The timber industry bitterly opposed the moratorium, stating that thousands of jobs would be lost if the northern spotted owl habitat were set aside.

The situation was more complex than simply jobs versus the environment, however. During the decade between 1977 and 1987, logging in Oregon's national forests increased by over 15%. At the same time, employment dropped by an estimated 12,000 jobs because of automation in the timber industry. In addition, the timber industry in the Pacific Northwest was not operating **sustainably**—that is, the industry removed trees faster than the forest could regenerate. If the industry had continued to log at its 1980s rates, most of the remaining old-growth forests would have disappeared within 20 years. Today timber is not as important to the economy of the Pacific Northwest, a change that began long before the controversy over the northern spotted owl erupted.

The 1994 **Northwest Forest Plan** provided federal aid to retrain some timber workers for other careers. State programs also helped reduce unemployment. For example, hundreds of former loggers were employed to restore watersheds and salmon habitats in the forests they used to harvest. As a result of the plan, logging was resumed on federal forests in Washington, Oregon, and northern California, but at only about one-fifth of 1980s level.

The plan, along with previous congressional and administrative actions, reserved about 75% of federal timberlands to safeguard watersheds and protect the northern spotted owl and several hundred other species.

Timber-cutting interests were not happy with the plan, and they tried unsuccessfully to revoke or revise the laws. Environmental groups were also unhappy, and in 1999 they sued the U.S. Forest Service and the Bureau of Land Management, the two agencies overseeing logging of old-growth forests on federal land, maintaining that these agencies had not adequately carried out the provisions of the Northwest Forest Plan. Specifically, the plan required the agencies to complete surveys of 77 local endangered and threatened species before granting the timber industry permission to log. The judge agreed that the agencies had failed to complete the surveys. The Forest Service conducted the surveys until 2004 when they discontinued the practice, arguing that they were an inconsequential bureaucratic hurdle. The issue was back in the courts in 2006.

In this chapter we continue our presentation of how environmental problems are addressed, which we began in Chapter 1 with the role of science in environmental decision making. We will first examine the environmental history of the United States, and then the roles of government, economics, and ethics in handling environmental issues.

World View and Closer to You...
Real news footage relating to environmental laws, economics and ethics around the world and in your own geographic region is available at www.wiley.com/college/raven by clicking on "World View and Closer to You."

A Brief Environmental History of the United States

LEARNING OBJECTIVES

- Briefly outline the environmental history of the United States.
- Describe the environmental contributions of the following people: George Perkins Marsh, Theodore Roosevelt, Gifford Pinchot, John Muir, Aldo Leopold, Wallace Stegner, Rachel Carson, and Paul Ehrlich.
- Distinguish between utilitarian conservationists and biocentric preservationists.

From the establishment of the first permanent English colony at Jamestown, Virginia, in 1607, the first two centuries of U.S. history were a time of widespread environmental destruction. Land, timber, wildlife, rich soil, clean water, and other resources were cheap and seemingly inexhaustible. The European settlers did not dream that the bountiful natural resources of North America would one day become scarce. During the 1700s and early 1800s, most Americans had a **frontier attitude,** a desire to conquer and exploit nature as quickly as possible. Concerns about the depletion and degradation of resources occasionally surfaced, but efforts to conserve were seldom made because the vastness of the continent made it seem that we would always have enough resources.

Protecting Forests

The great forests of the Northeast were leveled within a few generations, and shortly after the Civil War in the 1860s, loggers began deforesting the Midwest at an accelerated rate. Within 40 years they deforested an area the size of Europe, stripping Minnesota, Michigan, and Wisconsin of virgin forest (**Figure 2.1**). By 1897 the sawmills of Michigan had processed 160 billion board feet of white pine, leaving less than 6 billion board feet standing in the whole state.

During the 19th century, many U.S. naturalists began to voice concerns about conserving natural resources. **John James Audubon** (1785–1851) painted lifelike portraits of birds and other animals in their natural surroundings. His paintings, based on detailed field observations, aroused widespread public interest in the wildlife of North America. **Henry David Thoreau** (1817–1862), a prominent U.S. writer, lived for two years on the shore of Walden Pond near Concord, Massachusetts. There he observed nature and contemplated how people could economize and simplify their lives to live in harmony with the natural world. **George Perkins Marsh** (1801–1882) was a farmer, linguist, and diplomat at various times during his life. Today he is most remembered for his book *Man and Nature*, which recognized the interrelatedness of human and environmental systems and provided one of the first discussions of humans as agents of global environmental change. Marsh was widely traveled, and *Man and Nature* was based in part on his observations of environmental damage in areas as geographically separate as the Middle East and his native Vermont.

In 1875 a group of public-minded citizens formed the *American Forestry Association*, with the intent of influencing public opinion against the wholesale destruction of America's forests. Sixteen years later, in 1891, the **General Revision Act** gave the president the authority to establish forest reserves on federally owned land. Benjamin Harrison (1833–1901), Grover Cleveland (1837–1908), and **Theodore Roosevelt** (1858–1919) used this law to remove 17.4 million hectares (43 million acres) of forest, primarily in the West, from logging.

In 1907 angry Northwest congressmen pushed through a bill rescinding the president's powers to establish forest reserves. Theodore Roosevelt, an important contributor to the conservation movement, responded by designating 21 new national forests that totaled 6.5 million hectares (16 million acres). He then signed the bill into law that prevented him and future presidents from establishing additional forest reserves.

Figure 2.1   Logging Operations in 1884. This huge logjam occurred on the St. Croix River near Taylors Falls, Minnesota. (© Minnesota Historical Society/Corbis Images)

utilitarian conservationist: A person who values natural resources because of their usefulness for practical purposes but uses them sensibly and carefully.

biocentric preservationist: A person who believes in protecting nature because all forms of life deserve respect and consideration.

Roosevelt appointed **Gifford Pinchot** (1865–1946) the first head of the U.S. Forest Service. Both Roosevelt and Pinchot were utilitarian conservationists who viewed forests in terms of their usefulness for people—such as in providing jobs. Pinchot supported expanding the nation's forest reserves and managing forests scientifically, such as by harvesting trees only at the rate at which they re-grow. Today, national forests are managed for multiple uses, from biological habitats to recreation to timber harvest to cattle grazing.

Establishing and Protecting National Parks and Monuments

Congress established the world's first national park in 1872 after a party of Montana explorers reported on the natural beauty of the canyon and falls of the Yellowstone River; Yellowstone National Park now includes parts of Idaho, Montana, and Wyoming. In 1890 the *Yosemite National Park Bill* established the Yosemite and Sequoia National Parks in California, largely in response to the efforts of a single man, naturalist and writer **John Muir** (1838–1914) (**Figure 2.2**). Muir, a biocentric preservationist, founded the *Sierra Club*, a national conservation organization that is still active on a range of environmental issues.

In 1906 Congress passed the **Antiquities Act,** which authorized the president to set aside as national monuments sites, such as the Badlands in South Dakota, that had scientific, historic, or prehistoric importance. By 1916 there were 13 national parks and 20 national monuments, under the loose management of the U.S. Army. (Today there are 58 national parks and 73 national monuments under the management of the *National Park Service.*)

Some environmental battles involving the protection of national parks were lost. John Muir's Sierra Club fought such a battle with the city of San Francisco over its efforts to dam a river and form a reservoir in the Hetch Hetchy Valley (**Figure 2.3**), which lay within Yosemite National Park and was as beautiful as Yosemite Valley. In 1913 Congress voted to approve the dam.

The controversy generated a strong sentiment favoring better protection for national parks, and in 1916 Congress created the National Park Service to manage the national parks and monuments for the enjoyment of the public "without impairment." It was this clause that gave a different outcome to another battle, fought in the 1950s between conservationists and dam builders over the construction of a dam within Dinosaur National Monument. No one could deny that to

Figure 2.2 **President Theodore Roosevelt (left) and John Muir.** Photo was taken on Glacier Point above Yosemite Valley, California. (© Bettmann/Corbis Images)

Figure 2.3 **Hetch Hetchy Valley in Yosemite.** A view in Hetch Hetchy Valley (a) before and (b) after Congress approved a dam to supply water to San Francisco. (© National Archives, WallaceKleck/Terraphotographics/BPS)

drown the canyon with 400 feet of water would "impair" it. This victory for conservation established the "use without impairment" clause as the firm backbone of legal protection afforded U.S. national parks and monuments. Today, discussion is underway to restore Hetch Hetchy, which the State of California estimates would cost as much as $10 billion.

Conservation in the Mid-20th Century

During the Great Depression, the federal government financed many conservation projects to provide jobs for the unemployed. During his administration **Franklin Roosevelt** (1882–1945) established the Civilian Conservation Corps, which employed more than 175,000 men to plant trees, make paths and roads in national parks and forests, build dams to control flooding, and perform other activities to protect natural resources.

During the droughts of the 1930s, windstorms carried away much of the topsoil in parts of the Great Plains, forcing many farmers to abandon their farms and search for work elsewhere (see Chapter 15). The so-called *American Dust Bowl* alerted the United States to the need for soil conservation, and in 1935 President Roosevelt formed the *Soil Conservation Service*.

Aldo Leopold (1886–1948) was a wildlife biologist and environmental visionary who was influential in the conservation movement of the mid- to late 20th century. His textbook, *Game Management*, was published in 1933 and supported the passage of a 1937 act in which new taxes on sporting weapons and ammunition funded wildlife management and research. Leopold also wrote philosophically about humanity's relationship with nature and about the need to conserve wilderness areas in *A Sand County Almanac*, published in 1949. Leopold argued persuasively for a land ethic and the sacrifices such an ethic requires.

Aldo Leopold had a profound influence on many American thinkers and writers, including **Wallace Stegner** (1909–1993), who penned his famous "Wilderness Essay" in 1962. Stegner's essay, written to a commission conducting a national inventory of wilderness lands, helped create support for passage of the Wilderness Act of 1964. Stegner wrote:

> *Something will have gone out of us as a people if we ever let the remaining wilderness be destroyed; if we permit the last virgin forests to be turned into comic books and plastic cigarette cases; if we drive the few remaining members of the wild species into zoos or to extinction; if we pollute the last clean air and dirty the last clean streams and push our paved roads through the last of the silence, so that never again will Americans be free in their own country from the noise, the exhausts, the stinks of human and automotive waste.*
>
> *We simply need that wild country available to us, even if we never do more than drive to its edge and look in. For it can be a means of reassuring ourselves of our sanity as creatures, a part of the geography of hope.*

During the 1960s public concern about pollution and resource quality began to increase, in large part because of marine biologist **Rachel Carson** (1907–1964). Carson wrote about interrelationships among living organisms, including humans, and the natural environment (**Figure 2.4**). Her most famous work, *Silent Spring*, was published in 1962. In this work Carson wrote against the indiscriminate use of pesticides.

Silent Spring heightened public awareness and concern about the dangers of uncontrolled use of DDT and other pesticides, including the poisoning of birds and other wildlife and the contamination of human food supplies. Ultimately, this heightened public awareness led to restrictions on the use of certain pesticides. About this time the media began to increase its coverage of environmental incidents, such as hundreds of deaths in New York City from air pollution (1963); closed beaches and fish kills in Lake Erie from water pollution (1965); and detergent foam in a creek in Pennsylvania (1966).

In 1968, when the population of Earth was "only" 3.5 billion people (compared to 6.5 billion in 2006), ecologist **Paul Ehrlich** published *The Population Bomb*. In it he described the unavoidable environmental damage necessary for Earth to support such a huge population, including irreversible soil loss, depletion of groundwater, and loss of other living organisms. Ehrlich's book raised public awareness of the dangers of overpopulation and triggered debates on how to deal effectively with population issues.

Figure 2.4   Rachel Carson. Carson's book, *Silent Spring*, heralded the beginning of the environmental movement. Photographed in Maine in 1962. (Erich Hartmann/Magnum Photos, Inc.)

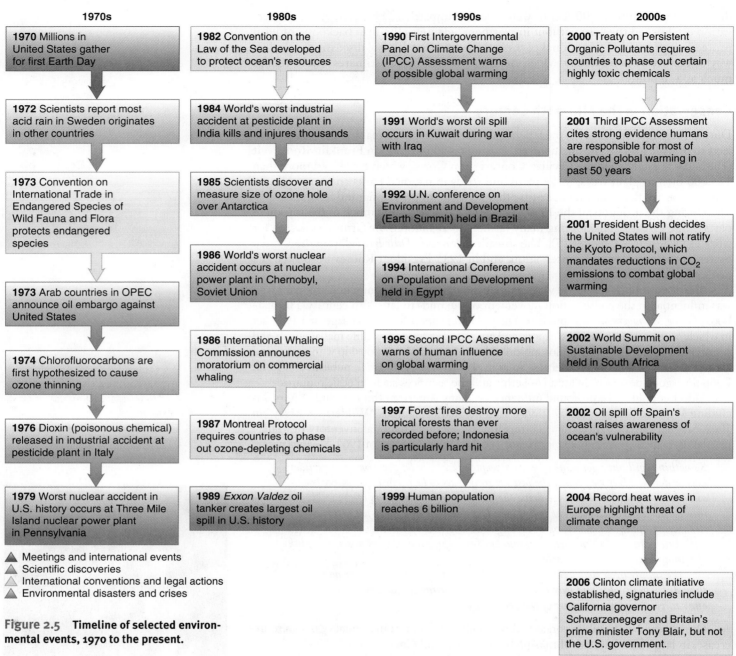

1970s	1980s	1990s	2000s

1970 Millions in United States gather for first Earth Day

1972 Scientists report most acid rain in Sweden originates in other countries

1973 Convention on International Trade in Endangered Species of Wild Fauna and Flora protects endangered species

1973 Arab countries in OPEC announce oil embargo against United States

1974 Chlorofluorocarbons are first hypothesized to cause ozone thinning

1976 Dioxin (poisonous chemical) released in industrial accident at pesticide plant in Italy

1979 Worst nuclear accident in U.S. history occurs at Three Mile Island nuclear power plant in Pennsylvania

1982 Convention on the Law of the Sea developed to protect ocean's resources

1984 World's worst industrial accident at pesticide plant in India kills and injures thousands

1985 Scientists discover and measure size of ozone hole over Antarctica

1986 World's worst nuclear accident occurs at nuclear power plant in Chernobyl, Soviet Union

1986 International Whaling Commission announces moratorium on commercial whaling

1987 Montreal Protocol requires countries to phase out ozone-depleting chemicals

1989 *Exxon Valdez* oil tanker creates largest oil spill in U.S. history

1990 First Intergovernmental Panel on Climate Change (IPCC) Assessment warns of possible global warming

1991 World's worst oil spill occurs in Kuwait during war with Iraq

1992 U.N. conference on Environment and Development (Earth Summit) held in Brazil

1994 International Conference on Population and Development held in Egypt

1995 Second IPCC Assessment warns of human influence on global warming

1997 Forest fires destroy more tropical forests than ever recorded before; Indonesia is particularly hard hit

1999 Human population reaches 6 billion

2000 Treaty on Persistent Organic Pollutants requires countries to phase out certain highly toxic chemicals

2001 Third IPCC Assessment cites strong evidence humans are responsible for most of observed global warming in past 50 years

2001 President Bush decides the United States will not ratify the Kyoto Protocol, which mandates reductions in CO_2 emissions to combat global warming

2002 World Summit on Sustainable Development held in South Africa

2002 Oil spill off Spain's coast raises awareness of ocean's vulnerability

2004 Record heat waves in Europe highlight threat of climate change

2006 Clinton climate initiative established, signatures include California governor Schwarzenegger and Britain's prime minister Tony Blair, but not the U.S. government.

▲ Meetings and international events
▲ Scientific discoveries
▲ International conventions and legal actions
▲ Environmental disasters and crises

Figure 2.5 Timeline of selected environmental events, 1970 to the present.

The Environmental Movement of the Late 20th Century

Until 1970 the voices of **environmentalists** were heard in the United States primarily through societies such as the *Sierra Club* and the *National Wildlife Federation.* There was no generally perceived environmental movement until the spring of 1970, when **Gaylord Nelson,** former senator of Wisconsin, urged Harvard graduate student **Denis Hayes** to organize the first nationally celebrated Earth Day. This event awakened U.S. environmental consciousness to population growth, overuse of resources, and pollution and degradation of the environment. On Earth Day 1970 an estimated 20 million people in the United States demonstrated their support of environmental quality by planting trees, cleaning roadsides and riverbanks, and marching in parades. (**Figure 2.5** presents a timeline of selected environmental events since Earth Day 1970.)

ENVIRONEWS

Religion and the Environment

A new degree of commitment to environmental issues has grown steadily within many religious faiths and organizations. In the past decade, the world's major religions have collaborated in several conferences on religious perspectives and the natural environment. Here are several examples:

a. From 1996 to 1998 Harvard University's Center for the Study of World Religions held a series of conferences on the ecological connections with the world's major religions (listed in declining number of adherents): Christianity, Islam, Hinduism, Confucianism, Buddhism, indigenous religions, Sikhism, Judaism, Spiritualism, Baha'i faith, Jainism, Shintoism, and Taoism.

b. In 2000 more than 1,000 religious leaders met at the U.N. Millennium World Peace Summit of Religious and Spiritual Leaders. Protecting the environment was a major topic of discussion.

c. In 2001 the U.N. Environment Program and the Islamic Republic of Iran sponsored an International Seminar on Religion, Culture, and Environment. They considered ways to counter environmental degradation.

d. In 2006 the Evangelical Climate Initiative, which includes traditionally conservative Evangelicals in the United States, identified global warming as an important religious issue requiring responsible stewardship and attention to future generations.

In the years that followed the first Earth Day, environmental awareness and the belief that individual actions could repair the damage humans were doing to Earth became a pervasive popular movement. Musicians such as Marvin Gaye, Joni Mitchell (a Canadian), and the group Alabama popularized environmental concerns. Many of the world's religions embraced environmental themes such as protecting endangered species.

By Earth Day 1990 the movement had spread around the world, signaling the rapid growth in environmental consciousness (**Figure 2.6**). An estimated 200 million people in 141 nations demonstrated to increase public awareness of the importance of individual efforts ("Think globally, act locally"). The theme of Earth Day 2000—"Clean Energy Now!"—reflected the dangers of global climate change and advised what individuals and communities could do: Replace fossil fuel energy sources, which produce greenhouse gases, with solar electricity, wind power, and the like. By 2000 many environmental activists had begun to think that individual actions, though collectively important, are not as important as pressuring governments and large corporations to make environmentally appropriate decisions.

Figure 2.6 Earth Day. Schoolchildren join with environmentalists to celebrate Earth Day 1990 in Hong Kong. (© Reuters Newmedia, Inc/Corbis Images)

REVIEW

1. Which U.S. president designated 21 national forests and then signed a law that prevented him and future presidents from doing just that?
2. What was the world's first national park?
3. Which occurred first in the U.S. environmental movement: Concerns about forest conservation or concerns about pollution?
4. How did the United States establish national parks? How are the parks protected?

U.S. Environmental Legislation

LEARNING OBJECTIVES

- Explain why the National Environmental Policy Act is the cornerstone of U.S. environmental law.
- Relate how environmental impact statements provide such powerful protection of the environment.
- Explain why industry groups often favor regulatory reform policies, while some environmental advocates focus on environmental justice.

By the late 1960s much of the U.S. public had become increasingly disenchanted with governmental secrecy, and many distrusted industry to work in the public interest. This broad social transformation, which included opposition to the Vietnam War and resistance to racist policies, was reflected in environmental attitudes as well. Galvanized by well-publicized ecological disasters, such as the 1969 oil spill off the coast of Santa Barbara, California, and by overwhelming public support for the Earth Day movement, in 1970 the **Environmental Protection Agency (EPA)** was formed, and the **National Environmental Policy Act (NEPA)** was signed into law. A key provision of NEPA required the federal government to consider the environmental impact of any proposed federal action, such as financing highway or dam construction. NEPA provides the basis for developing detailed **environmental impact statements (EISs)** to accompany every federal recommendation or legislative proposal. These EISs are supposed to help federal officials make informed decisions. Each EIS must include the following:

environmental impact statement: A document that summarizes the potential and expected adverse impacts on the environment associated with a project, as well as alternatives to the proposed project; typically mandated by law for public and/or private projects.

1. The nature of the proposal and why it is needed.
2. The environmental impacts of the proposal, including short-term and long-term effects and any adverse environmental effects if the proposal is implemented (**Figure 2.7**).
3. Alternatives to lessen the adverse effects of the proposal. This part of the EIS generally concentrates on ways to mitigate the impact of the project.

A required step in the EIS process is the solicitation of public comments, which generally provide a broader perspective on the proposal and its likely effects. NEPA established the **Council on Environmental Quality** to monitor the required EISs and report directly to the president. This council had no enforcement powers, and NEPA was originally considered innocuous, generally more a statement of good intentions than a regulatory policy. During the next few years, environmental activists took people, corporations, and the federal government to court to challenge their EISs and use them to block proposed development. The courts decreed that EISs must be substantial documents that thoroughly analyze the environmental consequences of anticipated projects on soil, water, and organisms. The courts also said that the public must have access to EISs. These rulings put sharp teeth into the law—particularly the provision for public scrutiny, which placed intense pressure on federal agencies to respect EIS findings.

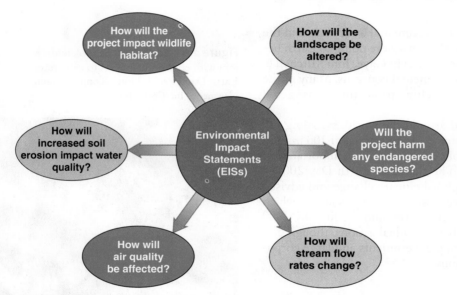

Figure 2.7 Environmental Impact Statements. These detailed statements help federal agencies and the public consider the environmental impacts of proposed activities. When anticipated impacts are likely to be high, there will be pressure to seek alternative actions.

NEPA revolutionized environmental protection in the United States. In addition to overseeing federal highway construction, flood and erosion control, military projects, and many other public works, federal agencies oversee nearly one-third of the land in the United States. Federally owned holdings include extensive fossil fuel and mineral reserves as well as millions of hectares of public grazing land and public forests, all subject to NEPA. Many states and local governments now require EISs for public (and sometimes private) projects, as do many foreign counties.

NEPA has evolved considerably through four decades of litigation and seven presidential administrations, but the fundamental effect of publicly available information remains. Although almost everyone agrees that NEPA has helped federal agencies reduce adverse environmental impacts of their activities and projects, it has its critics. Environmentalists complain that EISs are sometimes incomplete, include only alternatives that make a predetermined alternative look attractive, or are ignored when decisions are made. Other critics think the EISs delay important projects ("paralysis by analysis") because the EISs are too involved, take too long to prepare, and are often the targets of lawsuits.

Environmental Policy Since 1970

While legislation to manage many environmental problems existed before 1970, the regulatory system that exists today was largely put into place during the subsequent decade. Congress has passed many environmental laws that address a wide range of issues, such as endangered species, clean water, clean air, energy conservation, hazardous wastes, and pesticides (**Table 2.1**). These laws greatly increased federal regulation of pollution, creating a tough interlocking mesh of laws to improve environmental quality. Many environmental laws contain statutes that allow private citizens to take violators,

Table 2.1 Some Important Federal Environmental Legislation

General

Freedom of Information Act of 1966
National Environmental Policy Act of 1969
National Environmental Education Act of 1990

Conservation of Energy

Energy Policy and Conservation Act of 1975
Northwest Power Act of 1980
National Appliance Energy Conservation Act of 1987
Energy Policy Act of 1992

Conservation of Wildlife

Fish and Wildlife Act of 1956
Anadromous Fish Conservation Act of 1965
Fur Seal Act of 1966
National Wildlife Refuge System Act of 1966
Species Conservation Act of 1966
Marine Mammal Protection Act of 1972
Marine Protection, Research, and Sanctuaries Act of 1972
Endangered Species Act of 1973
Federal Noxious Weed Act of 1974
Magnuson Fishery Conservation and Management Act of 1976
Whale Conservation and Protection Study Act of 1976
Fish and Wildlife Improvement Act of 1978
Fish and Wildlife Conservation Act of 1980
Fur Seal Act Amendments of 1983
Wild Bird Conservation Act of 1992
National Invasive Species Act of 1996

Conservation of Land

General Revision Act of 1891
Taylor Grazing Act of 1934
Soil Conservation Act of 1935
Multiple Use Sustained Yield Act of 1960 (re: national forests)
Wilderness Act of 1964
Land and Water Conservation Fund Act of 1965
Wild and Scenic Rivers Act of 1968
National Trails System Act of 1968
Coastal Zone Management Act of 1972
National Reserves Management Act of 1974
Forest and Rangeland Renewable Resources Act of 1974
Federal Land Policy and Management Act of 1976
National Forest Management Act of 1976
Soil and Water Resources Conservation Act of 1977
Surface Mining Control and Reclamation Act of 1977
Public Rangelands Improvement Act of 1978
Antarctic Conservation Act of 1978
Endangered American Wilderness Act of 1978
Alaska National Interest Lands Act of 1980
Coastal Barrier Resources Act of 1982

Emergency Wetlands Resources Act of 1986
North American Wetlands Conservation Act of 1989
California Desert Protection Act of 1994
Farm Security and Rural Investment Act of 2002 (the latest version of the "farm bill," which has been amended and renamed every 5 years or so since the 1930s)

Air Quality and Noise Control

Noise Control Act of 1965
Clean Air Act of 1970
Quiet Communities Act of 1978
Asbestos Hazard and Emergency Response Act of 1986
Clean Air Act Amendments of 1990

Water Quality and Management

Refuse Act of 1899
Water Resources Research Act of 1964
Water Resources Planning Act of 1965
Clean Water Act of 1972
Ocean Dumping Act of 1972
Safe Drinking Water Act of 1974
National Ocean Pollution Planning Act of 1978
Water Resources Development Act of 1986
Great Lakes Toxic Substance Control Agreement of 1986
Water Quality Act of 1987 (amendment of Clean Water Act)
Ocean Dumping Ban Act of 1988
Oceans Act of 2000

Control of Pesticides

Food, Drug, and Cosmetics Act of 1938
Federal Insecticide, Fungicide, and Rodenticide Act of 1947
Food Quality Protection Act of 1996

Management of Solid and Hazardous Wastes

Solid Waste Disposal Act of 1965
Resource Recovery Act of 1970
Hazardous Materials Transportation Act of 1975
Toxic Substances Control Act of 1976
Resource Conservation and Recovery Act of 1976
Low-Level Radioactive Policy Act of 1980
Comprehensive Environmental Response, Compensation, and Liability ("Superfund") Act of 1980
Nuclear Waste Policy Act of 1982
Hazardous and Solid Waste Amendments of 1984
Superfund Amendments and Reauthorization Act of 1986
Medical Waste Tracking Act of 1988
Marine Plastic Pollution Control Act of 1987
Oil Pollution Act of 1990
Pollution Prevention Act of 1990
State or Regional Solid Waste Plans (RCRA Subtitle D) of 1991

whether they are private industries or government-owned facilities, to court for non-compliance. These citizen suits have contributed significantly in the enforcement of environmental legislation.

In the early 1980s President Reagan attempted to reverse the pro-environmental trend by appointing pro-business EPA Administrator Ann Gorsuch and leaving many top-level positions in the agency unfilled. Congressional and public backlash to this led to even more restrictive environmental laws. Gorsuch was replaced with William Ruckleshaus, who had been the first EPA administrator, and who was widely respected for having established the agency's credibility and authority. Ruckleshaus advocated a rational, organized approach to environmental regulation that would avoid the "pendulum" effect of stringent and then relaxed environmental regulation.

Since 1980 the rate of new major environmental legislation has slowed, but environmental policy continues to evolve through administrative actions and the judicial process. The Northwest Forest Plan described at the beginning of this chapter is one such example. Through the late 1980s and early 1990s, EPA and a number of states engaged in environmental prioritization exercises. These efforts, often called **Comparative Risk Analyses,** evaluate the health, economic, and ecosystem impacts of a range of environmental issues.

In 1994 President Clinton issued Executive Order 12898, requiring that all new environmental regulations take **environmental justice** issues into account. Many environmental advocates are concerned that much environmental degradation and pollution takes place in areas where other social problems are prevalent, including poverty, substandard housing, and higher levels of disease.

The EPA is usually given the job of translating environmental laws into specific regulations. Before the regulations become official, several rounds of public comments allow affected parties to present their views; the EPA is required to respond to all of these comments. Then the Office of Management and Budget, which oversees the federal budget, reviews the new regulations. Some regulations proposed by the EPA must be justified by **cost-benefit analysis,** while others (including the Clean Air Act) specifically prohibit cost-benefit analysis. Implementation and enforcement of environmental regulations often fall to state governments, which must send the EPA detailed plans showing how they plan to achieve regulatory goals and standards. Currently, the EPA oversees thousands of pages of environmental regulations that affect individuals, corporations, local communities, and states.

Environmental laws have not always worked as intended. The *Clean Air Act* of 1977 required coal-burning power plants to outfit their smokestacks with expensive "scrubbers" to remove sulfur dioxide from their emissions but made an exception for tall smokestacks (**Figure 2.8**). This loophole led directly to the proliferation of tall stacks that have since produced *acid rain* throughout the Northeast. The *Clean Air Act Amendments* of 1990, described in Chapter 20, go a long way toward closing this loophole.

In addition many environmental regulations are not enforced properly. According to Oliver Houck of the Tulane Law School, environmental laws such as the Clean Water Act have compliance rates of about 50%. Proper enforcement of existing rules and rigorous sentences and stiff fines for lawbreakers would improve the quality of our environment without the need for additional legislation.

Despite imperfections, U.S. environmental legislation has had overall positive effects. Since 1970

a. Twenty-three national parks have been established, and the National Wilderness Preservation System now totals more than 43 million hectares (106 million acres).

b. Millions of hectares of farmland particularly vulnerable to erosion have been withdrawn from production, reducing soil erosion in the United States by more than 60%.

c. Many previously endangered species are better off than they were in 1970, and the American alligator, California gray whale, and bald eagle have recovered enough to be removed from the endangered species list. (Dozens of other species, such as the manatee, ivory-billed woodpecker, and Kemp's sea turtle, have suffered further declines or extinction since 1970.)

environmental justice: Dealing with concerns that populations at high risk because of social or economic factors also face elevated impacts from environmental hazards.

Figure 2.8   Tall smokestacks. Such smokestacks, which emit sulfur dioxide from coal-burning power plants, were exempt from the requirement of pollution-control devices under the Clean Air Act of 1977. (Albert Copley/Visuals Unlimited)

Although we still have a long way to go, pollution-control efforts have been particularly successful. According to the *EPA's Draft Report on the Environment 2003*,

a. Since 1970, emissions of six important air pollutants have dropped by almost 25%.

b. Since 1990, levels of wet sulfate, a major component of acid rain, have dropped by 20 to 30%.

c. In 2002, 94% of the U.S. population received water from community water systems that meet health-based drinking-water standards, up from 79% in 1993.

d. Release of toxic chemicals into water and air from industrial sources has declined 48% since 1988.

e. Of the 1,498 contaminated sites listed on the Superfund National Priorities List in 2002, 846 are now cleaned up. This figure is up from 149 in 1992.

f. Fewer rivers and streams are in violation of water quality standards. (The number of fish-consumption advisories relating to specific toxins such as mercury or polychlorinated biphenyls [PCBs] has risen since the 1970s, but this may be partly the result of more consistent monitoring.)

However, not all environmental trends are positive. In some major urban areas, such as Houston and Las Vegas, air quality has begun to deteriorate since about 2000. While technology has reduced environmental impacts of industries and products, population increases have outpaced technological improvements (recall the *IPAT* model from Chapter 1). Future environmental improvement will require much more sophisticated strategies, involving technological and behavior changes.

In the 1960s and 1970s, pollution was often obvious—witness the Cuyahoga River in Cleveland, Ohio, which burst into flames and burned for eight days in 1969 as a result of oily pollutants floating on its surface. Burning rivers and smoking exhaust pipes present visible threats that encourage immediate attention with little regard to cost. However, the last decade has witnessed increased interest in **regulatory reform,** in which environmental, health, safety, and other regulations are selected based on cost-effectiveness. Many industry groups see this as an opportunity to encourage more economically efficient regulation, and they point out that it should lead to more lives saved.

Some economists and industries contend that many regulations make pollution abatement unduly expensive. Environmental, consumer, and worker safety advocates see regulatory reform as an attempt to reverse hard-won protections. According to EPA estimates, during the early 2000s the cost of complying with these federal regulations was $210 billion per year, which amounts to about 2.6% of the U.S. gross domestic product, yet a 2003 Office of Management and Budget review of U.S. regulations reported that the environmental benefits of these laws exceed the costs by several times. The appropriate role of economic analysis in environmental regulation is an important current debate.

REVIEW

1. Which law is the cornerstone of U.S. environmental law? Why?

2. Describe the process for generating environmental impact statements.

Economics and the Environment

LEARNING OBJECTIVES

- Distinguish among the following economic terms: marginal cost of pollution, marginal cost of abatement, optimum amount of pollution.

- Explain why economists prefer efficient solutions to environmental problems.

- Describe various approaches to pollution control, including command and control regulation, incentive-based regulation, and cost-effectiveness analysis.

- Give two reasons why the national income accounts are incomplete estimates of national economic performance.

Economics is the study of how people use their limited resources to try to satisfy unlimited wants. Economists combine theoretical assumptions about individual and institutional behavior with analytical tools that include developing hypotheses, testing **models,** and analyzing observations and data. They try to understand the consequences of the ways in which people, businesses, and governments allocate their resources.

Economists who work on environmental problems must take a systems perspective. For example, Nobel Prize-winning economist **Amartya Sen (Figure 2.9)** addresses environment, poverty, nationalism, gender issues, and governmental structure. This approach recognizes the complex interactions among environment, society, health, and well-being.

Economics, especially as applied to public policy, relies on several precepts. First, economics is utilitarian. This means that all good and services—including those provided by the environment—have value to people, and that those values can be converted into some common currency. Thus, in a market economy such as that of the United States, goods and services have dollar values, which are determined by the amount of money someone is willing to pay for them.

This leads to a second precept, that of the **rational actor**. Economists assume that all individuals know what goods and services are worth to them, and spend their limited resources (money and time) in such a way that provides them the most **utility**.

> **rational actor model:** In economics, the assumption that all individuals try to spend their limited resources in a fashion that maximizes their individual utilities.

> **utility:** An economic term referring to the benefit that an individual gets from some good or service. Rational actors try to maximize utility.

Third, in an ideal economy, resources will be allocated efficiently. **Efficiency** is an economics term used to describe getting the greatest amount of goods or services from a limited set of resources. For example, if business plan A will build 9 cars with a given amount of material and number of workers, while business plan B will build 10 equivalent cars with the same resources, business plan B is more efficient and will succeed. To economists, environmental problems arise when market failures occur in one of several ways, principal among which are inefficiency and externalities.

Externalities occur when the producer of a good or service does not have to pay the full costs of production. An example is a blacksmith shop next door to a drip-dry laundry. If soot from the blacksmith shop lands on the laundry, then the laundry owner will have to rewash the laundry and try to pass the additional cost to its customers. Here, the blacksmith shop is causing a negative externality on the laundry customers, who in effect pay part of the cost of the blacksmith operation.

> **externality:** In economics, the effect (usually negative) of a firm that does not have to pay all the costs associated with its production.

A simple solution—advocated by some economists—is to clearly define rights and ownership. If the blacksmith does not have the right to release soot, then the laundry owner can demand compensation for the added cost of business. The blacksmith owner then has several options—for example, to close the shop, find a way to control the soot, or pay the laundry owner for the extra cost of business. The efficient solution is the one that costs the least money for everyone. If closing the shop costs $10,000 per year, soot control equipment costs $200 per year, and paying to rewash laundry costs $500 per year, the blacksmith should choose to control soot. This same solution will occur if the blacksmith *does* have the right to release soot, since the laundry owner should be willing to pay the blacksmith $200 to install soot-controlling equipment, rather than spend $500 to rewash laundry.

Unfortunately, most environmental externalities are not as clearly defined as the laundry-blacksmith example. Usually, there are many polluters and many affected individuals. A small town may have 5,000 automobiles, each of which generates a different amount of pollution that affects each of 10,000 residents. Damage associated with the releases may depend on the time of day and part of town where pollution is generated. Even if we could assess those damages, it would be impossible for each driver to compensate each person impacted by the car's emissions. The problem becomes even more extreme for climate change, where there are billions of individuals both generating and being affected by greenhouse gasses.

Figure 2.9 **Nobel Laureate Amartya Sen.** Economist Amartya Sen's work includes critiques of traditional assumptions about economic valuation of the environment, recognizing the complex, systemic nature of environment, health, and well-being. (Prakash Singh/Agence France Presse/ NewsCom)

Several economic solutions exist for cases of multiple polluters, all of which rely on the idea that we can identify an efficient or **optimal amount of pollution.** At this optimum, the cost to society of having less pollution is offset by the benefits to society of the activity creating the pollution. To find this optimum requires that we identify the **marginal cost of pollution,** where marginal cost refers to the cost of a small additional amount of pollution. Determining the marginal cost of pollution involves assessing damage to health, property, agriculture, and aesthetics caused by the pollution (see Chapter 7 for a discussion of risk assessment). Pollution can also reduce **ecosystem services,** which are benefits to humans—including clean water and fresh air—provided by natural systems. Determining the marginal cost of pollution is generally not an easy process.

Similarly, the **marginal cost of abatement** is the cost associated with reducing (abating) a small additional amount of pollution. Untreated wastewater from a paper mill contains a variety of chemicals and suspended wood fibers. The cost of filtering the fibers is relatively low, requiring mechanical screening. The cost of removing inorganic chemicals, however, (see Chapter 22) may be quite high. The societal benefit of filtering is relatively high compared to the benefit of removing a small amount of a chemical. If we insist that only perfectly pure water comes out of the pulp mill, the cost of paper will be extremely high.

Figure 2.10 demonstrates how the marginal cost of pollution

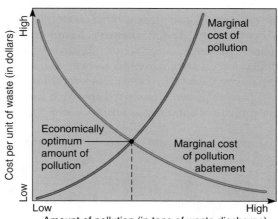

and the marginal cost of abatement lead to an optimal amount of pollution. The green line represents the marginal cost of abatement. When the amount of pollution being generated is high, abatement costs are relatively low. Similarly, when pollution concentration is low, the benefit of additional control is low, as represented by the red line. Pollution is at an optimal level when marginal costs of pollution and abatement are equal. When that is the case, the economic system is efficient, meaning that society would be worse off with either more *or less* pollution. If a pulp mill operating at that optimal level were to spend one additional dollar on control equipment, the value to society of the reduction in pollution would be less than one dollar.

It is unusual to find unregulated economic systems at the point of economic optimality. In unregulated markets, those who pollute often face only a fraction of the total costs of pollution. Thus the marginal cost of pollution faced by the polluter (private cost) is substantially lower than the marginal cost of pollution faced by society as a whole.

Figure 2.10   Economic optimality and pollution. This figure represents the economically optimal level of pollution in an efficient market. The upward-sloping curve (red line) represents the cost of damage associated with pollution at various levels. As pollution rises, the social cost (in terms of human health and a damaged environment) increases sharply. The downward-sloping curve (green line) represents the cost of reducing pollution to a lower, less-damaging level. The intersection of these two curves is the economically optimal point, where a shift in either direction (more or less pollution) will lead to lower total societal benefits.

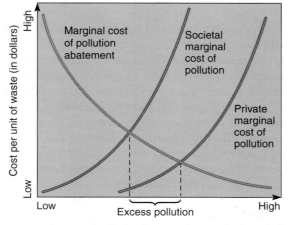

Figure 2.11   Inefficiency arising from different marginal costs. This figure represents the marginal costs of pollution faced by society (red line), the private marginal cost of pollution faced by the polluter (blue line), and the marginal cost of abatement (green line). The polluter will generate pollution at the point where private marginal cost of pollution intersects with the marginal cost of abatement. At this point, substantially more pollution is generated than is optimal from a societal perspective.

This is represented by **Figure 2.11**, which contains two marginal cost curves. The marginal cost curve on the left is the societal marginal cost of pollution, which includes that faced by the polluter. To the right of this is the private marginal cost of pollution curve, which is the fraction of the overall marginal cost paid by the polluter. Without some sort of regulation, the amount of pollution generated will be determined by the point at which the private marginal cost of pollution is equal to the marginal cost of abatement. Substantially more pollution may be generated at this intersection than is optimal from a societal perspective.

Strategies for Pollution Control

Economists favor market-based solutions to environmental externality problems. Historically, many environmental regulations have been **command and control** solutions. This means that the EPA or other government agency requires a particular piece of equipment be installed to limit emissions to water, air, or soil. Industries regulated in this fashion often object that they are thereby discouraged from developing lower-cost alternatives that would achieve the same level of pollution control for less money.

Command and control regulations may result in pollution levels that are lower or higher than the economically optimum level, and for this reason are not popular with economists. Consequently, most economists, whether they have progressive or conservative views, prefer **incentive-based** or cost-benefit-based regulation over command and control. Two major incentive-based approaches that have met with mixed policy success are environmental taxes and tradable permits. Cost-effectiveness analysis is a cost-benefit-based approach for prioritizing diverse environmental regulations.

A popular incentive-based approach for regulating pollution, particularly in Europe, involves imposing an **emission charge** on polluters. In effect, this charge is a tax on pollution. Economists propose such "green taxes" to correct what they perceive as distortions in the market caused by not including the external cost of driving an automobile or cutting down trees or polluting streams. The purpose of this tax is to force polluters to pay the full cost of pollution. If the tax is set at the correct level, the private marginal cost of pollution will be equivalent to the social cost of pollution. This is depicted in **Figure 2.12**.

Many European countries have restructured their taxes to take into account environmentally destructive products and activities. Germany increased taxes on gasoline, heating oil, and natural gas while simultaneously lowering its income tax. One result was an increase in carpooling. There were concerns, however, about whether Germany's energy-intensive industries could remain competitive with industries in countries without similar energy taxes. The Netherlands introduced a tax on natural gas, electricity, fuel oil, and heating oil that includes incentives for energy efficiency; income taxes were decreased to offset the tax burden on individuals. Electricity and fuel use has declined as a result. Finland implemented a carbon dioxide tax. Sweden introduced taxes on carbon and sulfur while simultaneously reducing income taxes. These charges increase the cost of polluting or of overusing natural resources. They also are usually "revenue neutral," since they are offset by rebates or reductions in other taxes.

If taxes are set at the correct level, users will react to the increase in cost resulting from emission charges by decreasing pollution or decreasing consumption. However, it can be very difficult to identify the correct level, and taxes on pollution are almost always set too low to have the desired effect on the behavior of people or companies. It is difficult to enact such taxes, especially in the United States, because people object to paying a tax on something they perceive as "free," and they tend to doubt that such taxes will be "revenue neutral."

While environmental taxes are designed to identify and replicate the social cost of pollution, **tradable permits** rely on identifying the optimal level of pollution. Government sets a cap on pollution and then issues a fixed number of **marketable waste-discharge permits,** allowing holders to emit a specified amount of a given pollutant, such as sulfur dioxide. A permit owner can decide whether to generate the pollution or

Figure 2.12   The corrective effect of "green taxes." This figure represents the marginal costs of pollution faced by society (red line), the private marginal cost of pollution faced by the polluter (blue line), and the marginal cost of abatement (green line). A tax is assessed on the polluter equal to the difference between the private marginal cost of pollution and the social marginal cost of pollution. As a result the entire private marginal cost curve is shifted upward. If the tax is designed well, the cost faced by the polluter will be identical to the marginal social cost of pollution. (Prakash Singh/Agence France Presse/NewsCom)

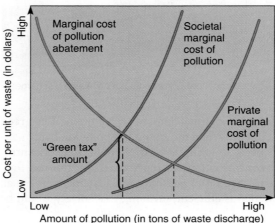

sell the permit, and ideally will choose the option that earns the most money. Once the market for permits has been established, industries that can easily reduce emissions will do so, and then sell their extra permits to industries that cannot.

Tradable permits can also be designed to reduce emissions over time. The Clean Air Act Amendments of 1990 included a plan to cut acid rain–causing sulfur dioxide emissions by issuing tradable permits for coal-burning electricity utilities. The amount of allowable pollution decreases with each trade. This approach has been effective. Sulfur dioxide emissions are being reduced ahead of schedule at about 50% of the original projected cost.

Cost-effectiveness analysis is an increasingly common regulatory tool. Rather than setting an optimum level of pollution, cost-effectiveness analysis asks "if we establish this regulation, how much will it cost to achieve some outcome?" where the outcome is lives saved or years of life saved. Thus, a regulatory action that costs $5,000 per saved life would be preferable to one that costs $10,000 per saved life. This approach could be used to compare a wide range of interventions, from banning pesticides to requiring catalytic converters on cars.

Critiques of Environmental Economics

There are two classes of critiques leveled at economic approaches to environmental regulation. First, it is difficult to assess the true costs of environmental damage by pollution and the cost of abatement. The impacts of pollution on people and nature are extremely uncertain. We often do not know how effective pesticides are at eliminating pests, or whether that pesticide harms humans. If we do not understand the economic benefits of ecosystem services, they may be undervalued in the assessment. Trying to place a value on quality of life or on damage to natural beauty can be highly controversial (**Figure 2.13**). As you will see in the next few chapters, the web of relationships within the environment is extremely intricate, and environmental systems may be more vulnerable to pollution damage than is initially obvious. When the optimal level of pollution is highly uncertain, too many or too few permits may be issued, and green taxes may be set too high or too low.

Second, we do not all agree that economics is an appropriate decision tool. The risks of unanticipated environmental catastrophe may not be taken into account, and dynamic changes over time may not be considered. Utilitarian economics does not account for fairness, so even if society were better off overall, some individuals may lose out. Many people distrust economic approaches, since they are so strongly associated with industry groups. Finally, economists assign prices based on what people can or do

Figure 2.13 Agriculture, timber, and scenic beauty at Coruña, Spain. The value of the environment derives from many factors that constitute a complex, interrelated system. These factors include aesthetics, recreation, agriculture, real estate, and ecosystem services. It can be difficult or impossible to assess the monetary value of all possible options for an area such as this one. (Juan Carlos Muñoz/Age Fotostock America)

pay for things. This means that the knowledge that a pristine stream or songbird exists will not have much impact in an economic assessment. Additionally, economics dictates that if it is inefficient to save a species, we should not do it. This idea clashes with many religious and individual beliefs.

Natural Resources, the Environment, and the National Income Accounts

Because much of our economic well-being flows from natural, rather than human-made, assets, we should include the use and misuse of natural resources and the environment in the national income accounts. **National income accounts** represent the total income of a nation for a given year. Two measures used in national income accounting are *gross domestic product (GDP)* and *net domestic product (NDP)*. Both GDP and NDP provide estimates of national economic performance used to make important policy decisions.

Unfortunately, current national income accounting practices are misleading and incomplete because they do not incorporate environmental factors. At least two important conceptual problems affect the way the national income accounts currently handle the economic use of natural resources and the environment. These problems involve costs and benefits of pollution control and depletion of **natural capital** (**Figure 2.14**).

natural capital: All of Earth's resources and processes that sustain living organisms, including humans.

Natural Resource Depletion   If a firm produces some product (output) but in the process wears out a portion of its plant and equipment, the firm's output is counted as part of GDP, but the depreciation of capital is subtracted in the calculation of NDP. Thus, NDP is a measure of the net production of the economy after a deduction for used-up capital. In contrast, when an oil company drains oil from an underground field, the value of the oil produced is counted as part of the nation's GDP, but no offsetting deduction to NDP is made to account for the nonrenewable resources used up.

The Cost and Benefits of Pollution Control at the National Level   Imagine that a company has the following choices: It can produce $100 million worth of output and, in the process, pollute the local river by dumping its wastes. Alternatively, by properly disposing of its wastes, it can avoid polluting but will only get $90 million of output. Under current national income accounting rules, if the firm chooses to pollute, its contribution to GDP is larger ($100 million rather than $90 million). National income accounts attach no explicit value to a clean river. In an ideal accounting system, the economic cost of environmental degradation is subtracted in the calcula-

Figure 2.14   Natural capital and the environment. Goods and services (products) and money (to pay for the products) flow between businesses (production) and consumers (consumption). Economies depend on natural capital to provide sources for raw materials and sinks for waste products.

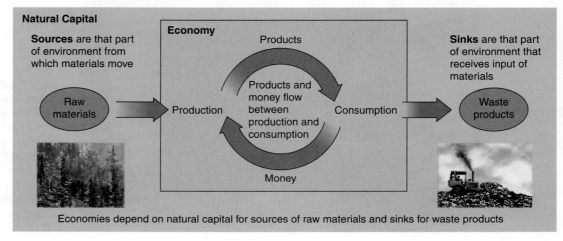

Economies depend on natural capital for sources of raw materials and sinks for waste products

tion of a firm's contribution to GDP, and activities that improve the environment—because they provide real economic benefits—are added to GDP. Estimates of environmental damage should be subtracted from GDP.

Discussing the national income accounting implications of resource depletion and pollution may seem to trivialize these important problems. However, because GDP and related statistics are used continually in policy analyses, abstract questions of measurement may turn out to have significant effects. Economic development experts have expressed concern that some poor countries, in attempting to raise their GDPs as quickly as possible, overexploit their natural resources and impair the environment. If "hidden" resource and environmental costs were explicitly incorporated into official measures of economic growth, policies that harm the environment might be modified. Similarly, in industrialized countries, political debates about the environment have sometimes emphasized the impact on GDP of proposed pollution-control measures, rather than the impact on overall economic welfare. Better accounting for environmental quality might refocus these debates on the more relevant question of whether, for any given environmental proposal, the benefits (economic and noneconomic) exceed the cost (**Figure 2.15**). One option is to replace GDP and NDP with a more comprehensive measure of national income accounting that includes estimates of both natural resource depletion and the environmental costs of economic activities.

One tool that may be used alongside the GDP is the **Environmental Performance Index (EPI),** which assesses a country's commitment to environmental and resource management. In the first assessment, completed in 2006, the United States ranked 28th out of 133 countries assessed, which is below the rankings of most Western European countries. Most African countries are ranked in the bottom half. **Table 2.2** presents the EPI rankings of some selected countries.

We have examined the roles of government and economics in addressing environmental concerns, particularly in the United States. Now let us consider environmental destruction in Central and Eastern Europe, an environmental problem closely connected to both governmental and economic policies.

"But why <u>save</u> the environment when it could be <u>invested</u>!?!"

Figure 2.15   An economic view of the environment? Does this cartoon fairly depict the economist's view of the environment? Why or why not? (Patrick Hardin/Cartoon Stock)

Table 2.2 Environmental Performance Index Scores And Rankings For Select Countries.

Country	EPI	Ranking (out of 133)
New Zealand	88.0	1
Sweden	87.8	2
Finland	87.0	3
Czech Rep.	86.0	4
United Kingdom	85.6	5
Canada	84.0	8
Japan	81.9	14
Costa Rica	81.6	15
Australia	80.1	20
United States	78.5	28
Mexico	64.8	66
China	56.2	94
India	47.7	118
Mali	33.9	130
Mauritania	32.0	131
Chad	30.5	132
Niger	25.7	133

Canada ranks highest among North American countries. The lowest-ranked countries are in Sub-Saharan Africa.
Source: www.yale.edu/epi/.

CASE IN POINT

Environmental Problems in Central and Eastern Europe

The fall of the Soviet Union and communist governments in Central and Eastern Europe during the late 1980s revealed a grim legacy of environmental destruction (**Figure 2.16**). Over the previous decades, the value of natural capital had been ignored. Water was so poisoned from raw sewage and chemicals that it could not be used for industrial purposes, let alone for drinking. Unidentified chemicals leaked out of dumpsites into the surrounding soil and water, while nearby, fruits and vegetables were grown in chemical-laced soil. Power plants emitted soot and sulfur dioxide into the air, producing a persistent chemical haze. Buildings and statues eroded, and entire forests died because of air pollution and acid rain. Crop yields fell despite intensive use of chemical pesticides and fertilizers. One of the most polluted areas in the world was the "Black Triangle," which consisted of the bordering regions of the former East Germany, northern Czech Republic, and southwest Poland.

Many Central and Eastern Europeans suffered from asthma, emphysema, chronic bronchitis, and other respiratory diseases as a result of breathing the filthy, acrid air. By the time most Polish children were 10 years old, they suffered from chronic respiratory diseases or heart problems. The levels of cancer, miscarriages, and birth defects were extremely high. Life expectancies are still lower in Eastern Europe than in other industrialized nations; in 2006 the average Eastern European lived to age 69, which was 10 years less than the average Western European.

The economic assumption behind communism was one of high production and economic self-sufficiency, regardless of damages to environment and natural resources, and so pollution in communist-controlled Europe went largely unchecked. Meeting industrial production quotas always took precedence over environmental concerns, even though production was not carried out for profit. It was acceptable to pollute because clean air, water, and soil were granted no economic value. The governments supported heavy industry—power plants, chemicals, metallurgy, and large machinery—at the expense of the more environmentally benign service industries. As a result, Central and Eastern Europe were over-industrialized with old plants that lacked pollution abatement equipment.

Communism did little to encourage resource conservation, which can curb pollution. Neither industries nor individuals had incentive to conserve energy because energy subsidies and lack of competition allowed power plants to provide energy at prices far below its actual cost.

Figure 2.16 Pollution problems in former communist countries. This coal mine on the Mius River in Ukraine produces pollutants that seep through piles of waste into soil, surface water, and groundwater. Note the scummy layer of chemical wastes covering the river water. Thousands of these sites exist in former communist countries, the result of rapid expansion of industrialization without regard for the environment. It will take decades for these sites to be cleaned up. Photographed in the late 1990s. (TASS/Sovfoto/Eastfoto)

ENVIRONEWS

How Green Is Your Campus?

No longer restricting their environmental education to the classroom, students at many colleges and universities are identifying environmental problems on their campuses, then mapping out and implementing solutions. The tremendous success of many of these efforts has required understanding the campus as a system, and demands the cooperation of faculty, administrators, staff, and students. Real-life environmental situations become educational opportunities. There are many notable examples including:

a. The Rebel Recycling program at the University of Nevada–Las Vegas, which accepts materials from off campus and collects reusable materials during dorm moveouts. The program was based on the senior thesis of an Environmental Studies Major who continues to run it.

b. The development of a "campus environmental audit" by students at the University of California–Los Angeles, now published and widely used by institutions across the United States.

c. Campus sustainability initiatives at dozens of campuses across the country.

d. The reduction of hazardous wastes in chemistry labs across the United States, a result of a student-faculty effort to develop "microscale" experiments at Bowdoin College in Maine.

Overall, student efforts nationally include recycling programs on 80% of campuses and a total annual savings of nearly $17 million from 23 campus conservation projects, ranging from transportation initiatives to energy and water conservation, reuse and recycling, and composting. Organizations such as the Campus Ecology program associated with the National Wildlife Federation, the Sierra Student Coalition, Focus the Nation, and the American Association of Sustainability in Higher Education, reflect an increasing level of environmental stewardship found at the nation's institutions of higher learning.

See www.nwf.org/campusecology/index.cfm or www.aashe.org.

More recently, while switching from communism to democracy with market economies, Central and Eastern European governments have faced the overwhelming responsibility of improving the environment. Although no political system in transition will make the environment its first priority, several have formulated environmental policies based on the experiences of the United States and Western Europe over the past several decades. However, experts predict it will take decades to clean up the pollution legacy of communism. How much will it cost? The figures are staggering. For example, improving the environment in what was formerly East Germany will cost up to $300 billion. This represents the loss of natural capital during the 1960s, 1970s, and 1980s. Eastern Europeans today are paying the costs of their parents' environmental neglect.

The environment of countries in the former Soviet bloc is slowly improving. Some, such as Hungary, Poland, and the Czech Republic, have been relatively successful in moving toward market economies. They generate enough money to invest in environmental cleanup. The Czech Republic ranks fourth in the EPI 2006 assessment, well ahead of the United States. Approximately 1.5% of the Czech Republic's GDP is invested in environmental protection. Air pollution (sulfur dioxide, nitrogen oxide, and particulates) is declining despite the Czech Republic's increased economic growth.

In other countries, such as Romania (which is ranked 90th in the EPI Assessment) and Bulgaria, the transition to a market economy has not been as smooth. Economic recovery has been slow in these countries, and severe budgetary problems have forced the environment to take a backseat to political and economic reform. ■

REVIEW

1. What is the marginal cost of pollution? the marginal cost of pollution abatement? the optimum amount of pollution?

2. What do economists mean by "efficient" regulation?

3. What is natural capital, and why is it often excluded from national income accounts?

Figure 2.17   Tomorrow's generation. The choices we make today will determine whether future generations will inherit a sustainable world. (Purestock/Superstock)

Environmental ethics: A field of applied ethics that considers the moral basis of environmental responsibility and the appropriate extent of this responsibility.

environmental worldview: A worldview that helps us make sense of how the environment works, our place in the environment, and right and wrong environmental behaviors.

Western worldview: An understanding of our place in the world based on human superiority and dominance over nature, the unrestricted use of natural resources, and increased economic growth to manage an expanding industrial base.

deep ecology worldview: An understanding of our place in the world based on harmony with nature, a spiritual respect for life, and the belief that humans and all other species have an equal worth.

 ## Environmental Ethics, Values, and Worldviews

LEARNING OBJECTIVES

- Define *environmental ethics.*
- Define *environmental worldview* and discuss distinguishing aspects of the Western and deep ecology worldviews.

We now shift our attention to the worldviews of different individuals and societies and how those worldviews affect our ability to understand and solve sustainability problems. **Ethics** is the branch of philosophy that is derived through the logical application of human values. These **values** are the principles that an individual or society considers important or worthwhile. Values are not static entities but change as societal, cultural, political, and economic priorities change. Ethics helps us determine which forms of conduct are morally acceptable or unacceptable, right or wrong. Ethics plays a role in whatever types of human activities involve intelligent judgment and voluntary action. When values conflict, ethics helps us choose which value is better, or worthier, than other values.

Environmental ethics examines moral values to determine how humans should relate to the natural environment. Environmental ethicists consider such questions as what role we should play in determining the fate of Earth's resources, including other species. Or how do we develop an environmental ethic that is acceptable in the short term for us as individuals but also in the long term for our species and the planet? These questions and others like them are difficult intellectual issues that involve political, economic, societal, and individual tradeoffs.

Environmental ethics considers not only the rights of people living today, both individually and collectively, but also the rights of future generations (**Figure 2.17**). This aspect of environmental ethics is critical because the impacts of today's activities and technologies are changing the environment. In some cases, these impacts may be felt for hundreds or even thousands of years. Addressing issues of environmental ethics puts us in a better position to use science, government policies, and economics for long-term environmental sustainability.

Human-Centered and Life-Centered Worldviews

Each of us has a particular **worldview**—a commonly shared perspective based on a collection of our basic values that helps us make sense of the world, understand our place and purpose in it, and determine right and wrong behaviors. These worldviews lead to behaviors and lifestyles that may or may not be compatible with environmental sustainability.

There are many worldviews; some worldviews share certain fundamental beliefs, whereas others are mutually exclusive. A worldview that is considered ethical in one society may be considered irresponsible or even sacrilegious in another. The following are two extreme, opposing **environmental worldviews**: The Western worldview and the deep ecology worldview. These two worldviews, admittedly broad generalizations, are near opposite ends of a spectrum of worldviews relevant to global sustainability problems, and each approaches environmental responsibility in a radically different way.

The traditional **Western worldview**, also known as the *expansionist worldview*, is *anthropocentric* (human-centered) and utilitarian. This perspective mirrors the beliefs of the 18th-century *frontier attitude*, a desire to conquer and exploit nature as quickly as possible. The Western worldview also advocates the inherent rights of individuals, accumulation of wealth, and unlimited consumption of goods and services to provide material comforts. According to the Western worldview, humans have a primary obligation to humans and are therefore responsible for managing natural resources to benefit human society. Thus, any concerns about the environment are derived from human interests.

The **deep ecology worldview** is a diverse set of viewpoints that dates from the 1970s and is based on the work of **Arne Naess,** a Norwegian philosopher, and others, including ecologist **Bill Devall** and philosopher **George Sessions.** The principles of deep ecology, as expressed by Arne Naess in *Ecology, Community, and Lifestyle,* include the following:

1. All life has intrinsic value. The value of nonhuman life forms is independent of the usefulness they may have for narrow human purposes.
2. Richness and diversity of life forms contribute to the flourishing of human and nonhuman life on Earth.
3. Humans have no right to reduce this richness and diversity except to satisfy vital needs.
4. Present human interference with the nonhuman world is excessive, and the situation is rapidly worsening.
5. The flourishing of human life and cultures is compatible with a substantial decrease in the human population. The flourishing of nonhuman life requires such a decrease.
6. Significant improvement of life conditions requires changes in economic, technological, and ideological structures.
7. The ideological change is mainly that of appreciating life quality rather than adhering to a high standard of living.
8. Those who subscribe to the foregoing points have an obligation to participate in the attempt to implement the necessary changes.

Compared to the Western worldview, the deep ecology worldview represents a radical shift in how humans relate themselves to the environment. The deep ecology worldview stresses that all forms of life have the right to exist, and humans are not different or separate from other organisms. Humans have an obligation to themselves and to the environment. The deep ecology worldview advocates sharply curbing human population growth. It does not favor returning to a society free of today's technological advances but instead proposes a significant rethinking of our use of current technologies and alternatives. It asks individuals and societies to share an inner spirituality connected to the natural world.

Ted Perry, a film scriptwriter, eloquently summarized the worldview that deep ecologists say must be embraced to help solve the many serious environmental problems of today. His words are based on notes of a speech reportedly delivered in 1854 by Chief Sealth, chief of the Suquamish tribe in the Pacific Northwest, in response to President Franklin Pierce's offer to buy their land and provide them with a reservation:

> *How can you buy [the land,] or [we] sell the sky ? This idea is strange to us. If we do not own the freshness of the air and the sparkle of the water, how can you buy them? This we know—the Earth does not belong to man, man belongs to the Earth. All things are connected like the blood which unites one family. Whatever befalls the Earth befalls the sons of the Earth. Man does not weave the web of life, he is merely a strand in it. Whatever he does to the web, he does to himself.*

Most people today do not fully embrace either the Western worldview or the deep ecology worldview. The Western worldview emphasizes the importance of humans as the overriding concern in the grand scheme of things. In contrast, the deep ecology worldview is *biocentric* (life-centered) and views humans as one species among others. The planet's natural resources could not support its more than 6 billion humans if each consumed the high level of goods and services sanctioned by the Western worldview. On the other hand, if all humans adhered completely to the tenets of deep ecology, it would entail giving up some of the material comforts and benefits of modern technology.

The world as envisioned by the deep ecology worldview could only support a fraction of the existing human population (recall the discussion of ecological footprints in Chapter 1). These worldviews, although not practical for widespread adoption, are useful to keep in mind as you examine various environmental issues in later chapters. In the meantime, think about your own worldview and discuss it with others. Listen carefully to their worldviews, which will probably be different from your own. Thinking leads to actions, and actions lead to consequences. What are the short-term and long-term consequences of your particular worldview? We must develop and incorporate a long-lasting, environmentally sensitive worldview into our culture if the environment is to be sustainable for us, for other living organisms, and for future generations.

REVIEW

1. What is environmental ethics?
2. What are worldviews? How do Western and deep ecology worldviews differ?

 REVIEW OF LEARNING OBJECTIVES WITH KEY TERMS

• Briefly outline the environmental history of the United States.

The first two centuries of U.S. history were a time of widespread environmental destruction. During the 1700s and early 1800s, most Americans had a desire to conquer and exploit nature as quickly as possible. During the 19th century, naturalists became concerned about conserving natural resources. The earliest conservation legislation revolved around protecting land—forests, parks, and monuments. By the late 20th century, environmental awareness had become a pervasive popular movement.

• Describe the environmental contributions of the following people: George Perkins Marsh, Theodore Roosevelt, Gifford Pinchot, John Muir, Aldo Leopold, Wallace Stegner, Rachel Carson, and Paul Ehrlich.

George Perkins Marsh wrote about humans as agents of global environmental change. **Theodore Roosevelt** appointed **Gifford Pinchot** as the first head of the U.S. Forest Service. Pinchot supported expanding the nation's forest reserves and managing forests scientifically. The Yosemite and Sequoia National Parks were established, largely in response to the efforts of naturalist **John Muir**. In *A Sand County Almanac*, **Aldo Leopold** wrote about humanity's relationship with nature. **Wallace Stegner** helped create support for passage of the Wilderness Act of 1964. **Rachel Carson** published *Silent Spring*, alerting the public to the dangers of uncontrolled pesticide use. **Paul Ehrlich** wrote *The Population Bomb*, which raised the public's awareness of the dangers of overpopulation.

• Distinguish between utilitarian conservationists and biocentric preservationists.

A **utilitarian conservationist** is a person who values natural resources because of their usefulness for practical purposes but uses them sensibly and carefully. A **biocentric preservationist** is a person who believes in protecting nature because all forms of life deserve respect and consideration.

• Explain why the National Environmental Policy Act is the cornerstone of U.S. Environmental Law.

The **National Environmental Policy Act (NEPA),** passed in 1970, stated that the federal government must consider the environmental impact of a proposed federal action, such as financing a highway or constructing a dam. NEPA established the **Council on Environmental Quality** to monitor required **environmental impact statements (EISs)** and report directly to the president.

• Relate how environmental impact statements provide for powerful protection of the environment.

By requiring EISs that are open to public scrutiny, NEPA initiated serious environmental protection in the United States. NEPA allows citizen suits, in which private citizens take violators, whether they are private industries or government-owned facilities, to court for noncompliance.

• Explain why industry groups favor regulatory reform policies, while some environmental advocates focus on environmental justice.

Regulatory reform includes a number of changes that might eliminate some of the least cost-effective regulations. Industry groups expect that dropping some of the most expensive regulations would save them considerable amounts of money. Environmental advocates are concerned that some of those may be regulations that protect already at-risk populations, which is an environmental justice issue.

• Distinguish among the following economic terms: marginal cost of pollution, marginal cost of abatement, and optimum amount of pollution.

From an economic point of view, the appropriate amount of pollution is a trade-off between harm to the environment and inhibition of development. The **marginal cost of pollution** is the added cost for all present and future members of society of an additional unit of pollution. The **marginal cost of abatement** is the added cost for all present and future members of society of reducing one unit of a given type of pollution. Economists think that the use of resources for pollution abatement should increase only until the cost of abatement equals the cost of the pollution damage. This results in the **optimum amount of pollution**—the amount of pollution that is economically efficient.

• Explain why economists prefer efficient solutions to environmental problems.

Efficient solutions are those where there is the greatest total social benefit. Solutions that are inefficient will spend more on abatement than the pollution costs, or will spend less on abatement than pollution reductions are worth. Economic solutions assume that **rational actors** try to maximize **utility** given limited resources.

• Describe various approaches to pollution control, including command and control regulation, incentive-based regulation , and cost-effectiveness analysis.

To control pollution, governments often use **command and control regulations,** which are pollution-control laws that require specific technologies. **Incentive-based regulations** are pollution-control laws that work by establishing emission targets and providing industries with incentives to reduce emissions. **Cost-effectiveness analysis** is an economic tool used to estimate costs associated with achieving some goal, such as saving a life.

• Give two reasons why the national income accounts are incomplete estimates of national economic performance.

National income accounts are a measure of the total income of a nation's goods and services for a given year. An **external cost** is a harmful environmental or social cost that is borne by people not directly involved in buying or selling a product. Currently, national income accounting does not include estimates of such external costs as depletion of **natural capital** and the environmental cost of economic activities. Planned economies, such as the former Soviet republics, did not treat the environment as having value, which led to substantial costs to current and future generations.

• Define *environmental ethics.*

Environmental ethics is a field of applied ethics that considers the moral basis of environmental responsibility and the appropriate extent of this responsibility. Environmental ethicists consider how humans should relate to the natural environment.

• Define *environmental worldview* and discuss distinguishing aspects of the Western and deep ecology worldviews.

An **environmental worldview** is a worldview that helps us make sense of how the environment works, our place in the environment, and right and wrong environmental behaviors. The **Western worldview** is an understanding of our place in the world based on human superiority and dominance over nature, the unrestricted use of natural resources, and increased economic growth to manage an expanding industrial base. The **deep ecology worldview** is an understanding of our place in the world based on harmony with nature, a spiritual respect for life, and the belief that humans and all other species have an equal worth.

Thinking About the Environment

1. Briefly describe each of these aspects of U.S. environmental history: protection of forests; establishment and protection of national parks and monuments; conservation in the mid-20th century; and the environmental movement of the late 20th century.

2. Describe the environmental contributions of two of the following: George Perkins Marsh, Theodore Roosevelt, Gifford Pinchot, John Muir, Aldo Leopold, Wallace Stegner, Rachel Carson, and Paul Ehrlich.

3. If you were a member of Congress, what legislation would you introduce to deal with each of these problems?

    a. Toxic materials from a major sanitary landfill are polluting some rural drinking water wells.

    b. Acid rain from a coal-burning power plant in a nearby state is harming the trees in your state. Loggers and foresters are upset.

    c. There is a high incidence of cancer in the area of your state where heavy industry is concentrated.

4. Would you expect environmental advocates to oppose the concept of efficiency? Why or why not?

5. Based on what you have learned in this chapter, do you think the economy is part of the environment, or is the environment part of the economy? Explain your answer.

6. Describe a case where command and control regulation would work better than incentive-based regulation.

7. Explain how environmental taxes can be revenue neutral.

8. Can economic approaches to environmental management adequately account for the complex interactive systems that make up the environment? Why or why not?

9. The Environmental Performance Index has detractors among both environmentalists and industry groups. Based on what you have learned in this chapter, suggest reasons for each.

10. Graph the marginal cost of pollution, the marginal cost of abatement, and the optimum amount of pollution.

11. Describe how environmental destruction in formerly communist countries relates to natural capital.

12. Determine whether each of these statements reflects the Western worldview, the deep ecology worldview, or both:

    a. Species exist for humans to use.

    b. All organisms, humans included, are interconnected and interdependent.

    c. There is a unity between humans and nature.

    d. Humans are a superior species capable of dominating other organisms.

    e. Humans should protect the environment.

    f. Nature should be used, not preserved.

    g. Economic growth will help Earth manage an expanding human population.

    h. Humans have the right to modify the environment to benefit society.

    i. All forms of life are intrinsically valuable and have the right to exist.

(© Steve Greenberg)

13. What is this cartoon trying to say? When do you think this cartoon was published?

Quantitative question relating to this chapter are on our Web site.

Take a Stand

Visit our Web site at www.wiley.com/college/raven (select Chapter 2 from the Table of Contents) for links to more information about the controversies surrounding the old-growth forests in the Pacific Northwest. Consider the opposing views of loggers and other rural people and environmentalists, and then debate the issues with your classmates. You will find tools to help you organize your research, analyze the data, think critically about the issues, and construct a well-considered argument. *Take a*

Stand activities can be done individually or as a team, as oral presentations, written exercises, or as Web-based (e-mail) assignments.

    Additional online materials relating to this chapter, including a Student Testing Section with study aids and self-tests, Environmental News, Activity Links, Environmental Investigations, and more, are also on our Web site.

Ecosystems and Energy

Cordgrass in a Chesapeake Bay salt marsh.
(Robert Noonan/Photo Researchers, Inc.)

The environment often contains a remarkable assortment of organisms that interact with one another and are interdependent in a variety of ways. Consider for a moment a salt marsh in the Chesapeake Bay on the East Coast of the United States. This bay is an **estuary**, a semi-enclosed body of water found where fresh water from a river drains into the ocean. The Chesapeake is one of the world's richest estuaries in terms of the amount of seafood harvested from it.

Estuaries are complex *systems* under the influence of tides and gradually change from unsalty fresh water to salty ocean water. In the Chesapeake Bay, this change results in three distinct communities: freshwater marshes at the head of the bay, brackish (moderately salty) marshes in the middle bay region, and salt marshes on the ocean side of the bay.

A Chesapeake Bay salt marsh consists of flooded meadows of cordgrass (*Spartina;* see photo). Few other plants are found because high salinity and twice-daily tidal inundations produce a challenging environment to which only a few plants have adapted. Nutrients such as nitrates and phosphates, much of them from treated sewage and agriculture,

drain into the marsh from the land and promote rapid growth of both cordgrass and microscopic algae (photosynthetic aquatic organisms) suspended in the water. These organisms are eaten directly by some animals, and when they die, their remains provide food for other salt marsh inhabitants.

A visitor to a salt marsh would observe two major kinds of animal life: insects and birds. Insects, particularly mosquitoes and horseflies, number in the millions. Birds nesting in the salt marsh include seaside sparrows, laughing gulls, and clapper rails. However, the salt marsh has numerous other organisms. Shrimps, lobsters, crabs, barnacles, worms, clams, and snails seek refuge in the water surrounding the cordgrass. Here these animals eat, hide from predators to avoid being eaten, and reproduce.

Chesapeake Bay marshes are an important nursery for numerous marine fishes—spotted sea trout, Atlantic croaker, striped bass, and bluefish, to name just a few. These fishes typically spawn (reproduce) in the open ocean, and the young then enter the estuary, where they grow into juveniles.

Almost no amphibians inhabit salt marshes— the salty water dries out their skin—but a few reptiles, such as the northern diamondback terrapin (a semi-aquatic turtle), have adapted.

The terrapin spends its time basking in the sun or swimming in the water searching for food—snails, crabs, worms, insects, and fish. Although a variety of snakes live in the dry areas adjacent to salt marshes, only the northern water snake, which preys on fish, is adapted to salty water.

The meadow vole is a small rodent that lives in the salt marsh. It constructs a nest of cordgrass on the ground above the high-tide zone. Meadow voles are excellent swimmers and scamper about the salt marsh day and night. Their diet consists mainly of insects and the leaves, stems, and roots of cordgrass.

Add to all these visible plants and animals the unseen microscopic world of the salt marsh, which includes countless numbers of protozoa, fungi, and bacteria, and you begin to appreciate the complexity of a salt marsh. In this chapter you begin your study of ecology, which is central to environmental science. Throughout the chapter, which focuses on energy flow, you will encounter many examples from the salt marsh community.

World View and Closer to You...
Real news footage relating to ecosystems and energy around the world and in your own geographic region is available at www.wiley.com/college/raven/ by clicking on "World View and Closer to You."

What Is Ecology?

LEARNING OBJECTIVES

- Define *ecology*.
- Distinguish among the following ecological levels: population, community, ecosystem, landscape, and biosphere.

Ernst Haeckel, a 19th-century scientist, developed the concept of ecology and named it—*eco* from the Greek word for "house" and *logy* from the Greek word for "study." Thus, ecology literally means "the study of one's house." The environment—one's house—consists of two parts, the **biotic** (living) environment, which includes all organisms, and the **abiotic** (nonliving, or physical) surroundings, which include living space, temperature, sunlight, soil, wind, and precipitation.

The focus of ecology is local or global, specific or generalized, depending on what questions the scientist is trying to answer. One ecologist might determine the temperature or light requirements of a single oak, another might study all the organisms that live in a forest where the oak is found, and another might examine how nutrients flow between the forest and surrounding communities.

Ecology is the broadest field within the biological sciences, and it is linked to every other biological discipline. The universality of ecology links subjects that are not traditionally part of biology. Geology and earth science are extremely important to ecology, especially when ecologists examine the physical environment of planet Earth. Chemistry and physics are also important; in this chapter, for example, you are studying chemistry when you read about photosynthesis and physics when you read about the laws of thermodynamics. Humans are biological organisms, and our activities have a bearing on ecology. Even economics and politics have profound ecological implications, as was presented in Chapter 2.

How does the field of ecology fit into the organization of the biological world? As you may know, one of the characteristics of life is its high degree of organization (**Figure 3.1**). Atoms are organized into molecules, which in turn are organized into cells. In multicellular organisms, cells are organized into tissues, tissues into organs such as a bone or stomach, organs into body systems such as the nervous system and digestive system, and body systems into individual organisms such as dogs and ferns.

Ecologists are most interested in the levels of biological organization that include or are above the level of the individual organism. Individuals of the same **species** occur in **populations**. A population ecologist might study a population of polar bears or a population of marsh grass. Populations, particularly human populations, are so important in environmental science that we devote three chapters, Chapters 8, 9, and 10, to their study. Species are considered further in Chapter 17.

Populations are organized into **communities**. Ecologists characterize communities by the number and kinds of species that live there, along with their relationships with one another. A community ecologist might study how organisms interact with one another—including feeding relationships (who eats whom)—in an alpine meadow community or in a coral reef community (**Figure 3.2**).

Ecosystem is a more inclusive term than *community*. An **ecosystem** includes all the biotic interactions of a community as well as the interactions between organisms and their abiotic environment. Like other *systems*, an ecosystem consists of multiple interacting parts that form a unified whole. An ecosystem is a system in which all of the biological, physical, and chemical components of an area form a complex, interacting network of energy flow and materials cycling. An ecosystem ecologist might examine how energy, nutrients, organic (carbon-containing) materials, and water affect the organisms living in a desert ecosystem or a coastal bay ecosystem.

The ultimate goal of ecosystem ecologists is to understand how ecosystems function. This is not a simple task, but it is important because ecosystem processes collectively regulate global cycles of water, carbon, nitrogen, phosphorus, and sulfur essential to the survival of humans and all other organisms. As humans increasingly alter ecosystems for

ecology: The study of systems that include interactions among organisms and between organisms and their abiotic environment.

species: A group of similar organisms whose members freely interbreed with one another in the wild to produce fertile offspring; members of one species generally do not interbreed with other species of organisms.

population: A group of organisms of the same species that live in the same area at the same time.

community: A natural association that consists of all the populations of different species that live and interact within an area at the same time.

ecosystem: A community and its physical environment.

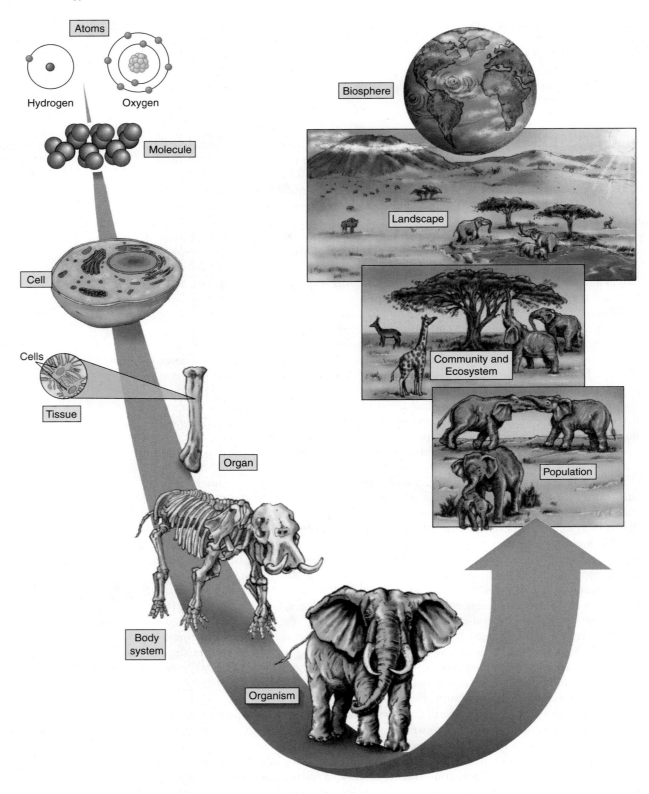

Figure 3.1 Levels of biological organization. Starting at the simplest level, atoms are organized into molecules, which are organized into cells. Cells are organized into tissues, tissues into organs, organs into body systems, and body systems into individual multicellular organisms. A group of individuals of the same species is a population. Populations of different species interact to form communities. A community and its abiotic environment are an ecosystem, whereas a region with several ecosystems is a landscape. The layer of Earth containing all living organisms comprises the biosphere.

Figure 3.2 Coral reef community. Coral reef communities have the greatest number of species and are the most complex aquatic community. A coral reef community in the Indian Ocean off the coast of Maldives is shown. Coral reefs worldwide are threatened by global climate change and ocean acidification (see Chapter 21). (Denise Tackett/Tom Stack & Associates)

their own uses, the natural functioning of ecosystems is changed, and we must determine whether these changes will affect the sustainability of our life-support system.

Landscape ecology is a subdiscipline of ecology that studies ecological processes that operate over large areas. Landscape ecologists examine the connections among ecosystems found in a particular region. Consider a simple landscape consisting of a forest ecosystem located adjacent to a pond ecosystem. One connection between these two ecosystems might be great blue herons, which eat fish, frogs, insects, crustaceans, and snakes along the shallow water of the pond but often build nests and raise their young in the secluded treetops of the nearby forest. Landscapes, then, are based on larger land areas that include several ecosystems.

The organisms of the biosphere—Earth's communities, ecosystems, and landscapes—depend on one another and on the other realms of Earth's physical environment: the atmosphere, hydrosphere, and lithosphere (**Figure 3.3**). The **atmosphere** is the gaseous envelope surrounding Earth; the **hydrosphere** is Earth's supply of water—liquid and frozen, fresh and salty; and the **lithosphere** is the soil and rock of Earth's crust. Ecologists who study the biosphere examine global interrelationships among Earth's atmosphere, land, water, and organisms.

The biosphere is filled with life. Where do these organisms get the energy to live? And how do they harness this energy? Let's examine the importance of energy to organisms, which survive only as long as the environment continuously supplies them with energy. You will revisit the importance of energy as it relates to human endeavors in many chapters throughout this text.

landscape: A region that includes several interacting ecosystems.

biosphere: The parts of Earth's atmosphere, ocean, land surface, and soil that contain all living organisms.

Figure 3.3 Earth's four realms.

(a) Earth's four realms, represented as intersecting circles, are a system of interrelated parts.

(b) In this scene, photographed at the northern tip of Palawan Island, Philippines, the atmosphere contains cumulus clouds, which indicate warm, moist air. The jagged spires of rock, formed from volcanic lava flows that have eroded over time, represent the lithosphere. The shallow water of the bay represents the hydrosphere. The biosphere includes the green vegetation and human settlement along the shore as well as the coral reefs visible as darker areas in the bay. (Jerry Alexander/Stone/Getty Images)

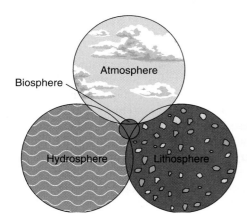

REVIEW

1. What is ecology?

2. What is the difference between a community and an ecosystem? Between an ecosystem and a landscape?

The Energy of Life

LEARNING OBJECTIVES

- Define *energy* and explain how it is related to work and to heat.
- Use examples to contrast potential energy and kinetic energy.
- State the first and second laws of thermodynamics, and discuss the implications of these laws as they relate to organisms.
- Write summary reactions for photosynthesis and cellular respiration, and contrast these two biological processes.

Energy is the capacity or ability to do work. In organisms, any biological work, such as growing, moving, reproducing, and maintaining and repairing damaged tissues, requires energy. Energy exists in several forms: chemical, radiant, thermal, mechanical, nuclear, and electrical. *Chemical energy* is energy stored in the bonds of molecules; for example, food contains chemical energy. *Radiant energy* is energy, such as radio waves, visible light, and X-rays, which is transmitted as electromagnetic waves. *Solar energy* is radiant energy from the sun; it includes ultraviolet radiation, visible light, and infrared radiation. *Thermal energy* is heat that flows from an object with a higher temperature (the heat source) to an object with a lower temperature (the heat sink). *Mechanical energy* is energy in the movement of matter. Atomic nuclei can be converted into *nuclear energy*. *Electrical energy* is energy that flows as charged particles. You will encounter these forms of energy throughout the text.

Biologists generally express energy in units of work (*kilojoules, kJ*) or units of heat (*kilocalories, kcal*). One kilocalorie, the energy required to raise the temperature of 1 kg of water by 1°C, equals 4.184 kJ. The kcal is the unit that nutritonists use to express the energy level of the foods we eat.

Energy can exist as stored energy—called **potential energy**—or as **kinetic energy,** the energy of motion (**Figure 3.4**). Think of potential energy as an arrow on a drawn bow, which equals the work the archer did when drawing the bow to its position. When the string is released, the bow's potential energy is converted to the arrow's kinetic energy of motion. Similarly, the cordgrass that a meadow vole eats has chemical potential energy in the bonds of its molecules; as molecular bonds are broken, this energy is converted to kinetic energy and heat as the meadow vole swims in the salt marsh. Thus, energy changes from one form to another.

The study of energy and its transformations is called **thermodynamics.** When considering thermodynamics, scientists use the word *system* to refer to a group of atoms, molecules, or objects being studied.[1] The rest of the universe other than the system is known as the *surroundings.* A **closed system** is self-contained and isolated—that is, it does not exchange energy with its surroundings (**Figure 3.5**). Closed systems are very rare in nature. In contrast, an **open system** exhibits an exchange of energy with its surroundings. This text discusses many kinds of open systems. For example, a city is an open system with an input of energy (as well as food, water, and consumer goods). Outputs

Figure 3.4   Potential and kinetic energy. Potential energy is stored in the drawn bow and is converted to kinetic energy as the arrow speeds toward its target.

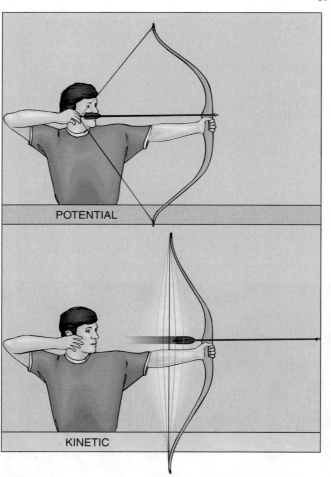

POTENTIAL

KINETIC

[1] The meaning of *system* in thermodynamics is different from the meaning of *system* in environmental science (a set of components that interact and function as a whole).

from a city system include energy (as well as manufactured goods, sewage, and solid waste). On a global scale, Earth is an open system dependent on a continual supply of energy from the sun.

Regardless of whether a system is open or closed, two laws about energy apply to all things in the universe: the first and second laws of thermodynamics.

The First Law of Thermodynamics

According to the first law of thermodynamics, an organism may absorb energy from its surroundings, or it may give up some energy into its surroundings, but the total energy content of the organism and its surroundings is always the same. As far as we know, the energy present in the universe at its formation, approximately 15 billion to 20 billion years ago, equals the amount of energy present in the universe today. This is all the energy that will ever be present in the universe. Similarly, the energy of any system and its surroundings is constant. A system may absorb energy from its surroundings, or it may give up some energy into its surroundings, but the total energy content of that system and its surroundings is always the same.

The first law of thermodynamics specifies that an organism cannot create the energy it requires to live. Instead, it must capture energy from the environment to use for biological work, a process involving the transformation of energy from one form to another. In photosynthesis, plants absorb the radiant energy of the sun and convert it into the chemical energy contained in the bonds of carbohydrate (sugar) molecules. Similarly, some of that chemical energy may later be transformed by an animal that eats the plant into the mechanical energy of muscle contraction, enabling it to walk, run, slither, fly, or swim (**Figure 3.6**).

The Second Law of Thermodynamics

As each energy transformation occurs, some energy is changed to heat that is released into the cooler surroundings. No other organism can ever reuse this energy for biological work; it is "lost" from the biological point of view. It is not really gone from a thermodynamic point of view because it still exists in the surrounding physical environment. The use of food to enable us to walk or run does not destroy the chemical energy once present in the food molecules. After we have performed the task of walking or running, the energy still exists in the surroundings as heat.

According to the second law of thermodynamics, the amount of usable energy available to do work in the universe decreases over time. The second law of thermodynamics is consistent with the first law; that is, the total amount of energy in the universe is not decreasing with time. However, the total amount of energy in the universe available to do work decreases over time.

Less usable energy is more diffuse, or disorganized. **Entropy** is a measure of this disorder or randomness; organized, usable energy has low entropy, whereas disorganized energy such as heat has high entropy. Entropy is continuously increasing in the universe in all natural processes. Billions of years from now, all energy may exist as heat uniformly distributed throughout the universe. If that happens, the universe as a closed system will cease to operate because no work will be possible. Everything will be at the same temperature, so there will be no way to convert the thermal energy of the universe into usable mechanical energy. Another way to explain the second law of thermodynamics, then, is that entropy, or disorder, in a system spontaneously increases over time. (The word *spontaneously* in this context means that entropy occurs naturally rather than that some external influence causes it.)

As a result of the second law of thermodynamics, no process requiring an energy conversion is ever 100% efficient because much of the energy is dispersed as heat, resulting in an increase in entropy. An automobile engine, which converts the chemical

Figure 3.5 Closed and open systems, with regard to energy.

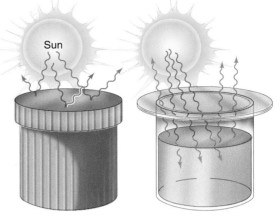

(a) **Closed system.** Energy is not exchanged between a closed system and its surroundings. A thermos bottle is an approximation of a closed system. Closed systems are very rare in nature.

(b) **Open system.** Energy is exchanged between an open system and its surroundings. Earth is an open system because it receives energy from the sun, and this energy eventually escapes Earth as it dissipates into space.

first law of thermodynamics: Energy cannot be created or destroyed, although it can change from one form to another.

second law of thermodynamics: When energy is converted from one form to another, some of it is degraded into heat, a less usable form that disperses into the environment.

Figure 3.6 Meadow vole. The chemical energy produced by photosynthesis and stored in stems and leaves of marsh vegetation is transferred to the meadow vole (*Microtus pennsylvanicus*) as it eats. (Andy Harmer/Photo Researchers, Inc.)

energy of gasoline to mechanical energy, is between 20% and 30% efficient. That is, only 20% to 30% of the original energy stored in the chemical bonds of the gasoline molecules is actually transformed into mechanical energy, or work. In our cells, energy use for metabolism is about 40% efficient, with the remaining energy given to the surroundings as heat.

Organisms are highly organized and at first glance appear to refute the second law of thermodynamics. As organisms grow and develop, they maintain a high level of order and do not become more disorganized. However, organisms maintain their degree of order over time only with the constant input of energy. That is why plants must photosynthesize and why animals must eat food.

Photosynthesis and Cellular Respiration

Photosynthesis is the biological process in which light energy from the sun is captured and transformed into the chemical energy of carbohydrate (sugar) molecules. Photosynthetic pigments such as *chlorophyll*, which gives plants their green color, absorb radiant energy. This energy is used to manufacture the carbohydrate glucose ($C_6H_{12}O_6$) from carbon dioxide (CO_2) and water (H_2O), with the liberation of oxygen (O_2).

Photosynthesis:

$$6CO_2 + 12H_2O + \text{radiant energy} \rightarrow C_6H_{12}O_6 + 6H_2O + 6O_2$$

The chemical equation for photosynthesis is read as follows: 6 molecules of carbon dioxide plus 12 molecules of water plus light energy are used to produce 1 molecule of glucose plus 6 molecules of water plus 6 molecules of oxygen. (See Appendix I for a review of basic chemistry.)

Plants, some bacteria, and algae perform photosynthesis, a process essential for almost all life. Photosynthesis provides these organisms with a ready supply of energy in carbohydrate molecules which they use as the need arises. The energy can also be transferred from one organism to another—for example, from plants to the organisms that eat plants. Oxygen, which many organisms require when they break down glucose or similar foods to obtain energy, is a by-product of photosynthesis.

The chemical energy that plants store in carbohydrates and other molecules is released within cells of plants, animals, or other organisms through **cellular respiration.** In *aerobic* cellular respiration, molecules such as glucose are broken down in the presence of oxygen and water into carbon dioxide and water, with the release of energy.

Aerobic cellular respiration:

$$C_6H_{12}O_6 + 6O_2 + 6H_2O \rightarrow 6CO_2 + 12H_2O + \text{energy}$$

Cellular respiration makes the chemical energy stored in glucose and other food molecules available to the cell for biological work, such as moving around, courting, and growing new cells and tissues. All organisms, including green plants, respire to obtain energy. Some organisms do not use oxygen for this process. *Anaerobic* bacteria that live in waterlogged soil, stagnant ponds, or animal intestines respire in the absence of oxygen.

CASE IN POINT

Life Without the Sun

The sun is the energy source for almost all ecosystems. A notable exception was discovered in 1977 when an oceanographic expedition aboard the submersible research craft *Alvin* studied the Galapagos Rift, a deep cleft in the ocean floor off the coast of Ecuador. The expedition revealed a series of **hydrothermal vents** on the floor of the deep ocean where seawater had penetrated and been heated by the hot rocks below. During its time within Earth, the water had been charged with inorganic compounds, including hydrogen sulfide (H_2S).

At the tremendous depth (greater than 2,500 m, or 8,200 ft) of the Galapagos Rift, there is no light for photosynthesis. But the hot springs support a rich ecosystem that contrasts with the surrounding "desert" of the deep-ocean floor. Giant, blood-red tube worms almost 3 m (10 ft) in length cluster in great numbers around the vents (**Figure 3.7**). Other animals around the hydrothermal vents include clams, crabs, barnacles, and mussels.

Scientists initially wondered what the ultimate source of energy for the species in this dark environment is. Most deep-sea ecosystems depend on the organic material that drifts down from surface waters; that is, they depend on energy derived from photosynthesis. But the Galapagos Rift ecosystem and other hydrothermal vent ecosystems are too densely clustered and too productive to depend on chance encounters with organic material from surface waters.

The base of the *food web* in these aquatic oases consists of certain bacteria that survive and multiply in water so hot (exceeding 200°C, or 392°F) that it would not remain in liquid form were it not under such extreme pressure. These bacteria function as producers, but they do not photosynthesize. Instead, they obtain energy and make carbohydrate molecules from inorganic raw materials by **chemosynthesis.** Chemosynthetic bacteria possess enzymes (organic catalysts) that cause the inorganic molecule hydrogen sulfide to react with oxygen, producing water and sulfur or sulfate. Such chemical reactions provide the energy to support these bacteria and other organisms in deep-ocean hydrothermal vents. Many of the Galapagos Rift animals consume the bacteria directly by filter feeding. Others, such as giant tubeworms, obtain energy from chemosynthetic bacteria that live symbiotically inside their bodies. ■

Figure 3.7 **Hydrothermal vent ecosystem.** Bacteria living in the tissues of these tube worms extract energy from hydrogen sulfide to manufacture organic compounds. These worms lack digestive systems and depend on the organic compounds the bacteria provide, along with materials filtered from the surrounding water. Also visible in the photograph are some filter-feeding clams (yellow) and a crab (white). (D. Foster, Science VU-WHOI/Visuals Unlimited)

REVIEW

1. Is water stored behind a dam an example of potential or kinetic energy? What would cause the water to convert to the other form of energy?

2. When coal is burned in a power plant, only 3% of the energy in the coal is converted into light in a light bulb. What happens to the other 97% of the energy? Explain your answer using the laws of thermodynamics.

3. Distinguish between photosynthesis and cellular respiration. Which organisms perform each process?

The Flow of Energy Through Ecosystems

LEARNING OBJECTIVES

- Define *energy flow, trophic level,* and *food web.*
- Summarize how energy flows through a food web, using producer, consumer, and decomposer in your explanation.
- Describe typical pyramids of numbers, biomass, and energy.
- Distinguish between gross primary productivity and net primary productivity, and discuss human impact on the latter.

With the exception of a few ecosystems such as hydrothermal vents, energy enters ecosystems as radiant energy (sunlight), some of which plants trap during photosynthesis. The energy, now in chemical form, is stored in the bonds of organic molecules such as glucose. To obtain energy, animals eat plants or eat animals that ate plants. All organisms—plants, animals, and microorganisms—respire to obtain some of the energy in organic molecules. When cellular respiration breaks these molecules apart, the energy becomes available for work such as repairing tissues, producing body heat, or reproducing. As the work is accomplished, the energy escapes the organism and dissipates into the environment as heat (recall the second law of thermodynamics). Ultimately, this heat radiates into space. Once an organism has used energy, it becomes unusable for all other organisms. The movement of energy just described is called **energy flow.**

energy flow: The passage of energy in a one-way direction through an ecosystem.

Figure 3.8 **Producers, consumers, and decomposers.** In photosynthesis, producers (green plants) use the energy from sunlight to make organic molecules from carbon dioxide and water. Consumers obtain energy when they eat producers or other consumers. For example, the caterpillar in the robin's beak ate plant leaves, and then the robin ate the caterpillar. Wastes and dead organic material from producers and consumers supply decomposers, such as bacteria and fungi, with energy. During every energy transaction, some energy is lost to biological systems as it disperses into the environment as heat.

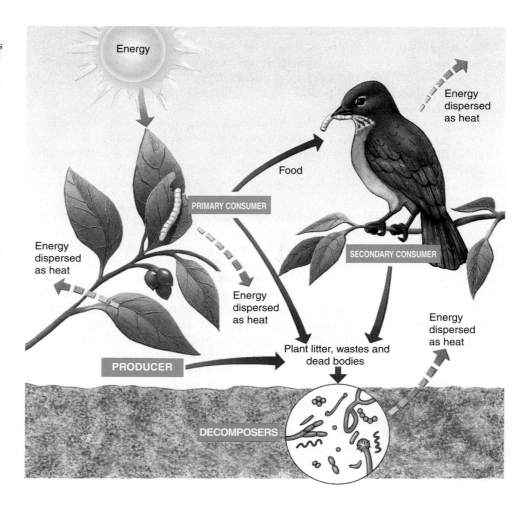

Producers, Consumers, and Decomposers

The organisms of an ecosystem are divided into three categories on the basis of how they obtain nourishment: producers, consumers, and decomposers (**Figure 3.8**). Virtually all ecosystems contain representatives of all three groups, which interact extensively, both directly and indirectly, with one another.

Producers, also called **autotrophs** (Greek *auto,* "self," and *tropho,* "nourishment"), manufacture organic molecules from simple inorganic substances, generally carbon dioxide and water, usually using the energy of sunlight. In other words, most producers perform the process of photosynthesis. Producers incorporate the chemicals they manufacture into their own bodies, becoming potential food resources for other organisms. Whereas plants are the most significant producers on land, algae and certain types of bacteria are important producers in aquatic environments. In the salt marsh ecosystem discussed in the chapter introduction, cordgrass, algae, and photosynthetic bacteria are important producers.

Animals are **consumers** that use the bodies of other organisms as a source of food energy and bodybuilding materials. Consumers are also called **heterotrophs** (Greek *heter,* "different," and *tropho,* "nourishment"). Consumers that eat producers are **primary consumers** or **herbivores** (plant eaters). Rabbits and deer are examples of primary consumers, as is the marsh periwinkle, a type of snail that feeds on algae in the salt marsh ecosystem.

Secondary consumers eat primary consumers, whereas **tertiary consumers** eat secondary consumers. Both secondary and tertiary consumers are flesh-eating **carnivores** that eat other animals. Lions, lizards, and spiders are examples of carnivores, as are the northern diamondback terrapin and the northern water snake in the salt marsh ecosystem. Other consumers, called **omnivores,** eat a variety of organisms, both plant

ENVIRONEWS

The Hippo-Tilapia Connection

It is common knowledge in Africa that where you find hippopotamuses, you always find lots of fish. In Africa's Lake Edward, located on the border of Uganda and Congo, an ongoing civil war has disrupted the food web that links hippopotamuses and tilapia fish. At night hippopotamuses wade ashore to feed on the surrounding vegetation. These animals then spend their days in the lake, digesting their food and depositing large quantities of dung into the water. (A hippo produces about 27 kilograms, or 60 pounds, of feces each day.) As the fecal material decomposes, it provides nutrients for algae, which reproduce in large numbers and are consumed by worms and insect larvae.

Tilapia eat the worms and larvae, and the fish in turn provide food for the local people.

The civil war that began in the Congo in the mid-1990s continues to affect Lake Edward despite a peace accord signed in 2003. Roving bands of armed ex-militia occupy the area near Lake Edward, and they use their automatic weapons to kill the hippos for food and money. Hippo numbers have declined by over 90% compared to the hippo population counted in Lake Edward in the 1970s. As a result of the precipitous decline in the hippo population, local fishermen are catching far fewer tilapia, a situation that threatens their livelihood. Congolese soldiers and U.N. peacekeepers are slowly disarming and removing the militia groups in hope of stabilizing the hippo population and preventing a serious humanitarian crisis in the area.

and animal. Bears, pigs, and humans are examples of omnivores; the meadow vole, which eats both insects and cordgrass in the salt marsh ecosystem, is an omnivore.

Some consumers, called **detritus feeders** or **detritivores,** consume **detritus,** organic matter that includes animal carcasses, leaf litter, and feces. Detritus feeders, such as snails, crabs, clams, and worms, are especially abundant in aquatic environments, where they burrow in the bottom muck and consume the organic matter that collects there. Marsh crabs are detritus feeders in the salt marsh ecosystem. Earthworms, termites, beetles, snails, and millipedes are terrestrial (land-dwelling) detritus feeders. An earthworm actually eats its way through the soil, digesting much of the organic matter contained there. Detritus feeders work with microbial decomposers to destroy dead organisms and waste products.

Decomposers, also called **saprotrophs** (Greek *sapro,* "rotten," and *tropho,* "nourishment"), are heterotrophs that break down dead organic material and use the decomposition products to supply themselves with energy. They typically release simple inorganic molecules, such as carbon dioxide and mineral salts, that producers can reuse. Bacteria and fungi are important decomposers. For example, during the decomposition of dead wood, sugar-metabolizing fungi first invade the wood and consume simple carbohydrates, such as glucose and maltose. When these carbohydrates are exhausted, other fungi, often aided by termites with symbiotic bacteria in their guts, complete the digestion of the wood by breaking down cellulose, the main carbohydrate of wood.

Ecosystems such as the Chesapeake Bay salt marsh contain a balanced system of producers, consumers, and decomposers, all of which have indispensable roles in ecosystems. Producers provide both food and oxygen for the rest of the community. Consumers play an important role by maintaining a balance between producers and decomposers. Detritus feeders and decomposers are necessary for the long-term survival of any ecosystem because, without them, dead organisms and waste products would accumulate indefinitely. Without microbial decomposers, important elements such as potassium, nitrogen, and phosphorus would remain permanently in dead organisms, unavailable for new generations of organisms.

The Path of Energy Flow: Who Eats Whom in Ecosystems

In an ecosystem, energy flow occurs in **food chains,** in which energy from food passes from one organism to the next in a sequence (**Figure 3.9**). Each level, or "link," in a food chain is a **trophic level** (recall the Greek *tropho* means nourishment). An organism is assigned a trophic level based on the number of energy transfer steps to that level.

trophic level: An organism's position in a food chain, which is determined by its feeding relationships.

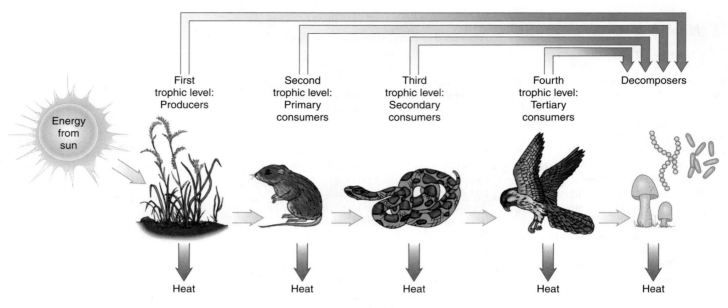

Figure 3.9 **Energy flow through a food chain.** Energy enters ecosystems from an external source (the sun), flows linearly—in a one-way direction—through ecosystems, and exits as heat loss. Much of the energy acquired by a given level of the food chain is used for cellular respiration and escapes into the surrounding environment as heat. This energy, as stipulated by the second law of thermodynamics, is unavailable to the next level of the food chain.

Producers (organisms that photosynthesize) form the first trophic level, primary consumers (herbivores) the second trophic level, secondary consumers (carnivores) the third trophic level, and so on. At every step in a food chain are decomposers, which respire organic molecules in the carcasses and body wastes of all members of the food chain.

Simple food chains rarely occur in nature because few organisms eat just one kind of organism. More typically, the flow of energy and materials through an ecosystem takes place in accordance with a range of food choices for each organism involved. In an ecosystem of average complexity, numerous alternative pathways are possible. A hawk eating a rabbit is a different energy pathway than a hawk eating a snake. A **food web** is a more realistic **model** of the flow of energy and materials through an ecosystem (**Figure 3.10**). A food web helps us to visualize feeding relationships that indicate how a community is organized.

The most important thing to remember about energy flow in ecosystems is that it is linear, or one way. Energy moves along a food chain or food web from one organism to the next as long as it has not been used for biological work. Once an organism has used energy, it is lost as heat and is unavailable for any other organism in the ecosystem.

food web: A representation of the interlocking food chains that connect all organisms in an ecosystem.

CASE IN POINT

How Humans Have Affected the Antarctic Food Web

Although the icy waters around Antarctica may seem an inhospitable environment, a complex food web is found there. The base of the food web consists of microscopic algae present in vast numbers in the well-lit, nutrient-rich water. A huge population of herbivores—tiny shrimplike **krill**—eat these marine algae (**Figure 3.11**). Krill, in turn, support a variety of larger animals. A major consumer of krill is the baleen whale, which filters krill out of the frigid water. Baleen whales include blue whales, humpback whales, and right whales. Squid and fishes also consume krill in great quantities. These, in turn, are eaten by other carnivores: toothed whales such as the sperm whale, elephant seals and leopard seals, king penguins and emperor penguins, and birds such as the albatross and petrel.

(1) Pitch pine	(5) Eastern chipmunk	(9) Red-tailed hawk	(13) American robin	(17) Worms and ants	(21) Insect larvae
(2) White oak	(6) Eastern cottontail	(10) Eastern bluebird	(14) Woodpecker	(18) Moths	(22) Insects
(3) Barred owl	(7) Red fox	(11) Red-winged blackbird	(15) Red clover	(19) Deer mouse	(23) Fungi
(4) Gray squirrel	(8) White-tailed deer	(12) Blackberry	(16) Bacteria and fungi	(20) Spiders	

Figure 3.10 **Food web at the edge of an eastern deciduous forest.** This food web is greatly simplified compared to what actually happens in nature. Groups of species are lumped into single categories such as "spiders" and "fungi," other species are not included, and many links in the web are not shown.

Figure 3.11 **Antarctic krill.** These tiny, shrimplike animals live in large swarms and eat photosynthetic algae in and around the pack ice. Whales, squid, and fishes consume vast numbers of krill. (Tom McHugh/Photo Researchers, Inc.)

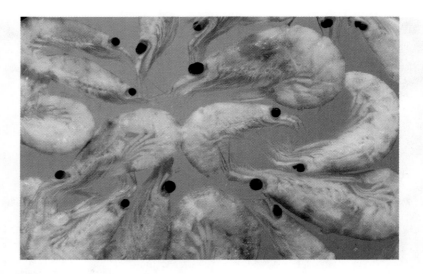

Humans have had an impact on the Antarctic food web as they have had on most other ecosystems. Before the advent of whaling, baleen whales consumed huge quantities of krill. Before a 1986 global ban on hunting large whales, whaling steadily reduced the number of large baleen whales in Antarctic waters. As a result of fewer whales eating krill, more krill became available for other krill-eating animals, whose populations increased.

Now that commercial whaling is regulated, it is hoped that the number of large baleen whales will slowly increase, and that appears to be the case for some species. The populations of most baleen whales in the Southern Hemisphere are still a fraction of their pre-whaling levels. It is not known whether baleen whales will return to their former position of dominance in terms of krill consumption in the food web. Biologists will monitor changes in the Antarctic food web as the whale populations recover.

Recently, a human-related change has developed in the atmosphere over Antarctica with the potential to cause far greater effects on the entire Antarctic food web— thinning of the ozone layer in the stratospheric region of the atmosphere. This ozone thinning allows more of the sun's ultraviolet radiation to penetrate to Earth's surface. Ultraviolet radiation contains more energy than visible light and can break the chemical bonds of some biologically important molecules, such as deoxyribonucleic acid (DNA). Scientists are concerned that ozone thinning over Antarctica may damage the algae that form the base of the food web in the Southern Ocean. Increased ultraviolet radiation is penetrating the surface waters around Antarctica, and algal productivity has declined, probably as a result of increased exposure to ultraviolet radiation. (The problem of stratospheric ozone depletion is discussed in detail in Chapter 20.)

Another human-induced change that may be responsible for declines in certain Antarctic populations is global warming. As the water has warmed in recent decades around Antarctica, less pack ice has formed during winter months. Large numbers of marine algae are found in and around the pack ice, providing a critical supply of food for the krill, which spawn in the area. Years with below-average pack ice cover mean fewer algae, which mean less krill spawning. Scientists have demonstrated that low krill abundance coincides with unsuccessful breeding seasons in penguins and fur seals, all of which struggle to find food during warmer winters. Scientists are concerned that global warming may continue to decrease the amount of pack ice, which will reverberate through the food web. (Global climate change, including the effect on Adélie penguins in Antarctica, is discussed in Chapter 21.)

To complicate matters, some commercial fishermen have started to harvest krill to make fishmeal for aquaculture industries (discussed in Chapter 19). Scientists worry that the human harvest of krill may endanger the many marine animals that depend on krill for food. ■

Ecological Pyramids

An important feature of energy flow is that most of the energy going from one trophic level to the next in a food chain or food web dissipates into the environment as a result of the second law of thermodynamics. **Ecological pyramids** often graphically represent the relative energy values of each trophic level. There are three main types of pyramids—a pyramid of numbers, a pyramid of biomass, and a pyramid of energy.

A **pyramid of numbers** shows the number of organisms at each trophic level in a given ecosystem, with greater numbers illustrated by a larger area for that section of the pyramid (**Figure 3.12**). In most pyramids of numbers, the organisms at the base of the food chain are the most abundant, and fewer organisms occupy each successive trophic level. In African grasslands the number of herbivores, such as zebras and wildebeests, is far greater than the number of carnivores, such as lions. Inverted pyramids of numbers, in which higher trophic levels have *more* organisms than lower trophic levels, are often observed among decomposers, parasites, tree-dwelling herbivorous insects, and similar organisms. One tree may provide food for thousands of leaf-eating insects, for example. Pyramids of numbers are of limited usefulness because they do not indicate the biomass of the organisms at each level, and they do not indicate the amount of energy transferred from one level to another.

A **pyramid of biomass** illustrates the total biomass at each successive trophic level. **Biomass** is a quantitative estimate of the total mass, or amount, of living material; it indicates the amount of fixed energy at a particular time. Biomass units of measure vary: Biomass is represented as total volume, as dry weight, or as live weight. Typically, pyramids of biomass illustrate a progressive reduction of biomass in succeeding trophic levels (**Figure 3.13**). For example, if one assumes there is about a 90% reduction of biomass for each trophic level, 10,000 kg of grass should support 1000 kg of grasshoppers, which in turn support 100 kg of toads. The 90% reduction in biomass is used for illustrative purposes only; actual field numbers for biomass reduction in nature vary widely. By this logic, however, the biomass of toad eaters such as snakes could be, at most, only about 10 kg. From this brief exercise, it is apparent that although carnivores do not eat vegetation, a great deal of vegetation is required to support them.

A **pyramid of energy** illustrates the energy content, often expressed as kilocalories per square meter per year, of the biomass of each trophic level (**Figure 3.14**). These pyramids always have large energy bases and get progressively smaller through succeeding trophic levels. Energy pyramids show that most energy dissipates into the environment when going from one trophic level to the next. Less energy reaches each successive trophic level from the level beneath it because organisms at the lower level use some energy to perform work, and some is lost. (Remember, because of the second law of thermodynamics, no biological process is ever 100% efficient.) Energy pyramids explain why there are so few trophic levels: Food webs are short because of the dramatic reduction in energy content at each trophic level. (See "You Can Make a Difference: Vegetarian Diets" in Chapter 19 for a discussion of how the eating habits of humans relate to food chains and trophic levels.)

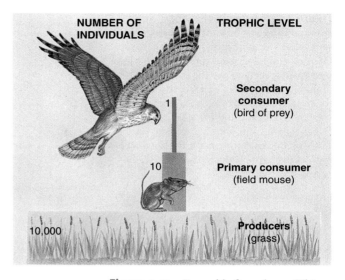

Figure 3.12   Pyramid of numbers. This pyramid is for a hypothetical area of temperate grassland; in this example, 10,000 grass plants support 10 mice, which support one bird of prey. Based on the number of organisms found at each trophic level, a pyramid of numbers is not as useful as other ecological pyramids. It provides no information about biomass differences or energy relationships between one trophic level and the next. (Note that decomposers are not shown.)

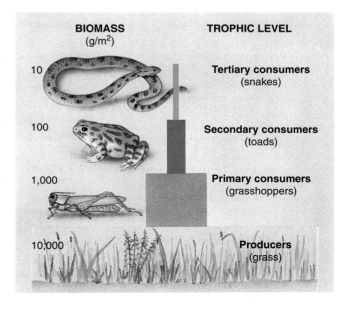

Figure 3.13   Pyramid of biomass. This pyramid is for a hypothetical area of temperate grassland. Based on the biomass at each trophic level, pyramids of biomass generally have a pyramid shape with a large base and progressively smaller areas for each succeeding trophic level. (Note that decomposers are not shown.)

Figure 3.14 **Pyramid of energy.** This pyramid indicates how much energy is present at each trophic level in a salt marsh in Georgia and how much is transferred to the next trophic level. Note the substantial loss of usable energy from one trophic level to the next; this loss is because of the energy used metabolically and given off as heat. (Note that decomposers are not shown. The 36,380 kcal/m²/year for the producers is gross primary productivity, or GPP, discussed shortly.)

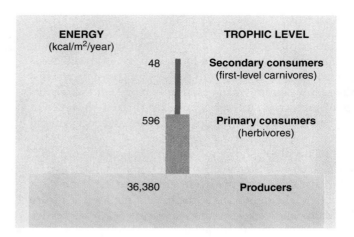

gross primary productivity (GPP): The total amount of photosynthetic energy that plants capture and assimilate in a given period.

net primary productivity (NPP): Productivity after respiration losses are subtracted.

Ecosystem Productivity

The **gross primary productivity (GPP)** of an ecosystem is the rate at which energy is captured during photosynthesis. (Gross and net primary productivities are referred to as *primary* because plants occupy the first trophic level in food webs.)

Of course, plants respire to provide energy for their own use, and this acts as a drain on photosynthesis. Energy in plant tissues after cellular respiration has occurred is **net primary productivity (NPP)**. That is, NPP is the amount of biomass found in excess of that broken down by a plant's cellular respiration. NPP represents the rate at which this organic matter is actually incorporated into plant tissues for growth.

Net primary productivity (plant growth per unit area per unit time)	=	**Gross primary productivity** (total photosynthesis per unit area per unit time)	−	**Plant cellular respiration** (per unit area per unit time)

Both GPP and NPP are expressed as energy per unit area per unit time (kilocalories of energy fixed by photosynthesis per square meter per year) or as dry weight (grams of carbon incorporated into tissue per square meter per year).

Only the energy represented by NPP is available as food for an ecosystem's consumers. Consumers use most of this energy for cellular respiration to contract muscles (obtaining food and avoiding predators) and to maintain and repair cells and tissues. Any energy that remains is used for growth and for production of young, collectively called *secondary productivity*. Any environmental factor that limits an ecosystem's primary productivity—an extended drought, for example—limits secondary productivity by its consumers.

Ecosystems differ strikingly in their productivities (**Figure 3.15**). On land, tropical rain forests have the highest NPP, probably because of their abundant rainfall, warm temperatures, and intense sunlight. As you might expect, tundra, with its harsh, cold winters, and deserts, with their lack of precipitation, are the least productive terrestrial ecosystems. Wetlands—swamps and marshes that connect terrestrial and aquatic environments—are extremely productive. The most productive aquatic ecosystems are algal beds, coral reefs, and estuaries. The lack of available nutrient minerals in some regions of the open ocean makes them extremely unproductive, equivalent to aquatic deserts. (Earth's major aquatic and terrestrial ecosystems are discussed in Chapter 6.)

Human impact on net primary productivity Humans consume far more of Earth's resources than any other of the millions of animal species (**Figure 3.16**). **Peter Vitousek** and colleagues at Stanford University calculated in 1986 how much of the global NPP is appropriated for the human economy and therefore not transferred to other organisms. When both direct and indirect human impacts are accounted for, Vitousek estimated that humans use 32% of the annual NPP of land-based ecosystems. This is a huge amount considering that we humans represent about 0.5% of the total biomass of all consumers on Earth. Essentially, humans' use of global productivity is competing with other species' needs for energy. Our use of so much of the world's productivity may contribute to the loss of many species, some potentially useful to humans, through extinction. At these levels of consumption of Earth's resources, human population growth becomes a serious threat to the planet's ability to support both its nonhuman and human occupants.

In 2001 **Stuart Rojstaczer** and colleagues at Duke University reexamined Vitousek's influential work. Rojstaczer used contemporary data sets, many of which are satellite-based and more accurate than the data Vitousek had for his groundbreaking research. Rojstaczer's mean value for his conservative estimate of annual land-based NPP appropriation by humans was 32%, like Vitousek's, although Rojstaczer arrived at that number using different calculations. Both Vitousek's and Rojstaczer's numbers are estimates, not actual values. However, the take-home message is simple: If we want our planet to operate sustainably, we must share terrestrial photosynthesis products—that is, NPP—with other organisms.

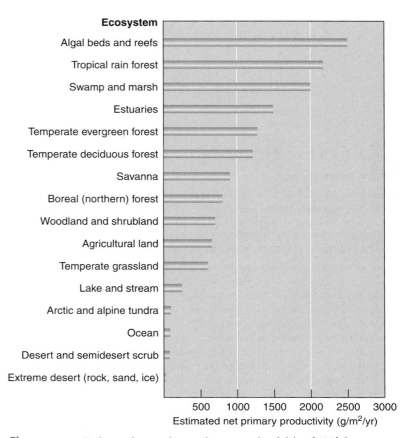

Figure 3.15 **Estimated annual net primary productivities (NPP) for selected ecosystems.** NPP is expressed as grams of dry matter per square meter per year (g/m²/yr). (After R. H. Whittaker)

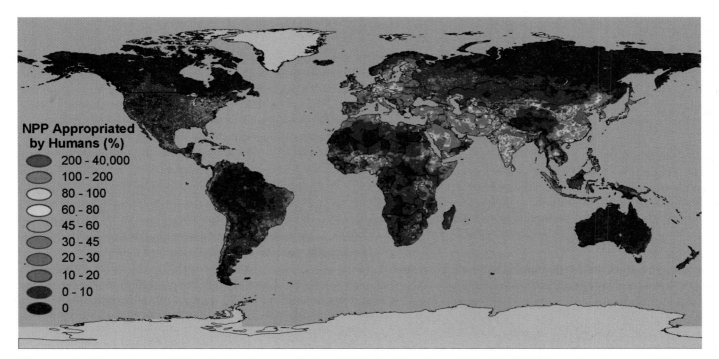

Figure 3.16 **Human appropriation of net primary production (NPP).** In this study, published in 2004, the authors integrated climate and satellite data to calculate Earth's net primary production (NPP). They then calculated the human share of NPP for various regions based on human consumption of foods and wood (for fuel and construction). In large urban areas where human appropriation is greater than 100%, food and wood from other areas supplement local NPP. (Marc L. Imhoff, Ph.D./ Biospheric Sciences Branch/NASA)

REVIEW

1. What are producers, consumers, and decomposers?

2. How does energy flow through a food web consisting of producers, consumers, and decomposers?

3. What is a pyramid of energy?

4. What is gross primary productivity? net primary productivity?

REVIEW OF LEARNING OBJECTIVES WITH KEY TERMS

• **Define** *ecology*.

Ecology is the study of the interactions among organisms and between organisms and their abiotic environment.

• **Distinguish among the following ecological levels: population, community, ecosystem, landscape, and biosphere.**

A **population** is a group of organisms of the same species that live in the same area at the same time. A **species** is a group of similar organisms whose members freely interbreed with one another in the wild to produce fertile offspring; members of one species generally do not interbreed with other species of organisms. A **community** is a natural association that consists of all the populations of different species that live and interact within an area at the same time. An **ecosystem** is a community and its physical environment. A **landscape** is a region that includes several interacting ecosystems. The **biosphere** is the parts of Earth's atmosphere, ocean, land surface, and soil that contain all living organisms.

• **Define** *energy* **and explain how it is related to work and to heat.**

Energy is the capacity to do work. Energy can be transformed from one form to another but is often measured as **heat**; the unit of heat is the kilocalorie (kcal).

• **Use examples to contrast potential energy and kinetic energy.**

Potential energy is stored energy; **kinetic energy** is energy of motion. Using a bow and arrow as an example, potential energy is stored in the drawn bow and is converted to kinetic energy as the string is released and the arrow speeds toward its target.

• **State the first and second laws of thermodynamics, and discuss the implications of these laws as they relate to organisms.**

According to the **first law of thermodynamics**, energy cannot be created or destroyed, although it can change from one form to another. According to the **second law of thermodynamics**, when energy is converted from one form to another, some of it is degraded into heat, a less usable form that disperses into the environment. The first law explains why organisms cannot produce energy but must continuously capture it from the surroundings. The second law explains why no process requiring energy is ever 100% efficient. In every energy transaction, some energy is dissipated as heat, which contributes to **entropy**.

• **Write summary reactions for photosynthesis and cellular respiration, and contrast these two biological processes.**

In photosynthesis:

$$6CO_2 + 12H_2O + \text{radiant energy} \rightarrow C_6H_{12}O_6 + 6H_2O + 6O_2$$

In cellular respiration:

$$C_6H_{12}O_6 + 6O_2 + 6H_2O \rightarrow 6CO_2 + 12H_2O + \text{energy}$$

Plants, algae, and some bacteria capture radiant energy during **photosynthesis** and incorporate some of it into carbohydrate molecules. All organisms obtain the energy in carbohydrate and other molecules by **cellular respiration,** in which molecules such as glucose are broken down with the release of energy.

• **Define** *energy flow, trophic level,* **and** *food web.*

Energy flow is the passage of energy in a one-way direction through an ecosystem. A **trophic level** is an organism's position in a food chain, which is determined by its feeding relationships. A **food web** is a representation of the interlocking food chains that connect all organisms in an ecosystem.

• **Summarize how energy flows through a food web, using producer, consumer, and decomposer in your explanation.**

Energy flow through an ecosystem is linear, from the sun to producer to consumer to decomposer. Much of this energy is converted to less usable heat as the energy moves from one organism to another, as stipulated in the second law of thermodynamics. **Producers** are the photosynthetic organisms (plants, algae, and some bacteria) that are potential food resources for other organisms. **Consumers,** which feed on other organisms, are almost exclusively animals. **Primary consumers** feed on plants; **secondary consumers** feed on primary consumers; **detritus feeders** feed on **detritus,** dead organic material. Microbial **decomposers** feed on the components of dead organisms and organic wastes, degrading them into simple inorganic materials that producers can then use to manufacture more organic material.

• **Describe typical pyramids of numbers, biomass, and energy.**

A **pyramid of numbers** shows the number of organisms at each successive trophic level. **Biomass** is a quantitative estimate of the total mass, or amount, of living material; it indicates the amount of fixed energy at a particular time. A **pyramid of biomass** illustrates the total biomass at each successive trophic level. A **pyramid of energy** illustrates the energy content of the biomass of each trophic level.

• **Distinguish between gross primary productivity and net primary productivity, and discuss human impact on the latter.**

Gross primary productivity (GPP) is the total amount of photosynthetic energy that plants capture and assimilate in a given period. **Net primary productivity (NPP)** is productivity after respiration losses are subtracted. Scientists have estimated how much of the global NPP is appropriated for the human economy and therefore not transferred to other organisms. When both direct and indirect human impacts are considered, humans are conservatively estimated to use 32% of the annual NPP of land-based ecosystems.

Thinking About the Environment

1. Draw a food web containing organisms found in a Chesapeake Bay salt marsh.

2. Distinguish among Earth's four realms: atmosphere, hydrosphere, lithosphere, and biosphere.

3. Which scientist—a population ecologist or a landscape ecologist—would be most likely to study broad-scale environmental issues and land management problems? Explain your answer.

4. What are two examples of resources that ecosystems require from the abiotic environment?

5. Give two examples of potential energy, and in each case tell how it is converted to kinetic energy.

6. How are the following forms of energy significant to organisms in ecosystems: (a) radiant energy, (b) mechanical energy, (c) chemical energy, (d) thermal energy?

7. State the first and second laws of thermodynamics.

8. How is the first law of thermodynamics related to the movement of an automobile?

9. Give an example of a natural process in which order becomes increasingly disordered.

10. Why do deep-sea organisms cluster around hydrothermal vents? What is their energy source?

11. Give an example of an herbivore, a carnivore, and an omnivore.

12. Why is the concept of a food web generally preferred over that of a food chain?

13. Could you construct a balanced ecosystem that contained only producers and consumers? only consumers and decomposers? only producers and decomposers? Explain the reasons for your answers.

14. How have humans affected the Antarctic food web?

15. Suggest a food chain with an inverted pyramid of numbers—that is, greater numbers of organisms at higher rather than at lower trophic levels.

16. Is it possible to have an inverted pyramid of energy? Why or why not?

17. Relate the pyramid of energy to the second law of thermodynamics.

18. Is this an example of an open system or a closed system? Explain your answer.

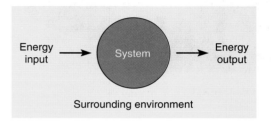

19. Is a rabbit an example of a closed system or an open system? Why?

20. Explain why a living system is more ordered than the abiotic environment.

Quantitative questions relating to this chapter are on our Web site.

Take a Stand

Visit our Web site at http://www.wiley.com/college/raven/ (select Chapter 3 from the Table of Contents) for links to more information about the environmental issues surrounding Antarc-tica. Consider the opposing views of whalers and environmentalists who oppose whaling, and debate the issues with your classmates. You will find tools to help you organize your research, analyze the data, think critically about the issues, and construct a well-considered argument. *Take a Stand* activi-ties can be done individually or as a team, as oral presentations, as written exercises, or as Web-based (e-mail) assignments.

Additional online materials relating to this chapter, including a Student Testing Section with study aids and self-tests, Environmental News, Activity Links, Environmental Investigations, and more, are also on our Web site.

4

Ecosystems and Living Organisms

Gray wolf. The reintroduction of gray wolves (*Canis lupus*) has caused many changes in plant and wildlife populations in Yellowstone National Park. (Courtesy Tracy Brooks, U.S. Fish & Wildlife Service)

Gray wolves originally ranged across North America from northern Mexico to Greenland, but they were trapped, poisoned, snared, and hunted to extinction in most places. To prevent wolves from attacking livestock, a federal program assisted ranchers in removing wolves from Yellowstone and the American Rocky Mountains in the 1930s. By 1960, the only wolves remaining in the lower 48 states were small populations in Minnesota. Removing wolves disrupted a balanced natural system.

Under the provisions of the Endangered Species Act (ESA), wolves were listed as endangered in 1974. Many scientists recommended reintroducing wolves, but this proposal was controversial and was not acted on for more than two decades. Beginning in 1995, the U.S. Fish and Wildlife Service (FWS) captured a small number of gray wolves in Canada and released them into Yellowstone National Park in Wyoming (see photograph). The population thrived and has increased to several hundred individuals in the Yellowstone area.

It will take decades for biologists to understand the effects of the wolf reintroduction on the Yellowstone ecosystem. Wolves prey on elk and occasionally on mule deer, moose, and bison. In some areas of Yellowstone, hunting by wolf packs has helped reduce Yellowstone's elk population, which was at an all-time high before the wolves returned. When elk numbers are not managed properly, they overgraze their habitat, and thousands starve during hard winters.

The reduction and redistribution of Yellowstone's elk population have relieved heavy grazing pressure on aspen, willow, cottonwood, and other plants, particularly at sites where elk cannot see approaching wolves or escape from them rapidly. As a result of a more lush and varied plant composition, herbivores such as beavers and snow hares have increased in number, which in turn supports small predators such as foxes, badgers, and martens.

Wolf packs have decimated some coyote populations, which has allowed populations of the coyotes' prey, such as ground squirrels, chipmunks, and pronghorn (coyotes prey on the fawns), to increase. Scavengers such as ravens, magpies, bald eagles, wolverines, and bears (both grizzlies and black bears) have benefited from dining on scraps from wolf kills. Ongoing research will continue to sort out the interactions among wolves and Yellowstone's other inhabitants.

The reintroduction of wolves was controversial. Ranchers and farmers in the area were against the reintroduction because their livelihood depends on livestock being safe from predators. To deal with this concern, ranchers can kill wolves that attack cattle and sheep, and federal officers can remove any wolf that threatens humans or livestock. To interject flexibility into the ESA, which forbids killing an endangered species, the Yellowstone wolf was declared an "experimental nonessential" species in the area. The Defenders of Wildlife, an environmental organization that works to protect wild plant and animal species in their natural environments, has a fund that reimburses ranchers for livestock lost to wolves.

Regardless of whether an ecosystem is as large as Yellowstone National Park or as small as a roadside drainage ditch, its organisms continually interact with one another and adapt to changes in the environment. This chapter is concerned with making sense of community structure and diversity by finding common patterns and processes in a wide variety of communities.

World View and Closer to You...
Real news footage relating to ecosystems and living organisms around the world and in your own geographic region is available at www.wiley.com/college/raven/ by clicking on "World View and Closer to You."

Evolution: How Populations Change over Time

LEARNING OBJECTIVES

- Define *evolution*.
- Explain the four premises of evolution by natural selection as proposed by Charles Darwin.
- Identify the three domains and six kingdoms of living organisms.

Where did the many species of plants, animals, fungi, and microorganisms in Yellowstone come from? They, like all Earth's species living today, are thought to have descended from earlier species by the process of **evolution**. The concept of evolution dates back to the time of Aristotle (384–322 BCE), but **Charles Darwin,** a 19th-century naturalist, proposed the mechanism of evolution that the scientific community still accepts today. As you will see, the environment plays a crucial role in Darwin's theory of natural selection as a mechanism of evolution.

It occurred to Darwin that, from one generation to the next, inherited traits favorable to survival in a given environment would be preserved, whereas unfavorable ones would be eliminated. The result would be **adaptation,** evolutionary modification that improves the chances of survival and reproductive success of the population in its environment.

> **evolution:** Cumulative genetic changes that occur over time in a population of organisms; evolution explains many patterns observed in the natural world.

Natural Selection

Darwin proposed the theory of evolution by **natural selection** in his monumental book, *The Origin of Species by Means of Natural Selection*, published in 1859. Since then, scientists have accumulated an enormous body of observations and experiments that support Darwin's theory. Although biologists still do not agree completely on some aspects of how evolutionary changes occur, the concept of evolution by natural selection is now well documented.

As a result of natural selection, the population changes over time; the frequency of favorable traits increases in successive generations, and unfavorable traits decrease or disappear. Evolution by natural selection consists of four observations about the natural world: overproduction, variation, limits on population growth, and differential reproductive success.

> **natural selection:** The process in which better-adapted individuals—those with a combination of genetic traits better suited to environmental conditions—are more likely to survive and reproduce, increasing their proportion in the population.

1. **Overproduction.** Each species produces more offspring than will survive to maturity. Natural populations have the reproductive potential to increase their numbers continuously over time. For example, if each breeding pair of elephants produces six offspring during a 90-year life span, in 750 years a single pair of elephants will have given rise to a population of 19 million! Yet elephants have not overrun the planet.

2. **Variation.** The individuals in a population exhibit variation. Each individual has a unique combination of traits, such as size, color, and ability to tolerate harsh environments. Some traits improve the chances of an individual's survival and reproductive success, whereas others do not. The variation necessary for evolution by natural selection must be inherited so that it can be passed to offspring.

3. **Limits on population growth, or a struggle for existence.** There is only so much food, water, light, growing space, and so on available to a population, and organisms compete with one another for the limited resources available to them. Not all of the offspring will survive to reproductive age because there are more individuals than the environment can support. Other limits on population growth include predators and diseases.

4. **Differential reproductive success.** Those individuals that possess the most favorable combination of characteristics (those that make individuals better adapted to their environment) are more likely to survive, reproduce, and pass their traits to the next generation. Offspring tend to resemble their parents

because the next generation obtains the parents' heritable traits. Reproduction is the key to natural selection: The best-adapted individuals reproduce most successfully, whereas less-fit individuals die prematurely or produce fewer or inferior offspring. In some cases, enough changes may accumulate over time in geographically separated populations (often with slightly different environments) to produce new species.

Charles Darwin formulated his ideas on natural selection when he was a ship's naturalist on a five-year voyage around the world. During an extended stay in the Galápagos Islands off the coast of Ecuador, he studied the plants and animals of each island, including 14 species of finches. Each finch species was specialized for a particular lifestyle different from those of the other species and different from finches on the South American mainland. The finch species, although similar in color and overall size, exhibit remarkable variation in the shape and size of their beaks, which are used to feed on a variety of foods (**Figure 4.1**). Darwin realized that the 14 species of Galápagos finches descended from a single common ancestor—one or a small population of finches that originally colonized the Galápagos from the South American

Figure 4.1 Darwin's finches. Note the various beak sizes and shapes, which are related to a species' diet. Since 1973 the long-term research of Peter and Rosemary Grant and others has verified and extended Darwin's observations on the evolution of Galápagos finches by natural selection.

mainland. Over many generations, the surviving finch populations underwent natural selection, making them better adapted to their environments, including feeding on specific food sources.

One premise on which Darwin based his theory of evolution by natural selection is that individuals transmit traits to the next generation. However, Darwin could not explain *how* this occurs or *why* individuals vary within a population. Beginning in the 1930s and 1940s, biologists combined the principles of genetics with Darwin's theory of natural selection. The result was a unified explanation of evolution known as the **modern synthesis.** (In this context, *synthesis* refers to combining parts of several previous theories to form a unified whole.) For example, the modern synthesis explains Darwin's observation of variation among offspring in terms of **mutation,** or changes in DNA. Mutations provide the genetic variability on which natural selection acts during evolution. The modern synthesis, which has dominated the thinking and research of many biologists, has resulted in many new discoveries that validate evolution by natural selection.

A vast body of evidence supports evolution, most of which is beyond the scope of this text. This evidence includes observations from the fossil record, comparative anatomy, biogeography (the study of the geographic locations of organisms), and molecular biology. In addition, evolutionary hypotheses are tested experimentally. On the basis of these kinds of evidence, biologists generally accept the principle of evolution by natural selection but try to better understand certain aspects of evolution. What is the role of chance in evolution? How rapidly do new species evolve? These questions have arisen in part from a reevaluation of the fossil record and in part from discoveries in molecular aspects of inheritance. Such critical analyses are an integral part of the scientific process because they stimulate additional observation and experimentation, along with reexamination of previous evidence. Science is an ongoing process, and information obtained in the future may require modifications to certain parts of the theory of evolution by natural selection.

Evolution of Biological Diversity: The Domains and Kingdoms of Life

Biologists arrange organisms into logical groups to try to make sense of the remarkable diversity of life that has evolved on Earth. For hundreds of years, biologists regarded organisms as falling into two broad categories—plants and animals. With the development of microscopes, however, it became increasingly obvious that many organisms did not fit well into either the plant kingdom or the animal kingdom. For example, bacteria have a *prokaryotic* cell structure: They lack organelles enclosed by membranes, including a nucleus. This feature, which separates bacteria from all other organisms, is far more fundamental than the differences between plants and animals, which have similar cell structures. Hence, it became clear that bacteria were neither plants nor animals.

These and other considerations eventually led to the three domain/six kingdom system of classification that many biologists use today (**Figure 4.2**). The prokaryotes fall into two groups that are sufficiently distinct from each other to be classified into two domains, **Archaea** and **Bacteria.** The archaea frequently live in oxygen-deficient environments and are often adapted to harsh conditions; these include hot springs, salt ponds, and hydrothermal vents (see Case in Point: Life without the Sun in Chapter 3). The thousands of remaining kinds of prokaryotes are collectively called bacteria. The eukaryotes, organisms with *eukaryotic* cells, are classified in domain **Eukarya.** Eukaryotic cells have a high degree of internal organization, containing nuclei, chloroplasts (in photosynthetic cells), and mitochondria.

Each of the six kingdoms is assigned to one of the three domains. Kingdom Archaea corresponds to domain Archaea, and kingdom Bacteria corresponds to the domain Bacteria. The remaining four kingdoms are classified in the domain Eukarya. Unicellular or relatively simple multicellular eukaryotes, such as algae, protozoa, slime molds, and water molds, are classified as members of the kingdom Protista. In addition to the kingdom Protista, there are three specialized groups of multicellular organisms—fungi, plants, and animals—that evolved independently from different groups of Protista. The kingdoms Fungi, Plantae, and Animalia differ from one another in— among other features—their types of nutrition. Fungi (molds and yeasts) secrete

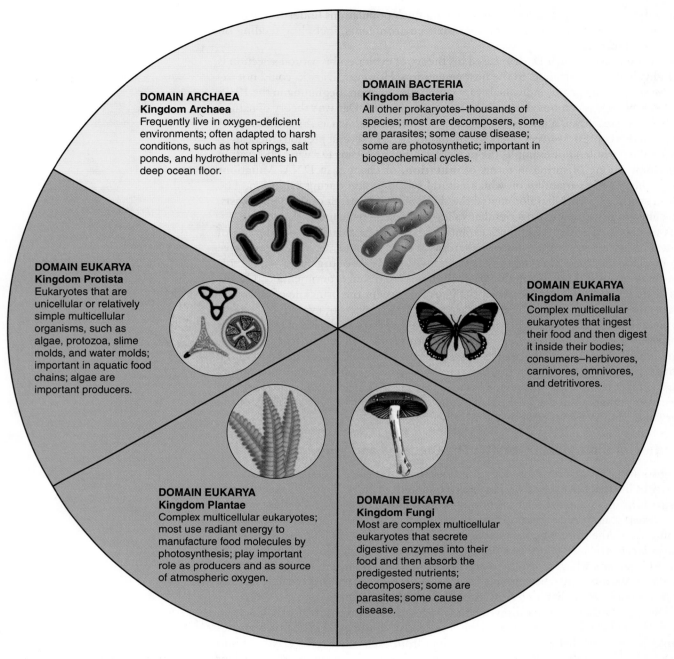

DOMAIN ARCHAEA
Kingdom Archaea
Frequently live in oxygen-deficient environments; often adapted to harsh conditions, such as hot springs, salt ponds, and hydrothermal vents in deep ocean floor.

DOMAIN BACTERIA
Kingdom Bacteria
All other prokaryotes–thousands of species; most are decomposers, some are parasites; some cause disease; some are photosynthetic; important in biogeochemical cycles.

DOMAIN EUKARYA
Kingdom Protista
Eukaryotes that are unicellular or relatively simple multicellular organisms, such as algae, protozoa, slime molds, and water molds; important in aquatic food chains; algae are important producers.

DOMAIN EUKARYA
Kingdom Animalia
Complex multicellular eukaryotes that ingest their food and then digest it inside their bodies; consumers–herbivores, carnivores, omnivores, and detritivores.

DOMAIN EUKARYA
Kingdom Plantae
Complex multicellular eukaryotes; most use radiant energy to manufacture food molecules by photosynthesis; play important role as producers and as source of atmospheric oxygen.

DOMAIN EUKARYA
Kingdom Fungi
Most are complex multicellular eukaryotes that secrete digestive enzymes into their food and then absorb the predigested nutrients; decomposers; some are parasites; some cause disease.

Figure 4.2   The three domain/six kingdom system of classification.

digestive enzymes into their food and then absorb the predigested nutrients. Plants use radiant energy to manufacture food molecules by photosynthesis. Animals ingest their food, then digest it inside their bodies.

Although the three domain/six-kingdom system is a definite improvement over the two-kingdom system, it is not perfect. Most of its problems concern the kingdom Protista, which includes some organisms that may be more closely related to members of other kingdoms than to certain other protists. For example, green algae are protists similar to plants but are not closely related to other protists, such as slime molds and brown algae.

REVIEW

1. What are Darwin's four premises of evolution by natural selection?

Biological Communities

LEARNING OBJECTIVE

• Define *ecological succession* and distinguish between primary and secondary succession.

The vast assemblage of organisms that are classified into the domains and kingdoms just discussed are organized into communities. The term *community* has a far broader sense in ecology than in everyday speech. For the biologist, a **community** is an association of different populations of organisms that live and interact in the same place at the same time.

The organisms in a community are interdependent in a variety of ways. Species compete with one another for food, water, living space, and other resources. (Used in this context, a *resource* is anything from the environment that meets a particular species' needs.) Some organisms kill and eat other organisms. Some species form intimate associations with one another, whereas other species seem only distantly connected. As discussed in Chapter 3, each organism plays one of three main roles in community life: producer, consumer, or decomposer. The unraveling of the many positive and negative, direct and indirect interactions of organisms living as a community is one of the goals of community ecologists.

Communities vary greatly in size, lack precise boundaries, and are rarely completely isolated. They interact with and influence one another in countless ways, even when the interaction is not readily apparent. Furthermore, communities are nested within one another like Chinese boxes—that is, there are communities within communities. A forest is a community, but so is a rotting log in that forest. Insects, plants, and fungi invade a fallen tree as it undergoes a series of decay steps (**Figure 4.3**). First, wood-boring insects and termites forge paths through the bark and wood. Later, other insects, plant roots, and fungi follow and enlarge these openings. Mosses and lichens that become established on the log's surface trap rainwater and extract nutrient minerals, and fungi and bacteria speed decay, providing nutrients for other inhabitants. As decay progresses, small mammals burrow into the wood and eat the fungi, insects, and plants.

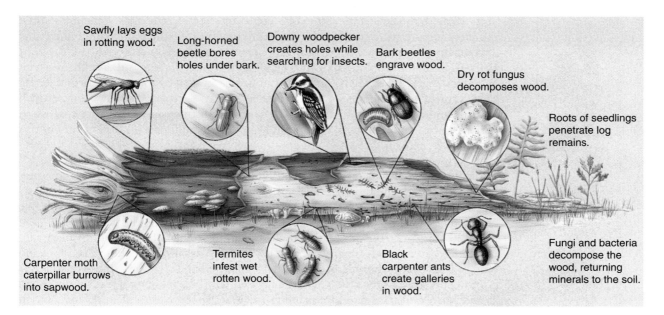

Figure 4.3   A rotting log community. Detritivores and decomposers consume a rotting log, which teems with a variety of fungi, animals, and plants.

Organisms exist in an abiotic (nonliving) environment that is as essential to their existence as are their interactions with one another. Minerals, air, water, and sunlight are just as much a part of a honeybee's environment as the flowers it pollinates and from which it takes nectar. A biological community and its abiotic environment comprise an **ecosystem** (Chapter 5 considers an ecosystem's abiotic environment).

Succession: How Communities Change over Time

A community develops gradually, through a sequence of species. The process of community development over time, which involves species in one stage being replaced by different species, is called **succession.** Certain organisms that initially colonize an area are replaced over time by others, which themselves are replaced much later by still others.

The actual mechanisms that underlie succession are not clear. In some cases, an earlier species modifies the environment in some way, thereby making it more suitable for a later species to colonize. It is also possible that earlier species exist because there is little competition from other species. Later, as more invasive species arrive, the original species are outcompeted and displaced.

Ecologists initially thought that succession inevitably led to a stable and persistent community, such as a forest, called a *climax community*. But more recently, the traditional view has fallen out of favor. The apparent stability of a "climax" forest is probably the result of how long trees live relative to the human life span. Mature "climax" communities are not in a state of permanent equilibrium but rather in a state of continual disturbance. Over time, a mature community changes in species composition and in the relative abundance of each species, despite its overall uniform appearance.

Succession is usually described in terms of the changes in the species composition of the plants growing in an area, although each stage of succession may have its own characteristic kinds of animals and other organisms. The time involved in ecological succession is on the scale of tens, hundreds, or thousands of years, not the millions of years involved in the evolutionary time scale.

Primary Succession

primary succession: The change in species composition over time in a previously uninhabited environment.

Primary succession is ecological succession that begins in an environment that has not been inhabited before. No soil exists when primary succession begins. Bare rock surfaces, such as recently formed volcanic lava and rock scraped clean by glaciers, are examples of sites where primary succession may take place (**Figure 4.4**).

Although the details vary from one site to another, on bare rock, lichens are often the most important element in the **pioneer** community—the initial community that develops during primary succession. Lichens secrete acids that help break the rock apart, beginning the process of soil formation. Over time, mosses and drought-resistant ferns may replace the lichen community, followed in turn by tough grasses and herbs. Once enough soil accumulates, grasses and herbs may be replaced by low shrubs, which in turn would be replaced by forest trees in several distinct stages. Primary succession on bare rock from a pioneer community to a forest community often occurs in this sequence: lichens → mosses → grasses → shrubs → trees.

Primary Succession on Krakatoa Primary succession takes hundreds or thousands of years to occur. The Indonesian island of Krakatoa has provided scientists with a perfect long-term study of primary succession in a tropical rain forest. In 1883, a volcanic eruption destroyed all life on the island. Biologists have surveyed Krakatoa in the more than 120 years since the devastation to document the return of life forms. Primary succession has been extremely slow, in part because of Krakatoa's isolation. Both Java and Sumatra are nearby, but many species have a limited ability to disperse over water. A portion of Krakatoa's forest might have only one-tenth the number of tree species found in undisturbed tropical rain forests on nearby islands. The lack of plant diversity has in turn limited the number of colonizing animal species. In a forested area of Krakatoa where zoologists would expect more than 100 butterfly species, there are only 2 species.

Figure 4.4 **Primary succession on glacial moraine.** During the past 200 years, glaciers have retreated in Glacier Bay, Alaska. Although these photos were not taken in the same location, they show some of the stages of primary succession on glacial moraine (rocks, gravel, and sand deposited by a glacier).

(a) After the glacier's retreat, lichens initially colonize the barren landscape, then mosses and small shrubs. (Wolfgang Kaehler)

Primary Succession on Sand dunes   Some lake and ocean shores have extensive sand dunes deposited by wind and water. At first these dunes are blown about by the wind. The sand dune environment is severe, with high temperatures during the day and low temperatures at night. The sand is also deficient in certain nutrient minerals needed by plants. As a result, few plants can tolerate the environmental conditions of a sand dune.

Henry Cowles developed the concept of succession in the 1880s when he studied succession on sand dunes around the shores of Lake Michigan, which has been gradually shrinking since the last ice age. The shrinking lake exposed new sand dunes that displayed a series of stages in the colonization of the land.

Grasses are common pioneer plants on sand dunes around the Great Lakes. As the grasses extend over the surface of a dune, their roots hold it in place, helping to stabilize it. At this point, mat-forming shrubs may invade, further stabilizing the dune. Later, shrubs are replaced by poplars (cottonwoods), which years later are replaced by pines and finally by oaks. Soil fertility remains low, and as a result, other forest trees rarely replace oaks. A summary of how primary succession on sand dunes around the Great Lakes might proceed is in this sequence: grasses → shrubs → poplars (cottonwoods) → pine trees → oak trees.

(b) At a later time, dwarf trees and shrubs colonize the area. (Glenn N. Oliver/Visuals Unlimited)

Secondary Succession

Secondary succession is ecological succession that begins in an environment following destruction of all or part of an earlier community. Abandoned farmland and open areas caused by a forest fire are common examples of sites where secondary succession occurs.

During the summer of 1988, wildfires burned approximately one-third of Yellowstone National Park. This natural disaster provided a chance for biologists to study secondary succession in areas that were once forests. After the conflagration, gray ash covered the forest floor, and most of the trees, though standing, were charred and dead. Secondary succession in Yellowstone has occurred rapidly. Less than one year later, in the spring of 1989, trout lily and other herbs had sprouted and covered much of the ground. Ten years after the fires, a young forest of knee-high to shoulder-high lodgepole pines had dominated the area, and Douglas fir seedlings were also present. Biologists continue to monitor the changes in Yellowstone as secondary succession proceeds.

Old Field Succession   Biologists have extensively studied secondary succession on abandoned farmland. It takes more than 100 years for secondary succession to occur at a single site, but a single researcher can study old field succession in its entirety by observing different sites undergoing succession for different periods. The biologist may examine county tax records to determine when each field was abandoned.

(c) Still later, spruces dominate the community. (Wolfgang Kaehler)

secondary succession: The change in species composition that takes place after some disturbance destroys the existing vegetation; soil is already present.

A predictable succession of communities colonizes abandoned farmland in North Carolina (**Figure 4.5**). The first year after cultivation ceases, crabgrass dominates. During the second year, horseweed, a larger plant that outgrows crabgrass, is the dominant species. Horseweed does not dominate more than one year because decaying horseweed roots inhibit the growth of young horseweed seedlings. In addition, horseweed does not compete well with other plants that become established in the third year. During the third year after the last cultivation, other weeds— broomsedge, ragweed, and aster—become established. Typically, broomsedge out-competes aster because broomsedge is drought-tolerant, whereas aster is not.

In years 5 to 15, the dominant plants in an abandoned field are pines such as short-leaf pine and loblolly pine. Through the buildup of soil litter, such as pine needles and branches, pines produce conditions that cause the earlier dominant plants to decline in importance. Over the next century or so, pines give up their dominance to hardwoods such as oaks. The replacement of pines by oaks depends primarily on the environmental changes produced by the pines. The pine litter causes soil changes, such as an increase in water-holding capacity, which young oak seedlings need to become established. In addition, hardwood seedlings are more tolerant of shade than are young pine seedlings. Secondary succession on abandoned farmland in the southeastern United States proceeds in this sequence: crabgrass → horseweed → broomsedge and other weeds → pine trees → hardwood trees.

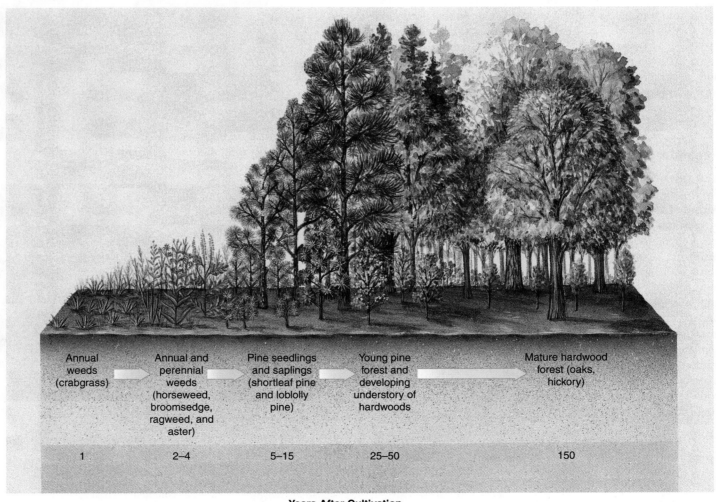

Years After Cultivation

Figure 4.5   Secondary succession on an abandoned field in North Carolina.

As secondary succession in North Carolina proceeds, a progression of animal life follows the changes in vegetation. Although a few animals—such as the short-tailed shrew—are found in all stages of abandoned farmland succession, most animals appear with certain stages and disappear with others. During the crabgrass and weed stages of secondary succession, the open fields support grasshoppers, meadow voles, cottontail rabbits, and birds such as grasshopper sparrows and meadowlarks. As young pine seedlings become established, animals of open fields give way to animals common in mixed herbaceous and shrubby habitats. Now white-tailed deer, white-footed mice, ruffed grouse, robins, and song sparrows are common, whereas grasshoppers, meadow mice, grasshopper sparrows, and meadowlarks disappear. As the pine seedlings grow into trees, animals of the forest replace those common in mixed herbaceous and shrubby habitats. Cottontail rabbits give way to red squirrels, and warblers and veeries replace ruffed grouse, robins, and song sparrows. Thus, each stage of succession supports its own characteristic animal life.

REVIEW

1. What is ecological succession?

2. How do primary and secondary succession differ?

Interactions among Organisms

LEARNING OBJECTIVES

- Define *symbiosis* and distinguish among mutualism, commensalism, and parasitism.
- Define *predation* and describe the effects of natural selection on predator-prey relationships.
- Define *competition* and distinguish between intraspecific and interspecific competition.

No organism exists independently of other organisms. The producers, consumers, and decomposers of an ecosystem interact with one another in a variety of ways, and each forms associations with other organisms. Three main types of interactions occur among species in an ecosystem: symbiosis, predation, and competition.

Symbiosis

In **symbiosis**, individuals of one species usually live in or on the individuals of another species. At least one of the species—and sometimes both—uses its partner's resources. The partners of a symbiotic relationship, called **symbionts,** may benefit from, be unaffected by, or be harmed by the relationship. Examples of symbiosis occur across all of the domains and kingdoms of life.

Symbiosis is the result of **coevolution,** the interdependent evolution of two interacting species. Flowering plants and their animal pollinators have a symbiotic relationship that is an excellent example of coevolution. Plants are rooted in the ground and lack the mobility that animals have when mating. Many flowering plants rely on animals to help them reproduce. Bees, beetles, hummingbirds, bats, and other animals transport the male reproductive units, called pollen grains, from one plant to another, in effect giving plants mobility. How has this come about?

During the millions of years these associations developed, flowering plants evolved several ways to attract animal pollinators. One of the rewards for the pollinator is food—nectar (a sugary solution) and pollen. Plants often produce food precisely adapted for one type of pollinator. The nectar of bee-pollinated flowers usually contains 30% to 35% sugar, the concentration bees need to make honey. Bees will not visit flowers with lower sugar concentrations in their nectar. Bees also use pollen to make beebread, a nutritious mixture of nectar and pollen that bees feed to their larvae.

symbiosis: Any intimate relationship or association between members of two or more species; includes mutualism, commensalism, and parasitism.

Plants possess a variety of ways to get the pollinator's attention, most involving colors and scents. Showy petals visually attract the pollinator much as a neon sign or golden arches attract a hungry person to a restaurant. Insects have a well-developed sense of smell, and many insect-pollinated flowers have strong scents that are also pleasant to humans. A few specialized kinds of flowers have unpleasant odors. The carrion plant produces flowers that mimic the smell of rotting flesh. Carrion flies move from one flower to another, looking for a place to deposit their eggs, and in the process pollen is transferred.

During the time plants were evolving specialized features to attract pollinators, the animal pollinators coevolved specialized body parts and behaviors to aid pollination and to obtain nectar and pollen as a reward. Coevolution is responsible for the hairy bodies of bumblebees, which catch and hold the sticky pollen for transport from one flower to another. The long, curved beaks of certain honeycreepers, Hawaiian birds that insert their beaks into tubular flowers to obtain nectar, are the result of coevolution (**Figure 4.6**).

Animal behavior has also coevolved. The flowers of certain orchids resemble female wasps in coloring and shape. The resemblance between one orchid species and female wasps is so strong that male wasps mount the flowers and attempt to copulate with them. During this misdirected activity, a pollen sac usually attaches to the back of the wasp. When the frustrated wasp departs and attempts to copulate with another orchid flower, pollen grains are transferred to the flower.

The thousands, or even millions, of symbiotic associations that result from coevolution fall into three categories: mutualism, commensalism, and parasitism.

mutualism: A symbiotic relationship in which both partners benefit.

commensalism: A type of symbiosis in which one organism benefits and the other one is neither harmed nor helped.

Mutualism: Sharing Benefits   In **mutualism**, different species living in close association provide benefits to each other. The interdependent association between nitrogen-fixing bacteria of the genus *Rhizobium* and legumes (plants such as peas, beans, and clover) is an example of mutualism. Nitrogen-fixing bacteria live in nodules in the roots of legumes and supply the plants with all of the nitrogen they need. The legumes supply sugar to their bacterial symbionts.

Another example of mutualism is the association between reef-building coral animals and microscopic algae. These symbiotic algae, called **zooxanthellae** (pronounced *zoh-zan-thel'ee*), live inside cells of the coral, where they photosynthesize and provide the animal with carbon and nitrogen compounds as well as oxygen. Zooxanthellae have a stimulatory effect on the growth of corals, causing calcium carbonate skeletons to form around their bodies much faster when the algae are present. The corals, in turn, supply their zooxanthellae with waste products such as ammonia, which the algae use to make nitrogen compounds for both partners.

Mycorrhizae (*my-kor-rye'zee*) are mutualistic associations between fungi and the roots of about 80% of all plants. The fungus, which grows around and into the root as well as into the surrounding soil, absorbs essential minerals, especially phosphorus, from the soil and provides them to the plant. In return, the plant provides the fungus with food produced by photosynthesis. Plants grow more vigorously in mycorrhizal relationships and tolerate *environmental stressors* such as drought and high soil temperatures better (**Figure 4.7**). Indeed, some plants cannot maintain themselves under natural conditions if the fungi with which they normally form mycorrhizae are not present.

Commensalism: Taking Without Harming   **Commensalism** is an association between two different species in which one benefits and the other is unaffected. One example of commensalism is the relationship between two kinds of insects: silverfish and army ants. Certain kinds of silverfish move along in permanent association with the marching columns of army ants and share the plentiful food caught in their raids. The army ants derive no apparent benefit or harm from the silverfish.

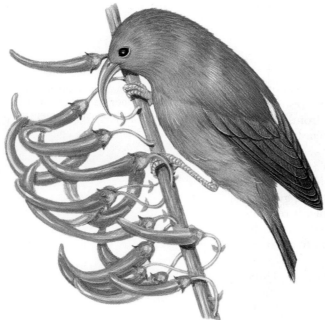

Figure 4.6   Coevolution. The gracefully curved bill of the 'I'iwi, one of the Hawaiian honeycreepers, lets it sip nectar from flowers of the lobelia. The 'I'iwi bill fits perfectly into the long, tubular lobelia flowers.

Figure 4.7 **Mutualism.** Western red cedar (*Thuja plicata*) seedlings respond to mycorrhizal fungi.

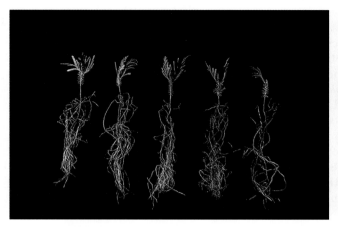

(a) These seedlings, which represent an experimental control, were grown in low phosphorus in the absence of the fungus. (Courtesy Randy Molina, U.S. Forest Service)

(b) Comparable seedlings were grown in low phosphorus with the fungus. Both roots and shoots are larger and developed more extensively in the presence of mycorrhizal fungi. (Courtesy Randy Molina, U.S. Forest Service)

Another example of commensalism is the relationship between a tropical tree and many **epiphytes,** smaller plants, such as mosses, orchids, and ferns that live attached to the bark of the tree's branches (**Figure 4.8**). The epiphyte anchors itself to the tree but does not obtain nutrients or water directly from the tree. Its location on the tree lets it obtain adequate light, water (as rainfall dripping down the branches), and required nutrient minerals (washed out of the tree's leaves by rainfall). Thus, the epiphyte benefits from the association, whereas the tree is apparently unaffected.

Parasitism: Taking at Another's Expense   In parasitism, one organism, the *parasite*, obtains nourishment from another organism, its *host*. Although a parasite may weaken its host, it rarely kills it. (A parasite would have a difficult life if it kept killing off its hosts!) Some parasites, such as ticks, live outside the host's body; other parasites, such as tapeworms, live within the host. Parasitism is a successful lifestyle; more than 100 parasites live in or on the human species alone!

parasitism: A symbiotic relationship in which one organism benefits and the other is adversely affected.

Figure 4.8 **Commensalism.** Epiphytes are small plants that grow attached to the branches and trunks of larger trees. Photo was taken in Costa Rica. (Michael Melford/National Geographic Society)

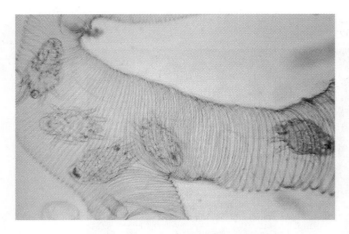

Figure 4.9 Parasitism. Tiny tracheal mites live in the breathing tubes of honeybees and suck their blood, weakening and eventually killing them. These mites have devastated many of North America's wild and domestic honeybee populations. (USDA, Agricultural Research Service)

predation: The consumption of one species (the prey) by another (the predator).

Since the 1980s, wild and domestic honeybees in the United States have been dying off. Although *habitat fragmentation* and pesticide use have contributed to the problem, tracheal mites (**Figure 4.9**) and larger varroa mites are a major reason for the honeybee decline. The number of commercial colonies has fallen about 50% during the past several decades. A major die-off across the United States was reported in 2006 to 2007; scientists had not yet determined the cause of the decline, known as *colony collapse disorder*, when we went to press.

In the United States honeybees annually pollinate more than 90 food, fiber, and seed crops valued at $14.6 billion and produce about $158 million of honey (2005 data), so their decline is a major threat to U.S. agriculture. Beekeepers in several states, in cooperation with the U.S. Department of Agriculture, are testing Russian honeybees for pollination and honey production. Preliminary results indicate that Russian bees are resistant to mites and tolerate cold winters better than American bees.

When a parasite causes disease and sometimes the death of a host, it is known as a **pathogen.** Crown gall disease, caused by a bacterium, occurs in many kinds of plants and results in millions of dollars' worth of damage each year to ornamental and agricultural plants. Crown gall bacteria, which live on detritus (organic debris) in the soil, enter plants through small wounds such as those caused by insects. They cause galls, or tumorlike growths, often at a plant's crown (between the stem and the roots, at or near the soil surface). Although plants seldom die from crown gall disease, they are weakened, grow more slowly, and often succumb to other pathogens.

Many parasites do not cause disease. Humans become infected with the beef tapeworm by eating undercooked beef infested with immature tapeworms. Once the tapeworm is inside the human digestive system, it attaches itself to the wall of the small intestine and grows rapidly by absorbing nutrients that pass through the small intestine. A single beef tapeworm that lives in a human digestive tract does not usually cause any noticeable symptoms, although some weight loss may be associated with a multiple infestation.

Predation

Predation includes both animals eating other animals (for example, herbivore-carnivore interactions) and animals eating plants (producer-herbivore interactions). Predation has resulted in an evolutionary "arms race," with the coevolution of predator strategies— more efficient ways to catch prey—and prey strategies—better ways to escape the predator. An efficient predator exerts a strong selective force on its prey, and over time, the prey species may evolve some sort of countermeasure that reduces the probability of being captured. The countermeasure that the prey acquires in turn may act as a strong selective force on the predator.

Adaptations related to predator-prey interactions include predator strategies (pursuit and ambush) and prey strategies (plant defenses and animal defenses). Keep in mind as you read these descriptions that such strategies are not "chosen" by the respective predators or prey. New traits arise randomly in a population as a result of changes in genetic material, or deoxyribonucleic acid (DNA). Some new traits may be beneficial, whereas others may be harmful or have no effect at all. As a result of natural selection, beneficial strategies, or traits, persist in a population because such characteristics make the individuals that possess them well suited to thrive and reproduce. In contrast, characteristics that make the individuals that possess them poorly suited to their environment tend to disappear in a population.

Pursuit and Ambush A day gecko sights a spider and pounces on it (**Figure 4.10**). Orcas (killer whales), which hunt in packs, often herd salmon or tuna into a cove so that they are easier to catch. Any trait that increases hunting efficiency, such as the speed of a gecko or the intelligence of orcas, favors predators that pursue their prey. Because these carnivores must process information quickly during the pursuit of prey, their brains are generally larger, relative to body size, than those of the prey they pursue.

Ambush is another effective way to catch prey. The goldenrod spider is the same color as the white or yellow flowers in which it hides. This camouflage prevents unwary insects that visit the flower for nectar from noticing the spider until it is too late. Predators that *attract* prey are particularly effective at ambushing. For example, a diverse group of deep-sea fishes called anglerfish possess rodlike luminescent lures close to their mouths to attract prey.

Plant Defenses Against Herbivores   Plants cannot escape predators by fleeing, but they possess adaptations that protect them from being eaten. The presence of spines, thorns, tough leathery leaves, or even thick wax on leaves discourages foraging herbivores from grazing. Other plants produce an array of protective chemicals that are unpalatable or even toxic to herbivores. The active ingredients in such plants as marijuana, opium poppy, tobacco, and peyote cactus may discourage the foraging of herbivores. For example, the nicotine found in tobacco is so effective at killing insects that it is a common ingredient in many commercial insecticides.

Milkweeds are an excellent example of the evolutionary arms race between plants and herbivores. Milkweeds produce alkaloids and cardiac glycosides, chemicals poisonous to all animals except for a small group of insects. During the course of evolution, these insects acquired the ability to either tolerate or metabolize the milkweed toxins. As a result, they can eat milkweeds without being poisoned. These insects avoid competition from other herbivorous insects because few other insects can tolerate milkweed toxins. Predators learn to avoid these insects, which accumulate the toxins in their tissues and are usually brightly colored.

Figure 4.10   Predation.   A Madagascar day gecko has just caught a spider. Unlike most geckos, the Madagascar day gecko is diurnal (active during the day). Native to Madagascar, the day gecko spends most of its time in trees, where it blends into the foliage. (Roy Toft/NG Image Collection)

Figure 4.11 Warning coloration. The poison arrow frog (*Dendrobates tinctorius*) advertises its poisonous nature with its conspicuous warning coloration to avoid would-be predators. Photo was taken in Guyana, South America. (Michael Fogden/Animals Animals)

competition: The interaction among organisms that vie for the same resources (such as food or living space) in an ecosystem.

Defensive Adaptations of Animals Many animals, such as woodchucks, flee from predators by running to their underground burrows. Others have mechanical defenses, such as the quills of a porcupine and the shell of a pond turtle. Some animals live in groups—a herd of antelope, colony of honeybees, school of anchovies, or flock of pigeons. This social behavior decreases the likelihood of a predator catching one of them unaware; the group has many eyes, ears, and noses watching, listening, and smelling for predators.

Chemical defenses are common among animal prey. The South American poison arrow frog has poison glands in its skin. Its bright **warning coloration** prompts avoidance by experienced predators (**Figure 4.11**). Snakes or other animals that have tried to eat a poisonous frog do not repeat their mistake! Other examples of warning coloration occur in the striped skunk, which sprays acrid chemicals from its anal glands, and the bombardier beetle, which sprays harsh chemicals at potential predators.

Some animals blend into their surroundings to hide from predators. The animal's behavior often enhances such camouflage. There are many examples of camouflage. Certain caterpillars resemble twigs so closely you would never guess they are animals until they move. Pipefish are almost perfectly camouflaged in green eelgrass. The pygmy seahorse resembles gorgonian coral so closely that the little seahorse was not discovered until 1970 when a zoologist found it in the coral he had placed in an aquarium (**Figure 4.12**). Evolution has preserved and accentuated such camouflage.

Competition

Competition occurs when two or more individuals attempt to use an essential common resource such as food, water, shelter, living space, or sunlight. Resources are often in limited supply in the environment, and their use by one individual decreases the amount available to others. If a tree in a dense forest grows taller than surrounding trees, it absorbs more of the incoming sunlight, and less sunlight is available for nearby trees. Competition occurs among individuals within a population (*intraspecific competition*) or between species (*interspecific competition*).

Ecologists traditionally assumed that competition is the most important determinant of both the number of species found in a community and the size of each population. Today, ecologists recognize that competition is only one of many interacting biotic and abiotic factors that affect community structure. Furthermore, competition is not always a straightforward, direct interaction. A variety of flowering plants live in a young pine forest and presumably compete with conifers for such resources as soil moisture and soil nutrient minerals. Their relationship is more than simple competition, however. The flowers produce nectar consumed by some insect species that also prey on needle-eating insects, thereby reducing the number of insects feeding on pines. It is therefore difficult to assess the overall effect of flowering plants on pines. If the flowering plants were removed from the community, would the pines grow faster because they were no longer competing for necessary resources? Or would the increased presence of needle-eating insects caused by fewer omnivorous insects inhibit pine growth?

Short-term experiments in which one competing plant species is removed from a forest community often have demonstrated improved growth for the remaining species. However, few studies have tested the long-term effects on forest species of the removal of one competing species. Long-term effects may be subtle, indirect, and difficult to assess. They may lower or negate the negative effects of competition for resources.

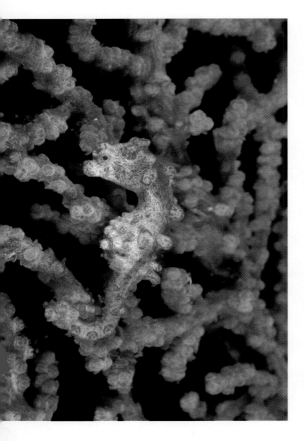

Figure 4.12 Camouflage. Pygmy seahorses (*Hippocampus bargibanti*), about the size of small fingernails, are virtually invisible in and around gorgonian coral (*Muricella*). Photographed off the coast of Sulawesi, Indonesia. (Chris Newbert/Minden Pictures, Inc.)

REVIEW

1. What is symbiosis? What are the three kinds of symbiosis?
2. What is predation? Describe how evolution has affected predator-prey relationships.
3. What is competition? What are the two main kinds of competition?

 The Ecological Niche

LEARNING OBJECTIVES

- Describe the factors that contribute to an organism's ecological niche and distinguish between fundamental niche and realized niche.
- Give several examples of limiting resources and discuss how they might affect an organism's ecological niche.
- Relate the concepts of competitive exclusion and resource partitioning.

You have examined some of the ways species interact to form interdependent relationships within a community. Now let us consider the way of life of a given species in its ecosystem. An ecological description of a species typically includes (1) whether it is a producer, consumer, or decomposer; (2) the kinds of symbiotic associations it forms; (3) whether it is a predator and/or prey; and (4) what species it competes with. Other details are needed, however, to provide a complete picture.

Every organism is thought to have its own role, or ecological niche, within the structure and function of an ecosystem. Because the ecological niche describes an organism's place and function within a complex system of biotic and abiotic factors, it is difficult to define precisely. An ecological niche takes into account all aspects of the organism's existence—all the physical, chemical, and biological factors an organism needs to survive, remain healthy, and reproduce. Among other things, the niche includes the local environment in which an organism lives—its **habitat.** An organism's niche also encompasses how the abiotic components of its environment, such as light, temperature, and moisture, interact with and influence it. A complete description of an organism's ecological niche involves numerous dimensions.

The ecological niche of an organism may be much broader potentially than it is in actuality. Put differently, an organism is potentially capable of using much more of its environment's resources or of living in a wider assortment of habitats than it actually does. The potential, idealized ecological niche of an organism is its **fundamental niche,** but various factors such as competition with other species usually exclude it from part of its fundamental niche. The lifestyle an organism actually pursues and the resources it actually uses make up its **realized niche.**

An example may clarify the distinction between fundamental and realized niches. The green anole, a lizard native to Florida and other southeastern states, perches on trees, shrubs, walls, or fences during the day and waits for insect and spider prey (**Figure 4.13a**). In the past, these little lizards were widespread in Florida. A number of years ago, a related species, the brown anole, was introduced from Cuba into southern Florida and quickly became common (**Figure 4.13b**). Suddenly the green anoles became rare, apparently driven out of their habitat by competition from the slightly larger brown lizards. Careful investigation disclosed that green anoles were still around. They were now confined largely to the vegetation in wetlands and to the foliated crowns of trees, where they were less obvious.

The habitat portion of the green anole's fundamental niche includes the trunks and crowns of trees, exterior house walls, and many other locations. Where they became established, brown anoles drove green anoles out from all but wetlands and tree crowns, so the green anole's realized niche became smaller as a result of competition (**Figure 4.13c, d**). Natural communities consist of numerous species, many of which compete to some extent, and the interactions among species produce the realized niche of each.

ecological niche: The totality of an organism's adaptations, its use of resources, and the lifestyle to which it is fitted.

Figure 4.13   Effect of competition on an organism's realized niche.

(a) The green anole (*Anolis carolinensis*) is native to Florida. (Ed Kanze/Dembinsky Photo Associates)

(b) The brown anole (*Anolis sagrei*) was introduced in Florida. (Robert Clay/Visuals Unlimited)

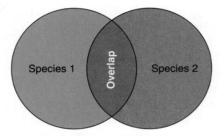

(c) The fundamental niches of the two lizards initially overlapped. Species 1 is the green anole, and species 2 is the brown anole

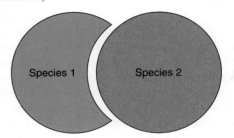

(d) The brown anole out-competed the green anole, restricting its niche.

Limiting Resources

What environmental resources help determine the niche of a species? An ecological niche is basically determined by all of a species' structural, physiological, and behavioral adaptations. Such adaptations determine the tolerance an organism has for environmental extremes. If any feature of its environment lies outside the bounds of its tolerance, then the organism cannot live there. Just as you would not expect to find a cactus living in a pond, you would not expect water lilies in a desert.

The environmental resources that determine an organism's ecological niche can be extremely difficult to identify. For this reason, the concept of ecological niche is largely abstract, although some of its dimensions can be experimentally determined. Any resource at a suboptimal level relative to an organism's need for it or at a level in excess of an organism's tolerance for it is a **limiting resource**.

Most limiting resources that scientists have investigated are simple variables such as the mineral content of soil, extremes of temperature, and amount of precipitation.

limiting resource: Any environmental resource that, because it is scarce or at unfavorable levels, restricts the ecological niche of an organism.

Figure 4.14   Limiting resource. An organism is limited by any environmental resource that exceeds its tolerance or is less than the required minimum.

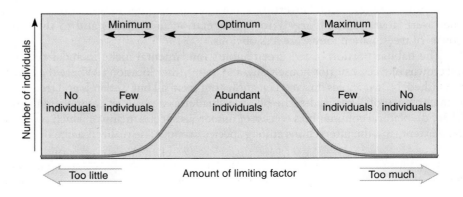

Figure 4.15   Interspecific competition.

Such investigations have disclosed that any resource that exceeds an organism's tolerance or is present in quantities smaller than the minimum required limits the occurrence of that organism in an ecosystem (**Figure 4.14**). By their interaction, such resources help define an organism's ecological niche.

Limiting resources often affect only one part of an organism's life cycle. Although adult blue crabs can live in brackish (slightly salty) water, they cannot become permanently established there because their larvae (immature forms) require water with a higher concentration of dissolved salt. Similarly, the ring-necked pheasant, a popular game bird, was introduced widely in North America but did not survive in the southern United States. The adult birds do well, but the eggs cannot develop properly in the warm southern temperatures.

Competitive Exclusion and Resource Partitioning

When two species are similar, as are the green and brown anoles, their fundamental niches may overlap. However, many ecologists think that no two species indefinitely occupy the same niche in the same community because **competitive exclusion** eventually occurs. In competitive exclusion, one species excludes another from a portion of a niche as a result of competition between species (interspecific competition). Although it is possible for species to compete for some necessary resource without being total competitors, two species with absolutely identical ecological niches cannot coexist. Coexistence *can* occur if the overlap in the two species' niches is reduced. In the lizard example, direct competition between the two species was reduced as the brown anole excluded the green anole from most of its former habitat until the only places open to it were wetland vegetation and tree crowns.

The initial evidence that competition between species determines an organism's realized niche came from a series of experiments conducted by the Russian biologist **G. F. Gause** in 1934. In one study, Gause grew populations of two species of *Paramecium* (a type of protist), *P. aurelia* and the larger *P. caudatum* (**Figure 4.15**). When grown in separate test tubes, the population of each species quickly increased to a high level and remained there for some time. When grown together, only *P. aurelia* thrived; *P. caudatum* dwindled and eventually died out. Under different environmental conditions, *P. caudatum* prevailed over *P. aurelia*. Gause concluded that one set of conditions favored one species, and a different set favored the other. Nonetheless, because both species were so similar, given time, one or the other would eventually triumph in a mixed culture.

Competition, then, has adverse effects on all species that use a limited resource and may result in competitive exclusion of one or more species. It follows that natural selection should favor those individuals of each species that avoid or at least reduce competition. In **resource partitioning,** coexisting species' niches differ from each other in one or more ways. Evidence of resource partitioning in animals is well documented and includes studies in tropical forests of Central and South America that demonstrate little overlap in the diets of fruit-eating birds, primates, and bats that coexist in the same habitat. Although fruits are the primary food for several hundred bird, primate, and bat species, the wide variety of fruits available has allowed fruit eaters to specialize, thereby reducing competition.

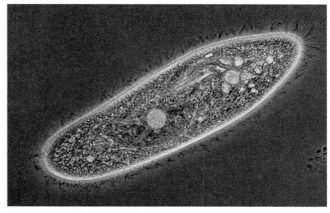

(a) Gause studied competition between two species of *Paramecium*, a unicellular protist. (Michael Abbey/Photo Researchers, Inc.)

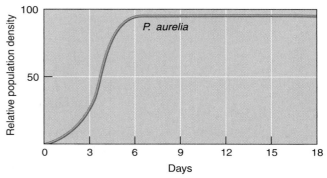

(b) How a population of *P. aurelia* grows in separate culture (in a single-species environment).

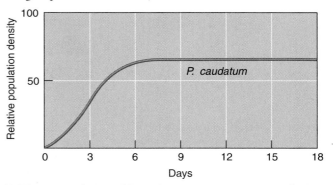

(c) How a population of *P. caudatum* grows in separate culture.

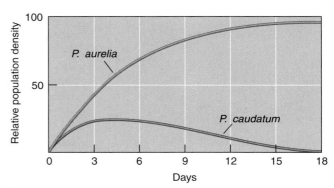

(d) How these two species grow in a mixed culture, in competition with each other. *P. aurelia* outcompetes *P. caudatum* and drives it to extinction. (b, c, d, adapted from G. F. Gause)

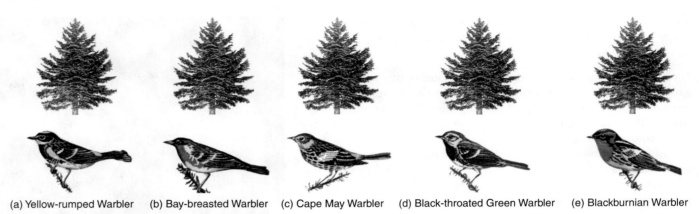

(a) Yellow-rumped Warbler (b) Bay-breasted Warbler (c) Cape May Warbler (d) Black-throated Green Warbler (e) Blackburnian Warbler

Figure 4.16 **Resource partitioning.** Robert MacArthur's study of five North American warbler species is a classic example of resource partioning. Although it initially appeared that their niches were nearly identical, MacArthur determined that individuals of each species spend most of their feeding time in different parts (brown areas) of the spruces and other conifer trees they frequent. They also move in different directions through the canopy, consume different combinations of insects, and nest at slightly different times. (Adapted from R. H. MacArthur)

Resource partitioning may include timing of feeding, location of feeding, nest sites, and other aspects of an organism's ecological niche. **Robert MacArthur**'s study of five North American warbler species is a classic example of resource partitioning (**Figure 4.16**).

REVIEW

1. What is an ecological niche?

2. What are limiting resources? How might a limiting resource affect a species' ecological niche?

3. What is the principle of competitive exclusion? resource partitioning?

 ## Keystone Species

LEARNING OBJECTIVE

• Define keystone species and discuss the wolf as a keystone species.

keystone species: A species, often a predator, that exerts a profound influence on a community in excess of that expected by its relative abundance.

Certain species are more crucial to the maintenance of their ecosystem than others. Such species, called **keystone species**, are vital in determining the nature and structure of the entire ecosystem—that is, its species composition and its ecosystem functioning. The fact that other species depend on or are greatly affected by the keystone species is revealed when the keystone species is removed. Keystone species are usually not the most abundant species in the ecosystem.

Identifying and protecting keystone species are crucial goals of conservation biologists because if a keystone species disappears from an ecosystem, many other organisms in that ecosystem may become more common, rare, or even disappear. One example of a keystone species is a top predator such as the gray wolf, which was discussed in the chapter introduction. Where wolves were hunted to extinction, the populations of deer, elk, and other herbivores increased explosively. As these herbivores overgrazed the vegetation, many plant species that could not tolerate such grazing pressure disappeared. Many smaller animals such as insects were lost from the ecosystem because the plants that they depended on for food were now less abundant. Thus, the disappearance of the wolf resulted in an ecosystem with considerably less biological diversity.

To date, few long-term studies have identified keystone species and determined the nature and magnitude of their effects on the ecosystems they inhabit. Additional studies are urgently needed to provide concrete information about the importance of keystone species in conservation biology.

REVIEW

1. What is a keystone species? Why is the wolf considered a keystone species?

Species Richness

LEARNING OBJECTIVES

- Describe factors associated with high species richness.
- Give several examples of ecosystem services.

Species richness varies greatly from one community to another. Tropical rain forests and coral reefs are examples of communities with extremely high species richness. In contrast, geographically isolated islands and mountaintops exhibit low species richness.

What determines the number of species in a community? Several factors appear to be significant: the abundance of potential ecological niches, closeness to the margins of adjacent communities, geographic isolation, dominance of one species over others, habitat stress, and geologic history.

Species richness is related to the abundance of potential ecological niches. An already complex community offers a greater variety of potential ecological niches than does a simple community. For example, in a study of chaparral habitats (shrubby and woody areas) in California, those with structurally complex vegetation provided birds with more kinds of food and hiding places than communities with low structural complexity (**Figure 4.17**).

Species richness is usually greater at the margins of adjacent communities than in their centers. This is because an **ecotone**—a transitional zone where two or more communities meet—contains all or most of the ecological niches of the adjacent communities as well as some niches unique to the ecotone. The change in species composition produced at ecotones is known as the **edge effect.**

Species richness is inversely related to the geographic isolation of a community. Isolated island communities are much less diverse than communities in similar environments found on continents. This is partly the result of the difficulty many species have in reaching and successfully colonizing the island (recall the discussion of primary succession on Krakatoa). Sometimes species become locally extinct as a result of random events, and in isolated environments such as islands or mountaintops, extinct species are not readily replaced. Isolated areas are often small and possess fewer potential ecological niches.

Species richness is reduced when any one species enjoys a decided position of dominance within a community because it may appropriate a disproportionate share of available resources, thus crowding out other species. Ecologist **James H. Brown** and colleagues of the University of New Mexico have studied species competition and diversity in long-term experiments conducted since 1977 in the Chihuahuan desert of southeastern Arizona. In one experiment, the scientists enclosed their study areas with fencing and then cut holes in the fencing to allow smaller rodents to come and go but to exclude the larger kangaroo rats. The removal of three dominant species, all kangaroo rats, from several plots resulted in an increased diversity of other rodent species. This increase was attributed to less competition for food and to an altered habitat because the abundance of grass species increased dramatically after the removal of the kangaroo rats.

Generally, species richness is inversely related to the environmental stress of a habitat. Only those species capable of tolerating extreme environmental conditions can live in an environmentally stressed community. The species richness of a highly polluted stream is low compared to that of a nearby pristine stream. Similarly, the species richness of high-latitude (further from the equator) communities exposed to harsh climates is less than that of lower-latitude (closer to the equator) communities with milder climates. Although the equatorial countries of Colombia, Ecuador, and Peru occupy only 2% of Earth's land, they contain an astonishing 45,000 native plant species. The

species richness: The number of species in a community.

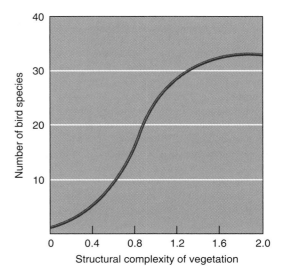

Figure 4.17   Effect of community complexity on species richness.   The structural complexity of chaparral vegetation (*x*-axis) in California is numerically assigned based on height and density of vegetation, from low complexity (very dry scrub) to high complexity (woodland). Note that species richness in birds increases as the vegetation becomes more structurally complex. (After M. L. Cody and J. M. Diamond)

continental United States and Canada, with a significantly larger land area, possess a total of 19,000 native plant species. Ecuador alone contains more than 1,300 native bird species, twice as many as the United States and Canada combined.

Geologic history greatly affects species richness. Tropical rain forests are probably old, stable communities that have undergone few climate changes in Earth's entire history. During this time, myriad species evolved in tropical rain forests, having experienced few or no abrupt climate changes that might have led to their extinction. In contrast, glaciers have repeatedly altered temperate and arctic regions during Earth's history. An area recently vacated by glaciers will have a low species richness because few species will as yet have had a chance to enter it and become established.

Species Richness, Ecosystem Services, and Community Stability

Ecologists and conservationists have long debated whether the extinction of species threatens the normal functioning and stability of ecosystems. This question is of great practical concern because ecosystems supply human societies with many environmental benefits (**Table 4.1**). Conservationists maintain that ecosystems with greater species richness better supply such ecosystem services than ecosystems with lower species richness.

Traditionally, most ecologists assumed that **community stability**—the ability of a community to withstand environmental disturbances—is a consequence of community complexity. That is, a community with considerable species richness may function better and be more stable than a community with less species richness. According to this view, the greater the species richness, the less critically important any single species should be. With many possible interactions within the community, it is unlikely that any single disturbance could affect enough components of the system to make a significant difference in its functioning. Evidence for this hypothesis includes the fact that destructive outbreaks of pests are more common in cultivated fields, which are low-diversity communities, than in natural communities with greater species richness.

Ongoing studies by **David Tilman** of the University of Minnesota and **John Downing** of the University of Iowa have strengthened the link between species richness and community stability. In their initial study, they established and monitored 207 plots of Minnesota grasslands for seven years. During the study, Minnesota's worst

ecosystem services: Important environmental benefits that ecosystems provide to people; include clean air to breathe, clean water to drink, and fertile soil in which to grow crops.

Table 4.1 Ecosystem Services

Ecosystem	Services Provided by Ecosystem
Forests	Purify air and water; produce and maintain soil; absorb carbon dioxide (carbon storage); provide wildlife habitat; provide humans with wood and recreation
Freshwater systems (rivers and streams, lakes, and groundwater)	Moderate water flow and mitigate floods; dilute and remove pollutants; provide wildlife habitat; provide humans with drinking and irrigation water, food, transportation corridors, electricity, and recreation
Grasslands	Purify air and water; produce and maintain soil; absorb carbon dioxide (carbon storage); provide wildlife habitat; provide humans with livestock and recreation
Coasts	Provide a buffer against storms; dilute and remove pollutants; provide wildlife habitat, including food and shelter for young marine species; provide humans with food, harbors and transportation routes, and recreation
Sustainable agricultural ecosystems*	Produce and maintain soil; absorb carbon dioxide (carbon storage); provide wildlife habitat for birds, insect pollinators, and soil organisms; provide humans with food and fiber crops

*Sustainable agricultural ecosystems are human-made and therefore inherently different from other ecosystems. Sustainable agriculture is discussed in Chapter 19.

drought in 50 years occurred (1987–1988). The biologists found that those plots with the greatest number of plant species lost less ground cover and recovered faster than did species-poor plots. Further research supported these conclusions and showed a similar effect of species richness on community stability during nondrought years. Similar work at grassland sites in Europe also supports the link between species richness and ecosystem functioning.

CASE IN POINT

Species Richness in Lake Victoria

Until relatively recently, the world's second largest freshwater lake, Lake Victoria in East Africa, was home to about 400 species of cichlids (pronounced *sik'lids*), small, colorful fishes. The cichlid species in Lake Victoria had remarkably different eating habits. Some grazed on algae; some consumed dead organic material at the bottom of the lake; and others ate insects, shrimp, or other cichlid species. These fishes thrived throughout the lake ecosystem and provided protein to the diets of 30 million humans living in the vicinity.

Today, the aquatic community in Lake Victoria is altered from what it was 50 years ago. More than half of the cichlids and other native fish species are now extinct. As a result of the disappearance of most of the algae-eating cichlids, the algal population has increased explosively. When these algae die, their decomposition uses up the dissolved oxygen in the water. The bottom zone of the lake, once filled with cichlids, is empty because it contains too little dissolved oxygen. Any fishes venturing into the oxygen-free zone suffocate. Local fishermen, who once caught and ate hundreds of types of fishes, now catch only a few types.

A major contributor to the destruction of Lake Victoria's ecological balance was the deliberate introduction of the Nile perch, a large and voracious predator, into the lake (**Figure 4.18**). Proponents of the introduction thought the successful establishment of the Nile perch would stimulate the local economy and help the fishermen. For many years, as its population slowly increased, the Nile perch did not have an appreciable effect on the lake. But by 1980, fishermen noticed they were harvesting increasing quantities of Nile perch and decreasing amounts of native fishes. By 1985, most of the annual catch was Nile perch, which was experiencing a population explosion fueled by an abundant food supply—the cichlids.

ENVIRONEWS

Otters in Trouble

Sea otters, a known keystone species off the Alaskan coast, play an important role: Otters feed on sea urchins, limiting the urchins from eating kelp, which allows the underwater kelp forests to thrive. Since the 1990s, scientists have observed an alarming decline in sea otter populations in western Alaska's Aleutian Islands that in turn poses wide-ranging threats to the coastal ecosystem there. The population of sea urchins in these areas is exploding, and kelp forests are being devastated. Strong evidence identifies orcas as the culprits. Recently, orcas began to prey on sea otters. (Orcas generally feed on sea lions, seals, and fishes.) In comparison, sea otters, the smallest marine mammal species, are more like a snack rather than a desirable meal. So why are the orcas now choosing sea otters? Biologists suggest it is because seal and sea lion populations have collapsed across the North Pacific.

In a scenario that is partly documented and partly speculative, the starting point of this chain of events is a drop in fish stocks, possibly caused by overfishing or climate change. With their food fish in decline, seal and sea lion populations have suffered, and orcas have looked elsewhere for food. Such a change in the orcas' feeding behavior has transformed the food web of kelp forests, putting orcas rather than otters at the top.

Figure 4.18   Nile perch. This fish, which was deliberately introduced into Lake Victoria, has caused ecological havoc. (Tom McHugh/Photo Researchers, Inc.)

Human-caused factors such as pollution are also responsible for the cichlids' disappearance. As nearby forests were cut for firewood to dry the large Nile perch, soil erosion made the water of Lake Victoria turbid (cloudy). Agricultural practices in the area contributed fertilizer as well as sediment pollution from soil erosion, and both of these pollutants increased the turbidity of the water. Thus, the introduction of Nile perch and other environmental changes drastically altered a delicately balanced system.

The Lake Victoria story is far from finished. In the early 2000s, conservation biologists noted that people were taking so many Nile perch from the lake that the perch population was declining. As a result, some of native cichlid populations that had been decimated by Nile perch began to recover. These changes are good news for the cichlids but bad news for the people who now depend on Nile perch for their livelihoods. Biologists are currently trying to determine whether there is a fishing level for Nile perch that is **sustainable** and that will also maintain species richness of native cichlids.   ■

REVIEW

1. What are two determinants of species richness? Give an example of each.
2. What are ecosystem services? Describe some ecosystem services a forest provides.

REVIEW OF LEARNING OBJECTIVES WITH KEY TERMS

• **Define *evolution*.**

Evolution is cumulative genetic changes that occur over time in a population of organisms; evolution explains many patterns observed in the natural world.

• **Explain the four premises of evolution by natural selection as proposed by Charles Darwin.**

Natural selection is the process in which better-adapted individuals—those with a combination of genetic traits better suited to environmental conditions—are more likely to survive and reproduce, increasing their proportion in the population. Natural selection, as envisioned by **Charles Darwin**, has four premises:

1. Each species produces more offspring than will survive to maturity.
2. The individuals in a population exhibit inheritable variation in their traits.
3. Organisms compete with one another for the resources needed to survive.
4. Those individuals with the most favorable combination of traits are most likely to survive and reproduce, passing their genetic characters on to the next generation.

• **Identify the three domains and six kingdoms of living organisms.**

Organisms are classified into three domains: **Archaea, Bacteria,** and **Eukarya;** and six kingdoms: Archaea, Bacteria, Protista (algae, protozoa, slime molds, and water molds), Fungi (molds and yeasts), Plantae, and Animalia.

• **Define *ecological succession* and distinguish between primary and secondary succession.**

Ecological succession is the orderly replacement of one community by another. **Primary succession** is the change in species composition over time in a previously uninhabited environment. **Secondary succession** is the change in species composition that takes place after some disturbance destroys the existing vegetation; soil is already present.

• **Define *symbiosis* and distinguish among mutualism, commensalism, and parasitism.**

Symbiosis, any intimate relationship or association between members of two or more species, includes mutualism, commensalism, and parasitism. **Mutualism** is a symbiotic relationship in which both partners benefit. **Commensalism** is a type of symbiosis in which one organism benefits and the other one is neither harmed nor helped. **Parasitism** is a symbiotic relationship in which one organism benefits and the other is adversely affected.

• **Define *predation* and describe the effects of natural selection on predator-prey relationships.**

Predation is the consumption of one species (the prey) by another (the predator). During **coevolution** between predator and prey, the predator evolves more efficient ways to catch prey, and the prey evolves better ways to escape the predator.

• **Define *competition* and distinguish between intraspecific and interspecific competition.**

Competition is the interaction among organisms that vie for the same resources (such as food or living space) in an ecosystem. Competition occurs among individuals within a population (intraspecific competition) or between species (interspecific competition).

• **Describe the factors that contribute to an organism's ecological niche and distinguish between fundamental niche and realized niche.**

An organism's **ecological niche** is the totality of its adaptations, its use of resources, and the lifestyle to which it is fitted. The ecological niche is the organism's place and role in a complex biotic and abiotic system. Organisms can potentially exploit more resources and play a broader role in the life of their community than they actually do. The potential ecological niche of an organism is its **fundamental niche,** whereas the niche an organism actually occupies is its **realized niche.**

• Give several examples of limiting resources and discuss how they might affect an organism's ecological niche.

A **limiting resource** is any environmental resource that, because it is scarce or at unfavorable levels, restricts the ecological niche of an organism. Examples of limiting resources include the mineral content of soil, temperature extremes, and amount of precipitation.

• Relate the concepts of competitive exclusion and resource partitioning.

Many ecologists think no two species occupy the same niche in the same community for an indefinite period. In **competitive exclusion,** one species excludes another as a result of competition for limited resources. Some species reduce competition by **resource partitioning,** in which they use resources differently.

• Define *keystone species* and discuss the wolf as a keystone species.

A **keystone species** is a species, often a predator, that exerts a profound influence on a community in excess of that expected by its relative abundance. One example of a keystone species is the gray wolf.

Where wolves were hunted to extinction, the populations of deer, elk, and other herbivores increased. As these herbivores overgrazed the vegetation, many plant species disappeared. Many smaller animals such as insects were lost because the plants they depended on for food were now less abundant. Thus, the disappearance of the wolf resulted in an ecosystem with considerably less biological diversity.

• Describe factors associated with high species richness.

Species richness is the number of species in a community. Species richness is often great when there are many potential ecological niches, when the area is at the margins of adjacent communities, when the community is not isolated or severely stressed, when one species does not dominate others, and when communities have a long history.

• Give several examples of ecosystem services.

Ecosystem services are important environmental benefits that ecosystems provide to people; include clean air to breathe, clean water to drink, and fertile soil in which to grow crops.

Thinking About the Environment

1. During mating season, male giraffes slam their necks together in fighting bouts to determine which male is stronger and can mate with females. Explain how long necks may have evolved under this scenario, using Darwin's theory of evolution by natural selection.

2. Describe an example of secondary succession. Begin your description with the specific disturbance that preceded it.

3. What type of symbiotic relationship—mutualism, commensalism, or predation—do you think exists between the pygmy seahorse and the gorgonian coral pictured in Figure 4.12? Explain your answer.

4. Biologists recognize that the three types of symbiosis are not always clearcut. For example, under certain conditions mutualism may become commensalism or even parasitism. What type of symbiosis is it if the fungi in mycorrhizae take so much of their host's food that the host cannot reproduce? Explain your answer.

5. How are symbiosis and predation related to the concept of energy flow through ecosystems that was covered in Chapter 3?

6. Why is a realized niche usually narrower, or more restricted, than a fundamental niche?

7. What portion of the human's fundamental niche are we occupying today? Do you think our realized niche has changed over the past 200 years? Why or why not?

8. Explain how fundamental and realized niches apply to the rodents in the Chihuahuan desert of Arizona.

9. What is the most likely limiting resource for plants and animals in deserts? Explain your answer.

10. Which study discussed in this chapter is a classic example of competitive exclusion?

11. Which study discussed in this chapter is a classic example of resource partitioning?

12. Some biologists think that protecting keystone species would help preserve biological diversity in an ecosystem. Explain.

13. Diagram the food web discussed in the Environews on the acorn-Lyme disease connection.

14. John Muir once said, "Tug at a single thing in nature, and you will find it connected to the universe." Explain this quote using the acorn-Lyme disease connection discussed in the chapter.

15. Draw a diagram of three concentric circles and label the circles to show the relationships among species, ecosystems, and communities. If you were adding symbiosis, predation, and competition to the simple system you have depicted, in which circle(s) would you place them?

Quantitative questions relating to this chapter are on our Web site.

Take a Stand

Visit our Web site at http://www.wiley.com/college/raven (select Chapter 4 from the Table of Contents) for links to more information about the controversies surrounding the reintroduction of wolves into Yellowstone National Park. Consider the opposing views of those who seek to protect the wolves and of ranchers who wish to protect their livestock, and debate the issues with your classmates. You will find tools to help you organize your research, analyze the data, think critically about the issues, and construct a well-considered argument. *Take a Stand* activities can be done individually or as a team, as oral presentations, written exercises, or Web-based (e-mail) assignments.

Additional online materials relating to this chapter, including a Student Testing Section with study aids and self-tests, Environmental News, Activity Links, Environmental Investigations, and more, are also on our Web site.

5

Ecosystems and the Physical Environment

Water surged through the 17th Street Canal, flooding homes in the aftermath of Hurricane Katrina. A study of subsidence in New Orleans three years before Hurricane Katrina suggests that the subsidence may have weakened levees and canals that subsequently failed. (Smiley N. Pool/Dallas Morning News/MCT/NewsCom)

Hurricane Katrina, which hit the north-central Gulf coast along Louisiana, Mississippi, and Alabama in August 2005, was one of the most devastating storms in U.S. history. It produced a storm surge that caused severe damage to the city of New Orleans as well as to other coastal cities and towns in the region. The high waters caused levees and canals to fail, flooding 80% of New Orleans and many nearby neighborhoods (see photograph).

Most people are aware of the catastrophic loss of life and property damage caused by Katrina. Here we focus on how humans have altered the geography and geology of the New Orleans area in ways that have impaired a balanced natural system and exacerbated the storm damage. Many of the world's cities located on river deltas are also vulnerable to storms and flooding.

The Mississippi River delta formed over millennia from sediments deposited at the mouth of the river. The location of New Orleans is

ideal for industry as well as for sea and river commerce. Over the years, engineers have constructed a system of canals to aid navigation and of levees to control flooding since the city is at or below sea level. As the city has grown, new development has taken place on wetlands—bayous, waterways, and marshes—that were drained and filled in.

Before their destruction, these coastal wetlands, which act like a sponge, provided some degree of protection against flooding from storm surges. We are not implying that had Louisiana's wetlands been intact, New Orleans would not have suffered any damage from a hurricane of Katrina's magnitude. However, had these wetlands been largely unaltered, they would have moderated the damage by absorbing much of the water from the storm surge. Wetlands that have been replaced by housing tracts, factories, and roads are less able to absorb excess water, which therefore spreads more rapidly and extensively.

Another reason that Katrina devastated New Orleans is that the city has been subsiding (sinking) for many years. The levees that were built to control occasional flooding also prevent the deposition of sediments that remain behind after floodwaters subside. (The sediments are now deposited in the Gulf of Mexico.) Under natural conditions, these sediments replenish and maintain the delta. The New Orleans area is also subsiding because its rich supply of un-

derground natural resources, such as groundwater, oil, and natural gas, are being extracted. As these resources are removed, the land compacts, lowering the city. New Orleans and nearby coastal areas are subsiding an average of 11 mm each year. At the same time, the sea level has been rising an average of 1 mm to 2.5 mm per year, in part because of human-induced changes in climate.

The destruction caused by Katrina will take many years to repair. City planners; engineers; representatives of local, state, and federal governments; and others are developing long-range plans to rebuild New Orleans. These plans range from a complete rebuilding of all damaged areas to a scaled-back restoration of areas that are less vulnerable to future hurricane damage.

This chapter examines the interactions and connections of Earth's ecosystems with the abiotic, physical environment on which all life depends. We focus on cycles of matter, solar radiation, the atmosphere, the ocean, weather and climate (including hurricanes), and internal planetary processes.

World View and Closer to You...
Real news footage relating to ecosystems and the physical environment around the world and in your own geographic region is available at www.wiley.com/college/raven/ by clicking on "World View and Closer to You."

The Cycling of Materials within Ecosystems

LEARNING OBJECTIVES

- Describe the main steps in each of these biogeochemical cycles: carbon, nitrogen, phosphorus, sulfur, and hydrologic cycles.
- Describe how humans have influenced the carbon, nitrogen, phosphorus, sulfur, and hydrologic cycles.

In Chapter 3, you learned that energy flows in one direction through an ecosystem. In contrast, matter, the material of which organisms are composed, moves through systems in numerous cycles from one part of an ecosystem to another—from one organism to another and from living organisms to the abiotic environment and back again (**Figure 5.1**). These **biogeochemical cycles** involve biological, geologic, and chemical interactions.

Five biogeochemical cycles of matter—carbon, nitrogen, phosphorus, sulfur, and water (hydrologic)—are representative of all biogeochemical cycles. These five cycles are particularly important to organisms because these materials make up the chemical compounds of cells. Carbon, nitrogen, and sulfur are elements that form gaseous compounds, whereas water is a compound that readily evaporates; these four cycles have components that move over large distances in the atmosphere with relative ease. Phosphorus does not form gaseous compounds, and as a result, only local cycling occurs easily. Humans are affecting all of these cycles on both local and global scales; we conclude this section with a case-in-point that discusses this important issue.

The Carbon Cycle

Proteins, carbohydrates, and other molecules essential to life contain carbon, so organisms must have carbon available to them. Carbon makes up approximately 0.04% of the atmosphere as a gas, carbon dioxide (CO_2). Carbon is present in the ocean in several forms: dissolved carbon dioxide—that is, carbonate (CO_3^{2-}) and bicarbonate (HCO_3^-); other forms of dissolved inorganic carbon; and dissolved organic carbon from decay processes. Carbon is also present in sedimentary rocks such as *limestone*, which consists primarily of calcium carbonate ($CaCO_3$). The global movement of carbon between organisms and the abiotic environment—including the atmosphere, ocean, and sedimentary rock—is known as the **carbon cycle** (**Figure 5.2**).

During **photosynthesis,** plants, algae, and certain bacteria remove CO_2 from the air and fix (incorporate) it into chemical compounds such as sugar. Plants use sugar to make other compounds. Thus, photosynthesis incorporates carbon from the abiotic environment into the biological compounds of producers. Those compounds are usually used as fuel for *cellular respiration* by the producer that made them, by a consumer that eats the producer, or by a decomposer that breaks down the remains of the producer or consumer. Thus, cellular respiration returns CO_2 to the atmosphere. A similar carbon cycle occurs in aquatic ecosystems between aquatic organisms and carbon dioxide dissolved in the water.

Sometimes the carbon in biological molecules is not recycled back to the abiotic environment for some time. A large amount of carbon is stored in the wood of trees, where it may stay for several hundred years or even longer. In addition, millions of years ago, vast coal beds formed from the bodies of ancient trees that did not decay fully before they were buried. Similarly, the organic compounds of unicellular marine organisms probably gave rise to the underground deposits of oil and natural gas that accumulated in the geologic past. Coal, oil, and natural gas, called **fossil fuels** because they formed from the

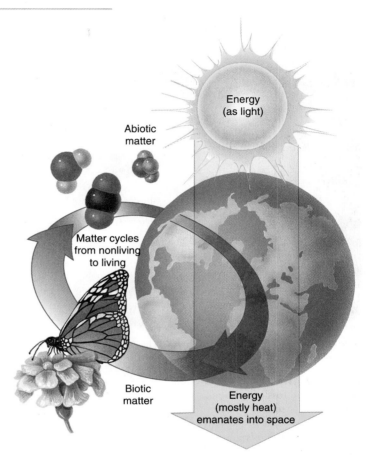

Figure 5.1 Energy and matter. Although energy flows one way through ecosystems, matter continually cycles from the abiotic to the biotic components of ecosystems and back again.

carbon cycle: The global circulation of carbon from the environment to living organisms and back to the environment.

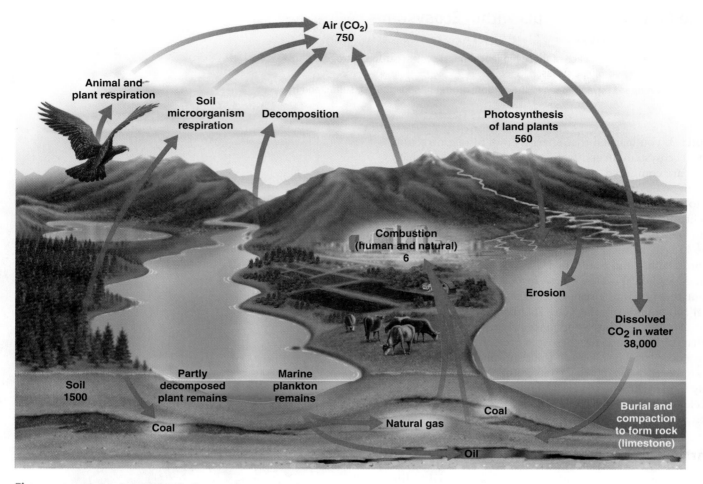

Figure 5.2 **The carbon cycle.** Sedimentary rocks and fossil fuels hold almost all of Earth's estimated 10^{23} g of carbon. The values shown for some of the active pools in the global carbon budget are expressed as 10^{15} g of carbon. For example, the soil contains an estimated 1500×10^{15} g of carbon. (Values from W. H. Schlesinger)

remains of ancient organisms, are vast deposits of carbon compounds, the end products of photosynthesis that occurred millions of years ago (see Chapter 11).

The carbon in coal, oil, natural gas, and wood can return to the atmosphere by burning, or **combustion.** In combustion, organic molecules are rapidly oxidized—combined with oxygen—and converted into CO_2 and water, with an accompanying release of heat and light.

The Carbon-Silicate Cycle On a geologic time scale involving millions of years, the carbon cycle interacts with the silicon cycle in the **carbon-silicate cycle.** The first step of the cycle involves chemical *weathering processes.* Atmospheric CO_2 dissolves in rainwater to form carbonic acid (H_2CO_3), which is a weak acid. As the slightly acidic rainwater moves through the soil, carbonic acid dissociates to form hydrogen ions (H^+) and bicarbonate ions (HCO_3^-). The hydrogen ions enter silicate-rich minerals such as feldspar and change its chemical composition, releasing calcium ions (Ca^{2+}). The calcium and bicarbonate ions wash into surface waters and eventually reach the ocean.

Microscopic marine organisms incorporate Ca^{2+} and HCO_3^- into their shells. When these organisms die, their shells sink to the ocean floor and are covered by sediments, forming carbonate deposits several kilometers thick. The deposits are eventually cemented to form the sedimentary rock limestone. Earth's crust is dynamically active, and over millions of years, sedimentary rock on the bottom of the sea floor may lift to form land surfaces. The summit of Mount Everest, for example, is composed of sedimentary rock. When the process of geologic uplift exposes limestone, it slowly erodes as a result of chemical and physical weathering processes, returning CO_2 to the water and atmosphere to participate in the carbon cycle once again.

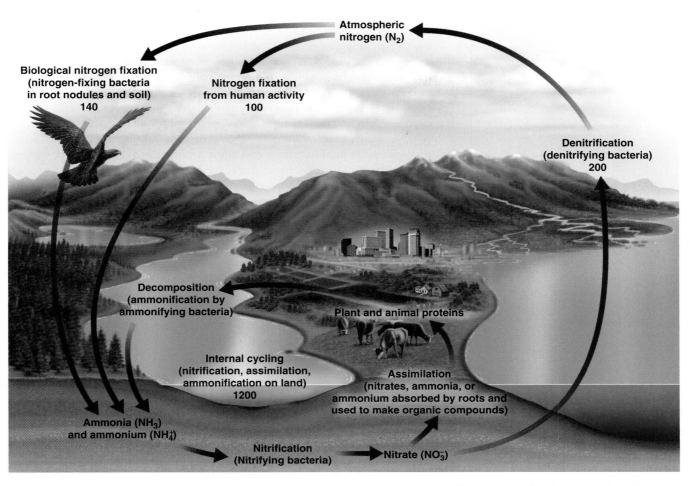

Figure 5.3 The nitrogen cycle. The atmosphere holds the largest pool of nitrogen, 3.9×10^{21} g. The values shown for some of the active pools in the global nitrogen budget are expressed as 10^{12} g of nitrogen per year. For example, each year humans fix an estimated 100×10^{12} g of nitrogen. (Values from W. H. Schlesinger)

Alternatively, the geologic process of *subduction* (discussed later in the chapter) buries the carbonate deposits. The increased heat and pressure of deep burial partially melts the sediments, releasing CO_2, which rises in volcanoes and is vented into the atmosphere.

The Nitrogen Cycle

Nitrogen is crucial for all organisms because it is an essential part of biological molecules such as proteins and nucleic acids (for example, DNA). At first glance it appears that a shortage of nitrogen for organisms is impossible. The atmosphere is 78% nitrogen gas (N_2), a two-atom molecule. But atmospheric nitrogen is so stable that it does not readily combine with other elements. Atmospheric nitrogen must first be broken apart before the nitrogen atoms can combine with other elements to form proteins and nucleic acids.

There are five steps in the **nitrogen cycle**, in which nitrogen cycles between the abiotic environment and organisms: nitrogen fixation, nitrification, assimilation, ammonification, and denitrification (**Figure 5.3**). Bacteria are exclusively involved in all of these steps except assimilation.

The first step in the nitrogen cycle, **nitrogen fixation,** is the conversion of gaseous nitrogen to ammonia (NH_3). The process gets its name from the fact that nitrogen is fixed into a form that organisms can use. Combustion, volcanic action, lightning discharges, and industrial processes, all of which supply enough energy to break apart atmospheric nitrogen, fix considerable nitrogen. Nitrogen-fixing bacteria, including cyanobacteria, carry out biological nitrogen fixation in soil and aquatic environments. Nitrogen-fixing bacteria employ the enzyme *nitrogenase* to split atmospheric nitrogen and combine the resulting nitrogen atoms with hydrogen. Nitrogenase functions only in the absence of oxygen, and the bacteria that use nitrogenase must insulate it from oxygen by some means.

nitrogen cycle: The global circulation of nitrogen from the environment to living organisms and back to the environment.

Some nitrogen-fixing bacteria live beneath layers of oxygen-excluding slime on the roots of certain plants. Other important nitrogen-fixing bacteria, *Rhizobium*, live inside special swellings, or nodules, on the roots of legumes such as beans or peas and some woody plants (See Chapter 4 discussion of mutualism). The relationship between *Rhizobium* and its host plants is mutualistic: The bacteria receive carbohydrates from the plant, and the plant receives nitrogen in a form it can use. In aquatic environments cyanobacteria perform most nitrogen fixation. Filamentous cyanobacteria have special oxygen-excluding cells that function as the sites of nitrogen fixation.

The conversion of ammonia (NH_3) or ammonium (NH_4^+, formed when water reacts with ammonia) to nitrate (NO_3^-) is **nitrification.** Soil bacteria perform nitrification, a two-step process. First, soil bacteria convert ammonia or ammonium to nitrite (NO_2^-). Then other soil bacteria oxidize nitrite to nitrate. The process of nitrification furnishes these bacteria, called nitrifying bacteria, with energy.

In **assimilation,** plant roots absorb nitrate (NO_3^-), ammonia (NH_3), or ammonium (NH_4^+)and incorporate the nitrogen of these molecules into plant proteins and nucleic acids. When animals consume plant tissues, they assimilate nitrogen by taking in plant nitrogen compounds (amino acids) and converting them to animal compounds (proteins).

Ammonification is the conversion of biological nitrogen compounds into ammonia (NH_3) and ammonium ions (NH_4^+). Ammonification begins when organisms produce nitrogen-containing waste products such as urea (in urine) and uric acid (in the wastes of birds). These substances, as well as the nitrogen compounds that occur in dead organisms, are decomposed, releasing the nitrogen into the abiotic environment as ammonia. The bacteria that perform this process both in the soil and in aquatic environments are called ammonifying bacteria. The ammonia produced by ammonification enters the nitrogen cycle and is once again available for the processes of nitrification and assimilation.

Denitrification is the reduction of nitrate (NO_3^-) to gaseous nitrogen. Denitrifying bacteria reverse the action of nitrogen-fixing and nitrifying bacteria by returning nitrogen to the atmosphere. Denitrifying bacteria prefer to live and grow where there is little or no free oxygen. For example, they are found deep in the soil near the water table, a nearly oxygen-free environment.

The Phosphorus Cycle

phosphorus cycle: The global circulation of phosphorus from the environment to living organisms and back to the environment.

Phosphorus does not form compounds in the gaseous phase and does not appreciably enter the atmosphere (except during dust storms). In the phosphorus cycle, phosphorus cycles from the land to sediments in the ocean and back to the land (**Figure 5.4**).

As water runs over apatite[1] and other minerals containing phosphorus, it gradually wears away the surface and carries off inorganic phosphate (PO_4^{3-}) molecules. The erosion of phosphorus-containing minerals releases phosphorus into the soil, where plant roots absorb it in the form of inorganic phosphates. Once in cells, phosphates are incorporated into biological molecules such as nucleic acids and ATP (adenosine triphosphate, an organic compound important in energy transfer reactions in cells). Animals obtain most of their required phosphate from the food they eat, although in some localities drinking water contains a substantial amount of inorganic phosphate. Phosphorus released by decomposers becomes part of the soil's pool of inorganic phosphate for plants to reuse. Like carbon, nitrogen, and other biogeochemical cycles, phosphorus moves through the food web as one organism consumes another.

Phosphorus cycles through aquatic communities in much the same way that it does through terrestrial communities. Dissolved phosphorus enters aquatic communities through absorption and assimilation by algae and plants, which are then consumed by plankton and larger organisms. A variety of fishes and mollusks eat these in turn. Ultimately, decomposers that break down wastes and dead organisms release inorganic phosphorus into the water, where it is available for aquatic producers to use again.

Phosphate can be lost from biological cycles. Streams and rivers carry some phosphate from the land to the ocean, where it can be deposited on the sea floor and remain

[1] Apatite is a group of phosphate-containing minerals that are often found as small green, yellow, or blue crystals in rocks.

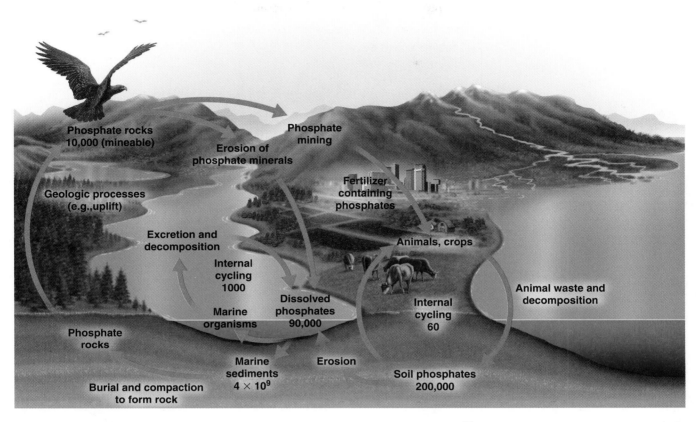

Figure 5.4 **The phosphorus cycle.** The values shown in the figure for the global phosphorus budget are expressed as 10^{12} g of phosphorus per year. For example, each year an estimated 60×10^{12} g of phosphorus cycles from the soil to terrestrial organisms and back again. (Values from W. H. Schlesinger)

for millions of years. The geologic process of uplift may someday expose these sea floor sediments as new land surfaces, from which phosphate will once again erode.

A small portion of the phosphate in the aquatic food web finds its way back to the land. Sea birds that eat fishes and other marine animals may defecate on land where they roost. Guano, the manure of sea birds, contains large amounts of phosphate and nitrate. Once on land, these minerals are available for the roots of plants to absorb. The phosphate contained in guano may enter terrestrial food webs in this way, although the amounts involved are quite small.

The Sulfur Cycle

Scientists are still piecing together how the global **sulfur cycle** works. Most sulfur is underground in sedimentary rocks and minerals (for example, gypsum and anhydrite), which over time erode to release sulfur-containing compounds into the ocean (**Figure 5.5**). Sulfur gases enter the atmosphere from natural sources in both the ocean and land. Sea spray delivers sulfates (SO_4^{2-}) into the air, as do forest fires and dust storms (desert soils are rich in calcium sulfate, $CaSO_4$). Volcanoes release both hydrogen sulfide (H_2S), a poisonous gas with a smell of rotten eggs, and sulfur oxides (SO_x). Sulfur oxides include sulfur dioxide (SO_2), a choking, acrid gas, and sulfur trioxide (SO_3).

Sulfur gases comprise a minor part of the atmosphere and are not long-lived because atmospheric sulfur compounds are reactive. Hydrogen sulfide reacts with oxygen to form sulfur oxides, and sulfur oxides (SO_x) react with water to form sulfuric acid (H_2SO_4). Although the total amount of sulfur compounds present in the atmosphere at any given time is relatively small, the total annual movement of sulfur to and from the atmosphere is substantial.

sulfur cycle: The global circulation of sulfur from the environment to living organisms and back to the environment.

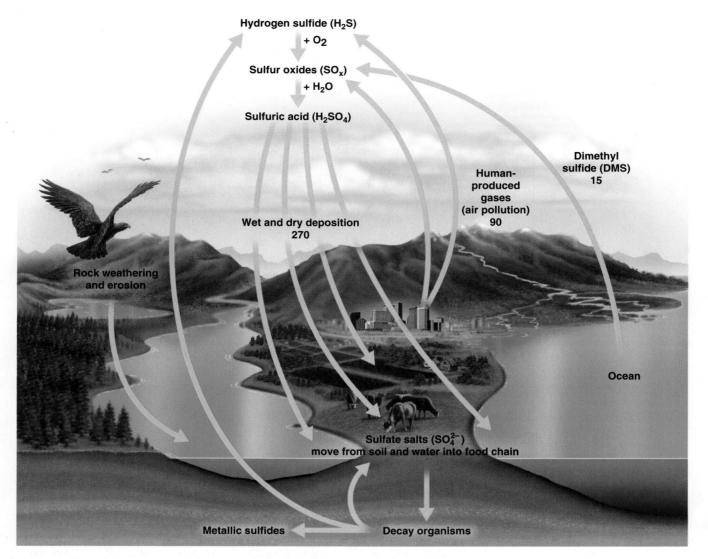

Hydrogen sulfide (H_2S)

+ O_2

Sulfur oxides (SO_x)

+ H_2O

Sulfuric acid (H_2SO_4)

Dimethyl sulfide (DMS)
15

Human-produced gases (air pollution)
90

Wet and dry deposition
270

Rock weathering and erosion

Ocean

Sulfate salts (SO_4^{2-})
move from soil and water into food chain

Metallic sulfides

Decay organisms

Figure 5.5   The sulfur cycle. The largest pool of sulfur on the planet is sedimentary rocks, which contain an estimated 7440×10^{18} g of sulfur. The second largest pool is the ocean, which contains 1280×10^{18} g of sulfur. The values shown in the figure for the global sulfur budget are expressed in units of 10^{12} g of sulfur per year. For example, the ocean emits an estimated 15×10^{12} g of sulfur per year as the gas dimethyl sulfide. (Values from W. H. Schlesinger)

A tiny fraction of global sulfur is present in living organisms, where it is an essential component of proteins. Plant roots absorb sulfate (SO_4^{2-}) and assimilate it by incorporating the sulfur into plant proteins. Animals assimilate sulfur when they consume plant proteins and convert them to animal proteins. In the ocean, certain marine algae release large amounts of a compound that bacteria convert to dimethyl sulfide, or DMS (with a chemical formula of CH_3SCH_3). DMS is released into the atmosphere, where it helps condense water into droplets in clouds and may affect weather and climate. In the atmosphere, DMS is converted to sulfate, most of which is deposited into the ocean.

As in the nitrogen cycle, bacteria drive the sulfur cycle. In freshwater wetlands, tidal flats, and flooded soils, which are oxygen-deficient, certain bacteria convert sulfates to hydrogen sulfide gas, which is released into the atmosphere, or to metallic sulfides, which are deposited as rock. In the absence of oxygen, other bacteria perform an ancient type of photosynthesis that uses hydrogen sulfide instead of water. Where oxygen is present, different bacteria oxidize sulfur compounds to sulfates.

The Hydrologic Cycle

Life would be impossible without water, which makes up a substantial part of the mass of most organisms. All life forms, from bacteria to plants and animals, use water as a medium for chemical reactions as well as for the transport of materials within and among cells.

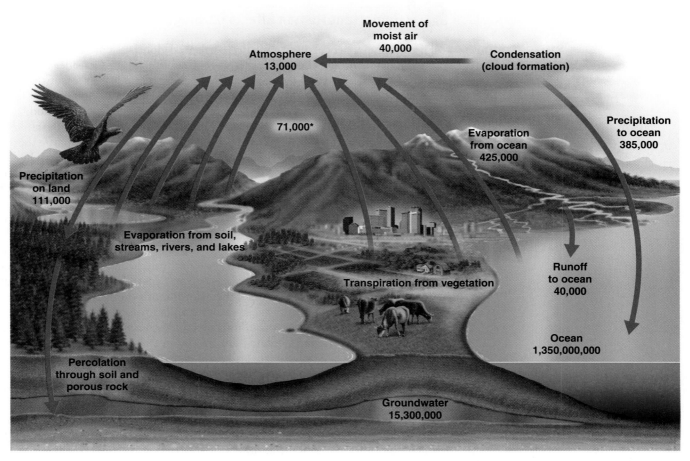

Figure 5.6   The hydrologic cycle. Estimated values for pools in the global water budget are expressed as km³, and the values for movements (associated with arrows) are in km³ per year. The starred value (71,000 km³ per year) includes both transpiration from plants and evaporation from soil, streams, rivers, and lakes. (Values from W. H. Schlesinger)

In the **hydrologic cycle**, water continuously circulates from the ocean to the atmosphere to the land and back to the ocean. It provides a renewable supply of purified water for terrestrial organisms. The hydrologic cycle results in a balance between water in the ocean, on the land, and in the atmosphere (**Figure 5.6**). Water moves from the atmosphere to the land and ocean in the form of precipitation—rain, snow, sleet, or hail. When water evaporates from the ocean surface and from soil, streams, rivers, and lakes on land, it forms clouds in the atmosphere. In addition, **transpiration,** the loss of water vapor from land plants, adds water to the atmosphere. Roughly 97% of the water a plant's roots absorb from the soil is transported to the leaves, where it is transpired.

Water may evaporate from land and reenter the atmosphere directly. Alternatively, it may flow in rivers and streams to coastal **estuaries** where fresh water meets the ocean. The movement of water from land to rivers, lakes, wetlands and, ultimately, the ocean is called **runoff,** and the area of land drained by runoff is a **watershed.** Water also percolates, or seeps, downward through the soil and rock to become **groundwater,** fresh water stored in underground caverns and porous layers of rock. Groundwater may reside in the ground for hundreds to many thousands of years, but eventually it supplies water to the soil, vegetation, streams and rivers, and the ocean.

Regardless of its physical form—solid, liquid, or vapor—or location, every molecule of water moves through the hydrologic cycle. Tremendous quantities of water are cycled annually between Earth and its atmosphere. The volume of water entering the atmosphere each year is about 389,500 km³ (95,000 mi³). Approximately three-fourths of this water reenters the ocean directly as precipitation; the remainder falls on land.

hydrologic cycle: The global circulation of water from the environment to living organisms and back to the environment.

Important Human Effects on Biogeochemical Cycles

Historically, biogeochemical cycles were balanced global systems. Human activities are increasingly disturbing the balance of biogeochemical cycles. Consider the carbon cycle. Since 1850, the advent of the Industrial Revolution, our industrial society has used a lot of energy, most of which we have obtained by burning increasing amounts of fossil fuels—coal, oil, and natural gas. This trend, along with a greater combustion of wood as a fuel and the burning of large sections of tropical forests, has released CO_2 into the atmosphere at a rate greater than the carbon cycle can handle. Numerous studies indicate that the rise of CO_2 in the atmosphere (see Figure 21.2) is causing human-induced global climate change. Global climate change is causing a rise in sea level, changes in precipitation patterns, the death of forests, the extinction of organisms, and problems for agriculture. It has begun to displace people from coastal areas; eventually, millions of people could be displaced.

Human activities have also disturbed the balance of the global nitrogen cycle. During the 20th century, humans more than doubled the amount of fixed nitrogen entering the global nitrogen cycle. (*Fixed nitrogen* refers to nitrogen chemically combined with hydrogen, oxygen, or carbon; we use fixed nitrogen for fertilizer.) Precipitation washes nitrogen fertilizer into rivers, lakes, and coastal areas where it stimulates the growth of algae. As these algae die, their decomposition by bacteria robs the water of dissolved oxygen, which in turn causes other aquatic organisms, including many fishes, to die of suffocation. An excess of nitrogen and other nutrients from fertilizer runoff has caused large, oxygen-depleted dead zones in almost 150 coastal areas around the world. (Chapter 22 discusses the dead zone in the Gulf of Mexico.) In addition, nitrates from fertilizer can leach (dissolve and wash down) through the soil and contaminate groundwater. Many people drink groundwater, and nitrate-contaminated groundwater is dangerous to drink, particularly for infants and small children.

Another human activity that affects the nitrogen cycle is the combustion of fossil fuels. When fossil fuels are burned—in automobiles, for example—the high temperatures of combustion convert some atmospheric nitrogen to **nitrogen oxides,** which produce **photochemical smog,** a mixture of air pollutants that injures plant tissues, irritates eyes, and causes respiratory problems. Nitrogen oxides react with water in the atmosphere to form acids that leave the atmosphere as **acid deposition** and cause the pH of surface waters (lakes and streams) and soils to decrease (see Chapter 20). Acid deposition is linked to declining animal populations in aquatic ecosystems and altered soil chemistry on land.

Humans affect the phosphorus cycle by accelerating the long-term loss of phosphorus from the land. For example, corn grown in Iowa, which contains phosphate absorbed from the soil, may fatten cattle in an Illinois feedlot. Part of the phosphate ends up in feedlot wastes, which may eventually wash into the Mississippi River. When people consume beef from the cattle, more of the phosphate ends up in human wastes that are flushed down toilets into sewer systems. Because sewage treatment rarely removes them, phosphates cause water quality problems in rivers, lakes, and coastal areas. For practical purposes, phosphorus that washes from the land into the ocean is permanently lost from the terrestrial phosphorus cycle (and from further human use), for it remains in the ocean for millions of years. Also, phosphorus is a limiting nutrient to plants and algae in certain aquatic ecosystems. Thus, the addition of excess phosphorus from fertilizer or sewage can contribute to enrichment of the water and lead to undesirable changes (recall the discussion of Lake Washington in Chapter 1).

Coal and, to a lesser extent, oil contain sulfur. When these fuels are burned in power plants, factories, and motor vehicles, sulfur dioxide, a major cause of acid deposition, is released into the atmosphere. Sulfur dioxide is also released during the smelting of sulfur-containing ores of such metals as copper, lead, and zinc. Pollution abatement, such as scrubbing smokestack gases to remove sulfur oxides, has reduced the amount of sulfur emissions in highly developed countries in recent years, but the global level continues to increase.

Some research suggests that air pollution may weaken the global hydrologic cycle. **Aerosols**—tiny particles of air pollution consisting mostly of sulfates, nitrates, carbon, mineral dusts, and smokestack ash—are produced largely from fossil fuel combustion and the burning of forests. Once in the atmosphere, aerosols enhance the scattering and absorption of sunlight in the atmosphere and cause brighter clouds to form. Both the clouds and the light-scattering effect in the atmosphere cause a warming of the atmosphere and a threefold reduction in the amount of solar radiation reaching Earth's surface, including the ocean. Clouds formed in aerosols are less likely to release their precipitation. As a result, scientists think aerosols may affect the availability and quality of water in some regions during the 21st century. Climate change caused by CO_2 is also altering the global hydrologic cycle by increasing glacial and polar ice-cap melting and by increasing evaporation in some areas. (We discuss all of these environmental issues in greater detail in later chapters.) ■

You have seen how living things depend on the abiotic environment to supply energy and essential materials (in biogeochemical cycles). Let us now consider five aspects of the physical environment that also affect organisms: solar radiation, the atmosphere, the ocean, weather and climate, and internal planetary processes.

REVIEW

1. What roles do photosynthesis, cellular respiration, and combustion play in the carbon cycle?

2. What are the five steps of the nitrogen cycle?

3. How does the phosphorus cycle differ from the carbon, nitrogen, and sulfur cycles?

4. What sulfur-containing gases are found in the atmosphere?

 ## Solar Radiation

LEARNING OBJECTIVE

• Summarize the effects of solar energy on Earth's temperature, including the influence of albedos of various surfaces.

The sun makes life on Earth possible. It warms the planet, including the atmosphere, to habitable temperatures. Without the sun's energy, the temperature would approach absolute zero ($-273°C$) and all water would freeze, even in the ocean. The sun powers the hydrologic cycle, carbon cycle, and other biogeochemical cycles and is the primary determinant of climate. Photosynthetic organisms capture the sun's energy and use it to make the food molecules required by almost all forms of life. Most of our fuels—wood, oil, coal, and natural gas—represent solar energy captured by photosynthetic organisms. Without the sun, almost all life would cease.

The sun's energy is the product of a massive nuclear fusion reaction and is emitted into space in the form of electromagnetic radiation—especially visible light and infrared and ultraviolet radiation, which are not visible to the human eye. Approximately one billionth of the total energy released by the sun strikes our atmosphere, and of this tiny trickle of energy, a minute part operates the biosphere.

Clouds and, to a lesser extent, surfaces (especially snow, ice, and the ocean) reflect about 31% of the solar radiation that falls on Earth (**Figure 5.7**). Glaciers and ice sheets have high **albedos** and reflect 80 to 90% of the sunlight hitting their surfaces. At the other extreme, asphalt pavement and buildings have low albedos and reflect 10% to 15%, whereas the ocean and forests reflect only about 5%.

albedo: The proportional reflectance of solar energy from Earth's surface, commonly expressed as a percentage.

As shown in Figure 5.7, the remaining 69% of the solar radiation that falls on Earth is absorbed and runs the hydrologic cycle, drives winds and ocean currents, powers photosynthesis, and warms the planet. Ultimately, all of this energy is lost through the continual radiation of long-wave infrared (heat) energy into space.

Temperature Changes with Latitude

The most significant local variation in Earth's temperature is produced because the sun's energy does not reach all places uniformly. A combination of Earth's roughly spherical shape and the tilt of its axis produces variation in the exposure of the surface to the sun's energy.

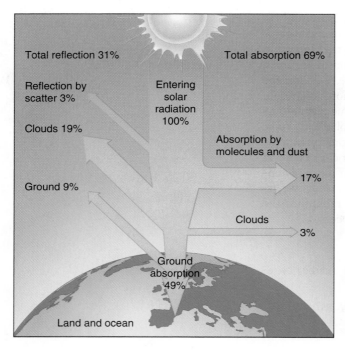

Figure 5.7 Fate of solar radiation that reaches Earth. Most of the sun's energy never reaches Earth. The solar energy that reaches Earth warms the planet's surface, drives the hydrologic cycle and other biogeochemical cycles, produces our climate, and powers almost all life through the process of photosynthesis. (Adapted from A. Strahler)

The principal effect of the tilt is on the angles at which the sun's rays strike different areas of the planet at any one time (**Figure 5.8**). On the average, the sun's rays hit vertically near the equator, making the energy more concentrated and producing higher temperatures. Near the poles, the sun's rays hit more obliquely, and as a result, their energy is spread over a larger surface area. Also, rays of light entering the atmosphere obliquely near the poles pass through a deeper envelope of air than does light entering near the equator. This causes more of the sun's energy to be scattered and reflected back to space, which in turn further lowers temperatures near the poles. Thus, solar energy that reaches polar regions is less concentrated, and temperatures are lower.

Temperature Changes with the Seasons

Seasons are determined primarily by Earth's inclination on its axis. Earth's inclination on its axis is 23.5 degrees from a line drawn perpendicular to the orbital plane. During half of the year (March 21 to September 22) the Northern Hemisphere tilts toward the sun, and during the other half (September 22 to March 21) it tilts away from the sun (**Figure 5.9**). The orientation of the Southern Hemisphere is just the opposite at these times. Summer in the Northern Hemisphere corresponds to winter in the Southern Hemisphere.

REVIEW

1. How does the sun affect temperature at different latitudes? Why?
2. What is albedo?

The Atmosphere

LEARNING OBJECTIVES

- Describe the five layers of Earth's atmosphere: troposphere, stratosphere, mesosphere, thermosphere, and exosphere.
- Discuss the roles of solar energy and the Coriolis effect in producing atmospheric circulation.
- Define *prevailing winds* and distinguish among polar easterlies, westerlies, and trade winds.

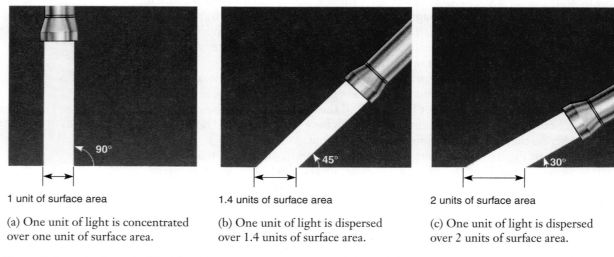

(a) One unit of light is concentrated over one unit of surface area.

(b) One unit of light is dispersed over 1.4 units of surface area.

(c) One unit of light is dispersed over 2 units of surface area.

Figure 5.8   Solar intensity and latitude. The angle at which the sun's rays strike Earth varies from one geographic location to another owing to Earth's spherical shape and its inclination on its axis. (a) Sunlight (represented by the flashlight) that shines vertically near the equator is concentrated on Earth's surface. (b, c) As one moves toward the poles, the light hits the surface more and more obliquely, spreading the same amount of radiation over larger and larger areas.

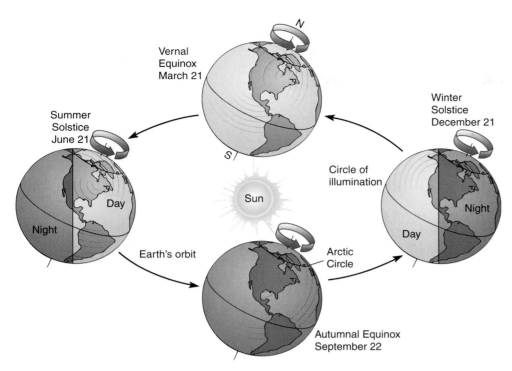

Figure 5.9 Progression of seasons. Earth's inclination on its axis remains the same as it travels around the sun. The sun's rays hit the Northern Hemisphere obliquely during its winter months and more directly during its summer. In the Southern Hemisphere, the sun's rays are oblique during its winter, which corresponds to the Northern Hemisphere's summer. At the equator, the sun's rays are approximately vertical on March 21 and September 22.

The atmosphere is an invisible layer of gases that envelops Earth. Oxygen (21%) and nitrogen (78%) are the predominant gases in the atmosphere, accounting for about 99% of dry air. Other gases, including argon, carbon dioxide, neon, and helium, make up the remaining 1%. In addition, water vapor and trace amounts of various air pollutants, such as methane, ozone, dust particles, microorganisms, and chlorofluorocarbons (CFCs) are present in the air. The atmosphere becomes less dense as it extends outward into space; as a result of gravity, most of the atmosphere's mass is found near Earth's surface (**Figure 5.10a**).

The atmosphere performs several ecologically important functions. It protects Earth's surface from most of the sun's ultraviolet radiation and X-rays as well as from lethal amounts of cosmic rays from space. Without this shielding by the atmosphere, most life would cease to exist. While the atmosphere protects Earth from high-energy radiation, it allows visible light and some infrared radiation to penetrate, and they warm the surface and the lower atmosphere. This interaction between the atmosphere and solar energy is responsible for weather and climate.

Organisms depend on the atmosphere, but they maintain and, in certain instances, modify its composition. Atmospheric oxygen is thought to have increased to its present level as a result of millions of years of photosynthesis. A balance between oxygen-producing photosynthesis and oxygen-using respiration maintains the current level of oxygen.

Layers of the Atmosphere

The atmosphere is composed of a series of five concentric layers—the troposphere, stratosphere, mesosphere, thermosphere, and exosphere (**Figure 5.10b**). These layers vary in altitude and temperature with latitude and season. The **troposphere** extends to a height of approximately 10 km (6.2 mi). The temperature of the troposphere decreases with increasing altitude about $-6°C$ ($-11°F$) for every kilometer. Weather, including turbulent wind, storms, and most clouds, occurs in the troposphere.

troposphere: The layer of the atmosphere closest to Earth's surface.

Figure 5.10   The atmosphere.

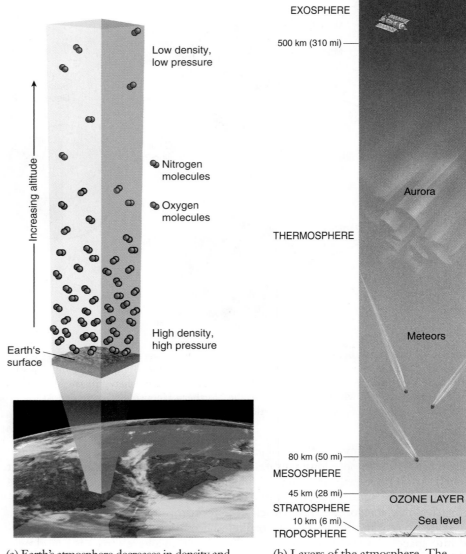

(a) Earth's atmosphere decreases in density and pressure with increasing altitude. The force of gravity is responsible for attracting more air molecules closer to Earth's surface.

(b) Layers of the atmosphere. The troposphere is closest to Earth's surface. The stratosphere is above the troposphere, followed by the mesosphere and thermosphere. The outermost layer, the exosphere, has no distinct boundary separating it from interplanetary space.

stratosphere: The layer of the atmosphere found directly above the troposphere.

In the next layer of atmosphere, the **stratosphere**, there is a steady wind but no turbulence. There is little water, and the temperature is more or less uniform (−45°C to −75°C) in the lower stratosphere; commercial jets fly here. The stratosphere extends from 10 km to 45 km (6.2 mi to 28 mi) above Earth's surface and contains a layer of ozone critical to life because it absorbs much of the sun's damaging ultraviolet radiation. The absorption of ultraviolet radiation by the ozone layer heats the air, and so temperature increases with increasing altitude in the stratosphere.

The **mesosphere,** the layer of atmosphere directly above the stratosphere, extends from 45 km to 80 km (28 mi to 50 mi) above Earth's surface. Temperatures drop steadily in the mesosphere to the lowest in the atmosphere—as low as −138°C.

The **thermosphere** extends from 80 km to 500 km (50 mi to 310 mi) and is very hot. Gases in the thin air of the thermosphere absorb X-rays and short-wave ultraviolet radiation. This absorption drives the few molecules present to great speeds, raising their temperature in the process to 1000°C or more. The aurora, a colorful display

Figure 5.11 **Atmospheric circulation.**

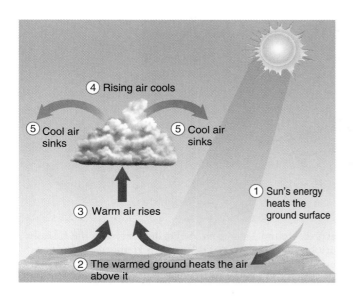

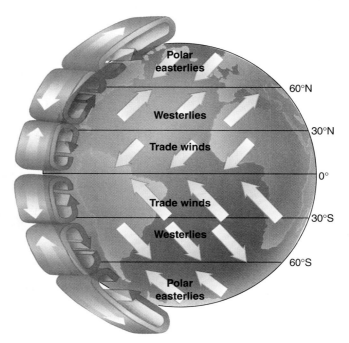

(a) In atmospheric convection, heating of the ground surface heats the air, producing an updraft of less dense, warm air. The convection process ultimately causes air currents that mix warmer and cooler parts of the atmosphere.

(b) Atmospheric circulation transports heat from the equator to the poles (left side of figure). The greatest solar energy input occurs at the equator, heating air most strongly in that area. The air rises, travels toward the poles, and cools in the process so that much of it descends again at around 30 degrees latitude in both hemispheres. At higher latitudes, the patterns of air circulation are more complex.

of lights in dark polar skies, is produced when charged particles from the sun hit oxygen or nitrogen molecules in the thermosphere. The thermosphere is important in long-distance communication because it reflects outgoing radio waves back to Earth without the aid of satellites.

The outermost layer of the atmosphere, the **exosphere,** begins about 500 km (310 mi) above Earth's surface. The exosphere continues to thin until it converges with interplanetary space.

Atmospheric Circulation

In large measure, differences in temperature caused by variations in the amount of solar energy reaching different locations on Earth drive the circulation of the atmosphere. The warm surface near the equator heats the air in contact with it, causing this air to expand and rise (**Figure 5.11**). As the warm air rises, it cools and then sinks again. Much of it recirculates almost immediately to the same areas it has left, but the remainder of the heated air splits and flows in two directions toward the poles. The air chills enough to sink to the surface at about 30 degrees north and south latitudes. This descending air splits and flows over the surface in two directions. Similar upward movements of warm air and its subsequent flow toward the poles occur at higher latitudes farther from the equator. At the poles, the cold polar air sinks and flows toward the lower latitudes, generally beneath the sheets of warm air that simultaneously flow toward the poles. The constant motion of air transfers heat from the equator toward the poles, and as the air returns, it cools the land over which it passes. This continuous turnover moderates temperatures over Earth's surface.

Surface Winds In addition to global circulation patterns, the atmosphere exhibits complex horizontal movements commonly called **winds.** The nature of wind, with its gusts, eddies, and lulls, is difficult to understand or predict. It results in part from differences in atmospheric pressure and from the Earth's rotation.

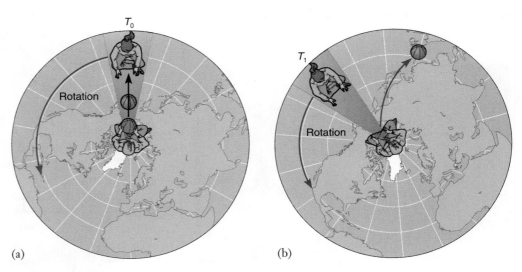

Figure 5.12 **Coriolis effect.** A merry-go-round (seen from above) demonstrates the Coriolis effect. To help visualize how the Coriolis effect operates, the Northern Hemisphere has been superimposed over the merry-go-round. The center of the merry-go-round corresponds to the North Pole, and the outer edge to the equator. Imagine you are sitting on the merry-go-round (in the center, at the North Pole), and a friend is also on the merry-go-round (sitting close to the equator); the merry-go-round is rotating counterclockwise. (a) At time zero, T_0, you throw a ball to your friend, but (b) by the time it would get to her, at time T_1, she is no longer in that position. Thus the ball appears to curve to the right instead of going straight. Winds curve to the right in the Northern Hemisphere and to the left in the Southern Hemisphere because of Earth's rotation.

Coriolis effect: The influence of Earth's rotation, which tends to turn fluids (air and water) toward the right in the Northern Hemisphere and toward the left in the Southern Hemisphere.

Atmospheric gases have weight and exert a pressure that is, at sea level, about 1,013 millibars (14.7 lb per in^2). Air pressure is variable, changing with altitude, temperature, and humidity. Winds tend to blow from areas of high atmospheric pressure to areas of low pressure, and the greater the difference between the high- and low-pressure areas, the stronger the wind.

Earth's rotation influences the direction of wind. Earth rotates from west to east, which causes the east-west movements of surface winds to deflect from their straight-line paths. The moving air swerves to the right of the direction in which it is traveling in the Northern Hemisphere and to the left of the direction in which it is traveling in the Southern Hemisphere. This tendency is the result of the **Coriolis effect**. In other words, the Coriolis effect deflects air currents in the direction of Earth's rotation. The Coriolis effect is greater at higher latitudes and negligible at the equator. Air moving eastward or westward at the equator is not deflected from its path.

To visualize the Coriolis effect, imagine you and a friend are sitting about 10 ft apart on a merry-go-round turning counterclockwise (**Figure 5.12**). Suppose you throw a ball directly to your friend. By the time the ball reaches the place where your friend was, he or she is no longer in that spot. From your vantage point, the ball will have swerved far to the right of your friend. This is how the Coriolis effect works in the Northern Hemisphere.

To visualize how the Coriolis effect works in the Southern Hemisphere, imagine you and your friend are sitting on the same merry-go-round, only this time it is moving clockwise. Now when you throw the ball, it will swerve far to the left of your friend (from your vantage point).

prevailing winds: Major surface winds that blow more or less continually.

The atmosphere has three **prevailing winds** (see Figure 5.11b). Prevailing winds that generally blow from the northeast near the North Pole or from the southeast near the South Pole are **polar easterlies.** Winds that generally blow in the midlatitudes from

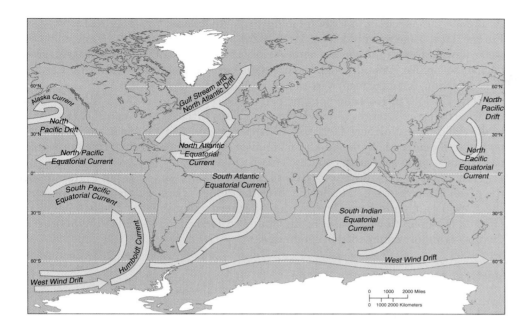

Figure 5.13 **Surface ocean currents.**
Winds largely cause the basic pattern of ocean currents. The main ocean current flow—clockwise in the Northern Hemisphere and counterclockwise in the Southern Hemisphere—results partly from the Coriolis effect.

the southwest in the Northern Hemisphere or from the northwest in the Southern Hemisphere are **westerlies.** Tropical winds that generally blow from the northeast in the Northern Hemisphere or from the southeast in the Southern Hemisphere are called **trade winds.**

REVIEW

1. What is the innermost layer of the atmosphere? Which layer of the atmosphere contains ozone that absorbs much of the sun's ultraviolet radiation?
2. What basic forces determine the circulation of the atmosphere?

The Global Ocean

LEARNING OBJECTIVES

- Discuss the roles of solar energy and the Coriolis effect in producing global water flow patterns, including gyres.
- Define El Niño-Southern Oscillation (ENSO) and La Niña and describe some of their effects.

The global ocean is a huge body of salt water that surrounds the continents and covers almost three-fourths of Earth's surface. It is a single, continuous body of water, but geographers divide it into four sections separated by the continents: the Pacific, Atlantic, Indian, and Arctic oceans. The Pacific Ocean is the largest. It covers one-third of Earth's surface and contains more than half of Earth's water.

Patterns of Circulation in the Ocean

The persistent prevailing winds blowing over the ocean produce surface-ocean water **currents (Figure 5.13).** The prevailing winds generate **gyres,** circular ocean currents. In the North Atlantic Ocean, the tropical trade winds tend to blow toward the west, whereas the westerlies in the midlatitudes blow toward the east. This helps establish a clockwise gyre in the North Atlantic. That is, the trade winds produce the westward North Atlantic Equatorial Current in the tropical North Atlantic Ocean. When this current reaches the North American continent, it is deflected northward, where the westerlies begin to influence it.

gyres: Large, circular ocean current systems that often encompass an entire ocean basin.

Figure 5.14   Ocean and landmasses in the Northern and Southern Hemispheres.

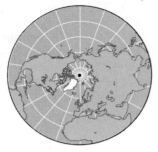

(a) The Northern Hemisphere as viewed from the North Pole.

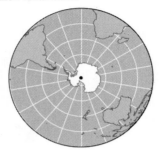

(b) The Southern Hemisphere as viewed from the South Pole. Ocean currents are freer to flow in a circumpolar manner in the Southern Hemisphere.

Figure 5.15   Ocean conveyor belt. This loop consists of both warm, shallow water and cold, deep water. The ocean conveyor belt is responsible for warming Europe and may affect global climate. (After W. S. Broecker.)

As a result, the current flows eastward in the midlatitudes until it reaches the landmass of Europe. Here some water is deflected toward the pole and some toward the equator. The water flowing toward the equator comes under the influence of trade winds again, producing the circular gyre. Although surface-ocean currents and winds tend to move in the same direction, there are many variations on this general rule.

The Coriolis effect influences the paths of surface-ocean currents just as it does the winds. Earth's rotation from west to east causes surface-ocean currents to swerve to the right in the Northern Hemisphere, helping establish the circular, clockwise pattern of water currents. In the Southern Hemisphere, ocean currents swerve to the left, thereby moving in a circular, counterclockwise pattern.

The position of landmasses affects ocean circulation. As you see in **Figure 5.14**, the ocean is not distributed uniformly over the globe: There is clearly more water in the Southern Hemisphere than in the Northern Hemisphere. The circumpolar (around the pole) flow of water in the Southern Hemisphere—called the Southern Ocean—is almost unimpeded by landmasses.

Vertical Mixing of Ocean Water

The varying **density** (mass per unit volume) of seawater affects deep-ocean currents. Cold, salty water is denser than warmer, less salty water. (The density of water increases with decreasing temperature down to 4°C.) Colder, salty ocean water sinks and flows under warmer, less salty water, generating currents far below the surface. Deep-ocean currents often travel in different directions and at different speeds than do surface currents, in part because the Coriolis effect is more pronounced at greater depths. **Figure 5.15** shows the present circulation of shallow and deep currents—the **ocean conveyor belt**—that moves cold, salty deep-sea water from higher to lower latitudes. Note that the Atlantic Ocean gets its cold, deep water from the Arctic Ocean, whereas the Pacific and Indian Oceans get theirs from the water surrounding Antarctica.

The ocean conveyor belt affects regional and possibly global climate. As the Gulf Stream and North Atlantic Drift (see Figure 5.13) push into the North Atlantic, they deliver an immense amount of heat from the tropics to Europe. As this shallow current transfers its heat to the atmosphere, the water becomes denser and sinks. The deep

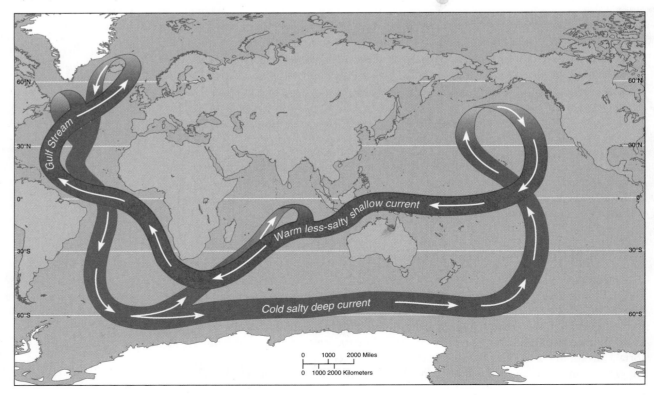

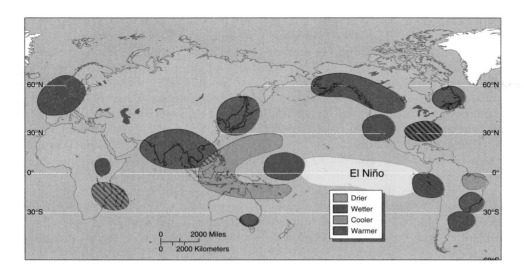

Figure 5.16 **Climate patterns associated with ENSO.** El Niño-Southern Oscillation (ENSO) events drastically alter the climate in many areas remote from the Pacific Ocean. As a result of ENSO, some areas are drier, some wetter, some cooler, and some warmer than usual. Typically, northern areas of the contiguous United States are warmer during the winter, whereas southern areas are cooler and wetter. (National Oceanic and Atmospheric Administration)

current flowing southward in the North Atlantic is, on average, 8°C (14°F) cooler than the shallow current flowing northward.

Evidence from seafloor sediments and Greenland ice indicates that the ocean conveyor belt is not unchanging but shifts from one equilibrium state to another in a relatively short period (a few years to a few decades). The present ocean conveyor belt reorganized between 11,000 and 12,000 years ago. During this period, heat transfer to the North Atlantic stopped, and both North America and Europe experienced conditions of intense cold. Global temperatures also dropped during this time. The exact causes and effects of such large shifts in climate are not currently known, but scientists are concerned that human activities may unintentionally affect the link between the ocean conveyor belt and global climate.

Ocean Interactions with the Atmosphere

The ocean and the atmosphere are strongly linked, with wind from the atmosphere affecting the ocean currents and heat from the ocean affecting atmospheric circulation. One of the best examples of the interaction between ocean and atmosphere is the **El Niño-Southern Oscillation (ENSO)** event, which is responsible for much of Earth's interannual (from one year to the next) climate variability. ENSO has global implications because it results in unusual weather in areas far from the tropical Pacific (**Figure 5.16**). Normally, westward-blowing trade winds restrict the warmest waters to the western Pacific near Australia. Every three to seven years, however, the trade winds weaken, and the warm mass of water expands eastward to South America, increasing surface temperatures in the East Pacific. Ocean currents, which normally flow westward in this area, slow down, stop altogether, or even reverse and go eastward. The name for this phenomenon, **El Niño** (Spanish, "the boy child"), refers to the Christ child because the warming usually reaches the fishing grounds off Peru just before Christmas. Most El Niños last from one to two years.

El Niño has a devastating effect on the fisheries off South America. Normally, the colder, nutrient-rich deep water[2] is about 40 m (130 ft) below the surface and **upwells** (comes to the surface) along the coast, partly in response to strong trade winds (**Figure 5.17**). During an El Niño event, the colder, nutrient-rich deep water is about 152 m (500 ft) below the surface in the Eastern Pacific, and the warmer surface temperatures and weak trade winds prevent upwelling. The lack of nutrients in the water results in a severe decrease in the populations of anchovies and many other marine fishes. During the 1982–1983 El Niño, one of the worst ever recorded, the anchovy population decreased by 99%. Other species such as shrimp and scallops thrive during an El Niño event.

[2] Why are deep waters nutrient-rich? As aquatic organisms die, their remains sink into the deep benthic environment; here they decay, releasing nutrients.

El Niño-Southern Oscillation (ENSO): A cycling of alternating warming and cooling of surface waters of the tropical eastern Pacific Ocean that affects both ocean and atmospheric circulation patterns.

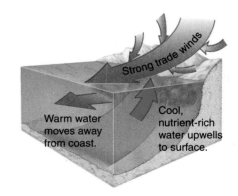

Figure 5.17 **Upwelling.** Coastal upwelling, where deeper waters come to the surface, occurs in the Pacific Ocean along the South American coast. Upwelling provides nutrients for microscopic algae, which in turn support a complex food web. Coastal upwelling weakens considerably during years with El Niño events, temporarily reducing fish populations.

ENSO alters global air currents, directing unusual weather to areas far from the tropical Pacific. The 1997–1998 ENSO, the strongest on record, caused more than 20,000 deaths and $33 billion in property damages worldwide. It resulted in heavy snows in parts of the western United States and ice storms in eastern Canada. This ENSO was responsible for torrential rains that flooded Peru, Ecuador, California, Arizona, and Western Europe and for droughts in Texas, Australia, and Indonesia. Indonesia was particularly hurt by an ENSO-related environmental crisis for which humans were mostly responsible: The 1997–1998 drought in Indonesia was the worst in 50 years. It exacerbated fires, many deliberately set by large, multinational companies to clear land for agriculture (such as rice and oil palm plantations). The fires burned out of control and destroyed an area as large as the state of New Jersey.

El Niño is not the only periodic ocean temperature event to affect the tropical Pacific Ocean. **La Niña** (Spanish, "the little girl") occurs when the surface water temperature in the eastern Pacific Ocean becomes unusually cool, and westbound trade winds become unusually strong. La Niña often occurs after an El Niño event and is considered part of the natural oscillation of ENSO. During the spring of 1998, the surface water of the Eastern Pacific cooled 6.7°C (12°F) in just 20 days. Like El Niño, La Niña affects weather patterns around the world, but its effects are more difficult to predict. In the contiguous United States, La Niña typically causes wetter than usual winters in the Pacific Northwest, warmer weather in the Southeast, and drought conditions in the Southwest. Atlantic hurricanes are stronger and more numerous during a La Niña event.

REVIEW

1. How are the sun's energy, prevailing winds, and surface ocean currents related?

2. What is the El Niño-Southern Oscillation (ENSO)? What are some of its global effects?

Weather and Climate

LEARNING OBJECTIVES

- Distinguish between weather and climate and give three causes of regional precipitation differences.
- Describe a rain shadow.
- Contrast tornadoes and tropical cyclones.

Weather refers to the conditions in the atmosphere at a given place and time; it includes temperature, atmospheric pressure, precipitation, cloudiness, humidity, and wind. Weather changes from one hour to the next and from one day to the next.

The two most important factors that determine an area's **climate** are temperature—both average temperature and temperature extremes—and precipitation—average precipitation, seasonal distribution, and variability. Other climate factors include wind, humidity, fog, and cloud cover. Depending on their layers, altitude, and density, clouds can absorb or reflect sunlight and can retain the planet's outgoing heat; details of the effects clouds have on climate have not yet been resolved. Lightning is an important aspect of climate in some areas because it starts fires.

Day-to-day variations, day-to-night variations, and seasonal variations in climate factors are important dimensions of climate that affect organisms. Latitude, elevation, topography, vegetation, distance from the ocean, and location on a continent or other landmass influence temperature, precipitation, and other aspects of climate. Unlike weather, which changes rapidly, climate usually changes slowly, over hundreds or thousands of years.

Earth has many climates, and because each is relatively constant for many years, organisms have adapted to them. The many kinds of organisms on Earth are here in part because of the large number of climates—from cold, snow-covered polar climates to tropical climates where it is hot and rains almost every day. A German botanist and

climate: The average weather conditions that occur in a place over a period of years.

climatologist, **Wladimir Köppen**, developed the most widely used system for classifying climates in the early part of the 20th century. He based his scheme on the observation that various types of vegetation are associated with different climates, particularly temperature and precipitation (see Chapter 6). **Figure 5.18** shows a world climate map modified from Köppen. Note that there are six climate zones—humid equatorial, dry, humid temperate, humid cold, cold polar, and highland climate—and that each is subdivided into climate types. For example, the three types of humid temperate climates are no dry season, dry winter, and dry summer.

Precipitation

Precipitation refers to any form of water, such as rain, snow, sleet, and hail, that falls from the atmosphere. Precipitation varies from one location to another and has a profound effect on the distribution and kinds of organisms present. One of the driest places on Earth is in the Atacama Desert in Chile, where the average annual rainfall is 0.05 cm (0.02 in). In contrast, Mount Waialeale in Hawaii, Earth's wettest spot, receives an average annual precipitation of 1,200 cm (472 in.).

Differences in precipitation depend on several factors. The heavy rainfall of some areas of the tropics results mainly from the equatorial uplift of moisture-laden air. High surface-water temperatures cause the evaporation of vast quantities of water from tropical parts of the ocean, and prevailing winds blow the resulting moist air over landmasses. Heating of the air over a land surface warmed by the sun causes moist air to rise. As it rises, the air cools, and its moisture-holding ability decreases (cool air holds less water vapor than warm air). When the air reaches its saturation point—when it cannot hold any additional water vapor—clouds form and water is released as precipitation. The air eventually returns to Earth on both sides of the equator near the Tropics of Cancer and Capricorn (latitudes 23.5 degrees north and 23.5 degrees south). By then most of its moisture has precipitated, and the dry air returns to the equator. This dry air makes little biological difference over the ocean, but its lack of moisture over land produces some of the great tropical deserts, such as the Sahara Desert.

Long journeys over landmasses dry the air. Near the windward (the side from which the wind blows) coasts of continents, rainfall is heavy. In the temperate areas, continental interiors are usually dry because they are far from the ocean that replenishes water in the air passing over it.

Rain Shadows   Mountains, which force air to rise, remove moisture from humid air. The air cools as it gains altitude, clouds form, and precipitation occurs, primarily on the windward slopes of the mountains. As the air mass moves down on the other side of the mountain, it is warmed, thereby lessening the chance of precipitation of any remaining moisture. This situation exists on the West Coast of North America, where precipitation falls on the western slopes of mountains close to the coast. The dry land on the side of the mountains away from the prevailing wind—in this case, east of the mountain range—is a **rain shadow** (**Figure 5.19**).

rain shadow: Dry conditions, often on a regional scale, that occur on the leeward side of a mountain barrier; the passage of moist air across the mountains removes most of the moisture.

Tornadoes

A **tornado,** or twister, is a powerful, rotating funnel of air associated with severe thunderstorms. Tornadoes form when a mass of cool, dry air collides with warm, humid air, producing a strong updraft of spinning air on the underside of a cloud. The spinning funnel becomes a tornado when it descends from the cloud and touches the ground. Wind velocity in a strong tornado may reach 480 km per hour (300 mi per hour). Tornadoes range from 1 m to 3.2 km (2 mi) in width. They last from several seconds to as long as 7 hours and travel along the ground from several meters to more than 320 km (200 mi).

On a local level, tornadoes have more concentrated energy than any other kind of storm. They can destroy buildings, bridges, and freight trains and even blow the water out of a river or small lake, leaving it empty. Tornadoes kill people: More than 10,000

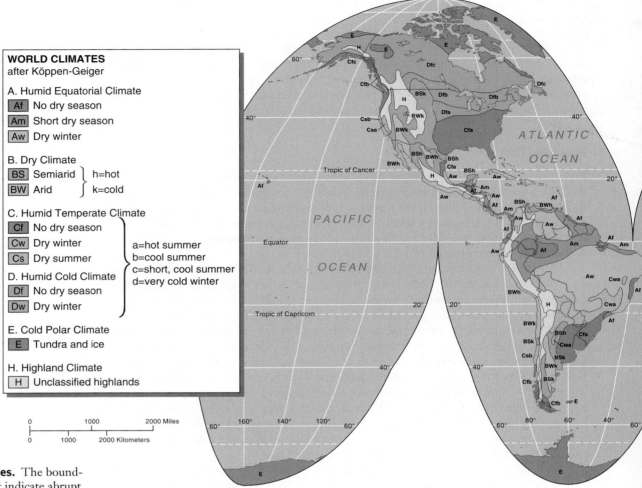

WORLD CLIMATES
after Köppen-Geiger

A. Humid Equatorial Climate
 Af No dry season
 Am Short dry season
 Aw Dry winter

B. Dry Climate
 BS Semiarid } h=hot
 BW Arid } k=cold

C. Humid Temperate Climate
 Cf No dry season
 Cw Dry winter
 Cs Dry summer

D. Humid Cold Climate
 Df No dry season
 Dw Dry winter

a=hot summer
b=cool summer
c=short, cool summer
d=very cold winter

E. Cold Polar Climate
 E Tundra and ice

H. Highland Climate
 H Unclassified highlands

0   1000   2000 Miles
0 1000 2000 Kilometers

Figure 5.18 **World climates.** The boundary lines on this map do not indicate abrupt changes in climate; instead, they are transition zones from one type of climate to another. In the next 50 years or so, the Köppen map will probably have to be revised to take into account global warming. (From H. J. deBlij and P. O. Muller.)

people in the United States died in tornadoes during the 20th century. Although tornadoes occur in other countries, the United States, known as the severe-storm capital of the world, has more tornadoes (typically about 1,000 per year) than anywhere else. They are most common in the spring months throughout the Great Plains and midwestern states (especially Texas, Oklahoma, and Kansas), as well as states along the Gulf of Mexico coast (especially Florida).

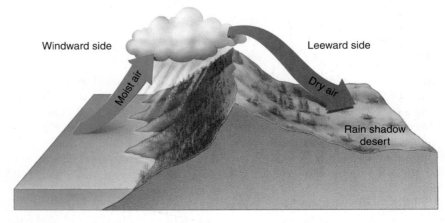

Windward side   Leeward side

Moist air   Dry air

Rain shadow desert

Figure 5.19 **Rain Shadow.** Prevailing winds blow warm, moist air from the windward side. Air cools as it rises, releasing precipitation so that dry air descends on the leeward side.

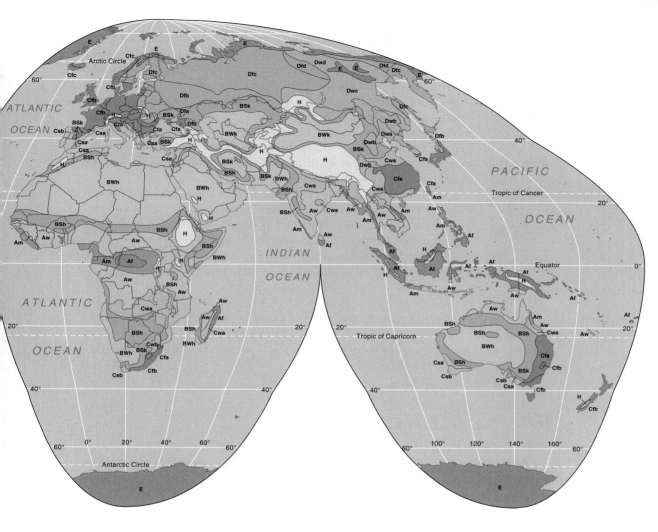

Tropical Cyclones

Tropical cyclones are giant, rotating tropical storms with winds of at least 119 km per hour (74 mi per hour); the most powerful have wind velocities greater than 250 km per hour (155 mi per hour). They form as strong winds pick up moisture over warm surface waters of the tropical ocean and start to spin as a result of Earth's rotation. The spinning causes an upward spiral of massive clouds as air is pulled upward. Known as *hurricanes* in the Atlantic, *typhoons* in the Pacific, and *cyclones* in the Indian Ocean, tropical cyclones are most common during summer and autumn months when ocean temperatures are warmest. With their spiral of clouds measuring about 800 km (500 mi) in diameter, tropical cyclones are easy to recognize in satellite photographs (**Figure 5.20**).

Tropical cyclones are destructive when they hit land, not so much from strong winds as from resultant storm surges, waves that rise as much as 7.5 m (25 ft) above the ocean surface (recall the discussion of Hurricane Katrina in the chapter introduction). Storm surges cause property damage and loss of life. Some hurricanes produce torrential rains. Hurricane Mitch, which hit the Atlantic coast of Central America in 1998, caused more than 10,000 deaths during the flooding and landslides that followed. Mitch, which also caused extensive damage in Honduras and Nicaragua, was the deadliest hurricane to occur in the Western Hemisphere in at least 200 years.

Figure 5.20 Hurricane Katrina.
This satellite image shows Hurricane Katrina as it struck Louisiana, Mississippi, and Alabama in August 2005. Note the hurricane's eye to the east of New Orleans. (GOES Project Science Office/NASA Visible Earth)

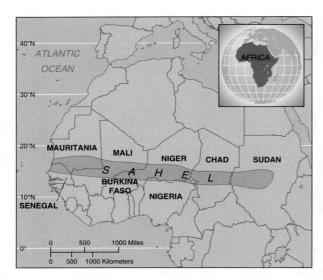

Figure 5.21 **The Sahel region in Africa.**
This area is a transition zone between the Sahara Desert to the north and the moist tropical rain forest to the south.

Some years produce more hurricanes than others. The 2005 hurricane season in the Atlantic Ocean was the most active on record, with 28 named tropical storms, 15 of which became hurricanes. (Tropical storms have wind speeds between 63 and 117 km per hour [39 and 73 mi per hour], whereas hurricanes have wind speeds of 119 km per hour [74 mi per hour] or greater.) Some of the factors that influence hurricane formation in the North Atlantic include precipitation in western Africa and water temperatures in the eastern Pacific Ocean. A wetter than usual rainy season in the western Sahel region of Africa (**Figure 5.21**) translates into more hurricanes, as does a dissipation of ENSO, which results in cooler water temperatures in the Pacific.

Since 1970, there has been an increase in the intensity of tropical cyclones. Several factors, such as global atmospheric patterns and vertical wind shear, interact in a complex way to produce tropical cyclone intensity. However, recent evidence suggests that sea-surface temperature is the most important factor, which implies that as the global climate warms (see Chapter 21), tropical cyclones may become more intense.

REVIEW

1. How do you distinguish between weather and climate? What are the two most important climate factors?

2. What are some of the environmental factors that produce areas of precipitation extremes, such as rain forests and deserts?

3. Distinguish between tornadoes and tropical cyclones.

Internal Planetary Processes

LEARNING OBJECTIVE

• Define *plate tectonics* and explain its relationship to earthquakes and volcanic eruptions.

plate tectonics: The study of the processes by which the lithospheric plates move over the asthenosphere.

Earth's outermost rigid rock layer (the **lithosphere**) is composed of seven large plates, plus a few smaller ones, that float on the **asthenosphere** (the region of the mantle where rocks become hot and soft) (**Figure 5.22**). The landmasses are situated on some of these plates. As the plates move horizontally across Earth's surface, the continents change their relative positions. **Plate tectonics** is the study of the dynamics of Earth's lithosphere—that is, the movement of these plates.

Any area where two plates meet—a **plate boundary**—is a site of intense geologic activity (**Figure 5.23**). Earthquakes and volcanoes are common in such a region. Both the San Francisco area, noted for its earthquakes, and the volcano Mount Saint Helens in Washington State are situated where two plates meet. Where landmasses are on the boundary between two plates, mountains may form. The Himalayas formed when the plate carrying India rammed into the plate carrying Asia. When two plates grind together, one of them sometimes descends under the other in the process of **subduction.** When two plates move apart, a ridge of molten rock from the mantle wells up between them; the ridge continually expands as the plates move farther apart. The Atlantic Ocean is growing as a result of the buildup of lava along the mid-Atlantic ridge, where two plates are separating.

Earthquakes

Forces inside Earth sometimes push and stretch rocks in the lithosphere. The rocks absorb this energy for a time, but eventually, as the energy accumulates, the stress is too great and the rocks suddenly shift or break. The energy—released as **seismic waves,** vibrations that spread through the rocks rapidly in all directions—causes one of the most powerful events in nature, an earthquake. Most earthquakes occur along **faults,** fractures where rock moves forward and backward, up and down, or from side to side. Fault zones are often found at plate boundaries. For example, the Pakistan earthquake that killed at least 75,000 people on October 8, 2005, occurred along boundary of the Eurasian plate and the Indian-Australian plate.

The site where an earthquake begins, often far below the surface, is the **focus.** Directly above the focus, at Earth's surface, is the earthquake's **epicenter.** When seismic waves reach the surface, they cause the ground to shake. Buildings and bridges collapse, and roads break. One of the instruments used to measure seismic waves is a *seismograph*, which helps seismologists (scientists who study earthquakes) determine where an earthquake started, how strong it was, and how long it lasted. In 1935, **Charles Richter,** a California seismologist, invented the *Richter scale*, a measure of the magnitude of energy released by an earthquake. Each unit on the Richter scale represents about 30 times more released energy than the unit immediately below it. As an example, a magnitude 8 earthquake is 30 times more powerful than a magnitude 7 earthquake and 900 times more powerful than a magnitude 6 earthquake. The Richter scale makes it easy to compare earthquakes, but it tends to underestimate the energy of large quakes.

Although the public is familiar with the Richter scale, seismologists typically do not use it. There are several ways to measure the magnitude of an earthquake, just as there are several ways to measure the size of a person (for example, height, weight, and amount of body fat). Most seismologists use a more precise scale, the *moment magnitude scale*, to measure earthquakes, especially those larger than magnitude 6.5 on the Richter scale. The moment magnitude scale calculates the total energy that a quake releases.

Seismologists record more than 1 million earthquakes each year. Some of these are major, but most are too small to be felt, equivalent to readings of about 2 on the Richter scale. A magnitude 5 earthquake usually causes property damage. On average, about every five years, a great earthquake occurs with a reading of 8 or higher. Such quakes usually cause massive property destruction and kill large numbers of people. Few people die directly because of the seismic waves; most die because of building collapses or fires started by ruptured gas lines.

Landslides and tsunamis are some of the side effects of earthquakes. A **landslide** is an avalanche of rock, soil, and other debris that slides swiftly down a mountainside. A 1970 earthquake in Peru resulted in a landslide that buried the town of Yungay and killed 17,000 people.

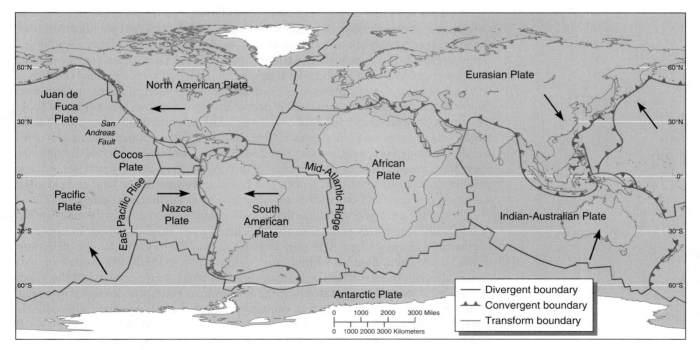

Figure 5.22 Plates and plate boundary locations. There are seven major independent plates that move horizontally across Earth's surface: African, Eurasian, Indian-Australian, Antarctic, Pacific, North American, and South American. Arrows show the directions of plate movements. The three types of plate boundaries are explained in Figure 5.23.

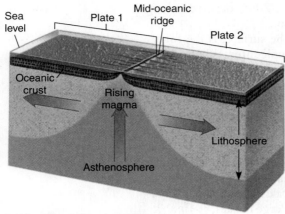

(a) **Divergent plate boundary**
Two plates move apart at a divergent plate boundary.

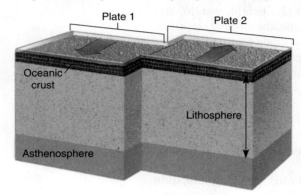

(c) **Transform plate boundary**

(b) **Convergent plate boundary**
When two plates collide at a convergent plate boundary
in the sea floor, subduction may occur. Convergent collision can also form a mountain range (not shown).

At a transform plate boundary, plates move horizontally
in opposite but parallel directions. On land, such a
boundary is often evident as a long, thin valley due to
erosion along the fault.

Figure 5.23   Plate boundaries. All three types of plate boundaries occur in the ocean and on land.

A **tsunami**, a giant sea wave caused by an underwater earthquake, volcanic eruption,
or landslide, sweeps through the water at more than 750 km (450 mi) per hour. Although
a tsunami may be only about 1 m high in deep-ocean water, it can build to a wall of water 30 m (100 ft) high—as high as a 10-story building—when it comes ashore, often far
from where the original earthquake triggered it. Tsunamis have caused thousands of
deaths. Colliding tectonic plates in the Indian Ocean triggered tsunamis on December
26, 2004, that killed more than 230,000 people in South Asia and Africa (**Figure 5.24**).
(About 300,000 were reported dead or missing.) The magnitude 9.3 earthquake was the
largest since an earthquake in Alaska in 1964. Not only did the tsunamis cause loss of life
and destruction of property, but they resulted in widespread environmental damage. Salt
water that moved inland as far as 3 km (1.9 mi) polluted soil and groundwater. Oil and
gasoline from overturned cars, trucks, and boats contaminated the land and poisoned
wildlife. Coral reefs and other offshore habitats were also damaged or destroyed.

One of the most geologically active places in North America is California's San Andreas Fault, which runs parallel to the California coast from the Mexican border to
northern California, a length of more than 1,100 km (700 mi). The Pacific plate (west
of the San Andreas Fault) is sliding northward relative to the North American plate
(east of the San Andreas Fault) at a rate of about 3.5 cm (1.4 in) per year. In 1906, much
of San Francisco, which is located near the San Andreas Fault, was destroyed by a magnitude 8.3 earthquake and the fire, caused by ruptured gas lines, that followed it. In
1989, a magnitude 6.9 earthquake along the San Andreas Fault in Loma Prieta, about
90 km south of San Francisco, killed 67 people and caused an estimated $6 billion in
damage to the San Francisco Bay area.

Not all earthquakes occur at plate boundaries. Some occur on smaller faults that
crisscross the large plates—the major earthquakes that damaged Northridge, Califor-

Figure 5.24 **Indian Ocean tsunami.**
Waves crash onto the shore of a Thailand town. Across the region, the December 2004 tsunamis generated high waves. (David Rydevik, Stockholm, Sweden)

nia, in 1994 and Kobe, Japan, in 1995, for example. Such earthquakes pose a seismic hazard that is difficult to evaluate because major quakes occur along a given small fault only every 1,000 to 5,000 years, in contrast to a larger fault line, which may have a major quake every century or so.

Volcanoes

The movement of tectonic plates on the hot, soft rock of the asthenosphere causes most volcanic activity. In places, the rock reaches the melting point, forming pockets of molten rock, or **magma.** When one plate slides under or away from an adjacent plate, magma may rise to the surface, often forming volcanoes. Magma that reaches the surface is called **lava.**

Volcanoes occur at three locations: subduction zones, spreading centers, and above hot spots. Subduction zones around the Pacific Basin have given rise to hundreds of volcanoes around Asia and the Americas in the region known as the "ring of fire." Plates that spread apart also form volcanoes. Iceland is a volcanic island that formed along the mid-Atlantic ridge. The Hawaiian Islands have a volcanic origin, but they did not form at plate boundaries. This chain of volcanic islands is thought to have formed as the Pacific plate moved over a **hot spot,** a rising plume of magma that flowed from deep within Earth's rocky mantle through an opening in the crust.

The largest volcanic eruption in the 20th century occurred in 1991 when Mount Pinatubo in the Philippines exploded. Despite the evacuation of more than 200,000 people, 338 deaths occurred, mostly from the collapse of buildings under the thick layer of wet ash that blanketed the area. The volcanic cloud produced when Mount Pinatubo erupted extended upward some 48 km (30 mi). We are used to hearing about human activities affecting climate, but many significant natural phenomena, including volcanoes, affect global climate. The magma and ash ejected into the atmosphere when Mount Pinatubo erupted blocked much of the sun's warmth and caused a slight cooling of global temperatures for a year or so.

REVIEW

1. What are tectonic plates and plate boundaries?

2. Where are earthquakes and volcanoes commonly located, and why?

ENVIRONEWS

Mangroves and the Asian Tsunami Disaster

It is not often that one hears of forests saving human lives, but that was the case during the December 2004 tsunami that swept across the Indian Ocean. The World Conservation Union (IUCN)* examined the death toll in two comparable coastal Sri Lankan villages. Extensive mangrove forests lined the coast near one village, whereas the mangroves had been removed near the other village. (Mangroves are often cut down to build tourist resorts or aquaculture facilities for shrimp farming.) The village with intact mangrove vegetation recorded 2 deaths from the tsunami, whereas the other settlement had almost 6,000 deaths. One of the many valuable ecosystem services that mangroves provide is to act as a barrier for storm surges and tsunamis. Based on this information, many Asian countries are replanting deforested coastal areas with mangroves where it is possible to do so.

* Formerly called the International Union for Conservation of Nature and Natural Resources, the World Conservation Union still goes by the acronym IUCN.

REVIEW OF LEARNING OBJECTIVES WITH KEY TERMS

- **Describe the main steps in each of these biogeochemical cycles: carbon, nitrogen, phosphorus, sulfur, and hydrologic cycles.**

1. The **carbon cycle** is the global circulation of carbon from the environment to living organisms and back to the environment. Carbon enters plants, algae, and cyanobacteria as carbon dioxide (CO_2), which is incorporated into organic molecules by photosynthesis. Cellular respiration by plants, by animals that eat plants, and by decomposers returns CO_2 to the atmosphere, making it available for producers again. **Combustion** and weathering also return CO_2 to the atmosphere.

2. The **nitrogen cycle** is the global circulation of nitrogen from the environment to living organisms and back to the environment. **Nitrogen fixation** is the conversion of nitrogen gas to ammonia. **Nitrification** is the conversion of ammonia or ammonium to nitrate. **Assimilation** is the biological conversion of nitrates, ammonia, or ammonium into proteins and other nitrogen-containing compounds by plants; the conversion of plant proteins into animal proteins is also part of assimilation. **Ammonification** is the conversion of organic nitrogen to ammonia and ammonium ions. **Denitrification** converts nitrate to nitrogen gas.

3. The **phosphorus cycle** is the global circulation of phosphorus from the environment to living organisms and back to the environment. This cycle has no biologically important gaseous compounds. Phosphorus erodes from rock as inorganic phosphates and plants absorb it from the soil. Animals obtain phosphorus from their diets, and decomposers release inorganic phosphate into the environment.

4. The **sulfur cycle** is the global circulation of sulfur from the environment to living organisms and back to the environment. In the cycle, most sulfur occurs as rocks or as sulfur dissolved in the ocean. Sulfur-containing gases, which include hydrogen sufide, sulfur oxides, and dimethyl sulfide (DMS), comprise a minor part of the atmosphere and are not long lived. A tiny fraction of sulfur is present in the proteins of living organisms. Bacteria drive the sulfur cycle.

5. The **hydrologic cycle** is the global circulation of water from the environment to living organisms and back to the environment. This cycle, which continuously renews the supply of water essential to life, involves an exchange of water among the land, the atmosphere, and organisms. Water enters the atmosphere by evaporation and **transpiration** and leaves the atmosphere as precipitation. On land, water filters through the ground or runs off to lakes, rivers, and the ocean. **Groundwater** is stored in underground caverns and porous layers of rock. The movement of surface water from land to rivers, lakes, wetlands, and ultimately the ocean is called **runoff.**

- **Describe how humans have influenced the carbon, nitrogen, phosphorus, sulfur, and hydrologic cycles.**

The level of atmospheric CO_2 increased dramatically during the last half of the 20th century and the first years of the 21st century; this increase is causing human-induced global climate change. Humans have more than doubled the amount of fixed nitrogen entering the global nitrogen cycle; the excess nitrogen is contributing to water quality problems, air pollution, and acid deposition. Phosphorus can be lost from terrestrial cycles for millions of years when it washes into the ocean and is deposited on the sea floor. Human activities have greatly increased sulfur emissions, which contribute to air pollution and acid deposition. **Aerosols,** tiny particles of air pollution produced from fossil fuel combustion and the burning of forests, enhance the scattering and absorption of sunlight in the atmosphere and cause brighter clouds to form. Scientists think that aerosols may weaken the hydrologic cycle.

- **Summarize the effects of solar energy on Earth's temperature, including the influence of albedos of various surfaces.**

Sunlight is the primary (almost sole) source of energy available to the biosphere. Of the solar energy that reaches Earth, 31% is immediately reflected away and the remaining 69% is absorbed. **Albedo** is the proportional reflectance of solar energy from Earth's surface, commonly expressed as a percentage. Glaciers and ice sheets have high albedos, and the ocean and forests have low albedos. Ultimately, all absorbed solar energy is radiated into space as infrared (heat) radiation. A combination of Earth's roughly spherical shape and the tilt of its axis concentrates solar energy at the equator and dilutes solar energy at the poles. Seasons are determined primarily by the inclination of Earth's axis.

- **Describe the five layers of Earth's atmosphere: troposphere, stratosphere, mesosphere, thermosphere, and exosphere.**

The **troposphere** is the layer of the atmosphere closest to Earth's surface; weather occurs in the troposphere. The **stratosphere,** found directly above the troposphere, contains a layer of ozone that absorbs much of the sun's damaging ultraviolet radiation. The **mesosphere,** found directly above the stratosphere, has the lowest temperatures in the atmosphere. The **thermosphere** has steadily rising temperatures because the air molecules absorb high-energy X-rays and short-wave ultraviolet radiation. The outermost layer of the atmosphere, the **exosphere**, converges with interplanetary space.

- **Discuss the roles of solar energy and the Coriolis effect in producing atmospheric circulation.**

Variations in the amount of solar energy reaching different places on Earth largely drive atmospheric circulation. Atmospheric heat transfer from the equator to the poles produces a movement of warm air toward the poles and a movement of cool air toward the equator, moderating the climate. In addition to these global circulation patterns, the atmosphere exhibits **winds,** complex horizontal movements that result in part from differences in atmospheric pressure and from the Coriolis effect. The **Coriolis effect** is the influence of Earth's rotation, which tends to deflect fluids (air and water) toward the right in the Northern Hemisphere and toward the left in the Southern Hemisphere.

- **Define *prevailing winds* and distinguish among polar easterlies, westerlies, and trade winds.**

Prevailing winds are major surface winds that blow more or less continually. Prevailing winds that blow from the northeast near the North Pole or the southeast near the South Pole are known as **polar easterlies.** Winds that blow in the midlatitudes from the southwest in the Northern Hemisphere or the northwest in the Southern Hemisphere are called **westerlies.** Tropical winds that blow from the northeast in the Northern Hemisphere or the southeast in the Southern Hemisphere are known as **trade winds.**

- **Discuss the roles of solar energy and the Coriolis effect in producing global water flow patterns, including gyres.**

Surface-ocean **currents** result largely from prevailing winds, which in turn are generated from solar energy. Other factors that contribute to ocean currents include the Coriolis effect, the position of landmasses, and the varying **density** of water. **Gyres** are large, circular ocean current systems that often encompass an entire ocean basin. Deep-ocean currents often travel in different directions and at different speeds than do surface currents. The present circulation of shallow and deep currents, known informally as the **ocean conveyor belt**, affects regional and possibly global climate.

- **Define El Niño-Southern Oscillation (ENSO) and La Niña and describe some of their effects.**

El Niño-Southern Oscillation (ENSO) is a periodic, large-scale warming of surface waters of the tropical eastern Pacific Ocean that temporarily alters both ocean and atmospheric circulation patterns. ENSO results in unusual weather in areas far from the tropical Pacific. During a **La Niña** event, surface water in the eastern Pacific becomes unusually cool.

- **Distinguish between weather and climate and give three causes of regional precipitation differences.**

Weather is the conditions in the atmosphere at a given place and time, whereas **climate** is the average weather conditions that occur in a place over a period of years. Temperature (both average temperature and temperature extremes) and precipitation (average precipitation, seasonal distribution, and variability) largely determine an area's climate. Precipitation is greatest where warm air passes over the ocean, absorbing moisture, and is then cooled, such as when mountains force humid air upward.

- **Describe a rain shadow.**

A **rain shadow** refers to dry conditions, often on a regional scale, that occur on the leeward side of a mountain barrier; the passage of moist air across the mountains removes most of the moisture from the air.

- **Contrast tornadoes and tropical cyclones.**

A **tornado** is a powerful, rotating funnel of air associated with severe thunderstorms. A **tropical cyclone** is a giant, rotating tropical storm with high winds. Tropical cyclones are called hurricanes in the Atlantic, typhoons in the Pacific, and cyclones in the Indian Ocean.

- **Define *plate tectonics* and explain its relationship to earthquakes and volcanic eruptions.**

Plate tectonics is the study of the processes by which the lithospheric plates move over the asthenosphere. Earth's lithosphere (outermost rock layer) consists of seven large plates and a few smaller ones. As the plates move horizontally, the continents change their relative positions. **Plate boundaries** are sites of intense geologic activity, such as mountain building, volcanoes, and earthquakes.

Thinking About the Environment

1. Scientists produce tentative conclusions based on a degree of uncertainty, whereas politicians and others making public policies prefer to deal in absolutes. How does this dichotomy relate to Hurricane Katrina and the restoration of New Orleans?

2. What is a biogeochemical cycle? Why is the cycling of matter essential to the continuance of life?

3. Describe how organisms participate in each of these biogeochemical cycles: carbon, nitrogen, phosphorus, and sulfur.

4. How are photosynthesis and cellular respiration involved in the carbon cycle?

5. What is the basic flow path of the nitrogen cycle?

6. A geologist or physical geographer would describe the phosphorus cycle as a "sedimentary pathway." Based on what you have learned about the phosphorus cycle in this chapter, what do you think that means?

7. How have global air temperatures changed in the recent past? How is this change related to the carbon cycle?

8. Diagram the hydrologic cycle.

9. What are the two lower layers of the atmosphere? Cite at least two differences between them.

10. Describe the general directions of atmospheric circulation.

11. How do ocean currents affect climate on land?

12. How do atmospheric and oceanic circulations transport heat toward the poles?

13. Relate the locations of earthquakes to plate tectonics.

14. Evaluate the area where you live with respect to natural dangers. Is there a threat of possible earthquakes, volcanic eruptions, hurricanes, tornadoes, or tsunamis?

15. The system encompassing Earth's global mean surface temperature can be diagrammed as follows:

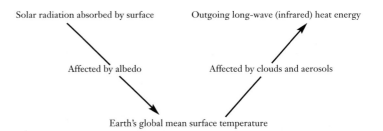

Please explain each part of this system.

16. Examine the following changes that have been identified in the arctic hydrologic system in the past few decades. Predict the effect of these changes on the salinity in the North Atlantic Ocean.

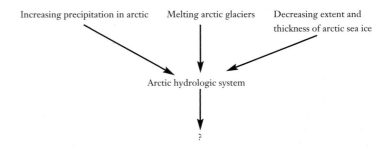

Quantitative questions relating to this chapter are on our Web site.

Take a Stand

Visit our Web site at http://www.wiley.com/college/raven (select Chapter 5 from the Table of Contents) for links to more information about human-induced changes in the global nitrogen cycle. Consider the opposing views of farmers who use nitrogen fertilizers and ecologists who study the effects of excess nitrate on water quality and coastal fisheries, and debate the issues with your classmates. You will find tools to help you organize your research, analyze the data, think critically about the issues, and construct a well-considered argument.

Take a Stand activities can be done individually or as a team, as oral presentations, written exercises, or Web-based (e-mail) assignments.

Additional online materials relating to this chapter, including a Student Testing Section with study aids and self-tests, Environmental News, Activity Links, Environmental Investigations, and more, are also on our Web site.

Wildfire. (Courtesy John McColgan, Alaska Fire Service/BLM)

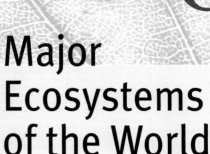

6

Major Ecosystems of the World

Fires started by lightning (wildfires) are an integral environmental force in many areas. Those areas most prone to wildfires have wet seasons followed by lengthy dry seasons. Vegetation grows and accumulates during the wet season and then dries out enough during the dry season to burn easily. Lightning ignites the dry organic material, and wind spreads the fire. At the peak of a wildfire season in the American West, where extensive lands are prone to fire, hundreds of new wildfires can break out each day.

Fires were part of the natural environment long before humans appeared, and plants in many terrestrial ecosystems that we discuss in this chapter have adapted to fire. African savannas, California chaparrals, North American grasslands, and pine forests of the southern United States are some fire-adapted ecosystems. For example, fire helps maintain grasses as the dominant vegetation in grasslands by removing fire-sensitive hardwood trees. (Grasses adapted to wildfire have underground stems and buds unaffected by a fire.)

Recent evidence suggests that climate change is causing more frequent and more intense wildfires in the western United States. It ap-

pears that the wildfire season is longer and drier due to an increase in spring and summer temperatures causing mountain snows to melt more quickly. (The greatest increase in wildfires has occurred in snow-dominated mountains.) Thus, climate change plays an important role in current wildfire trends.

Humans sometimes try to prevent fires. If fire is excluded from a fire-adapted ecosystem, organic litter (such as brush and slender trees) accumulates. When a fire does occur, it burns hotter and climbs into the upper canopies of trees. Called *crown fires*, these fires shoot flames high into the air, and wind pushes the flames to nearby trees. Many recent fires in the western United States have been caused in part by a century of suppressing fires in the region.

Humans sometimes conduct *prescribed burns*, in which the organic litter is deliberately burned under controlled conditions before it accumulates to dangerous levels. Prescribed burns also suppress fire-sensitive trees, thereby maintaining the natural fire-adapted ecosystem. However, prescribed burns do not always reduce fire risk. Fire management experts agree that prescribed burns reduce the damage but do not prevent fires. There will always be wildfires.

In this chapter, we consider Earth's major ecosystems, both aquatic ecosystems and those on land. As you read the descriptions of the plants and other organisms living in these ecosystems, consider how the physical environment, such as presence or absence of wildfires, influences each ecosystem.

Earth's Major Biomes

LEARNING OBJECTIVES

- Define biome and briefly describe the nine major terrestrial biomes: tundra, boreal forest, temperate rain forest, temperate deciduous forest, temperate grassland, chaparral, desert, savanna, and tropical rain forest.
- Relate at least one human effect on each of the biomes discussed.
- Explain the similarities and the changes in vegetation observed with increasing elevation and increasing latitude.

biome: A large, relatively distinct terrestrial region with a similar climate soil, plants, and animals, regardless of where it occurs in the world.

Earth has many **climates** based primarily on temperature and precipitation differences. Characteristic organisms have adapted to each climate. A **biome** is quite large in area and encompasses many interacting ecosystems (**Figure 6.1**; also see Figure 15.8 for the world's major soil types). In terrestrial ecology, a biome is considered the next level of ecological organization above those of community, ecosystem, and landscape.

Near the poles, temperature is generally the overriding climate factor, whereas in temperate and tropical regions, precipitation becomes more significant than temperature (**Figure 6.2**). Light is relatively plentiful in biomes, except in certain environments such as the rainforest floor. Other abiotic factors to which certain biomes are sensitive include temperature extremes as well as rapid temperature changes, fires, floods, droughts, and strong winds.

We now consider nine major biomes and how humans are affecting them: tundra, boreal forest, temperate rain forest, temperate deciduous forest, temperate grassland, chaparral, desert, savanna, and tropical rain forest.

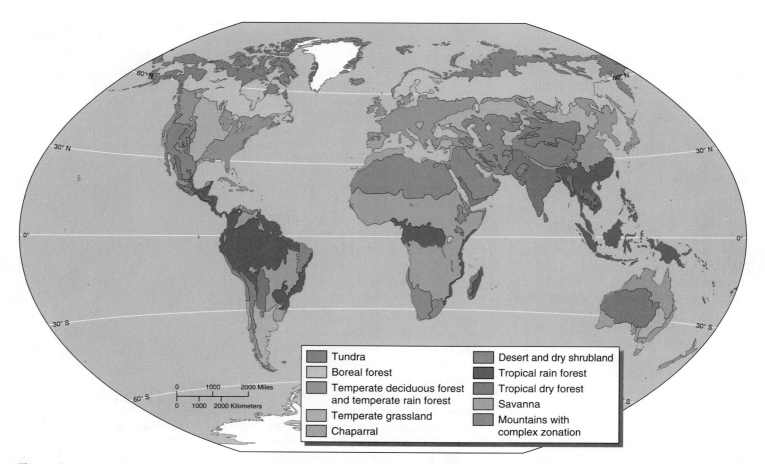

Tundra	Desert and dry shrubland
Boreal forest	Tropical rain forest
Temperate deciduous forest and temperate rain forest	Tropical dry forest
Temperate grassland	Savanna
Chaparral	Mountains with complex zonation

Figure 6.1 **Distribution of the world's terrestrial biomes.** Although sharp boundaries are shown in this highly simplified map, biomes actually blend together at their boundaries. Biomes generally correspond to the climate zones shown in Figure 5.18. (Adapted from World Wildlife Fund)

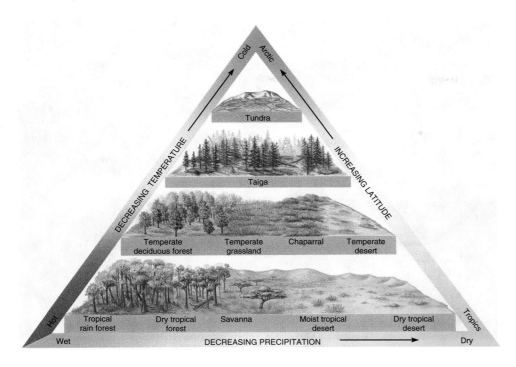

Figure 6.2 Temperature and precipitation.
The biomes are distributed primarily in accordance with two climate factors, temperature and precipitation. In the higher latitudes, temperature is the more important of the two. In temperate and tropical zones, precipitation is a significant determinant of community composition. (Adapted from L. Holdridge)

Tundra: Cold Boggy Plains of the Far North

Tundra (also called **arctic tundra**) occurs in the extreme northern latitudes wherever the snow melts seasonally (**Figure 6.3**). The Southern Hemisphere has no equivalent of the arctic tundra because it has no land in the corresponding latitudes. **Alpine tundra** is a similar ecosystem located in the higher elevations of mountains, above the tree line.

Arctic tundra has long, harsh winters and short summers. Although the growing season, with its warmer temperatures, is short (from 50 to 160 days depending on location), the days are long. Above the Arctic Circle, the sun does not set at all for many days in midsummer, although the amount of light at midnight is one-tenth that at noon. There is little precipitation (10 to 25 cm, or 4 to 10 in., per year) over much of the tundra, with most of it falling during the summer months.

Most tundra soils are geologically young because they formed when glaciers retreated after the last Ice Age. (Glacier ice, which occupied about 29% of Earth's land during the last Ice Age, began retreating about 17,000 years ago. Today, glacier ice occupies about 10% of the land.) These soils are usually nutrient-poor and have little organic litter such as dead leaves and stems, animal droppings, and remains of organisms. Although the soil melts at the surface during the summer, tundra has a layer of **permafrost,** permanently frozen ground that varies in depth and thickness. Permafrost is most extensive in northern Canada and Siberia. The thawed upper zone of soil is usually waterlogged during the summer because permafrost interferes with drainage. Permafrost limits the depth roots penetrate, thereby preventing the establishment of most woody species. The limited precipitation in combination with low temperatures, flat topography (surface features), and the permafrost layer produces a landscape of broad, shallow lakes and ponds, sluggish streams, and bogs.

Tundra has low **species richness** (the number of different species) and low **primary productivity** (the rate at which energy is accumulated; see section on ecosystem productivity in Chapter 3). Few plant species occur, but individual species often exist in great numbers. Mosses, lichens (such as reindeer moss), grasses, and grasslike sedges dominate tundra. No readily recognizable trees or shrubs grow except in sheltered locations, although dwarf willows, dwarf birches, and other dwarf trees are common. As a rule, tundra plants seldom grow taller than 30 cm (12 in.).

tundra: The treeless biome in the far north that consists of boggy plains covered by lichens and small plants such as mosses; has harsh, very cold winters and extremely short summers.

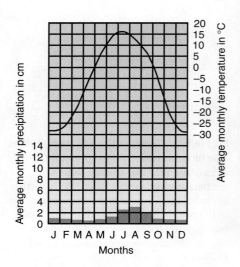

Figure 6.3 Arctic tundra. A caribou buck stands in the Alaskan tundra. Only small, hardy plants grow in the northernmost biome that encircles the Arctic Ocean. Climate graph shows monthly temperature (line graph at top) and precipitation data (bar graph at bottom) for Fort Yukon, Alaska. Data for all climate graphs are from www.worldclimate.com. (Warren Garst/Tom Stack & Associates)

boreal forest: A region of coniferous forest (such as pine, spruce, and fir) in the Northern Hemisphere; located just south of the tundra.

The year-round animal life of the tundra includes lemmings, voles, weasels, arctic foxes, snowshoe hares, ptarmigan, snowy owls, and musk oxen. In the summer, caribou migrate north to the tundra to graze on sedges, grasses, and dwarf willow. Dozens of bird species migrate north in summer to nest and feed on abundant insects. Mosquitoes, blackflies, and deerflies survive the winter as eggs or pupae, and adults occur in great numbers during summer. There are no reptiles or amphibians except for wood frogs, which are sometimes found in tundra ponds.

Tundra regenerates slowly after it has been disturbed. Even hikers can cause damage. Oil and natural gas exploration and military use have caused long-lasting injury, likely to persist for hundreds of years, to large portions of the arctic tundra (see Chapter 11, "Case In Point: The Arctic National Wildlife Refuge").

Boreal Forests: Conifer Forests of the North

Just south of the tundra is the **boreal forest** (also called **taiga**, pronounced tie′guh). The boreal forest stretches across North America and Eurasia, covering approximately 11% of Earth's land (**Figure 6.4**). A biome comparable to the boreal forest is not found in the Southern Hemisphere. Winters are extremely cold and severe, although not as harsh as in the tundra. The growing season of the boreal forest is somewhat longer than that of the tundra. Boreal forest receives little precipitation, perhaps 50 cm (20 in.) per year, and its soil is typically acidic and mineral-poor, with a deep layer of partly decomposed pine and spruce needles at the surface. Permafrost is patchy and, where found, is often deep under the soil. Boreal forest has numerous ponds and lakes in water-filled depressions that were dug by grinding ice sheets during the last Ice Age.

Black and white spruces, balsam fir, eastern larch, and other conifers (cone-bearing evergreens) dominate the boreal forest, although deciduous trees (trees that shed their leaves in autumn), such as aspen and birch, may form striking stands. Conifers have many drought-resistant adaptations, such as needlelike leaves with a minimal surface area for water loss. Such an adaptation lets conifers withstand the "drought" of the northern winter months when roots cannot absorb water because the ground is frozen. Being evergreen, conifers resume photosynthesis as soon as warmer temperatures return.

The animal life of the boreal forest consists of some larger species such as caribou, which migrate from the tundra to the boreal forest for winter; wolves; bears; and moose. Most mammals are medium-sized to small, including rodents, rabbits, and fur-bearing predators such as lynx, sable, and mink. Most species of birds are abundant in the summer but migrate to warmer climates for winter. Wildlife ecologists estimate that one of every three birds in the United States and Canada spends its breeding season in the boreal forests of North America. Insects are abundant, but there are few amphibians and reptiles except in the southern boreal forest.

Most of the boreal forest is not well suited to agriculture because of its short growing season and mineral-poor soil. However, the boreal forest yields lumber, pulpwood for paper products, animal furs, and other forest products. Currently, boreal forest is the world's primary source of industrial wood and wood fiber, and extensive logging of certain boreal forests has occurred. Mining, drilling for gas and oil, and farming have also contributed to loss of boreal forest. (See Chapter 18 for a discussion of deforestation of this biome.)

Temperate Rain Forest: Lush Temperate Forests

A coniferous **temperate rain forest** occurs on the northwest coast of North America. Similar vegetation exists in southeastern Australia and in southern South America. Annual precipitation in this biome is high, more than 127 cm (50 in.), and is augmented by condensation of water from dense coastal fogs. The proximity of temperate rain forest to the coastline moderates the temperature so the seasonal fluctuation is narrow; winters are mild and summers are cool. Temperate rain forest has relatively nutrient-poor soil, although its organic content may be high. Cool temperatures slow the activity of bacterial and fungal decomposers. Needles and large fallen branches and trunks accumulate on the ground as litter that takes many years to decay and release nutrient minerals to the soil.

The dominant plants in the North American temperate rain forest are large evergreen trees such as western hemlock, Douglas fir, western red cedar, Sitka spruce, and western arborvitae (**Figure 6.5**). Temperate rain forests are rich in epiphytic vegetation—smaller plants that grow on the trunks and branches of large trees. Epiphytes in this biome are mainly mosses, club mosses, lichens, and ferns, all of which also carpet the ground. Deciduous shrubs such as vine maple grow wherever a break in the overlying canopy occurs. Squirrels, wood rats, mule deer, elk, numerous bird species, and several species of amphibians and reptiles are common temperate rainforest animals.

Temperate rain forest is a rich wood producer, supplying us with lumber and pulpwood. It is also one of the world's most complex ecosystems in terms of species richness. We must avoid overharvesting the original old-growth (never logged) forest, because such an ecosystem takes hundreds of years to develop. When the logging industry harvests old-growth forest, it typically replants the area with a **monoculture** (a single species) of trees that it harvests in 40- to 100-year cycles. The old-growth forest ecosystem, once harvested, never has a chance to redevelop. A small fraction of the original old-growth temperate rain forest in Washington, Oregon, and northern California remains untouched. These stable forest ecosystems provide biological habitats for many species, including about 40 endangered and threatened species. The issues surrounding old-growth forests of the Pacific Northwest were explored in the Chapter 2 introduction.

Temperate Deciduous Forest: Broad-Leaved Trees That Shed Their Leaves

Seasonality (hot summers and cold winters) is characteristic of the **temperate deciduous forest**, which occurs in temperate areas where precipitation ranges from about 75 to 150 cm (30 to 60 in.) annually. Typically, the soil of a temperate deciduous forest

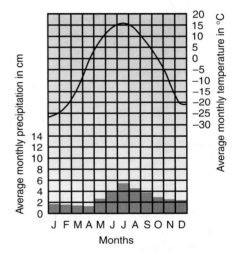

Figure 6.4 **Boreal forest.** These coniferous forests occur in cold regions of the Northern Hemisphere adjacent to the tundra. Photographed in Yukon Territory, Canada. Climate graph shows monthly temperature and precipitation data for Fort Smith, Northwest Territories, Canada. (Beth Davidow/Visuals Unlimited)

temperate rain forest: A coniferous biome with cool weather, dense fog, and high precipitation.

temperate deciduous forest: A forest biome that occurs in temperate areas with a moderate amount of precipitation.

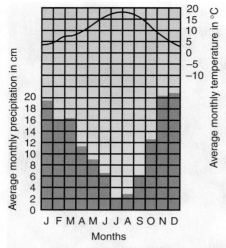

Figure 6.5 **Temperate rain forest.** This temperate biome has large amounts of precipitation, particularly during winter months. Photographed in Olympic National Park in Washington State. Climate graph shows monthly temperature and precipitation data for Estacada, Oregon. (David Muench Photography)

consists of a topsoil rich in organic material and a deep, clay-rich lower layer. As organic materials decay, mineral ions are released. Ions not absorbed by tree roots **leach** (filter) into the clay, where they may be retained.

The temperate deciduous forests of the northeastern and mideastern United States are dominated by broad-leaved hardwood trees, such as oak, hickory, maple, and beech, that lose their foliage annually (**Figure 6.6**). In the southern areas of the temperate deciduous forest, the number of broad-leaved evergreen trees, such as magnolia, increases. The trees of the temperate deciduous forest form a dense canopy that overlies saplings and shrubs.

Figure 6.6 **Temperate deciduous forest.** The broad-leaf trees that dominate this biome are deciduous and will shed their leaves before winter. Photographed during autumn in Pennsylvania. Climate graph shows monthly temperature and precipitation data for Nashville, Tennessee. (Barbara Miller/Biological Photo Service)

Temperate deciduous forests originally contained a variety of large mammals, such as puma, wolves, and bison, which are now absent, plus deer, bears, and many small mammals and birds. In Europe and North America, logging and land clearing for farms, tree plantations, and cities have removed much of the original temperate deciduous forest. Where it has regenerated, temperate deciduous forest is often in a seminatural state modified by recreation, livestock foraging, timber harvest, and other uses. Many forest organisms have successfully become reestablished in these returning forests.

Worldwide, deciduous forests were among the first biomes converted to agricultural use. In Europe and Asia, many soils that originally supported deciduous forests have been culti-

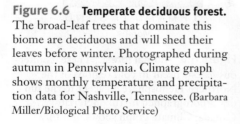

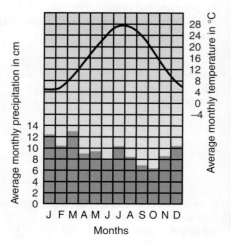

vated by traditional agricultural methods for thousands of years without a substantial loss in fertility. During the 20th century, intensive agricultural practices were widely adopted; these, along with overgrazing and deforestation, have contributed to the degradation of some agricultural lands. Most damage to farmland has happened since the end of World War II. We say more about humans and our effects on land degradation in later chapters.

Grasslands: Temperate Seas of Grass

Summers are hot, winters are cold, and rainfall is often uncertain in **temperate grasslands**. Annual precipitation averages 25 to 75 cm (10 to 30 in.). In grasslands with less precipitation, nutrient minerals tend to accumulate in a well-defined layer just below the topsoil. These nutrient minerals tend to wash out of the soil in areas with more precipitation. Grassland soil has considerable organic material because the aerial portions of many grasses die off each winter and contribute to the organic content of the soil, whereas the roots and rhizomes (underground stems) survive. The roots and rhizomes eventually die and add to the soil's organic material. Many grasses are sod formers—that is, their roots and rhizomes form a thick, continuous underground mat.

Moist temperate grasslands, or *tallgrass prairies*, occur in the United States in parts of Illinois, Iowa, Minnesota, Nebraska, Kansas, and other midwestern states (**Figure 6.7**). Although few trees grow except near rivers and streams, grasses grow in great profusion in the deep, rich soil. Periodic wildfires help maintain grasses as the dominant vegetation in grasslands. Several species of grasses that, under favorable conditions, grew as tall as a person on horseback dominated tallgrass prairies. The land was covered with herds of grazing animals, such as pronghorn elk and bison. According to the American Prairie Foundation, fewer than 7000 wild (genetically pure) bison remain today, compared with an estimated 30 million to 60 million bison that roamed the prairies 200 years ago. The principal predators were wolves, although in sparser, drier areas coyotes took their place. Smaller animals included prairie dogs and their predators (foxes, black-footed ferrets, and various birds of prey), grouse, reptiles such as snakes and lizards, and great numbers of insects.

Shortgrass prairies are temperate grasslands that receive less precipitation than the moist temperate grasslands just described but more precipitation than deserts. In the United States, shortgrass prairies occur in the eastern half of Montana, the western half of South Dakota, and parts of other midwestern states. Grasses that grow knee high or lower dominate shortgrass prairies. The plants grow in less abundance than in moister grasslands, and occasionally some bare soil is exposed. Native grasses of shortgrass prairies are drought-resistant.

The North American grassland, particularly the tallgrass prairie, was well suited to agriculture. More than 90% has vanished under the plow, and the remaining prairie is so fragmented that almost nowhere can you see what European settlers saw when they migrated into the Midwest. Today, the tallgrass prairie is considered North America's rarest biome. It is not surprising that the North American Midwest, the Ukraine, and other moist temperate grasslands became the breadbaskets of the world, because they provide ideal growing conditions for crops such as corn and wheat, which are also grasses.

Chaparral: Thickets of Evergreen Shrubs and Small Trees

Some hilly temperate environments have mild winters with abundant rainfall combined with dry summers. Such **mediterranean climates,** as they are called, occur in

temperate grasslands: A grassland with hot summers, cold winters, and less rainfall than the temperate deciduous forest biome.

Figure 6.7   Temperate grassland. This tallgrass prairie contains a profusion of grasses and other herbaceous flowering plants. Photographed in Iowa on one of the virgin tracts of tallgrass prairie owned by the Nature Conservancy. Climate graph shows monthly temperature and precipitation data for Lawrence, Kansas. (Annie Griffiths Belt/DRK Photo)

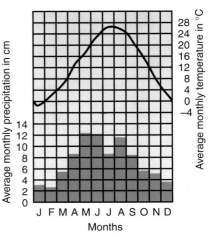

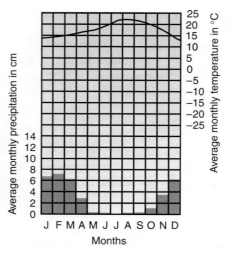

Figure 6.8 **Chaparral.** Chaparral vegetation consists mainly of drought-resistant evergreen shrubs and small trees. Chaparral has hot, dry summers and mild, rainy winters. Photographed in the Santa Monica Mountains, California. Climate graph shows monthly temperature and precipitation data for Culver City, California. (John Cunningham/Visuals Unlimited)

chaparral: A biome with mild, moist winters and hot, dry summers; vegetation is typically small-leaved evergreen shrubs and small trees.

desert: A biome in which the lack of precipitation limits plant growth; deserts are found in both temperate and subtropical regions.

the area around the Mediterranean Sea as well as in the North American Southwest, southwestern and southern Australia, central Chile, and southwestern South Africa. In mountain slopes of Southern California, this mediterranean-type community is known as **chaparral**. Chaparral soil is thin and often not fertile. Frequent fires occur naturally in this environment, particularly in late summer and autumn.

Chaparral vegetation looks strikingly similar in different areas of the world, even though the individual species are not the same. Chaparral usually has a dense growth of evergreen shrubs but may contain short, drought-resistant pine or scrub oak trees that grow 1 to 3 m (3.3 to 9.8 ft) tall (**Figure 6.8**). During the rainy winter season, the environment is lush and green, but the plants lie dormant during the hot, dry summer. Trees and shrubs often have hard, small, leathery leaves that resist water loss. Many plants are also fire-adapted and grow best in the months following a fire. Such growth is possible because fire releases nutrient minerals from aerial parts of the plants that burned. Fire does not kill the underground parts and seeds of many plants, and with the new availability of essential nutrient minerals, the plants sprout vigorously during winter rains. Mule deer, wood rats, chipmunks, lizards, and many species of birds are common animals of the chaparral.

The fires that occur at irregular intervals in California chaparral are quite costly to humans when they consume expensive homes built on the hilly chaparral landscape. Unfortunately, efforts to prevent the naturally occurring fires sometimes backfire. Denser, thicker vegetation tends to accumulate over several years; then, when a fire does occur, it is much more severe. Removing the chaparral vegetation, whose roots hold the soil in place, causes problems; witness the mudslides that sometimes occur during winter rains in these areas.

Deserts: Arid Life Zones

Deserts are dry areas found in both temperate (*cold deserts*) and subtropical regions (*warm deserts*). The low water vapor content of the desert atmosphere results in daily temperature extremes of heat and cold, and a major change in temperature occurs in a

ENVIRONEWS

Using Goats to Fight Fires

California has about 6000 wildfires each year, and they are becoming increasingly expensive and dangerous to manage because so many people are building homes and living in fire-vulnerable chaparral. For one thing, the topography is so steep that firefighters often cannot use mechanized equipment and must be transported to fires by helicopters. Fearing that prescribed burns will get out of control, local governments are increasingly employing an effective, low-tech method to reduce the fuel load: During the six-month fire season, goats are clearing hills around Oakland, Berkeley, Monterey, and Malibu.

A herd of 350 goats can denude an entire acre of heavy brush in about a day, but their use entails a lot of advance organization and support. Before goats clear hazardous dry fuels from surrounding hillsides, botanists must walk the terrain to put fences around any small trees and rare or endangered plants; fencing keeps the goats from eating those plants. A portable home site for goatherds is then installed, as are electric fencing and water troughs for the goats. The goatherds typically use dogs to help herd the goats.

Goats are an excellent tool for fire management because they preferentially browse woody shrubs and thick undergrowth—the exact fuel that causes disastrous fires. Fires that occur in areas after goats have browsed there are much easier to contain.

single 24-hour period. Deserts vary depending on the amount of precipitation they receive, which is generally less than 25 cm (10 in.) per year. A few deserts, such as the African Namib Desert and the Atacama Desert of northern Chile and Peru, are so dry that virtually no plant life occurs in them. As a result of sparse vegetation, desert soil is low in organic material but is often high in mineral content, particularly the salts sodium chloride (NaCl), calcium carbonate ($CaCO_3$), and calcium sulfate ($CaSO_4$). In some regions, such as areas of Utah and Nevada, the concentration of certain soil minerals reaches toxic levels for many plants.

Plant cover is so sparse in deserts that much of the soil is exposed. Both perennials (plants that live for more than two years) and annuals (plants that complete their life cycles in one growing season) occur in deserts. However, annuals are common only after rainfall. Plants in North American deserts include cacti, yuccas, Joshua trees, and sagebrushes (**Figure 6.9**). Desert plants tend to have reduced leaves or no leaves, an adaptation that conserves water. In cacti such as the giant saguaro, the stem, which expands accordion-style to store water, carries out photosynthesis; the leaves are modified into spines, which discourage herbivores. Other desert plants shed their leaves for most of the year, growing only during the brief moist season. Many desert plants are provided with spines, thorns, or toxins to resist the heavy grazing pressure often experienced in this food- and water-deficient environment.

Desert animals tend to be small. During the heat of the day, they remain under cover or return to shelter periodically, whereas at night they come out to forage or hunt. In addition to desert-adapted insects, there are a few desert-adapted amphibians (frogs and toads) and many specialized desert reptiles, such as the desert tortoise, desert iguana, Gila monster, and Mojave rattlesnake. Desert mammals include rodents such as gerbils and jerboas in African and Asian deserts and kangaroo rats in North American deserts. There are also mule deer and jackrabbits in these deserts, oryxes in African deserts, and kangaroos in Australian deserts. Carnivores such as the African fennec fox and some birds of prey, especially owls, live on the rodents and jackrabbits. During the driest months of the year, many desert insects, amphibians, reptiles, and mammals tunnel underground, where they remain inactive.

Humans have altered North American deserts in several ways. Off-road vehicles damage desert vegetation, which sometimes takes years to recover. In the early 1940s, military exercises in California's Mojave Desert left tracks that are still visible today. People who drive across the desert in four-wheel-drive vehicles inflict environmental damage. When the top layer of desert soil is disturbed, erosion occurs more readily,

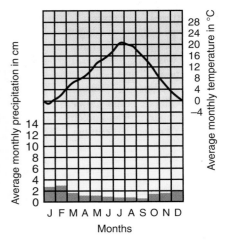

Figure 6.9 Desert. Inhabitants of deserts are strikingly adapted to the demands of their environment. The moister deserts of North America contain large columnar cacti such as the giant saguaro (*Carnegiea gigantea*), which grows 15 to 18 m (50 to 60 ft) tall. The smaller cacti, called chollas (*Opuntia* sp.), have a dense covering of barbed spines. Photographed in the Sonoran Desert in southwestern Arizona and southeastern California. Climate graph shows monthly temperature and precipitation data for Reno, Nevada. (Willard Clay/Dembinsky Photo Associates)

and less vegetation grows to support native animals. Certain cacti and desert tortoises are rare as a result of poaching. Houses, factories, and farms built in desert areas require vast quantities of water, which is imported from distant areas. Increased groundwater consumption by many desert cities has caused groundwater levels to drop. Aquifer depletion in U.S. deserts is particularly critical in southern Arizona and southwestern New Mexico. (See Chapter 14 for a discussion of water problems in this region.)

Savanna: Tropical Grasslands

savanna: A tropical grassland with widely scattered trees or clumps of trees.

Savanna occurs in areas of low rainfall or seasonal rainfall with prolonged dry periods (**Figure 6.10**). The temperatures in tropical savannas vary little throughout the year, and seasons are regulated by precipitation, not by temperature as they are in temperate grasslands. Annual precipitation is 76 to 150 cm (30 to 60 in.). Savanna soil is somewhat low in essential nutrient minerals, in part because it is strongly leached—that is, many nutrient minerals have filtered out of the topsoil. Savanna soil is often rich in aluminum, which resists leaching, and in places the aluminum reaches levels toxic to many plants. Although the African savanna is best known, savanna also occurs in South America and northern Australia.

Savanna has wide expanses of grasses interrupted by occasional trees such as acacia, which bristle with thorns that provide protection against herbivores. Both trees and grasses have fire-adapted features, such as extensive underground root systems, that let them survive seasonal droughts as well as periodic fires.

Spectacular herds of hoofed mammals—such as wildebeest, antelope, giraffe, zebra, and elephants—occur in the African savanna. Large predators, such as lions and hyenas, kill and scavenge the herds. In areas of seasonally varying rainfall, the herds and their predators may migrate annually.

Savannas are rapidly being converted into rangeland for cattle and other domesticated animals, which are replacing the big herds of wild animals. Half of the Cerrado (savanna) in central Brazil has been converted to cropland and pastures since 1970. The problem is more acute in Africa because it has the most rapidly growing human population of any continent. In some places, severe overgrazing and harvesting of trees for firewood have converted savanna to desert, a process called **desertification.**

Figure 6.10 Savanna. Tropical grasslands such as this one, with widely scattered Acacia trees, support large herds of grazing animals and their predators. These are swiftly vanishing under pressure from pastoral and agricultural land use. Photographed in Tanzania. Climate graph shows monthly temperature and precipitation data for Lusaka, Zambia. (Carlyn Iverson)

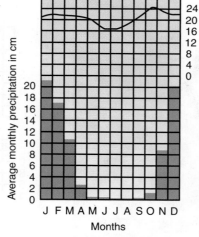

Tropical Rain Forests: Lush Equatorial Forests

Tropical rain forests occur where temperatures are warm throughout the year and precipitation occurs almost daily. The annual precipitation of a tropical rain forest is typically from 200 to 450 cm (80 to 180 in.). Much of this precipitation comes from locally recycled water that enters the atmosphere by transpiration (loss of water vapor from plants) of the forest's own trees.

tropical rain forest: A lush, species-rich forest biome that occurs where the climate is warm and moist throughout the year.

Tropical rain forest commonly occurs in areas with ancient, highly weathered, mineral-poor soil. Little organic matter accumulates in such soils because bacteria, fungi, and detritus-feeding ants and termites decompose organic litter quite rapidly. Roots and mycorrhizae quickly absorb nutrient minerals from the decomposing material. Thus, nutrient minerals of tropical rain forests are tied up in the vegetation rather

Figure 6.11   Tropical rain forest. A broad view of tropical rainforest vegetation along a riverbank in Southeast Asia. Except at riverbanks, tropical rain forests have a closed canopy that admits little light to the rainforest floor. Climate graph shows monthly temperature and precipitation data for Belem, Brazil. (Frans Lanting/Minden Pictures, Inc.)

than the soil. Tropical rain forests are found in Central and South America, Africa, and Southeast Asia (see Figure 18.10).

Tropical rain forests are very productive—that is, the plants capture a lot of energy by photosynthesis. Despite the scarcity of nutrient minerals in the soil, abundant solar energy and precipitation stimulate high productivity.

Of all the biomes, the tropical rainforest is unexcelled in species richness and variety. No single species dominates this biome. A person can travel hundreds of meters without encountering two individuals of the same tree species. Local factors, such as varying soil fertility and topography, affect the composition of rainforest species. Valleys have more plant diversity than hills, for example.

The trees of tropical rain forests are typically evergreen flowering plants (**Figure 6.11**). Their roots are often shallow and concentrated near the surface in a mat. The root mat catches and absorbs almost all nutrient minerals released from leaves and litter by decay processes. Swollen bases or braces called *buttresses* hold the trees upright and aid in the extensive distribution of the shallow roots.

A fully developed tropical rain forest has at least three distinct stories, or layers, of vegetation. The topmost story consists of the crowns of occasional, very tall trees, some 50 m (164 ft) or more in height, which are exposed to direct sunlight. The middle story reaches a height of 30 to 40 m (100 to 130 ft) and forms a continuous canopy of leaves that lets in little sunlight to support the sparse understory. Only 2% to 3% of the light bathing the forest canopy reaches the forest understory. Smaller plants specialized for life in the shade, as well as the seedlings of taller trees, comprise the understory. The vegetation of tropical rain forests is not dense at ground level except near stream banks or where a fallen tree has opened the canopy.

Tropical rainforest trees support extensive epiphytic communities of plants such as ferns, mosses, orchids, and bromeliads. Epiphytes grow in crotches of branches, on bark, or even on the leaves of their hosts, but they use their host trees primarily for physical support, not for nourishment.

Little light penetrates to the understory, and many plants living there are adapted to climb already established host trees rather than to invest their meager photosynthetic resources in building the cellulose tissues of their own trunks. Lianas (woody tropical vines), some as thick as a human thigh, twist up through the branches of the huge rainforest trees. Once in the canopy, lianas grow from the upper branches of one forest tree to another, connecting the tops of the trees and providing a walkway for many of the canopy's residents. They and herbaceous vines provide nectar and fruit for many tree-dwelling animals.

Not counting bacteria and other soil-dwelling organisms, about 90% of tropical rainforest organisms live in the upper canopy. Rainforest animals include the most abundant and varied insects, reptiles, and amphibians on Earth. The birds, often brilliantly colored, are varied, with some specialized to consume fruit (parrots, for example) and others to consume nectar (hummingbirds and sunbirds, for example). Most rainforest mammals, such as sloths and monkeys, live only in the trees and rarely climb to the ground, although some large, ground-dwelling mammals, including elephants, are also found in rain forests.

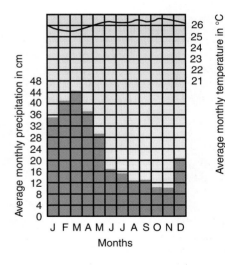

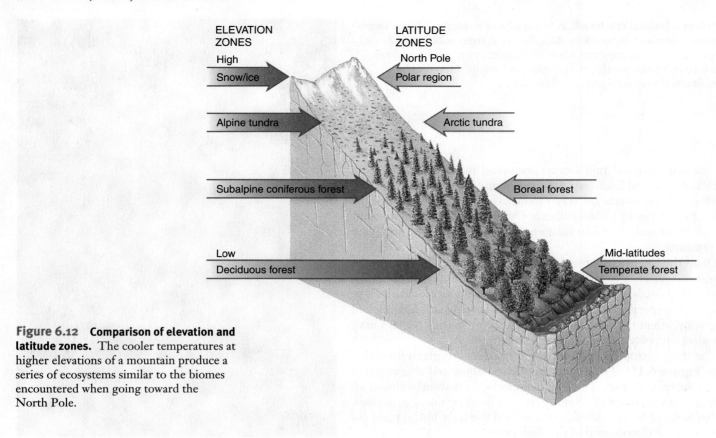

ELEVATION ZONES
High
Snow/ice
Alpine tundra
Subalpine coniferous forest
Low
Deciduous forest

LATITUDE ZONES
North Pole
Polar region
Arctic tundra
Boreal forest
Mid-latitudes
Temperate forest

Figure 6.12 Comparison of elevation and latitude zones. The cooler temperatures at higher elevations of a mountain produce a series of ecosystems similar to the biomes encountered when going toward the North Pole.

Human population growth and industrial expansion in tropical countries may spell the end of tropical rain forests during the 21st century. Biologists know many rainforest organisms will become extinct before they are even identified and scientifically described. (See Chapter 18 for more discussion of the ecological impacts of tropical rainforest destruction.)

Vertical Zonation: The Distribution of Vegetation on Mountains

Hiking up a mountain is similar to traveling toward the North Pole with respect to the major ecosystems encountered (**Figure 6.12**). This elevation-latitude similarity occurs because as one climbs a mountain, the temperature drops just as it does when one travels north. The types of organisms living on the mountain change as the temperature changes.

The base of a mountain in Colorado, for example, is covered by deciduous trees, which shed their leaves every autumn. At higher elevations, where the climate is colder and more severe, one might find a coniferous *subalpine forest*, which resembles the northern boreal forest. Higher still, where the climate is quite cold, *alpine tundra* occurs, with vegetation composed of grasses, sedges, and small tufted plants; it is called alpine tundra to distinguish it from arctic tundra. The top of the mountain might have a permanent ice or snow cap, similar to that in the nearly lifeless polar land areas.

Important environmental differences between high elevations and high latitudes affect the types of organisms found in each place. Alpine tundra typically lacks permafrost and receives more precipitation than arctic tundra. High elevations of temperate mountains do not have the great extremes in day length associated with the changing seasons in biomes at high latitudes. Furthermore, the intensity of solar radiation is greater at high elevations than at high latitudes. At high elevations, the sun's rays pass through less atmosphere, which results in a greater exposure to ultraviolet (UV) radiation than occurs at high latitudes. (The mountain atmosphere filters out less UV.)

1. What is a biome? What two climate factors are most important in determining an area's characteristic biome?
2. What climate and soil factors produce each of the major terrestrial biomes?
3. How does vegetation change with increasing elevation and latitude?

 ## Aquatic Ecosystems

LEARNING OBJECTIVES

- Summarize the important environmental factors that affect aquatic ecosystems.
- Briefly describe the eight aquatic ecosystems: flowing-water ecosystem, standing-water ecosystem, freshwater wetlands, estuary, intertidal zone, benthic environment, neritic province, and oceanic province.
- Relate at least one human effect on each of the aquatic ecosystems discussed.
- Discuss national marine sanctuaries.

Not surprisingly, aquatic life zones differ from terrestrial life zones in almost all respects. Recall that in biomes, temperature and precipitation are the major determinants of plant and animal inhabitants, and light is relatively plentiful, except in certain habitats such as the rainforest floor. Other environmental factors are significant in aquatic ecosystems. Generally speaking, temperature is somewhat less important in watery environments because the water itself tends to moderate temperature. Water is obviously not an important limiting factor in aquatic ecosystems.

The most fundamental division in aquatic ecology is probably between freshwater and saltwater environments. **Salinity**, the concentration of dissolved salts such as sodium chloride (NaCl) in a body of water, affects the kinds of organisms present in aquatic ecosystems, as does the amount of dissolved oxygen. Water greatly interferes with the penetration of light. Floating photosynthetic organisms remain near the water's surface, and vegetation attached to the bottom grows only in shallow water. In addition, low levels of essential nutrient minerals limit the number and distribution of organisms in certain aquatic environments. Other abiotic determinants of species composition in aquatic ecosystems include temperature, pH, and presence or absence of waves and currents.

Aquatic ecosystems contain three main ecological categories of organisms: free-floating plankton, strongly swimming nekton, and bottom-dwelling benthos. **Plankton** are usually small or microscopic organisms that are relatively feeble swimmers. For the most part, they are carried about at the mercy of currents and waves. They cannot swim far horizontally, but some species are capable of large daily vertical migrations and are found at different depths of water depending on the time of day and the season.

Plankton are generally subdivided into two major categories, phytoplankton and zooplankton. **Phytoplankton** are free-floating photosynthetic algae and cyanobacteria that form the base of most aquatic food webs. **Zooplankton** are nonphotosynthetic organisms that include protozoa (animal-like protists), tiny shrimplike crustaceans, and the larval (immature) stages of many animals. In aquatic food webs, zooplankton feed on algae and cyanobacteria and are in turn consumed by newly hatched fish and other small aquatic organisms.

Nekton are larger, more strongly swimming organisms such as fishes, turtles, and whales. **Benthos** are bottom-dwelling organisms that fix themselves to one spot (sponges, oysters, and barnacles), burrow into the sand (worms, clams, and sea cucumbers), or simply walk about on the bottom (crawfish, aquatic insect larvae, and brittle stars).

Freshwater Ecosystems

Freshwater ecosystems include rivers and streams (flowing-water ecosystems), lakes and ponds (standing-water ecosystems), and marshes and swamps (freshwater wetlands). Specific abiotic conditions and characteristic organisms distinguish each freshwater

Figure 6.13 Features of a typical river.

The river begins at a **source**, often high in the mountains and fed by melting snows or glaciers.

Headwater streams flow downstream rapidly, often over rocks (as **rapids**) or bluffs (as **waterfalls**).

Along the way, smaller **tributaries** feed into the river, adding to its flow.

The **flood plain** is the relatively flat area on either side of the river that is subject to flooding.

As the river's course levels out, the river flows more slowly and winds from side to side, forming bends called **meanders.**

Near the ocean, the river may form a **salt marsh** where fresh water from the river and salt water from the ocean mix.

The **delta** is a fertile, low-lying plain at the river's **mouth** that forms from sediments deposited by the slow-moving river as it empties into the ocean.

Ocean

ecosystem. Although freshwater ecosystems occupy a relatively small portion (about 2%) of Earth's surface, they have an important role in the hydrologic cycle: They assist in recycling precipitation that flows as surface runoff to the ocean (see hydrologic cycle in Chapter 5). Large bodies of fresh water moderate daily and seasonal temperature fluctuations on nearby land. Freshwater habitats provide homes for many species.

Rivers and Streams: Flowing-Water Ecosystems   Various conditions exist along the length of a river or stream (**Figure 6.13**). The nature of a **flowing-water ecosystem** changes greatly between its source (where it begins) and its mouth (where it empties into another body of water). Surrounding forest shades certain parts of the stream, whereas other parts are exposed to direct sunlight. *Headwater streams* (the small streams that are the sources of a river) are usually shallow, cool, swiftly flowing, and highly oxygenated. In contrast, rivers downstream from the headwaters are wider and deeper, cloudy (they contain suspended particulates), not as cool, slower flowing, and less oxygenated. Along parts of a river or stream, groundwater wells up through sediments on the bottom. This local input of water moderates the water temperature so that summer temperatures are cooler and winter temperatures are warmer than in adjacent parts of the flowing-water ecosystem.

The organisms in flowing-water ecosystems vary greatly from one stream to another, depending primarily on the strength of the current. In streams with fast currents, the inhabitants may have adaptations such as hooks or suckers to attach themselves to rocks so that they are not swept away. The larvae of blackflies attach themselves with a suction disk located on the end of their abdomen. Some stream inhabitants, such as

flowing-water ecosystem: A freshwater ecosystem such as a river or stream in which the water flows in a current.

immature water-penny beetles, may have flattened bodies to slip under or between rocks. The water-penny beetle larva (immature form) gets its common name from its flattened, nearly circular shape. Alternatively, inhabitants such as fish are streamlined and muscular enough to swim in the current. Organisms in large, slow-moving streams and rivers do not need such adaptations, although they are typically streamlined like most aquatic organisms to lessen resistance when moving through water.

Unlike other freshwater ecosystems, streams and rivers depend on the land for much of their energy. In headwater streams, almost all of the energy input comes from detritus, such as dead leaves carried from the land into streams and rivers by wind or surface runoff. Downstream, rivers contain more producers and are less dependent on detritus as a source of energy than are headwaters. The concept of a river system as a single ecosystem with a gradient in physical features from headwaters to mouth is known as the *river continuum concept*. This gradient results in predictable changes in the organisms inhabiting different parts of the river system.

Human activities have several adverse impacts on rivers and streams, including water pollution and the effects of dams built to contain or divert the water of rivers or streams. Pollution alters the physical environment and changes the biotic component downstream from the pollution source. Uncontrolled pollution threatens not only wildlife habitat but also our water supply and commercial and recreational fisheries. A dam causes water to back up, flooding large areas of land and forming a reservoir, which destroys terrestrial habitat. Below the dam, the once-powerful river is often reduced to a relative trickle, which alters water temperature, sediment transport, and delta replenishment and prevents fish migrations. (See Chapter 13 for more discussion of the environmental effects of dams.)

Lakes and Ponds: Standing-Water Ecosystems   Zonation characterizes **standing-water ecosystems**. A large lake has three zones: the littoral, limnetic, and profundal zones (**Figure 6.14**). The **littoral zone** is a shallow-water area along the shore of a lake or pond where light reaches the bottom. Emergent vegetation, such as cattails and bur reeds, as well as several deeper-dwelling aquatic plants and algae, live in the littoral zone. The littoral zone is the most productive section of the lake (photosynthesis is greatest here), in part because it receives nutrient inputs from surrounding land that stimulate the growth of plants and algae. Animals of the littoral zone include frogs and their tadpoles; turtles; worms; crayfish and other crustaceans; insect larvae; and many fishes, such as perch, carp, and bass. Surface dwellers such as water striders and whirligig beetles are found in the quieter areas.

The **limnetic zone** is the open water beyond the littoral zone—that is, away from the shore; it extends down as far as sunlight penetrates to permit photosynthesis. The main

standing-water ecosystem: A body of fresh water that is surrounded by land and that does not flow; a lake or a pond.

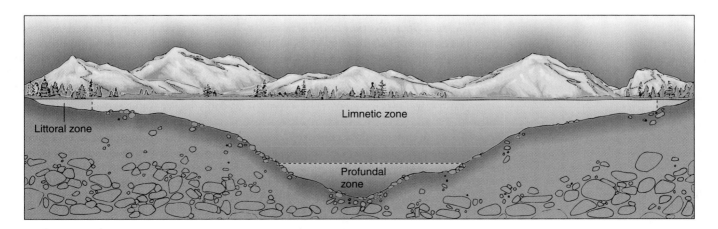

Figure 6.14   Zonation in a large lake. The littoral zone is the shallow-water area around the lake's edge. The limnetic zone is the open, sunlit water away from the shore. The profundal zone, under the limnetic zone, is below where light penetrates.

Figure 6.15 **Thermal stratification in a temperate lake.**

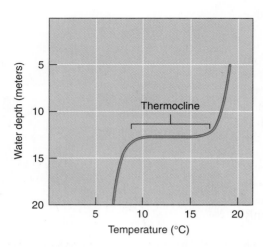

(a) Temperature varies by depth during the summer. There is an abrupt temperature transition, the thermocline, between the upper warm layer and the bottom cold layer.

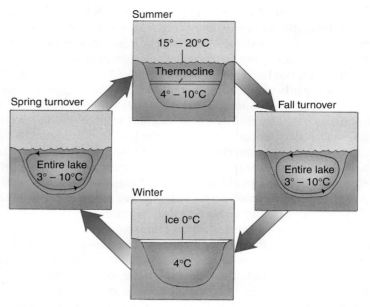

(b) During fall and spring turnovers, a mixing of upper and lower layers of water brings oxygen to the oxygen-depleted depths of the lake and nutrient minerals to the mineral-deficient surface water. During summer months, a layer of warm water develops over cooler, deeper water; no turnover occurs to supply oxygen to deeper water, where the oxygen level drops.

organisms of the limnetic zone are microscopic phytoplankton and zooplankton. Larger fishes spend most of their time in the limnetic zone, although they may visit the littoral zone to feed and reproduce. Owing to its depth, less vegetation grows here than in the littoral zone.

The deepest zone, the **profundal zone,** is beneath the limnetic zone of a large lake; smaller lakes and ponds typically lack a profundal zone. Light does not penetrate effectively to this depth, and plants and algae do not live here. Food drifts into the profundal zone from the littoral and limnetic zones. Bacteria decompose dead organisms and other organic material that reaches the profundal zone, using up oxygen and liberating the nutrient minerals contained in the organic material. These nutrient minerals are not effectively recycled because there are no producers to absorb them and incorporate them into the food web. As a result, the profundal zone is both mineral-rich and anaerobic (without oxygen), with few organisms other than anaerobic bacteria occupying it.

Thermal Stratification and Turnover in Temperate Lakes. The marked layering of large temperate lakes caused by how far light penetrates is accentuated by **thermal stratification,** in which the temperature changes sharply with depth. Thermal stratification occurs because the summer sunlight penetrates and warms surface waters, making them less dense. (The density of water is greatest at 4°C. Both above and below this temperature, water is less dense.) In the summer, cool, denser water remains at the lake bottom, separated from the warm, less dense water above by an abrupt temperature transition, the **thermocline (Figure 6.15a).** Seasonal distribution of temperature and oxygen (more oxygen dissolves in water at cooler temperatures) affects the distribution of fish in the lake.

In temperate lakes, falling temperatures in autumn cause **fall turnover**, a mixing of the layers of lake water (**Figure 6.15b**). (Such turnovers are not common in the tropics because there is little seasonal temperature variation there.) As the surface water cools, its density increases, and eventually it displaces the less dense, warmer, mineral-rich water beneath. The warmer water then rises to the surface where it, in turn, cools and sinks. This process of cooling and sinking continues until the lake reaches a uniform temperature throughout.

When winter comes, the surface water may cool below 4°C, its temperature of greatest density, and if it is cold enough, ice forms. Ice forms at 0°C and is less dense than cold water. Thus, ice forms on the surface, and the water on the lake bottom is warmer than the ice on the surface.

In the spring, **spring turnover** occurs as ice melts and the surface water reaches 4°C. Surface water again sinks to the bottom, and bottom water returns to the surface, resulting in a mixing of the layers. As summer arrives, thermal stratification occurs once again. The mixing of deeper, nutrient-rich water with surface, nutrient-poor water during the fall and spring turnovers brings essential nutrient minerals to the surface and oxygenated water to the bottom. The sudden presence of large amounts of essential nutrient minerals in surface waters encourages the growth of large algal and cyanobacterial populations, which form temporary **blooms** (population explosions) in the fall and spring. (Harmful algal blooms such as red tides are discussed in Chapters 7 and 22.)

Marshes and Swamps: Freshwater Wetlands   Freshwater wetlands include marshes, in which grasslike plants dominate, and swamps, in which woody trees or shrubs dominate (**Figure 6.16**). Freshwater wetlands include hardwood bottomland forests (lowlands along streams and rivers that are periodically flooded), prairie potholes (small, shallow ponds that formed when glacial ice melted at the end of the last Ice Age), and peat moss bogs (peat-accumulating wetlands where sphagnum moss dominates). Wetland soils are waterlogged and anaerobic for variable periods. Most wetland soils are rich in accumulated organic materials, in part because anaerobic conditions discourage decomposition.

Wetland plants are highly productive and provide enough food to support a variety of organisms. Wetlands are valued as a wildlife habitat for migratory waterfowl and many other bird species, beaver, otters, muskrats, and game fishes. Wetlands provide natural flood control because they are holding areas for excess water when rivers flood their banks. The floodwater stored in wetlands then drains slowly back into the rivers, providing a steady flow of water throughout the year. Wetlands serve as groundwater recharging areas. One of their most important roles is to help cleanse water by trapping and holding pollutants in the flooded soil. These important environmental functions are called **ecosystem services.**

At one time wetlands were considered wastelands, areas to fill in or drain so that farms, housing developments, and industrial plants

freshwater wetlands: Lands that shallow fresh water covers for at least part of the year; wetlands have a characteristic soil and water-tolerant vegetation.

Figure 6.16   Freshwater swamp. Freshwater swamps are inland areas permanently saturated or covered by water and dominated by trees, such as bald cypress (*Taxodium distichum*). A floating carpet of tiny aquatic plants covers the water surface. Photographed in northeast Texas. (Gregory G. Dimijian/Photo Researchers, Inc.)

E N V I R O N E W S

The Chesapeake Bay Program

The Chesapeake Bay system is the largest estuary in the United States. Until relatively recently, it was also the most productive, annually yielding tons of seafood such as oysters, blue crabs, shad, striped bass, and white perch. The Chesapeake Bay watershed is home to more than 17 million people. The bay has suffered in recent years from deteriorating water quality due to pollution associated with the motor vehicles, farms, industries, and homes of all these people. Consider nitrogen pollution, which empties into the bay from agricultural runoff, sewage treatment plants, electric power companies, and motor vehicles. Most of this pollution remains in the bay, causing excessive algal growth that results

in a much lower level of dissolved oxygen in the water. Oxygen-deprived water suffocates oysters, crabs, and other aquatic organisms. Other problems for the bay include loss of habitat, overfishing, and sediment pollution (from soil particles that enter the water).

The long process of restoring and protecting the bay and its resources is underway. The **Chesapeake Bay Program** is a partnership of federal and state agencies with goals such as sound land use, restoration of water quality and habitats, and management of fisheries. The state has reduced some of the pollution load of both nutrients and sediment going into the bay and restricted some coastal development. Many problems remain, and the Chesapeake Bay Program is committed to the preservation and restoration of the bay and its watershed. (See the Chapter 3 introduction for a discussion of wildlife in the Chesapeake Bay.)

could be built. Wetlands, as breeding places for mosquitoes, were viewed as a menace to public health. Today, the crucial ecosystem services that wetlands provide are widely recognized, and wetlands have some legal protection, although they are still threatened by agriculture, pollution, dam construction, and urban and suburban development. In many parts of the United States, we continue to lose wetlands (see Chapter 18).

Estuaries: Where Fresh Water and Salt Water Meet

Several ecosystems may occur where the ocean meets the land: a rocky shore, a sandy beach, an intertidal mud flat, or a tidal **estuary**. Water levels in an estuary rise and fall with the tides, whereas salinity fluctuates with tidal cycles, the time of year, and precipitation. Salinity changes gradually within the estuary, from unsalty fresh water at the river entrance to salty ocean water at the mouth of the estuary. Estuarine organisms must tolerate the significant daily, seasonal, and annual variations in temperature, salinity, and depth of light penetration.

Estuaries are among the most fertile ecosystems in the world, often having a much greater productivity than either the adjacent ocean or the fresh water upriver. This high productivity is brought about by four factors. (1) Nutrients are transported from the land into rivers and creeks that flow into the estuary. (2) Tidal action promotes a rapid circulation of nutrients and helps remove waste products. (3) A high level of light penetrates the shallow water. (4) The presence of many plants provides an extensive photosynthetic carpet and mechanically traps detritus, forming the base of a detritus food web. Many species, including commercially important fishes and shellfish, spend their larval stages in estuaries among the protective tangle of decaying plants.

Temperate estuaries usually contain **salt marshes,** shallow wetlands dominated by salt-tolerant grasses (see the photograph in the Chapter 3 introduction). Salt marshes have often appeared as worthless, empty stretches of land to uninformed people. As a result, they have been used as dumps and become severely polluted or have been filled with dredged bottom material to form artificial land for residential and industrial development. A large part of the estuarine environment has been lost in this way, along with many of its ecosystem services, such as biological habitats, sediment and pollution trapping, groundwater supply, and storm buffering. (Salt marshes absorb much of the energy of a storm surge and thereby prevent flood damage elsewhere.) According to the Environmental Protection Agency, estuaries in the United States are in fair to poor condition, with the greatest problems in northeastern and Gulf estuaries.

Mangrove forests, the tropical equivalent of salt marshes, cover perhaps 70% of tropical coastlines (**Figure 6.17**). Like salt marshes, mangrove forests provide valuable ecosystem services. Their interlacing roots are breeding grounds and nurseries for several commercially important fishes and shellfish, such as blue crabs, shrimp, mullet, and spotted sea trout. For example, biologists studied the number of commercially fished yellowtail snapper on coral reefs adjacent to neighboring mangrove-rich and mangrove-poor areas. In 2004, they reported that the snapper biomass on reefs near mangrove-rich areas was two times greater than the biomass near mangrove-poor areas. (Recall from Chapter 3 that *biomass* is an estimate of the mass, or amount, of living material.)

Mangrove branches are nesting sites for many species of birds, such as pelicans, herons, egrets, and roseate spoonbills. Mangrove roots stabilize the submerged soil, thereby preventing coastal erosion, and provide a barrier against the ocean during storms such as hurricanes. Mangroves may be even more effective than concrete sea walls in dissipating wave energy and controlling floodwater from tropical storms. Unfortunately, mangroves are under assault from coastal development, unsustainable logging, and aquaculture (see Figure 19.15). Some countries, such as the Philippines, Bangladesh, and Guinea-Bissau, have lost 70% or more of their mangrove forests.

estuary: A coastal body of water, partly surrounded by land, with access to the open ocean and a large supply of fresh water from a river.

Figure 6.17   Mangroves. Red mangroves (*Rhizophora mangle*) have stiltlike roots that support the tree. These roots grow into deeper water as well as into mudflats exposed by low tide. Many animals live in the root systems of mangrove forests. Photographed at low tide along the coast of Florida, near Miami. (Patti Murray/Animals Animals/Earth Scenes)

Marine Ecosystems

Although lakes and the ocean are comparable in many ways, there are many physical differences. The depths of even the deepest lakes do not approach those of the ocean trenches, with areas that extend more than 6 km (3.6 mi) below the sunlit surface. Tides and currents exert a profound influence on the ocean. Gravitational pulls of both the sun and the moon usually produce two high tides and two low tides each day along the ocean's coastlines, but the height of those tides varies with the season, local topography, and phases of the moon (a full moon causes the highest tides).

The immense marine environment is subdivided into several life zones: the intertidal zone, where organisms live along the shore, between high and low tides; the benthic environment, where organisms live on or under the seafloor; and the pelagic environment, where organisms live in the water (**Figure 6.18**). The pelagic environment is in turn divided into two provinces based on depth: the neritic and oceanic provinces. The neritic province consists of shallow waters close to shore, and the oceanic province comprises most of the ocean.

The Intertidal Zone: Transition Between Land and Ocean   Although the high levels of light and nutrients, together with an abundance of oxygen, make the **intertidal zone** a biologically productive habitat, it is a stressful one. If an intertidal beach is sandy, inhabitants must contend with a constantly shifting environment that threatens to engulf them and gives them scant protection against wave action. Consequently, most sand-dwelling organisms, such as mole crabs, are continuous and active burrowers. They usually do not have notable adaptations to survive drying out or exposure because they follow the tides up and down the beach.

A rocky shore provides fine anchorage for seaweeds and marine animals but is exposed to wave action when immersed during high tides and to drying and temperature changes when exposed to air during low tides (**Figure 6.19**). A typical rocky-shore inhabitant has some way of sealing in moisture, perhaps by closing its shell, if it has one, plus a powerful means of anchoring itself to the rocks. Mussels have tough, threadlike anchors secreted by a gland in the foot, and barnacles have special glands that secrete a tightly bonding glue that hardens under water. Rocky-shore intertidal algae, such as rockweed (*Fucus*), usually have thick, gummy coats, which dry out slowly when exposed to air, and flexible bodies not easily broken by wave action.

intertidal zone: The area of shore line between low and high tides.

Figure 6.18   Zonation in the ocean. The ocean has three main life zones: the intertidal zone, the benthic environment, and the pelagic environment. The pelagic environment consists of the neritic and oceanic provinces. The neritic province overlies the ocean floor from the shoreline to a depth of 200 m. The oceanic province overlies the ocean floor at depths greater than 200 m. The ocean floor is not a flat expanse; it consists of mountains, valleys, canyons, seamounts, ridges, and trenches. (The slopes of the ocean floor are not as steep as shown; they are exaggerated to save space.)

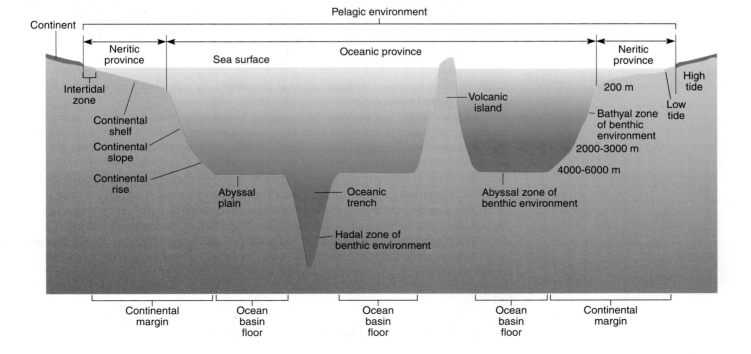

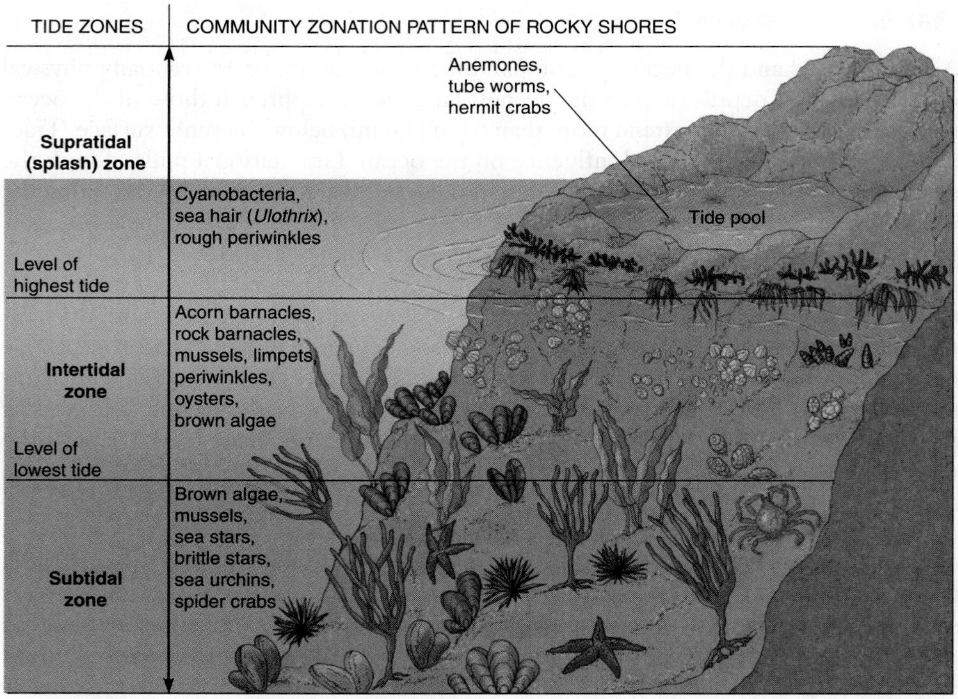

TIDE ZONES | COMMUNITY ZONATION PATTERN OF ROCKY SHORES

Anemones, tube worms, hermit crabs

Supratidal (splash) zone

Cyanobacteria, sea hair (*Ulothrix*), rough periwinkles

Tide pool

Level of highest tide

Intertidal zone

Acorn barnacles, rock barnacles, mussels, limpets, periwinkles, oysters, brown algae

Level of lowest tide

Subtidal zone

Brown algae, mussels, sea stars, brittle stars, sea urchins, spider crabs

Figure 6.19   Zonation along a rocky shore. Three zones are shown: The supratidal (splash) zone, the intertidal zone, and the subtidal zone (part of the benthic environment). Representative organisms are given for each of these zones. Although exposed to wave action when submerged and to drying and seasonal heat and freezing when exposed to the air, the intertidal zone is crowded with many species. The subtidal zone is rarely exposed to the atmosphere except during spring tides. A greater diversity of organisms lives here than at higher levels on the rocky shore.

benthic environment: The ocean floor, which extends from the intertidal zone to the deep ocean trenches.

Some rocky-shore organisms hide in burrows or under rocks or crevices at low tide, and some small crabs run about the splash line, following it up and down the beach.

The Benthic Environment: Seagrass Beds, Kelp Forests, and Coral Reefs   Most of the **benthic environment** consists of sediments (mostly sand and mud) where many animals, such as worms and clams, burrow. Bacteria are common in marine sediments and have even been reported in ocean sediments more than 500 m (1625 ft) below the ocean floor at several sites in the Pacific Ocean.

The deeper parts of the benthic environment are divided into three zones; they are, from shallowest to deepest, the bathyal, abyssal, and hadal zones. The *bathyal benthic zone* is the benthic environment that extends from a depth of 200 to 4000 m (650 ft to 2.5 mi). The *abyssal benthic zone* extends from a depth of 4000 m to a depth of 6000 m (2.5 to 3.7 mi), whereas the *hadal benthic zone* extends from 6000 m to the bottom of the deepest trenches. ("Case in Point: Life Without the Sun" in Chapter 3 examines the unusual organisms in the deeper part of the benthic environment.) Here we describe shallow benthic communities that are particularly productive—seagrass beds, kelp forests, and coral reefs.

Seagrass Beds. **Sea grasses** are flowering plants adapted to complete submersion in salty ocean water. They only occur in shallow water, to depths of 10 m (33 ft) where they receive enough light to photosynthesize efficiently. Extensive beds of sea grasses occur in quiet temperate, subtropical, and tropical waters. Eelgrass is the most widely distributed sea grass along the coasts of North America; the largest eelgrass bed in the world is in Izembek Lagoon on the Alaska Peninsula. The most common sea grasses in the Caribbean Sea are manatee grass and turtle grass (**Figure 6.20**). Sea grasses have a high primary productivity and are ecologically important in shallow marine areas. Their roots and rhizomes help stabilize the sediments, reducing surface erosion. Sea grasses provide food and habitat for many marine organisms. In temperate waters, ducks and geese eat sea grasses, whereas in tropical waters, manatees, green turtles, parrot fish, sturgeon fish, and sea urchins eat them. These herbivores consume only about 5% of the sea grasses. The remaining 95% eventually enters the detritus food web and is decomposed when the sea grasses die. The decomposing bacteria are in turn consumed by a variety of animals such as mud shrimp, lugworms, and mullet (a type of fish).

Kelp Forests. **Kelps**, which may reach lengths of 60 m (200 ft), are the largest brown algae (**Figure 6.21**). Kelps are common in cooler temperate marine waters of both Northern and Southern Hemispheres. They are especially abundant in relatively shallow waters (depths of about 25 m, or 82 ft) along rocky coastlines. Kelps are photosynthetic and are the primary food producers for the kelp "forest" ecosystem. Kelp forests provide habitats for many marine animals. Tube worms, sponges, sea cucumbers, clams, crabs, fishes, and sea otters find refuge in the algal fronds. Some animals eat the fronds, but kelps are mainly consumed in the detritus food web. Bacteria that decompose the kelp remains provide food for sponges, tunicates, worms, clams, and snails. The diversity of life supported by kelp beds almost rivals that found in coral reefs. (In Chapter 4 the Environews on population declines in sea otters discusses recent changes in the kelp forest ecosystem.)

Coral Reefs. **Coral reefs**, built from accumulated layers of calcium carbonate ($CaCO_3$), are found in warm (usually greater than 21°C), shallow seawater. The living portions of coral reefs must grow in shallow waters where light penetrates. Some coral reefs are composed of red coralline algae that require light for photosynthesis. Most coral reefs consist of colonies of millions of tiny coral animals, which require light for the large number of **zooxanthellae,** symbiotic algae that live and photosynthesize in their tissues. In addition to obtaining food from the zooxanthellae that live inside them, coral animals capture food at night with stinging tentacles that paralyze zooplankton and small animals that drift nearby. The waters where coral reefs are found are often poor in nutrients, but other factors are favorable for a high productivity, including the presence of zooxanthellae, a favorable temperature, and year-round sunlight.

Coral reefs grow slowly, as coral organisms build on the calcium carbonate remains of countless organisms before them. On the basis of their structure and underlying geologic features, there are three kinds of coral reefs: fringing reefs, atolls, and barrier reefs (**Figure 6.22**). The most common type of coral reef is the fringing reef. A **fringing reef** is directly attached to the shore of a volcanic island or continent and has no lagoon associated with it. An **atoll** is a circular coral reef that surrounds a central lagoon of quiet water. An atoll forms on top of the cone of a submerged volcano island. More than 300 atolls are found in the Pacific and Indian oceans, whereas the

Figure 6.20   Seagrass bed. A fisherman holds a fish in shallow water near turtle grass. Turtle grass (*Thalassia testudinum*) and other sea grasses form underwater meadows that are ecologically important for shelter and food for many organisms. Photographed off the coast of Grand Bahama Island. (© Doug Wilson/Corbis)

Figure 6.21   Kelp forest. These underwater "forests" are ecologically important because they support many kinds of aquatic organisms. Photographed off the coast of California. (Norbert Wu/Peter Arnold, Inc.)

Figure 6.22 **Types of coral reefs.**

(a) Fringing reef

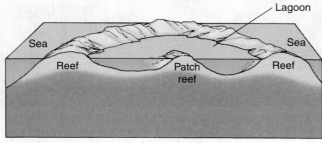

(b) Atoll

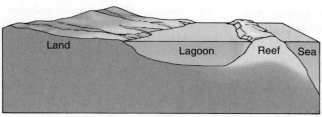

(c) Barrier reef

Atlantic Ocean, which is geologically different from the other two ocean basins, has few atolls. A lagoon of open water separates a **barrier reef** from the nearby land. The world's largest barrier reef is the Great Barrier Reef, which is nearly 2000 km (more than 1200 mi) in length and up to 100 km (62 mi) across. It extends along the northeastern coast of Australia. The second largest barrier reef is the Mesoamerican Reef in the Caribbean Sea off the coast of Belize, Mexico, and Honduras.

Coral reef ecosystems are the most diverse of all marine environments (see Figure 3.2). They contain thousands of species of fishes and invertebrates, such as giant clams, snails, sea urchins, sea stars, sponges, flatworms, brittle stars, sea fans, shrimp, and spiny lobsters. The Great Barrier Reef occupies only 0.1% of the ocean's surface, but 8% of the world's fish species live there. Some fishes are brightly colored to advertise they are poisonous.[1] The conspicuous lion fish, for example, defends itself from potential predators with numerous spiny projections that contain a powerful poison. The multitude of relationships and interactions that occur at coral reefs is comparable only to tropical rain forests among terrestrial ecosystems. As in the rain forest, competition is intense, particularly for light and space to grow.

Many unusual relationships occur at coral reefs. Certain tiny fishes swim over and even inside the mouths of larger fishes to remove potentially harmful parasites. Fishes sometimes line up at these cleaning stations, waiting their turn to be serviced. As another example, *Podillopora* corals have a remarkable defense mechanism that protects them from coral browsers such as the crown-of-thorns sea star. Tiny crabs live among the branches of the coral. When a sea star crawls onto the coral, the crabs swarm over it, nipping off its tube feet and causing it to retreat. Sometimes population explosions of crown-of-thorns sea stars have overrun coral reefs, causing extensive devastation of the coral. Some research suggests that the cause of these outbreaks is high levels of nutrient-rich runoff from farms and cities. The runoff stimulates the growth of phytoplankton, an important food source for sea star larvae.

Coral reefs are ecologically important because they both provide a habitat for many kinds of marine organisms and protect coastlines from shoreline erosion. They provide humans with seafood, pharmaceuticals, and recreational/tourism dollars. Although coral formations are important ecosystems, they are being degraded and destroyed. Of 109 countries with large reef formations, 90 are damaging them. Approximately one-fourth of the world's coral reefs are seriously degraded and at high risk. A 2002 report of a five-year census of global U.S. reefs, including those of former U.S. territories (Micronesia, the Marshall Islands, and Palau), substantiates the decline.

Human Impacts on Coral Reefs. Globally, many coral reefs are at risk. In some areas, silt washing downstream from clearcut inland forests has smothered reefs under a layer of sediment. In addition to pollution from coastal runoff, overfishing, fishing with dynamite or cyanide, disease, and coral bleaching are serious threats. Land reclamation, tourism, oil spills, boat groundings, anchor draggings, hurricane damage, ocean dumping, and the mining of corals for building material also take a toll.

Since the late 1980s, corals in the tropical Atlantic and Pacific have suffered extensive bleaching, in which stressed corals expel their zooxanthellae, becoming pale or white in color. Scientists are beginning to understand what causes bleaching, which may result in death of the colony. They suspect several environmental stressors, the

---

[1] Not all poisonous fishes are brightly colored, and not all brightly colored fishes are poisonous.

most likely being warmer seawater temperatures (water only about 1 degree above average can lead to bleaching). Many scientists attribute recent record sea temperatures and the large die-off of corals to El Niño effects, global climate change, or a combination of the two. Other potential stressors are pollution and coral diseases.

Although many coral reefs have not recovered from bleaching, some have. The coral-zooxanthellae relationship is highly complex and flexible. Research indicates that corals may lose up to 75% of their zooxanthellae without harming the reef. Corals may hold a "secret reserve" of zooxanthellae, not immediately apparent, that allows them to recover when bleached. In addition, corals take in any of several zooxanthella species, perhaps allowing one zooxanthella species to rescue coral when another abandons it. (Some species of zooxanthellae are more stress-resistant than others.)

Prospects for coral reefs remain uncertain; the incidence of coral diseases is on the rise, and estimates of the percentage of reefs beyond recovery continue to climb. Coral reefs off southeastern Asia, which contain the most species of all coral reefs, are the most threatened of any region.

The Pelagic Environment: The Vast Marine System

The **pelagic environment** consists of all of the ocean water, from the shoreline down to the deepest ocean trenches. It is subdivided into several life zones based on depth and degree of light penetration. The upper reaches of the pelagic environment comprise the **euphotic zone,** which extends from the surface to a maximum depth of 150 m (488 ft) in the clearest open ocean water. Sufficient light penetrates the euphotic zone to support photosynthesis. Large numbers of phytoplankton, particularly diatoms in cooler waters and dinoflagellates in warmer waters, produce food by photosynthesis and are the base of food webs.

The two main divisions of the pelagic environment are the neritic and oceanic provinces.

The Neritic Province. Organisms that live in the neritic province, the water that overlies the continental shelf, are all floaters or swimmers. Zooplankton (including tiny crustaceans, jellyfish, comb jellies, protists such as foraminiferans, and larvae of barnacles, sea urchins, worms, and crabs) feed on phytoplankton in the euphotic zone. Zooplankton are consumed by plankton-eating nekton such as herring, sardines, squid, baleen whales, and manta rays. These in turn become prey for carnivorous nekton such as sharks, porpoises, toothed whales, and tuna (**Figure 6.23**). Nekton are mostly confined to the shallower neritic waters (less than 60 m, or 195 ft, deep), near their food.

neritic province: The part of the pelagic environment that overlies the ocean floor from the shoreline to a depth of 200 m (650 ft).

Figure 6.23   Carnivorous nekton in the neritic province. Shown is a small school of Pacific bluefin tuna (*Thunnus thynnus*). (Richard Herrmann/Visuals Unlimited)

oceanic province: The part of the pelagic environment that overlies the ocean floor at depths greater than 200 m (650 ft).

The Oceanic Province. The oceanic province is the largest marine environment, comprising about 75% of the ocean's water; the oceanic province is the open ocean that does not overlie the continental shelf. Most of the oceanic province is loosely described as the "deep sea." (The average depth of the ocean is 4 km, more than 2 mi.) All but the surface waters of the oceanic province have cold temperatures, high hydrostatic pressure, and an absence of sunlight. These environmental conditions are uniform throughout the year.

Most organisms of the deep waters of the oceanic province depend on **marine snow,** organic debris that drifts down into their habitat from the upper, lighted regions of the oceanic province. Organisms of this little-known realm are filter feeders, scavengers, or predators. Many are invertebrates, some of which attain great sizes. The giant squid measures up to 18 m (59 ft) in length, including its tentacles. Fishes of the deep waters of the oceanic province are strikingly adapted to darkness and scarcity of food. An organism that encounters food infrequently must eat as much as possible when food is available. Adapted to drifting or slow swimming, animals of the oceanic province often have reduced bone and muscle mass. Many of these animals have light-producing organs to locate one another for mating or food capture. The dragonfish has a pocket of red light shining from beneath each eye. Because other species living in the ocean's depths cannot see red light, the dragonfish detects organisms in its surroundings without being seen. (Most animals in the ocean depths produce a blue-green light.)

national marine sanctuary: A marine ecosystem set aside to minimize human impacts and protect unique natural resources and historic sites.

National Marine Sanctuaries   The United States has national marine sanctuaries along the Atlantic, Pacific, and Gulf of Mexico coasts. These sanctuaries include kelp forests off the coast of California, coral reefs in the Florida Keys, fishing grounds along the continental shelf, and deep submarine canyons, as well as shipwrecks and other sites of historic value (see Figure 18.2).

The *National Marine Sanctuary Program*, which is part of the National Oceanic and Atmospheric Administration, administers the sanctuaries. Like many federal lands, they are managed for multiple purposes, including conservation, recreation, education, mining of some resources, scientific research, and ship salvaging. Commercial fishing is permitted in most of them, although several "no-take" zones exist where all fishing and collecting of biological resources is banned. Several recent studies have shown that these no-take zones for fish promote population increases of individual species as well as species richness (the no-take reserves typically contain 20% to 30% more species than nearby fished areas). The reserves have a spillover effect—that is, they boost populations of fish surrounding their borders. Fishing groups and some politicians are opposed to no-take reserves because they threaten public access to the seas.

The National Marine Sanctuary system includes 14 sites—13 national marine sanctuaries and one national monument. In 2006, President George W. Bush established the world's largest protected marine area when he designated the Northwestern Hawaiian Islands and surrounding waters—an area almost as large as California—as a national monument. Such a designation provides permanent funding to manage and preserve the area. This protected area is home to thousands of species of seabirds, fishes, marine mammals, and coral reef colonies.

Human Impacts on the Ocean   The ocean is so vast that it is hard to visualize how human activities could harm it. Such is the case, however. The development of resorts, cities, industries, and agriculture along coasts alters or destroys many coastal ecosystems, including mangrove forests, salt marshes, seagrass beds, and coral reefs. Coastal and marine ecosystems receive pollution from land, from rivers emptying into the ocean, and from atmospheric contaminants that enter the ocean via precipitation. Disease-causing viruses and bacteria from human sewage contaminate seafood, such as shellfish, and pose an increasing threat to public health. Millions of tons of trash, including plastic, fishing nets, and packaging materials, find their way into coastal and marine ecosystems (**Figure 6.24**). Some of this trash entangles and kills marine organisms. Less visible contaminants of the ocean include fertilizers, pesticides, heavy metals, and synthetic chemicals from agriculture and industry.

Offshore mining and oil drilling pollute the neritic province with oil and other contaminants. Millions of ships dump oily ballast and other wastes overboard in the neritic and oceanic provinces. Fishing is highly mechanized, and new technologies detect and remove every single fish in a targeted area of the ocean (see Figure 19.14). Scallop dredges and shrimp trawls are dragged across the benthic environment, destroying entire communities with a single swipe.

Conservation groups and government agencies have studied these problems for many years and have made numerous recommendations to protect and manage the ocean's resources. For example, from 2002 to 2003, the Pew Oceans Commission, composed of scientists, economists, fishermen, and other experts, verified the seriousness of ocean problems in a series of seven studies (**Figure 6.25**).

The 2004 report by the U.S. Commission on Ocean Policy, the first comprehensive review of federal ocean policy in 35 years, recommended improving the ocean and coasts in three main ways:

- **Create a new ocean policy to improve decision making**. Currently, an array of agencies and committees manages U.S. waters, and their respective goals often conflict. The Commission recommends that the National Oceanic and Atmospheric Administration (NOAA) be reorganized and strengthened. Later, other federal ocean programs would be consolidated under NOAA. In the future, a new agency would be created to manage air, land, and water resources together.

TWENTY-FIRST CENTURY HERMIT CRABS

Figure 6.24   Ocean pollution. What is the serious message behind this cartoon? (Chris Madden/Inkline Press)

**Nonpoint Source Pollution
(runoff from land)**
Example: Agricultural runoff (fertilizers,
pesticides, and livestock wastes)
pollutes water.

Invasive Species
Example: Release of ships' ballast water,
which contains foreign crabs, mussels,
worms, and fishes.

Overfishing
Example: The populations of many
commercial fish species are severely
depleted.

Bycatch
Example: Fishermen unintentionally
kill dolphins, sea turtles, and sea birds.

Aquaculture
Example: Produces wastes that can pollute
ocean water and harm marine organisms;
requires wild fish to feed farmed fish.

Point Source Pollution
Example: Passenger cruise ships dump
sewage, shower and sink water, and oily
bilge water.

Coastal Development
Example: Developers destroy important
coastal habitat, such as salt marshes and
mangrove swamps.

Habitat Destruction
Example: Trawl nets (fishing equipment
pulled along the ocean floor) destroy habitat.

Climate Change
Example: Coral reefs are particularly
vulnerable to increasing temperatures
and ocean acidification.

Figure 6.25 **An overview of major threats to the ocean.** (Based on S. R. Palumbi)

- **Strengthen science and generate information for decision makers.** There is a critical need for high-quality research on how marine ecosystems function and how human activities affect them. Currently, ocean research makes up 3.5% of federal research spending, and the Commission recommends doubling this amount.

- **Enhance ocean education to instill citizens with a stewardship ethic.** Environmental education should be part of the curriculum and should include a strong marine component. According to the Commission, ocean education programs should result in "lifelong learning, an adequate and diverse workforce, informed decision makers, improved science literacy, and a sense of stewardship for ocean and coastal resources."

The actions just listed will require billions of dollars and years of effort to be realized. Like most countries, the United States recognizes the importance of the ocean to life on this planet. However, it remains to be seen if the United States will make a strong commitment of resources to protect and manage the ocean effectively.

1. What environmental factors are most important in determining the kinds of organisms found in aquatic environments?

2. How do you distinguish between freshwater wetlands and estuaries? between flowing-water and standing-water ecosystems?

3. What are the four main marine environments?

4. Which marine environment is transitional between land and ocean?

5. How do the neritic and oceanic provinces differ?

6. What are four purposes of national marine sanctuaries?

 ## Interaction of Life Zones and Humans

LEARNING OBJECTIVE

• Outline the environmental history of the Florida Everglades.

Although we have discussed terrestrial and aquatic life zones as discrete entities, none of them exists in isolation. When parts of the Amazon rain forest flood annually, fishes leave the stream beds and swim all over the forest floor, where they play a role in dispersing the seeds of many species of plants. And in the Antarctic, where waters are much more productive than land areas, the many seabirds and seals form a link between marine and terrestrial environments. Although the ocean supports these animals, their waste products, cast-off feathers, and the like, when deposited on land, support whatever lichens, insects, and microorganisms live there.

Some inhabitants of terrestrial and aquatic life zones cover great distances—in the case of migratory fishes and birds, even global distances. Many young albacore tuna migrate from the California coast across the Pacific Ocean to Japan. Some species of flycatchers spend their summers in Canada and the United States and their winters in Central and South America. Like the flycatchers, many other migratory birds commonly spend critical parts of their life cycles in different countries, which makes their conservation difficult. It does little good, for instance, to protect a songbird in one country if the inhabitants of the next put it in the cooking pot as soon as it arrives. Such large-scale interactions among various ecosystems make ecological concepts difficult for people to grasp and nations to apply.

Throughout this chapter, you have seen how human activities are altering and in some cases destroying biomes and aquatic ecosystems. The Everglades is an excellent example of a severely altered ecosystem that, depending on human policies and intervention, may be partially restored during the first half of the 21st century.

 CASE IN POINT

The Everglades

The Everglades in the southernmost part of Florida is a vast expanse of predominantly sawgrass wetlands dotted with small islands of trees. At one time 80 km (50 mi) wide, 160 km (100 mi) long, and 15 cm to 0.9 m (6 in. to 3 ft) deep, the "river of grass" drifted south in a slow-moving sheet of fresh water from Lake Okeechobee to Florida Bay (**Figure 6.26**). The Everglades is a haven for wildlife such as alligators, snakes, panthers, otters, raccoons, and thousands of wading birds and birds of prey. The region's natural wonders were popularized in an environmental classic, *The Everglades: River of Grass*, written by **Marjory Stoneman Douglas** in 1947. The southernmost part of the Everglades is now protected as a national park.

The Everglades today is about half its original size, and it has many serious environmental problems. Most water bird populations are down 90% in recent decades, and the area is now home to 50 endangered or threatened species. Basically, two problems override all others in the Everglades today—it receives too little water, and the

Figure 6.26   Florida Everglades.

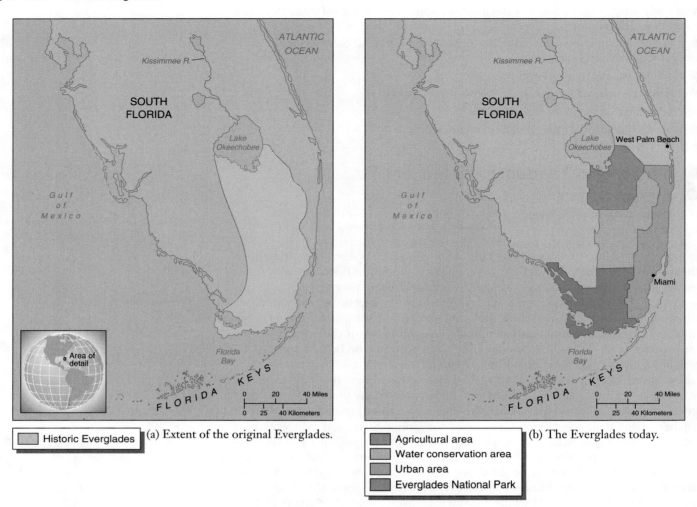

Historic Everglades   (a) Extent of the original Everglades.

Agricultural area
Water conservation area
Urban area
Everglades National Park   (b) The Everglades today.

water it receives is polluted with nutrient minerals from agricultural runoff. (In its natural state, the Everglades was nutrient-poor.)

During heavy rains, Lake Okeechobee historically flooded its banks, creating wetlands that provided biological habitat and helped recharge the Everglades. However, when a hurricane hit the lake in 1928 and many people died, the Army Corps of Engineers built the Hoover Dike along the eastern part of the lake. The Hoover Dike, completed in 1932, stopped the flooding but prevented the water in Lake Okeechobee from recharging the Everglades. Four canals built by the Everglades Drainage District effectively drained 214,000 hectares (530,000 acres) south of Lake Okeechobee, which was converted to farmland. Fertilizers and pesticides used here make their way to the Everglades, where they alter native plant communities. Phosphorus is particularly harmful because it encourages the growth of nonnative cattails that overrun the native sawgrasses and disrupt the flow of water.

After several tropical storms caused flooding damage in South Florida in 1947, the Army Corps of Engineers constructed an extensive system of canals, levees, and pump stations to prevent flooding, provide drainage, and supply water to South Florida. These structures, which divert excess water to the Atlantic Ocean rather than the Everglades, stopped the periodic floods in South Florida. The drier lands encouraged accelerated urban growth, particularly along the East Coast, and the expansion of agriculture into the southernmost parts of the Everglades. Thus, more than 70 years of engineering projects have reduced the quantity of water flowing into the Everglades, and the water that does enter is polluted from agricultural runoff. Urbanization has caused loss of habitat, contributing to the Everglades' problems.

In 1996, the state of Florida and the U.S. government began planning a massive restoration project to undo some of the damage from decades of human interference. The project will take about 30 years and cost $8 billion. The plan has three parts:

1. Sugar farmers will clean up their runoff so that the amount of phosphorus entering the Everglades is reduced.

2. Some agricultural land located at the southern end of the Everglades Agricultural Area below Lake Okeechobee will be bought and converted into marshes that will filter and further clean the agricultural runoff of the remaining farmland before it reaches the Everglades.

3. The U.S. Army Corps of Engineers and the South Florida Water Management District will reengineer the area's entire system of canals, levees, and pumps to restore a more natural flow of water to the Everglades. Roads in the area must be raised so that the water can flow under them.

The **Comprehensive Everglades Restoration Plan** will eventually supply clean, fresh water to the Everglades and to the fast-growing population of South Florida. (The population in South Florida is currently 6 million and is projected to be 15 million by 2050.) The plan involves drilling several hundred wells into the Florida Aquifer, a porous layer of limestone about 305 m (1000 ft) underground. The wells, located near Lake Okeechobee, would pump excess water produced during the rainy season underground, where it would be stored until needed during dry spells.

Some environmentalists and scientists have criticized the plan, particularly the significant uncertainties of using aquifer storage. Obviously, scientists must do a great deal of research as the plan goes forward. Many questions remain about how best to restore a more natural water flow, repel invasions of foreign species, and reestablish native species. ∎

REVIEW

1. How did the Florida Everglades get so degraded?
2. What are some of the challenges associated with restoration of the Florida Everglades?

REVIEW OF LEARNING OBJECTIVES WITH KEY TERMS

• Define *biome* and briefly describe the nine major terrestrial biomes: tundra, boreal forest, temperate rain forest, temperate deciduous forest, temperate grassland, chaparral, desert, savanna, and tropical rain forest.

A **biome** is a large, relatively distinct terrestrial region with a similar climate, soil, plants, and animals, regardless of where it occurs in the world. **Tundra** is the treeless biome in the far north that consists of boggy plains covered by lichens and small plants such as mosses; tundra has harsh, very cold winters and extremely short summers. **Boreal forest** is a region of coniferous forest (such as pine, spruce, and fir) in the Northern Hemisphere; it is located south of the tundra. **Temperate rain forest** is a coniferous biome with cool weather, dense fog, and high precipitation. **Temperate deciduous forest** is a forest biome that occurs in temperate areas with a moderate amount of precipitation. **Temperate grassland** is grassland with hot summers, cold winters, and less rainfall than the temperate deciduous forest biome. **Chaparral** is a biome with mild, moist winters and hot, dry summers; vegetation is typically small-leaved evergreen shrubs and small trees. **Desert** is a biome in which the lack of precipitation limits plant growth; deserts are found in both temperate and subtropical regions. **Savanna** is tropical grassland with widely scattered trees or clumps of trees. **Tropical rain forest** is a lush, species-rich forest biome that occurs where the climate is very warm and moist throughout the year.

• Relate at least one human effect on each of the biomes discussed.

Oil exploration and military use have caused long-lasting injury to large areas of the arctic tundra. Boreal forest is the world's primary source of industrial wood and wood fiber, and extensive logging has occurred. Temperate rain forest supplies lumber and pulpwood, and very little of original old-growth forest remains untouched. Many temperate deciduous forests have been converted to agricultural lands, and some have become degraded by intensive agricultural practices. Most temperate grasslands have become agricultural lands; the remaining grassland is often fragmented into small patches. The hilly chaparral on which homes have been built are subject to fires and mudslides. Human developments in desert areas require vast quantities of water, which is imported from distant areas; increased groundwater consumption by many desert cities has caused groundwater levels to drop. Savannas are rapidly being converted to rangeland for domesticated animals, which are replacing the big herds of wild animals. Human population growth and industrial expansion are rapidly destroying tropical rain forests, and many rainforest organisms are losing habitat.

• Explain the similarities and the changes in vegetation observed with increasing elevation and increasing latitude.

Similar ecosystems are encountered in climbing a mountain (increasing elevation) and traveling to the North Pole (increasing latitude).

This elevation-latitude similarity occurs because the temperature drops as one climbs a mountain, just as it does when one travels north.

- **Summarize the important environmental factors that affect aquatic ecosystems.**

In aquatic ecosystems, important environmental factors include **salinity**, amount of dissolved oxygen, and availability of light for photosynthesis.

- **Briefly describe the eight aquatic ecosystems: flowing-water ecosystem, standing-water ecosystem, freshwater wetlands, estuary, intertidal zone, benthic environment, neritic province, and oceanic province.**

A **flowing-water ecosystem** is a freshwater ecosystem such as a river or stream in which the water flows in a current. A **standing-water ecosystem** is a body of fresh water that is surrounded by land and that does not flow (a lake or a pond). **Freshwater wetlands** are lands that shallow fresh water covers for at least part of the year; wetlands have a characteristic soil and water-tolerant vegetation. An **estuary** is a coastal body of water, partly surrounded by land, with access to the open ocean and a large supply of fresh water from a river. Four important marine environments are the intertidal zone, benthic environment, neritic province, and oceanic province. The **intertidal zone** is the area of shoreline between low and high tides. The **benthic environment** is the ocean floor, which extends from the intertidal zone to the deep-ocean trenches. The **pelagic environment** consists of all of the ocean water, from the shoreline down to the deepest ocean trenches. The two main divisions of the pelagic environment are the neritic and oceanic provinces. The **neritic province** is the part of the pelagic environment that overlies the ocean floor from the shoreline to a depth of 200 m. The **oceanic province** is the part of the pelagic environment that overlies the ocean floor at depths greater than 200 m.

- **Relate at least one human effect on each of the aquatic ecosystems discussed.**

All aquatic ecosystems are subject to human-produced pollution, such as sewage discharge, agricultural runoff, oil spills, and trash. The building of dams has adverse environmental effects on rivers and streams. Many wetlands have been filled in or drained so that farms, housing developments, and industrial plants could be built. Estuaries, including mangrove forests, are under assault from coastal development, unsustainable logging, and aquaculture. Coral reefs in the benthic environment are suffering from silt from clearcut forests, high salinity from the diversion of fresh water, tourism, and the mining of corals for building material. Millions of ships dump oily ballast and other wastes overboard in the neritic and oceanic provinces. New fishing technologies remove every single fish in a targeted area of the ocean.

- **Discuss national marine sanctuaries.**

A **national marine sanctuary** is a marine ecosystem set aside to minimize human impacts and protect unique natural resources and historic sites. The National Oceanic and Atmospheric Administration (NOAA) administers the sanctuaries for multiple purposes, including conservation, recreation, education, mining, scientific research, ship salvaging, and commercial fishing.

- **Outline the environmental history of the Florida Everglades.**

The Everglades in the southernmost part of Florida is a vast expanse of sawgrass wetlands dotted with small islands of trees. The Everglades today is about half its original size and has many serious environmental problems. Most water bird populations are down 90% in recent decades. Engineering projects have reduced the quantity of water flowing into the Everglades, and the water that does enter is polluted from agricultural runoff. Urbanization has caused loss of habitat. Although the Everglades will never return completely to its original natural condition, it can be partially restored. **The Comprehensive Everglades Restoration Plan** will eventually supply fresh water to the Everglades and to the fast-growing population of South Florida.

Thinking About the Environment

1. Offer a possible reason why the tundra has such a low species richness.

2. Relate the tundra's temperature and precipitation to its biotic characteristics.

3. Describe representative organisms of the forest biomes discussed in the text: boreal forest, temperate deciduous forest, temperate rain forest, and tropical rain forest.

4. In which biome do you live? If your biome does not match the description given in this book, how do you explain the discrepancy?

5. Which biomes are best suited for agriculture? Explain why each of the biomes you did not specify is less suitable for agriculture.

6. What human activities are harmful to deserts? to grasslands? to forests?

7. Which biome do you think is in greatest immediate danger from human activities? Why?

8. What would happen to the organisms in a river with a fast current if a dam were built? Explain your answer.

9. Explain the role of freshwater wetlands in water purification.

10. If you were to find yourself on a boat in the Chesapeake Bay, what aquatic ecosystem would you be in? What ecosystem would you be in if you were in the middle of Everglades National Park?

11. Which aquatic ecosystem is often compared to a tropical rain forest? Why?

12. Explain how coral reefs are vulnerable to human activities.

13. What is the largest marine environment, and what are some of its features?

14. Briefly discuss two major human-caused problems in the ocean.

15. A 2003 study reports that in recent years tropical ocean waters have become saltier, whereas polar ocean waters have become less salty. What do you think is a possible explanation for these changes?

16. In modeling the Chesapeake Bay system, scientists consider the following health indicators for the bay, which are organized into three categories. Pick two human-caused problems and explain how each affects the biotic and abiotic components of the Chesapeake Bay system.

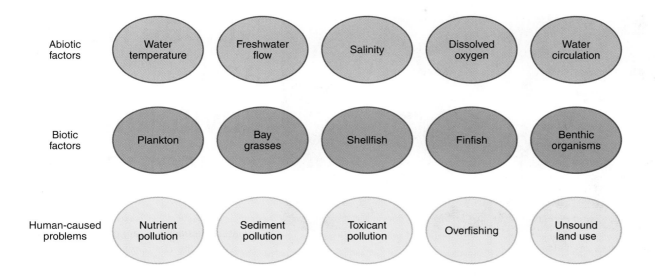

Quantitative questions relating to this chapter are on our Web site.

Take a Stand

Visit our Web site at http://www.wiley.com/college/raven (select Chapter 6 from the Table of Contents) for links to more information about the restoration of the Florida Everglades. Consider the opposing views of sugar farmers and conservationists, and debate the issue with your classmates. You will find tools to help you organize your research, analyze the data, think critically about the issues, and construct a well-considered argument. *Take a Stand* activities can be done individually or as a team, as oral presentations, written exercises, or Web-based (e-mail) assignments.

Additional online materials relating to this chapter, including a Student Testing Section with study aids and self-tests, Environmental News, Activity Links, Environmental Investigations, and more, are also on our Web site.

Human Health and Environmental Toxicology

Cars line up to drop off children at an elementary school. More children are driven to school today than in the past, when most children walked or rode bicycles. Seemingly simple lifestyle choices, such as driving children to school rather than having them walk, can have significant impacts on their health. (David Young-Wolff/Visual Unlimited)

Often, when people think about environment and health, they focus on things like pollution and pesticides. Certainly, toxic substances can sicken people, and there's a need to reduce or eliminate them. However, in the United States, pollution is only one of many aspects of the relationship between healthy people and a healthy environment. How individuals and societies interact with the environment has substantial implications for lifelong physical and mental health. Thinking about humans and the environment functioning as a system allows us to understand where and how changes in the system can improve human and environmental health.

For example, today, many children in suburban America are driven to school, whereas several decades ago they would have walked or bicycled (see photograph). While this change might seem minor, it represents a dramatic change in how children interact with their environment. They are establishing behaviors that will last a lifetime, with long-term health implications. Effects include reduced

exercise, increased pollution and accident risks, and the loss of interaction with the natural and physical world.

Consider a typical U.S. suburban school, with a thousand students. Even if only one-third of students are driven to school—the fraction can be much higher—as many as three hundred cars, mini-vans, SUVs, and pickup trucks may be driving to the school, all at the same time. The large numbers of vehicles present a physical threat to those children (and their parents, siblings, or grandparents) who do walk, especially when drivers are distracted by passengers, other drivers, and cell phones. The vehicles also present a health threat in the form of hundreds of gasoline-fueled engines idling and accelerating near occupied playgrounds. On days with slow winds, air pollution can be significantly elevated in areas where children are playing before their school day begins.

Children driven to school might otherwise be walking a mile or so each day. They miss out on light physical activity that is not made up in other ways. Childhood inactivity is linked to the increase in adult obesity. Being overweight is associated with elevated blood pressure, high cholesterol, and a low level of high-density lipoproteins (HDLs, the "good" cholesterol). These conditions harm the blood vessels that supply the brain, heart, and kidneys, thereby increasing the risk of stroke, heart disease, and kidney failure. Type 2 diabetes, a dangerous medical condition, is also

related to obesity, as are osteoarthritis and cancers of the colon, gallbladder, prostate, kidney, cervix, ovaries, and uterus.

Finally, in being driven to school on a regular basis, many children miss out on one of the few opportunities they have to interact with their environments. Children may experience daily and seasonal changes in weather for, at most, minutes each day. Small animals like insects, spiders, and snails go unnoticed, as do the details of trees and grasses.

Over the past century, human life spans and general health have improved markedly. This has come from improved medicine and nutrition, a better understanding of causes and treatment of disease, and reductions in poverty. However, the current generation of U.S. children may be the first in decades to face *lower* life expectancies than their parents. Understanding the systemic relationship between human health and the natural, physical, and built environment is critical to a future of improved health.

In this chapter, we explore the factors that underlie health and disease in highly developed and developing countries. In future chapters, we will explore a range of related environment–human health interactions.

World View and Closer to You...
Real news footage relating to health and toxicology around the world and in your own geographic region is available at www.wiley.com./college/raven by clicking on "World View and Closer to You."

Human Health

Two indicators of human health in a given country are *life expectancy* (how long people are expected to live) and *infant mortality* (how many infants die before the age of one). These health indicators vividly demonstrate the contrasts in health among different nations. In a highly developed country such as Japan, the average life expectancy for a woman is 86 and for a man 79. The infant mortality rate in Japan is 2.8, which means that almost three infants under age one die per 1000 live births. In contrast, in Zambia, a less developed country in East Africa, the average life expectancy is 37 for women and 38 for men. The infant mortality rate in Zambia is 92, which means that 92 out of 1000 Zambian infants die before reaching the age of one.

Why is there such a difference between these two countries? A Japanese child receives good health care, including vaccinations, and has sufficient nutrition for growth and development. The average Japanese woman gives birth to one or two children, and, during her pregnancies, can eat well, rest when necessary, and receive medical attention as needed. As Japanese people age, they may develop the chronic degenerative diseases associated with aging, but they have access to quality health care, medications, and rehabilitation services. The average person in Japan spends $550 (in U.S. dollars) on medicines each year.

The picture is very different in Zambia. Many Zambian children receive few immunizations or adequate nutrition for normal growth and development; many children are underweight. The average Zambian woman is likely to be in poorer health due to childhood factors. She may give birth to five or six children. Because she does not receive prenatal care, she has an increased risk of death during childbirth. Moreover, one of her children will probably die in infancy. The average Zambian spends about U.S. $5 on medicines each year. Those Zambians who survive middle age often die prematurely because they lack treatment for the chronic degenerative diseases associated with aging.

Differences in health and health care between highly developed and developing nations highlight the effects of different lifestyles and levels of poverty. In developing countries, there are about 170 million children who are underweight, and 3 million of these will die this year (**Figure 7.1**). Meanwhile, more than one billion people are overweight worldwide, and 300 million people are obese. In the highly developed countries of North America and Europe, about 500,000 people will die this year from obesity-related diseases.

Figure 7.1 Underweight infant in Bangladesh. This baby, just born in the less developed country of Bangladesh, is underweight (five and a half pounds) and therefore at greater risk of dying in infancy. The infant mortality rate in Bangladesh is 65. (Karen Kasmauski/National Geographic Society)

Health Issues in Highly Developed Countries

By many measures, the health of people in the United States and other highly developed nations is good. Improved sanitation during the 20th century reduced many diseases, such as typhoid, cholera, and diarrhea, that previously made people ill or killed them. Many childhood diseases, such as measles, polio, and mumps, have been conquered. In 1900, the average life expectancy for Americans was 51 years for women and 48 years for men. Today, the average life expectancy is 80 for women and 75 for men.

In 1900, the three leading causes of death in the United States were pneumonia and influenza, tuberculosis, and gastritis and colitis (diarrhea); these are infectious diseases caused by microorganisms. Currently, the three leading causes of death in the United States are cardiovascular diseases (of the heart and blood vessels), cancer, and chronic obstructive pulmonary disease (of the lungs); these diseases are noninfectious chronic health problems, and many are associated with aging. A significant fraction of premature deaths in the United States is caused in part by individual lifestyle habits involving poor diet, lack of exercise, and smoking.

While undernutrition is uncommon in the United States, two related types of malnourishment represent a growing health threat. Many Americans consume inadequate amounts of vitamins, fiber, and essential nutrients. In addition, typical American diets include too many total calories—overnutrition—resulting in a significant obesity problem (see Chapter 19).

Healthcare professionals use the **body mass index (BMI)** to determine whether a person is overweight or obese. To calculate your BMI, multiply your weight by 740, then divide that number by your height in inches, squared. For example, if you are 5 feet tall and weigh 130 pounds, multiply 130 by 740 = 96,200. Five feet equals 60 inches, and the square of 60 is 3600. Now divide 96,200 by 3600 = 26.7.

- If your BMI = 18.5 or less, you are considered underweight.
- If your BMI is from 18.5 to 24.9, you are at a healthy weight.
- If your BMI is between 25 and 29.9, you are overweight.
- If your BMI is 30 or more, you are considered obese.

In this example, the person whose BMI is 26.7 is considered overweight but not obese.

Health Issues in Developing Nations

There is good news and bad news regarding the health and well-being of people in developing nations. Gradual improvements in sanitation and drinking water supplies in moderately developed countries are reducing the incidence of diarrheal diseases such as cholera. Mass immunization programs have eliminated smallpox and reduced the risk of polio, yellow fever, measles, and diphtheria in most countries (see Meeting the Challenge: Global Polio Eradication). Despite these gains, malnutrition, unsafe water, poor sanitation, and air pollution still prevail in many less developed countries (**Figure 7.2**). In addition, research suggests that many unhealthy lifestyles associated with developed countries, including smoking, obesity, and diabetes, are increasingly common in less developed countries.

Although overall life expectancy has increased to 65 years in developing countries, 19 of the very poorest developing countries have life expectancies of 45 years or less. Except for Afghanistan, all of these countries are in Africa. HIV/AIDS has reduced life expectancies by more than 20 years in the African countries of Botswana, Lesotho, Swaziland, and Zimbabwe.

Figure 7.2   Workers burning trash in Asab, Eritrea. In many developing countries, sorting and burning garbage with little or no protective equipment is common. Workers risk infection, injury, and toxic exposures. (Robert Caputo, Aurora Photos)

MEETING THE CHALLENGE

Global Polio Eradication

During the 20th century, significant progress was made in eradicating certain infectious diseases caused by viruses. Smallpox was eliminated worldwide, and polio (more correctly called *poliomyelitis*) was eliminated in North and South America, Europe, Australia, and much of Asia.

Children under the age of five are the main victims of polio, which is spread by contaminated drinking water. The poliovirus attacks the central nervous system (brain and spinal cord), causing paralysis. Death sometimes occurs when the muscles that control breathing become paralyzed. Polio vaccines, first developed in the 1950s, are usually administered orally (by mouth), and a series of four doses is required to provide complete protection to infants. Supplementary doses are given to children under five years of age.

In 2006, there were a total of 1997 polio cases worldwide. Most of these cases occurred in the four countries in which polio is currently endemic (constantly present): Nigeria, India, Pakistan, and Afghanistan. Nigeria is the country with the most cases of polio—1123 in 2006.

In 1988, polio was endemic in more than 125 countries. Since 1988, the World Health Organization (WHO) has conducted a vigorous campaign to eliminate polio worldwide. For example, a single case of polio reported in Egypt triggered a campaign to vaccinate at least one million children there. India is also making significant progress; after an epidemic in 2002, 100 million children were vaccinated!

In Africa, health experts conduct an annual synchronized campaign to vaccinate children during the peak season for polio. However, in 2003, there was a major setback in efforts to eliminate polio in Africa when the Nigerian state of Kano suspended vaccinations over concerns of its safety. An outbreak of polio occurred in Nigeria, reinfecting areas that were previously polio-free. Polio also spread from northern Nigeria to other African countries, including Sudan, Central African Republic, Niger, Chad, Mali, Ivory Coast, Cameroon, Burkina Faso, Guinea, Benin, Botswana, Egypt, and Ethiopia, causing the worst epidemic in years.

In the fall of 2004, vaccinations resumed in Kano after health officials found a new supplier of the vaccine. Health officials are also conducting a synchronized campaign in nearby countries in an effort to contain the virus in Nigeria until it can be eradicated completely. However, the quickness with which polio rebounded after suspension of vaccinations vividly demonstrated that serious efforts at both the national and international levels are needed to make polio the second disease to be globally eradicated.

According to the World Health Organization, 18% of the 57 million deaths that occur worldwide each year are children less than five years of age. Child mortality is particularly serious in Africa, where 14 countries have higher levels of child mortality today than they did in 1990. Leading causes of death in children in developing countries include malnutrition, lower respiratory tract infections, diarrheal diseases, and malaria. In sub-Saharan Africa, HIV/AIDS is a significant cause of death for many young children.

Emerging and Reemerging Diseases

At one time, it was mistakenly thought that infectious diseases had either been conquered or were on the way to being conquered. We now know this is not true. *Emerging diseases* are infectious diseases that were not previously found in humans; emerging diseases typically jump from an animal host to the human species. Acquired immune deficiency syndrome (AIDS) is the most serious emerging disease: About 20 million people have died from HIV/AIDS since 1981, and 38 million people currently live with the disease. Epidemiologists think the HIV virus jumped from nonhuman primates to humans about 60 to 70 years ago, perhaps when humans were unintentionally exposed to contaminated blood when butchering or eating chimpanzees for food.

Other emerging diseases include Lyme disease, West Nile virus, Creutzfeld-Jakob disease (the human equivalent of mad cow disease), severe acute respiratory syndrome (SARS), Ebola virus, and monkeypox. In addition, new strains of influenza arise each year, some far more deadly than others. To successfully contain an emerging disease, epidemiologists must recognize the symptoms, identify the disease-causing agent, and inform public health authorities. In turn, public health officials must isolate patients with those symptoms and track down all people who have come into contact with the patients. Meanwhile, medical researchers develop treatment strategies and try to determine and eliminate the origins of the disease.

Reemerging diseases are infectious diseases that existed in the past but for a variety of reasons are increasing in incidence or in geographic range. The most serious reemerging disease is tuberculosis. People with HIV/AIDS are more susceptible to tuberculosis because their immune systems are compromised. Thus, the HIV/AIDS pandemic has helped fuel the tuberculosis outbreak. Another factor responsible for an increased incidence of tuberculosis is the evolution of antibiotic-resistant strains of the bacterium that causes tuberculosis; we must use antibiotics more judiciously to slow the inevitable increase in antibiotic-resistant bacteria. Tuberculosis is associated with poverty, and the increased incidence of urban poverty has also contributed to the emergence of tuberculosis. Other important reemerging diseases include yellow fever, malaria, and dengue fever.

Health experts have determined the main factors involved in the emergence or reemergence of infectious diseases. Some of the most important are:

- Evolution in the infectious organisms so they can move from animal to human hosts.
- Evolution of antibiotic resistance in the infectious organisms.
- Urbanization, associated with overcrowding and poor sanitation.
- A growing population of elderly people who are more susceptible to infection.
- Pollution, environmental degradation, and changing weather patterns.
- Growth in international travel and commerce.
- Poverty and social inequality.

CASE IN POINT

Influenza Pandemics Past and Future

The influenza virus has been a threat to human health for centuries. Every year, one or more new strains of the virus appear, and quickly spread around the globe during the "flu season," which usually runs from late fall through the winter. In a typical year, anywhere from 5–20% of the U.S. population contracts the flu, with symptoms ranging from mild headaches to severe muscle aches, digestive and breathing problems, and high fever. Typically, about 36,000 people in the United States die from the flu each year.

In contrast, a particularly virulent strain during the 1918–1919 flu season may have killed more than 850,000 in the United States, or nearly 1% of the population of 100 million people. The 1918–1919 flu season corresponded with the end of the first World War, a time already characterized by hardships including hunger and a lack of medicines, as well as a time of unprecedented international travel. The 1918 flu was unusual in that, unlike most strains, it was at least as potent to young healthy people as to the very young, elderly, and infirm.

While individuals can be vaccinated, successful vaccination requires that researchers predict which influenza strains will appear, successfully prepare enough vaccine for everyone who needs it, and then distribute the vaccine to all those people. In recent years, an increasing fraction of those 65 or older have been getting the vaccine (65% in 2003, the latest year on record). However, less than 20% of those 18–49 years old were vaccinated each year from 1989 through 2003, which means that a strain similar to that of the 1918–1919 season could be devastating. In addition, such a strain could easily spread around the world via air travel—a scenario envisioned by Paul Ehrlich in his 1968 book *The Population Bomb*.

Another recent concern is that of a pandemic of avian influenza, or bird flu. A **pandemic** is a disease that reaches nearly every part of of the world, and has the potential to infect almost every person. Avian influenza is a strain that appears commonly in birds (**Figure 7.3**). It tends to be very difficult for humans to contract, since it

Figure 7.3 Dead swans. Wild swan deaths provided early evidence that avian influenza had migrated from Asia to Europe. Photographed in 2006. (Carsten Koall/Getty Images)

usually is transferred from birds to humans, but not from human to human. It is, however, extremely potent once contracted, with a very high fatality rate. A major concern for epidemiologists is the possibility of a strain that is easily transferred from human to human—a change that might involve a single genetic mutation. An avian flu pandemic could kill millions, or even billions, of people within a single year.

Understanding and controlling an avian influenza pandemic requires an understanding of the environment that allows the virus to survive and travel, as well as the cooperation of many governments and individuals. The virus often originates in areas with large numbers of domestic birds, especially chickens, that are raised in small cages. In the past several years, large numbers of domestic poultry have been killed and burned to prevent or stop a disease outbreak. Avian flu has become endemic in domestic and wild birds in Asia, Europe and Northern Africa, and may have reached the United States by the time this book is published. See www.pandemicflu.gov for information on how to be prepared for a flu pandemic. ■

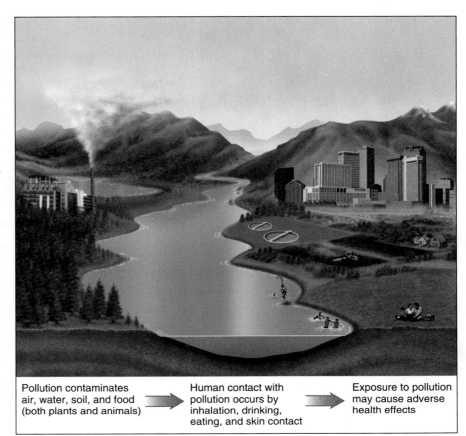

| Pollution contaminates air, water, soil, and food (both plants and animals) | → | Human contact with pollution occurs by inhalation, drinking, eating, and skin contact | → | Exposure to pollution may cause adverse health effects |

Figure 7.4 **Pathways of pollution in the environment.**

REVIEW

1. What is the average life expectancy for a woman in Japan? a woman in Zambia? Explain the difference.

2. What are three health issues for children in developing countries?

Environmental Pollution and Disease

LEARNING OBJECTIVES

- Summarize the problems associated with chemicals that exhibit persistence, bioaccumulation, and biological magnification in the environment.
- Define *endocrine disrupter*.
- Briefly describe some of the data suggesting that certain chemicals used by humans may also function as endocrine disrupters in animals, including humans.

It is often difficult to establish a direct relationship between environmental pollution and disease (**Figure 7.4**). The relationship is fairly clear for certain pollutants, such as the link between radon and lung cancer (see Chapter 20), or between lead and disorders of the nervous system (see Chapter 22). However, the evidence is less definite for many pollutants, and scientists can only suggest that there is an "association" between the pollutant and a specific illness. One reason it is so difficult to establish a direct cause and effect is that other factors—such as a person's genetic makeup, diet, level of exercise, and smoking habits—complicate the picture. These factors are often difficult to quantify.

Other complications include the fact that certain segments of society may be more susceptible to adverse health effects from an environmental pollutant. Children are particularly sensitive to pollution, as will be discussed later in this chapter. Other groups at higher risk from pollution include the elderly and those people with chronic diseases or compromised immune systems (such as people with HIV/AIDS or those taking chemotherapy to treat cancer). Also, people living in poor neighborhoods may be exposed to greater levels of pollution. We say more about the heath effects of specific pollutants on these various groups throughout the text.

(a) A bald eagle feeds its chick. Most people know that eagles were on the brink of extinction a few decades ago. However, most don't understand the scientific research behind the eagles' comeback. (Roy Toft/National Geographic Society)

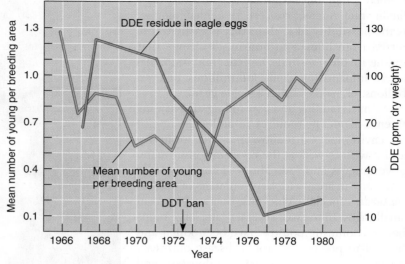

*DDT is converted to DDE in the birds' bodies

(b) A comparison of the number of successful bald eagle offspring with the level of DDT residues in their eggs. DDT was banned in 1972. Note that reproductive success improved after DDT levels decreased. (DDE is a derivative of DDT.) (J. W. Grier)

Figure 7.5 **Effect of DDT on birds.**

We now examine the characteristics of certain environmental pollutants that make them particularly dangerous to both the environment and human health. First, we consider toxic chemicals that persist and accumulate in the environment and magnify their concentration in the food web. Then, we examine a diverse group of pollutants that may affect the body's endocrine system, which produces *hormones* to regulate many aspects of body function.

Persistence, Bioaccumulation, and Biological Magnification of Environmental Contaminants

Some toxic substances exhibit persistence, bioaccumulation, and biological magnification. These substances include certain pesticides (such as DDT, or dichlorodiphenyl-trichloroethane), radioactive isotopes, heavy metals (such as lead and mercury), flame retardants (for example, PBDEs, or polybrominated diphenyl ethers), and industrial chemicals (such as dioxins and PCBs, or polychlorinated biphenyls).

The effects of the pesticide DDT on many bird species first demonstrated the problems with these chemicals. Falcons, pelicans, bald eagles, ospreys, and many other birds are sensitive to traces of DDT in their tissues. A substantial body of scientific evidence indicates that one of the effects of DDT on these birds is that they lay eggs with extremely thin, fragile shells that usually break during incubation, causing the chicks' deaths. After 1972, the year DDT was banned in the United States, the reproductive success of many birds improved (**Figure 7.5**).

The impact of DDT on birds is the result of three characteristics of DDT: its persistence, bioaccumulation, and biological magnification. Some pesticides, particularly chlorinated hydrocarbons such as DDT, take many years to be broken down into less toxic forms. The **persistence** of synthetic pesticides is a result of their novel (not found in nature) chemical structures. Natural decomposers such as bacteria have not yet evolved ways to degrade many synthetic pesticides, so they accumulate in the environment and in the food web.

persistence: A characteristic of certain chemicals that are extremely stable and may take many years to be broken down into simpler forms by natural processes.

When a pesticide is not metabolized (broken down) or excreted by an organism, it is simply stored, usually in fatty tissues. Over time, the organism may **bioaccumulate**, or **bioconcentrate**, high concentrations of the pesticide.

Organisms at higher levels on food webs tend to have greater concentrations of bioaccumulated pesticide stored in their bodies than those lower on food webs. This increase as the pesticide passes through successive levels of the food web is known as **biological magnification** or **biological amplification.**

As an example of the concentrating characteristic of persistent pesticides, consider a food chain studied in a Long Island salt marsh that was sprayed with DDT over a period of years for mosquito control: algae and plankton → shrimp → American eel → Atlantic needlefish → ring-billed gull (**Figure 7.6**). The concentration of DDT in water was extremely dilute, on the order of 0.00005 parts per million (ppm). The algae and other plankton contained a greater concentration of DDT, 0.04 ppm. Each shrimp grazing on the plankton concentrated the pesticides in its tissues to 0.16 ppm. Eels that ate shrimp laced with pesticide had a pesticide level of 0.28 ppm, and needlefish that ate eels contained 2.07 ppm of DDT. The top carnivores, ring-billed gulls, had a DDT level of 75.5 ppm from eating contaminated fishes. Although this example involves a bird at the top of the food chain, all top carnivores, from fishes to humans, are at risk from biological magnification. Because of this risk, currently approved pesticides in the United States have been tested to ensure they do not persist and accumulate in the environment beyond a level predetermined to be safe.

Endocrine Disrupters

Mounting evidence suggests that dozens of industrial and agricultural chemicals—many of which exhibit persistence, bioaccumulation, and biological magnification—are also **endocrine disrupters**. These chemicals, many of which are no longer used in the United States, include chlorine-containing industrial compounds known as PCBs and dioxins; the heavy metals lead and mercury; some pesticides such as DDT, kepone, dieldrin, chlordane, and endosulfan; flame retardants (PBDEs, or polybrominated diphenyl ethers); and certain plastics and plastic additives such as phthalates.

Hormones are chemical messengers produced by organisms in minute quantities to regulate their growth, reproduction, and other important biological functions. Some endocrine disrupters mimic the *estrogens*, a class of female sex hormones, and send false signals to the body that interfere with the normal functioning of the reproductive system. Because both males and females of humans and many other species produce estrogen, endocrine disrupters that mimic estrogen can affect both sexes. Additional endocrine disrupters interfere with the endocrine system by mimicking hormones other than estrogen, such as *androgens* (male hormones such as testosterone) and *thyroid hormones*. Like hormones, endocrine disrupters are active at very low concentrations and, as a result, may cause significant health effects at low doses.

Many endocrine disrupters appear to alter reproductive development in males and females of various animal species. Accumulating evidence indicates that fishes, frogs, birds, reptiles such as turtles and alligators, mammals such as polar bears and otters, and other animals exposed to these environmental pollutants exhibit reproductive disorders and are often left sterile.

A chemical spill in 1980 contaminated Lake Apopka, Florida's third largest lake, with DDT and other agricultural chemicals with known estrogenic properties. Male alligators living in Lake Apopka in the years following the spill had low levels of testosterone (an androgen) and elevated levels of estrogen. Their reproductive organs were

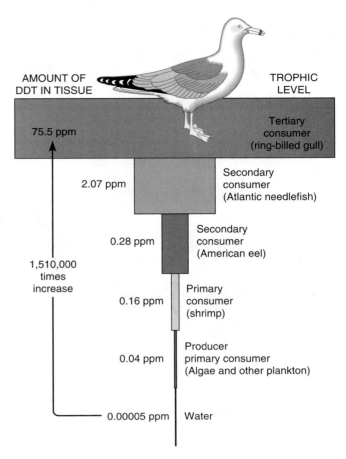

Figure 7.6   Biological magnification of DDT in a Long Island salt marsh. Note how the level of DDT, expressed as parts per million, increased in the tissues of various organisms as DDT moved through the food chain from producers to consumers. The ring-billed gull at the top of the food chain had approximately 1.5 million times more DDT in its tissues than the concentration of DDT in the water. (G. M. Woodwell, C. F. Worster, and A. R. Isaacson.)

bioaccumulation: The buildup of a persistent toxic substance, such as certain pesticides, in an organism's body, often in fatty tissues.

biological magnification: The increased concentration of toxic chemicals, such as PCBs, heavy metals, and certain pesticides, in the tissues of organisms that are at higher levels in food webs.

endocrine disrupter: A chemical that mimics or interferes with the actions of the endocrine system in humans and wildlife.

(a) A young American alligator (*Alligator mississippiensis*) hatches from eggs taken by University of Florida researchers from Lake Apopka, Florida. Many of the young alligators that hatch have abnormalities in their reproductive systems. This young alligator may not leave any offspring. (Stuart Bauer)

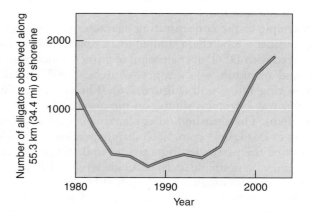

(b) Population of juvenile alligators in Lake Apopka after a chemical spill in 1980. (The chemical spill occurred just prior to the 1980 surveys). (A. R. Woodward)

Figure 7.7 **Lake Apopka alligators.**

often feminized or abnormally small. The mortality rate for eggs in this lake was extremely high, which reduced the alligator population for many years (**Figure 7.7**).

Humans may be equally at risk from endocrine disrupters, as the number of reproductive disorders, infertility cases, and hormonally related cancers (such as testicular cancer and breast cancer) appears to be increasing. More than 60 studies since 1938 have reported that sperm counts in a total of nearly 15,000 men from many nations, including the United States, dropped by more than 50% between 1940 and 1990. Although only one sperm is needed to fertilize an egg, infertility increases as sperm counts decline. Scientists do not know if there is a link between this apparent decline and environmental factors, however.

Certain *phthalates*, ingredients of cosmetics, fragrances, nail polish, medications, and common plastics used in a variety of food packaging, toys, and household products, have been implicated in birth defects and reproductive abnormalities. Animal studies in which rats are fed large doses of phthalates have shown that these compounds can damage fetal development, particularly of reproductive organs. In humans, there may be a correlation between certain phthalates and the increase since the 1970s in the incidence of hypospadias, a birth defect in which the urethral opening (passageway for urine) is on the underside of the penis instead of at its tip. Certain phthalates have also been linked to an increased observance of premature breast development in young girls, typically between 6 and 24 months of age. Other phthalates are of negligible concern, according to the National Toxicology Program's Center for the Evaluation of Risks to Human Reproduction.

In 2001, the Centers for Disease Control and Prevention (CDC) reported that they tested a diverse sample of the civilian U.S. population for the presence of 27 different environmental chemicals—heavy metals, nicotine from tobacco smoke, pesticides, and plastics, many of which are thought to be endocrine disrupters. Nearly every chemical was detected in every participant's body at higher levels than previously guessed. Because scientists had never actually measured 24 of the 28 chemicals in humans before, the study could not say if the levels of various environmental chemicals are increasing or decreasing. However, the study provides a good baseline of exposure of the U.S. population to these compounds against which future studies (the CDC plans to continue the tests) can be compared. The study is also a first step toward determining whether any of these chemicals is the hidden cause of modern illnesses or simply a benign byproduct of modern society.

Definite links between environmental endocrine disrupters and human health problems cannot be made at this time because of the limited number of human studies. Human exposure to endocrine-disrupting chemicals needs to be quantified so that we know exactly how much of these chemicals affects various communities. Complicating such assessments is the fact that humans are also exposed to natural hormone-mimicking substances in the plants we eat. Soy-based foods such as bean curd and soymilk, for example, contain natural estrogens.

Congress amended the Food Quality Protection Act and the Safe Drinking Water Act in 1996 to require the U.S. Environmental Protection Agency (EPA) to develop a plan and establish priorities to test thousands of chemicals for their potential to disrupt the endocrine system. In the first round of testing, chemicals are tested to see if they interact with any of five different *endocrine receptors*. (The body has specially shaped receptors on and in cells to which specific hormones attach. Once a hormone attaches to a receptor, it triggers other changes within the cell.) Chemicals testing positive—that is, binding to one or more types of receptors—are subjected to an extensive battery of tests to determine what specific damages, if any, they cause to reproduction and other biological functions. These tests, which may take decades to complete, should reveal the level of human and animal exposure to endocrine disrupters and the effects of this exposure. (The effects of toxic chemicals, including several endocrine disrupters, are discussed throughout the text.)

REVIEW

1. What are the differences among persistence, bioaccumulation, and biological magnification? How are these chemical characteristics interrelated?
2. How did the 1980 chemical spill in Lake Apopka affect alligators?

Determining Health Effects of Environmental Pollution

LEARNING OBJECTIVES

- Define *toxicant* and distinguish between acute and chronic toxicities.
- Describe how a dose-response curve helps determine the health effects of environmental pollutants.
- Discuss pesticide risks to children.
- Explain the relative advantages and disadvantages of toxicology and epidemiology.

The human body is exposed to many kinds of chemicals in the environment. Both natural and synthetic chemicals are in the air we breathe, the water we drink, and the food we eat. *All* chemicals, even "safe" chemicals such as sodium chloride (table salt), are toxic if exposure is high enough. A one-year-old child will die from ingesting about two tablespoons of table salt; table salt is also harmful to people with heart or kidney disease.

The study of **toxicants**, or toxic chemicals, is called **toxicology**. It encompasses the effects of toxicants on living organisms and how they cause toxicity, as well as ways to prevent or minimize adverse effects, such as developing appropriate handling or exposure guidelines.

The effects of toxicants following exposure can be immediate (acute toxicity) or prolonged (chronic toxicity). **Acute toxicity**, which ranges from dizziness and nausea to death, occurs immediately to within several days following a single exposure. In comparison, **chronic toxicity** generally produces damage to vital organs, such as the kidneys or liver, following a long-term, low-level exposure to chemicals. Toxicologists know far less about chronic toxicity than they do about acute toxicity, in part because the symptoms of chronic toxicity often mimic those of other chronic diseases.

toxicant: A chemical with adverse human health effects.

acute toxicity: Adverse effects that occur within a short period after exposure to a toxicant.

chronic toxicity: Adverse effects that occur some time after exposure to a toxicant, or after extended exposure to the toxicant.

Table 7.1 LD$_{50}$ Values for Selected Chemicals

Chemical	LD$_{50}$ (mg/kg)*
Aspirin	1750.0
Ethanol	1000.0
Morphine	500.0
Caffeine	200.0
Heroin	150.0
Lead	20.0
Cocaine	17.5
Sodium cyanide	10.0
Nicotine	2.0
Strychnine	0.8

*Administered orally to rats.
Source: M. D. Josten and J. L. Wood

We measure toxicity by the extent to which adverse effects are produced by various doses of a toxicant. A **dose** of a toxicant is the amount that enters the body of an exposed organism. The **response** is the type and amount of damage that exposure to a particular dose causes. A dose may cause death (*lethal dose*) or may cause harm but not death (*sublethal dose*). Lethal doses, usually expressed in milligrams of toxicant per kilogram of body weight, vary depending on the organism's age, sex, health, metabolism, genetic makeup, and how the dose was administered (all at once or over a period of time). Lethal doses in humans are known for many toxicants because of records of homicides and accidental poisonings.

One way to determine acute toxicity is to administer various doses to populations of laboratory animals, measure the responses, and use these data to predict the chemical effects on humans. The dose lethal to 50% of a population of test animals is the **lethal dose-50%, or LD$_{50}$.** It is usually reported in milligrams of chemical toxicant per kilogram of body weight. There is an inverse relationship between the LD$_{50}$ and the acute toxicity of a chemical: The smaller the LD$_{50}$, the more toxic the chemical, and, conversely, the greater the LD$_{50}$, the less toxic the chemical (**Table 7.1**). The LD$_{50}$ is determined for all new synthetic chemicals—thousands are produced each year—as a way of estimating their toxic potential. It is generally assumed that a chemical with a low LD$_{50}$ for several species of test animals is toxic in humans.

The **effective dose-50%, or ED$_{50}$,** is used for a wide range of biological responses, such as stunted development in the offspring of a pregnant animal, reduced enzyme activity, or onset of hair loss. The ED$_{50}$ causes 50% of a population to exhibit whatever response is under study.

A **dose-response curve** shows the effect of different doses on a population of test organisms (**Figure 7.8**). Scientists first test the effects of high doses and then work their way down to a **threshold** level, the maximum dose with no measurable effect (or, alternatively, the minimum dose with a measurable effect). It is often assumed that doses lower than the threshold level will not have an effect on the organism and are safe (but see the EnviroNews).

Figure 7.8 **Dose-response curves.**

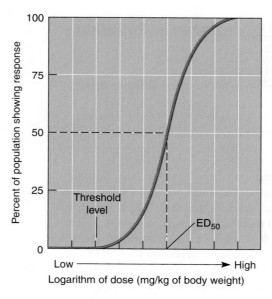

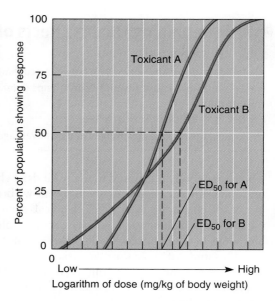

(a) This hypothetical dose-response curve demonstrates two assumptions of classical toxicology: First, the biological response increases as the dose is increased; second, there is a safe dose—a level of the toxicant at which no response occurs. Harmful responses occur only above a certain threshold level.

(b) Dose-response curves for two hypothetical toxicants, A and B. In this example, toxicant A has a lower effective dose-50% (ED$_{50}$) than toxicant B. At lower doses, toxicant B is more toxic than toxicant A.

E N V I R O N E W S

Rethinking the Dose-Response Relationship

Many chemicals are essential to humans and animals in small amounts, but toxic at higher levels. Vitamin D is a good example: We know that the human body requires vitamin D to properly absorb calcium and phosphorus. Too much or too little vitamin D can cause a variety of symptoms, including digestive complications. Thus the "dose-response" relationship between vitamin D and health is U-shaped. While the biological effects of chemicals at low concentrations are difficult to assess, some evidence suggests that a variety of chemicals, including some pesticides and trace metals like cadmium, may be healthful in small doses, but dangerous at higher levels.

This effect, called **hormesis** (blue line), is depicted in the figure. In this hypothetical data set, animals exposed to relatively low levels of a chemical were less likely to have tumors than those in the control group, while those at higher levels had clearly elevated levels. It is possible that there is a threshold (that is, below some concentration, the chemical does not cause cancer) and that the chemical causes tumors at low levels, but not enough to show up in a test of only a few animals. It is also possible that the chemical has a **hormetic** effect, and somehow suppresses tumors.

Some evidence suggests that ionizing radiation—the sort of radiation associated with nuclear waste and radon—may have a hormetic effect. However, most radiation experts conclude that while hormesis is possible, it is more likely that radiation causes cancer even at very low doses.

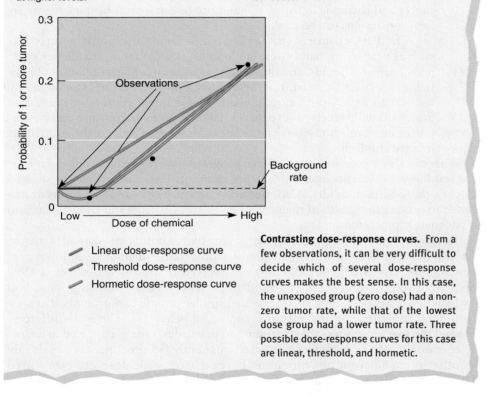

- Linear dose-response curve
- Threshold dose-response curve
- Hormetic dose-response curve

Contrasting dose-response curves. From a few observations, it can be very difficult to decide which of several dose-response curves makes the best sense. In this case, the unexposed group (zero dose) had a non-zero tumor rate, while that of the lowest dose group had a lower tumor rate. Three possible dose-response curves for this case are linear, threshold, and hormetic.

One complication of toxicology is that each individual's genes largely determine that person's response to a specific toxicant. The National Institute of Environmental Health Science has identified several hundred *environmental susceptibility genes*. Subtle differences in these genes affect how the body metabolizes toxicants, making them more or less toxic. Other gene variations allow certain toxicants to bind strongly—or less so—to the genetic molecule DNA. (Generally, when a toxicant binds to our DNA, it is a bad thing.) Genetic variation is one of the most important factors that determine why some people develop lung cancer after years of smoking, but others do not. For example, researchers have identified variations in the P450 gene (which codes for an enzyme) that determine how the body metabolizes some of the cancer-causing chemicals in tobacco smoke.

**Drawings of a person
(by 4-year-olds)**

Foothills   Valley

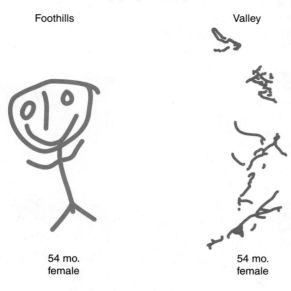

54 mo.   54 mo.
female   female

(a) Most preschoolers who received little pesticide exposure were able to draw a recognizable stick figure.

(b) Most preschoolers who lived in an area where agricultural pesticides were widely used could draw only meaningless lines and circles.

Figure 7.9   Effect of pesticide exposure on preschoolers. A study in Sonora, Mexico, found that Yaqui Indian preschoolers varied in their motor skills on the basis of the degree of pesticide exposure. The children were asked to draw a person. Two representative pieces of art are shown; both are drawn by 4 1/2-year-old girls. (E. A. Guilette et al.)

Children and Chemical Exposure

Children are more susceptible to most chemicals than are adults because their bodies are still developing and are not as effective in dealing with toxicants. Children are also more susceptible to chemicals because they weigh substantially less than adults. Consider a toxicant with an LD_{50} of 100 mg/kg. A potentially lethal dose for a child who weighs 11.3 kg (25 lb) is $100 \times 11.3 = 1130$ mg, equal to a scant 1/4 teaspoon if the chemical is a liquid. In comparison, the potentially lethal dose for an adult who weighs 68 kg (150 lb) is 6800 mg, or slightly less than 2 teaspoons. Thus, we must protect children from exposure to environmental chemicals because harmful doses are smaller for children than for adults.

Pesticides and Children   In recent years, increased attention has been paid to the health effects of household pesticides on children because it appears that household pesticides are a greater threat to children than to adults. For one thing, children tend to play on floors and lawns, where they are exposed to greater concentrations of pesticide residues. Also, children, especially when very young, are more likely to put items in their mouths. Several preliminary studies suggest that exposure to household pesticides may cause brain cancer and leukemia in children, but more research must be done before any firm conclusions can be made.

The EPA estimates that 84% of U.S. homes use pesticide products, such as pest strips, bait boxes, bug bombs, flea collars, pesticide pet shampoos, aerosols, liquids, and dusts. Several thousand different household pesticides are manufactured, and these contain over 300 active ingredients and more than 2500 inert ingredients. Poison control centers in the United States annually receive more than 130,000 reports of exposure and possible poisoning from household pesticides. More than half of these incidents involve children.

There is also concern about children's ingestion of pesticide residues on food. It is not known if infants and children are more, or less, susceptible to pesticide residues on food than adults. Also, because current pesticide regulations are intended to protect the health of the general population, infants and children may not be adequately protected.

Research supports an emerging hypothesis that exposure to pesticides may affect the development of intelligence and motor skills of infants, toddlers, and preschoolers. One study, published in *Environmental Health Perspectives* in 1998, compared two groups of rural Yaqui Indian preschoolers. These two groups, both of which live in northwestern Mexico, shared similar genetic backgrounds, diets, water mineral contents, cultural patterns, and social behaviors. The main difference between the two groups was their exposure to pesticides: One group lived in a farming community where pesticides were used frequently (45 times per crop cycle) and the other in an adjacent nonagricultural area where pesticides were rarely used. When asked to draw a person, most of the 17 children from the low-pesticide area drew recognizable stick figures, whereas most of the 34 children from the high-pesticide area drew meaningless lines and circles (**Figure 7.9**). Additional tests of simple mental and physical skills revealed similar striking differences between the two groups of children.

Identifying Cancer-Causing Substances

Traditionally, cancer was the principal disease evaluated in toxicology because many people were concerned about cancer-inducing chemicals in the environment and because cancer is so feared. Environmental contaminants are linked to several other serious diseases, such as birth defects, damage to the immune response, reproductive problems, and damage to the nervous system or other body systems. Although cancer is not the only disease caused or aggravated by toxicants, we focus here on risk

Table 7.2 Hypothetical Data Set for Animals Exposed to a Chemical

Number of animals in test	Number of animals with cancer	Dose (mg/kg/day)	Probability of cancer*
50	0	0.0	0
50	2	5.0	0.04
50	6	10	0.12
50	22	20	0.44

*The probability of getting cancer at a given dose is the number of animals with cancer at that dose level divided by the total number exposed at that dose level.

assessment as it relates to cancer. Noncancer hazards, such as diseases of the liver, kidneys, or nervous system, are assessed in ways similar to cancer risk assessment.

Toxicology and epidemiology are the two most common methods for determining whether a chemical causes cancer. Toxicologists expose laboratory animals such as rats to varying doses of the chemical and see whether they develop cancer. Epidemiologists look at historical exposures of humans to the same chemical, and see whether exposed groups show increased cancer rates. Each of these methods has advantages and disadvantages, and even when used together can provide only rough estimates of a chemical's carcinogenic potential.

Toxicology has the advantage that doses can be measured and administered in very precise amounts. Usually, two, three, or four groups of animals are exposed to different amounts or doses, including a "control group" that is not exposed. At the end of the experiment (about two years for mice and rats), the animals are dissected, and the ratio of animals with tumors to animals without in each group is recorded. This is used to determine a dose-response curve for the chemical. Doses are converted from animal dose to "equivalent human dose" by comparing body weight and metabolism rates.

Several uncertainties limit comparisons of animal laboratory studies to humans. First, the dose levels in the experiments are typically much, much larger than those faced by humans in the environment, so extrapolation from high-dose effects to low-dose effects may be inaccurate. **Extrapolation** is estimating the expected effects at some dose of interest from the effects at known doses. Second, humans and laboratory animals may process the chemicals in different ways. Third, laboratory animals only live about 2 years, while human life expectancy is around 70–80 years. Finally, human exposures are much more sporadic, and humans are exposed to a variety of chemicals that may amplify or offset each other. **Table 7.2** and **Figure 7.10** depict a toxicological data set and its associated dose-response curve. Extrapolation from high to low dose is highly uncertain.

toxicology: The study of the effects of toxic chemicals on human health.

epidemiology: The study of the effects of toxic chemicals and diseases on human populations.

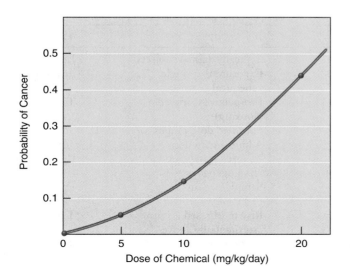

Figure 7.10   Dose-response curve associated with hypothetical data set. Table 7.2 contains a data set in which animals were exposed to four different dose levels of a toxicant. The probability of getting cancer is plotted against the dose levels, and a dose-response curve is fitted to these data points. This dose-response curve suggests that we would expect some increased cancer risk at any dose level, no matter how small.

Epidemiological studies have the advantage that they look at people who were actually exposed to the chemical. Ideally, a **cohort** or group of individuals who were exposed to the chemical is compared to an otherwise similar group who were not exposed. For example, a cohort exposed to benzene in a variety of industrial settings clearly had higher levels of leukemia than did similar groups who were not exposed.

Epidemiology has several limitations. First, it can be very difficult to reconstruct, or estimate, historical doses. Second, there may be confounding factors—for example, industry workers may have also been exposed to some other chemical that was not recorded. Third, the individuals in an industrial setting—in this case, healthy males between 18 and 60 years of age—may respond differently to the chemical than would others, such as children or pregnant women.

Epidemiology may be more representative than toxicology, but toxicology can be more precise. Ideally, epidemiological data and toxicological data can be combined to provide a clearer picture of the causes of cancer. In the case of benzene, animal tests confirm the epidemiological finding that benzene is a carcinogen, but do not improve our understanding of benzene's potency. **Table 7.3** compares the advantages and disadvantages of epidemiology and toxicology.

Chemical Mixtures

Humans are frequently exposed to various combinations of chemical compounds. Cigarette smoke contains a mixture of chemicals, as does automobile exhaust. The vast majority of toxicology studies are performed on single chemicals rather than chemical mixtures, however, and for good reason. Mixtures of chemicals interact in a variety of ways, increasing the level of complexity in risk assessment. Moreover, there are too many chemical mixtures to evaluate them all.

Chemical mixtures interact by additivity, synergy, or antagonism. When a chemical mixture is **additive**, the effect is exactly what one would expect, given the individual effects of each component of the mixture. If a chemical with a toxicity level of 1 is mixed with a different chemical with a toxicity level of 1, the combined effect of exposure to the mixture is 2. A **synergistic** chemical mixture has a greater combined effect than expected; two chemicals, each with a toxicity level of 1, might have a combined toxicity of 3. An **antagonistic** interaction in a chemical mixture results in a smaller combined effect than expected; for example, the combined effect of two chemicals, each with toxicity levels of 1, might be 1.3.

If toxicological studies of chemical mixtures are lacking, how do scientists assign the effects of chemical mixtures? Risk assessors typically assign risk values to mixtures

Table 7.3 Comparison of Advantages and Disadvantages of Toxicological and Epidemiological Studies

Epidemiology	Toxicology	Advantage
Human subjects	Typically animal subjects	Epidemiology
Exposure to multiple chemicals	Exposure to a single chemical	Toxicology
Retrospective (backward looking)	Prospective (forward looking)	Toxicology
Arbitrary dose ranges	Specified dose ranges	Toxicology
Estimated doses	Administered doses	Toxicology
Exposed group genetically diverse	Exposed group genetically homogeneous	Epidemiology
Sample sizes of 100 to 10,000	Sample size of 10 to 100	Epidemiology
Risk to exposed group near or slightly above background rate	Risk to exposed group substantially above background rate	Toxicology

by additivity—adding the known effects of each compound in the mixture. Such an approach usually underestimates but may sometimes overestimate the risk involved, but it is the best approach currently available. The alternative—waiting for years or decades until numerous studies are designed, funded, and completed—is unreasonable.

REVIEW

1. What are toxicology and epidemiology, and how do they differ?

2. Why are children particularly susceptible to environmental contaminants such as pesticides?

Ecotoxicology: Toxicant Effects on Communities and Ecosystems

LEARNING OBJECTIVES

- Discuss how discovery of the effects of DDT on birds led to the replacement of the dilution paradigm with the boomerang paradigm.
- Define *ecotoxicology* and explain why knowledge of ecotoxicology is essential to human well-being.

People used to think—and indeed some still do—that "the solution to pollution is dilution." This so-called dilution **paradigm** meant that you could discard pollution into the environment and it would be diluted sufficiently to cause no harm. We know today that the dilution paradigm is not generally valid. Our text is full of examples of the failure of the dilution paradigm. For example, recall from Chapter 1 how dumping treated sewage into Lake Washington caused a major water pollution problem. A more serious example, from the perspective of human well-being, is *Love Canal*, a small community in New York State contaminated by toxic waste dumped into a pit near their homes by a chemical company (see Chapter 24). To be fair, the chemical company dumped wastes at a time (1942–1953) when the dilution paradigm was still widely accepted. However, failure of the dilution paradigm brought scant consolation to the people living in Love Canal. They had to abandon their homes, and many have had health problems since leaving the contaminated site.

paradigm: A generally accepted understanding of how some aspect of the world works.

Today virtually all environmental scientists have rejected the dilution paradigm in favor of the boomerang paradigm: "What you throw away can come back and hurt you." The boomerang paradigm was adopted during the latter half of the 20th century after several well-publicized events captured the public's attention. Notable among these events was the discovery that the pesticide DDT was accumulating in birds at the top of the food web. The implications were clear that not only was DDT an unacceptable threat to ecosystem health but also potentially to human health as well.

As a result of the environmental impacts of DDT and the many other environmental problems that have arisen since DDT, a new scientific field—**ecotoxicology**—was born. Ecotoxicology, also called *environmental toxicology*, is an extension of the field of toxicology, which is human-oriented. However, ecotoxicology is human-oriented in the sense that humans produce the contaminants that adversely affect the environment.

ecotoxicology: The study of contaminants in the biosphere, including their harmful effects on ecosystems.

The scope of ecotoxicology is broad—from molecular interactions in the cells of individual organisms to effects on populations (e.g., local extinctions), communities and ecosystems (e.g., loss of species richness), and the biosphere (e.g., global climate change). Expanding knowledge in ecotoxicology indicates many examples of linkages between human health and the health of natural systems.

Ecotoxicology helps policymakers determine the costs and benefits of the many industrial and technological "advances" that affect us and the ecosystems upon which we depend. However, most environmental regulations are currently based on data for single species. Scientists have only begun to collect data to determine the environmental status of populations, communities, ecosystems, and higher levels of natural systems. Obtaining this higher-level information is complicated because (1) natural systems are exposed to many **environmental stressors** (changes that tax the

ENVIRONEWS

No Quick Fix for the Salton Sea

The Salton Sea of southern California holds many dilemmas for scientists, conservationists, and policymakers. It was accidentally created in 1905, when engineers were attempting to run irrigation canals from the Colorado River, and the entire river emptied into a shallow basin for 16 months. The 984-km² sea provides an unusual habitat—an inland marine environment. The sea's salinity level is 25% higher than the ocean's because it is fed by salty irrigation runoff. Only certain hardy fish species, stocked beginning in the 1950s, tolerate these stressful high-saline conditions. These fishes have experienced massive die-offs in recent years, probably from pollution and the high levels of nutrients from agricultural runoff.

As a desert oasis, the Salton Sea hosts millions of wintering birds and "stopover" migrants, including several threatened or endangered species. These birds feed on the fishes. Prime location notwithstanding, many consider the Salton Sea a poor bird habitat, or even haz-ardous to birds' health. Hundreds of thousands of birds have died there since the early 1990s, apparently from such diseases as botulism, avian cholera, and Newcastle virus. Scientists suspect a yet unproven link between the environment's high salinity and the animals' susceptibility to disease.

Some scientists think that the sea should cycle naturally, even if it reaches salinities that eliminate fishes. Others insist on reducing the salinity to preserve the wildlife habitat at all costs, especially in the face of so much wetland destruction elsewhere. To complicate matters further, the federal government requires California to reduce its use of water from the Colorado River. California had always used more water from the river than an agreement with other states allowed. To get the water it needs, California may have to reuse the agricultural runoff that provides the main source of water for the Salton Sea. If so, the Salton Sea would dry up and be largely gone by 2030. Any water remaining would be too salty to support fish or other forms of life, so the huge bird populations that rely on the Salton Sea would have no place to go.

environment); (2) natural systems must be evaluated for an extended period to establish important trends; and (3) the results must be clear enough for policymakers and the public to evaluate.

CASE IN POINT

The Ocean and Human Health

The ocean is important to us as a source of both food and natural chemical compounds that could benefit human health. These chemicals include, but are not limited by, novel pharmaceuticals, nutritional supplements, agricultural pesticides, and cosmetics. The discovery and development of beneficial marine compounds is in its infancy. Sponges, corals, mollusks, marine algae, and marine bacteria are some of the ocean's organisms that have the potential to provide these natural chemicals. In addition to serving as a source of food and other products, the ocean absorbs many wastes from human-dominated land areas.

Negative Health Impacts of Marine Microorganisms Marine microorganisms occur naturally in every part of the ocean environment. On average, one million bacteria and 10 million viruses occur in each milliliter of seawater! These organisms usually perform their ecological roles (such as decomposition) without causing major problems to humans. However, human activities now have a measurable effect on the ocean, including increases in land-based nutrient runoff and pollution and a small rise in ocean temperatures. These changes are causing an increase in the number and distribution of disease-causing microorganisms, particularly bacteria and viruses that pose a significant health threat to humans. (Recall the discussion of the boomerang paradigm.) For example, humans can become ill—or even die, if their immune systems are compromised—from drinking water or eating fish or shellfish (such as mussels or clams) contaminated with disease-causing microorganisms associated with human sewage or livestock wastes. Symptoms include stomach pain, diarrhea, headache, vomiting, and fever. Additional research is needed to study links between human-produced pollution and the growth of disease-causing microorganisms in the ocean system.

Scientists have observed an association in Bangladesh between outbreaks of cholera, caused by a waterborne bacterium, and increases in the surface temperature of coastal waters in the Bay of Bengal. The association is an indirect one. Apparently, the increased temperatures encourage the growth of microscopic plankton, which in turn produce ideal growing conditions for the cholera bacterium. Tides carry these contaminated waters into rivers that provide drinking water, increasing the risk of cholera in the local population.

Certain species of harmful algae sometimes grow in large concentrations called *algal blooms*. When some pigmented marine algae experience blooms, their great abundance frequently colors the water orange, red, or brown (**Figure 7.11**). Known as **red tides,** these blooms may cause serious environmental harm and threaten the health of humans and animals. Some of the algal species that form red tides produce toxins that attack the nervous systems of fishes, leading to massive fish kills. Water birds such as cormorants suffer and sometimes die when they eat the contaminated fishes. The toxins also work their way up the food web to marine mammals and people. In 1997, more than 100 monk seals, one-third of that endangered species' total population, died from algal toxin poisoning off the West African coast. Humans may also suffer if they consume algal toxins, which often bioaccumulate in shellfish or fishes. Even nontoxic algal species may wreak havoc when they bloom, as they shade aquatic vegetation and upset food web dynamics.

No one knows what triggers red tides, which are becoming more common and more severe, but many scientists think the blame lies with coastal pollution. Wastewater and agricultural runoff to coastal areas contain increasingly larger quantities of nitrogen and phosphorus, two nutrients that stimulate algal growth. Changes in ocean temperatures, such as those attributed to global warming, may also trigger algal blooms. In addition, a possible connection exists between red tide outbreaks in Florida's coastal waters and the arrival of dust clouds from Africa. These dust clouds, which sometimes blow across the Atlantic Ocean, enrich the water with iron and appear to trigger algal blooms.

Because we do not know what causes red tides, no control measures are in place to prevent the blooms or to end them when they occur. However, newer technologies like satellite monitoring and weather-tracking systems allow better prediction of conditions likely to stimulate blooms. ■

Figure 7.11   Red tide. The presence of billions of toxic algae color the water. Photographed in the Gulf of Carpentaria, Australia. (Bill Bachman/Photo Researchers, Inc.)

REVIEW

1. What environmental catastrophe was largely responsible for replacement of the dilution paradigm by the boomerang paradigm?
2. What is ecotoxicology?

Decision Making and Uncertainty: Assessment of Risks

LEARNING OBJECTIVES

- Define *risk* and explain how risk assessment helps determine adverse health effects.
- Discuss the precautionary principle as it relates to the introduction of new technologies or products.
- Define *ecological risk assessment*.

Throughout this chapter, we have discussed various **risks** to human health and the environment. Each of us makes many risk management decisions every day, most of which are based on intuition, habit, and experience. We make these decisions effectively— after all, most of us make it through each day without getting injured or killed.

risk: The probability that a particular adverse effect will result from some exposure or condition.

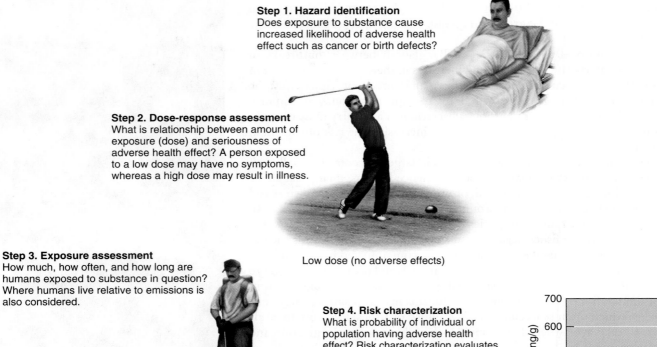

Step 1. Hazard identification
Does exposure to substance cause increased likelihood of adverse health effect such as cancer or birth defects?

Step 2. Dose-response assessment
What is relationship between amount of exposure (dose) and seriousness of adverse health effect? A person exposed to a low dose may have no symptoms, whereas a high dose may result in illness.

Low dose (no adverse effects)

Step 3. Exposure assessment
How much, how often, and how long are humans exposed to substance in question? Where humans live relative to emissions is also considered.

Agricultural workers have a greater exposure to chemicals such as pesticides

Step 4. Risk characterization
What is probability of individual or population having adverse health effect? Risk characterization evaluates data from dose-response assessment and exposure assessment (steps 2 and 3). Risk characterization indicates that Mexican-Americans, many of whom are agricultural workers, are more vulnerable to pesticide exposure than other groups (see graph).

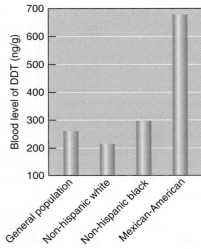

Figure 7.12 **The four steps of a risk assessment for adverse health effects.**

However, environmental and health decisions often impact many individuals, and the best choices cannot always be made on an intuitive level. *Risk analysis* is a tool used to organize how we think about complex environmental systems. When we think about risks from a systems perspective, we can decide whether it's most effective to

- change our activities to avoid particular risks altogether,
- limit the extent to which a particular **hazard** can come in contact with us,
- limit the extent to which the hazard can harm us, or
- provide some sort of offset or compensation for being harmed by the hazard.

hazard: A condition that has the potential to cause harm.

For example, consider the risk of injury from car accidents. We can redesign cities to allow people to work near where they live. We can establish rules on how fast people drive, or which side of the street they drive on. We can require seatbelts and airbags so that people who are in accidents don't get hurt as badly. Finally, we can purchase insurance to cover expenses, lost time, and suffering. Risk analysis also allows us to think about the tradeoffs between different activities. For example, we could bicycle instead of driving. This would take more time, and we might be at higher risk in an accident, but we would also get some health benefits from the exercise.

Table 7.4 Probability of Death by Selected Causes for a U.S. Citizen

Cause of Death*	One-year Odds**	Lifetime Odds**
Cardiovascular disease (1999)	1 in 300 (3.3×10^{-3})	1 in 4 (2.5×10^{-1})
Cancer, all types (1999)	1 in 510 (2.0×10^{-3})	1 in 7 (1.4×10^{-1})
Motor vehicle accidents (2000)	1 in 6,700 (1.5×10^{-4})	1 in 88 (1.1×10^{-2})
Suicide (1998)	1 in 9,200 (1.1×10^{-4})	1 in 120 (8.3×10^{-3})
Homicide (2000)	1 in 18,000 (5.6×10^{-5})	1 in 240 (4.2×10^{-3})
Killed on the job (2000)	1 in 48,000 (2.1×10^{-5})	1 in 620 (1.6×10^{-3})
Drowning in bathtub (1998)	1 in 840,000 (1.1×10^{-6})	1 in 11,000 (9.1×10^{-5})
Tornado (2000)	1 in 3,000,000 (3.3×10^{-7})	1 in 39,000 (2.6×10^{-5})
Commercial aircraft (2001)	1 in 3,100,000 (3.2×10^{-7})	1 in 40,000 (2.5×10^{-5})
Hornet, wasp, or bee sting (1998)	1 in 6,100,000 (1.6×10^{-7})	1 in 80,000 (1.3×10^{-5})

* Data are for the year noted in parentheses and are the most recent data available.
** Probability of risk is in parentheses.
Source: D. Ropeik and G. Gray

Risk management is the process of identifying, assessing, and reducing risks. The four steps involved in **risk assessment** for adverse health effects are summarized in **Figure 7.12.** Risk assessment involves using statistical methods to quantify the risks of a particular action so that they can be compared and contrasted with other risks.

Risk is calculated as the probability that some negative effect or event will occur. These are reported as fractions, and can range from 0 (certain not to occur) to 1 (certain to occur). For example, according to the American Cancer Society, in 2002 about 170,000 Americans who smoked died of cancer. This translates into a probability of risk of 0.00059 (that is, 5.9×10^{-4}). (See **Table 7.4** for probabilities of death by selected causes. Note that these are population estimates: None of them applies to any individual.)

Risk assessment is used in several ways for environmental regulation. A regulatory agency may establish a "maximum risk" standard. For example, the EPA may decide that people should face no more than one-in-one million additional risk of cancer from trichloroethylene (TCE, a common contaminant) in municipal drinking water. If we know the **cancer potency** of TCE, and we know how much water the average person drinks, we can calculate the maximum allowable concentration of TCE.

Alternatively, risk managers may be concerned about the expected risk associated with an existing or historical exposure. For example, a company might need to know how much money to set aside to pay medical costs associated with construction workers who were exposed to asbestos. It may be impossible to tell which individuals will contract asbestosis (a lung disease associated with chronic asbestos exposure), but we can estimate the number of diseases among a large group of individuals.

Many of our decisions about risks have far more to do with our trust in individuals and institutions who manage the risks than with the calculated values of those risks. For example, many people distrust the nuclear industry, which means it doesn't matter how accurately they can calculate the risks of a nuclear accident. We worry less about risks that we feel we can control (such as driving or eating) than those outside our control (such as pesticide contamination). We also worry more about things we dread (dying from cancer) than those we don't (bicycling). Effective risk management, then, cannot be based on calculated risks alone, but must also account for intuition, trust, and social conditions.

cancer potency: An estimate of the expected increase in cancer associated with a unit increase in exposure to a chemical.

Cost–Benefit Analysis of Risks

Risk assessment is also an important input to *cost–benefit analysis* (see Chapter 2). In a cost–benefit analysis, the estimated cost of some regulation to reduce risk is compared with potential benefits associated with that risk reduction. Cost–benefit analysis is an important mechanism to help decision makers formulate environmental legislation, but it is only as good as the data and assumptions on which it is based. Corporate estimates of the cost to control pollution are often many times higher than the actual cost turns out to be. During the debate over phasing out leaded gasoline in 1971, the oil industry predicted that the cost during the transition would be $7 billion per year, but the actual cost was less than $500 million per year.

Despite the wide range that often occurs between projected and actual costs, the cost portion of cost–benefit analysis is often easier to determine than are the health and environmental benefits. The cost of installing air pollution control devices at factories is relatively easy to estimate, but how does one put a price tag on the benefits of a reduction in air pollution? What is the value of reducing respiratory problems in children and the elderly, two groups susceptible to air pollution? How much is clean air worth?

Another problem with cost–benefit analysis is that the risk assessments on which such analyses are based are far from perfect. Even the best risk assessments are based on assumptions that, if changed, could substantially alter the estimated risk. Risk assessment is an uncertain science.

To summarize, cost–benefit analyses and risk assessments are useful in evaluating and solving environmental problems, but decision makers must recognize the limitations of these methods when developing new government regulations.

The Precautionary Principle

precautionary principle: The idea that no action should be taken or product introduced when the science is inconclusive but unknown risks may exist.

You have probably heard the expression "An ounce of prevention is worth a pound of cure." This statement is the heart of a policy—the **precautionary principle**—advocated by many politicians and environmental activists. According to the precautionary principle, when a new technology or chemical product is suspected of threatening health or the environment, precautionary measures should be taken, even if there is uncertainty about the scope of danger.

The precautionary principle may also be applied to existing technologies when new evidence suggests they are more dangerous than originally thought. When observations and experiments suggested that chlorofluorocarbons (CFCs) harm the ozone layer in the stratosphere, the precautionary principle led to these compounds being phased out. Studies made after the phase-out supported this step.

To many people, the precautionary principle is common sense, given that science and risk assessment often cannot provide definitive answers to policymakers. The precautionary principle puts the burden of proof on the developers of the new technology or substance, who must demonstrate safety beyond a reasonable doubt.

While the precautionary principle has been incorporated into decisions in the European Union, the United States, and elsewhere, it has many detractors. Some scientists think that it challenges the role of science and endorses making decisions without the input of science. Advocates of cost–benefit analysis note that the precautionary principle may be extremely expensive, or may cause us to hold off on new technologies that are much safer than those already in place.

Finally, some critics contend that the precautionary principle's imprecise definition can reduce trade and limit technological innovations. For example, several European countries made precautionary decisions to ban beef from the United States and Canada because these countries use growth hormones to make the cattle grow faster. Europeans contend that the growth hormone might harm humans eating the beef, but the ban, in effect since 1989, is widely viewed as protecting their own beef industry. Another international controversy in which the precautionary principle is involved is the introduction of genetically modified foods (discussed in Chapter 19).

Ecological Risk Assessment

The EPA and other federal and state environmental monitoring groups are increasingly trying to evaluate ecosystem health using risk methods developed for assessing human health. Detailed guidelines exist for performing **ecological risk assessments.**

The field of ecotoxicology, discussed earlier, is directly concerned with ecological risk assessments. Like human health risk assessment, ecological risk assessment involves hazard identification, dose-response assessment, exposure assessment, and risk characterization (see Figure 7.12).

Such analyses are difficult because effects may occur on a wide scale, from individual animals or plants in a local area to ecological communities across a large region. Given the hazards and exposure levels of human-induced environmental stressors, ecological effects range from good to bad, or from acceptable to unacceptable. Using scientific knowledge in environmental decision making is filled with uncertainty because many ecological effects are incompletely understood or difficult to measure. A real need exists to quantify risk to the environmental system and to develop strategies to cope with the uncertainty.

The EPA is using ecological risk assessment to tackle environmental problems such as the cumulative effects of many natural and human-induced stressors on various species in the Snake River ecosystem in southern Idaho. The Snake River provides irrigation water for agriculture, and dams harness the water to generate electricity (**Figure 7.13**). These and other land use practices in the **watershed** (the area of land a river drains) have resulted in a reduced river flow, elevated water temperature, and nutrient enrichment. Algae and aquatic weeds now grow in great profusion, and many fish and aquatic invertebrates are severely reduced in number. Ecological risk assessment is helping the EPA and other federal agencies, regional groups, state agencies, Native American tribes, local groups, and private individuals set priorities in order to meet their common goal of managing and protecting the biological communities in the Snake River watershed system.

Figure 7.13 **Hell's Canyon Dam on the Snake River, Idaho.** Ecological risk assessment of the Snake River ecosystem will help sustainably manage the river and its watershed, which are suffering from a variety of human-induced stressors, including dams. (Harald Sund/The Image Bank/Getty Images)

A Balanced Perspective on Risks

Some threats to our health, particularly from toxic chemicals in the environment, make big news. Other threats, while larger from a risk assessment perspective, go unnoticed. Media reports of risk events and situations are constrained by the need to entertain as well as inform, and most reporters are not trained in science or risk assessment. Consequently, the risks that get the most attention in the media may not be as detrimental as others.

This does not mean that we should ignore chemicals that humans introduce into the environment. Nor does it mean we should discount stories the news media sometimes sensationalize. These stories serve an important role in getting the regulatory wheels of the government moving to protect us as much as possible from the dangers of our technological and industrialized world. They reflect distrust in industry and government to manage risks, and thereby suggest opportunities to improve those institutions.

Most people do not expect no-risk foods, no-risk water, or no-risk anything else. Risk is inherent in all our actions and in everything in our environment. However, it is often helpful to have information about the risks we face. We should not ignore small risks just because larger ones exist. We must have an adequate understanding of the nature and size of risks before deciding what actions are appropriate to avoid them.

REVIEW

1. What is risk assessment?
2. What is the precautionary principle?
3. What information does a cost–benefit analysis provide decision makers?

REVIEW OF LEARNING OBJECTIVES WITH KEY TERMS

• **Contrast health issues in highly developed and developing countries.**

Cardiovascular diseases, cancer, and chronic obstructive pulmonary disease are health problems in the United States and other highly developed nations; many of these diseases are chronic health problems associated with aging and are caused in part by lifestyle choices involving diet, exercise, and smoking. Child mortality is particularly serious in developing countries, where leading causes of death in children include malnutrition, diarrheal diseases, and malaria.

• **Summarize the problems associated with chemicals that exhibit persistence, bioaccumulation, and biological magnification in the environment.**

Chemicals that exhibit **persistence** are extremely stable and may take many years to be broken down into simpler forms by natural processes. **Bioaccumulation** is the buildup of a persistent toxic substance, such as certain pesticides, in an organism's body, often in fatty tissues. **Biological magnification** is the increased concentration of toxic chemicals, such as PCBs, heavy metals, and certain pesticides, in the tissues of organisms at higher levels in food webs.

• **Define *endocrine disrupter*.**

An **endocrine disrupter** is a chemical that mimics or interferes with the actions of the endocrine system in humans and wildlife.

• **Briefly describe some of the data suggesting that certain chemicals used by humans may also function as endocrine disrupters in animals, including humans.**

A chemical spill in 1980 contaminated Lake Apopka, Florida, with DDT and other agricultural chemicals with estrogenic properties. Male alligators living in Lake Apopka in the years following the spill had low levels of testosterone (an androgen) and elevated levels of estrogen. Their reproductive organs were often feminized or abnormally small, and the mortality rate for eggs was extremely high. Humans may be equally at risk from endocrine disrupters, as the number of reproductive disorders, infertility cases, and hormonally related cancers (such as testicular cancer and breast cancer) appears to be increasing.

• **Define *toxicant* and distinguish between acute and chronic toxicities.**

A **toxicant** is a chemical with adverse human health effects. **Acute toxicity** refers to adverse effects that occur within a short period after exposure to a toxicant. **Chronic toxicity** refers to adverse effects that occur some time after exposure to a toxicant, or after extended exposure to a toxicant.

• **Describe how a dose–response curve helps determine the health effects of environmental pollutants.**

A **dose–response curve** is a graph that shows the effect of different doses on a population of test organisms. Scientists first test the effects of high doses and then work their way down to a **threshold** level.

• **Discuss pesticide risks to children.**

More than half of all reports of exposure and possible poisoning from household pesticides involve children. Several studies suggest that exposure to household pesticides may cause brain cancer and leukemia in children, but more research must be done before any firm conclu-

sions can be made. Some research suggests that exposure to pesticides may affect the development of intelligence and motor skills of infants, toddlers, and preschoolers.

• **Explain the relative advantages and disadvantages of toxicology and epidemiology.**

Toxicology is the study of the effects of toxic chemicals on human health, whereas **epidemiology** is the study of the effects of toxic chemicals and diseases on human populations. Toxicology can provide very specific information about effects of chemicals, and can do so using animals and even tissues or cells, without having to worry about humans being impacted. However, epidemiology, while often less precise, may more realistically reflect exposure conditions.

• **Discuss how discovery of the effects of DDT on birds led to the replacement of the dilution paradigm with the boomerang paradigm.**

A **paradigm** is a generally accepted understanding of how some aspect of the world works. According to the dilution paradigm, "the solution to pollution is dilution." Environmental scientists reject the dilution paradigm in favor of the boomerang paradigm, which is "what you throw away can come back and hurt you." The boomerang paradigm was adopted after the discovery that the pesticide DDT was accumulating in birds at the top of the food web. The implications were that DDT represented an unacceptable threat to ecosystem health and human health.

• **Define *ecotoxicology* and explain why knowledge of ecotoxicology is essential to human well-being.**

Ecotoxicology is the study of contaminants in the biosphere, including their harmful effects on ecosystems. Ecotoxicology helps policymakers determine the costs and benefits of the many industrial and technological "advances" that affect us and the ecosystems on which we depend.

• **Define *risk* and explain how risk assessment helps determine adverse health effects.**

A **risk** is the probability that a particular adverse effect will result from some exposure or condition. **Risk assessment** is the process of estimating those probabilities and consequences. Risk assessments, when properly performed, provide information that is useful to determining whether and how to reduce a particular risk. The process of identifying potential risks, soliciting risk assessments, and making appropriate decisions is known as risk management.

• **Discuss the precautionary principle as it relates to the introduction of new technologies or products.**

The **precautionary principle** is the idea that no action should be taken or product introduced when the science is inconclusive but unknown risks may exist. A new technology or chemical product suspected of threatening human health or the environment should not be introduced until it is demonstrated that the risks are small and that the benefits outweigh the risks.

• **Define *ecological risk assessment*.**

Ecological risk assessment is the process by which the ecological consequences of human activities are estimated. Ecotoxicology is directly concerned with ecological risk assessments.

Thinking About the Environment

1. What are the three leading causes of death in the United States? How are they related to lifestyle choices?

2. Why are public health researchers concerned about "exporting" health problems associated with developed countries to less developed countries?

3. A researcher who studies obesity once described the cause of obesity as a combination of "computer chips and potato chips." Explain what he meant.

4. What was the first viral disease to be globally eradicated? What disease do health officials hope will be eradicated soon?

5. What is a "pandemic," and why does the potential for pandemics trouble many public health officials?

6. Distinguish among persistence, bioaccumulation, and biological magnification.

7. How do acute and chronic toxicity differ?

8. What is a dose–response curve? What can it tell us about effects at low doses if experimental information is on high doses?

9. Examine this cartoon criticizing cutbacks in pollution testing in the mid-1990s. What does this cartoon suggest about the relative value of toxicological information and epidemiological testing? Do you think this is a reasonable comparison? Explain your answer.

10. Describe three ways that chemical mixtures can interact. Which of these is usually assigned when the effects of chemical mixtures are unknown?

11. Describe the common methods for determining whether a chemical causes cancer.

12. Distinguish between the dilution paradigm and the boomerang paradigm.

13. Select one of the two choices to complete the following sentence, and then explain your choice: The absence of certainty about the health effects of an environmental pollutant (is/is not) synonymous with the absence of risk.

14. Why might industrial interests be more strongly opposed to the precautionary principle than are environmental advocacy organizations?

15. How are risk assessments for human health and ecological risk assessments for environmental health alike? How do they differ?

16. Why is knowledge of ecotoxicology essential to human well-being? Why must we think in terms of systems and interactions in order to understand both human and environmental health?

Quantitative questions relating to this chapter are on our Web site.

Toles © 1995 The Washington Post. Reprinted with permission of Universal Press Syndicate. All rights reserved./Universal Press Syndicate

Take a Stand

Visit our Web site at http://www.wiley.com/college/raven (select Chapter 7 from the Table of Contents) for links to more information about red tides. You will find tools to help you organize your research, analyze the data, think critically about the issues, and construct a well-considered argument. *Take a Stand* activities can be done individually or as a team, as oral presentations, written exercises, or Web-based (e-mail) assignments.

Additional online materials relating to this chapter, including a Student Testing section with study aids and self-tests, Environmental News, Activity Links, Environmental Investigations, and more, are also on our Web site.

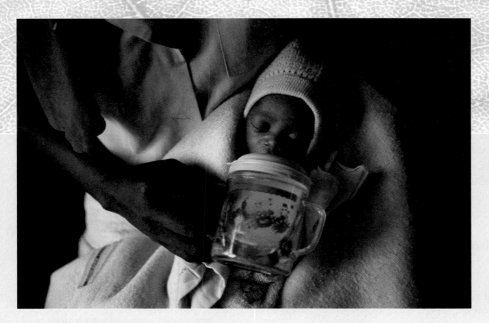

An AIDS-infected mother feeds her baby.
The Botswana government provides her
with formula to keep her baby healthy. (The
virus can pass from mother to child in breast
milk.) (Koren Kasmauski/National Geographic
Society)

8

Population Change

Africa has the most rapidly growing population of all the continents. Most of Africa's population is concentrated in sub-Saharan Africa (that part of Africa located south of the Sahara Desert). Experts currently predict that the sub-Saharan population of 767 million will almost double to 1.4 billion by 2030.

During the 1990s and early 2000s, many population experts lowered their estimates of population growth rates in sub-Saharan Africa. These lower projections were not based on declining birth rates. Tragically, the estimates were lowered because acquired immune deficiency syndrome (AIDS), caused by the human immune deficiency virus (HIV), is ravaging many countries in the sub-Saharan region. The U.N. Program on AIDS estimates that 26 million of the world's 40 million people who are now infected with HIV/AIDS live in sub-Saharan Africa.[1] An estimated 7.2% of the adult sub-Saharan population, ages 15 to 49, are infected (see photograph). Swaziland in southern Africa has the highest prevalence

of HIV/AIDS in the world: More than 38% of the adult population in this country is infected. In Africa, women now make up 57% of the infected population, but they have little access to prevention information or effective treatments. Even more tragic is that at the end of 2003, about 12 million sub-Saharan African children living in the region had lost one or both parents to AIDS. By 2010, about 18 million children in this region will probably be orphaned. Many of these AIDS orphans do not receive adequate education, healthcare, or nutrition.

In developing countries, including those in sub-Saharan Africa, almost all people infected with HIV/AIDS die, usually within 10 years of infection, because they cannot afford the high cost of potent antiviral drugs. Countries in sub-Saharan Africa are experiencing high death rates from HIV/AIDS, now the leading cause of death in Africa. In 2005, 2.4 million people died of AIDS in sub-Saharan Africa. An estimated 3.1 million people died worldwide of AIDS that year.

The high mortality from HIV/AIDS has caused life expectancies to decline in many African countries. In Botswana, for example, the average life expectancy declined from a high of about 60 years in the late 1980s to 34 years

in 2006. Africa's AIDS crisis has many repercussions beyond a reduced life expectancy. The HIV/AIDS epidemic threatens the economic stability and social support networks of the region and has overwhelmed healthcare systems. Hunger has significantly increased in some countries, owing to the loss of HIV/AIDS-infected agricultural workers; by 2020, the United Nations estimates that AIDS will have killed at least 20% of agricultural workers in southern Africa. Labor shortages are common, and foreign investments have slowed because many investors are wary of the high rate of infection among workers.

Outside of Africa the HIV/AIDS epidemic continues to grow, particularly in developing countries that have not mounted aggressive education campaigns. In the United States, Canada, and Western Europe, after steady declines in the late 1990s, infection rates are again increasing. Many people have become complacent as a result of improved availability of treatment programs, and key prevention messages are not reaching at-risk populations. Regardless of improved treatment programs, the "safe sex" message must continue to be communicated.

The HIV/AIDS epidemic is but one factor affecting the human population. In this chapter, we focus on the dynamics of population change characteristic of all organisms and then describe the current state of the human population.

[1] Each HIV or AIDS estimate represents the midpoint of a range.

World View and Closer to You...
Real news footage relating to population change around the world and in your own geographic region is available at www.wiley.com/college/raven by clicking on "World View and Closer to You."

Principles of Population Ecology

LEARNING OBJECTIVES

- Define *population* and *population ecology.*
- Define *growth rate (r)* and explain the four factors that produce changes in population size.
- Use intrinsic rate of increase, exponential population growth, and carrying capacity to explain the differences between J-shaped and S-shaped growth curves.

Because the human population is central to so many environmental problems and their solutions, you need to understand how populations increase or decrease. Studying populations of other species provides insights into the biological principles that affect human population changes.

Individuals of a given species are part of a larger organization—a **population**. Populations exhibit characteristics distinct from those of the individuals that comprise them. Some of the features characteristic of populations but not of individuals are population density, birth and death rates, growth rates, and age structure.

Population ecology is the branch of biology that deals with the numbers of a particular species found in an area and how and why those numbers change (or remain fixed) over time. Population ecologists try to determine the population processes common to all populations. They study how a population responds to its environment, such as how individuals in a population compete for food or other resources, and how predation, disease, and other environmental pressures affect the population. A population, whether of bacteria or maples or giraffes, cannot increase indefinitely because of such environmental pressures.

Additional aspects of populations important to environmental science are their reproductive success or failure (that is, extinction) and how populations affect the normal functioning of communities and ecosystems. Scientists in applied disciplines, such as forestry, agronomy (crop science), and wildlife management, must understand population ecology to effectively manage populations of economic importance, such as forest trees, field crops, game animals, and fishes. An understanding of the population dynamics of endangered species plays a key role in efforts to prevent their slide to extinction. Knowing the population dynamics of pest species helps in efforts to prevent their increase to levels that cause significant economic or health impacts.

population: A group of individuals of the same species that live in the same geographic area at the same time.

Population Density

By itself, the size of a population tells us relatively little. Population size is meaningful only when the boundaries of the population are defined. Consider the difference between 1000 mice in 100 hectares (250 acres) and 1000 mice in 1 hectare (2.5 acres). Often a population is too large to study in its entirety. Such a population is examined by sampling a part of it and then expressing the population in terms of density. Examples include the number of dandelions per square meter of lawn, the number of water fleas per liter of pond water, and the number of cabbage aphids per square centimeter of cabbage leaf. **Population density**, then, is the number of individuals of a species per unit of area or volume at a given time.

For a given species, different environments support varying population densities. Density also varies in a single habitat (local environment) from season to season or year to year. As an example, consider red grouse populations in northwest Scotland at two locations only 2.5 km (1.6 mi) apart. At one location the population density of these birds remained stable during a three-year period, but at the other site it almost doubled in the first two years and then declined to its initial density in the third year. The reason was likely a difference in habitat. The area where the population density increased had been experimentally burned, and young heather shoots produced during the two years following the burn provided nutritious food for the red grouse. External factors in the environment, then, determine population density to a large extent.

Figure 8.1 Factors that affect population size.

Increases population: Decreases population:

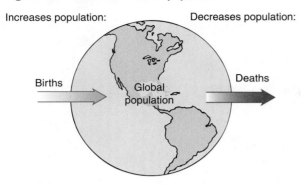

(a) On a global scale, the change in a population is due to the number of births and deaths.

Increases population: Decreases population:

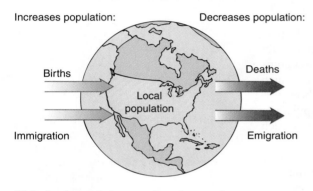

(b) In local populations, such as the population of the United States, the number of births, deaths, immigrants, and emigrants affect population size.

growth rate (r): The rate of change of a population's size, expressed in % per year.

intrinsic rate of increase: The exponential growth of a population that occurs under ideal conditions.

How Do Populations Change in Size?

Populations of organisms, whether they are sunflowers, eagles, or humans, change over time. On a global scale, this change is due to two factors: the rate at which individuals produce offspring (the birth rate) and the rate at which organisms die (the death rate) (**Figure 8.1a**). In humans, the **birth rate (b)** is usually expressed as the number of births per 1000 people per year, and the **death rate (d)** as the number of deaths per 1000 people per year.

The **growth rate (r)** of a population is equal to the birth rate (b) minus the death rate (d). Growth rate is also called **natural increase** in human populations.

$$r = b - d$$

As an example, consider a hypothetical human population of 10,000 in which there are 200 births per year (that is, by convention, 20 births per 1000 people) and 100 deaths per year (that is, 10 deaths per 1000 people).

$$r = \underbrace{20/1000}_{b} - \underbrace{10/1000}_{d}$$

$$r = 0.02 - 0.01 = 0.01, \text{ or } 1\% \text{ per year}$$

If organisms in the population are born faster than they die, r is a positive value, and population size increases. If organisms in the population die faster than they are born, r is a negative value, and population size decreases. If r is equal to zero, births and deaths match, and population size is stationary despite continued reproduction and death.

In addition to birth and death rates, **dispersal,** or movement from one region or country to another, is considered when changes in populations on a local scale are examined. There are two types of dispersal: **immigration (i),** in which individuals enter a population and increase its size, and **emigration (e),** in which individuals leave a population and decrease its size. The growth rate of a local population must take into account birth rate (b), death rate (d), immigration (i), and emigration (e) (**Figure 8.1b**). The growth rate equals (the birth rate minus the death rate) plus (immigration minus emigration):

$$r = (b - d) + (i - e)$$

For example, the growth rate of a population of 10,000 that has 100 births (by convention, 10 per 1000), 50 deaths (5 per 1000), 10 immigrants (1 per 1000), and 100 emigrants (10 per 1000) in a given year is calculated as follows:

$$r = (\underbrace{10/1000}_{b} - \underbrace{5/1000}_{d}) + (\underbrace{1/1000}_{i} - \underbrace{10/1000}_{e})$$

$$r = (0.010 - 0.005) + (0.001 - 0.010)$$

$$r = 0.005 - 0.009 = -0.004, \text{ or } -0.4\% \text{ per year}$$

Maximum Population Growth

The maximum rate that a population could increase under ideal conditions is its **intrinsic rate of increase** (also called *biotic potential*). Different species have different intrinsic rates of increase. Several factors influence a particular species' intrinsic rate of increase. These include the age that reproduction begins, the fraction of the life span during which an individual can reproduce, the number of reproductive periods per lifetime, and the number of offspring produced during each period of reproduction. These factors, called *life history characteristics*, determine whether a particular species has a large or a small intrinsic rate of increase.

Generally, larger organisms, such as blue whales and elephants, have the smallest intrinsic rates of increase, whereas microorganisms have the greatest intrinsic rates of increase. Under ideal conditions (that is, an environment with unlimited resources), certain bacteria reproduce by dividing in half every 30 minutes. At this rate of growth, a single bacterium would increase to a population of more than 1 million in just 10 hours (**Figure 8.2a**), and the population from a single individual would exceed 1 billion in 15 hours! If you plot the population number versus time, the graph has a J shape characteristic of **exponential population growth** (**Figure 8.2b**). When a population grows exponentially, the larger the population gets, the faster it grows.

An everyday example of exponential growth is a savings account in which a fixed percentage of interest—say, 2%—is accumulating (i.e., compounding). Assuming you do not deposit or withdraw any money, the amount that your money grows starts increasing slowly and proceeds faster and faster over time, as the balance increases. Exponential population growth works the same way. A small population that is growing exponentially increases at a slow rate initially, but growth proceeds faster and faster as the population increases.

Regardless of the species, whenever the population is growing at its intrinsic rate of increase, population size plotted versus time gives a curve of the same shape. The only variable is time. It may take longer for a lowland gorilla population than for a bacterial population to reach a certain size (because gorillas do not reproduce as rapidly as bacteria), but both populations will always increase exponentially as long as their growth rates remain constant.

Environmental Resistance and Carrying Capacity

Certain populations may exhibit exponential population growth for a short period. Exponential population growth has been experimentally demonstrated in bacterial and protist cultures and in certain insects. However, organisms cannot reproduce indefinitely at their intrinsic rates of increase because the environment sets limits, collectively called **environmental resistance.** Environmental resistance includes such unfavorable environmental conditions as the limited availability of food, water, shelter, and other essential resources (resulting in increased competition), as well as limits imposed by disease and predation.

Using the earlier example, we find that bacteria would never reproduce unchecked for an indefinite period because they would run out of food and living space, and poisonous body wastes would accumulate in their vicinity. With crowding, bacteria would become more susceptible to parasites (high population densities facilitate the spread of infectious organisms such as viruses among individuals) and predators (high population densities increase the likelihood of a predator catching an individual). As the environment deteriorates, their birth rate (*b*) would decline and their death rate (*d*) would increase. The environmental conditions might worsen to a point where *d* would exceed *b*, and the population would decrease. The number of individuals in a population, then, is controlled by the ability of the environment to support it. As the number of individuals in a population increases, so does environmental resistance, which acts to limit population growth. Environmental resistance is an excellent example of a **negative feedback mechanism,** in which a change in some condition triggers a response that counteracts, or reverses, the changed condition.

Over longer periods, the rate of population growth (*r*) may decrease to nearly zero. This leveling out occurs at or near the **carrying capacity (*K*)**, the limit of the environment's ability to support a population. In nature, the carrying capacity is dynamic and changes in response to environmental changes. An extended drought, for example, could decrease the amount of vegetation growing in an area, and this change, in turn, would lower the carrying capacity for deer and other herbivores in the environment.

Figure 8.2   Exponential population growth.

Time (hours)	Number of bacteria
0	1
0.5	2
1.0	4
1.5	8
2.0	16
2.5	32
3.0	64
3.5	128
4.0	256
4.5	512
5.0	1,024
5.5	2,048
6.0	4,096
6.5	8,192
7.0	16,384
7.5	32,768
8.0	65,536
8.5	131,072
9.0	262,144
9.5	524,288
10.0	1,048,576

(a) When bacteria divide at a constant rate, their numbers increase exponentially. This set of figures assumes a zero death rate, but even if a certain percentage of each generation of bacteria died, exponential population growth would still occur; it would just take longer to reach the high numbers.

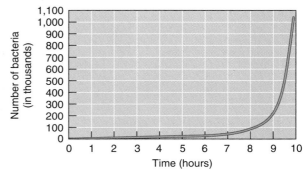

(b) When these data are graphed, the curve of exponential population growth has a characteristic J shape.

exponential population growth: The accelerating population growth that occurs when optimal conditions allow a constant reproductive rate over a period of time.

carrying capacity (*K*): The maximum number of individuals of a given species that a particular environment can support for an indefinite period, assuming there are no changes in the environment.

G. F. Gause, a Russian ecologist who conducted experiments during the 1930s, grew a population of a single species, *Paramecium caudatum*, in a test tube (**Figure 8.3a**). He supplied a limited amount of food (bacteria) daily and replenished the media occasionally to eliminate the buildup of metabolic wastes. Under these conditions, the population of *P. caudatum* increased exponentially at first, but then their growth rate declined to zero, and the population size leveled off.

When a population affected by environmental resistance is graphed over a long period (**Figure 8.3b**), the curve has the characteristic S shape of **logistic population growth.** The curve shows an approximate exponential increase initially (note the curve's J shape at the start, when environmental resistance is low), followed by a leveling out as the carrying capacity of the environment is approached. In logistic population growth, the rate of population growth is proportional to the amount of existing resources, and competition leads to limited population growth. Although logistic population growth is an oversimplification of how most populations change over time, it fits some populations studied in the laboratory, as well as a few studied in nature.

A population rarely stabilizes at K (carrying capacity), as shown in **Figure 8.3**, but may temporarily rise higher than K. It will then drop back to, or below, the carrying capacity. Sometimes a population that overshoots K will experience a **population crash,** an abrupt decline from high to low population density. Such an abrupt change is commonly observed in bacterial cultures, zooplankton, and other populations whose resources are exhausted.

The availability of winter forage largely determines the carrying capacity for reindeer, which live in cold northern habitats. In 1910, a small herd of 26 reindeer was introduced on one of the Pribilof Islands of Alaska (**Figure 8.4**). The herd's population increased exponentially for about 25 years until there were approximately 2000 reindeer, many more than the island could support, particularly in winter. The reindeer overgrazed the vegetation until the plant life was almost wiped out. Then, in slightly over a decade, as reindeer died from starvation, the number of reindeer plunged to 8, about one-third the size of the original introduced population and less than 1% of the population at its peak. Recovery of arctic and subarctic vegetation after overgrazing by reindeer takes 15 to 20 years, and during that time the carrying capacity for reindeer is greatly reduced.

REVIEW

1. What is the effect of each of the following on population size: birth rate, death rate, immigration, and emigration?

2. How do intrinsic rate of increase and carrying capacity produce the J-shaped and S-shaped population growth curves?

Figure 8.3   Logistic population growth.

(a) *Paramecium*, a unicellular protist. (Michael Abbey/Photo Researchers, Inc.)

(b) In many laboratory studies, including Gause's investigation with *Paramecium caudatum*, population growth increases approximately exponentially when the population is low but slows as the carrying capacity of the environment is approached. This produces a curve with a characteristic S shape.

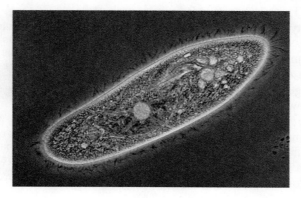

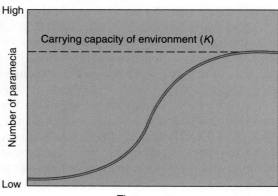

Factors That Affect Population Size

LEARNING OBJECTIVES

- Distinguish between density-dependent and density-independent factors that affect population size and give examples of each.
- Describe some of the density-dependent factors that may affect boom-or-bust population cycles.

Certain natural mechanisms influence population size. These mechanisms fall into two categories: density-dependent factors and density-independent factors. These two sets of factors vary in importance from one species to another, and in most cases they probably interact simultaneously to determine the size of a population.

Density-Dependent Factors

Certain environmental factors such as predation have a greater influence on a population when its density is greater. For example, if a pond has a dense population of frogs, predators such as water birds tend to congregate and eat more of the frogs than if the pond were sparsely populated by frogs. If a change in population density alters how an environmental factor affects the population, then the environmental factor is called a **density-dependent factor**. As population density increases, density-dependent factors tend to slow population growth by causing an increase in death rate and/or a decrease in birth rate. The effect of these density-dependent factors on population growth increases as the population density increases—that is, density-dependent factors affect a larger proportion, not just a larger number, of the population. Density-dependent factors also enhance population growth when population density declines, by decreasing death rate and/or increasing birth rate. Thus, density-dependent factors tend to keep a population at a relatively constant size near the carrying capacity of the environment. (Of course, the environment continually changes, and these changes continually affect the size of the carrying capacity.)

Predation, disease, and competition are examples of density-dependent factors. As the density of a population increases, predators are more likely to find an individual of a given prey species. When population density is high, the members of a population encounter one another more frequently, and the chance of their transmitting infectious disease organisms increases. As population density increases, so does competition for resources such as living space, food, cover, water, minerals, and sunlight. The opposite effects occur when the density of a population decreases. Predators are less likely to encounter individual prey, parasites are less likely to be transmitted from one host to another, and competition among members of the population for resources such as living space and food declines.

Most studies of density dependence are conducted in laboratory settings where all density-dependent (and density-independent) factors except one are controlled experimentally. Populations in natural settings are exposed to a complex set of variables that continually change, and it is difficult to evaluate the relative effects of density-dependent and density-independent factors.

Ecologists from the University of California, Davis, noted that few spiders occur on tropical islands inhabited by lizards, whereas more spiders and more species of spiders are found on lizard-free islands. Deciding to study these observations experimentally, the researchers selected 12 tiny Caribbean islands, 4 with lizards and 8 without; all contained web-building spiders. They introduced a small population of lizards onto four of the lizard-free islands. After seven years, spider population densities were higher in the lizard-free islands

Figure 8.4 **A population crash.**

(a) A herd of reindeer (*Rangifer tarandus*) on one of the Pribilof Islands in the Bering Sea, off the coast of Alaska. (Yvona Momatiuk and John Eastcott/Photo Researchers, Inc.)

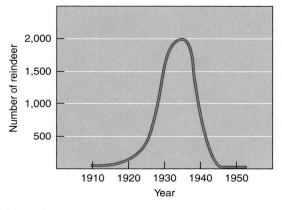

(b) The reindeer experienced rapid population growth followed by a sharp decline when the excess of reindeer damaged the environment. (After V. C. Scheffer)

density-dependent factor: An environmental factor whose effects on a population change as population density changes.

Figure 8.5 Lemming population oscillations.

(a) The brown lemming (*Lemmus trimucronatus*) lives in the arctic tundra. (Tom McHugh/Photo Researchers, Inc.)

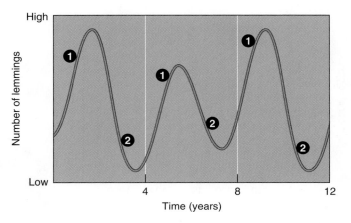

(b) This hypothetical diagram shows the cyclic nature of lemming population oscillations, which are not well understood. At the parts of the curve labeled 1, the population is increasing, and density-dependent factors are increasingly severe; as a result, the population peaks and begins to decline. At the parts of the curve labeled 2, the population is declining, and density-dependent factors are increasingly relaxed. As a result, the population bottoms out and begins to increase.

than in islands with lizards. Moreover, the islands without lizards had more species of spiders. You might conclude that lizards control spider populations. But even in this relatively simple experiment, the results are explained by a combination of two density-dependent factors, predation (lizards eat spiders) and competition (lizards compete with spiders for insect prey—that is, both spiders and lizards eat insects). In this experiment, the effects of the two density-dependent factors in determining spider population size cannot be evaluated separately. Additional experiments are needed to determine the relative roles of competition and predation.

Density Dependence and Boom-or-Bust Population Cycles

Lemmings are small rodents that live in the colder parts of the Northern Hemisphere. Lemming populations undergo dramatic increases in their population, followed by crashes over fairly regular time intervals—every three to four years (**Figure 8.5**). This cyclic fluctuation in abundance is often described as a boom-or-bust cycle. Other species such as snowshoe hares and red grouse also undergo cyclic population fluctuations.

The actual causes of such population oscillations are poorly understood, but many hypotheses involve density-dependent factors. The population density of lemming predators, such as weasels, arctic foxes, and jaegers (arctic birds that eat lemmings), may increase in response to the increasing density of prey. As more predators consume the abundant prey, the prey population declines. Another possibility is that a huge prey population overwhelms the food supply. Recent studies on lemming population cycles suggest that lemming populations crash because they eat all the vegetation in the area, not because predators eat them. It appears that prey populations may or may not be controlled by predators, whereas predator populations are most likely held in check by the availability of prey.

Parasites may influence cyclic fluctuations. Red grouse populations that live in the wild or in managed wildlife preserves exhibit population oscillations. Studies suggest that reproduction in red grouse is related to the presence of parasitic roundworms living in the birds' intestines. When adult birds are infected with worms, they do not breed successfully, and the population crashes. Biologists eliminated population oscillations in several red grouse populations after catching the birds and treating them with a medicine that ejects the worms from their bodies.

CASE IN POINT

Predator-Prey Dynamics on Isle Royale

During the early 1900s, a small herd of moose wandered across the ice of frozen Lake Superior to an island, Isle Royale. In the ensuing years, until 1949, moose became successfully established, although the population experienced oscillations. Beginning in 1949, a few Canadian wolves wandered across the frozen lake and discovered abundant moose prey on the island. The wolves remained and became established. Since 1958, wildlife biologists have studied the population dynamics of moose and wolves on Isle Royale.

Moose are the wolf's primary winter prey on Isle Royale. Wolves hunt in packs, encircling a moose and trying to get it to run so that they can attack it from behind. (The moose is the wolf's largest, most dangerous prey. A standing moose is more dangerous than a running one because when standing, it can kick and slash its attackers with its hooves.)

Wildlife biologists, such as **Rolf Peterson** of Michigan Technological University, have studied the effects of both density-dependent and density-independent factors on the Isle Royale populations of moose and wolves. They found that the two populations fluctuated over the years (**Figure 8.6**). Generally, as the population of wolves decreased, the population of moose increased. The reverse was also true: As the population of wolves increased, the population of moose decreased. Although the overall effect of wolves was to reduce the population of moose, the wolves did not eliminate moose from the island. Studies indicate that wolves primarily feed on the old and young in the moose population. Healthy moose in their peak reproductive years are not eaten.

During the 1980s and early 1990s, the wolf population plunged from 50 animals in 1980 to a low of 12 animals in 1989, possibly as a result of a deadly disease, canine parvovirus. Analysis of their blood revealed the presence of antibodies to canine parvovirus, confirming that the wolves had been exposed.

As expected, the moose population increased as the wolf population declined; in 1995, there were more than 2400 moose on Isle Royale, many more than the island's vegetation could support. The moose overgrazed the island, particularly mountain ash and aspen, their preferred food. Lack of food in combination with a particularly bad winter (1995–1996) caused hundreds of moose to die; only 500 moose were counted in the 1997 survey.

More recently, the wolf population has increased. In the 2003–2004 season, for example, each of the island's three territorial packs raised several pups, resulting in a total population of 29 wolves in 2004. The moose population declined from an estimated 900 animals in 2003 to an estimated 750 animals in 2004. This change was partly in response to heavy predation pressure by wolves on both calves and adults.

Other factors also interacted and influenced the moose population decline. Infestations of blood-sucking ticks weakened the moose during winter and spring months, affecting both their survival and their reproductive success. (In particularly severe cases, a moose can be so infested with ticks that it must replace the equivalent of its entire blood supply in a few weeks!)

For reasons that are not completely understood, the tick population is greater when either spring or fall months are warmer than usual. Isle Royale experienced unusually high temperatures and drought from 1998 to 2002. Thus, warmer temperatures indirectly hurt the moose population by providing ideal conditions for the tick population. Warmer temperatures also impacted the moose directly: Moose do not have sweat glands and are therefore physiologically stressed during warm summer months. (Moose, which range across much of Canada and Alaska, commonly do not live south of Isle Royale.) Scientists hypothesize that global climate change may be at least partially responsible for these unusual weather conditions. ■

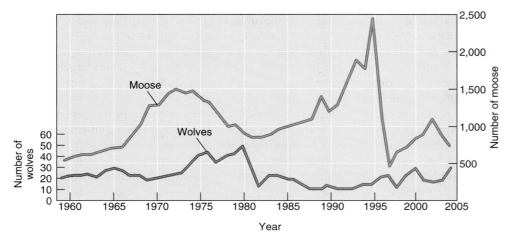

Figure 8.6   Wolf and moose populations on Isle Royale, 1958–2004. Data from aerial surveys indicate that there were 29 wolves and 750 moose in 2004. (After Rolf O. Peterson)

density-independent factor: An environmental factor that affects the size of a population but is not influenced by changes in population density.

Density-Independent Factors

Density-independent factors are typically abiotic. Random weather events that reduce population size serve as density-independent factors. A killing frost, severe blizzard, hurricane, or fire may cause extreme and irregular reductions in a population regardless of its size and thus are largely density independent.

Consider a density-independent factor that influences mosquito populations in arctic environments. These insects produce several generations per summer and achieve high population densities by the end of the season. A shortage of food is not a limiting factor for mosquitoes, nor is there any shortage of ponds in which to breed. Instead, winter puts a stop to the skyrocketing mosquito population. Not a single adult mosquito survives winter, and the entire population grows afresh the next summer from the few eggs and hibernating larvae that survive. The timing and severity of winter weather is a density-independent factor that affects arctic mosquito populations.

Density-dependent and density-independent factors are often interrelated. Social animals often resist dangerous weather conditions by their collective behavior, as in the case of sheep huddling together in a snowstorm. In this case, the greater the population density, the better their ability to resist the environmental stress of a density-independent event (the snowstorm).

REVIEW

1. What are three examples of density-dependent factors that affect population growth? What are three examples of density-independent factors?

2. What is a boom-or-bust population cycle? What density-dependent factors may influence such cyclic population oscillations?

 ## Reproductive Strategies

LEARNING OBJECTIVE

• Define survivorship and describe type I, type II, and type III survivorship curves.

Each species has a lifestyle uniquely adapted to its reproductive patterns. Many years pass before a young magnolia tree flowers and produces seeds, whereas wheat grows from seed, flowers, and dies in a single season. A mating pair of black-browed albatrosses produces a single chick every year, whereas a mating pair of gray-headed albatrosses produces a single chick biennially (every other year). Biologists try to understand the adaptive consequences of these various *life history strategies*.

Imagine an organism possessing the "perfect" life history strategy that ensures continual reproduction at the maximum intrinsic rate of increase. In other words, this hypothetical organism produces the maximum number of offspring, and all of these offspring survive to reproduce. Such an organism would have to reach reproductive maturity immediately after it was born so that it could begin reproducing at an early age. It would reproduce frequently throughout its long life and produce large numbers of offspring each time. Furthermore, it would have to provide care for all of its young to ensure their survival.

In nature, such an organism does not exist because if an organism were to put all its energy into reproduction, it could not expend any energy toward ensuring its own survival. Animals use energy to hunt for food, and plants use energy to grow taller than surrounding plants (to obtain adequate sunlight). Nature, then, requires organisms to make tradeoffs in the expenditure of energy. Successful individuals must do what is required to survive as individuals *and* as populations (by reproducing). If they allocate all their energy for reproduction, none is available for the survival of the individual, and the individual dies. If they allocate all their energy for the individual, none is available for reproduction, and there are no further generations.

Each species has its own life history strategy—its own reproductive characteristics, body size, habitat requirements, migration patterns, and behaviors—that represents a series of tradeoffs reflecting this energy compromise. Although many life history strategies exist, some ecologists recognize two extremes with respect to reproductive characteristics, *r*-selected species and *K*-selected species. As you read the following descriptions of *r* selection and *K* selection, keep in mind that these concepts, though useful, over-simplify most life histories. Many species possess a combination of *r*-selected and *K*-selected traits, as well as traits that are neither *r*-selected nor *K*-selected.

Populations described by *r selection* have traits that contribute to a high population growth rate. Recall that *r* designates the growth rate. Such organisms have a high *r* and are called *r strategists* or *r-selected species*. Small body size, early maturity, short life span, large broods, and little or no parental care are typical of many *r* strategists, which are usually opportunists found in variable, temporary, or unpredictable environments where the probability of long-term survival is low. Some of the best examples of *r* strategists are insects such as mosquitoes and common weeds such as the dandelion.

In populations described by *K selection*, traits maximize the chance of surviving in an environment where the number of individuals (*N*) is near the carrying capacity (*K*) of the environment. These organisms, called *K strategists* or *K-selected species*, do not produce large numbers of offspring. They characteristically have long life spans with slow development, late reproduction, large body size, and low reproductive rate. Redwood trees are classified as *K* strategists. Animals that are *K* strategists typically invest in the parental care of their young. *K* strategists are found in relatively constant or stable environments, where they have a high competitive ability.

Tawny owls are *K* strategists that pair-bond for life, with both members of a pair living and hunting in separate, well-defined territories. Their reproduction is regulated in accordance with resources, especially the food supply, in their territories. In an average year, 30% of the birds do not breed. If food supplies are more limited than initial conditions had indicated, many of those that breed fail to incubate their eggs. Rarely do the owls lay the maximum number of eggs they are physiologically capable of laying, and breeding is often delayed until late in the season, when the rodent populations on which they depend have become large. Thus, the behavior of tawny owls ensures better reproductive success of the individual and leads to a stable population at or near the carrying capacity of the environment. Starvation, an indication that the tawny owl population has exceeded the carrying capacity, rarely occurs.

Survivorship

Ecologists construct **life tables** for plants and animals that show the likelihood of survival for individuals at different times during their lives. Insurance companies originally developed life tables to determine how much policies should cost; life tables show the relationship between a client's age and the likelihood the client will survive to pay enough insurance premiums to cover the cost of the policy.

Survivorship is the proportion of newborn individuals that are alive at a given age. **Figure 8.7** is a graph of the three main survivorship curves recognized by ecologists. In type I survivorship, as exemplified by humans and elephants, the young (that is, pre-reproductive individuals) and those at reproductive age have a high probability of living. The probability of survival decreases more rapidly with increasing age, and deaths are concentrated later in life.

In type III survivorship, the probability of death is greatest early in life, and those individuals that avoid early death subsequently have a high probability of survival. In animals, type III survivorship is characteristic of many fish species and oysters. Young oysters have three free-swimming larval stages before adulthood, when they settle down and secrete a shell. These larvae are extremely vulnerable to predation, and few survive to adulthood.

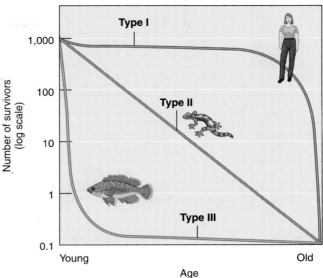

Figure 8.7 Survivorship. These generalized survivorship curves represent the ideal survivorships of species in which death is greatest in old age (type I), spread evenly across all age groups (type II), and greatest among the young (type III).

survivorship: The probability a given individual in a population will survive to a particular age.

In type II survivorship, intermediate between types I and III, the probability of survival does not change with age. The probability of death is likely across all age groups, resulting in a linear decline in survivorship. This constancy probably results from essentially random events that cause death with little age bias. This relationship between age and survivorship is relatively rare; some lizards have a type II survivorship.

The three survivorship curves are generalizations, and few populations exactly fit one of the three. Some species have one type of survivorship curve early in life and another type as adults. Herring gulls have a type III survivorship curve early in life and a type II curve as adults (**Figure 8.8**). With most, death occurs almost immediately after hatching, despite the parent bird providing the chicks with protection and care. Herring gull chicks die from predation or attack by other herring gulls, inclement weather, infectious disease, or starvation following the death of the parent. Once the chicks become independent, their survivorship increases dramatically, and death occurs at about the same rate throughout their remaining lives. As a result, few or no herring gulls die from the degenerative diseases of "old age" that cause death in most humans.

REVIEW

1. What are the three main survivorship curves?

The Human Population

LEARNING OBJECTIVES

- Define *demography* and summarize the history of human population growth.
- Identify Thomas Malthus, relate his ideas on human population growth, and explain why he may or may not be correct.
- Explain why it is impossible to answer precisely how many people Earth can support—that is, Earth's carrying capacity for humans.

Figure 8.8   **Survivorship for a herring gull population.**

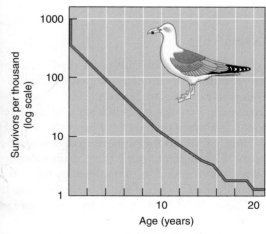

(a) You would be lucky to find this young herring gull (*Larus argentatus*) huddling in its nest of beach grass. Young gulls are hard to see because their mottled brownish color blends into the grasses. (James L. Amos/National Geographic Society)

(b) Despite their protective coloration, most gull chicks die at an early age. This survivorship curve reveals type III survivorship as chicks and type II survivorship as adults. (After R. A. Paynter, Jr.)

Now that you have examined some of the basic concepts of population ecology, let us apply those concepts to the human population. **Demography** is the science of population structure and growth. The application of population statistics is called **demographics.**

Examine **Figure 8.9**, which shows the world increase in population since 1800. Now reexamine Figure 8.2b and compare the two curves. The characteristic J curve of exponential population growth shown in Figure 8.9 reflects the decreasing amount of time it has taken to add each additional billion people to our numbers. It took thousands of years for the human population to reach 1 billion, a milestone that occurred around 1800. It took 130 years to reach 2 billion (in 1930), 30 years to reach 3 billion (in 1960), 15 years to reach 4 billion (in 1975), 12 years to reach 5 billion (in 1987), and 12 years to reach 6 billion (in 1999). The human population in 2006 was 6.6 billion (**Figure 8.10**).

One of the first to recognize that the human population cannot increase indefinitely was **Thomas Malthus,** a British economist (1766–1834). He pointed out that human population growth is not always desirable—a view contrary to the beliefs of his day and to those of many people even today—and that the human population can increase faster than its food supply. The inevitable consequences of population growth, he maintained, are famine, disease, and war. Since Malthus's time, the human population has grown from about 1 billion to more than 6 billion. At first glance, it appears that Malthus was wrong. Our population has grown so dramatically because scientific advances have allowed food production to keep pace with population growth. Malthus's ideas may ultimately prove correct because we do not know if our increase in food production is *sustainable*. Have we achieved this increase in food production at the environmental cost of reducing the ability of the land to meet the needs of future populations?

Current Population Numbers

Our world population, which was 6.6 billion in 2006, increased by about 78 million from 2005 to 2006. This increase is not due to an increase in the birth rate (*b*). In fact, the world birth rate has declined during the past 200 years. Instead, the increase in population is due to a dramatic decrease in the death rate (*d*), which has occurred primarily because greater food production, better medical care, and improvements in water quality and sanitation practices have increased the life expectancies for a great majority of the global population.

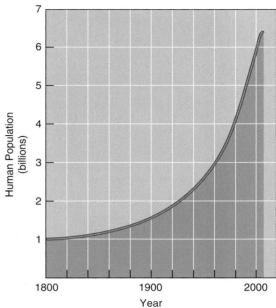

Figure 8.9   Human population numbers, 1800 to present. The human population has been increasing exponentially. Population experts predict that the population will level out during the 21st century, possibly forming the S curve observed in certain other species. (Population Reference Bureau)

demography: The applied branch of sociology that deals with population statistics and provides information on the populations of various countries or groups of people.

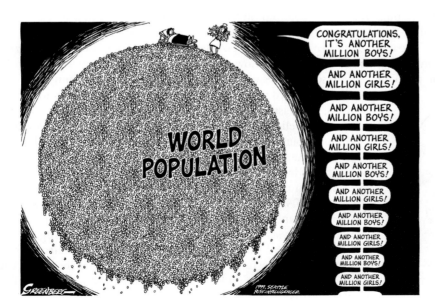

Figure 8.10   World population growth. Do you think the human population can continue to increase indefinitely? Why or why not? (© Steve Greenberg)

Consider Mexico. From about 1920 to 2000, the death rate in Mexico fell from approximately 40 to 5 per 1000 individuals, whereas the birth rate dropped from approximately 40 to 26 per 1000 individuals (**Figure 8.11**). Mexico's birth rate currently remains greater than its death rate, resulting in an annual growth rate (*r*) of 1.7% in 2006:

$$r = b - d$$
$$r = 22/1000 - 5/1000$$
$$r = 0.022 - 0.005 = 0.017, \text{ or } 1.7\% \text{ per year}$$

In comparison, the annual growth rate in the United States is 0.6%.

Projecting Future Population Numbers

Although our numbers continue to increase, the world growth rate (*r*) has declined over the past several years, from a peak of 2.2% per year in the mid-1960s to 1.2% per year in 2006. Population experts at the United Nations and the World Bank have projected that the growth rate will continue to decrease slowly until zero population growth is attained. At that time, exponential growth of the human population will end, and the J curve may be replaced by the S curve. It is projected that **zero population growth**—when the birth rate equals the death rate—will occur toward the end of the 21st century.

The United Nations periodically publishes population projections for the 21st century. U.N. figures in their 2004 revision forecast that the human population will be between 7.7 billion (their "low" projection) and 10.6 billion (their "high" projection) in the year 2050, with 9.1 billion thought "most likely" (**Figure 8.12**). These estimates take into account both lower projected fertility levels and higher projected mortality from HIV/AIDS. Lower fertility levels translate into an aging population; the percentage of the world population over the age of 65 years was about 7% in 2006 and is expected to rise. (Population aging is discussed later in this chapter.)

Such population projections are "what if" exercises: Given certain assumptions about future tendencies in the birth rate, death rate, and migration, an area's population can be calculated for a given number of years into the future. Population projections must be interpreted with care because they vary depending on what assumptions are made. In projecting that the world population will be 7.7 billion (their low projection) in the year 2050, U.N. population experts assume that the average number of children born to each woman in all countries will have declined to 1.5 in the 21st century.

In 2006, the average number of children born to each woman on Earth was 2.7. If the decline to 1.5 does not occur, our population could be significantly higher. If the average number of children born to each woman declines to only 2.5 instead of 1.5, the 2050 population will be 10.6 billion (the U.N. high projection). Small differences in fertility, then, produce large differences in population forecasts.

The main unknown factor in any population growth scenario is Earth's *carrying capacity*. Most published estimates of how many people Earth can support range from 4 billion to 16 billion. These estimates vary widely depending on what assumptions are made about standard of living, resource consumption, technological innovations, and waste generation. If we want all people to have a high level of material well-being equivalent to the lifestyles common in highly developed countries, then Earth will clearly support far fewer humans than if everyone lives just above the subsistence level. Earth's carrying capacity for humans is not decided simply by environmental constraints. Human choices and values must be factored into the assessment.

In 2004, **Jeroen Van Den Bergh** and **Piet Rietveld**, environmental economists in the Netherlands, performed a detailed statistical analysis of 69 recent studies of Earth's carrying capacity for humans. Based on current technology, they estimate that 7.7 billion is the upper limit of human population that the world can support. The medium and high U.N. population projections for 2050 exceed this value.

Figure 8.11 Birth and death rates in Mexico, 1900–2000. Both birth and death rates generally declined in Mexico during the 20th century. The death rate declined much more than the birth rate, so Mexico experienced a high growth rate. (The Mexican Revolution caused the high death rate prior to 1925.) (Population Reference Bureau)

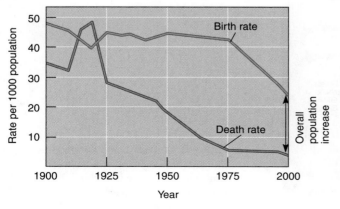

It is not clear what will happen to the human population when the carrying capacity is approached. Optimists suggest that the human population will stabilize because of a decrease in the birth rate. Some experts take a more pessimistic view and predict that the widespread degradation of our environment caused by our ever-expanding numbers will make Earth uninhabitable for humans as well as for other species. These experts contend that a massive wave of human suffering and death will occur. This view does not mean that we will become extinct as a species but that there will be severe hardships for many people. Some experts think the human population has already exceeded the carrying capacity of the environment, a potentially dangerous situation that threatens our long-term survival as a species.

REVIEW

1. Describe human population growth for the past 200 years.

2. Who was Thomas Malthus, and what were his views on human population growth?

3. When determining Earth's carrying capacity for humans, why is it not enough to just consider human numbers? What else must be considered?

 ## Demographics of Countries

LEARNING OBJECTIVES

• Explain how highly developed and developing countries differ in population characteristics such as infant mortality rate, total fertility rate, and age structure.

• Explain how population growth momentum works.

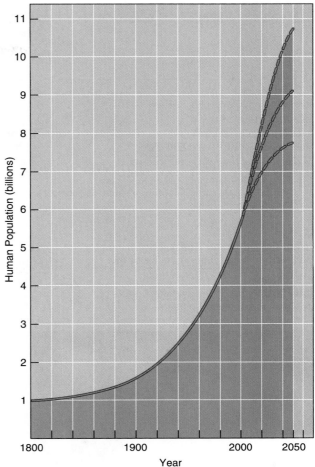

Figure 8.12 **Population projections to 2050.** In 2004, the United Nations made three projections, each based on different fertility rates. The medium projection was 9.1 billion. (World Population Prospects: The 2004 Revision)

Whereas world population figures illustrate overall trends, they do not describe other important aspects of the human population story, such as population differences from country to country (**Table 8.1**). As you may recall from Chapter 1, not all countries have the same rate of population increase. Countries are classified into two groups—highly developed and developing—depending on growth rates, degree of industrialization, and relative prosperity (**Table 8.2**). **Highly developed countries** (also called *developed countries*), such as the United States, Canada, France, Germany, Sweden, Australia, and Japan, have low rates of population growth and are highly industrialized relative to the rest of the world. Highly developed countries have the lowest birth rates in the world. Indeed, some countries such as Germany have birth rates just below those needed to sustain their populations and are declining slightly in numbers.

Table 8.1 The World's 10 Most Populous Countries

Country	2006 Population (in millions)*	Population Density (per mi²)
China	1311.4	355
India	1121.8	884
United States	299.1	80
Indonesia	225.5	307
Brazil	186.8	57
Pakistan	165.8	539
Bangladesh	146.6	2637
Russia	142.3	22
Nigeria	134.5	377
Japan	127.8	876

* These figures are from mid-2006. At the end of 2006, the United States reached a population milestone of 300 million people.

Table 8.2 Comparison of 2006 Population Data in Developed and Developing Countries

	Developed	Developing	
	(Highly Developed) United States	(Moderately Developed) Brazil	(Less Developed) Ethiopia
Fertility rate	2.0	2.3	5.4
Projected population change, 2006–2050*	+40%	+39%	+94%
Infant mortality rate	6.7 per 1000	27 per 1000	77 per 1000
Life expectancy at birth	78 years	72 years	49 years
Per capita GNI PPP (2002; U.S. $)**	$41,950	$8,230	$1,000
Women using modern contraception	68%	70%	14%

* Includes fertility, mortality, and migration estimates.
** GNI PPP = gross national income in purchasing power parity.

infant mortality rate: The number of infant deaths (under age 1) per 1000 live births.

total fertility rate: The average number of children born to each woman.

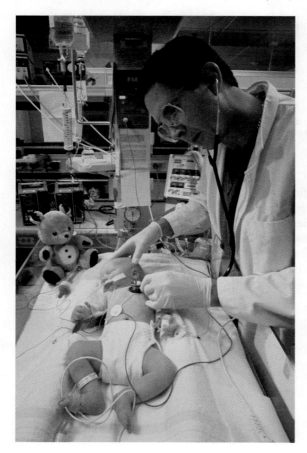

Figure 8.13 **A doctor examines a premature baby in the intensive care unit.** Medical advances have contributed to a decline in infant deaths, particularly in highly developed countries. Photographed in Dennison, Texas. (Stewart Cohen/Stone/Getty Images)

Highly developed countries have **low infant mortality rates** (**Figure 8.13**). The infant mortality rate of the United States was 6.7 in 2006, compared with a world rate of 52. Highly developed countries also have longer life expectancies (78 years in the United States versus 67 years worldwide) and high average per capita GNI PPPs ($41,950 in the United States versus $9,190 worldwide). Per capita GNI PPP is gross national income (GNI) in purchasing power parity (PPP) divided by midyear population. It indicates the amount of goods and services an average citizen of a particular country could buy in the United States.

Developing countries fall into two subcategories, moderately developed and less developed. Mexico, Turkey, Thailand, and most South American nations are examples of **moderately developed countries.** Their birth rates and infant mortality rates are higher than those of highly developed countries, but they are declining. Moderately developed countries have a medium level of industrialization, and their average per capita GNI PPPs are lower than those of highly developed countries. **Less developed countries (LDCs)** include Bangladesh, Niger, Ethiopia, Laos, and Cambodia. These countries have the highest birth rates, the highest infant mortality rates, the shortest life expectancies, and the lowest average per capita GNI PPPs in the world.

One way to express the population growth of a country is to determine its **doubling time (t_d),** the amount of time it would take for its population to double in size, assuming its current growth rate does not change. A simplified formula for doubling time is $t_d = 70/r$, sometimes called the *rule of 70*. The actual formula involves calculus and is beyond the scope of this text. A country's doubling time usually identifies it as a highly, moderately, or less developed country: The shorter the doubling time, the less developed the country. In 2006, the doubling time was 29 years for Ethiopia ($r = 2.4$), 54 years for Turkey ($r = 1.3$), 117 years for the United States ($r = 0.6$), and 350 years for Denmark ($r = 0.2$).

Replacement-level fertility is the number of children a couple must produce to "replace" themselves. Replacement-level fertility is usually given as 2.1 children. The number is greater than 2.0 because some infants and children die before they reach reproductive age. Worldwide, the **total fertility rate** is currently 2.7, well above the replacement level.

Demographic Stages

Based on the 1945 work of Princeton demographer **Frank Notestein,** demographers recognize four stages, based on their observations of Europe as it became industrialized and urbanized (**Figure 8.14**). These stages converted Europe from relatively high birth and death rates to relatively low birth and death rates. To date, all highly developed and moderately developed countries with more advanced economies have gone through this progression, or **demographic transition.** Demographers generally assume that the same demographic transition will occur in less developed countries as they become industrialized.

In the first stage—the **preindustrial stage**—birth and death rates are high, and population grows at a modest rate. Although women have many children, the infant mortality rate is high. Intermittent famines, plagues, and wars also increase the death rate, so the population grows slowly or temporarily declines. If we use Finland to demonstrate the four demographic stages, we can say that Finland was in the first demographic stage from the time of its first human settlements until the late 1700s.

As a result of improved healthcare and more reliable food and water supplies that accompany the beginning of an industrial society, the second demographic stage, called the **transitional stage,** has a lowered death rate. The population grows rapidly because the birth rate is still high. Finland in the mid-1800s was in the second demographic stage.

The third demographic stage, the **industrial stage,** is characterized by a decline in birth rate and takes place at some point during the industrialization process. The decline in birth rate slows population growth despite a relatively low death rate. Finland experienced this stage in the early 1900s.

Low birth and death rates characterize the fourth demographic stage, called the **postindustrial stage.** In heavily industrialized countries, people are better educated and more affluent; they tend to desire smaller families, and they take steps to limit family size. The population grows slowly or not at all in the fourth demographic stage.

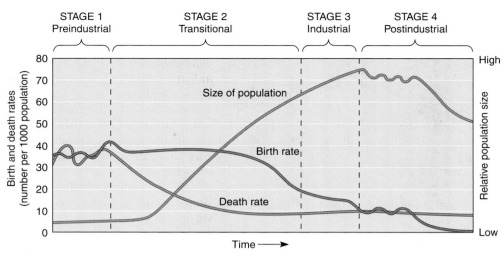

Figure 8.14 **Demographic transition.** The demographic transition consists of four demographic stages through which a population progresses as its society becomes industrialized. Note that the death rate declines first, followed by a decline in the birth rate.

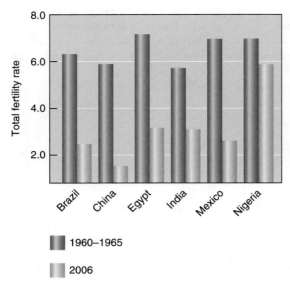

■ 1960–1965

■ 2006

Figure 8.15 **Fertility changes in selected developing countries.** (Population Reference Bureau)

age structure: The number and proportion of people at each age in a population.

population growth momentum: The potential for future increases or decreases in a population based on the present age structure.

This is the situation in such highly developed countries and groups of countries as the United States, Canada, Australia, Japan, and Europe, including Finland.

Why has the population stabilized in highly developed countries in the fourth demographic stage? The decline in birth rate is associated with an improvement in living standards, although it is not known whether improved socioeconomic conditions have caused a decrease in birth rate, or a decrease in birth rate has caused improved socioeconomic conditions. Perhaps both are true. Another reason for the decline in birth rate in highly developed countries is the increased availability of family planning services. Other socioeconomic factors that influence birth rate are increased education, particularly of women, and urbanization of society. We consider these factors in greater detail in Chapters 9 and 10.

Once a country reaches the fourth demographic stage, is it correct to assume that it will continue to have a low birth rate indefinitely? The answer is we do not know. Low birth rates may be a permanent response to the socioeconomic factors of an industrialized, urbanized society. On the other hand, low birth rates may be a response to the changing roles of women in highly developed countries. Unforeseen changes in the socioeconomic status of women and men in the future may change birth rates. No one knows for sure.

The population in many developing countries is approaching stabilization. See **Figure 8.15** and note the general decline in total fertility rates in selected developing countries from the 1960s to 2006. The total fertility rate in developing countries decreased from an average of 6.1 children per woman in 1970 to 2.7 in 2006.

Although the fertility rates in these countries have declined, most still exceed replacement-level fertility. Consequently, the populations in these countries are still increasing. Even when fertility rates equal replacement-level fertility, population growth will still continue for some time. To understand why this is so, let us examine the age structure of various countries.

Age Structure of Countries

To predict the future growth of a population, you must know its **age structure**, or distribution of people by age. The number of males and number of females at each age, from birth to death, are represented in an age structure diagram (**Figure 8.16**).

The overall shape of an age structure diagram indicates whether the population is increasing, stable, or shrinking. The age structure diagram of a country with a high growth rate, based on a high fertility rate—for example, Nigeria or Bolivia—is shaped like a pyramid (**Figure 8.17a**). The probability of future population growth is great because the largest percentage of the population is in the prereproductive age group (0 to 14 years of age). A positive **population growth momentum** exists because when

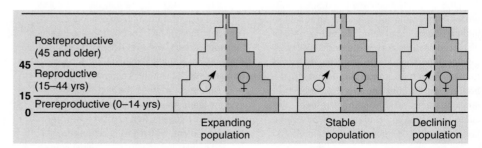

Figure 8.16 **Generalized age structure diagrams.** These diagrams show an expanding population, a stable population, and a population decreasing in size. Each diagram is divided vertically in half, the left side representing the males in a population and the right side the females. The bottom third of each diagram represents prereproductive humans (between 0 and 14 years of age); the middle third, reproductive humans (15 to 44 years); and the top third, postreproductive humans (45 years and older). The widths of these segments are proportional to the population sizes—a broader width implies a larger population.

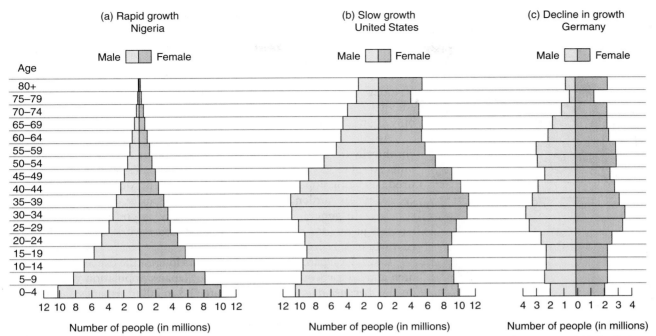

Figure 8.17   Age structure diagrams. Shown are countries with (a) fast (Nigeria), (b) slow (United States), and (c) declining (Germany) population growth. These age structure diagrams indicate that less developed countries such as Nigeria have a much higher percentage of young people than highly developed countries. As a result, less developed countries are projected to have greater population growth than highly developed countries. (Population Reference Bureau)

all these children mature, they will become the parents of the next generation, and this group of parents will be larger than the previous group. Even if the fertility rate of such a country has declined to replacement level (that is, couples are having smaller families than their parents did), the population will continue to grow for some time. Population growth momentum, which can be positive or negative, explains how the present age distribution affects the future growth of a population.

In contrast, the more tapered bases of the age structure diagrams of countries with slowly growing, stable, or declining populations indicate that a smaller proportion of the population will become the parents of the next generation (**Figure 8.17b and c**). The age structure diagram of a stable population, one that is neither growing nor shrinking, demonstrates that the numbers of people at prereproductive and reproductive ages are approximately the same. A larger percentage of the population is older— that is, postreproductive—than in a rapidly increasing population. Many countries in Europe have stable populations. In a population shrinking in size, the prereproductive age group is *smaller* than either the reproductive or postreproductive group. Russia, Ukraine, and Germany are examples of countries with slowly shrinking populations.

Worldwide, 29% of the human population is under age 15 (**Figure 8.18**). When these people enter their reproductive years, they have the potential to cause a large increase in the growth rate. Even if the birth rate does not increase, the growth rate will increase simply because there are more people reproducing.

Since 1950, most of the world population increase has occurred in developing countries—as a result of the younger age structure and the higher-than-replacement-level fertility rates of their populations. In 1950, 67% of the world's population was in developing countries in Africa, Asia (minus Japan), and Latin America. After 1950 the world's population more than doubled, but most of that growth occurred in developing countries. As a reflection of this fact, in 2006, the number of people in developing countries had increased to 81% of the world population. Most of the population increase during the 21st century will occur in developing countries, largely the result of their younger age structures. Because these countries are not industrialized, they are least able to support such growth.

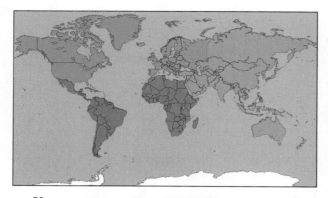

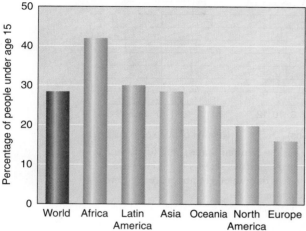

Figure 8.18   Percentages of the population under age 15 for various regions in 2006. The higher this percentage, the greater the potential for population growth when people in this group reach their reproductive years. (Note: The bars are color-coded to match the corresponding geographic locations on the map.) (Population Reference Bureau)

Age Structure: Effects of an Aging Population   Declining fertility rates have profound social and economic implications because as fertility rates drop, the percentage of the population that is elderly increases. For example, although the United States has about the same number of young people today as it had during the baby boom years following World War II, today's children make up a lower percentage of the overall population. About 36% of the U.S. population was younger than 15 in 1960, whereas only about 20% are younger than 15 today.

An aging population has a higher percentage of people who are chronically ill or disabled, and these people require more healthcare and other social services. Because the elderly produce less wealth (most are retired), an aging population reduces a country's productive workforce, increases its tax burden, and strains its social security, health, and pension systems. (In general, these systems are "pay-as-you-go," with current workers paying for retirees' benefits, and future workers paying for current workers' retirement benefits.)

Consider Japan, which has the world's longest life expectancy—79 for men and 86 for women. By 2007 or so, Japan's population will begin to decline because of its low fertility rate (1.3 in 2006). Japanese leaders are concerned that there will not be enough young people to support Japan's growing elderly population. To address some of the problems in its workforce, Japan is starting to offer incentives to the elderly to work longer before retiring (and before receiving retirement benefits), and many Japanese are happy to continue working, either full time or part time, for a longer period.

The Russian Federation is faced with a similar challenge. Its fertility rate is low (1.3 in 2006), and its population is declining. The U.N. population division predicts that Russia's population in 2050 will have declined to 112 million (its 2006 population was 142.3 million). Like many countries with declining populations, Russia is offering incentives to young couples who choose to have children; these include assisting with maternity leave and helping to pay for childcare when the mother returns to the workplace. Many demographers think that this approach will not affect the birth rate appreciably.

Also, given that the world population continues to increase, providing incentives in countries with aging populations to encourage women to have more children does not seem the way to go. After all, if a given country raises its birth rate, it is not only increasing the world population but also contributing to the aging problem down the road: More babies today means more elderly tomorrow. Similarly, filling the workforce by encouraging young people from developing countries to migrate only postpones the aging problem: More young immigrants today means more old people tomorrow.

Changes in a population's age structure can affect many social problems (such as crime) that at first glance do not appear to be directly related. Sociologists have observed that in an aging population the crime rate may decline. Most crimes involve young adults between the ages of 18 and 24, and in an aging population, these young adults represent a smaller proportion of the population. In the late 1990s, for example, the United States experienced a reduction in the rate of violent crimes, and the aging of the U.S. population was cited as at least one factor. (Like all social problems, the causes of crime are complex.)

Thus, the aging of populations offers a mixed bag of benefits and problems. No country has been faced with an aging population before now, and we do not know how aging populations will function. Despite the uncertainties, most policy analysts think that countries with higher proportions of elderly will probably have to increase the age of retirement (through incentives that encourage people to work after age 65) and decrease benefits for the elderly. Also, because benefits for the elderly will most certainly decline over time, experts recommend that young people begin saving aggressively for their retirements *early* in their careers instead of after their children have grown.

1. What is infant mortality rate? How does it affect life expectancy?
2. What is the difference between the total fertility rate and replacement-level fertility?
3. If all the women in the world suddenly started bearing children at replacement-level fertility rates, would the population stop increasing immediately? Why or why not?

Demographics of the United States

LEARNING OBJECTIVE

• Briefly describe the Immigration Reform and Control Act (IRCA).

The United States has the largest population of all highly developed countries, and compared to the global population, U.S. citizens have a high level of material well-being. However, you could argue persuasively that the United States is one of the most overpopulated countries in the world because of overconsumption (see Chapter 1). Moreover, the population of the United States continues to grow significantly, in part because people from developing countries immigrate to the United States to try for a better life. According to the U.S. government and other organizations such as the Pew Hispanic Center, international immigration is now responsible for at least 50% of U.S. population growth. Assuming recent immigration levels continue, the U.N. Population Division estimates that the U.S. population will reach 395 million by 2050. In comparison, the mid-2006 U.S. population was 299.1 million.

Population experts and environmental scientists try to predict what the increasing population size and high consumption level of the United States will mean to future generations, both in the United States and worldwide. Should the United States have a formal population policy that stipulates desirable family size and economic prosperity? Should we—or can we—control immigration? Is it ethical to curtail the individual rights of the present generation to assure a better future for generations that will follow us?

Currently, the United States does not have a formal population policy. The closest it came to such a policy occurred under President Nixon's leadership, when a commission was established to examine U.S. population growth. In 1972, when the U.S. population was 210 million, the commission concluded that the United States would gain no substantial benefits from continued population growth. The commission recommended that the United States should try to stabilize its population. Despite this recommendation, the U.S. population has continued to increase.

The United States has one of the highest rates of population increase of all the highly developed countries. The U.S. population increased 15 million in the five years from 2001 to 2006. This translates into a 0.6% annual increase in 2006, which is greater than that in most highly developed countries. As a comparison, the 2006 annual increase in Western Europe was 0.1%. These annual increases take only birth and death rates into account; immigration is not considered.

Among all countries, the United States has the largest number of immigrants, an estimated 35 million. Since 1992, the United States has accepted about one million legal immigrants annually. The number of unauthorized foreigners who gain access to the United States and are not deported is not known with any certainty, but the U.S. Department of Homeland Security (DHS) estimates that it is about 525,000 per year. DHS estimates that the total population of unauthorized migrants living in the United States in 2005 was 11 million.

CASE IN POINT

U.S. Immigration

Human migration across international borders is a worldwide phenomenon that has escalated dramatically during the past few decades. There are many reasons for the increase in international migration. People migrate in search of jobs or an improved standard of living (the most important reason); to escape war or persecution for their race,

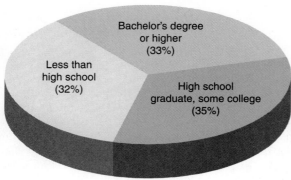

Figure 8.19   Education levels of recently arrived foreign-born Americans. Data were assembled in 2005, based on immigrants who entered the United States after 1999. (U.S. Census Bureau)

religion, nationality, or political opinions; or to join other family members who have already migrated.

The increase in the human population is an underlying factor that has contributed to the current surge in migration, which population experts predict will continue to increase during the next 30 years. Deteriorating environmental conditions brought on by population growth may contribute to international migration. Population growth is also responsible for the addition of millions of people to the workforce each year; many countries' economies do not create enough new jobs to accommodate these people.

History of Immigration in the United States   Prior to 1875, the United States had no immigration laws and no such thing as unauthorized migrants. In 1875, Congress passed a law denying convicts and prostitutes entrance to the United States. In 1882, the Chinese Exclusion Act was passed, and in 1891, the Bureau of Immigration was established. Thus, a policy of selective exclusion was officially established and began to shape the population of this country.

During the early 20th century, Congress set numerical restrictions on immigration, including quotas allowing only a certain number of people from each foreign country to immigrate. With these stronger laws, the influx of unauthorized migrants began to increase. U.S. immigration policy relaxed during World War II, when labor shortages made it possible for workers from other countries to gain temporary residence in the United States.

In 1952, the **Immigration and Nationality Act** was passed, and although it has been revised since then (it is now called the **Immigration Reform and Control Act, or IRCA**), it is still the basic immigration law in effect. Amendments passed in 1965 abolished national quotas and gave three groups of people priority when immigrating to the United States: those with family members already living in the United States, those who can fill vacant jobs (employment-based preference), and those who are refugees seeking asylum.

In the past, waves of immigrants entering the United States were racially and ethnically homogeneous. In contrast, typical immigrants to the United States today are a more diverse group, racially, ethnically, and culturally. Currently, the top five countries from which legal U.S. immigrants migrate are Mexico, the Philippines, Vietnam, the Dominican Republic, and China. About one-third of these new arrivals are poor and have few skills, but the rest are high school or college graduates, many with desirable skills (**Figure 8.19**).

Immigration is a controversial environmental issue. Some environmentalists think the population of the United States is already too large. According to this view, the presence of legal immigrants and unauthorized migrants is undesirable because it contributes greatly to pollution and resource depletion as these people adopt the affluent, high-consumption lifestyles of "typical" U.S. citizens. People with this view say the United States must reduce both birth rates and immigration rates to bring about sustainability.

Other environmentalists point out that the United States is not the only country to take in immigrants. They say that it is morally unacceptable to deny people a chance for a better life and that the United States can absorb the environmental consequences of immigrants, who often have simpler lifestyles than most U.S. citizens (at least for the first generation). Immigrants tend to have smaller families than if they had stayed in their countries of origin, so immigration reduces the rate of global population growth. Furthermore, immigrants provide U.S. environmental groups with a much-needed global consciousness, a "we're in this together" viewpoint that is easy to overlook when simply considering U.S. environmental issues. ∎

REVIEW

1. Should the United States increase or decrease the number of legal immigrants? Present arguments in favor of both sides.

REVIEW OF LEARNING OBJECTIVES WITH KEY TERMS

● **Define *population* and *population ecology*.**

A **population** is a group of individuals of the same species that live in the same geographic area at the same time. **Population ecology** is the branch of biology that deals with the numbers of a particular species found in an area and how and why those numbers change (or remain fixed) over time.

● **Define *growth rate* (r) and explain the four factors that produce changes in population size.**

Growth rate (*r*) is the rate of change of a population's size, expressed in % per year. On a global scale (when **dispersal** is not a factor), growth rate (*r*) is due to **birth rate** (*b*) and **death rate** (*d*): $r = b - d$. **Emigration** (*e*), the number of individuals leaving an area, and **immigration** (*i*), the number of individuals entering an area, affect a local population's size and growth rate. For a local population (where dispersal is a factor), $r = (b - d) + (i - e)$.

● **Use intrinsic rate of increase, exponential population growth, and carrying capacity to explain the differences between J-shaped and S-shaped growth curves.**

Intrinsic rate of increase is the exponential growth of a population that occurs under ideal conditions. **Exponential population growth** is the accelerating population growth that occurs when optimal conditions allow a constant reproductive rate over a period of time. The **carrying capacity (*K*)** is the maximum number of individuals of a given species that a particular environment can support for an indefinite period, assuming there are no changes in the environment. Although populations with a constant reproductive rate exhibit exponential population growth for limited periods (the J curve), eventually the growth rate decreases to around zero or becomes negative. The S curve shows an initial lag phase (when the population is small), followed by an exponential phase, followed by a leveling phase as the carrying capacity of the environment is reached. The S curve is an oversimplification of how most populations change over time.

● **Distinguish between density-dependent and density-independent factors that affect population size and give examples of each.**

A **density-dependent factor** is an environmental factor whose effects on a population change as population density changes. Predation, disease, and competition are examples. A **density-independent factor** is an environmental factor that affects the size of a population but is not influenced by changes in population density. Hurricanes and fires are examples.

● **Describe some of the density-dependent factors that may affect boom-or-bust population cycles.**

Cyclic population oscillations are poorly understood but may involve density-dependent factors such as predation, competition for food, and parasites. Examples of animals with such boom-or-bust cycles include lemmings, snowshoe hares, and red grouse.

● **Define survivorship and describe type I, type II, and type III survivorship curves.**

Survivorship is the probability that a given individual in a population will survive to a particular age. There are three general types of survivorship curves. In type I survivorship, death is greatest in old age. In type III survivorship, death is greatest among the young. In type II survivorship, death is spread evenly across all age groups.

● **Define *demography* and summarize the history of human population growth.**

Demography is the applied branch of sociology that deals with population statistics and provides information on the populations of various countries or groups of people. It took thousands of years for the human population to reach 1 billion (around 1800). It took 130 years to reach 2 billion (in 1930), 30 years to reach 3 billion (in 1960), 15 years to reach 4 billion (in 1975), 12 years to reach 5 billion (in 1987), and 12 years to reach 6 billion (in 1999). Although our numbers continue to increase, the world growth rate (*r*) declined from a peak of 2.2% per year in the mid-1960s to 1.3% per year in 2004.

● **Identify Thomas Malthus, relate his ideas on human population growth, and explain why he may or may not be correct.**

Thomas Malthus, a 19th-century British economist, pointed out that the human population can increase faster than its food supply, resulting in famine, disease, and war. Some people think Malthus was wrong because our population has grown to more than 6 billion; scientific advances have allowed food production to keep pace with population growth. But Malthus may ultimately be correct because we do not know if our increase in food production is sustainable.

● **Explain why it is impossible to answer precisely how many people Earth can support—that is, Earth's carrying capacity for humans.**

Estimates of how many people Earth can support vary widely depending on what assumptions are made about standard of living, resource consumption, technological innovations, and waste generation. If we want all people to have a high level of material well-being equivalent to the lifestyles common in highly developed countries, then Earth will clearly support far fewer humans than if everyone lives just above the subsistence level.

● **Explain how highly developed and developing countries differ in population characteristics such as infant mortality rate, total fertility rate, and age structure.**

Infant mortality rate is the number of infant deaths (under age 1) per 1000 live births. **Total fertility rate** is the average number of children born to each woman. **Age structure** is the number and proportion of people at each age in a population. Highly developed countries have the lowest infant mortality rates, lowest total fertility rates, and oldest age structure. Developing countries have the highest infant mortality rates, highest total fertility rates, and youngest age structure.

● **Explain how population growth momentum works.**

Population growth momentum is the potential for future increases or decreases in a population based on the present age structure. A country can have replacement-level fertility and still experience population growth if the largest percentage of the population is in the prereproductive years; when all these children mature, they will become the parents of the next generation, and this group of parents will be larger than the previous group.

● **Briefly describe the Immigration Reform and Control Act (IRCA).**

Immigration has a greater effect on population size in the United States than in many other countries. The **Immigration Reform and Control Act (IRCA)** is the basic immigration law in effect in the United States. It gives three groups priority when migrating to the United States: those with family members living in the United States, those who can fill vacant jobs, and refugees seeking asylum.

Thinking About the Environment

1. What are some of the factors that have contributed to the huge increase in the incidence of AIDS in sub-Saharan Africa?

2. What is population density?

3. How are these factors related in determining the growth rate: birth rate, death rate, immigration, and emigration?

4. If a population (or bank account) is growing by the same percentage every year, what type of growth does it exhibit?

5. Draw a graph to represent the long-term growth of a population of bacteria cultured in a test tube containing a nutrient medium that is replenished. Now draw a graph to represent the growth of bacteria in a test tube when the nutrient medium is not replenished. Explain the difference.

6. What is environmental resistance?

7. In the Caribbean study involving spiders and lizards, was density dependence or density independence more significant in determining spider populations? Explain your answer.

8. Describe the general interactions between wolf and moose populations on Isle Royale.

9. Give an example of a *K*-selected species. What features are characteristic of *K*-selected species?

10. How do survivorship curves relate to *r* selection and *K* selection in animals?

11. Explain how the spread of human diseases such as the Black Death, tuberculosis, and AIDS is related to high population densities found in urban environments. Are such diseases density dependent or density independent?

12. What is zero population growth? When do population experts think the human population will have zero population growth?

13. What is infant mortality rate? Which group of countries has the highest infant mortality rates? Which group has the lowest?

14. What is total fertility rate? Which group of countries has the highest total fertility rates? Which group has the lowest?

15. Examine Figure 8.11, which shows the birth and death rates of Mexico during the 20th century. Now compare it to Figure 8.14, which shows the demographic transition. What stage is Mexico in? Is Mexico at the beginning, middle, or end of this demographic stage?

16. Russia and Japan are entering a period of negative population growth momentum. Explain what this means.

17. Which population is more likely to have a positive population growth momentum, one with a young age structure or one with an old age structure? Which is more likely to have a negative population growth momentum? Explain your answers.

18. Do you think the United States should change its immigration policy? Why or why not?

19. Should the average citizen in the United States be concerned about the rapid increase in world population? Why or why not?

20. Explain how the loss of a predator from a small island will affect not only the population of the prey species, but also of the other species in that system.

Quantitative questions relating to this chapter are on our Web site.

WORLD POPULATION DATA SHEET ASSIGNMENT

Use the Population Reference Bureau's *World Population Data Sheet* from inside the back cover to answer the following questions.

1. What is the estimated world population in billions?

2. What is the current birth rate for the world?

3. Which continent has the highest birth rate, and what is this rate?

4. Which country or countries has/have the highest birth rate, and what is this rate?

5. Which continent has the lowest birth rate, and what is this rate?

6. Which country or countries has/have the lowest birth rate, and what is this rate?

7. What is the current death rate for the world?

8. Which country or countries has/have the highest death rate, and what is this rate?

9. Which country or countries does/do not have a positive natural increase (i.e., is/are either stable or decreasing in population)?

10. Which country or countries has/have the highest infant mortality rate, and what is this rate?

11. Which country or countries has/have the lowest infant mortality rate, and what is this rate?

12. Which country or countries has/have the highest total fertility rate, and what is this rate?

13. Which country or countries has/have the lowest total fertility rate, and what is this rate?

14. Which country or countries has/have the highest percentage of the population under age 15, and what is this percentage?

15. Which country or countries has/have the highest percentage of the population over age 65, and what is this percentage?

16. Which country or countries has/have the longest life expectancy, and what is this figure?

17. Which country or countries has/have the shortest life expectancy, and what is this figure?

18. Which country or countries has/have the highest percentage of the population living in urban areas, and what is this percentage?

19. Which country or countries has/have the highest per capita GNI PPP, and what is this figure?

20. Which country or countries has/have the lowest per-capita GNI PPP, and what is this figure?

Take a Stand

Visit our Web site at http://www.wiley.com/college/raven (select Chapter 8 from the Table of Contents) for links to more information about the controversies surrounding U.S. immigration policies. Consider the opposing views of supporters and opponents of current immigration policies, and debate the issues with your classmates. You will find tools to help you organize your research, analyze the data, think critically about the issues, and construct a well-considered argument. *Take a Stand* activities can be done individually or as a team,
as oral presentations, written exercises, or Web-based (e-mail) assignments.

Additional online materials relating to this chapter, including a Student Testing Section with study aids and self-tests, Environmental News, Activity Links, Environmental Investigations, and more, are also on our Web site.

Addressing Population Issues

Family planning. Women at the Dakalaia Health Clinic in Egypt learn about family planning and birth control. (Donna DeCesare)

Like many developing countries, Egypt has experienced an enormous population increase in recent decades. Its population of 75.4 million has more than tripled since 1950, and because most of the land in Egypt is uninhabitable desert, the majority of Egyptians live in crowded strips along the Nile River. Yet, despite its population growth, Egypt's **total fertility rate** (TFR), which is the average number of children born to each woman, declined to 3.1 in 2006, down from 7.0 during the 1960s.

Egypt's successful national family planning program, established in 1965, became particularly effective in the late 1980s when its administrators decided to communicate through religion and the media. (At that time, Egypt's TFR was 5.3.) Teams of religious leaders and healthcare workers started holding neighborhood meetings to educate people about birth control. Today about 57% of married Egyptian women use modern methods of contraception (see photograph).

Traditionally, Islamic leaders opposed any form of birth control as an impediment against God's wish to create life. However, some Islamic scholars have reinterpreted the Prophet Mohammed's words as sanctioning birth control because it contributes to the social welfare of Muslims.

Mass media entered the campaign in the 1990s with a popular television dramatic series, *And the Nile Flows On,* which promoted the desirability of marrying at a later age and the benefits of small families. The series also addressed other culturally sensitive topics, such as women's rights, childhood marriages and early pregnancy, village morality, and religious support for family planning. Mass media reach large audiences and have the potential to alter human behavior. The Egyptian government conducted a survey after the series ran and found that almost 60% of viewers who saw the series said they would definitely use the services of a family planning clinic versus 29% of viewers who did not see the series.

As you can see, the communications media have contributed to a changing cultural climate in Egypt by challenging long-held assumptions about reproduction and the role of women. Thus, it is not surprising that the national education system in Egypt provides equal educational opportunities to both males and females. (Many developing countries do not provide equal educational opportunities for young women.) Families, particularly those living in Cairo and other urban areas, encourage their daughters to delay marriage and childbirth until they have completed their education, including advanced studies such as medical school.

St. Lucia, a small island nation in the Caribbean, and Kenya in East Africa are examples of other developing countries with radio and television series that educate people about family planning, contraceptive use, and sexually transmitted diseases such as acquired immune deficiency syndrome (AIDS). Partly in response to such programs, St. Lucia's TFR declined from 3.8 in 1990 to 2.2 in 2006, whereas Kenya's TFR declined from 6.7 in 1990 to 4.9 in 2006.

Many other countries are experiencing fertility declines, but we are far from achieving population stabilization. In this chapter you will examine how rapid population growth aggravates chronic hunger, underdevelopment, and poverty. You will consider factors that influence the total fertility rate, including cultural traditions, the social and economic status of women, and the availability of family planning services. If population growth is slowed, the world will be in a better position to tackle many of its other serious social and environmental problems.

World View and Closer to You...
Real news footage relating to population issues around the world and in your own geographic region is available at www.wiley.com/college/raven by clicking on "World View and Closer to You."

Population and Quality of Life

LEARNING OBJECTIVES

- Relate carrying capacity to agricultural productivity.
- Define *food insecurity* and relate human population to chronic hunger.
- Describe the relationship between economic development and population growth.

Most people would agree that all people in all countries should have access to the basic requirements of life: a balanced diet, clean water, decent shelter, and adequate clothing. As the 21st century proceeds, it will become increasingly difficult to meet these basic needs, especially in countries that have not achieved population stabilization. About 81% of the world's population lives in less developed countries. If their rate of population growth continues, many of these countries will double their populations by the year 2050.

Many demographers consider rapid population growth in the poorest countries of the world to be the most urgent global population problem. It is likely that the social, political, and economic problems resulting from continued population growth in these countries will affect other countries that have already achieved stabilized populations and high standards of living (see Chapter 8 for a discussion of immigrants and refugees). For these reasons, population growth is of concern to the entire world community, regardless of where it is occurring.

As our numbers increase during the 21st century, environmental degradation, hunger, persistent poverty, economic stagnation, urban deterioration, and health issues will continue to challenge us (**Figure 9.1**). Already, the need for food for the increasing numbers of people living in environmentally fragile arid lands, such as parts of sub-Saharan Africa, has led to overuse of the land for grazing and crop production. (*Sub-Saharan Africa* refers to all African countries located south of the Sahara Desert.) When land overuse occurs in combination with an extended drought, these formerly productive lands may decline in agricultural productivity—that is, their **carrying capacity** may decrease. Although it is possible to reclaim such arid lands, the large number of people and their animal herds trying to live off the land make reclamation efforts difficult. (Carrying capacity as it relates to environmental resistance was discussed in Chapter 8.)

When the carrying capacity of the environment is reached, the population may stabilize or crash because of a decrease in the birth rate, an increase in the death rate, or a combination of both. No one knows if Earth can sustainably support the 9.1 billion people that the United Nations projects for 2050 (see Figure 8.12). It is not even clear that Earth can sustainably support the 6.6 billion people we currently have. We cannot quantify the carrying capacity for humans in any meaningful way, in part because our impact on natural resources and the environment involves more than the number of humans.

To estimate carrying capacity for humans, we must make certain assumptions about our *quality* of life. Do we assume that everyone in the world should have the same standard of living as average U.S. citizens currently do? If so, then Earth would support a fraction of the humans it could support if everyone in the world had only the barest minimum of food, clothing, and shelter. Also, we do not know what future technology might be developed that would completely alter Earth's sustainable population size. We may have already reached or overextended our carrying capacity, and the numerous environmental problems we are experiencing will cause the world population increase to come to a halt or even decline precipitously.

On a national level, developing countries have the largest rates of population increase and often have the fewest resources to support their growing numbers. If a

carrying capacity: The maximum number of individuals of a given species that a particular environment can support for an indefinite period, assuming there are no changes in the environment.

Figure 9.1 Poverty and environmental degradation in the developing world. These children, who are too poor to attend school, collect waste for recycling. Photographed in Manila Bay, Philippines. (Hartmmut Schwarzback/UNEP/Still Pictures/Peter Arnold, Inc.)

country is to support its human population, it must have either the agricultural land to raise enough food for those people or enough of other natural resources, such as minerals or oil, to provide buying power to purchase food.

Population and Chronic Hunger

Food security is the condition in which people do not live in hunger or fear of starvation. Many of the world's people—more than 800 million—do not have food security. These people do not get enough food to thrive; in certain areas of the world, people, especially children, still starve to death (**Figure 9.2**). According to the U.N. Food and Agriculture Organization (FAO), 86 countries are considered low income and food deficient. South Asia and sub-Saharan Africa are the two regions of the world with the greatest **food insecurity** (**Figure 9.3**). People with food insecurity always live under the threat of starvation. Worldwide, the FAO estimates that as many as 2 billion people face food insecurity intermittently as a result of poverty, drought, or civil strife.

Most of the millions of people who die each year from hunger are not victims of *famine*. Famines are usually attributed to bad weather (droughts or floods), insect outbreaks, armed conflict (which causes a breakdown in social and political institutions), or some other disaster. Famines, which generally receive widespread press coverage, account for between 5% and 10% of the world's hungry people in a given year. The remaining 90% to 95%, which represent people with chronic hunger, receive little attention in the world news.

The effects of chronic hunger are insidious. Chronic hunger saps a person's strength and weakens the immune system, leading to greater vulnerability to illness and disease. For example, diarrhea, acute respiratory illness, malaria, and measles are the four biggest killers of children. Studies indicate that each of these illnesses is far more likely to kill children who are malnourished than those who are well nourished.

The cause of chronic hunger is not simply the failure of food production to keep pace with population growth. Agriculture currently produces enough food for everyone to have enough if it were distributed evenly. Experts agree that chronic hunger, population, poverty, and environmental problems are interrelated, but they have reached no consensus as to the most effective way to stop chronic hunger.

food insecurity: The condition in which people live with chronic hunger and malnutrition.

Figure 9.2 **Hunger.** A young Sudanese boy collapses from hunger at a food distribution center run by a medical charity. Overpopulation is not the only reason there is not enough food for all the world's peoples. In Sudan, the underlying cause is continuing civil strife. Photographed in 2005. (Antony Njuguna/NewsCom)

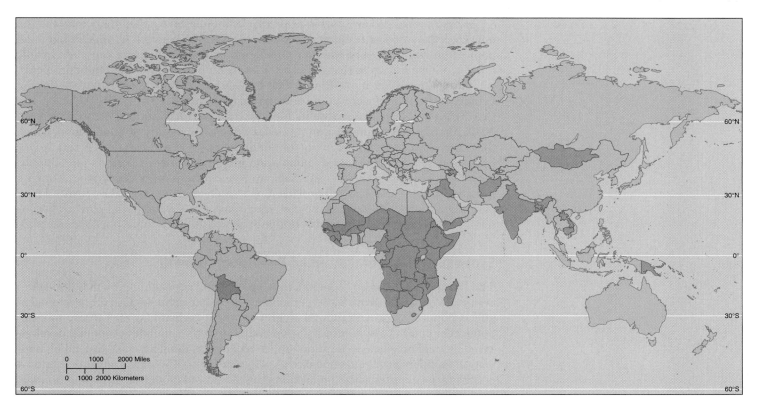

Figure 9.3 **The correlation between hunger and child mortality.** Chronic hunger and malnutrition, which are concentrated in sub-Saharan Africa and south Asia, are linked to an increased mortality for infants and children. In the countries shaded orange, more than 20% of the population are undernourished, and 75 or more of every 1000 children die before the age of five. (Based on U.N. FAO)

Those who assert that population growth is the root cause of the world's food problem point out that countries with some of the highest total fertility rates also have the greatest food shortages. Some of the more extreme members of this group argue that it is imperative to reduce population growth, even through drastic measures such as establishing world population quotas. Under such a system, a country that exceeded its assigned population size would not be eligible for relief from the international community during times of food shortages.

Many politicians and economists think the best way to tackle world food problems is to promote the **economic development** of countries that do not produce adequate food supplies for their people. They presume that economic development would provide the appropriate technology for the people living in those countries to increase their food production or their food-purchasing ability. Once a country becomes more developed, its total fertility rate should decline, helping to lessen the population problem. (Recall the Chapter 8 discussion of declining fertility rates in highly developed countries.)

A third group of people maintains that neither controlling population growth nor enhancing economic development will solve world food problems. They argue that the inequitable distribution of resources is the primary cause of chronic hunger. According to this view, there are enough resources, land, and technologies to produce food for all humans, but people on the lower end of the socioeconomic scale in many countries do not have access to the resources they need to support themselves. The more than 800 million people who are undernourished have neither land nor money, and so they cannot grow or afford to buy food. (Poverty and hunger are revisited in Chapters 19 and 25.)

economic development: An expansion in a government's economy, viewed by many as the best way to raise the standard of living.

These differing viewpoints indicate that economic development, poverty, and the uneven distribution of resources affect the relationship between chronic hunger and population. Regardless of whether population growth is the main cause of chronic hunger, all experts agree that population pressures exacerbate world food problems.

Consider sub-Saharan Africa, which is plagued with food insecurity and periodic famine. If this region were to continue its current rate of growth, the FAO estimates that Africa will have to increase its food production 300% to feed its population in 2050. Alternatively, many African countries will have to import additional food, which they currently cannot afford to pay for out of their own export earnings. A real danger exists that sub-Saharan Africa's population growth will overwhelm its food supply. Slowing population growth in sub-Saharan countries and other developing countries in Latin America and Asia will provide time to increase food production, expand economic development (providing the purchasing power to buy food), and potentially conserve natural resources (which will not be exploited as much to pay for food imports).

Economic Effects of Continued Population Growth

The relationship between economic development and population growth is difficult to evaluate. Population growth affects economic development *and* economic development affects population growth, but the degree to which each affects the other is unclear. Some economists have argued that population growth stimulates economic development and technological innovation. Other economists think a rapidly expanding population hampers developmental efforts. Most major technological advances are now occurring in countries where population growth is low to moderate, an observation that supports the latter point of view.

The National Research Council[1] examined whether large increases in population are a deterrent to economic development. They took into account the interactions among global problems such as underdevelopment, hunger, poverty, environmental problems, and population growth. While concluding that population stabilization alone will not eliminate other world problems, the panel determined that for most of the developing world, economic development would profit from slower population growth. Population stabilization will not guarantee higher living standards but will probably promote economic development, which in turn will raise the standard of living.

In the 1990s, the U.S. National Academy of Sciences took a stronger stance and issued a statement with the Royal Society of London that if population growth continues at currently predicted levels, much of the world will experience irreversible environmental degradation and continued poverty.

Debt in Developing Countries If a country is to raise its standard of living, its economic growth must be greater than its population growth. If a population doubles every 40 years, then its economic goods and services must more than double during that time. Until recently, many developing nations realized economic growth despite increases in population, largely because of financial assistance in the form of loans from banks and governments of highly developed nations, or from multilateral institutions such as the World Bank and the International Monetary Fund (IMF). This money was used to build roads, power plants, and schools, and to finance other economic development projects.

It is increasingly difficult for many developing countries to continue raising their standards of living, because the tremendous debts they have accumulated, while funding past economic development, preclude future loans. For example, sub-Saharan African countries are overwhelmed by a massive foreign debt, which the World Bank estimates accounts for 70% of the entire region's output in goods and services. This debt burden severely impedes economic development because to ensure there is

---

[1] The National Research Council is a private, nonprofit society of distinguished scholars. It was organized by the National Academy of Sciences to advise the U.S. government on complex issues in science and technology.

enough money for debt repayments, many countries are forced to spend less for basic health and education services.

REVIEW

1. How is human population growth related to chronic hunger?
2. How is human population growth related to economic development?

 ## Reducing the Total Fertility Rate

LEARNING OBJECTIVES

- Define *culture* and explain how total fertility rate and cultural values are related.
- Define *gender inequality* and relate the social and economic status of women to total fertility rate.
- Explain how the availability of family planning services affects total fertility rate.

Dispersal, moving from one place to another, used to be a solution for overpopulation, but it is not today. As a species, we have expanded our range throughout Earth, and few habitable areas remain that have the resources to adequately support a major increase in human population. Nor is increasing the death rate an acceptable means of regulating population size. Clearly, the way to control our expanding population is to reduce the number of births. Cultural traditions, women's social and economic status, and family planning all influence the total fertility rate.

culture: The ideas and customs of a group of people at a given period; culture, which is passed from generation to generation, evolves over time.

Culture and Fertility

The values and norms of a society—what is considered right and important and what is expected of a person—are all part of a society's **culture**. A society's culture exerts a powerful influence over individuals by controlling their behaviors.

Gender is an important part of culture. Different societies have different gender expectations—that is, varying roles men and women are expected to fill. In parts of Latin America men do the agricultural work, whereas in sub-Saharan Africa this is the woman's job. With respect to fertility and culture, a couple is expected to have the number of children determined by the cultural traditions of their society.

High total fertility rates are traditional in many cultures. The motivations for having many babies vary from culture to culture, but overall a major reason for high total fertility rates is high infant and child mortality rates. For a society to endure, it must produce enough children who can survive to reproductive age. If infant and child mortality rates are high, total fertility rates must be high to compensate. Although world infant and child mortality rates are decreasing, it will take longer for culturally embedded fertility levels to decline. Parents must have enough confidence that the children they already have will survive before they stop having additional babies. Another reason for the lag in fertility decline is cultural: Changing anything traditional, including large family size, usually takes a long time.

Higher total fertility rates in some developing countries are due to the important economic and societal roles of children. In some societies, children usually work in family enterprises such as farming or commerce, contributing to the family's livelihood (**Figure 9.4**). When these children become adults, they provide support for their aging parents.

The International Labor Organization (ILO) estimates that, worldwide, about 250 million children between the ages of 5 and 14 work; at least 120 million children work full time. (The ILO does not

Figure 9.4   Children at work. These children are gathering firewood near the Ethiopian village of Meshal. In developing countries, total fertility rates are high partly because children contribute to the family by working. Young children gather firewood, carry water, and work in the fields with their parents. Older children often have wage-producing jobs that add to the family income. (Mike Goldwater/Alamy Images)

count household chores as labor.) Almost all of these children live in developing countries. Many child laborers do hazardous work, such as mining and construction; these child laborers often suffer from chronic health problems caused by the dangerous, unhealthy conditions they are exposed to. Children who work full time do not have childhoods, nor do they receive any education.

In contrast, children in highly developed countries have less value as a source of labor because they attend school and because less human labor is required in an industrialized society. Furthermore, highly developed countries provide many economic and social services for the elderly, so the burden of their care does not fall entirely on their offspring.

Many cultures place a higher value on male children than on female children. In these societies, a woman who bears many sons achieves a high status; the social pressure to have male children keeps the total fertility rate high. For example, the Hindu religion in India traditionally required that parents be buried alongside their son, and having a son still has deep cultural significance in India today.

Religious values are another aspect of culture that affects total fertility rates. Several studies done in the United States point to differences in TFRs among Catholics, Protestants, and Jews. In general, Catholic women have a higher TFR than either Protestant or Jewish women, and women who do not follow any religion have the lowest TFR of all. However, the observed differences in TFRs may not be the result of religious differences alone. Other variables, such as ethnicity (certain religions are associated with particular ethnic groups) and residence (certain religions are associated with urban or with rural living), complicate any generalizations.

The Social and Economic Status of Women

gender inequality: The social construct that results in women not having the same rights, opportunities, or privileges as men.

Gender inequality exists in most societies, although the extent of the gap between men and women varies in different cultures. Gender disparities include the lower political, social, economic, and health status of women compared to men. For example, more women than men live in poverty, particularly in developing countries. In most countries, women are not guaranteed equality in legal rights, education, employment and earnings, or political participation.

Because sons are more highly valued than daughters, girls are often kept at home to work rather than being sent to school. In most developing countries, a higher percentage of women are illiterate than men (**Figure 9.5**). However, definite progress has been made in recent years in increasing literacy in both women and men and in narrowing the gender gap. Fewer young women and men are illiterate than older women and men within a given country.

Fewer women than men attend secondary school (high school). In some African countries only 2% to 5% of girls are enrolled in secondary school. Worldwide, some 90 million girls are not given the opportunity to receive a primary (elementary school) education. Laws, customs, and lack of education often limit women to low-skilled, low-paying jobs. In such societies, marriage is usually the only way for a woman to achieve social influence and economic security.

The single most important factor affecting high total fertility rates may be the low status of women in many societies. A significant way to address population growth, then, is to improve the social and economic status of women (see "Meeting the Challenge: Microcredit Programs"). We will say more about this later, but for now we will examine how marriage age and educational opportunities, especially for women, affect fertility.

Marriage Age and Fertility The total fertility rate is affected by the average age women marry, which is determined by the laws and customs of the society in which they live. Women who marry are more apt to bear children than women who do not marry, and the earlier a woman marries, the more children she is likely to have.

The percentage of women who marry and the average age at marriage vary widely among different societies, but there is generally a correlation between marriage age

Figure 9.5 Percent illiteracy of men and women in selected developing countries, 2002. A higher percentage of women than men are illiterate. (World Resources)

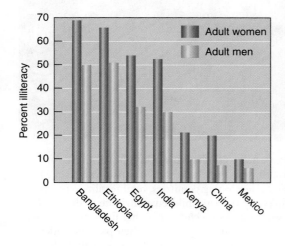

MEETING THE CHALLENGE

Microcredit Programs

"The poor stay poor, not because they're lazy but because they have no access to capital." This statement by Peruvian economist **Hernando de Soto** provides the philosophical basis for **microcredit**, financial services used to empower women and attack poverty. Microcredit consists of small loans (*microloans* of $50 to $500) extended to very poor people to help them establish self-employment projects that generate income. These people lack the assets to qualify for loans from commercial banks.

Microcredit is a flexible program that works at the grassroots level, based on local conditions and needs. The poor have used these loans for a variety of projects. Some have purchased used sewing machines to make clothing faster than sewing by hand. Others have opened small grocery stores after purchasing used refrigerators to store food so that it does not spoil. Still others have used the money to bake bread, weave mats, or raise chickens (see photograph).

The **Foundation for International Community Assistance (FINCA)** is a nonprofit agency that administers a global network of microcredit banks. FINCA uses *village banking*, in which a group of very poor neighbors guarantees one another's loans, administers group lending and savings activities, and provides mutual support. Thus, village banks give autonomy to local people.

FINCA primarily targets women, and for good reason, because an estimated 70% of the world's poorest people are women. Also, the number of households in which women provide the sole support for their children is growing worldwide. FINCA believes that the best way to alleviate the effects of poverty and hunger on children is to provide their mothers with a means of self-employment. A woman's status

in the community is raised as she begins earning income from her business.

The loans help people rise above poverty because they can generate more income and even save money for the future. Many women who borrow from FINCA spend their earnings to improve the nutrition and health of their children. For these women the next priority after improved nutrition and health is education for their children.

Microcredit was first tried in Bangladesh in the 1970s where it has helped thousands of urban and rural Bangladeshis start successful businesses. No collateral is required for these loans; the only requirement for a loan is that a person is poor. **Muhammad Yunus,** the Bangladeshi who founded the *Grameen Bank* and pioneered microcredit, was awarded a Nobel Peace Prize in 2006.

FINCA currently offers loans in 21 countries around the world, such as Armenia, Ecuador, Guatemala, Kyrzygstan, Malawi, Nicaragua, and Uganda. Other microcredit programs are offered through CARE, Catholic Relief Services, Freedom from Hunger, and Save the Children.

Microcredit is not charity—it is investment. The loans must be paid back with interest. The institutions that provide the loans also offer technical assistance in business, bookkeeping, and savings. Globally, 97% of all microloans are repaid on time; this repayment rate is as good or better than that in most commercial banks. When a loan is repaid, the money becomes available for investment in another starting business. After a person starting a small business pays off the loan, she or he can qualify for another loan if it is needed to expand the business.

Because of the time and labor involved in tracking many small loans, microcredit banks seldom make money; they rely on international aid from governments, corporations, religious organizations, and individuals. However, some banks have become financially self-sufficient.

Microcredit has helped about 92 million of the world's poor so far. But the need remains great: An estimated 200 million more of the world's poor could benefit from microcredit programs.

Microcredit. This Bangladeshi woman feeds chickens at her poultry farm. She received her first microcredit loan to buy a few chickens 30 years ago and has built the farm into a thriving business. Photographed in 2006. (Rafiqur Rahman/Reuters/NewsCom)

and total fertility rate. Consider Sri Lanka and Bangladesh, two developing countries in South Central Asia. In Sri Lanka the average age at marriage is 25, and the average number of children born per woman is 2.0, which means it will take 54 years for the population to double (at its current rate of natural increase). In contrast, in Bangladesh the average age at marriage is 17, the average number of children born per woman is 3.3, and the doubling time is 37 years. The difference in TFRs and doubling times for these two countries is not greater because 47% of Bangladeshi women in the 15- to 49-year age group now use modern contraceptive methods; this percentage is up from 36% just 10 years ago.

Figure 9.6 **Education and fertility.**

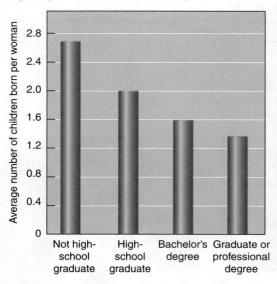

(a) The amount of education a woman receives affects the total number of children she has (TFR). The graph shows TFRs for 35- to 44-year-old women in the United States by level of education. (U.S. Census Bureau)

(b) Teen mothers gather during lunch at their high school in Nebraska. Because these young women are continuing their education, they will likely have fewer children than if they had dropped out of school. (Karen Kasmauski/National Geographic Society)

Educational Opportunities and Fertility   In nearly all societies, women with more education tend to marry later and have fewer children. **Figure 9.6** shows the total fertility rates of women in the United States with different education levels. Providing women with educational opportunities delays their first childbirth, thereby reducing the number of childbearing years and increasing the amount of time between generations. Education opens the door to greater career opportunities and may change women's lifetime aspirations. In the United States, it is not uncommon for a woman to give birth to her first child in her thirties or forties, after she has established a career.

Studies in dozens of countries show a strong correlation between the average amount of education women receive and the total fertility rate. For example, women in Botswana with a secondary education have an average of 3.1 children each, women with a primary education have 5.1 children each, and women with no formal education have 5.9 children each.

Education increases the probability that women will know how to control their fertility, and it provides them with knowledge to improve the health of their families, which results in a decrease in infant and child mortality. A study in Kenya showed that 10.9% of children born to women with no education died by age 5, as compared with 7.2% of children born to women with a primary education and 6.4% of children born to women with a secondary education. Education increases women's options, providing ways of achieving status besides having babies.

Education may have an indirect effect on total fertility rate as well. Children who are educated have a greater chance of improving their living standards, partly because they have more employment opportunities. Parents who recognize this may be more willing to invest in the education of a few children than in the birth of many children whom they cannot afford to educate. The ability of better-educated people to earn more money may be one of the reasons smaller family size is associated with increased family income.

Family Planning Services

family planning services: Services that enable men and women to limit family size, safeguard individual health rights, and improve the quality of life for themselves and their children.

Socioeconomic factors may encourage people to want smaller families, but reduction in fertility will not become a reality without the availability of **family planning services**. Traditionally, family planning services have focused on maternal and child health, including prenatal care to help prevent infant and maternal death or disability. However, because of gender inequality and cultural constraints, many women cannot ensure their own reproductive health without the support of their partners. Polls of women in developing countries reveal that many who say they do not want additional children still fail to practice any form of birth control. When asked why they do not use birth control, these women frequently respond that their husbands or in-laws want additional children. Thus, in many countries men make reproductive decisions such as whether or not to use contraceptives, whether or not to have a child, and even whether or not a woman can get health care. Increasingly, family planning services are offering reproductive services to men, such as information on sexuality, contraception, sexually transmitted diseases, and parenting.

The governments of most countries recognize the importance of educating people about basic maternal and child health care. Developing countries that have had success in significantly lowering total fertility rates credit many of these results to effective family planning programs. Prenatal care and proper birth spacing make women healthier. In turn, healthier women give birth to healthier babies, leading to fewer infant deaths. The percentage of women using family planning services to limit family size has increased from less than 10% in the 1960s to more than 50% today. However, the actual *number* of women who are not using family planning has increased owing to population growth.

Family planning services provide information on reproductive physiology and contraceptives, as well as the actual contraceptive devices, to those who wish to control the number of children they produce or to space their children's births. Family planning programs are most effective when they are designed with sensitivity to local social and cultural beliefs. Family planning services do not try to force people to limit their family sizes but rather attempt to convince people that small families (and the contraceptives that promote small families) are acceptable and desirable (**Figure 9.7a**).

Contraceptive use is strongly linked to lower total fertility rates (**Figure 9.7b**). In highly developed countries, where total fertility rates are at replacement levels or lower, the percentage of married women of reproductive age who use contraceptives is often greater than 70%. Fertility declines have occurred in developing countries where contraceptives are readily available. For example, research has shown that 90% of the decrease in fertility in 31 developing countries was a direct result of the increased knowledge and availability of contraceptives. Since the 1970s, use of contraceptives in East Asia and many areas of Latin America has increased significantly, and these regions have experienced a corresponding decline in birth rate. In areas where contraceptive use remained low, such as parts of Africa, there was little or no decline in birth rate.

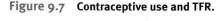

Figure 9.7 Contraceptive use and TFR.

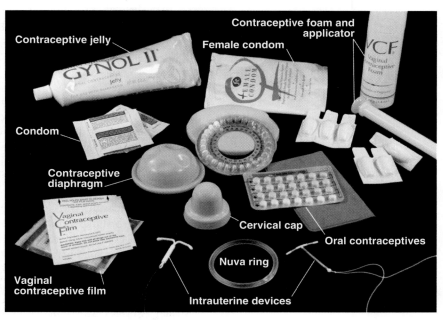

(a) Some commonly used contraceptives. (Custom Medical Stock Photo, Inc.)

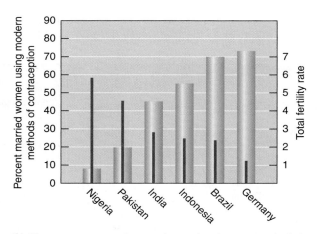

(b) Greater contraceptive use (green bars) among married women of reproductive age correlates with a lower fertility rate (red bars). Data are for 2006. (Population Reference Bureau)

REVIEW

1. How are high TFRs related to the economic roles of children in certain cultures?
2. What appears to be the single most important factor affecting high TFRs?
3. How do health and family planning services influence TFR?

Government Policies and Fertility

LEARNING OBJECTIVES

- Compare how the governments of China, India, Mexico, and Nigeria have tried to slow human population growth.
- Define *population growth momentum* and explain its role in Mexico's and Nigeria's population growth.
- Describe population concerns in Europe, including the views of pronatalists.
- Describe at least four of the Millennium Development Goals that came out of the U.N. Millennium Summit.

The involvement of governments in childbearing and child rearing is well established. Laws determine the minimum age people may marry and the amount of compulsory education. Governments may allot portions of their budgets to family planning services, education, health care, old-age security, or incentives for smaller or larger family size. The tax structure, including additional charges or allowances based on family size, influences fertility.

In recent years, the governments of at least 78 developing countries in Africa, Asia, Latin America, and the Caribbean have recognized they must limit population growth. They have formulated policies, such as economic rewards and penalties, to achieve this goal. Most countries sponsor family planning projects, many of which are integrated with health care, education, economic development, and efforts to improve women's status. The U.N. Fund for Population Activities supports many of these activities.

Many developing countries have initiated population control measures. Let us now examine those measures in China, India (the world's two most populous nations), Nigeria, and Mexico. We then examine population concerns in Europe as an interesting comparison.

China's Controversial Family Planning Policy

China, with a mid-2006 population of 1.3 billion, has the largest population in the world, and its population exceeds the combined populations of all highly developed countries. Recognizing that its rate of population growth had to decrease or the quality of life for everyone in China would be compromised, the Chinese government in 1971 began to pursue birth control seriously. It urged couples to marry later, increase spacing between children, and limit the number of children to two.

In 1979, China instigated an aggressive plan to push China into the third demographic stage. (As discussed in Chapter 8, the third demographic stage is marked by a decline in the birth rate along with a relatively low death rate.) Announcements were made for incentives to promote later marriages and one-child families (**Figure 9.8**). Local jurisdictions were assigned the task of reaching this goal. A couple who signed a pledge to limit themselves to a single child might be eligible for such incentives as medical care and schooling for the child, cash bonuses, preferential housing, and retirement funds. Penalties were instituted, including fines and the surrender of all of these privileges if a second child was born.

China's aggressive plan brought about the most rapid and drastic reduction in fertility in the world, from 5.8 births per woman in 1970 to 2.1 births per woman in 1981. However, it was controversial and unpopular because it compromised individual freedom of choice. In some instances, social pressures from the community induced women who were pregnant with a second child to get an abortion.

Figure 9.8 China's one-child family policy. A billboard campaign in China promotes the one-child family. Note that the child is a male rather than a female. More recent ads have featured happy parents with a single female child to subtly counteract the traditional preference for boys. (E. F. Anderson/Visuals Unlimited)

Moreover, based on the disproportionate number of male versus female babies reported born in recent years, it is suspected that many expectant parents abort the female fetus. In addition, hundreds of thousands of newborn baby girls were killed or abandoned by parents who, required to conform to the one-baby policy, wanted a boy. In China, sons carry on the family name and traditionally provide old-age security for their parents. Thus, sons are valued more highly than daughters are. Demographers project that by the middle of the 21st century, marriageable males will outnumber marriageable females by 1 million.

The one-child family policy is now relaxed in rural China, where 63% of all Chinese live. China's recent population control program has relied on education, publicity campaigns, and fewer penalties to achieve its goals. China trains population specialists at institutions such as the Nanjing College for Family Planning Administrators. In addition, thousands of secondary school teachers integrate population education into the curriculum. China's TFR in 2006 was 1.6, and the Chinese government continues its policy of strict population control.

India's Severe Population Pressures

India is the world's second most populous nation, with a mid-2006 population of 1.1 billion. In the 1950s, it became the first country to establish government-sponsored family planning. Unlike China, India did not experience immediate results from its efforts to control population growth, in part because of the diverse cultures, religions, and customs in different regions of the country. Indians speak 14 main languages and more than 700 other languages and dialects, which makes communicating a broad program of family planning education difficult. Like China's, the Indian culture is biased toward male children. Husbands often desert wives who do not produce sons. Custom dictates that when a husband dies, the sons rather than the daughters support their mother. These and other factors promote a "large family" psychology in Indian society.

In 1976, the Indian government became more aggressive. It introduced incentives to control population growth and controversial programs of compulsory sterilization in several states. If a man had three or more living children, he was compelled to obtain a vasectomy. Compulsory sterilization was a failure; it had little effect on the birth rate and was exceedingly unpopular. Even after compulsory sterilization was abandoned, population experts continued to criticize India's family planning policy because it still put too much emphasis on sterilizations, which are almost exclusively used by mature couples who have already had several children and are seeking a long-term birth control procedure.

In recent years, India's family planning policies have taken a different approach—to educate young couples, who must delay childbirth and space their young if the TFR is going to reach its target of 2.1. India has attempted to integrate economic development and family planning projects. Adult literacy and population education programs have been combined. Multimedia advertisements and education promote voluntary birth control, and contraceptives are more available. India has emphasized improving health services to lower infant and child mortality rates and reduce the rate of female infanticide. These changes have had an effect: India's TFR declined from 5.3 in 1980 to 2.9 in 2006.

Despite these relatively modest gains, India's total fertility rate remains above replacement level, and the nation is facing severe problems. Population pressure has caused the deterioration of India's environment in the past few decades, and 80% of Indians live below the official poverty level (less than $2 a day). Population experts predict that India's population will soon exceed China's. These numbers will exacerbate India's poverty, environmental degradation, and economic underdevelopment.

Mexico's Young Age Structure

Mexico, with a mid-2006 population of 108.3 million, is the second most populous nation in Latin America. (Brazil, with a population of 186.8 million, is the most populous.) Mexico has a tremendous potential for growth because 32% of its population is

population growth momentum: The potential for future increases or decreases in a population based on the present age structure.

less than 15 years of age. Even with a low birth rate, positive **population growth momentum** will cause the population to continue to increase in the future because of the large number of young women having babies.

Traditionally, the Mexican government supported rapid population growth, but in the late 1960s the government became alarmed at how rapidly the population was increasing. In 1974, the Mexican government instigated several measures to reduce population growth, such as educational reform, family planning, and health care. The time was right for such changes, as many Mexican women were already ignoring the decrees of the pro-growth government and the Roman Catholic Church and purchasing contraceptives on the black market. Mexico has had great success in reducing its fertility level, from 6.7 births per woman in 1970 to 2.4 births per woman in 2006.

Mexico's goal includes both population stabilization and balanced regional development. Its urban population makes up 75% of its total population, and most of these people live in Mexico City. Although Mexico is largely urbanized compared with other developing countries, its urban-based industrial economy has been unable to absorb the large number of people in the workforce. (According to the Mexico office of the International Labor Organization, about 1.3 million new workers join Mexico's labor force each year.) Unemployment in Mexico is high, and many Mexicans have migrated, both legally and illegally, to the United States.

Mexico's recent efforts at population control include multimedia campaigns. As in Egypt, popular television and radio shows carry family planning messages such as "Small families live better." Booklets on family planning are distributed, and population education is being integrated into the public school curriculum. Social workers receive training in family planning as part of their education.

The Population Challenge in Nigeria

Nigeria is part of the sub-Saharan region with the world's most rapid population growth. It has the largest population of any African country: In mid-2006, its population was 134.5 million. Nigeria's TFR today (5.9) is basically what it was in the early 1980s—6.0 births per woman. Nigeria has a huge reproductive potential because 43% of the population is less than 15 years of age. Today only about 8% of married women use a modern method of contraception. Furthermore, less than half of all Nigerian women in their reproductive years (ages 15 to 49) have any knowledge of contraception. The average life expectancy in Nigeria is 44 years, which is low in part because of high infant and child mortality rates.

The Nigerian government recognizes that it is more likely to attain its economic goals if its rate of population growth decreases. Nigeria has developed a national population policy that integrates population and economic development projects. The plan involves improving health care, including training nurses and other healthcare professionals. Population education is being used to encourage later marriages (currently, half of all women are married by age 17) and birth spacing.

Population Concerns in Europe

Population has stabilized in Europe, where most countries have lower-than-replacement-level total fertility rates; several countries have even experienced a slight decline in population. As a result of fewer births, the proportion of elderly people in the European population is increasing (see the section on aging populations in Chapter 8). Population control has become a controversial issue in Europe, with two opposing viewpoints emerging.

pronatalist: A person who is in favor of population growth.

Pronatalists think that declining birth rates threaten the vitality of their region. They are concerned that the decrease in population might result in a loss of economic growth. (In contrast, a *neo-Malthusian* maintains that a rapidly expanding population hampers economic growth.) Pronatalists presume that their countries' positions in the world will weaken and that immigrants from non-European countries will dilute their cultural identity. (Declining birth rates and increased immigration to Europe have resulted in increased diversity. The current wave of immigrants to Europe is primarily from North Africa and the Middle East.) Pronatalists are also concerned that the large number of European elderly will overwhelm pension and old age security systems unless a larger workforce is available to contribute to those systems. The most outspoken pronatalists assert that women should marry young and have many children for the good of society. Pronatalists favor government policies that provide incentives for larger families (such as paid maternity and paternity leaves, easily available child care, and monetary "baby bonuses") and penalties for smaller families.

The opponents of pronatalists do not view European society as declining. Instead, they contend that European influence is increasing because they judge power in economic rather than population terms. Generally, they are not as opposed to immigration, and they maintain that Europe should not be making a concerted effort to increase its total fertility rate when overpopulation is such a serious problem in much of the world. Furthermore, they point out that technological innovations have eliminated many jobs in Europe, and the consequent unemployment would only be made worse by an increase in the labor force caused by a rise in birth rate. Opponents of pronatalists question whether the elderly are a burden to society (**Figure 9.9**). They argue that the elderly are not the only ones the workforce supports; the cost of providing for the young, especially for many years of education, cancels out any perceived benefit from an increased birth rate. Further, they point out that an increase in the birth rate now will lead to an increase in the number of elderly in a few decades.

Figure 9.9 Elderly Europeans on a bicycle excursion. Photographed in France. (David R. Frazier Photolibrary, Inc./Alamy Images)

CASE IN POINT

The Millennium Development Goals

In September 2000, 189 heads of state met at the U.N. Millennium Summit to address how to meet the needs of the hundreds of millions of people living in abject poverty. This gathering adopted the U.N. Millennium Declaration, which ultimately committed their countries to a global partnership with a concrete plan of action, known as the **Millennium Development Goals (MDGs)**, with a target date of 2015. The MDGs were assembled by task forces consisting of more than 250 international experts, including scientists, government policymakers, and representatives from nongovernmental organizations (NGOs), the World Bank, and the United Nations. As you will see, the MDGs are directly and indirectly related to the population issues discussed in this chapter. Although these MDGs are lofty, each one is attainable; each has measurable indicators that can be used to monitor our progress.

- **Goal 1: Eradicate extreme hunger and poverty.** One target is to halve the number of people living on an income of less than $1 a day by 2015. A second target is to halve the proportion of people who suffer from hunger.
- **Goal 2: Achieve universal primary education.** All boys and girls, regardless of where they live, should enroll in and complete a primary education (grades 1 through 5).
- **Goal 3: Promote gender equality and empower women.** This goal is also based on education—by 2015, gender disparity should be eliminated in all levels of education.
- **Goal 4: Reduce child mortality.** The target is to reduce the mortality rate of infants and children under five years of age by two-thirds.
- **Goal 5: Improve maternal health.** By 2015, the maternal mortality ratio should be reduced by two-thirds.
- **Goal 6: Combat HIV/AIDS, malaria, and other diseases.** Halt and begin to reduce the spread of HIV/AIDS, malaria, and other important diseases.

E N V I R O N E W S

The Millennium Villages Project

All too often, international aid to impoverished people does not reach them, or the aid is piecemeal and provides a temporary, band-aid approach that quickly loses its effectiveness after the money is spent. The Millennium Villages Project (MVP) is a different approach to fighting poverty. National and local African governments, the Earth Institute at Columbia University, various aid groups, and other participants work with selected communities in rural Africa to help them achieve the Millennium Development Goals. Currently, 12 sub-Saharan villages are participating. Each of these villages, which are located in Kenya, Ethiopia, Ghana, Malawi, Mali, Nigeria, Senegal, Rwanda, Tanzania, and Uganda, represents one of the 12 agro-ecological zones and farming systems in Africa. The villages are all located in areas of food insecurity: At least 20% of all children five years or younger are underweight.

The MVP is unique in that it involves a co-ordinated set of interventions (strategies) in agriculture, nutrition, health, education, energy, water, and the environment. First, a detailed assessment of current conditions is prepared, both to determine what interventions are most needed and to provide a baseline against which future assessments can monitor the effectiveness of the program. The project involves community-led economic development and is based on the premise that rural African communities can improve their standard of living if they are taught and provided with technologies that simultaneously improve agricultural productivity, health, and education. The program provides assistance for at least 5 years, during which time management of improvements are transitioned from the aid groups to the villagers themselves.

- **Goal 7: Ensure environmental sustainability.** This is a multi-part goal that includes reversing the loss of natural resources, halving the number of people without sustainable access to safe drinking water and basic sanitation, and improving the lives of at least 100 million people living in slums (see Chapters 10 and 25).

- **Goal 8: Develop a global partnership for economic development**. The financial aspects of the MDGs are encompassed in this goal. Among the seven targets are providing measures for debt relief for the external debts in developing countries, creating productive jobs for young people, and providing access to affordable medicines in developing countries.

Highly developed countries have committed to provide developing countries with more foreign aid, debt relief, and fair trade. Developing countries have committed to govern better and invest more money in health care and education. Overall, progress toward these goals has been limited, particularly in sub-Saharan Africa. Asian countries have made the most progress toward the MDGs, although hundreds of millions of people there still live in extreme poverty.

These goals will not be reached quickly. As then U.N. Secretary-General **Kofi A. Annan** said, "It takes time to train the teachers, nurses, and engineers; to build the roads, schools, and hospitals; to grow the small and large businesses able to create jobs and income needed." But having lofty goals like these is a beginning that should encourage us all to do our part in improving the human condition during the 21st century. (Chapter 25 is organized around a similar plan of action that has five goals.) ◼

REVIEW

1. What are the successes and failures of China, India, Mexico, and Nigeria in slowing human population growth?

2. What is a pronatalist?

3. What are Millennium Development Goals?

Achieving Population Stabilization

LEARNING OBJECTIVES

- Outline how governments can help achieve global population stabilization.
- Explain how individuals can adopt voluntary simplicity to mitigate the effects of population growth.

In this chapter you have considered how population issues exacerbate global problems, including chronic hunger, poverty, and economic underdevelopment. Population stabilization is critically important if we are to effectively tackle these serious problems. But what kinds of policies will help achieve population stabilization?

Developing countries should increase the amount of money allotted to public health (to reduce infant and child mortality rates) and family planning services, particularly the dissemination of information on affordable, safe, and effective methods of birth control. Governments should take steps to increase the average level of education, especially of women, and women must be given more employment opportunities. National population policies will not work without cultural awareness, respect for individual religious beliefs, and local community involvement and acceptance; community leaders can provide feedback on the effectiveness of programs.

Highly developed countries can help by providing financial support for the U.N. Fund for Population Activities, which supports international family planning efforts, and by supporting research on new birth control methods. Providing funding for economic development projects that increase the incomes of poor people in developing countries would help alleviate the hunger and poverty associated with population growth.

Most important, highly developed nations must face their own population problems, particularly the environmental costs of consumption overpopulation by affluent people (see Chapter 1). Policies should be formulated that support reducing the use of resources, increasing the reuse and recycling of materials, and expunging the throwaway mentality. These policies will show that highly developed countries are serious about the issues of overpopulation at home as well as abroad.

On a personal level, individuals in highly developed countries should examine their own consumptive practices and consider taking steps to reduce them. Individual efforts to reduce material consumption—sometimes called **voluntary simplicity**—are effective for their collective effects and because they may influence additional people to reduce unnecessary consumption. Moving from a lifestyle based on accumulation of wealth and the things money can buy to a lifestyle of voluntary simplicity is called *downshifting*.

Voluntary simplicity is not just living frugally. Instead, it involves desiring less of what money can buy. Each person who scales back his or her consumption benefits the environment (see discussion of ecological footprints in Chapter 1). As a result of needing less money, many people who downshift can afford to work fewer hours. People who advocate voluntary simplicity maintain that their lifestyle gives them less "things" but more time—time to do volunteer work, experience the natural world, and enjoy relationships (**Figure 9.10**).

Before purchasing a product, individuals should ask if it is really needed. A material lifestyle—two or three cars in the garage, steaks on the grill, a television in every room, and similar luxury items—may provide short-term satisfaction. Such a lifestyle is not as fulfilling in the long term as learning about the world, interacting with others in meaningful ways, and contributing to the betterment of family and community. In the end, a rich life is measured not by what was owned but by what was done for others.

Figure 9.10 People picnic and enjoy a free outdoor concert. In voluntary simplicity, quality of life is a better measure of success than wealth and material possessions. Photographed in Madison, Wisconsin. (Zane Williams/Stone/Getty Images)

voluntary simplicity: A way of life that involves wanting and spending less.

REVIEW

1. How can governments in developing countries achieve a decrease in population growth? How can highly developed countries help?

REVIEW OF LEARNING OBJECTIVES WITH KEY TERMS

• **Relate carrying capacity to agricultural productivity.**

Carrying capacity is the maximum number of individuals of a given species that a particular environment can support for an indefinite period, assuming there are no changes in the environment. When land overuse occurs, often in combination with an extended drought, formerly productive lands may decline in agricultural productivity—that is, their carrying capacity may decrease.

• **Define *food insecurity* and relate human population to chronic hunger.**

Food insecurity is the condition in which people live with chronic hunger and malnutrition. The rapid increase in population exacerbates many human problems, including hunger. The countries with the greatest food shortages have some of the highest **total fertility rates (TFRs)**.

• **Describe the relationship between economic development and population growth.**

Economic development is an expansion in a government's economy, viewed by many as the best way to raise the standard of living. Most economists think that slowing population growth promotes economic development.

• **Define *culture* and explain how total fertility rate and cultural values are related.**

Culture comprises the ideas and customs of a group of people at a given period; culture, which is passed from generation to generation, evolves over time. The relationship between TFR and culture is complex. A combination of four factors is primarily responsible for high TFRs: high infant and child mortality rates, the important economic and societal roles of children in some cultures, the low status of women in many societies, and a lack of health and family planning services. Culture influences all of these factors.

• **Define *gender inequality* and relate the social and economic status of women to total fertility rate.**

Gender inequality is the social construct that results in women not having the same rights, opportunities, or privileges as men. The single most important factor affecting high TFRs is the low status of women in many societies.

• **Explain how the availability of family planning services affects total fertility rate.**

Family planning services are services that enable men and women to limit family size, safeguard individual health rights, and improve the quality of life for themselves and their children. TFRs tend to decrease where family planning services are available.

• **Compare how the governments of China, India, Mexico, and Nigeria have tried to slow human population growth.**

The governments of many developing countries are trying to limit population growth. In 1979 China began a coercive, one-child family policy to reduce TFR. The successful but unpopular program is now relaxed in rural China. Education and publicity campaigns are used today. India's government has sponsored family planning since the 1950s.

In 1976, India introduced compulsory sterilization in several states, but this policy failed. Education and publicity campaigns are the focus of its family planning efforts today. Mexico's government has sponsored family planning since 1974. Mexico's TFR continues to drop, largely in response to education, improved health care, economic development, and publicity campaigns. Nigeria put a national population policy into place that includes education, improved health care, and economic development. These efforts have been largely ineffective, and Nigeria's TFR remains almost as high today as it was in the early 1980s.

• **Define *population growth momentum* and explain its role in Mexico's and Nigeria's population growth.**

Population growth momentum is the potential for future increases or decreases in a population based on the present age structure. Even with lower birth rates, Mexico's and Nigeria's positive population growth momentums will cause their populations to increase because of the large number of young women having babies.

• **Describe population concerns in Europe, including the views of pronatalists.**

A **pronatalist** is a person who is in favor of population growth. Pronatalists think the vitality of Europe is at risk because of declining birth rates. Other Europeans contend that European economic influence is increasing and that Europe should not make a concerted effort to increase its total fertility rate when overpopulation is such a serious global problem.

• **Describe at least four of the Millenium Development Goals that came out of the U.N. Millenium Summit.**

The U.N. Millennium Summit formed a global partnership with a plan of action, known as the **Millennium Development Goals (MDGs)**, with a target date of 2015. The goals are to: eradicate extreme hunger and poverty; achieve universal primary education; promote gender equality and empower women; reduce child mortality; improve maternal health; combat HIV/AIDS, malaria, and other diseases; ensure environmental sustainability; and develop a global partnership for economic development.

• **Outline how governments can help achieve global population stabilization.**

Developing countries should increase the amount of money allotted to public health and family planning services. Developing countries should also take steps to increase the average level of education, especially of women. Highly developed countries should provide financial support for the U.N. Fund for Population Activities, which supports international family planning efforts, and for research on new birth control methods.

• **Explain how individuals can adopt voluntary simplicity to mitigate the effects of population growth.**

Voluntary simplicity is a way of life that involves wanting and spending less. Each person who scales back unnecessary consumption lessens the effects of population growth.

Thinking About the Environment

1. What is food insecurity? What regions of the world have the greatest food insecurity?

2. What are two different views of the relationship between chronic hunger, population growth, and economic development?

3. How does poverty cause hunger? How does hunger cause poverty?

4. How do cultural values affect fertility rate?

5. What is the relationship between fertility rate and marriage age? between fertility and educational opportunities for women?

6. How does microcredit help poor people establish thriving small businesses? Why is microcredit targeted primarily for women?

7. What is family planning? Is family planning effective in reducing fertility rates?

8. Health experts are concerned that the surplus of men in China will lead to an increase in AIDS, particularly in cities. Explain the link.

9. Why do Mexico and Nigeria have positive population growth momentums?

10. Who are pronatalists? neo-Malthusians? Are you a pronatalist or a neo-Malthusian? Why?

11. Explain the rationale behind this statement: It is better for highly developed countries to spend millions of dollars on family planning in developing countries now than to have to spend billions of dollars on relief efforts later.

12. What are four Millennium Development Goals that the nations of the world have aspired to? What is the target date for these goals?

13. How can governments in highly developed nations help achieve population stability?

14. What is voluntary simplicity? How is it related to resource consumption?

15. Discuss this statement: The current human population crisis causes or exacerbates all environmental problems.

16. Discuss some of the ethical issues associated with overpopulation. Is it ethical to have more than two children? Is it ethical to consume so much in the way of material possessions? Is it ethical to try to influence a couple's decision about family size?

17. How is carrying capacity of the Earth system related to agricultural productivity?

Quantitative questions relating to this chapter are on our Web site.

Take a Stand

Visit our Web site at http://www.wiley.com/college/raven (select Chapter 9 from the Table of Contents) for links to more information about the controversy of debt in developing countries. Consider the opposing views of proponents and opponents of the Highly Indebted Poor Countries (HIPC) initiative, started by the World Bank and the International Monetary Fund (IMF) to reduce the debt burden of developing countries, and debate the issues with your classmates. You will find tools to help you organize your research, analyze the data, think critically about the issues, and construct a well-considered argument. *Take a Stand* activities can be done individually or as a team, as oral presentations, written exercises, or Web-based (e-mail) assignments.

Additional online materials relating to this chapter, including a Student Testing section with study aids and self-tests, Environmental News, Activity Links, Environmental Investigations, and more, are also on our Web site.

10

The Urban World

Aerial view of Chicago. Lake Michigan is in the background. Photographed in 2005. (Picture Arts/NewsCom)

Situated along Lake Michigan and near the St. Lawrence Seaway and the Mississippi River, the city of Chicago is a major commercial and industrial center in the United States. With a population in 2005 of 8.8 million, Chicago is the third largest city in the United States and the 24th largest in the world. Chicago is still growing: The U.N. Population Division projects that Chicago's population in 2015 will be 9.5 million.

Chicago's early history would not have suggested its remarkable growth into a major commercial and cultural force. In 1795, Indians living in the area ceded land to the U.S. government for the erection of Fort Dearborn. A settlement of civilians grew around the fort. After the town of Chicago was laid out in 1830, it began to grow rapidly, partly because of the westward migration of settlers. By 1850, more than 29,000 people lived there.

Cities do not exist entirely by themselves; they depend on the surrounding natural environment. Chicago's proximity to abundant natural resources enabled it to grow. We have already mentioned Chicago's location near major waterways, which helped Chicago

become part of the expanding network of U.S. urban trade. But what products did the city trade? The forests of Michigan and Wisconsin provided timber, the lands in Illinois provided coal, and the fertile grasslands of central Illinois provided forage for cattle and a rich soil for growing grains. By the 1850s, Chicago was a major railway center, and the combination of shipping by rail and boat made the city one the nation's largest trading centers.

Today dozens of suburban communities extend beyond the city's limits. However, many people choose to live in Chicago's downtown because of the many amenities there. Chicago's economy has done well in recent years, crime has declined, and the city offers numerous restaurants, theaters, museums, and shops. Thousands of trees have been planted, and new community gardens and parks have been added.

Chicago's population, like that of other urban areas, contains many people of foreign descent. Asians and Hispanics currently dominate the city's new immigrant population. Some of these immigrants are highly educated, bilingual professionals, whereas others are poorly educated and have little personal wealth. Minority groups are a growing political force, particularly African Americans and Hispanics.

Although Chicago has much to offer, it faces many of the same problems that beset other

major cities. These include school funding issues, urban renewal forcing the poor out of neighborhoods, suburban sprawl eating up nearby wetlands, and traffic congestion.

In this chapter we examine urban populations and trends so that we can get a better understanding of the dynamic, densely populated settlements known as *cities*. On a global scale, people are increasingly concentrating in cities. If this inundation continues as expected, cities will be home to two-thirds of all humans by 2025. Many cities, particularly those in the developing world, will not be able to provide their rapidly expanding populations with basic services, such as clean water, adequate housing, and new schools. Many urban dwellers in Calcutta and other large cities in the developing world are already suffering terrible deprivations, including poverty, malnutrition, and illiteracy. The continued rapid growth of cities has the potential to cause additional human suffer-ing and poverty as well as environmental problems.

World View and Closer to You...
Real news footage relating to urbanization around the world and in your own geographic region is available at www.wiley.com/college/raven by clicking on "World View and Closer to You."

Population and Urbanization

LEARNING OBJECTIVES

- Define *urbanization* and describe trends in the distribution of people in rural and urban areas.
- Distinguish between megacities and urban agglomerations.
- Describe some of the problems associated with the rapid growth rates of large urban areas.

Throughout human history, sociologists say that three urban revolutions have transformed human society. During the first urban revolution, from about 8000 to 2000 B.C.E., people moved into cities for the first time. The second urban revolution occurred from around 1700 to 1950. Modern cities developed and gained prominence as commerce replaced farming as the main way to make a living. As cities became wealthier, *urban migration* increased, with more people flocking to cities to take advantage of opportunities that the city-based economy provided. Thus, cities experienced a dramatic increase in population.

Today, a third urban revolution is underway. This revolution differs from the prior two in that the greatest population growth is taking place in cities in developing countries, not in the already established centers of commerce located in the highly developed countries. Also, the scale of urban migration today makes the earlier two revolutions pale in comparison. According to current projections, by 2007 or 2008, half of the world's population will live in urban areas for the first time in history (**Figure 10.1**).

The geographic distribution of people in rural areas, towns, and cities significantly influences the social, environmental, and economic aspects of population growth. When Europeans first settled in North America, the majority of the population consisted of farmers in rural areas. Today, approximately 21% of the people in the United States live in rural areas, and 79% of the U.S. population live in cities. Urbanization involves the movement of people from rural to urban areas as well as the transformation of rural areas into urban areas.

How many people does it take to make an urban area or city? The answer varies from country to country; it can be anything from 100 homes clustered in one place to a population of 50,000 residents. A population of 250 qualifies as a city in Denmark, whereas in Greece a city has a population of 10,000 or more. According to the U.S. Bureau of the Census, a location with 2500 or more people qualifies as an urban area.

One important distinction between rural and urban areas is not how many people live there but how people make a living. Most people residing in a rural area have occupations that involve harvesting natural resources—such as fishing, logging, and farming. In urban areas, most people have jobs that are not directly connected with natural resources.

Cities have grown at the expense of rural populations for several reasons. With advances in agriculture, including the increased mechanization of farms, fewer farmers support an increased number of people. Poor people also move to cities because of **land tenure** issues: In many developing countries, a few wealthy people own most of the land, to which poor farmers are denied access. Consequently, people in rural settings have fewer employment opportunities. Cities have traditionally provided more jobs because cities are the sites of industry, economic development, educational and cultural opportunities, and technology advancements—all of which generate income.

Figure 10.1   The worldwide shift from rural to urban areas, 1950-2030. In 2007 or 2008, a significant milestone will be reached, as 50% of the world's population will live in urban areas for the first time in history. (From World Urbanization Prospects: The 2003 Revision)

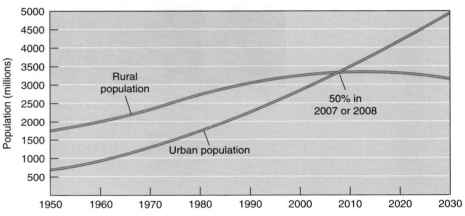

urbanization: The process in which people increasingly move from rural areas to densely populated cities.

Figure 10.2   Vancouver, Canada. This city is Canada's most important port city on the Pacific coast. Noted for its lovely parks and gardens, Vancouver has mild weather with a wet winter season. The Greater Vancouver region has a population of 2.2 million, including a large immigrant population, particularly from China, India, the Philippines, Taiwan, and South Korea. (© Gunter Marx Photography/Corbis Images)

Characteristics of the Urban Population

Every city is unique in terms of its size, climate, culture, and economic development (**Figure 10.2**). Although there is no such thing as a typical city, certain traits are common to urban populations in general. One of the basic characteristics of city populations is their far greater heterogeneity with respect to race, ethnicity, religion, and socioeconomic status than populations in rural areas. People living in urban areas are generally younger than those in the surrounding countryside. The young age structure of cities is due not to a higher birth rate but to the influx of many young adults from rural areas.

Urban and rural areas often have different proportions of males and females. Cities in developing nations tend to have more males. In cities in Africa, for example, males migrate to the city in search of employment, whereas females tend to remain in the country to care for the farm and their children. Cities in highly developed countries often have a higher ratio of females to males. Women in rural areas often have little chance of employment after they graduate from high school, and so they move to urban areas.

Urbanization Trends

Urbanization is a worldwide phenomenon. According to the Population Reference Bureau (PRB), 48% of the world population currently lives in urban areas. (The PRB considers towns with populations of 2000 or greater as urban.) The percentage of people living in cities compared with rural settings currently is greater in highly developed countries than in developing countries. In 2006, urban inhabitants comprised 77% of the total population of highly developed countries but only 41% of the total population of developing countries.

Although proportionately more people still live in rural settings in developing countries, urbanization is increasing rapidly. Currently, most urban growth in the world

Table 10.1 The World's 10 Largest Cities

1975	2005	2015
Tokyo, Japan, 26.6*	Tokyo, Japan, 35.2	Tokyo, Japan, 35.5
New York, USA 15.9	Mexico City, Mexico, 19.4	Mumbai (Bombay), India, 21.9
Mexico City, Mexico, 10.7	New York, USA, 18.7	Mexico City, Mexico, 21.6
Osaka-Kobe, Japan, 9.8	São Paulo, Brazil, 18.3	São Paulo, Brazil, 20.5
São Paulo, Brazil, 9.6	Mumbai (Bombay), India, 18.2	New York, USA, 19.9
Los Angeles, USA, 8.9	Delhi, India, 15.0	Delhi, India, 18.6
Buenos Aires, Argentina, 8.7	Shanghai, China, 14.5	Shanghai, China, 17.2
Paris, France, 8.6	Calcutta, India, 14.3	Calcutta, India, 17.0
Calcutta, India, 7.9	Jakarta, Indonesia, 13.2	Dhaka, Bangladesh, 16.8
Moscow, Russian Federation, 7.6	Buenos Aires, Argentina, 12.6	Jakarta, Indonesia, 16.8

*Population in millions.
Source: United Nations Population Division, Urban Agglomerations, 2005

is occurring in developing countries, whereas highly developed countries are experiencing little urban growth. As a result of the greater urban growth of developing nations, most of the world's largest cities are in developing countries. In 1975, four of the world's 10 largest cities were in developing countries: Mexico City, São Paulo, Buenos Aires, and Calcutta. In 2005, eight of the world's 10 largest cities were in developing countries: Mexico City, São Paulo, Mumbai (Bombay), Delhi, Shanghai, Calcutta, Jakarta, and Buenos Aires (**Table 10.1**).

According to the United Nations, almost 400 cities worldwide have a population of at least one million inhabitants, and 284 of these cities are in developing countries. The number and size of **megacities** has also increased. In some places, separate urban areas have merged into **urban agglomerations** (**Figure 10.3**). An example is the Tokyo-Yokohama-Osaka-Kobe agglomeration in Japan, which is home to about 50 million people. However, according to the U.N. Population Division, most of the world's urban population still lives in small or medium-sized cities with populations of less than one million.

megacities: Cities with more than 10 million inhabitants.

urban agglomeration: An urbanized core region that consists of several adjacent cities or megacities and their surrounding developed suburbs.

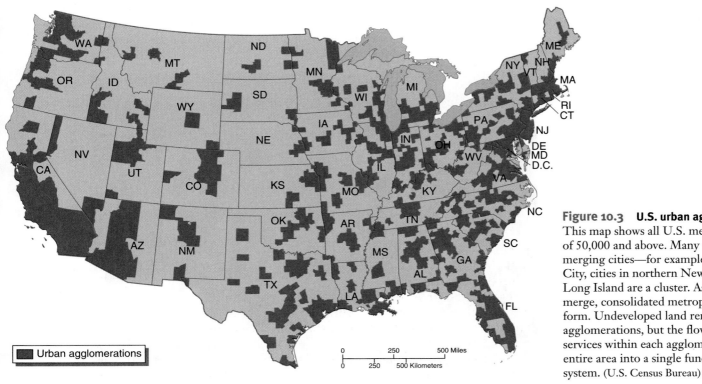

Urban agglomerations

Figure 10.3   U.S. urban agglomerations. This map shows all U.S. metropolitan areas of 50,000 and above. Many are clusters of merging cities—for example, New York City, cities in northern New Jersey, and Long Island are a cluster. As adjacent cities merge, consolidated metropolitan areas form. Undeveloped land remains in these agglomerations, but the flow of goods and services within each agglomeration links the entire area into a single functional system. (U.S. Census Bureau)

Mexico City illustrates the rapidity of urban growth in Latin America, which has a much larger urban population than is found in North America. According to the U.N. Population Division, Mexico City had a population of 2.9 million in 1950 and was the world's 19th largest city. By 2005, its population had increased to 19.4 million, and it was the world's second largest city, surpassed only by Tokyo. Mexico City's population continues to increase by more than 950 new immigrants *each day* from economically stagnated rural areas.

Africa now has a larger urban population than North America. Lagos, Nigeria, is an example of an African city with dramatic growth. In just 30 years, from 1975 to 2005, the population of Lagos grew from 1.9 million to 10.9 million. The United Nations Population Division projects that the population of Lagos will be 16.1 million in 2015, which will make it the world's 11th largest city.

Almost half of the world's 100 largest cities are found in Asia. Mumbai (Bombay) is representative of rapid urban growth in Asia. Its population increased from 7.1 million in 1975, when it was the world's 15th largest city, to 18.2 million in 2005, when it was the fifth largest. By 2015, Mumbai's population is projected to be 21.9 million, which will make it the second largest city in the world.

Urbanization is increasing in highly developed nations, too, but at a much slower rate. Consider the United States as representative of highly developed nations. Here, most of the migration to cities occurred during the past 150 years, when an increased need for industrial labor coincided with a decreased need for agricultural labor. The growth of U.S. cities over such a long period was typically slow enough to allow important city services such as water purification, sewage treatment, education, and adequate housing to keep pace with the influx of people from rural areas.

In contrast, the recent faster pace of urban growth in developing nations has outstripped the limited capacity of many cities to provide basic services. It has overwhelmed their economic growth, although cities still offer more job opportunities than rural areas. Consequently, cities in developing nations generally face more serious challenges than cities in highly developed countries. These challenges include poverty; exceptionally high unemployment; heavy pollution; and inadequate or nonexistent water, sewage, and waste disposal (see Figure 25.10 of Jakarta, Indonesia). Rapid urban growth strains school, medical, and transportation systems.

Substandard housing (slums and squatter settlements) is a critical issue (**Figure 10.4**). Squatters illegally occupy the land they build on and cannot obtain city services such as clean water, sewage treatment, garbage collection, paved roads, or police and fire protection. Many squatter settlements are built on land unsuitable for housing, such as

Figure 10.4 Squatter settlement. The substandard, poor-quality housing in this section of Port au Prince, Haiti, is common in cities of developing nations. These houses are built out of whatever materials their owners salvage and patch together. (© Les Stone/Corbis Sygma)

Figure 10.5   Homelessness. Entire families of desperately poor "pavement people" eat, sleep, and raise their families on a street in Calcutta, India. Homelessness, a serious problem in the cities of highly developed countries, is an even greater problem in cities of developing nations. (Deshakalyan Chowdhury/AFP/Getty Images)

high hills where mudslides are a danger during periods of rainfall. Squatters have an insecure existence because they are always facing the risk of eviction. According to the United Nations, squatter settlements house about one-third of the entire urban population in most developing countries.

Cities in highly developed and developing countries share some problems, such as homelessness. Every country, even a highly developed country such as the United States, has city-dwelling people who lack shelter. Although estimates vary widely, most urban scholars estimate that the United States has a total of between 300,000 and 500,000 homeless people on any given night. Urban problems such as homelessness are usually more pronounced in the cities of developing nations. In Calcutta, India, perhaps 250,000 homeless people sleep in the streets each night (**Figure 10.5**).

REVIEW

1. What is urbanization?

2. What is a megacity? an urban agglomeration?

3. What parts of the world have the fastest current rates of urbanization?

The City as an Ecosystem

LEARNING OBJECTIVES

- Explain how cities are analyzed from an ecosystem perspective.
- Describe brownfields.
- Distinguish between urban heat islands and dust domes.
- Define *compact development.*

Many urban sociologists use an ecosystem approach to better understand how cities function and how they change over time. Recall from Chapter 3 that an **ecosystem** is an interacting system that encompasses a biological community and its nonliving, physical environment. Urban ecologists study urban trends and patterns in the context of four variables: population, organization, environment, and technology; they use the acronym POET to refer to these variables.

Population refers to the number of people; the factors that change this number (births, deaths, immigration, and emigration); and the composition of the city by age, sex, and ethnicity. *Organization* is the social structure of the city, including its economic policies, method of government, and social hierarchy. *Environment* considers both the natural environment, such as if the city is situated near a river or in a desert, and the city's physical infrastructure, including its roads, bridges, and buildings (**Figure 10.6**). Environment also includes changes in the natural environment caused by humans—air and water pollution, for example. *Technology* refers to human inventions that directly affect the urban environment. Examples of technology include aqueducts, which carry water long distances to cities in arid environments, and air conditioning, which allows people to live in comfort in hot, humid cities.

Figure 10.6 **Satellite photographs of land use patterns.** Urban land use patterns exhibit different configurations. Note the amount of paved roads and parking lots in certain of these landscapes.

(a) Full view of Park City, Utah. All other views are enlarged and cropped from this shot. (IKONOS Satellite Images by Space Imaging)

(b) Mixed use (commercial, industrial, and residential). Note the multifamily residential on left side of figure.

(c) Single-family residential.

(d) Undeveloped open space.

The four variables (POET) do not function independently of one another; they are interrelated and interact much like the parts of natural ecosystems. For example energy is needed for all parts of POET. Well-designed urban systems can have a substantial effect on the energy efficiency of the entire system, including how much energy is required to keep it functioning properly.

Phoenix, Arizona: Long-Term Study of an Urban Ecosystem

The National Science Foundation has several Long-Term Ecological Research (LTER) sites to gather extensive data on various ecosystems, such as deserts, mountains, lakes, and forests. Two LTER choices—the urban settings of Baltimore and Phoenix—challenge the conventional approach to ecology. With the majority of Americans now living in cities, researchers faced a void when considering the effects of humans on their urban environment. As with all LTER projects, the efforts in Baltimore and Phoenix are assessing ecosystem health over a long period. The urban research focuses on the ecological effects of human settlement, rather than the interactions among the humans themselves.

Many research questions about cities as ecological systems are the same as those for other LTER sites, such as changes in plant and animal populations and the effects of major disturbances such as fire, drought, or hurricanes. For the urban programs, the approach is more complicated because the flow of water, energy, and resources into and out of the sites is linked to the flow of money and the human population (**Figure 10.7**). Potential relationships may exist between political power and environmental quality of specific neighborhoods. Paving might greatly alter water runoff and subsequent water quality. Preliminary findings from Phoenix suggest a surprising diversity of organisms there, though different from those in surrounding rural environments.

Researchers are entering new territory to answer broad ecological system questions in urban settings. The knowledge gained in urban ecology could increase public awareness and eventually influence policy decisions.

Environmental Problems Associated with Urban Areas

The concept of the city as an ecosystem would be incomplete without considering the effects the city has on its natural environment. Growing urban areas affect land use patterns and destroy or fragment wildlife habitat by suburban development that encroaches into former forest, wetlands, desert, or agricultural land in rural areas. For example, large portions of Chicago, Boston, and New Orleans are former wetlands. Most cities have blocks and blocks of **brownfields**. Meanwhile, the suburbs continue to expand outward, swallowing natural areas and farmland.

Reuse of brownfields is complicated because many have environmental contaminants that must be cleaned up before redevelopment can proceed. Nonetheless,

brownfield: An urban area of abandoned, vacant factories, warehouses, and residential sites that may be contaminated from past uses.

Natural capital (inputs)		Products and wastes (outputs)
Energy (fuel) Clean water Clean air Food Raw and refined materials for construction and industry Business and consumer products		Waste heat, greenhouse gases Waste water, water pollution Air pollution Solid waste Goods, services

Figure 10.7   The city as a dynamic system. Cities have much in common with natural ecosystems. Both are open systems. The human population in an urban environment requires inputs from the surrounding countryside and produces outputs that flow into surrounding areas. Not shown in this figure is the internal cycling of materials and energy within the urban system.

brownfields represent an important potential land resource. Pittsburgh, Pennsylvania, is best known for its redevelopment of brownfields that were once steel mills. Residential and commercial sites now occupy several of these former brownfields.

Cities affect water flow by covering the rainfall-absorbing soil with buildings and paved roads. Storm sewage systems are built to handle the runoff from rainfall, which is polluted with organic wastes (garbage, animal droppings, and such), motor oil, lawn fertilizers, and heavy metals. In most cities across the United States, urban runoff is cleaned up in sewage treatment plants before being discharged into nearby waterways. In many cities, however, high levels of precipitation can overwhelm the sewage treatment plant and result in the release of untreated urban runoff. When this occurs, the polluted runoff contaminates bodies of water far beyond the boundaries of the city.

Most workers in U.S. cities have to commute dozens of miles through traffic-congested streets, from suburbs where they live to downtown areas where they work. Automobiles are a necessity to accomplish everyday chores because development is so spread out in the suburbs. This heavy dependence on motor vehicles as our primary means of transportation increases air pollution and causes other environmental problems.

The high density of automobiles, factories, and commercial enterprises in urban areas causes a buildup of airborne emissions, including particulate matter (dust), sulfur oxides, carbon oxides, nitrogen oxides, and volatile organic compounds. Urban areas in developing nations have the worst air pollution in the world. In Mexico City, the air is so polluted that schoolchildren are not permitted to play outside during much of the school year. Although progress has been made in reducing air pollution in highly developed nations, the atmosphere in many of their cities often contains higher levels of pollutants than are acceptable based on health standards.

urban heat island: Local heat buildup in an area of high population density.

Streets, rooftops, and parking lots in areas of high population density absorb solar radiation during the day and radiate heat into the atmosphere at night. Heat released by human activities such as fuel combustion is also highly concentrated in cities. The air in urban areas is therefore warmer than the air in the surrounding suburban and rural areas and is known as an **urban heat island** (**Figure 10.8**). Urban heat islands affect local air currents and weather conditions, particularly by increasing the number of thunderstorms over the city (or downwind from it) during summer months. The uplift of warm air over the city produces a low-pressure cell that draws in cooler air from the surroundings. As the heated air rises, it cools, causing water vapor to condense into clouds and producing thunderstorms.

dust dome: A dome of heated air that surrounds an urban area and contains a lot of air pollution.

Urban heat islands also contribute to the buildup of pollutants, especially particulate matter, in the form of **dust domes** over cities (**Figure 10.9a**). Pollutants concentrate in a dust dome because convection (that is, the vertical motion of warmer air) lifts pollutants into the air, where they remain because of somewhat stable air masses produced by the urban heat island. If wind speeds increase, the dust dome moves downwind from the city, and the polluted air spreads over rural areas (**Figure 10.9b**).

Noise Pollution **Sound** is caused by vibrations in the air (or some other medium) that reach the ears and stimulate a sensation of hearing. Sound is called **noise pollution** when it becomes loud or disagreeable, particularly when it results in physiological or psychological harm.

Most of the noise that travels in the atmosphere is of human origin. Vehicles, from trains to trucks to powerboats, produce a great deal of noise. Power lawn mowers, jets flying overhead, chain saws, jackhammers, cars with stereos booming, and heavy traffic are just a few examples of the outside noise that assails our ears. Indoors, dishwashers, trash compactors, washing machines, televisions, and stereos add to the din.

Prolonged exposure to noise damages hearing. In addition to hearing loss, noise increases the heart rate, dilates the pupils, and causes muscle contractions. Evidence exists that prolonged exposure to high levels of noise causes a permanent constriction of blood vessels, which can increase the blood pressure, thereby contributing to heart disease. Other physiological effects associated with noise pollution include migraine headaches, nausea, dizziness, and gastric ulcers. Noise pollution also causes psychological stress.

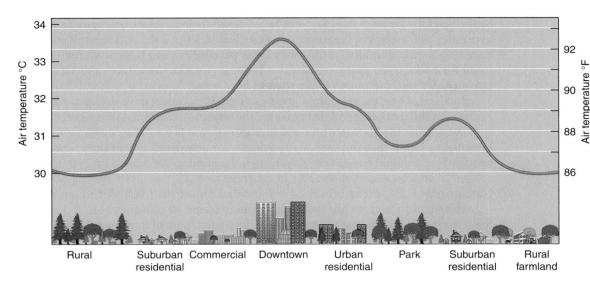

Figure 10.8   Urban heat island. This figure shows how temperatures might vary on a summer afternoon. The city stands out as a heat island against the surrounding rural areas.

Obviously, producing less noise can reduce noise pollution. This can be accomplished in a variety of ways, from restricting the use of sirens and horns on busy city streets to engineering motorcycles, vacuum cleaners, jackhammers, and other noisy devices so that they produce less noise. The engineering approach is technologically feasible but is often avoided because consumers associate loud noise with greater power. Putting up shields between the noise producer and the hearer can also help control noise pollution. One example of a sound shield is the noise barriers erected along heavily traveled highways.

Environmental Benefits of Urbanization

Our previous discussion suggests that the concentration of people into cities has an overall harmful effect on the environment. Yet urbanization has the potential to provide tangible environmental benefits that in many cases outweigh the negative aspects. A well-planned city actually benefits the environment by reducing pollution and preserving rural areas. A solution to urban growth is **compact development**, which uses

compact development: The design of cities in which tall, multiple-unit residential buildings are close to shopping and jobs, and all are connected by public transportation.

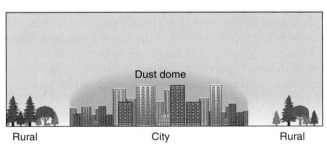

(a) A dust dome of pollutants forms over a city when the air is somewhat calm and stable.

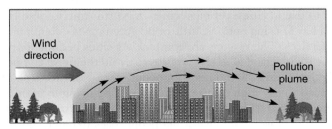

(b) When wind speeds increase, the pollutants move downwind from the city.

Figure 10.9   Dust dome.

land efficiently. Dependence on motor vehicles and their associated pollution is reduced as people walk, cycle, or take public transit such as buses or light rails to work and shop. Because compact development requires fewer parking lots and highways, there is more room for parks, open space, housing, and businesses. Compact development makes a city more livable, and more people want to live there.

Portland, Oregon, provides a good example of compact development. Although Portland is still grappling with many issues, the city government has developed effective land use policies that dictate where and how growth will occur. The city looks inward to brownfields rather than outward to the suburbs for new development sites. Since 1975, Portland's population has grown 50%, from 0.9 million to 1.8 million (in 2005), yet the urbanized area increased about 2%. In contrast, from 1975 to 2005 the population of Chicago grew 22%, yet its urbanized area increased more than 50% due to sprawl.

Although the automobile is still the primary means of transportation in Portland, the city's public transportation system is an important part of its regional master plan. Public transportation incorporates light-rail lines, bus routes (many of which have buses arriving every 15 minutes), bicycle lanes, and walkways in addition to the automobile. Employers are encouraged to provide bus passes to their employees instead of paying for parking. The emphasis on public transportation has encouraged commercial and residential growth along light rail and bus stops instead of in suburbs.

REVIEW

1. What are brownfields?

2. Why are cities associated with urban heat islands and dust domes?

3. What is compact development?

Urban Land Use Planning

LEARNING OBJECTIVES

- Discuss the use of zoning in land use planning.
- Relate how a city's transportation infrastructure affects urban development.
- Define *suburban sprawl* and discuss a problem caused or exacerbated by sprawl.

Land use in many cities is based on economic concerns. Taxes pay for the city's infrastructure—its roads, schools, water treatment plants, prisons, and garbage trucks. The city's center—the *central business district*—typically pays the highest taxes. Surrounding residential properties pay lower taxes than the central business district, but the taxes are still high. Thus, many residential buildings near the central business district are high-rises filled with small apartments or condominiums; the collective tax for the property is high, but individual taxes are more reasonable. Circling the residential properties—and farther from the city's center—are land-intensive businesses that require lower taxes: golf courses, cemeteries, water treatment plants, sanitary landfills, and such. Parks and other open spaces are interspersed among the various land uses. People living in the suburbs, often far from the central business district, pay less in terms of taxes but more in terms of transportation costs.

High taxes near a city's central business district mean that only more affluent people can afford to live in cities; yet most cities in North America also have poor neighborhoods with few housing options, little or no green space, often inferior schools, and fewer public services. The reasons are complex. As cities became more industrialized, the more affluent citizens fled to the suburbs to avoid the noise and pollution, leaving the poor in the inner cities. Although poor people cannot afford to pay high taxes, they also cannot afford to pay high transportation costs. By living in high-density housing, the tax burden is shared with many families, mitigating the overall cost to individuals. **Gentrification**, the movement of wealthier people back to older, run-down homes that have been renovated, sometimes displaces the urban poor who can no longer afford to live in the neighborhood.

Social scientists have examined factors that influence urban development. For example, **David Harvey,** an English geographer, did a detailed analysis of Baltimore, Maryland, in the 1970s. He divided Baltimore's real estate into various areas based on income and ethnicity. Harvey demonstrated that financial institutions and government agencies did not deal consistently with different neighborhoods. For example, banks were less likely to lend money to the inner-city poor for housing than to those living in more affluent neighborhoods. Thus, the investment capitalists' discrimination in the housing market affected the dynamics of buying and selling. Harvey concluded that real estate investors and government programs largely determined if a neighborhood remained viable or decayed and was eventually abandoned.

Who makes decisions that determine how a city grows and allocates its land? Both political and economic factors influence land use planning. Cities do not exist as separate entities; they are part of larger political organizations, including counties, states or provinces, and countries, all of which affect urban development. Real estate investment in property has a profound effect on where commercial and residential development occurs. Economic institutions, such as banks and multinational corporations, also influence land use in cities.

> **land use planning:** The process of deciding the best uses for undeveloped land in a given area.

Cities regulate land use mainly through *zoning*, in which the city is divided into **use zones,** areas restricted to specific land uses, such as commercial, residential, or industrial (see Figure 10.6). These categories are often subdivided. For example, residential use zones may designate single-family residences or multifamily residences (apartments). Property owners can develop their properties as long as they meet the zoning ordinances in which the property is located. Often these rules are very specific, regulating building height, how the building is situated on the property, and what the property can be used for. Zoning has largely resulted in separate industrial parks, shopping centers, and residential districts.

Transportation and Urban Development

Transportation and land use are inextricably linked because as cities grow, they expand along public transportation routes. The kinds of transportation available at a particular period in history affect a city's spatial structure. **Figure 10.10** shows the expansion of a hypothetical city along the eastern coast of North America. During the 1700s to the 1850s, transportation in the city was limited to walking, horse-drawn carriages, and ships (Figure 10.10a).

Technological advances enabled fixed transportation routes (railroads and electric street trolleys) to spread out of the city from the central business district during the 1870s to 1910s (Figure 10.10b). People could move beyond the city limits quickly and

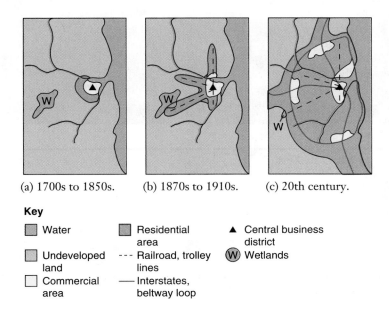

(a) 1700s to 1850s.   (b) 1870s to 1910s.   (c) 20th century.

Key

- ▨ Water
- ▨ Undeveloped land
- ☐ Commercial area
- ▨ Residential area
- - - - Railroad, trolley lines
- —— Interstates, beltway loop
- ▲ Central business district
- Ⓦ Wetlands

Figure 10.10   The relationship between transportation and urban spatial form.
Cities grow outward along transportation routes. During the 20th century, the automobile dramatically expanded suburbanization along roads and highways built to accommodate automobiles.

inexpensively. Real estate developers began building housing tracts in areas that were at one time inaccessible. The first suburbs clustered around railway stations, so that the city's outward growth assumed a shape reminiscent of a sea star. Still, the city was relatively contained; as an example, the average commute was only about 1.5 miles in the 1920s.

Cars and trucks forever changed the city, increasing its spatial scale. With the advent of the automobile era during the 20th century, roads expanded the access to undeveloped areas between the "arms" of prior metropolitan development, filling in the wild spaces (Figure 10.10c). Construction of the interstate highway system and outer-city beltway "loops," beginning in the 1950s, encouraged development even farther from the city's central business district. Today, many people live in suburbs far from their place of employment, and daily commutes of 20 or more miles each way are commonplace.

Suburban Sprawl

suburban sprawl: A patchwork of vacant and developed tracts around the edges of cities; contains a low population density.

Urbanization and its accompanying suburban sprawl affect land use and have generated a new series of concerns (**Figure 10.11**). Prior to World War II, jobs and homes were concentrated in cities, but during the 1940s and 1950s, jobs and homes began to move from urban centers to the suburbs. New housing and industrial and office parks were built on rural land surrounding the city, along with a suburban infrastructure that included new roads, schools, and the like.

Further development extended the edges of the suburbs, cutting deeper and deeper into the surrounding rural land and causing environmental problems such as loss of wetlands, loss of biological habitat, air pollution, and water pollution. Meanwhile, peo-

Figure 10.11 **Suburban sprawl.**
Commercial and residential areas are separated in suburban sprawl, resulting in a greater dependence on automobiles.

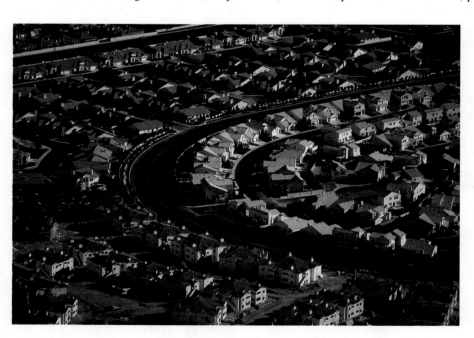

(a) Suburban Sprawl near Las Vegas, Nevada. These homes are not located within walking distance of stores and restaurants. (Peter Essick/Aurora Photos)

(b) What point is the cartoonist making about suburban sprawl? (Pardon My Planet: © Vic Lee /Reprinted with special permission of King Features Syndicate)

ple who remained in the city and older suburbs found themselves the victims of declining property values and increasing isolation from suburban jobs. This pattern of land use, which has intensified in recent years, has increased the economic disparity between older neighborhoods and newer suburbs. In the past few decades, the rate of land development around most U.S. cities has exceeded that of population growth.

There is an urgent need for regional planning involving government and business leaders, environmentalists, inner-city advocates, suburbanites, and farmers to determine where new development should take place and where it should not. Also, most metropolitan areas need to make more efficient use of land that has already been developed, including central urban areas. The Greater Atlanta area is an excellent example of the rapid spread of urban sprawl. During the 1990s and early 2000s, tract houses, strip malls, business parks, and access roads replaced an average of 500 acres of surrounding farms and forests each week. The Greater Atlanta area, with a 2005 population of 4.3 million people, extends more than 175 km (110 mi) across, almost twice the distance it was in 1990. Atlanta now has 1750 km^2 (700 mi^2) of sprawl.

U.S. voters have grown increasingly concerned about the unrestricted growth of suburban sprawl. At least 11 states now have comprehensive, statewide growth-management laws. Maryland, for example, has a smart growth plan that protects open space in highly developed areas while promoting growth in areas that could be helped by growth. **Smart growth** is an urban planning and transportation strategy that mixes land uses (commercial, manufacturing, entertainment, and a range of housing types).

Smart growth incorporates compact development, creates communities in which it is easy to walk from one place to another, and preserves open space, farmlands, and important environmental areas. Because people live near jobs and shopping, the need for a continually expanding highway system is lessened. Smart growth involves a long-range "look at the big picture" instead of developing individual tracts of land as they become available. Arlington, Virginia, and Minneapolis–St. Paul, Minnesota, are two examples of cities that have incorporated smart growth policies.

REVIEW

1. How does zoning help regulate urban land use?
2. How are transportation and land use linked?
3. What is suburban sprawl?

Making Cities More Sustainable

LEARNING OBJECTIVES

- List at least five characteristics of an ideal sustainable city.
- Explain how city planners have incorporated environmental sustainability into the design of Curitiba, Brazil.

Most environmental scientists think that, on balance, increased urbanization is better for the environment than having the same number of people living in rural areas spread across the landscape. However, the challenge is to make these cities more sustainable through better design. This challenge is particularly difficult because the rapid urban growth evident in many cities is overburdening the existing infrastructure, leading to a lack of urban services and environmental degradation.

Imagine you are commissioned to design a **sustainable city** for 100,000 people. What features would you want it to have? To design a sustainable city, it is first necessary to think about the city as a system. Changes in one variable, such as increasing population density, always impact all other parts of the city system. Greater population density, for example, means a greater demand for housing, public transit and/or parking lots, and energy use for heating and cooling.

You should make sure that the city has clear, cohesive urban policies that enable the government infrastructure to manage it effectively. This is essential because poor planning, corruption, and ineffective urban governance exacerbate many of the

sustainable city: A city with a livable environment, a strong economy, and a social and cultural sense of community; sustainable cities enhance the well-being of current and future generations of urban dwellers.

problems associated with urbanization. According to social scientists, the most effective urban, regional, and national governments are democratic and participatory, in which local citizens are encouraged to work together to address local problems.

Cities have the potential to produce low levels of energy consumption, resource use, and wastes. Your sustainable city should be designed to reduce energy consumption by using energy and other resources efficiently. It should have codes that require buildings, motor vehicles, and appliances to be energy-efficient (see Meeting the Challenge: Green Architecture). Your sustainable city would make use of solar and other forms of renewable energy as much as possible.

Your sustainable city should be designed to reduce pollution and wastes by reusing and recycling materials in the waste stream. In your city, much of the municipal solid waste—paper, plastics, aluminum cans, and such—would be recycled, thereby reducing consumption of virgin materials. Yard wastes would be composted and used to enrich the soil in public places. Sewage (wastewater) treatment in your city would involve the use of living plants in large tanks or, even better, in marshes that could also provide wildlife habitat.

Your sustainable city should be designed with large areas of green space that provides habitat for wildlife, thereby supporting biological diversity. The green space would also provide areas of recreation for the city's inhabitants; people are more likely to engage in recreation if there is local green space than if there is not. Humans are a part of the natural world, and it is easy for urban dwellers to forget this fact. Urban living can lead to both a skewed perception of the importance of the natural world to our survival and the misperception that we can use engineering and technology to solve all environmental issues. Having parks and other open spaces within a city encourages residents to spend more time outside, which promotes awareness of the natural world. In addition, substantial evidence indicates that interaction with open space is linked to health; the beneficial effects of spending time in green space are both physical and psychological.

Your design for a sustainable city should it people-centered, not automobile-centered (**Figure 10.12**; also see Figure 25-11). It should be designed to allow people to move about the city by walking or bicycling or, for longer distances, mass transit. The use of automobiles would be limited (perhaps by closing certain streets to motor vehicles), and public transport would be readily available, clean, and inexpensive. Restricting automobile use would also reduce fossil fuel consumption and motor vehicle-generated air pollution.

The people living in your city should grow some of their food—for example, in rooftop gardens, window boxes, greenhouses, and community gardens. Rooftop gardens are particularly popular because they cannot be vandalized. Currently, urban farmers supply about 15% of the food consumed in cities worldwide, and that number is projected to increase. Berlin, Germany, has more than 80,000 urban farmers.

Compact development should be a prominent characteristic of your sustainable city. Abandoned lots and brownfields would be cleaned up and redeveloped, thereby reducing the spread of the city into nearby rural areas—farms, wetlands, and forests.

No city exists that incorporates all the sustainable features we have just discussed. However, many cities around the world provide unique examples of specific sustain-

Figure 10.12 **Copenhagen, Denmark, a people-centered city.** In Nuhavn, a public area in the city of Copenhagen, curbside parking was replaced with bicycle paths and sidewalks, creating a people-centered place. (Robert Harding World Imagery/Getty Images, Inc.)

ability initiatives. Portland, Oregon (discussed earlier in the chapter), is an example of a North American city that exhibits some features of sustainability. Many cities in developing nations are also making progress. One of these is Curitiba, Brazil.

CASE IN POINT

Curitiba, Brazil

Livable cities are not restricted to highly developed countries. Curitiba, a city of more than 2.9 million people in Brazil, provides a good example of compact development in a moderately developed country. Curitiba's city officials and planners have had notable successes in public transportation, traffic management, land use planning, waste reduction and recycling, and community livability.

The city developed an inexpensive, efficient mass transit system that uses clean, modern buses that run in high-speed bus lanes (**Figure 10.13**). High-density development was largely restricted to areas along the bus lines, encouraging population growth where public transportation was already available. About two million people use Curitiba's mass transportation system each day. Since 1975 Curitiba's population has more than tripled, yet traffic has declined. Curitiba has less traffic congestion and significantly cleaner air, both of which are major goals of compact development. Instead of streets crowded with vehicular traffic, the center of Curitiba is a *calcadao*, or "big sidewalk," that consists of 49 downtown blocks of pedestrian walkways connected to bus stations, parks, and bicycle paths.

Curitiba was the first city in Brazil to use a special low-polluting fuel that contains a mixture of diesel fuel, alcohol, and soybean extract. In addition to burning cleanly, this fuel provides economic benefits for people in rural areas who grow the soybeans and grain (used to make the alcohol). It is estimated that 50,000 jobs are created to provide each billion liters of alcohol.

Over several decades, Curitiba purchased and converted flood-prone properties along rivers in the city to a series of interconnected parks crisscrossed by bicycle paths. This move reduced flood damage and increased the per capita amount of "green space"

Figure 10.13   Curitiba, Brazil.

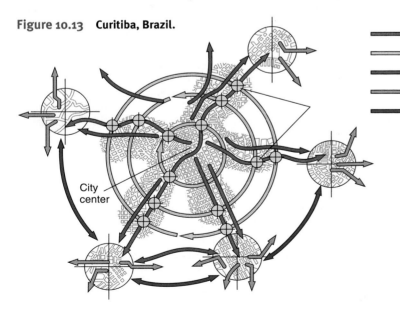

Express routes
Interdistrict routes
Direct routes
Feeder bus routes
Workers' routes

City center

(a) Curitiba's bus network, arranged like the spokes of a wheel, has concentrated development along the bus lanes, saving much of the surrounding countryside from development.

(b) Shown is a tube shelter with departing bus riders. Passengers pay their fares in advance in the tube shelter and then step directly onto the bus when it arrives (the bus and tube floors are at the same level). (H. John Maier, Jr./Time Life Pictures/Getty Images)

MEETING THE CHALLENGE

Green Architecture

The Adam Joseph Lewis Center for Environmental Studies at Oberlin College near Cleveland, Ohio, represents a vision implemented by dedicated students, faculty members, architects, and many others (see figure). Dedicated in 2000, the building is an award-winning example of **green architecture**, which encompasses environmental considerations such as energy conservation, improved indoor air quality, water conservation, and recycled or reused building materials. The Lewis Center is designed to work like a natural ecosystem such as a forest or pond—that is, it obtains its energy from sunlight and produces few wastes that cannot be recycled or reused.

An advanced-design, earth-coupled heat pump, which extracts heat from 24 geothermal

wells in the ground, supplies some of the energy for heating and cooling the building. On the roof, an array of photovoltaic (PV) cells collects the sun's energy to meet most of the building's electricity needs. On sunny days, the PV cells produce more electricity than the building uses, so the excess is sold to the Ohio power grid. On cloudy winter days, the building buys the extra energy it needs from the grid.

Energy is saved by the triple-paned windows, which allow sunlight to enter the building but insulate against heat loss during winter and heat gain during summer. Classrooms in the building have motion sensors that detect when people are not in the room, triggering the energy-efficient lighting fixtures and ventilation systems to shut down.

The building's wastewater from sinks and toilets is cleansed and recycled by a so-called

living machine. This organic system consists of a series of tanks that contain bacteria, aquatic organisms, and plants. These organisms interact to remove organic wastes, nutrients such as nitrogen and phosphorus, and disease-causing organisms from the water as it moves through the tanks. After it has been cleaned, the water is stored and reused in the building's toilets.

The acoustic panels in the auditorium are made from agricultural straw wastes, and the computer table in the resource center is constructed from a bowling alley lane from the old campus bowling alley. All new wood used in the building came from nearby forests that are certified to produce sustainably harvested wood. All the steel frames used in the building are composed of recycled steel. The carpets in the classrooms are installed in squares that are easy to remove. As the carpet wears out, it is not disposed of in a landfill. Instead, the squares are returned to the manufacturer, where they are recycled into "new" carpet.

The landscape surrounding the building is carefully designed to highlight a variety of natural and agricultural ecosystems. Because most of Oberlin College's students come from urban areas, these landscaped grounds provide an outdoor classroom that exposes students to the natural environment. Much of the area on which Oberlin College was built originally consisted of marshes and deciduous forests. A wetland that provides a haven for frogs, insects such as dragonflies, and songbirds has been constructed along one side of the building. Native forest trees are planted along another side of the building. Orchards of apple and pear trees, gardens, and a greenhouse are also part of the landscape. Near the orchard is a cistern that collects precipitation runoff from the roof for reuse.

Green architecture. The Adam Joseph Lewis Center for Environmental Studies at Oberlin College is an excellent example of ecological design. It uses about 20% of the energy that a new building typically uses. (Barney Taxel, Cleveland, Ohio)

from 0.5 m² in 1950 to 50 m² today, a significant accomplishment considering Curitiba's rapid population growth during the same period.

Another example of Curitiba's creativity is its labor-intensive Garbage Purchase program, in which poor people exchange filled garbage bags for bus tokens, surplus food (eggs, butter, rice, and beans), or school notebooks. This program encourages garbage pickup from the unplanned squatter settlements (where garbage trucks cannot drive) that surround the city. Curitiba supplies more services to these unplanned settlements than most cities do. It tries to provide water, sewer, and bus service to them; the bus service allows the settlers to seek employment in the city.

These changes did not happen overnight. Like Curitiba, most cities can be carefully reshaped over several decades to make better use of space and to reduce dependence on motor vehicles. City planners and local and regional governments are increasingly adopting measures to provide the benefits of sustainable development in the future. ■

REVIEW

1. What features does a sustainable city possess?

2. Why is Curitiba, Brazil, a model of sustainability in the developing world?

REVIEW OF LEARNING OBJECTIVES WITH KEY TERMS

- **Define *urbanization* and describe trends in the distribution of people in rural and urban areas.**

Urbanization is the process in which people increasingly move from rural areas to densely populated cities. As a nation develops economically, the proportion of the population living in cities increases. In developing nations, most people live in rural settings, but their rates of urbanization are rapidly increasing.

- **Distinguish between megacities and urban agglomerations.**

Megacities are cities with more than 10 million inhabitants. In some places, separate urban areas have merged into an **urban agglomeration**, which is an urbanized core region that consists of several adjacent cities or megacities and their surrounding developed suburbs; an example is the Tokyo-Yokohama-Osaka-Kobe agglomeration in Japan, which is home to about 50 million people.

- **Describe some of the problems associated with the rapid growth rates of large urban areas.**

Rapid urban growth often outstrips the capacity of many cities to provide basic services. Challenges include substandard housing; poverty; high unemployment; pollution; and inadequate or nonexistent water, sewage, and waste disposal. Rapid urban growth also strains school, medical, and transportation systems. Cities in developing nations are generally faced with more serious challenges than are cities in highly developed countries.

- **Explain how cities are analyzed from an ecosystem perspective.**

Urban ecologists study urban trends and patterns in the context of four variables: population, organization, environment, and technology. Population refers to the number of people; the factors that change this number; and the composition of the city by age, sex, and ethnicity. Organization is the social structure of the city, including its economic policies, method of government, and social hierarchy. Environment considers both the natural environment and the city's physical infrastructure, including its roads, bridges, and buildings. Technology refers to human inventions that directly affect the urban environment. These four variables are interrelated.

- **Describe brownfields.**

A **brownfield** is an urban area of abandoned, vacant factories, warehouses, and residential sites that may be contaminated from past uses.

- **Distinguish between urban heat islands and dust domes.**

An **urban heat island** is local heat buildup in an area of high population density. A **dust dome** is a dome of heated air that surrounds an urban area and contains a lot of air pollution.

- **Define *compact development*.**

Compact development is the design of cities in which tall, multiple-unit residential buildings are close to shopping and jobs, and all are connected by public transportation. Portland, Oregon, is a good example of compact development in a highly developed country.

- **Discuss the use of zoning in land use planning.**

Land use planning is the process of deciding the best uses for undeveloped land in a given area. The main way that cities regulate land use is by zoning, in which the city is divided into **use zones**, areas restricted to specific land uses, such as commercial, residential, or industrial. Zoning has largely resulted in separate industrial parks, shopping centers, apartment districts, and the like.

- **Relate how a city's transportation infrastructure affects urban development.**

Transportation and land use are inextricably linked, because as cities grow, they expand along public transportation routes. The kinds of transportation available at a particular period affect a city's spatial structure. Technological advances enabled fixed transportation routes (railroads and electric street trolleys) to spread from the central business district during the 1870s to 1910s. Later, cars and trucks increased the city's spatial scale. The interstate highway system and outer-city beltway "loops" encouraged development even further from the city's central business district. Today, many people live in suburbs far from their place of employment.

- **Define *suburban sprawl* and discuss a problem caused or exacerbated by sprawl.**

Suburban sprawl is a patchwork of vacant and developed tracts around the edges of cities; sprawl contains a low population density. Sprawl cuts into the surrounding rural land and causes environmental problems such as loss of wetlands, loss of biological habitat, air pollution, and water pollution.

- **List at least five characteristics of an ideal sustainable city.**

A **sustainable city** is a city with a livable environment, a strong economy, and a social and cultural sense of community; sustainable cities enhance the well-being of current and future generations of urban dwellers. A sustainable city has clear, cohesive urban policies that enable the government infrastructure to manage it effectively. A sustainable city uses energy and other resources efficiently and makes use of renewable energy as much as possible. A sustainable city reduces pollution and wastes by reusing and recycling materials. A sustainable city has large areas of green space. A sustainable city is people-centered, not automobile-centered.

- **Explain how city planners have incorporated environmental sustainability into the design of Curitiba, Brazil.**

Curitiba developed an inexpensive, efficient mass transit system that uses clean, modern buses that run in high-speed bus lanes. High-density development was largely restricted to areas along the bus lines. Curitiba uses a special low-polluting fuel that contains a mixture of diesel fuel, alcohol, and soybean extract. Over several decades, Curitiba purchased and converted flood-prone properties along rivers in the city to a series of interconnected parks crisscrossed by bicycle paths. This move reduced flood damage and increased the per capita amount of "green space."

Thinking About the Environment

1. What is urbanization?

2. Which countries are the most urbanized? the least urbanized? What is the urbanization trend today in largely rural nations?

3. What is an urban agglomeration? Give an example.

4. What are some of the problems brought on by rapid urban growth in developing countries?

5. What are the four variables that urban ecologists study?

6. Suggest a reason why many squatter settlements are built on floodplains or steep slopes.

7. Generally, the higher a country's level of urbanization, the lower its level of poverty. Suggest a possible explanation for this observation.

8. What is a brownfield?

9. How has transportation affected the spatial structure of cities?

10. How can land use planning promote compact development?

11. What is an urban heat island? a dust dome?

12. How has Curitiba, Brazil, been designed to incorporate sustainability?

13. Why is good governance so important in increasing sustainability in cities?

14. Why are the U.S. urban agglomerations shown in Figure 10.3 each considered a functional system?

Quantitative questions relating to this chapter are on our Web site.

Take a Stand

Visit our Web site at http://www.wiley.com/college/raven (select Chapter 10 from the Table of Contents) for links to more information about suburban sprawl in U.S. cities. Select a specific U.S. city and debate the issues with your classmates. You will find tools to help you organize your research, analyze the data, think critically about the issues, and construct a well-considered argument. *Take a Stand* activities can be done individually or as a team, as oral presentations, written exercises, or Web-based (e-mail) assignments.

Additional online materials relating to this chapter, including a Student Testing Section with study aids and self-tests, Environmental News, Activity Links, Environmental Investigations, and more, are also on our Web site.

11

Fossil Fuels

An oil refinery burns in Novi Sad, Serbia.
Pollution from oil comes not only from automobile and truck emissions, but also at each stage of production: pumping, shipping, processing, storage, and delivery. This includes oil spilled in water and on land, and both day-to-day and emergency conditions at refineries, as depicted here. Oil refineries are attractive targets in military actions, such as the one that set this refinery ablaze during the 1999 conflict in Kosovo. (Desmond Boylan/Reuters Photo Archive/NewsCom)

In his January 2006 State of the Union address, President George W. Bush warned that the United States is "addicted to oil." This echoes similar claims made over the past several decades, not just by environmentalists but by business leaders, academics, and world leaders. Oil is cheap (relative to most other energy sources), easy to transport and use, and extremely versatile. Oil provides a range of goods and services. Not only is it the world's largest single source of energy, it is also one of three main sources of useful material—the other two being organic materials like wood and cotton, and minerals like sand and steel. Oil also seems to be fairly abundant, with large deposits found on every continent except Antarctica. So why the concern with this "addiction"?

First, the United States imports more than two-thirds of the oil that we use, and much of the oil we import comes from a small number of countries. Part of the concern, then, is that we have to rely on other countries to maintain our current lifestyle. Threats to oil supply are often interpreted as national threats. Second, like most addictive substances, oil can have substantial side effects. Oil drilling is a hazardous, energy intensive venture, as are transportation and refining (see photograph). Burning oil (usually as diesel or gasoline) releases a variety of pollutants that have effects from local to global scale. And, as mentioned in Chapter 7, reliance on oil for transportation can lead to health problems including obesity and heart disease.

Finally, it is not clear how long our addiction can continue. Some estimates of available oil suggest that we have reached "Peak Oil." Peak Oil is the point in time at which the maximum amount of oil is being pumped from underground. This amount depends on several factors—including cost of pumping from various locations and the total amount of oil available. However, once we pass the peak, the amount of oil available worldwide will go down each year, while price goes up. Several of the world's largest oil fields passed their peak productivity in 2005 and 2006, and are now in what is called "depletion," where each year less oil can be pumped out. The peak for oil fields in the United States occurred in the early 1970's.

The analogy of "addiction" is not lightly chosen. Addicts do not necessarily act prudently when their supplies are threatened, and are not known to be rational about the risks posed by their addictions. Weaning from an addiction can be difficult, and typically requires intervention. Quitting abruptly can be difficult, or even traumatic; quitting oil tomorrow would grind all of our transportation, agriculture, construction, manufacturing, food production, heating, lighting, and computing—in short, our entire economy—to a halt.

In this chapter you will examine the various fossil fuels—nonrenewable fuels such as petroleum, natural gas, coal, and synfuels—and the environmental problems associated with their production and use. The chapter concludes with a discussion of the current U.S. National Energy Policy, which was formulated because a systems perspective shows that our dependence on fossil fuels cannot continue. The following two chapters look at alternatives to fossil fuels: nuclear (which is currently in a distant fourth place to oil, natural gas, and coal) in Chapter 12, and alternative sources—including energy conservation and efficiency—in Chapter 13.

World View and Closer to You...
Real news footage relating to fossil fuels around the world and in your own geographic region is available at www.wiley.com/college/raven by clicking on "World View and Closer to You."

Energy Sources and Consumption

LEARNING OBJECTIVES

- Define *energy density* and *energy efficiency* and explain their importance.
- Compare per capita energy consumption in highly developed and developing countries.

Everything humans do requires energy. We use energy to move things and build things, and to heat, cool, and illuminate our living and work spaces. We use energy to plant, water, harvest, process, ship, and store food. Energy is required to capture energy—to drill for and pump oil, to mine coal and uranium, to build solar panels, and to run wind turbines.

Just a few hundred years ago, almost all of the energy used by people was derived from agriculture (including wood, dung, and peat), wind, or water. Energy sources were local, and individuals were limited by the **energy density** of those sources. The discovery of fire and the domestication of animals significantly increased humans' abilities to manipulate their environments, but these advances were minor in comparison to the control of heat from fossil fuels and nuclear energy. Electricity allows the concentration of large amounts of very useful energy from a wide range of sources. However, the concentration of energy has also led to a concentration of wastes associated with energy, including heat and a range of pollutants.

A conspicuous difference in per capita energy consumption exists between highly developed and developing nations (**Figure 11.1**). Highly developed nations consume much more energy per person than developing nations. Although only 20% of the world's population lived in highly developed countries in 2000, these people used 60% of the commercial energy consumed worldwide.[1] The average person in the United States used approximately ten times more energy than did the average Nigerian. In addition, about 60% of energy used in Nigeria was for residential uses, compared to only about 20% in the United States.

A comparison of energy requirements for food production clearly illustrates the energy consumption differences between developing and highly developed countries. Farmers in developing nations rely on their own physical energy or the energy of animals to plow and tend fields. In contrast, agriculture in highly developed countries involves many energy-consuming machines, such as tractors, automatic loaders, and combines. Additional energy is required to produce the fertilizers and pesticides widely used in industrialized agriculture. The larger energy input is one reason the agricultural productivity of highly developed countries is greater than that of developing countries.

World energy consumption has increased every year since 1982, with most of the increase occurring in developing countries. From 2003 to 2004, for example, energy consumption increased worldwide by about 2.5%, and most of this increase came from China and India. One of the goals of developing countries is to improve their standard of living. One way to achieve this goal is through economic development, a process usually accompanied by a rise in per capita energy consumption. In addition, as you learned in Chapter 8, the human population continues to increase, and most of this growth will occur in developing countries. In contrast, the population in highly developed nations is more stable, and many energy experts think those nations' per capita energy consumption may be at or near a saturation maximum. Additional energy demands may be met

energy density: The amount of energy contained within a given volume or mass of an energy source. Gasoline has a higher energy density than does dry wood, which in turn has a higher energy density than wet wood.

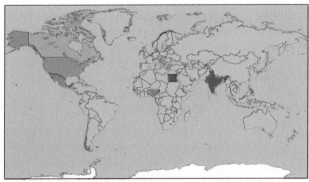

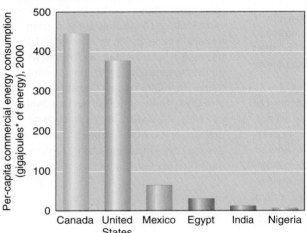

*1 gigajoule = 1 billion joules

Figure 11.1 **Annual per capita commercial energy consumption in selected countries, 2000.** Energy consumption per person in highly developed nations is much greater than it is in developing countries. (The map is color-coded with the bar graph.)

[1]Unless noted otherwise, all energy facts cited in this chapter were obtained from the Energy Information Administration (EIA), the statistical agency of the U.S. Department of Energy (DOE). Because EIA and DOE weren't created until 1975, pre-1975 data cannot be verified and aren't provided in this chapter.

by increased **energy efficiency**, the use of less energy to accomplish the same task, of such items as appliances, automobiles, and home insulation (see Chapter 13). Energy efficiency ranges from 0 to 100%; use of natural gas for heating has an efficiency of close to 100%, while the efficiency of burning natural gas to generate electricity has a maximum efficiency of about 60%.

Figure 11.2 shows how energy is used in the United States. Industry, which encompasses the production of chemicals, minerals, food, and additional energy resources, accounts for about 42% of the energy we consume. Another 33% of our consumed energy is used to make buildings comfortable with heating, air conditioning, lighting, and hot water. The remaining 25% of energy we consume provides transportation, primarily for motor vehicles.

energy efficiency: A measure of the fraction of energy used relative to the total energy available in a given source.

REVIEW

1. How does per capita energy consumption compare in highly developed and developing countries?
2. Why are fuels with higher energy density more useful?

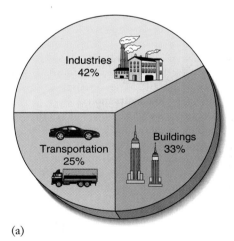
(a)

Figure 11.2   Energy consumption in the United States.
(a) Overall, industries consume 42%, buildings 33%, and transportation 25%. (b), (c), and (d) Breakdown of energy consumption by (b) industries, (c) buildings, and (d) transportation.

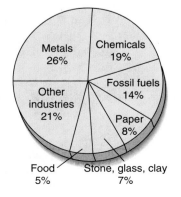
(b)

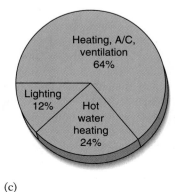
(c)

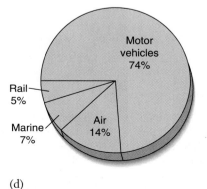
(d)

Fossil Fuels

- Define *fossil fuel*, and distinguish among coal, oil, and natural gas.
- Describe the processes that formed coal, oil, and natural gas.
- Explain the role of photosynthesis in the creation of fossil fuels.

fossil fuel: Combustible deposits in Earth's crust, composed of the remnants (fossils) of prehistoric organisms that existed millions of years ago. Coal, oil (petroleum), and natural gas are the three types of fossil fuel.

Energy is obtained from a variety of sources, including **fossil fuels**, nuclear reactors (see Chapter 12), biomass, solar and other alternative energy sources (see Chapter 13). Today, fossil fuels—*coal, oil,* and *natural gas*—supply most of the energy required in North America. A fossil fuel is composed of the remnants of organisms, compressed in an oxygen-free environment. Fossil fuels resulted from photosynthesis (Chapter 3) that captured solar energy millions of years ago.

Fossil fuels are **nonrenewable resources**—Earth's crust has a finite, or limited, supply of them, and that supply is depleted by use. Although natural processes are still forming coal and other fossil fuels, they are forming too slowly (on a scale of millions of years) to replace the fossil fuel reserves we are using. Fossil fuel formation does not keep pace with use, and as fossil fuels are used up, we will have to switch to other forms of energy.

How Fossil Fuels Formed

coal: A black, combustible solid composed mainly of carbon, water, and trace elements found in Earth's crust; formed from the remains of ancient plants that lived millions of years ago.

Three hundred million years ago, the climate of much of Earth was mild and warm, and atmospheric carbon dioxide levels were higher. Vast swamps were filled with plant species that have long since become extinct. Many of these plants—horsetails, ferns, and club mosses—were as large as trees (**Figure 11.3**).

Plants in most environments decay rapidly after death, owing to the activities of decomposers such as bacteria and fungi. As the ancient swamp plants died, either from old age or from storm damage, they fell into the swamp and were covered by water. Their watery grave prevented the plants from decomposing much; wood-rotting fungi cannot act on plant material where oxygen is absent, and anaerobic bacteria, which thrive in oxygen-deficient environments, do not decompose wood rapidly. Over time, more and more dead plants piled up. As a result of periodic changes in sea level, sediment (mineral particles deposited by gravity) accumulated, forming layers that covered the plant material. Aeons passed, and the heat and pressure that accompanied burial converted the nondecomposed plant material into a carbon-rich rock called **coal**, and the layers of sediment into sedimentary rock. Much later, geologic upheavals raised these layers so that they were nearer Earth's surface.

Figure 11.3 Reconstruction of a Carboniferous swamp. The plants of the Carboniferous period, 360 million to 286 million years ago, included giant ferns, horsetails, and club mosses that formed our present-day coal deposits. (No. GEO85638c, Field Museum of Natural History)

Oil formed when large numbers of microscopic aquatic organisms died and settled in the sediments. As these organisms accumulated, their decomposition depleted the small amount of oxygen present in the sediments. The resultant oxygen-deficient environment prevented further decomposition. Over time, the dead remains were covered and buried deeper in the sediments. Although we do not know the basic chemical reactions that produce oil, the heat and pressure caused by burial presumably aided in the conversion of these remains to the mixture of **hydrocarbons** (molecules containing carbon and hydrogen) known as oil.

Natural gas, composed primarily of the simplest hydrocarbon, **methane,** formed in essentially the same way as oil, only at higher temperatures, typically greater than 100°C. Over millions of years, as the remains of organisms were converted to oil or natural gas, the sediments covering them were transformed into sedimentary rock.

In one sense, burning of fossil fuels represents the completion of the carbon cycle, part of a natural system. Normally, solar energy and carbon dioxide are captured through photosynthesis stored for weeks or years, then consumed and released. In the case of fossil fuels, however, the energy and carbon were stored up over millions of years, but are being released over about a century.

oil: A thick, yellow to black, flammable liquid hydrocarbon mixture found in Earth's crust; formed from the remains of ancient microscopic aquatic organisms.

natural gas: A mixture of energy-rich gaseous hydrocarbons (primarily methane) that occurs, often with oil deposits, in Earth's crust.

REVIEW

1. What are fossil fuels?
2. How are coal, oil, and natural gas formed?

Coal

LEARNING OBJECTIVES

- Distinguish between surface mining and subsurface mining.
- Discuss the advantages and disadvantages of coal.
- Summarize the environmental problems associated with using coal, including acid mine drainage, climate change, and acid deposition.
- Explain how resource recovery and fluidized-bed combustion can make coal a cleaner fuel.

Although coal was used as a fuel for centuries, not until the 18th century did it begin to replace wood as the dominant fuel in the Western world. Since then, coal has had a significant impact on human history. It was coal that powered the steam engine and supplied the energy for the Industrial Revolution, which began in the mid-18th century. Today utility companies use coal to produce electricity, and heavy industries use coal for steel production. Coal consumption has surged in recent years, particularly in the rapidly growing economies of China and India, both of which have large coal reserves.

Coal occurs in different grades, largely as a result of the varying amounts of heat and pressure it was exposed to during formation. Coal exposed to high heat and pressure during its formation is drier, is more compact (and therefore harder), and has a higher heating value (that is, a higher energy density). Lignite, subbituminous coal, bituminous coal, and anthracite are the four most common grades of coal (**Table 11.1**).

Table 11.1 A Comparision of Different Kinds of Coal

Type of Coal	Color	Water Content (%)	Relative Sulfur Content	Carbon Content (%)	Average Heat Value (BTU/pound)	2006 Cost at Mine for 2000 lbs of Cool
Lignite	Dark brown	45	Medium	30	6,000	$13.49
Subbituminous coal	Dull black	20–30	Low	40	9,000	$8.68
Bituminous coal	Black	5–15	High	50–70	13,000	$36.80
Anthracite	Black	4	Low	90	14,000	$41.00

Sources: EIA, U.S. Department of Energy, and USGS

Figure 11.4 **Burning coal seam in Black Dragon Mountain, Mongolia.** One problem with mining coal near the surface is that it is highly flammable—a spark can set an entire seam burning, and in some cases can be impossible to extinguish. This burning seam in Black Dragon Mountain in Mongolia releases huge amounts of pollutants into the atmosphere. (Elleringmann/Laif/Aurora Photos)

Lignite is a soft coal, brown or brown-black in color with a soft, woody texture. It is moist and produces little heat compared with other types of coal. Lignite is often used to fuel electric power plants. Sizable deposits of it are found in the western states, and the largest producer of lignite in the United States is North Dakota.

Subbituminous coal is a grade of coal intermediate between lignite and bituminous. Like lignite, subbituminous coal has a relatively low heat value and sulfur content. Many coal-fired electric power plants in the United States burn subbituminous coal because its sulfur content is low. It is found primarily in Alaska and a few western states, such as Montana and Wyoming.

Bituminous coal, the most common type, is also called **soft coal,** even though it is harder than lignite and subbituminous coal. Bituminous coal is dull to bright black with dull bands. Much bituminous coal contains sulfur, a chemical element that causes severe environmental problems (discussed shortly) when the coal is burned in the absence of pollution-control equipment. Nevertheless, electric power plants use bituminous coal extensively because it produces a lot of heat. In the United States, bituminous coal deposits are found in the Appalachian region, near the Great Lakes, in the Mississippi Valley, and in central Texas.

The highest grade of coal, **anthracite** or **hard coal,** was exposed to extremely high temperatures during its formation. It is a dark, brilliant black and burns most cleanly—it produces the fewest pollutants per unit of heat released—of all the types of coal because it is not contaminated by large amounts of sulfur. Anthracite has the highest heat-producing capacity of any grade of coal. Anthracite deposits in the United States are largely depleted; most of the remaining deposits are east of the Mississippi River, particularly in Pennsylvania. Coal is usually found in seams, underground layers that vary from 2.5 cm (1 in.) to more than 30 m (100 ft) in thickness. Geologists think that most, if not all, major coal deposits have been identified. Scientists working with coal are therefore concerned less about finding new deposits than about the safety and environmental problems associated with coal, such as the burning coal seam in **Figure 11.4**.

Coal Reserves

Coal, the most abundant fossil fuel in the world, is found primarily in the Northern Hemisphere (**Figure 11.5**). The largest coal deposits are in the United States, Russia, China, Australia, India, Germany, and South Africa. The United States has 25% of the world's coal supply in its massive deposits. According to the World Resources Institute, known world coal reserves could last more than 200 years at the present rate of consumption. Coal resources currently too expensive to develop have the potential to provide enough coal to last for 1000 or more years (at current consumption rates). For example, some coal deposits are buried more than 5000 feet inside Earth's crust. Drilling a shaft that deep would cost considerably more than the current price of coal would justify.

Coal Mining

The two basic types of coal mines are surface and subsurface (underground) mines. The type of mine chosen depends on surface contours and on the location of the coal bed relative to the surface. If the coal bed

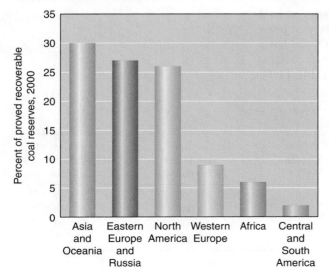

Figure 11.5 **Distribution of coal deposits.** Data are presented as percentages of the 2000 estimated recoverable reserves—that is, of coal known to exist that can be recovered under present economic conditions with existing technologies. The majority of the world's coal deposits are located in the Northern Hemisphere. (The map is color-coded with the bar graph.)

is within 30 m (100 ft) or so of the surface, **surface mining** is usually done (**Figure 11.6**). In one type of surface mining, **strip mining,** a trench is dug to extract the coal, which is scraped out of the ground and loaded into railroad cars or trucks. Then a new trench is dug parallel to the old one, and the *overburden* from the new trench is put into the old trench, creating a **spoil bank,** a hill of loose rock. Digging the trenches involves using bulldozers, giant power shovels, and wheel excavators to remove the ground covering the coal seam. Surface mining is used to obtain approximately 60% of the coal mined in the United States.

When the coal is deeper in the ground or runs deep into the ground from an outcrop on a hillside, it is mined underground. **Subsurface mining** accounts for approximately 40% of the coal mined in the United States.

Surface mining has several advantages over subsurface mining: It is usually less expensive and safer for miners, and it generally allows a more complete removal of coal from the ground. However, surface mining disrupts the land much more extensively than subsurface mining and has the potential to cause several serious environmental problems.

Safety Problems Associated with Coal

Although we usually focus on the environmental problems caused by mining and burning coal, there are significant human safety and health risks in the mining process itself. Underground mining is a hazardous occupation. According to the DOE, during the 20th century more than 90,000 American coal miners died in mining accidents, although the number of deaths per year declined significantly in the latter part of the century. Miners have an increased risk of cancer and **black lung disease,** a condition in which the lungs are coated with inhaled coal dust, and the exchange of oxygen between the lungs and blood is severely restricted. It is estimated that these diseases are responsible for the deaths of at least 2000 miners in the United States each year.

Environmental Impacts of the Mining Process

Coal mining, especially surface mining, has substantial effects on the environment. Prior to passage of the 1977 **Surface Mining Control and Reclamation Act (SMCRA),** abandoned surface coal mines were usually left as large open pits or trenches. *Highwalls,* cliffs of excavated rock, some more than 30 m (100 ft) high, were left exposed. Acid and toxic mineral drainage from such mines, along with the removal of topsoil, which was buried or washed away by erosion, prevented most plants from naturally recolonizing the land. Streams were polluted with sediment and **acid mine drainage,** produced when rainwater seeps through iron sulfide minerals exposed in mine wastes. Dangerous landslides occurred on hills unstable from the lack of vegetation.

Surface-mined land can be restored to prevent such degradation and to make the land productive for other purposes, although restoration is expensive and technically challenging. The SMCRA requires coal companies to restore areas that have been surface mined, beginning in 1977. The SMCRA protects the environment by requiring permits and inspections of active coal mining operations and reclamation sites. The SMCRA prohibits coal mining in sensitive areas such as national parks, wildlife refuges, wild and scenic rivers, and sites listed on the National Register of Historic Places. In addition, the SMCRA stipulates that surface-mined land abandoned prior to 1977 (more than 0.4 million hectares, or 1 million acres) should gradually be restored, using money from a tax that coal companies pay on currently mined coal. According to the U.S. Office of Surface Mining, more than $1.5 billion has been spent reclaiming the most dangerous abandoned mine lands; approximately two-thirds of this money was spent in four states—Pennsylvania, Kentucky, West Virginia, and Wyoming. However, so many abandoned mines exist that it is doubtful they can all be restored.

One of the most land-destructive types of surface mining is mountaintop removal. A **dragline,** a huge shovel with a 20-story-high arm, takes enormous chunks out of a mountain, eventually removing the entire mountaintop to reach the coal located below. According to Environmental Media Services, mountaintop removal has leveled between 15 and 25% of the mountaintops in southern West Virginia. The valleys and

surface mining: The extraction of mineral and energy resources near Earth's surface by first removing the soil, subsoil, and overlying rock strata (i.e., the overburden).

subsurface mining: The extraction of mineral and energy resources from deep underground deposits.

Figure 11.6 Surface coal mining. The overlying vegetation, soil, and rock are stripped away, and then the coal is extracted out of the ground. Photographed near Douglas, Wyoming. (Kristin Finnegan Photograhy)

acid mine drainage: Pollution caused when sulfuric acid and dangerous dissolved materials such as lead, arsenic, and cadmium wash from coal and metal mines into nearby lakes and streams.

streams between the mountains are gone as well, filled with debris from the mountaintops. Mountaintop removal is also occurring in parts of Kentucky, Pennsylvania, Tennessee, and Virginia. The SMRCA specifically exempts mountaintop removal. In 1977, when the SMCRA was passed, existing technology for mountaintop removal could take out the top seam of coal along a mountain ridge. Today, mountaintop removal takes out as many as 16 seams of coal, from the ridge to the base of the mountain (**Figure 11.7**). In 2002 a federal judge upheld a 1999 court decision to limit mountaintop removal because disposing of mountaintop waste rock in valleys violates the Clean Water Act (the disposal buries streams).

Environmental Impacts of Burning Coal

The equilibrium between CO_2 in the atmosphere, CO_2 dissolved in the ocean, and CO_2 in organic matter changes over long periods—thousands or millions of years. Over the past century, however, we have been releasing so much CO_2 into the atmosphere through consumption of fossil fuels that Earth's CO_2 equilibrium has been disrupted. Global temperature has been affected because atmospheric CO_2 prevents heat from escaping from the planet. An increase of a few degrees in global temperature caused by higher levels of CO_2 and other greenhouse gases may not seem serious at first glance, but a closer look reveals that such an increase is likely to be quite harmful.

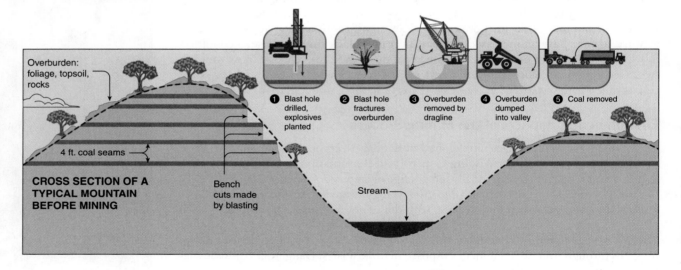

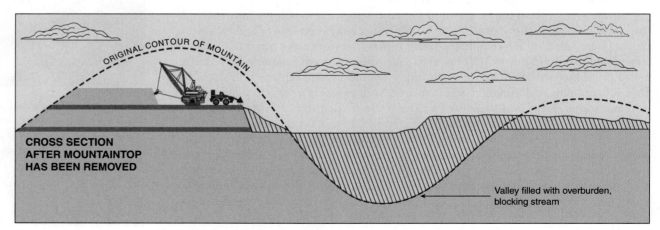

Figure 11.7   Removing coal by leveling a mountain. When coal is removed from mountains in Kentucky, the geography of a region can change. The top of a mountain above a coal seam is scraped off and dumped in the adjacent valley. Next, the coal seam is removed. If another seam lies lower in the mountain, the process is repeated. Streams are buried, lakes created and destroyed, and large amounts of sediment washed downstream as bare earth is eroded.

A global rise in temperature has caused polar and glacial ice to melt, raised sea levels, and will lead to flooding of coastal areas. This will increase coastal erosion and put many coastal buildings (and the people who live in them) at higher risk from violent storms. Other serious environmental consequences of global climate change are considered in Chapter 21. Burning coal causes a more severe CO_2 problem than burning other fossil fuels because coal burning releases more CO_2 per unit of heat produced.

Coal burning generally contributes more air pollutants (including CO_2) than does burning either oil or natural gas. Coal often contains mercury that is released into the atmosphere during combustion. This mercury moves readily from the atmosphere to water and land, and accumulates in the environment where it harms humans as well as wildlife. (See Chapter 22 for a discussion of the human health hazards of elevated levels of mercury.) In the United States, coal-burning electric power plants currently produce one-third of all airborne mercury emissions.

Much bituminous coal contains sulfur and nitrogen that, when burned, are released into the atmosphere as sulfur oxides (SO_2 and SO_3) and nitrogen oxides (NO, NO_2, and N_2O). Sulfur oxides and the nitrogen oxides NO and NO_2 form acids when they react with water (See Chapter 20). These reactions result in **acid deposition**. The combustion of coal is responsible for acid deposition, which is particularly prevalent downwind from coal-burning electric power plants. Normal rain is slightly acidic (pH 5.6), but in some areas acid precipitation has a pH of 2.1, equivalent to that of lemon juice. (See Appendix I for a review of pH.) Acidification of lakes and streams has resulted in the decline of aquatic animal populations and is linked to some of the forest decline documented worldwide (**Figure 11.8**). Acid precipitation and forest decline are discussed in greater detail in Chapter 20.

Although it is relatively easy to identify and measure pollutants such as sulfur oxides in the atmosphere, it is more difficult to trace their exact origins. Air currents transport and disperse air pollutants, which are often altered as they react chemically with other pollutants in the air. Even so, some nations clearly suffer the damage of acid deposition caused by air pollutants produced in other countries, and as a result acid deposition is an international issue.

acid deposition: A type of air pollution in which acid falls from the atmosphere to the surface as precipitation (acid precipitation) or as dry acid particles.

Figure 11.8 Dead trees enveloped in an acid fog. Forest decline was first documented in Germany and Eastern Europe. More recently, it has been observed in eastern North America, particularly at higher elevations. Acid deposition contributes to forest decline. Photographed on Mount Mitchell, North Carolina. (John Shaw Photography)

E N V I R O N E W S

ON CAMPUS

Reducing Fossil Fuel Demand

An increasing number of campuses—from small, private colleges to large public universities—have begun to move away from historically high-energy demand to more sustainable approaches.

Middlebury College in Vermont has gone well beyond now-common practices such as efficient lighting and recycling. The campus has made a commitment to local, sustainably harvested wood (to heat increasingly efficient buildings) as well as policies that reduce paper use, waste generation, and motor vehicle travel. Recently, they committed to offsetting the CO_2 emissions at their campus-run ski area, in part out of concern that climate change would dramatically affect the ski season!

Arizona State University (ASU), with over 60,000 students in a desert environment (compared to just over 2000 at Middlebury, in the often frigid Northeast), provides another example. ASU established a campus-wide commitment to sustainability that includes a major effort to facilitate alternative transportation around campus. One component is free bus passes for students between campus and many nearby locations. This led to over 100,000 student bus trips in a single month. The campus recycling program includes batteries, shoes, and cell phones, in addition to traditional paper, plastic, glass, and metals. Sustainability is integrated across the campus, from purchasing to mail services to buildings. In 2006, ASU established several academic degrees in sustainability, to promote students thinking about the environment from an interdisciplinary systems perspective.

In another innovative approach to avoiding fossil fuels, students at Middle Tennessee State University and Tennessee Tech ran a successful campaign to support energy conservation and alternative energy sources. In referendums on both campuses, more than 90% of students voting agreed to increase their annual fees by $16. This amount will be used to fund energy conservation measures around campus and purchase green energy (see Chapter 13).

In all of these cases, schools are taking a systems perspective, which leads them to alter their traditional habits and practices.

Making Coal a Cleaner Fuel

It is possible to reduce sulfur emissions associated with the combustion of coal by installing **scrubbers**, or desulfurization systems, to clean the power plants' exhaust. As the polluted air passes through a scrubber, chemicals in the scrubber react with the pollution and cause it to precipitate (settle) out. Modern scrubbers remove 98% of the sulfur and 99% of the particulate matter in smokestacks. Desulfurization systems are expensive; they cost about $50 to $80 per installed kilowatt, or about 10 to 15% of the construction costs of a coal-fired electric power plant.

In lime scrubbers, a chemical spray of water and lime neutralizes acidic gases such as sulfur dioxide, which remain behind as a calcium sulfate sludge that becomes a disposal problem (see Figure 20.11d). A large power plant may produce enough sludge annually to cover 2.6 km^2 (1 mi^2) of land 0.3 m (1 ft) deep. Although many power plants currently dispose of the sludge in landfills, some have found markets for the material. In **resource recovery**, the sludge is treated as a marketable product rather than as a polluted emission. Some utilities have begun selling calcium sulfate from scrubber sludge to wallboard manufacturers. (Wallboard is traditionally manufactured from gypsum, a mineral composed of calcium sulfate.) Other companies are using fly ash, the ash from the chimney flues, to make a lightweight concrete that could substitute for wood in the building industry. Some farmers have started applying one type of sludge (calcium sulfate) as a soil conditioner. Plants grow better because calcium sulfate neutralizes acids in some soils and increases the water-holding capacity of the soil. (The calcium sulfate acts like a sponge.)

The **Clean Air Act Amendments of 1990** required the nation's 111 dirtiest coal-burning power plants to cut sulfur dioxide emissions. Compliance resulted in a total annual decrease of 3.8 million metric tons nationwide. This represented a significant portion of the total amount of sulfur dioxide emitted in the United States each year (15.7 million metric tons in 1993, before reductions mandated by the Clean Air Act Amendments went into effect). In the second phase of the Clean Air Act Amendments, more than 200 additional power plants made SO$_2$ cuts by the year 2000, resulting in a total annual decrease of 10 million metric tons nationwide. A nationwide cap on SO$_2$ emissions from coal-burning power plants was imposed after 2000. Utilities also cut nitrogen oxide emissions by 2.6 million tons per year, out of 7.2 million tons per year total.

Clean coal technologies are new methods being developed for burning coal that will not contaminate the atmosphere with sulfur oxides and will significantly reduce nitrogen oxide production. Clean coal technologies include fluidized-bed combustion and coal gasification and liquefaction (considered shortly, in the discussion of synfuels). However, these technologies have little impact on reducing CO$_2$ emissions.

Fluidized-bed combustion mixes crushed coal with particles of limestone in a strong air current during combustion (**Figure 11.9**). Fluidized-bed combustion takes place at a lower temperature than regular coal burning, and fewer nitrogen oxides are produced. (Higher temperatures cause atmospheric nitrogen and oxygen to combine, forming nitrogen oxides.) Because the sulfur in coal reacts with the calcium in limestone to form calcium sulfate, which then precipitates out, sulfur is removed from the coal during the burning process, so scrubbers are not needed to remove it after combustion.

In the United States several large power plants are testing fluidized-bed combustion, and a few small power plants that use this technology are already in commercial operation. The Clean Air Act Amendments of 1990 provide incentives for utility companies to convert to clean coal technologies, such as fluidized-bed combustion. The cost of installing fluidized-bed combustion compares favorably with the cost of installing desulfurization systems.

Fluidized-bed combustion is more efficient than traditional coal burning—that is, it produces more heat from a given amount of coal—and therefore reduces CO$_2$ emissions per unit of electricity produced. If improvements of this technology were developed and adopted widely by coal-burning power plants, fluidized-bed combustion could significantly reduce the amount of CO$_2$ released into the atmosphere. Pressurized fluidized-bed combustion is being developed as a way to reduce CO$_2$ and nitrogen and

resource recovery: The process of removing any material—sulfur or metals, for example—from polluted emissions or solid waste and selling it as a marketable product.

fluidized-bed combustion: A clean-coal technology in which crushed coal is mixed with limestone to neutralize the acidic sulfur compounds produced during combustion.

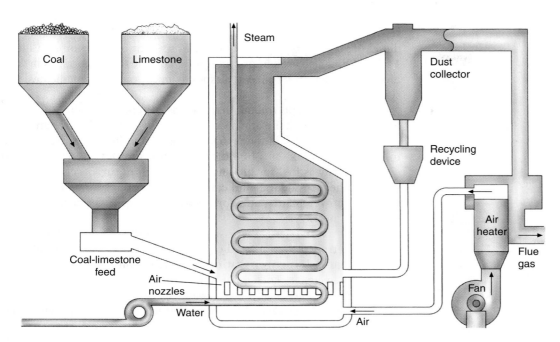

Figure 11.9 Fluidized-bed combustion of coal. Crushed coal and limestone are suspended in air. As the coal burns, limestone neutralizes most of the sulfur dioxide in the coal. The heat generated during combustion converts water to steam, which powers various industrial processes.

sulfur oxides. By operating fluidized-bed combustion under high pressure, complete combustion of coal occurs at low temperatures. Sulfur emissions are removed as calcium sulfate, and few nitrogen oxides form because of the low temperatures. Pressurized fluidized-bed combustion is more expensive than regular fluidized-bed combustion, because it requires a costly pressurized vessel.

REVIEW

1. What are the advantages and disadvantages of using coal as an energy resource?
2. Which type of coal mining—surface or subsurface mining—is more land-intensive?
3. What are acid mine drainage and acid deposition?
4. What are the environmental benefits of resource recovery? of fluidized-bed combustion?

Oil and Natural Gas

LEARNING OBJECTIVES

- Define *structural trap* and give two examples.
- Explain what is meant by "Peak Oil," and why it might concern us.
- Discuss the environmental problems of using oil and natural gas.
- Summarize the continuing controversy surrounding the Arctic National Wildlife Refuge.

Although coal was the most important energy source in the United States during the early 1900s, oil and natural gas became increasingly important, beginning in the 1940s. This change occurred largely because oil and natural gas are easier to transport and because they burn cleaner than coal. In 2005, oil and natural gas supplied approximately 63% of the energy used in the United States. In comparison, other U.S. energy sources included coal (23%), nuclear power (8.1%), and hydropower (3.1%). Globally in 2004, oil and natural gas provided 60.6% of the world's energy. In comparison, other major energy sources included coal (25.5%), hydroelectric power (6.3%), and nuclear power (6.1%) (**Figure 11.10**).

Figure 11.10 World commercial energy sources, 2004. Note the overwhelming importance of oil, coal, and natural gas as commercial energy sources. "Alternatives" include geothermal, solar, wind, and wood.

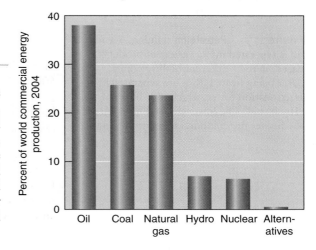

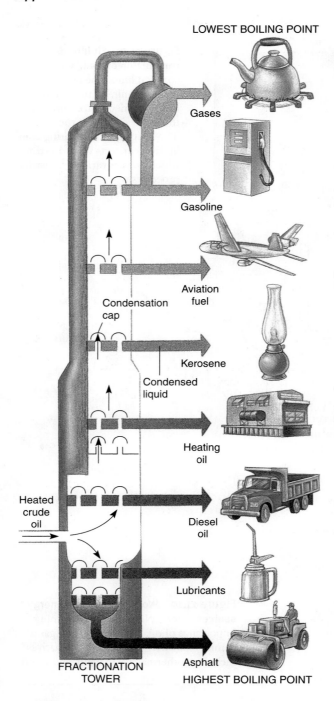

LOWEST BOILING POINT

Gases

Gasoline

Aviation
fuel

Condensation
cap

Kerosene

Condensed
liquid

Heating
oil

Heated
crude
oil

Diesel
oil

Lubricants

FRACTIONATION
TOWER

Asphalt

HIGHEST BOILING POINT

Figure 11.11   Petroleum refining. Crude oil is separated into a variety of products based on their different boiling points. After being heated, they are separated in a fractionation tower, which may be 30 m (100 ft) tall. The lower the boiling point, the higher the compounds rise in the tower.

Petroleum, or **crude oil,** is a liquid composed of hundreds of hydrocarbon compounds. During petroleum refining, the compounds are separated into different products—such as gases, gasoline, heating oil, diesel oil, and asphalt—based on their different boiling points (**Figure 11.11**). Oil is also used to produce **petrochemicals,** compounds in such diverse products as fertilizers, plastics, paints, pesticides, medicines, and synthetic fibers.

In contrast to petroleum, natural gas contains only a few different hydrocarbons: methane and smaller amounts of ethane, propane, and butane. Propane and butane are separated from the natural gas, stored in pressurized tanks as a liquid called **liquefied petroleum gas,** and used primarily as fuel for heating and cooking in rural areas. Methane is used to heat residential and commercial buildings, to generate electricity in power plants, and for a variety of purposes in the organic chemistry industry.

Use of natural gas is increasing in three main areas—generation of electricity, transportation, and commercial cooling. One example of a systems approach is **cogeneration,** in which natural gas is used to produce both electricity and steam; the heat of the exhaust gases provides the energy to make steam for water and space or industrial heating (see Figure 13.17). Cogeneration systems that use natural gas provide relatively clean and efficient electricity.

Natural gas as a fuel for trucks, buses, and automobiles offers significant environmental advantages over gasoline or diesel: Natural gas vehicles emit 80 to 93% fewer hydrocarbons, 90% less carbon monoxide, 90% fewer toxic emissions, and almost no soot. As of 2000, about 100,000 vehicles that use compressed natural gas for fuel were in use in the United States. Most of these are fleet vehicles. The city of Los Angeles has the largest fleet of natural gas-powered transit buses in North America.

Natural gas efficiently fuels residential and commercial air-cooling systems. One example is the use of natural gas in a desiccant-based (air-drying) cooling system, which is ideal for supermarkets, where humidity control is as important as temperature control. Restaurants are also important users of natural gas-powered desiccant-based cooling systems.

The main disadvantage of natural gas is that deposits are often located far from where the energy is used. Because it is a gas and is less dense than a liquid, natural gas costs four times more to transport through pipelines than crude oil. To transport natural gas over long distances, it must first be compressed to form **liquefied natural gas (LNG),** then carried on specially constructed refrigerated ships (**Figure 11.12**).

After LNG arrives at its destination, it must be returned to the gaseous state at regasification plants before being piped to where it will be used. Currently, the United States has only four such plants, which severely restricts the importation of natural gas from other countries. American energy companies claim that the United States needs at least 40 regasification plants to keep costs down for natural gas and to meet increasing demands. The potential market for LNG is huge: Imports increased from 500 to 800 cubic feet between 2003 and 2006, and the industry would like to see at least 10 more terminals in the United States by 2020. However, port communities resist building new plants because, while unlikely to occur, an explosion could cause damage over a large geographic area, causing burns to people located as much as a mile away.

Exploration for Oil and Natural Gas

Geologic exploration is continually under way in search of new oil and natural gas deposits, usually found together under one or more layers of rock. (Recall that oil and natural gas tend to migrate upward until they reach an impermeable rock layer.)

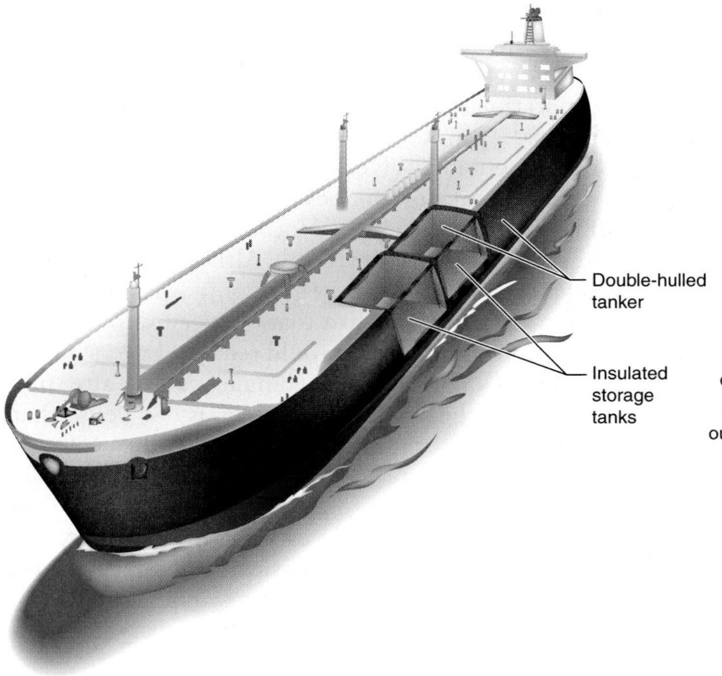

Figure 11.12 **Liquefied natural gas (LNG) ship and tank.** Natural gas liquefies at very low temperatures, and so is transported and stored below -150°C (-260°F).

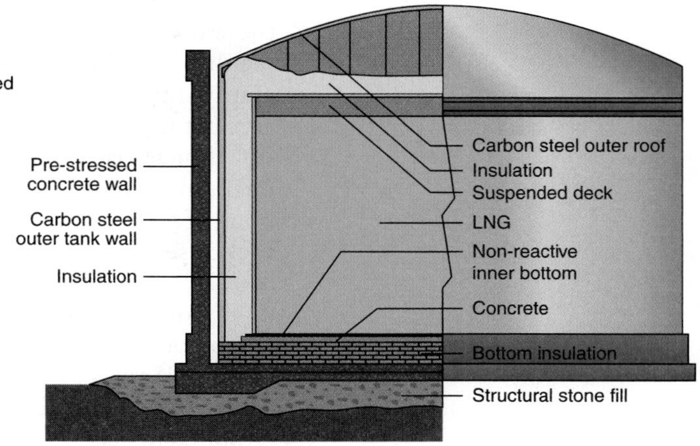

Double-hulled tanker

Insulated storage tanks

Pre-stressed concrete wall

Carbon steel outer tank wall

Insulation

Carbon steel outer roof

Insulation

Suspended deck

LNG

Non-reactive inner bottom

Concrete

Bottom insulation

Structural stone fill

(a) LNG is shipped in tankers that carry it in large, cylindrical tanks. Ships like this one arrive daily at ports in Japan.

(b) On land, LNG is stored in double-walled insulated tanks. Major LNG accidents are unusual: As LNG heats, it forms a vapor that can burn at a high temperature, but the vapor becomes lighter than air and disperses above −105°C (−160°F).

Oil and natural gas deposits are usually discovered indirectly by the detection of **structural traps** (**Figure 11.13**).

Plate tectonic movements sometimes cause the upward folding of sedimentary rock **strata** (layers). Sometimes the strata that arch upward include both porous and impermeable rock. If impermeable layers overlie porous layers, any oil or natural gas present from a source rock such as shale may work its way up through the porous rock to accumulate under the impermeable layer.

structural traps: Underground geologic structures that tend to trap any oil or natural gas if it is present.

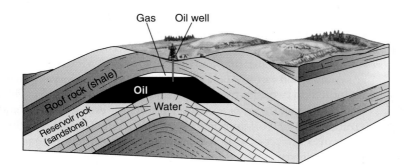

Gas   Oil well

Roof rock (shale)

Oil

Water

Reservoir rock (sandstone)

Figure 11.13 **Structural trap.** The most important of several kinds of structural traps is shown. These traps form when sedimentary rock strata buckle, or fold upward. Oil and natural gas seep through porous reservoir rock such as sandstone and collect under nonporous layers such as a roof of shale. Natural gas accumulates on top of the oil, which in turn floats on groundwater.

Many important oil and natural gas deposits (for example, oil deposits known to exist in the Gulf of Mexico) are found in association with **salt domes,** underground columns of salt. Salt domes develop when extensive salt deposits form at Earth's surface because of the evaporation of water. All surface water contains dissolved salts. The salts dissolved in ocean water are so concentrated they can be tasted, but even fresh water contains some dissolved material. If a body of water lacks a passage to the ocean, as an inland lake often does, the salt concentration in the water gradually increases. (The Great Salt Lake in North America is an example of a salty inland body of water that formed in this way. Although three rivers empty into the Great Salt Lake, water escapes from the lake only by evaporation, accounting for its high salinity—four times higher than that of ocean water.)

If such a lake were to dry up, a massive salt deposit would remain. Layers of sediment may eventually cover such deposits and convert to sedimentary rock after millions of years. The rock layers settle, and the salt deposit, which is less dense than rock, rises in a column—a salt dome. The ascending salt dome, together with the rock layers that buckle over it, provides a trap for oil or natural gas.

Geologists use a variety of techniques to identify structural traps that might contain oil or natural gas. One method is to drill test holes in the surface and obtain rock samples. Another method is to produce an explosion at the surface and measure the echoes of sound waves that bounce off rock layers under the surface. These data are interpreted to determine whether structural traps are present. However, many structural traps do not contain oil or natural gas.

Three-dimensional seismology is a new technology that maps oil fields three-dimensionally, enabling geologists to have a higher rate of success when drilling. Another new technology that improves oil recovery is horizontal drilling. Traditional oil wells are vertical and cannot veer off to follow the contours of underground formations that contain oil. Wells dug with horizontal drilling follow contours, and they generally yield three to five times as much oil as vertical wells.

Even with the new technologies, searching for oil and natural gas is expensive. It costs millions of dollars for the basic geologic analyses to find structural traps. And once oil or natural gas is located, drilling and operating the wells cost additional millions.

Reserves of Oil and Natural Gas

Although oil and natural gas deposits exist on every continent, their distribution is uneven, and a large share of total oil deposits is clustered relatively close together. Enormous oil fields containing more than half of the world's total estimated reserves are situated in the Persian Gulf region, which includes Iran, Iraq, Kuwait, Oman, Qatar, Saudi Arabia, Syria, United Arab Emirates, and Yemen (**Figure 11.14**). In addition, major oil fields are known to exist in Venezuela, Mexico, Russia, Kazakhstan, Libya, and the United States (in Alaska and the Gulf of Mexico).

Almost half of the world's proved recoverable reserves of natural gas are located in two countries, Russia and Iran (**Figure 11.15**). The United States has more deposits of natural gas than Western Europe, and use of natural gas is more common in North America. Canada and the United States also extract **coal bed methane,** a form of natural gas associated with coal deposits.

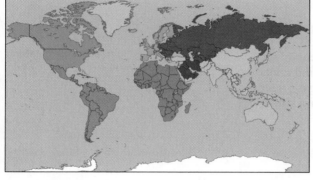

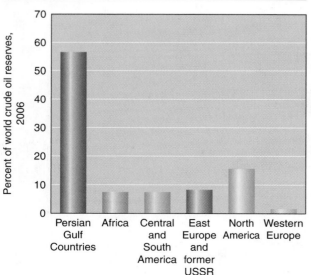

Figure 11.14   Distribution of oil deposits. The Persian Gulf region contains huge oil deposits in a relatively small area, whereas other regions have few. Data are presented as regional percentages of the 2006 world estimate of crude oil reserves. (The map is color-coded with the bar graph.)

Fossil fuel resources (Figures 11.5, 11.14, and 11.15) are concentrated in only a few countries. Africa and South America—with the exception of a few countries—have few fossil fuel resources. Their only options for expanding energy use will be to purchase fossil fuels from other countries or develop alternative energy resources. Either of these approaches requires substantial financial capital—another resource generally lacking in developing countries. Thus, worldwide access to energy is a significant equity issue.

It is unlikely that major new oil fields will be discovered in the continental United States, where production peaked over three decades ago. Since the 1980s, the success rate of searches for new oil fields has declined, as has the amount of exploration.

Large oil deposits probably exist under the **continental shelves,** the relatively flat underwater areas that surround continents, and in deepwater areas adjacent to the continental shelves. Despite problems such as storms at sea and the potential for major oil spills, many countries engage in offshore drilling for this oil. New technologies, such as platforms the size of football fields, enable oil companies to drill several thousand feet for oil, making seafloor oil fields once considered inaccessible open for tapping. As many as 18 billion barrels (756 billion gallons) of oil and natural gas may exist in the deep water of the Gulf of Mexico, just off the continental shelf from Texas to Alabama. Continental shelves off the coasts of western Africa and Brazil are also promising. The oil industry is currently developing remote-controlled robots that can install and maintain underwater equipment and pipelines. Environmentalists generally oppose opening the outer continental shelves for oil and natural gas exploration because of the threat a major oil spill would pose to marine and coastal environments. Coastal industries, including fishing and tourism, also oppose oil and natural gas exploration in these areas.

How Long Will Oil and Natural Gas Supplies Last?   It is difficult to project when the world will run out of oil and natural gas, but by some estimates the peak level of oil production has passed, and global resources are in decline. We do not know how many additional oil and natural gas reserves will be discovered, nor do we know if or when technological breakthroughs will allow us to extract more fuel from each deposit. The answer to how long these fuels will last also depends on whether world consumption of oil and natural gas increases, remains the same, or decreases. Economic factors influence oil and natural gas availability and consumption. As reserves are exhausted, prices increase, which drives down consumption and stimulates greater energy efficiency, the search for additional deposits, and the use of alternative energy sources.

Despite adequate oil supplies for the near future, in the long term, we will need other resources. Gasoline has been readily available and inexpensive in the United States for most of the past century. An exception was the brief disruption caused when the Organization of Petroleum Exporting Countries (OPEC) limited global oil supply in the early 1970s. During the next several decades, however, the remaining oil will become more difficult and expensive to obtain (**Figure 11.16**). Most experts think we will begin to have serious problems with oil supplies sometime during the 21st century.

Some experts think that global oil production has already reached **Peak Oil,** the point at which the oil is being withdrawn at the highest possible rate. About 80% of current production comes from oil fields discovered before 1973, and most of these fields have started to decline in production. These analysts say the world must move quickly to develop alternative energy sources because the global demand for energy will only continue to increase even as production declines.

Industry analysts tend to be more optimistic. They think that improving technology will allow us to extract more oil out of old oil fields. (Currently, about 60% is left because it is too expensive to remove using current technology.) New technologies may help us obtain oil from fields formerly unreachable (such as beneath deep-ocean waters). Improved technology may allow us to produce oil from natural gas, coal, and

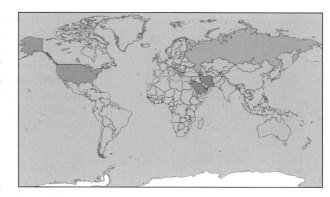

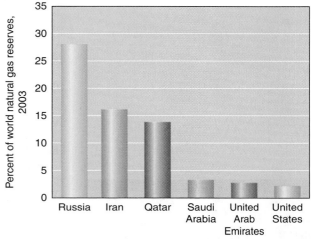

Figure 11.15   **Six countries with the greatest natural gas deposits.** Data are presented as percentages of the 2003 world estimate of natural gas reserves. Note that Russia and Iran together possess 43% of the world's natural gas deposits. (The map is color-coded with the bar graph. Qatar and United Arab Emirates, which are not shown on the map, are tiny countries east of Saudi Arabia.)

Peak Oil: Also known as "Hubberts Peak," after the U.S. geologist who came up with the idea, is the point at which global oil production has reached a maximum rate; by some estimates, Peak Oil is already past.

synfuels (discussed later in the chapter). Even so, the most optimistic predictions are for Peak Oil to occur at around 2035.

Natural gas is more plentiful than oil. Experts estimate that readily recoverable reserves of natural gas, if converted into a liquid fuel, would be equivalent to between 500 billion and 770 billion barrels of crude oil, enough to keep production rising for at least 10 years after conventional supplies of petroleum have begun to decline. However, if the global use of natural gas continues to increase as it has in recent years, then its life supply will be shorter than current projections predict.

Global Oil Demand and Supply

One difficult aspect of the oil market is that the world's major oil producers are not its major oil consumers. In 2006 North America and Western Europe consumed 50.0% of the world's total petroleum, yet these same countries produced only 22.9% of the world's crude oil. In contrast, the Persian Gulf region consumed 5.9% of the world's petroleum and produced 27.9 of the world's crude oil.

The United States currently imports more than two-thirds of its oil, which produced a 2006 balance of payments deficit of $270 billion. This oil bill is expected to increase. The U.S. Department of Energy and other knowledgeable experts project that the United States will be importing almost 100% of its oil by 2015.

The imbalance between oil consumers and oil producers will probably worsen in the future because the Persian Gulf region has much higher proven reserves than other countries. At current rates of production, North America's oil reserves will run out decades before those of the Persian Gulf nations, which have 65% of the known world oil reserves and may produce oil at current rates for perhaps a century. This dependence of the United States and other countries on Middle Eastern oil has potential international security implications as well as economic impacts.

Environmental Impacts of Oil and Natural Gas

Two sets of environmental problems are associated with the use of oil and natural gas: the problems that result from burning the fuels (combustion) and the problems involved in obtaining them (production and transport). We have already mentioned the CO_2 emissions that are a direct result of the combustion of fossil fuels. As with coal, the burning of oil and natural gas produces CO_2. Every gallon of gasoline you burn in your automobile releases an estimated 9 kg (20 lb) of CO_2 into the atmosphere. As CO_2 accumulates in the atmosphere, it insulates the planet, preventing planetary heat from radiating back into space. The global climate is warming more rapidly now than it did during any of the warming periods following the ice ages, and the environmental impact of rapid global climate change could be catastrophic.

Another negative environmental impact of burning oil is *acid deposition*. Although oil does not produce appreciable amounts of sulfur oxides, it does produce nitrogen oxides, mainly through gasoline combustion in automobiles, which contribute approximately half the nitrogen oxides released into the atmosphere. (Coal combustion is responsible for the other half.) Nitrogen oxides contribute to acid deposition and, along with unburned gasoline vapors, the formation of photochemical smog. Poorly tuned engines and diesel-burning vehicles also contribute particulate matter—small particles that are inhaled and cause lung damage and disease (see Chapter 20).

The burning of natural gas, on the other hand, does not pollute the atmosphere as much as the burning of oil. Natural gas is a relatively clean, efficient source of energy that contains almost no sulfur, a contributor to acid deposition. In addition, natural gas produces far less CO_2, fewer hydrocarbons, and almost no particulate matter, as compared to oil and coal.

One of the concerns in oil and natural gas production is the environmental damage that may occur during their transport, often over long distances by pipelines or ocean tankers. A serious spill along the route creates an environmental crisis, particularly in aquatic ecosystems, where the oil slick can travel. For example, one of the worst oil spills

Figure 11.16 Long lines at a filling station, 1973. In 1973 the United States had a short-lived but traumatic gasoline shortage, resulting from an embargo imposed by OPEC. Currently, while gasoline prices can vary greatly, it is always available at a price. How long this ready availability will continue is uncertain; lines like this could be commonplace in the future. (© Tony Korody/Corbis Sygma)

in Europe's history occurred in 2002 when a storm caused the oil tanker *Prestige* to break up off the coast of Spain. Thousands of tons of oil spewed into the sea, contaminating hundreds of kilometers of coastline and bringing the large fishing industry there to a halt.

The Largest Oil Spill in the United States In 1989 the supertanker *Exxon Valdez* hit Bligh Reef and spilled 260,000 barrels (10.9 million gallons) of crude oil into Prince William Sound along the coast of Alaska, creating the largest oil spill in U.S. history. As it spread, the black, tarry gunk eventually covered thousands of square kilometers of water (**Figure 11.17**) and contaminated hundreds of kilometers of shoreline. According to the U.S. Fish and Wildlife Service and the Alaska Department of Environmental Conservation, more than 30,000 birds (sea ducks, loons, cormorants, bald eagles, and other species) and between 3500 and 5500 sea otters died as a result of the spill. The area's killer whale and harbor seal populations declined, and salmon migration was disrupted. Throughout the area, there was no fishing season that year.

Within hours of the spill, scientists began to arrive on the scene to advise both Exxon Corporation and the government on the best way to try to contain and clean up the spill. But it took much longer for any real action to occur. Eventually, nearly 12,000 workers took part in the cleanup; their activities included mechanized steam cleaning and rinsing, which killed additional shoreline organisms such as barnacles, clams, mussels, eelgrass, and rockweed.

In late 1989, Exxon declared the cleanup "complete." But it left behind contaminated shorelines, particularly rocky coasts, marshes, and mudflats; continued damage to some species of birds (such as the common loon and harlequin duck), fishes (such as murrelet

Figure 11.17 *Exxon Valdez* **oil spill, 1989.**

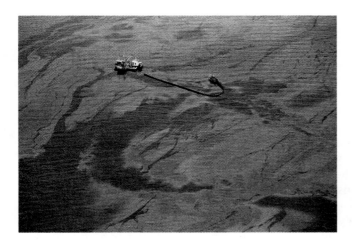

(a) An aerial view of the massive oil slick at the southwest end of Prince William Sound. (Chris Wilkins/AFP/Getty Images)

(b) Workers try to clean the rocky shoreline of Eleanor Island, Alaska, several months after the spill. (Charles Mason/Black Star)

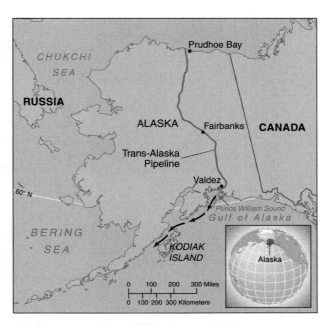

(c) The extent of the spill (black arrows). Water currents caused it to spread rapidly for hundreds of kilometers. Countless animals, such as sea otters and ocean birds, died.

and rockfish), and mammals (such as the harbor seal); and a reduced commercial salmon catch, among other problems. Multiple studies are still being conducted to determine the status of the cleanup. In 1991 Exxon agreed to pay Alaska a settlement of $1 billion, and in 1994, a jury awarded 34,000 fishermen and other Alaskans an additional $5 billion in punitive damages. Exxon is appealing the punitive damages, and additional lawsuits will take years to settle. The final cost to Exxon may exceed $10 billion.

One positive outcome of the disaster was passage of the **Oil Pollution Act** of 1990. This legislation establishes liability for damages to natural resources resulting from a catastrophic oil spill, including a trust fund that pays to clean up spills when the responsible party cannot; a tax on oil provides money for the trust fund. The Oil Pollution Act requires double hulls on all oil tankers that enter U.S. waters by 2015. Had the Exxon Valdez possessed a double hull, the disaster may not have occurred because only the outer hull might have broken.

The Largest Global Oil Spill   The world's most massive oil spill occurred in 1991 during the Persian Gulf War, when about 6 million barrels (250 million gallons) of crude oil—more than 20 times the amount of the *Exxon Valdez* spill—were deliberately dumped into the Persian Gulf. Many oil wells were set on fire, and lakes of oil spilled into the desert around the burning oil wells. Cleanup efforts along the coastline and in the desert were initially hampered by the war. In 2001 Kuwait began a massive remediation project to clean up its oil-contaminated desert. Progress is slow, and it may take a century or more for the area to completely recover.

CASE IN POINT

The Arctic National Wildlife Refuge

The proposed opening of the **Arctic National Wildlife Refuge** to oil exploration has been a major environment-versus-economy conflict off and on since 1980. On one side are those who seek to protect rare and fragile natural environments; on the other side are those whose higher priority is the development of some of the last major U.S. oil supplies.

The refuge, called "America's Serengeti," is home to many animal species, including polar bears, arctic foxes, peregrine falcons, musk oxen, Dall sheep, wolverines, and snow geese. It is the calving area for a large migrating herd of caribou: The Porcupine caribou herd contains more than 150,000 head. Dominant plants in this coastal plain of tundra include mosses, lichens, sedges, grasses, dwarf shrubs, and small herbs. Under a thin upper layer of soil is the permafrost layer, which contains permanently frozen water. Although it is biologically rich, the tundra is an extremely fragile ecosystem, in part because of its harsh climate. The organisms living here have adapted to their environment, but any additional stress has the potential to harm or even kill them. Thus, arctic organisms are particularly vulnerable to human activities.

History of the Arctic National Wildlife Refuge   In 1960 Congress declared a section of northeastern Alaska protected because of its distinctive wildlife. In 1980 Congress expanded this wilderness area to form the Arctic National Wildlife Refuge (**Figure 11.18**). The Department of the Interior was given permission to determine the potential for oil discoveries in the area, but exploration and development could proceed only with congressional approval.

Pressure to open the refuge to oil development subsided for about five years following the Alaskan oil spill, when public sentiments were strongly against oil companies. In the mid-1990s, pro-development interests became more vocal, partly because in 1994, for the first time in its history, the United States imported more than half the oil it used. Although the Department of the Interior concluded that oil drilling in the refuge would harm the area's ecosystem, both the Senate and the House of Representatives passed measures to allow it. (President Clinton vetoed the bill.)

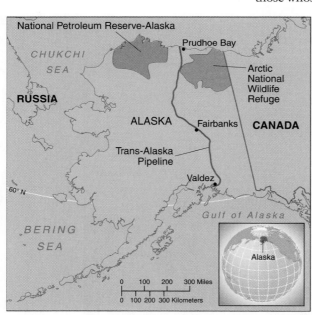

Figure 11.18   Arctic National Wildlife Refuge. Located in the northeastern part of Alaska, the refuge is situated close to the Trans-Alaska Pipeline, which begins at Prudhoe Bay and extends south to Valdez. The National Petroleum Reserve–Alaska is also shown.

In 2001, President George W. Bush announced his support for opening the refuge to oil drilling as part of his comprehensive energy policy (discussed later in the chapter). After a contentious debate in Congress, in 2005 the Senate voted against opening the refuge for oil development. Despite this setback, President Bush has announced that developing oil resources in Alaska is a priority during his second term.

Support for and Opposition to Oil Exploration in the Refuge   Supporters cite economic considerations as the main reason for drilling for oil in the refuge. The United States is spending a large proportion of its energy budget to purchase foreign oil. Development of domestic oil would improve the balance of trade and make us less dependent on foreign countries for our oil.

The oil companies are eager to develop this particular site because it is near Prudhoe Bay, where large oil deposits are already being tapped. (To date, Prudhoe Bay has produced about 14 billion barrels of crude oil.) Prudhoe Bay has a sprawling industrial complex to support oil production, including roads, pipelines, gravel pads, and storage tanks. The Prudhoe Bay oil deposits peaked in production in 1985 and have declined in productivity since then. As a result, the oil industry is looking for sites that can use the infrastructure already in place.

Conservationists think oil exploration poses permanent threats to the delicate balance of nature in the Alaskan wilderness, in exchange for a temporary oil supply. They reason that the money spent drilling for oil would be better used for research into alternative, renewable energy sources and energy conservation—a more permanent solution to the energy problem. Studies, such as one by the U.S. Fish and Wildlife Service, document considerable habitat damage and declining numbers of wolves and bears in the Prudhoe Bay area. (Top predators are usually more susceptible to environmental disruption than are organisms occupying lower positions in a food web.) Because it is not financially practical to restore developed areas in the Arctic to their natural states, development in the Arctic causes permanent changes in the natural environment. ■

REVIEW

1. What are two examples of structural traps?

2. What are three environmental problems associated with using oil and natural gas as energy resources?

3. What is the controversy surrounding the Arctic National Wildlife Refuge?

Synfuels and Other Potential Fossil Fuel Resources

LEARNING OBJECTIVES

- Define *synfuel* and distinguish among tar sands, oil shales, gas hydrates, liquid coal, and coal gas.
- Briefly consider the environmental implications of using synfuels.

Synthetic fuels, or **synfuels**, are fuels that are similar or identical to the chemical composition of oil or natural gas. Synfuels include tar sands, oil shales, gas hydrates, liquefied coal, and coal gas. Although synfuels are more expensive to produce than fossil fuels, they may become more important as fossil fuel reserves decline.

Tar sands, or **oil sands,** are underground sand deposits permeated with *bitumen*, a thick, asphalt-like oil. The bitumen in tar sands deep in the ground cannot be pumped out unless it is heated underground with steam to make it more fluid. If tar sands are close to Earth's surface, they are surface-mined. Once bitumen is obtained from tar sands, it must be refined like crude oil. World tar sand reserves are estimated to contain half again as much fuel as world oil reserves. Major tar sands are found in Venezuela and in Alberta, Canada, where an estimated 300 billion barrels of oil occur in tar sands; this estimate is greater than the estimated oil reserves in Saudi Arabia. Canadian mines are currently producing almost 300 million barrels of oil a year from oil sands.

synfuel: A liquid or gaseous fuel that is synthesized from coal and other naturally occurring resources and used in place of oil or natural gas.

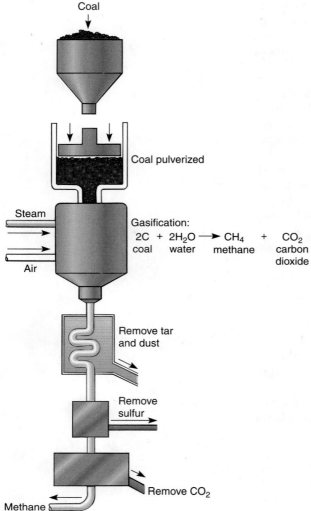

Figure 11.19   Goal gasification. Shown is one method of coal gasification, in which the combustible gas methane is generated from coal. To follow the steps in coal gasification, start at the top of the figure and work your way down.

Western American pioneers discovered "oily rocks" when their rock hearths caught fire and burned. **Oil shales** are sedimentary rocks containing a mixture of hydrocarbons known collectively as *kerogen*. Oil shales are crushed and heated to yield their oil, and the kerogen must be refined after it is mined. It is not yet cost-efficient to process oil shales because the mining and refinement require a great deal of energy. Large oil shale deposits are located in Australia, Estonia, Brazil, Sweden, the United States, and China. Wyoming, Utah, and Colorado have the largest deposits in the United States. Like tar sands, oil shale reserves may contain half again as much fuel as world oil reserves.

Gas hydrates, also called **methane hydrates**, are reserves of ice-encrusted natural gas located deep underground in porous rock. Massive deposits have been identified in the arctic tundra, deep under the permafrost, and in the deep-ocean sediments of the continental slope and ocean floor. Until recently, the U.S. oil industry was not particularly interested in extracting natural gas from gas hydrates because of the expense involved. Several U.S. oil companies are currently developing methods to extract gas hydrates. Countries with lots of gas hydrates (for example, Russia) or with few conventional fossil fuel deposits (for example, India and Japan) have established national gas hydrates programs. A 2003 pilot program conducted by an international consortium in northwest Canada showed that obtaining energy from gas hydrates is feasible. All they say they need is a gas pipeline that extends into the Canadian Arctic to transport the methane. Despite their optimism, most analysts say that gas hydrates will never become a major energy source because most deposits are too small to be removed economically.

A nonalcohol liquid fuel similar to oil can be produced from coal. The liquid fuel, which is cleaned before burning, is less polluting than solid coal. This process, called **coal liquefaction**, was developed before World War II, but its expense prevented it from replacing gasoline production. Technological improvements have lowered the cost of coal liquefaction. While it is still not cost-competitive with gasoline, there is a major push in the United States to increase coal liquifaction in the near future.

Another synfuel is a gaseous product of coal. **Coal gas** has been produced since the 19th century. As a matter of fact, it was the major fuel used for lighting and heating in American homes until oil and natural gas replaced it during the 20th century. **Coal gasification** is production of the combustible gas methane from coal by reacting it with air and steam (**Figure 11.19**). Several demonstration power plants that convert coal into gas have been constructed in the United States. One advantage of coal gas over solid coal is that coal gas burns almost as cleanly as natural gas. Scrubbers are not needed when coal gas is burned because sulfur is removed during coal gasification. Like other synfuels, coal gas is currently more expensive to produce than fossil fuels.

Environmental Impacts of Synfuels

Although synfuels are promising energy sources, they have many of the same undesirable effects as fossil fuels. Their combustion releases enormous quantities of CO_2 and other pollutants into the atmosphere, thereby contributing to global warming and air pollution. Some synfuels, such as coal gas, require large amounts of water during production and are of limited usefulness in arid areas, where water shortages are already commonplace. Also, enormously large areas of land would have to be surface mined to recover the fuel in tar sands and oil shales.

REVIEW

1. What are tar sands and oil shales?

2. What are liquid coal and coal gas?

3. How do the environmental problems associated with the use of synfuels compare to those of coal, oil, and natural gas?

The U.S. Energy Strategy

LEARNING OBJECTIVES

- Relate three reasons the United States needs a comprehensive national energy strategy.
- Explain how the government affects energy prices through the use of subsidies.
- Briefly describe the National Energy Policy of the George W. Bush administration.

The United States has a comprehensive energy policy for several reasons: (1) the supply of fossil fuels is limited; (2) the production, transport, and use of fossil fuels pollute the environment; and (3) our heavy dependence on foreign oil makes us economically vulnerable. Because of the complex nature of energy issues, any policy our political leaders adopt must have many approaches. Although there is no way to completely eliminate our vulnerability to disruptions in foreign oil supplies and to oil price increases, a comprehensive energy policy lessens the effects of such events. Such a strategy should provide a secure supply of energy, encourage us to use less energy, and protect the environment. However, as noted in the introductory section, changing our energy habits can be difficult. The following elements should be included in a comprehensive national energy policy.

Objective 1: Increase Energy Efficiency and Conservation

Over the past several decades, energy efficiency in the United States has improved significantly, but despite gains, the United States uses more energy than any other country. There is room for great improvement on all fronts, from individuals conserving heating oil by weatherproofing their homes, to groups of commuters conserving gasoline by carpooling, to corporations developing more energy-efficient products. The automobile industry could be required to increase the average new-car gasoline mileage, for example. Average fuel economy of motor vehicles in the United States has been diminishing; at the same time it is going up in other parts of the world. In large part, this is due to a switch to more and larger light trucks and SUVs.

Reinstating the 55-mile-per-hour speed limit in areas where the limit is now as high as 75 mph would conserve gasoline. Fuel consumption increases approximately 50% if a car is driven at 75 mph rather than at 55 mph. Fuel consumption increases 30% at 65 mph rather than 55 mph. In addition, federal financial support for transportation could shift from highway construction to public transportation. We say more about energy conservation in Chapter 13.

One way to encourage energy conservation is to eliminate **subsidies**, which keep energy prices artificially low because the government thinks they are beneficial to the economy. When prices reflect the true costs of energy, including the environmental costs incurred by its production, transport, and use, energy is used more efficiently. Gasoline prices in the United States do not reflect the true cost of gasoline and are unrealistically low, even though gasoline in early 2007 was as expensive as it has been since the early 1980s. During 2006, Western Europeans paid about 2.6 times more for gasoline than U.S. citizens did (**Table 11.2**). The price of gasoline affects the level of gasoline consumption: Lower prices encourage greater consumption. Over the next few years, a more realistic price for gasoline—including a tax to offset externalities (Chapter 2) could be introduced to encourage people to buy fuel-efficient automobiles, carpool, and use public transportation (**Table 11.3**).

subsidy: A form of government support (such as public financing or tax breaks) given to a business or institution to promote that group's activity.

Table 11.2 A Comparison of Gasoline Prices in Selected Countries (Including Taxes)

Country	Premium Gasoline Price (Dollars per Gallon, September 2006)
United States	$2.25
Canada	2.94
Japan	4.49
France	5.54
Germany	5.78
Italy	5.84
United Kingdom	6.13

Source: Energy Information Administration

Table 11.3 A Comparison of Energy Input for Different Kinds of Transportation

Method of Transportation	Energy Input (in BTUs)* per Person, per Mile
Automobile (driver only)	6530
Rail	3534
Carpool	2230
Vanpool	1094
Bus	939

*BTU stands for *British thermal unit*, an energy unit equivalent to 252 calories or 1054 joules.

Objective 2: Secure Future Fossil Fuel Energy Supplies

A comprehensive national energy strategy could include the environmentally sound and responsible development of domestically produced fossil fuels, especially natural gas. There are two types of opposition to this element of a national energy strategy; one is economic and the other is environmental. Some think it is better to deplete foreign oil reserves while prices are reasonable and save domestic supplies for the future. Most economists argue against this view because of the U.S. trade deficit. We do not currently finance our oil imports by exporting goods and services of equal value. Many environmentalists oppose the development and increased use of domestic fossil fuels, largely because of the environmental problems already discussed.

Everyone, environmentalists included, recognizes the need for a dependable energy supply. Securing a future supply of fossil fuels, whether domestic or foreign, is a *temporary* strategy because fossil fuels are nonrenewable resources that will eventually be depleted, regardless of how efficient our use or how much we conserve. Having a secure energy supply for the short term, however, will allow us to develop alternative energy sources for the long term.

Objective 3: Develop Alternative Energy Sources

We must expand research and development for all possible alternatives to fossil fuels, especially renewable energy sources such as solar and wind energy. Our long-term energy goal should shift us to energy sources that are less harmful to the environment. Who should pay for the research costs of improving energy conservation and developing alternative forms of energy? The answer is that we all should share in these costs because we will all share in the benefits. A gasoline tax has been suggested as a means of financing programs to achieve a sustainable energy future. Some policymakers have suggested a tax of as much as 50 cents per gallon.

YOU CAN MAKE A DIFFERENCE

Getting Around Town

Can you imagine getting around town without a car? How would you get to class, the grocery store, the laundromat? Hopping in your car for every errand seems like the natural thing to do. According to the American Automobile Association, American motorists drive an average of 10,100 miles annually and burn 507 gallons of gasoline in the process.

From the production of gasoline to the disposal of old automobiles, the car has a significant negative impact on the environment. Smog and global climate change are just two of the problems caused by gasoline combustion. Vehicle exhaust, acid deposition, and chronic low-level exposure to toxins are all health threats to car owners and to those who live in areas with a high density of cars. Dumping of engine oil, fumes from the burning of tires and batteries, and automobile junkyards threaten both our health and our environment.

There is something you can do about it. Granted, you may not be able to give up your car entirely, but you can cut down on its use wherever possible. For example, try the following:

1. Ride the bus or the train whenever possible. Think about how jammed the road would be if all 50 people on a bus were driving individual cars!

2. Consider whether you must drive to accomplish a task. Sometimes a phone call can substitute for a trip in the automobile.

3. Carpool to class, to work, to the grocery store, to social events. One car on the road is better than three or four.

4. Buy a good bicycle; it is less expensive than a car to buy and maintain, and it is great for local transportation. It is also good exercise.

5. Walk to class or work if you live within a mile or so. You must allow yourself a little extra time, but once you get into the habit, it is easy. Walking is good exercise, too.

6. Modify your driving habits to save gasoline. Minimize braking, and do not let your engine idle for more than one minute. Keep your car well tuned, replace air and oil filters often, and keep your tires inflated at the recommended pressure. Remove any unnecessary weight from your car. All of these measures help boost gasoline mileage.

7. When you purchase a motor vehicle, use the EPA's official mileage ratings (http://www.fueleconomy.gov) as one of the criteria to help you select a model. For example, using the average 2006 price for a gallon of regular gasoline, you will spend $876 to drive 15,000 miles in the 2007 Toyota Prius, which gets 55 miles per gallon for highway driving. Driving the same distance in a 2007 Chevrolet Suburban, which gets 19 miles per gallon for highway driving, will cost you $3018. (We say more about fuel-efficient, less polluting cars in Chapter 20.)

Objective 4: Meet the First Three Objectives without Further Damage to the Environment

The environmental costs of using a particular energy source must be weighed against its benefits when it is considered as a practical component of an energy policy. Domestic supplies of fossil fuels that are developed with as much attention to the environment as possible will help reduce our dependence on foreign oil. One suggestion is to add a 5-cent tax on each barrel of domestically produced oil to establish a reclamation fund for some of the environmental damage caused by mining and production of oil and natural gas.

How Politics Influences the National Energy Policy

The nation's energy policy reflects the varied political views of the president, Congress, and the American public. It is subject to major changes with every change in the U.S. presidential office. The current National Energy Policy, developed by President George W. Bush, is summarized in **Table 11.4**. It has five components: (1) modernize conservation; (2) modernize our energy infrastructure; (3) increase energy supplies; (4) accelerate the protection and improvement of the environment; and (5) increase our nation's energy security. However, the **Energy Policy Act** of 2005 focused largely on supporting energy research for fossil fuels, and subsidies continue, prompted in part by high gasoline prices in 2006. Taking a systems perspective—one that accounts for long-term cause and effect relationships—might lead to an energy policy that provides a secure and sustainable energy future for the United States and the world.

REVIEW

1. What are subsidies? How do they affect energy prices and use?
2. What are the main points in President George W. Bush's National Energy Policy?

Table 11.4 Highlights of President George W. Bush's National Energy Policy

Modernize Conservation

1. Increase funding for renewable energy efficiency research and development programs.
2. Create an income tax credit for people who purchase who purchase hybrid and fuel cell vehicles.
3. Extend the Energy Star efficiency program to include schools, retail buildings, healthcare facilities, and homes, and extend the Energy Star labeling program to additional products and appliances.
4. Fund the Intelligent Transportation Systems program, the Fuel Cell Powered Transit Bus program, and the Clean Buses program.
5. Review the Corporate Average Fuel Economy (CAFE) standards.

Modernize Our Energy Infrastructure

1. Expedite permits for energy-related projects on a national basis.
2. Grant authority for rights-of-way for electricity transmission lines, to create a national transmission grid.
3. Enact electricity legislation that encourages new electricity generation.
4. Improve the reliability of the interstate transmission system.

Increase Energy Supplies

1. Open the Arctic National Wildlife Refuge (ANWR) to exploration and production. Examine the potential of oil and natural gas development on other federal lands.
2. Earmark $1.2 billion from the leasing of ANWR to fund research into renewable energy resources.
3. Fund research on clean coal technology and on electricity from biomass cofired with coal.
4. Provide for the safe expansion of nuclear energy by establishing a national repository for nuclear waste and by streamlining the licensing of nuclear power plants.

Accelerate the Protection and Improvement of the Environment

1. Enact legislation to establish a flexible, market-based program to reduce and cap emissions of sulfur dioxide, nitrogen oxides, and mercury from electric power generators.
2. Earmark royalties from oil and gas exploration in ANWR to fund land conservation efforts.

Increase Our Nation's Energy Security

1. Prepare for potential energy-related emergencies.
2. Expand cross-border energy investments, oil and gas pipeline, and electricity grid connections with Canada and Mexico.
3. Expedite permits for a gas pipeline route from Alaska to the lower 48 states.

REVIEW OF LEARNING OBJECTIVES WITH KEY TERMS

• Define *energy density* and *energy efficiency* and explain their importance.

Energy density is the amount of energy that can be extracted from a given volume or mass of fuel. High energy density fuels are typically easier to transport and use than are low energy density fuels. **Energy efficiency** refers to the fraction of total energy in a fuel that is used. More efficient processes require less fuel.

• Compare per capita energy consumption in highly developed and developing countries.

Global energy consumption is increasing each year. Most of the increase is occurring in developing countries, which use more energy as they improve their standard of living. Highly developed nations consume much more energy per person than developing countries.

• Define *fossil fuel*, and distinguish among coal, oil, and natural gas.

Fossil fuels are combustible deposits in Earth's crust, composed of the remnants of prehistoric organisms that existed millions of years ago. Fossil fuels are **nonrenewable resources**; Earth has a finite supply of fossil fuels that are depleted by use. **Coal** is a black combustible solid formed from the remains of ancient plants that lived millions of years ago. **Oil** is a thick, yellow to black, flammable liquid hydrocarbon mixture. **Natural gas** is a mixture of gaseous hydrocarbons (primarily methane) that often occurs with oil deposits.

• Describe the processes that formed coal, oil, and natural gas.

Coal was formed when partially decomposed plant material was exposed to heat and pressure for aeons. Oil and natural gas formed when countless microscopic aquatic organisms died and settled in oxygen-deficient sediments.

• Explain the role of photosynthesis in the creation of fossil fuels.

Photosynthesis is the fundamental process whereby energy from the sun is incorporated into living organisms. This means that the energy in fossil fuels was captured through photosynthesis millions of years ago.

• Distinguish between surface mining and subsurface mining.

Surface mining is the extraction of mineral and energy resources near Earth's surface by first removing the soil, subsoil, and overlying rock **strata**. **Subsurface mining** is the extraction of mineral and energy resources from deep underground deposits. In the United States surface mining accounts for 60% of the coal mined, and subsurface mining accounts for 40%.

• Discuss the advantages and disadvantages of coal.

Coal is present in greater quantities than oil or natural gas (known world reserves could last more than 200 years at the present rate of consumption), but its use has a greater potential to harm the environment.

• Summarize the environmental problems associated with using coal, including acid mine drainage, climate change, and acid deposition.

Many environmental problems are associated with coal mining and combustion. Surface mining destroys existing vegetation and topsoil. As with all fossil fuels, the combustion of coal produces several pollutants—in particular, large amounts of CO_2, a greenhouse gas that prevents heat from escaping from Earth. Coal produces more CO_2 emissions per unit of heat than do other fossil fuels. **Acid mine drainage** is pollution caused when sulfuric acid and dangerous dissolved materials such as lead, arsenic, and cadmium wash from coal and metal mines into nearby lakes and streams. Burning soft coals that contain sulfur contributes to **acid deposition**, a type of air pollution in which acid falls from the atmosphere to the surface as precipitation (acid precipitation) or as dry acid particles.

• Explain how resource recovery and fluidized-bed combustion can make coal a cleaner fuel.

The installation of **scrubbers** (desulfurization systems) in smokestacks results in the production of a sulfur-containing sludge, which must be disposed of or sold. **Resource recovery** is the process of removing any material—sulfur in this case—from polluted emissions (such as sludge) and selling it as a marketable product. **Fluidized-bed combustion** is a clean-coal technology in which crushed coal is mixed with limestone to neutralize the acidic sulfur compounds produced during combustion.

• Define *structural trap* and give two examples.

Structural traps are underground geologic structures that tend to trap any oil or natural gas if it is present. Structural traps include upward foldings of rock strata and **salt domes** (underground columns of salt).

• Explain what is meant by "Peak Oil" and why it might concern us.

Peak Oil is the point at which global oil production has reached a maximum rate. Once that peak is passed, less and less oil will be removed each year. By some estimates we are just about at that peak, which means that oil supply will be dropping while demand rises.

• Discuss the environmental problems of using oil and natural gas.

Many environmental problems are associated with obtaining and burning oil and, to a lesser extent, natural gas. Oil exploration and extraction are a threat to environmentally sensitive areas. An accidental oil spill during transport, often over long distances by pipelines or ocean tankers, creates an environmental crisis. CO_2 emissions released when oil and natural gas are burned may contribute to global climate warming. Production of nitrogen oxides when oil is burned contributes to acid deposition.

• Summarize the continuing controversy surrounding the arctic national wildlife refuge.

Supporters of drilling in the **Arctic National Wildlife Refuge** say that development of domestic oil would improve the balance of trade and make us less dependent on foreign countries for our oil. Conservationists think oil exploration poses permanent threats to the delicate balance of nature in the Alaskan wilderness, in exchange for a temporary (and probably relatively small) oil supply.

• Define *synfuel* and distinguish among tar sands, oil shales, gas hydrates, liquid coal, and coal gas.

Synfuel is a liquid or gaseous fuel that is synthesized from coal and other naturally occurring resources and used in place of oil or natural gas. **Tar sands** are underground sand deposits permeated with bitumen, a thick, asphalt-like oil. **Oil shales** are sedimentary rocks containing a mixture of hydrocarbons known collectively as kerogen. **Gas hydrates** are reserves of ice-encrusted natural gas located deep underground in porous rock. Coal liquid is a nonalcohol liquid fuel similar to oil that can be produced from coal by the process of **coal liquefaction**. Another synfuel, **coal gas**, is a gaseous product of coal.

• Briefly consider the environmental implications of using synfuels.

Synfuels have many of the same undesirable effects as fossil fuels. Their combustion releases enormous quantities of CO_2 and other pollutants into the atmosphere, thereby contributing to global warming and air pollution. Some synfuels, such as coal gas, require large amounts of water during production and are of limited usefulness in arid areas. Enormously large areas of land would have to be surface mined to recover the fuel in tar sands and oil shales.

• **Relate three reasons the United States needs a comprehensive national energy strategy.**

The United States needs a comprehensive energy policy because (1) the supply of fossil fuels is limited; (2) the production, transport, and use of fossil fuels pollute the environment; and (3) our heavy dependence on foreign oil makes us economically vulnerable.

• **Explain how the government affects energy prices through the use of subsidies.**

A **subsidy** is a form of government support (such as public financing or tax breaks) given to a business or institution to promote that group's

activity. The U.S. government subsidizes energy prices. For example, gasoline prices in the United States do not reflect the true cost of gasoline and are unrealistically low.

• **Briefly describe the National Energy Policy of the George W. Bush administration.**

President George W. Bush's National Energy Policy focuses primarily on research and development of fossil fuel resources. It also includes modernizing conservation and our energy infrastructure, increasing energy supplies, and further developing domestic energy resources.

Thinking About the Environment

1. The Industrial Revolution may have been concentrated in the Northern Hemisphere because coal is located there. What is the relationship between coal and the Industrial Revolution?

2. Few countries in Africa have significant amounts of coal, oil, and natural gas resources. What does this suggest about opportunities for financial development in those countries?

3. Does thinking of "oil addiction" literally, as depicted in the cartoon, provide a useful way to get away from this addiction? Why or why not?

(Bizarro © Dan Piraro/Reprinted with special permission of King Features Syndicate)

4. How does U.S. dependence on foreign oil affect our energy security?

5. On the basis of what you have learned about coal, oil, and natural gas, which fossil fuel do you think the United States should exploit in the short term (during the next 20 years)? Explain your rationale.

6. Explain why the U.S. Department of Energy describes coal as a "true measure of the energy strength of the United States." Is this also true of China? India? Why or why not?

7. In your estimation, which fossil fuel has the greatest potential for the 21st century? Why?

8. Which of the negative environmental impacts associated with fossil fuels is most serious? Why?

9. Which major consumer of oil is most vulnerable to disruption in the event of another energy crisis: electric power generation, motor vehicles, heating and air conditioning, or industry? Why?

10. What are the implications of "Peak Oil" on future global energy supplies?

11. Do you think oil drilling should be permitted in the Arctic National Wildlife Refuge? Why or why not?

12. Some environmental analysts think that the latest war in Iraq was related in part to gaining control over the supply of Iraqi oil. Do you think this is plausible? Explain why or why not.

13. Why does the United States currently have a growing shortage of natural gas?

14. What are the five kinds of synfuels? Why are they not being used more extensively?

15. What is resource recovery?

16. Distinguish among fluidized-bed combustion, coal liquefaction, and coal gasification.

17. Give three reasons why the United States needs a comprehensive national energy policy.

18. Fossil fuels are "non-renewable" resources. Why is this a problem from a systems perspective? (*Hint:* irreversible change disrupts system stability.)

Quantitative questions relating to this chapter are on our Web site.

Take a Stand

Visit our Web site at http://www.wiley.com/college/raven (select Chapter 11 from the Table of Contents) for links to more information about the controversies surrounding the Arctic National Wildlife Refuge. Consider the opposing views of those who seek to protect rare and fragile natural environments and those who wish to develop the last major oil supplies in the United States, and debate the issues with your classmates. You will find tools to help you organize your research, analyze the data, think critically about the issues, and construct a well-considered argument. *Take a Stand* activities can be done individually or as a team, as oral presentations, written exercises, or Web-based (e-mail) assignments.

Additional online materials relating to this chapter, including a Student Testing Section with study aids and self-tests, Environmental News, Activity Links, Environmental Investigations, and more, are also on our Web site.

Nuclear Energy

Nuclear Power Plant at Salem, New Jersey.
There are three nuclear reactors here.
Nuclear power meets about one-third of
New Jersey's electricity needs. (Courtesy
Public Service Electric & Gas Co.)

When nuclear energy was discovered during the 20th century, human history was forever changed. Lauded as a potential source of energy for humanity, nuclear energy was also denounced as a weapon that could destroy human civilization. This dichotomous view continues to permeate issues involving nuclear energy today. The main question we address throughout this chapter is what role nuclear energy will have in providing electric energy in the 21st century. To consider this question effectively, we begin with a brief history of the scientific development of nuclear energy.

Nuclear reactions have the potential to release a vast amount of energy, primarily as heat. In this process, a small amount of the mass of an atom is transformed into a large amount of energy. In 1905 **Albert Einstein** first hypothesized that mass and energy are related in his now-famous equation, $E = mc^2$, in which energy (E) is equal to mass (m) times the speed of light (c) squared. This built on the work of French physicist **Henri Becquerel,** who in 1896 dis-

covered that uranium-containing minerals spontaneously and continually give off invisible rays of energy—that is, radiation. Beginning in 1898, British physicist **Ernest Rutherford** conducted a series of experiments that determined that radiation consists of high-energy particles. In 1911, his work and the work of others led Rutherford to the conclusion that each atom contains a nucleus where most of its mass is located.

Scientists began to study what happens when the nuclei of atoms are bombarded with high-energy particles. In 1919 Rutherford bombarded nitrogen nuclei with alpha particles (positively charged radiation), changing the nitrogen into oxygen in the process! Attempts to bombard heavier atoms with alpha particles failed because the larger nuclei repelled the positively charged particles. In 1938 two German scientists, **Otto Hahn** and **Fritz Strassmann,** succeeded in bombarding uranium with neutrons, which are atomic particles without a charge. During this reaction, some of the uranium nuclei split into two smaller nuclei of barium and krypton. Additional neutrons were also emitted, along with energy.

Scientists quickly realized the implications of splitting atomic nuclei. If the emitted neutrons could be made to bombard other uranium nuclei, a chain reaction would start. Using Albert Einstein's equation, scientists determined that 0.45 kg (1 lb) of uranium

could release as much energy as 7300 metric tons (8000 tons) of TNT, enough to make a powerful bomb. The development of nuclear weapons was the first application of nuclear energy attempted by several nations because, beginning in 1939, Europe was at war. Einstein fled to the United States to avoid persecution by the Nazis during World War II, and he informed President Roosevelt that the Germans were working on a nuclear bomb. Roosevelt established the top-secret Manhattan Project to build such a bomb. The scientists began to experiment with chain reactions and to produce enough enriched uranium and plutonium for a bomb. In July 1945, the world's first atomic bomb was exploded in the desert at Alamogordo, New Mexico. In August 1945, the United States destroyed much of Hiroshima and Nagasaki with the only two atomic bombs ever used as weapons.

Following World War II, the United States, the Soviet Union, the United Kingdom, France, and Canada began to design nuclear reactors to harvest nuclear energy for peaceful purposes (see figure). The world's first large-scale nuclear power plant opened at Calder Hall in England in 1956, whereas the first commercial nuclear power plant in the United States opened at Shippingport, Pennsylvania, in 1957, and Canada's opened in 1962 at Rolphton, Ontario. Today, there are 440 commercial nuclear reactors in 31 countries around the world, with more in various stages of planning and construction.

 World View and Closer to You...
Real news footage relating to nuclear power around the world and in your own geographic region is available at www.wiley.com/college/raven by clicking on "World View and Closer to You."

Introduction to Nuclear Processes

LEARNING OBJECTIVES

- Distinguish between nuclear energy and chemical energy.
- Contrast fission and fusion.
- Define *radioactive decay*.

As a way to obtain energy, nuclear processes are fundamentally different from the combustion that produces energy from fossil fuels. Combustion is a chemical reaction. In ordinary chemical reactions, atoms of one element do not change into atoms of another element, nor does any of their mass (matter) change into energy. The energy released in combustion and other chemical reactions comes from changes in the chemical bonds that hold the atoms together. Chemical bonds are associations between electrons, and ordinary chemical reactions involve the rearrangement of electrons (see Appendix I).

In contrast, **nuclear energy** involves changes in the *nuclei* of atoms; small amounts of matter from the nucleus are converted into large amounts of energy. Two different nuclear reactions release energy: fission and fusion. In **fission**, the process nuclear power plants use, larger atoms of certain elements are split into two smaller atoms of different elements. In **fusion**, the process that powers the sun and other stars, two smaller atoms are combined to make one larger atom of a different element. In both fission and fusion, the mass of the end product(s) is less than the mass of the starting material(s) because a small quantity of the starting material is converted to energy.

Nuclear reactions produce 100,000 times more energy per atom than is available from a chemical bond between two atoms. In nuclear bombs, energy from many atomic fissions is released all at once, producing a tremendous surge of energy that destroys everything in its vicinity. When nuclear energy is used to generate electricity, the nuclear reaction is controlled to produce smaller amounts of energy in the form of heat, which is then converted to electricity.

Atoms and Radioactivity

All atoms are composed of positively charged protons, negatively charged electrons, and electrically neutral neutrons (**Figure 12.1**). Protons and neutrons, which have approximately the same mass, are clustered in the center of the atom, making up its nucleus. Electrons, which possess little mass in comparison with protons and neutrons, orbit around the nucleus in distinct regions. Electrically neutral atoms possess identical numbers of positively charged protons and negatively charged electrons.

The **atomic mass** of an element is equal to the sum of protons and neutrons in the nucleus. Each element has a characteristic **atomic number,** or number of protons per atom. In contrast, the number of neutrons in each atom of a given element may vary, resulting in atoms of one element with different atomic masses. Forms of a single element that differ in atomic mass are known as **isotopes.** For example, normal hydrogen, the lightest element, contains one proton and no neutrons in the nucleus of each atom. The two isotopes of hydrogen are **deuterium,** which contains one proton and one neutron per nucleus, and **tritium,** which contains one proton and two neutrons per nucleus. Many isotopes are stable, but some are unstable; the unstable ones, called **radioisotopes,** are radioactive because they spontaneously emit **radiation,** a form of energy consisting of particles. The only radioisotope of hydrogen is tritium.

As a radioactive element emits radiation, its nucleus changes into the nucleus of a different, more stable element. This process is called **radioactive decay**. For example, the radioactive nucleus of one isotope of uranium, U-235, decays over time into lead (Pb-207). (Uranium-

nuclear energy: The energy released by nuclear fission or fusion.

fission: The splitting of an atomic nucleus into two smaller fragments, accompanied by the release of a large amount of energy.

fusion: The joining of two lightweight atomic nuclei into a single, heavier nucleus, accompanied by the release of a large amount of energy.

radioactive decay: The emission of energetic particles or rays from unstable atomic nuclei; includes positively charged alpha particles, negatively charged beta particles, and high-energy, electromagnetic gamma rays.

Figure 12.1   Atomic Structure. Atoms contain a nucleus made of positively charged particles (protons) and particles with no charge (neutrons). Circling the nucleus is a "cloud" of small negatively charged electrons.

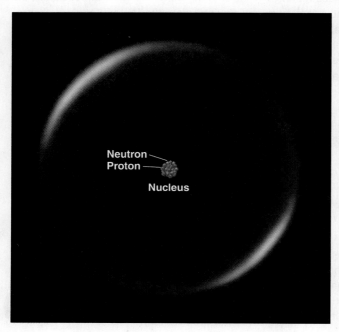

235 is an isotope of uranium with an atomic mass of 235.) Each radioisotope has its own characteristic rate of decay. The time required for one-half of the total amount of a radioactive substance to change into a different material is called its **radioactive half-life**. There is enormous variation in the half-lives of different radioisotopes (**Table 12.1**). The half-life of iodine (I-132) is only 2.4 hours, the half-life of tritium is 12.3 years, and the half-life of an isotope of uranium (U-234) is 250,000 years.

REVIEW

1. What is the difference between chemical energy and nuclear energy?
2. What is radioactive decay?

Nuclear Fission

LEARNING OBJECTIVES

- Describe the nuclear fuel cycle, including the process of enrichment.
- Define *nuclear reactor* and describe a typical nuclear power reactor.
- Contrast conventional nuclear fission, breeder nuclear fission, and mixed oxide fuel fission and describe spent fuel.

Uranium ore, the mineral fuel used in conventional nuclear power plants, is a nonrenewable resource present in limited amounts in sedimentary rock in Earth's crust. The steps, from mining to disposal, of the uranium fuel used in nuclear power plants are collectively called the **nuclear fuel cycle** (**Figure 12.2**).

Substantial deposits of uranium are found in Australia (20.4% of known world reserves), Kazakhstan (18.2%), the United States (10.6%), Canada (9.9%), and South

Table 12.1 Some Common Radioactive Isotopes Associated with the Fission of Uranium

Radioisotope	Half-Life (years)
Iodine-131	0.02 (8.1 days)
Xenon-133	0.04 (15.3 days)
Cerium-144	0.80
Ruthenium-106	1.00
Krypton-85	10.40
Tritium	12.30
Strontium-90	28.00
Cesium-137	30.00
Radium-226	1,600.00
Plutonium-240	6,600.00
Plutonium-239	24,400.00
Neptunium-237	2,130,000.00

nuclear fuel cycle: The processes involved in producing the fuel used in nuclear reactors and in disposing of radioactive (nuclear) wastes.

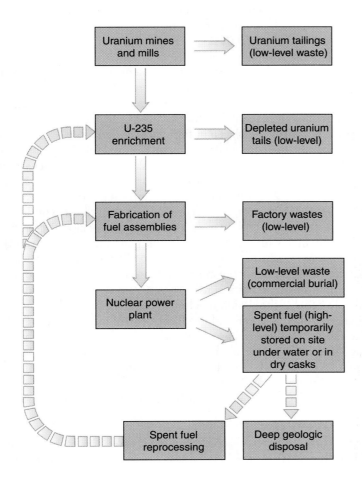

Figure 12.2 Nuclear fuel cycle, from mining to disposal. The left side of the figure shows how mined uranium becomes fuel for nuclear power plants. The right side shows radioactive wastes that must be handled and disposed of during the cycle. Broken lines indicate steps that are not currently occurring. In 1976 the United States stopped reprocessing (reusing spent fuel) for economic and political reasons. Currently, Japan, France, Russia, the United Kingdom, and India are the only countries reprocessing spent fuel for reuse as nuclear fuel. Deep geologic disposal of spent fuel is currently under study in several countries, including the United States.

Figure 12.3 **Uranium fuel.**

(a) Uranium dioxide pellets, held in a gloved hand, contain about 3% uranium-235, the fission fuel in a nuclear reactor. Each pellet contains the energy equivalent of one ton of coal. (Courtesy Westinghouse Electric Corp., Commercial Nuclear Fuel Division)

(b) The uranium pellets are loaded into long fuel rods, which are grouped into square fuel assemblies (shown). (Courtesy Westinghouse Electric Company)

enrichment: The process by which uranium ore is refined after mining to increase the concentration of fissionable U-235.

nuclear reactor: A device that initiates and maintains a controlled nuclear fission chain reaction to produce energy for electricity.

Africa (8.9%).[1] In the United States, uranium is found in Wyoming, Texas, Colorado, New Mexico, and Utah. Uranium ore contains three isotopes: U-238 (99.28%), U-235 (0.71%), and U-234 (less than 0.01%). Because U-235, the isotope used in conventional fission reactions, is such a minor part (less than 1%) of uranium ore, uranium must be refined after mining to increase the concentration of U-235 to at least 3%. This refining process, called **enrichment**, is energy-intensive.

After enrichment, the uranium fuel used in a **nuclear reactor** is processed into small pellets of uranium dioxide; each pellet contains the energy equivalent of a ton of coal (**Figure 12.3a**). The pellets are placed in **fuel rods,** closed pipes often as long as 3.7 m (12 ft). The fuel rods are then grouped into square **fuel assemblies,** generally of 200 rods each (**Figure 12.3b**). A typical nuclear reactor contains 150 to 250 fuel assemblies.

In nuclear fission, U-235 is bombarded with neutrons (**Figure 12.4**). When the nucleus of an atom of U-235 is struck by and absorbs a neutron, it becomes unstable and splits into two smaller atoms, each approximately half the size of the original uranium atom. In the fission process, two or three neutrons are also ejected from the uranium atom. They collide with other U-235 atoms, generating a chain reaction as those atoms are split and more neutrons are released to collide with additional U-235 atoms.

The fission of U-235 releases an enormous amount of heat, used to transform water into steam. The steam, in turn, is used to generate electricity. Production of electricity is possible because the fission reaction is controlled. Recall that nuclear bombs make use of uncontrolled fission reactions. If the control mechanism in a nuclear power plant were to fail, a bomblike nuclear explosion could not take place because nuclear fuel only has 3-5% U-235, whereas bomb-grade material contains at least 20%—and is usually about 85 to 90%—U-235. In the highly unlikely event of an uncontrolled fission reaction, an immense ammount of heat could be generated. However, the reactor vessel and massive concrete containment building are designed to contain the heat, along with the attendant radioactivity.

[1] Unless otherwise noted, all energy facts cited in this chapter were obtained from the Energy Information Administration (EIA), the statistical agency of the U.S. Department of Energy (DOE).

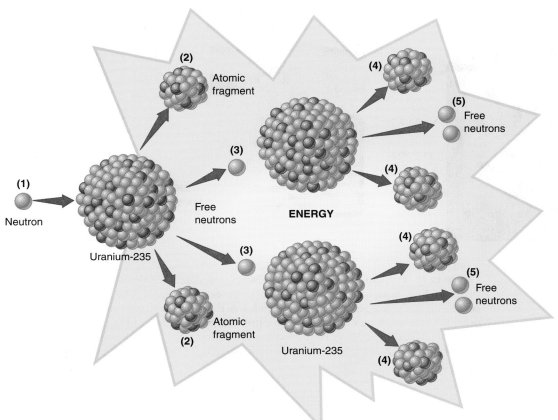

Figure 12.4   Nuclear fission.
Starting at the left side of the figure, neutron bombardment (1) of a uranium-235 (U-235) nucleus causes it to split into two smaller radioactive atomic fragments (2) and several free neutrons (3). The free neutrons bombard nearby U-235 nuclei, causing them to split (4) and release still more free neutrons (5) in a chain reaction. Many different pairs of radioactive atomic fragments are produced during the fission of U-235.

How Electricity Is Produced from Conventional Nuclear Fission

A typical nuclear power plant has four main parts: the reactor core, the steam generator, the turbine, and the condenser (**Figure 12.5**). Fission occurs in the **reactor core,** and the heat produced by nuclear fission is used to produce steam from liquid water in the **steam generator.** The **turbine** uses the steam to generate electricity, and the **condenser** cools the steam, converting it back to a liquid.

The **reactor core** contains the fuel assemblies. Above each fuel assembly is a **control rod** made of a special metal alloy that absorbs neutrons. The plant operator signals the control rod to move either up out of or down into the fuel assembly. If the control rod is out of the fuel assembly, free neutrons collide with uranium atoms in the fuel rods, and fission takes place. If the control rod is completely lowered into the fuel assembly, it absorbs the free neutrons, and fission of uranium no longer occurs. By exactly controlling the placement of the control rods, the plant operator produces the exact amount of fission required.

A typical nuclear power plant has three water circuits. The **primary water circuit** (orange circuit in Figure 12.5) heats water, using the energy produced by the fission reaction. This circuit is a closed system that circulates water under high pressure through the reactor core, where it is heated to about 293°C (560°F). This superheated water cannot expand to become steam because it is under high pressure, and so it remains in a liquid state.

From the reactor core, the hot water circulates to the steam generator, where it boils water held in a **secondary water circuit** (blue circuit in Figure 12.5), converting the water to steam. The pressurized steam goes to and turns the turbine, which in turn spins a generator to produce electricity. After it has turned the turbine, the steam in the secondary water circuit goes to a condenser, where it is converted to a liquid again. The cooling is necessary to obtain the pressure differential that helps turn the turbine blades.

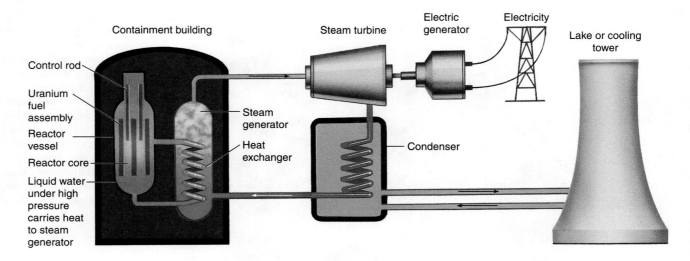

Figure 12.5   Pressurized water reactor. Fission of uranium-235 that occurs in the reactor vessel produces heat, used to produce steam in the steam generator. The steam drives a turbine to generate electricity. The steam then leaves the turbine and is pumped through a condenser before returning to the steam generator. Pumping hot water from the condenser to a lake or massive cooling tower controls excess heat. After it is cooled, the water is pumped back to the condenser. Approximately two-thirds of all nuclear power plants in the United States are of this type.

A **tertiary water circuit** (green circuit in Figure 12.5) provides cool water to the condenser, which cools the spent steam in the secondary water circuit. As the water in the tertiary water circuit is heated, it moves from the condenser to a lake or **cooling tower**, where it is cooled before circulating back to the condenser.

A huge, steel reactor vessel surrounds the reactor core where fission occurs. The **reactor vessel** is a safety feature designed to prevent the accidental release of radiation into the environment. The reactor vessel and the steam generator are placed in a **containment building**, an additional line of defense against accidental radiation leaks. Containment buildings have steel-reinforced concrete walls about 1 to 1.5 m (3 to 5 ft) thick and are built to withstand severe earthquakes and the high winds of hurricanes and tornadoes.

Breeder Reactors and Mixed Oxide Fuel (MOX) for Nuclear Fission

breeder nuclear fission: A type of nuclear fission in which nonfissionable U-238 is converted into fissionable Pu-239.

Uranium ore is mostly U-238, which is not fissionable and is a waste product of conventional nuclear fission. In breeder nuclear fission, however, U-238 is converted to plutonium, Pu-239, a human-made isotope that is fissionable. Some of the neutrons emitted in breeder nuclear fission produce additional plutonium from U-238 (**Figure 12.6**). A breeder reactor thus makes more fissionable fuel than it uses. The fuel is then **reprocessed** to concentrate the Pu-239 for use as fuel.

Because it can use U-238, plutonium-based breeder fission can generate much larger quantities of energy from uranium ore than nuclear fission using U-235. When one adds the uranium reserves in the ground to existing radioactive waste stockpiles, breeder fission has the potential to supply the entire country with electric energy for several centuries.

Although breeder fission sounds promising, there are both safety and weapons proliferations concerns. Breeder fission reactors use liquid sodium rather than water as a coolant. Sodium is a highly reactive metal that reacts explosively with water and burns spontaneously in air at the high temperatures maintained in the breeder reactor. Loss of this coolant would result in an uncontrolled reaction which would almost certainly rip open the containment building, releasing radioactive materials into the atmosphere.

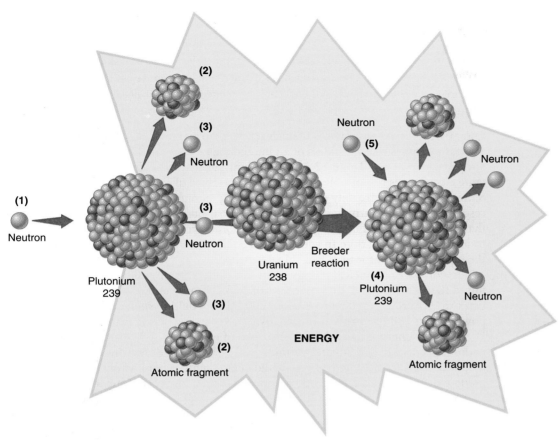

Figure 12.6 **Breeder nuclear fission.** A neutron (1) from a previous fission reaction bombards plutonium-239 (Pu-239), causing it to split into smaller radioactive atomic fragments (2). Additional neutrons (3) are ejected in the process, some of which may collide with uranium-238 to form Pu-239 (4). Neutron bombardment (5) of this Pu-239 molecule causes it to split, and the process continues.

In addition, because it is chemically different than uranium, plutonium is much easier to process into a weapons grade material. As additional countries develop nuclear power, breeder reactors will allow them to secretly produce bomb-quality plutonium more easily than by processing uranium. The plutonium used for India's first nuclear test bomb in 1974 came from its "civilian" breeder reactor development program. The United States performed the first breeder reactor experiments but abandoned its breeder reactor development program in 1977, during the administration of President Carter, a decision supported by every subsequent U.S. President.

More common, particularly in Europe, are reactors that use **mixed oxide fuel** (**MOX**). For MOX reactors, **spent fuel** from standard uranium-based reactors is reprocessed, and the extracted plutonium and U-235 are used as fuel. This option can also be used to mix some of the plutonium from nuclear weapons with uranium, and use it to generate electricity. MOX is now used in about 30 European reactors, and will probably be used in Japanese reactors in the near future. In contrast, only about three breeder reactors operate worldwide.

mixed oxide fuel: A reactor fuel that contains a combination of uranium oxide and plutonium oxide. The plutonium can come from reprocessed spent fuel or from other plutonium stockpiles, including dismantled weapons.

spent fuel: The used fuel elements that were irradiated in a nuclear reactor.

REVIEW

1. What is the nuclear fuel cycle?

2. How does a nuclear reactor produce electricity?

3. What is breeder nuclear fission?

4. What is mixed oxide fuel?

Pros and Cons of Nuclear Energy

LEARNING OBJECTIVES

- Discuss the pros and cons of electric power produced by nuclear energy versus coal.
- Describe the relationship between nuclear power and greenhouse gas reduction.
- Explain why nuclear power does not have much effect on U.S. oil needs.

One of the reasons proponents of nuclear energy argue for the widespread adoption of nuclear energy is that it has less of an immediate environmental impact than fossil fuels, particularly coal (**Table 12.2**). They point out that the combustion of coal to generate electricity is responsible for more than one-third of the air pollution in the United States. Coal is an extremely dirty fuel, especially because we have used up most of our reserves of cleaner-burning coal. Today most coal-burning power plants burn soft coal that produces sulfur-containing emissions that interact with moisture in the atmosphere to form acid precipitation. In addition, the combustion of coal releases carbon dioxide, the biggest contributor to global climate change.

In comparison, nuclear energy emits few pollutants into the atmosphere. According to the EIA, 70.3 million fewer tons of sulfur dioxide and 35.6 fewer tons of nitrogen oxides were emitted from 1973 to 2001 than would have been if existing nuclear power plants had been replaced with fossil fuel-burning power plants.

Nuclear energy is a carbon-free source of electricity—that is, it provides power without producing the climate-warming gas, carbon dioxide. At the very least, proponents say nuclear-generated power should be used until other carbon-free energy technologies, such as solar power, become more efficient and cost-competitive.

Although gasoline remains the major energy source for transportation, it is possible that in the future this will be replaced by a combination of hydrogen- and electric-powered vehicles (Chapter 13). Nuclear power plants currently produce electricity (which can be used, albeit inefficiently, to produce hydrogen), but in the future, it may be possible to generate hydrogen gas directly in specially designed nuclear power plants.

However, nuclear energy generates radioactive waste, such as *spent fuel*. Nuclear power plants also produce other radioactive wastes, such as radioactive coolant fluids and gases in the reactor. Spent fuel and other radioactive wastes are highly radioactive

Table 12.2 Comparison of Environmental Impacts of 1000-MWe Coal and Nuclear Power Plants

Impact	Coal	Nuclear (Conventional Fission)
Land use	17,000 acres	1900 acres
Daily fuel requirement	9000 tons/day	3 kg/day
Availability of fuel, based on present economics	A few hundred years	100 years, maybe longer (much longer with breeder fission)
Air pollution	Moderate to severe, depending on pollution controls	Low
Climate change risk (carbon dioxide emissions)	Severe	Relatively small
Radioactive emissions, routine	1 curie	28,000 curies
Water pollution	Often severe at mines	Potentially severe at nuclear waste disposal sites
Risk from catastrophic accidents	Short-term local risk	Long-term risk over large areas
Link to nuclear weapons	No	Yes
Annual occupational deaths	0.5 to 5	0.1 to 1
Certainty about risks	Well known	Highly uncertain

*Impacts include extraction, processing, transportation, and conversion. Assumes coal is strip-mined.
(A 1000-MWe utility, at a 60% load factor, produces enough electricity for a city of 1 million people.)

and dangerous. The extreme health and environmental hazards caused by this waste require special measures for its storage and disposal.

Nuclear power is also not climate neutral, since a number of steps along the way, from mining to processing to disposal, require substantial amounts of gasoline and diesel. This means that nuclear energy indirectly contributes to the greenhouse effect—about 2 to 6 grams of carbon per kilowatt hour, about two orders of magnitude lower than fossil fuels. In addition, replacing only 10% of the current U.S. fossil fuel use would require doubling the number of nuclear power plants, an expensive and long-term proposition.

Is Electricity Produced by Nuclear Energy Cheap?

While nuclear power advocates present many reasons why we don't have more nuclear power, the main hurdle for nuclear power is cost. It is more expensive to produce electricity from nuclear power than from coal, hydroelectric, or natural gas. In 2007, 104 nuclear power plants supplied the United States with about 20% of its electricity. U.S. government *subsidies* over the past fifty years, such as funding for research and development and for public management of the production and disposal of nuclear materials, has made nuclear power relatively affordable. During the 1960s and 1970s, many energy analysts predicted that the cost of nuclear power would be cheaper than fossil fuels by the year 2000. Generally speaking, this did not happen.

Proponents of nuclear energy usually point to France, which obtains 78% of its electricity from nuclear energy because electricity generated by French nuclear power plants is 27% less expensive than that generated from coal. The lower cost of nuclear power is partly explained because France uses the same design for all its nuclear power plants. Nuclear energy proponents believe that the United States could achieve the same economic efficiency. Opponents of nuclear power dismiss the example of France because the French government, like the governments of other countries that use nuclear power, heavily subsidizes the nuclear industry. Among other things, the French government funds research and development, radioactive waste disposal, and insurance coverage against accidents in its nuclear power plants.

The true costs of nuclear energy, like those of coal and other fossil fuels, are not always obvious in utility bills, whether in France or in the United States. Energy analysts estimate that U.S. tax dollars provide about $9.6 billion in nuclear energy subsidies annually. In an increasingly competitive power market, the generation of electricity using nuclear energy is expensive when all costs, including those subsidized by the government, are taken into account.

The Cost of Building a Nuclear Power Plant

In the United States, no nuclear power plants have been ordered since 1976. Two reasons electric utilities will not commit to building new nuclear power plants are their high cost and very long cost-recovery time. Traditionally designed nuclear power plants are large and take years to plan and build. In comparison, electric plants powered by fossil fuels, especially natural gas, are smaller and less expensive to build. The regulatory process required to obtain permits to build a nuclear power plant is cumbersome and adds to the expense.

Although the initial cost estimates for building a nuclear power plant are high, the actual costs are usually much higher than the forecasts. Cost overruns, borne by the utility and its customers, occur partly because the slow permitting process makes construction fall far behind schedule. Consider the Seabrook nuclear reactor in Seabrook, New Hampshire. Seabrook obtained its operating license from the Nuclear Regulatory Commission (NRC) in 1990, after numerous delays had put its construction 11 years behind schedule. The plant cost more than $6 billion, which was 12 times the original estimate.

Nuclear Power and Electrical Power Deregulation   Before the late 1990s, the price of electricity was averaged so that the higher cost of nuclear power-produced electricity was partly funded by conventionally produced electricity. During the late 1990s, some state governments deregulated the electricity market, which opened electricity to competition by giving businesses and residential customers choices about where to buy their electricity. In addition, as a result of deregulation, power companies began to purchase power from the least expensive sources.

In the deregulated market, the low operating costs of some existing nuclear power plants continued to make them attractive sources of electric power. Other nuclear power plants, particularly older ones, had much higher fuel and labor costs, and price competition led to the closing of certain nuclear power plants that produced expensive electricity.

In the early 2000s, amid widely publicized electricity shortages in several states, the market price of electricity soared, and nuclear power became more attractive economically. Power companies began to think about increasing the amount of electricity they generate by nuclear power. Several power companies asked the NRC if they could increase electricity production at nuclear power plants currently in operation. The NRC began to study the feasibility of reopening previously closed nuclear power plants. Currently, fixing many technical and safety problems in existing plants is economically possible.

Can Nuclear Energy Decrease Our Reliance on Foreign Oil?

Recently, the United States has become increasingly concerned about our reliance on foreign oil (see Chapter 11 Introduction). Some supporters of nuclear energy assert that our dependence on foreign oil would lessen if all oil-burning power plants were converted to nuclear plants. This claim is not as convincing as it seems because oil generates only about 3% of the electricity in the United States; we rely on oil primarily for transportation, for heavy machinery, and for heating buildings. Thus, replacing electricity generated by oil with electricity generated by nuclear energy would do little to lessen our dependence on foreign oil because nuclear energy cannot currently replace oil-intensive uses.

Technological advances could change nuclear power's potential contribution in the future. If electric heat pumps and electric motor vehicles become more common in the future, nuclear power plants have the potential to heat buildings and power automobiles and thus could decrease our reliance on foreign oil.

REVIEW

1. In generating electricity, how do the environmental effects of coal combustion and conventional nuclear fission compare?
2. Is electricity produced by nuclear energy inexpensive? Explain your answer.
3. Would expanding nuclear power in the United States reduce greenhouse gases and/or decrease reliance on oil?

 ## Safety Issues in Nuclear Power Plants

LEARNING OBJECTIVES

- Describe the nuclear power plant accidents at Three Mile Island and Chornobyl.
- Discuss the link between nuclear energy and nuclear weapons.

Although conventional nuclear power plants cannot explode like atomic bombs, accidents can happen in which dangerous levels of radiation are released into the environment and result in human casualties. At high temperatures the metal encasing the uranium fuel melts, releasing radiation; this is known as a **meltdown.** Also, the water used in a nuclear reactor to transfer heat can boil away during an accident, contaminating the atmosphere with radioactivity.

The probability that a major accident will occur is considered low by the nuclear industry, but public perception of the risk is high for several reasons. Nuclear power risks are involuntary and potentially catastrophic. In addition, many people are distrustful of the nuclear industry. The consequences of an accident are drastic and life threatening, both immediately and long after the accident has occurred. We now consider two major accidents, one in the United States (Three Mile Island) and the other at Chornobyl in Ukraine.

Three Mile Island

The most serious nuclear reactor accident in the United States occurred in 1979 at the Three Mile Island power plant in Pennsylvania, the result of the intersection of human and design errors. A 50% meltdown of the reactor core took place. Had there been a complete meltdown of the fuel assembly, dangerous radioactivity would have been emitted into the surrounding countryside. Fortunately, the containment building kept almost all the radioactivity released by the core material from escaping. Although a small amount of radiation entered the environment, there were no substantial environmental damages and no known human casualties. A study conducted within a 10-mile radius around the plant 10 years after the accident concluded that cancer rates were in the normal range and that no association could be made between cancer rates and radiation emissions from the accident. Numerous other studies have failed to link abnormal health problems (other than increased stress) to the accident.

Three Mile Island elevated public apprehension about nuclear power. It took 12 years and $1 billion to repair and reopen Three Mile Island. Although the reactor involved in the accident at Three Mile Island was destroyed, a second reactor that was undamaged during the accident is currently in operation. In the aftermath of the accident, public wariness prompted construction delays and cancellations of several new nuclear power plants across the United States.

On the positive side, the accident at Three Mile Island reduced the complacency that was commonplace in the nuclear industry. New safety regulations were put in place, including more frequent safety inspections, new risk assessments, and improved emergency and evacuation plans for nuclear power plants and surrounding communities. The Institute of Nuclear Power Operations was created to promote safety and improve the training of nuclear operators. The meaning of the accident continues to be debated. Nuclear power proponents argue that since there was no radiation release, it is a success story. Opponents counter that luck was the only reason that there was not a more serious result, and that in any case, the public had been promised that events like that at Three Mile Island would never occur.

Chornobyl

The worst accident ever to occur at a nuclear power plant took place in 1986 at the Chornobyl plant, located in the former Soviet Union in what is now Ukraine. One or possibly two explosions ripped apart a nuclear reactor and expelled large quantities of radioactive material into the atmosphere (**Figure 12.7**). The effects of this accident were not confined to the area immediately surrounding the power plant: Significant amounts of radioisotopes quickly spread across large portions of Europe. The Chornobyl accident affected and will continue to affect many nations.

The first task faced after the accident was to contain the fire that had broken out after the explosion and prevent it from spreading to other reactors at the power plant. Local firefighters, many of whom later died from exposure to the high levels of radiation, battled courageously to contain the fire. In addition, 116,000 people who lived within a 30-km (18.5-mi) radius around the plant were quickly evacuated and resettled. Ultimately, more than 170,000 people had to permanently abandon their homes.

Once the danger from the explosion and fire had passed, the radioactivity at the power station had to be cleaned up and contained so that it would not spread. Dressed in protective clothing, workers

Figure 12.7   Chornobyl. The arrow indicates the site of the explosion. The upper part of the reactor was completely destroyed. (Novosti)

were transported to the site in radiation-proof vehicles; initially the radioactivity was so high they could stay in the area for only a few minutes at a time. There are few photographs of the cleanup because the radiation quickly ruined the camera film. After the initial cleanup, the damaged reactor building was encased in 300,000 tons of concrete. Then the surrounding countryside had to be decontaminated. Highly radioactive soil was removed, and buildings and roads were scrubbed down to remove the radioactive dust that settled on them.

Although cleanup in the immediate vicinity of Chornobyl is finished, the people in Ukraine face many long-term problems. Much of the farmland and forests are so contaminated they cannot be used for more than a century. Loss of agricultural production is one of the largest costs for the local economy. Inhabitants in many areas of Ukraine still cannot drink the water or consume locally produced milk, meat, fish, fruits, or vegetables. Mothers do not nurse their babies because their milk is contaminated by radioactivity. The Ukrainian Scientific Center for Radiation Medicine has continually monitored the health of approximately 80,000 Chornobyl patients.

A 2003 report by the Ukrainian Intelligence Agency found a range of factors contributed to the accident. These included violations of safety rules, flawed design, inferior construction, and operator errors. The main design flaw was the lack of a containment building and instability at low power. This type of reactor—an RBMK reactor—is not used commercially in either North America or Western Europe because nuclear engineers consider it too unsafe. As of 2004, Russia and Lithuania still have 14 RBMK reactors in operation, although they have made safety improvements in them since the Chornobyl explosion.

Human error also contributed significantly. Many of the Chornobyl plant operators lacked scientific or technical understanding of the plant they were operating, and they made several major mistakes in response to the initial problem. As a result of the disaster at Chornobyl, the Soviet government developed a retraining program for operators at all their nuclear power plants, and safety features were added to existing re-

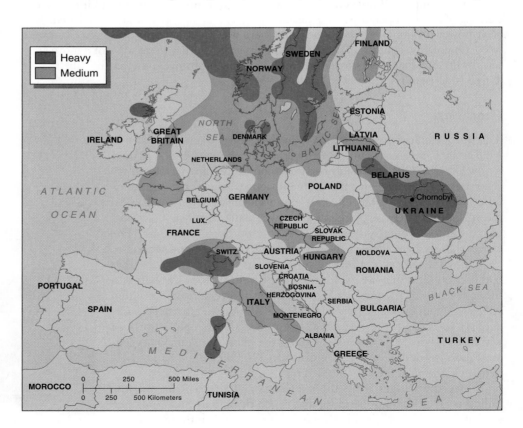

Figure 12.8 Radioactive fallout from Chornobyl. Parts of Ukraine, Belarus, Lithuania, Sweden, Norway, France, Italy, and Switzerland received heavy radioactive fallout from the accident.

actors. Nonetheless, nuclear power plants require many operators, and human error will continue to be a concern. Hiring highly qualified operators means more expensive and less competitive power.

One of the disquieting consequences of Chornobyl was the unpredictable course taken by spreading radiation (**Figure 12.8**). Chornobyl's radiation cloud unevenly dumped radioactive fallout over some areas of Europe and Asia, leaving other areas relatively untouched and making it difficult to plan emergency responses for a possible future nuclear accident.

As the health effects of the accident at Chornobyl are monitored over the years, the death toll is rising. Twenty years after the accident, the number of attributable deaths is highly disputed, with estimates ranging from around 10,000 to over 100,000. Because the level of screening and record keeping has increased since the accident, it is difficult to assess which deaths were caused by Chornobyl. In addition, estimates can vary because of uncertainty about the effects of small doses of radiation. While there is disagreement, the most recent Biological Effects of Ionizing Radiation (BEIR VII) expert panel concluded that even very small doses of **ionizing radiation** can be harmful.

Nearly 400,000 adults and more than one million children currently receive government aid for health problems related to Chornobyl. In addition, many people in nearby countries received dangerous doses of radiation. The frequency of birth defects and reduced cognitive function in newborns has increased in Belarus, Ukraine, and certain other parts of Europe. There has also been an increased incidence of leukemia in infants and of thyroid cancer and immune abnormalities in children exposed to radioactive iodine from the Chornobyl fallout (**Figure 12.9**). Breast cancer, stomach cancer, and other organ cancers are not expected to occur in large numbers until 20 years or more after the accident. Many of those living under the cloud of Chornobyl suffer from debilitating mental stress, anxiety, and depression, and their psychological injuries are still being assessed.

Figure 12.9 Suspected Chornobyl Victims, April 2006. These Ukrainian children, born since the 1986 Chornobyl accident, are thought to have illnesses caused by exposure to radiation released by the accident. The poster they are looking at commemorates the 20th anniversary of the accident. A significant (25-fold) increase in thyroid cancer in children and adolescents occurred within a few years after the accident. Fortunately, if the thyroid gland is removed, patients with thyroid cancer have high survival rates, although they require thyroid hormone replacement therapy for the rest of their lives. (Adalberto Roque/AFP/Getty Images)

The world has learned much from this nuclear disaster. Most countries now take nuclear power more seriously, hoping to prevent accidents. Safety features found in North American and Western European reactors are being incorporated into new nuclear power plants around the world. Nuclear engineers learned a great deal from the cleanup and entombment of Chornobyl; this knowledge will be useful when old nuclear power plants are dismantled. Doctors learned more about effective treatment of people exposed to massive doses of radiation. And ecologists have learned both about how nuclear materials move through the food chain, and more generally about how natural systems change when large numbers of humans abruptly move away from an area.

The Link Between Nuclear Energy and Nuclear Weapons

Fission is involved in both the production of electricity by nuclear energy and the destructive power of nuclear weapons. Uranium-235 and plutonium-239 are the two fuels commonly used in atomic fission weapons. Plutonium is intentionally produced in breeder reactors, and can be extracted from the spent fuel from conventional fission reactors.

Thirty-one countries currently use nuclear energy to generate electricity, and that number will certainly grow. The possession of nuclear power plants gives these countries access to the fuel needed for nuclear weapons (by reprocessing spent fuel). Many world leaders are concerned about the consequences of terrorist groups and countries of concern building nuclear weapons. Proliferation concerns cause many people and countries to shun nuclear energy, particularly breeder fission, and to seek alternatives that are not so intimately connected with nuclear weapons.

There are several hundred metric tons of weapons-grade plutonium worldwide, much of it surplus from scrapped nuclear weapons. Hundreds of additional metric tons of plutonium wastes exist from commercial reactors; this amount increases each year. Storing plutonium is a security nightmare because it takes only several kilograms to make a nuclear bomb as powerful as the ones that destroyed the Japanese cities of Nagasaki and Hiroshima in World War II. However, since the 9/11 terrorist attack, the security of nuclear power plants has been increased substantially to reduce the chance that terrorist groups could steal plutonium and enriched uranium and use them to make nuclear weapons or a "dirty bomb" that spreads radiation.

Of special concern to international security is Russia's political and economic instability. According to the EIA, the former Soviet Union has enough highly enriched uranium and plutonium stored at about 40 different sites to make 40,000 nuclear bombs! The U.S. Department of Energy and the International Atomic Energy Agency work with Russia's Atomic Energy Ministry to maintain nuclear security.

The United States plans to get rid of more than 50 tons of surplus plutonium from the dismantling of its nuclear warheads. After reviewing dozens of possible solutions, the Clinton administration decided to pursue a $2 billion program that includes two options. One, plutonium will be combined with highly radioactive waste and then vitrified into glass logs (discussed shortly). Two, some plutonium will be converted to MOX, and then "burned" as a fuel in commercial nuclear power reactors. Both options would make the plutonium so radioactive that it could not be safely handled to construct a bomb. The National Academy of Sciences and many nuclear scientists and arms-control experts endorsed the two-option Clinton plan, although all agreed there is no perfect solution. The George W. Bush administration is considering a revival of breeder reactors as well as burning plutonium in MOX. At the time this book went to press, neither of these has been done.

REVIEW

1. What caused the nuclear power plant accident at Three Mile Island in Pennsylvania?

2. What were some of the short-term effects of the nuclear power plant accident at Chornobyl in Ukraine? What are some long-term effects?

3. What is the link between nuclear energy that produces electricity and nuclear weapons?

Radioactive Wastes

LEARNING OBJECTIVES

- Distinguish between low-level and high-level radioactive wastes.
- Relate the pros and cons of permanent storage of high-level radioactive wastes at Yucca Mountain.
- Explain what happens to nuclear power plants after they are closed.

Radioactive wastes are classified as either low-level or high-level. Produced by nuclear power plants, university research labs, nuclear medicine departments in hospitals, and industries, **low-level radioactive wastes** include glassware, tools, paper, clothing, and other items contaminated by radioactivity. The **Low-Level Radioactive Waste Policy Act,** passed in 1980, specified that all states are responsible for the waste they generate, and it encouraged states to develop facilities to handle low-level radioactive wastes by 1996. Three sites currently accept waste for the entire country—Washington State, South Carolina, and Utah. New technologies to compact low-level radioactive waste have dramatically reduced the volume being disposed of in these nuclear dumps.

Examples of **high-level radioactive wastes** produced during nuclear fission include the reactor metals (fuel rods and assemblies), coolant fluids, and air or other gases found in the reactor. High-level radioactive wastes are also generated during the reprocessing of spent fuel. Produced by nuclear power plants and nuclear weapons facilities, high-level radioactive wastes are among the most dangerous human-made hazardous wastes.

Fuel rods absorb neutrons, thereby forming radioisotopes, and can be used for only about three years before becoming highly radioactive spent fuel. In 2002 the United States had 46,268 metric tons of spent fuel stored temporarily at more than 100 nuclear power plants around the country **(Table 12.3)**. Worldwide, about 10,000 metric tons of spent fuel are produced each year. As the radioisotopes in spent fuel decay, they produce considerable heat, are extremely toxic to organisms, and remain radioactive for thousands of years. Their dangerous level of radioactivity requires special handling. Secure storage of these materials must be guaranteed for thousands of years, until they decay sufficiently to be safe.

It is likely that nuclear power generation will increase substantially in the upcoming decades. In the long term, most waste experts conclude that deep geologic burial is the best solution. In the short term, however, the nearest date for opening a site is at least a decade away. Further, there are concerns that, in the next few decades, we may want to reprocess the spent fuel to reuse its plutonium and uranium in breeder reactors. Nonetheless, a safe, cost-effective solution for the problem of radioactive waste is a priority for the nuclear energy industry and regulators.

For permanent disposal, high-level radioactive wastes must be stored in an isolated area where there is minimal possibility they can contaminate the environment. The storage site must have geologic stability and little or no water flowing nearby, which might transport the waste away from its original site. In 2004, U.S. Federal Courts decided that any permanent burial site must meet EPA standards for the next one million years—an increase from the previous 10,000 year standard. (The National Academy of Sciences has said that radiation leaks could be highest hundreds of thousands of years from now.) Guaranteeing safety over a million-year period stretches the credibility of any scientific assessment.

What are the best sites for the long-term storage of high-level radioactive wastes? Many scientists recommend storing the wastes in stable rock formations deep in the ground. Another suggestion for the long-term storage of radioactive wastes is above-ground mausoleums built in remote locations. If we build mausoleums, we cannot simply

low-level radioactive wastes: Radioactive solids, liquids, or gases that give off small amounts of ionizing radiation.

high-level radioactive wastes: Radioactive solids, liquids, or gases that give off large amounts of ionizing radiation.

Table 12.3 Nuclear Power Plants and High-Level Radioactive Waste in Selected Countries

Country	Number of Reactors, 2006	Spent Fuel Inventories, 2000*
Argentina	2	2,480
Canada	20	27,860
France	59	30,480
Japan	55	17,450
South Africa	2	540
South Korea	20	4,780
Sweden	10	4,130
United Kingdom	19	41,430
United States	104	42,710

* Metric tons of radioactive waste.
Sources: International Atomic Energy Agency and Energy Information Administration

ENVIRONEWS

Public and Expert Attitudes Toward Nuclear Energy

Nuclear power advocates and opponents are engaged in a vicious circle that challenges the future of nuclear power. The **NIMBY**, or "not in my backyard," **response** commonly follows a proposal for a nuclear power plant or a radioactive waste disposal site. The NIMBY response is prevalent in nuclear energy issues partly because, despite the experts' assurances that a site will be safe, no one can guarantee complete safety, with no possibility of an accident. (Recall from the Chapter 7 discussion of risk assessment that nothing is risk-free.)

A sister response to NIMBY is the **NIMTOO response**, which stands for "not in my term of office." Politicians who wish to get reelected are sensitive to their constituents' concerns and are not likely to support the construction of a nuclear power plant or radioactive waste disposal site in their districts.

The NIMBY and NIMTOO responses of opponents have parallel responses in the "irrational public" belief held by many pro nuclear advocates and experts, and the DAD ("decide, announce, defend") approach taken by industry and government. Historic exposures from nuclear testing, misrepresentations of nuclear power safety that became known in the 1970s, and the "impossible" event at Three Mile Island have generated substantial distrust of both the nuclear industry and the governmental regulators. This distrust is amplified by interactions with experts, who often have little understanding of the nature of public resistance or training in effective risk communication. Experts who encounter this distrust misinterpret it as irrationality, "emotion," or ignorance.

Unfortunately, the driving forces behind NIMBY, NIMTOO, DAD, and distrust are difficult to resolve. Nuclear power proponents, believing that the public is uninformed, react with campaigns explaining how "safe" a proposal is. The public feels insulted, which leads to further distrust, and dismissal of expert arguments. . . which the experts again misinterpret as irrationality and anti-scientific attitudes. The vicious circle continues.

Consider the disposal of radioactive waste. There is universal agreement that we must safely isolate radioactive waste until it decays enough to cause little danger. But when it comes to identifying a specific location for the waste, there is opposition, since people object to transportation or storage near where they work and live. "Experts" are called out to explain to the public how safe the waste disposal and transportation will be. The already wary public finds these experts condescending and dismissive, and distrusts them even more.

Most people agree that our generation has the responsibility to dispose of radioactive waste generated by the nuclear power plants we have already built. Breaking the vicious circle between nuclear advocates who chronically misunderstand the nature of public opposition and the resistance of many people to anything nuclear will be essential for a successful nuclear waste management plan.

store the wastes and forget about them. Mausoleums will have to have adequate security to guarantee their safety. Other long-term possibilities include storage in Antarctic ice sheets and burial in the seabed (beneath the ocean floor). International agreements currently prohibit ocean disposal of both low-level and high-level radioactive wastes because ocean disposal has the potential to harm the marine environment.

Meanwhile, radioactive wastes continue to accumulate. In the United States, there are dozens of "temporary" sites where domestic radioactive wastes have been stored for decades. Most commercially operated nuclear power plants store their spent fuel in huge indoor pools of water on-site. None of these plants was designed for long-term storage of spent fuel. Because they have nowhere to send their spent fuel when on-site storage areas are full, these plants must either expand their on-site storage—an expensive proposition because of legal battles that usually occur before expansion is granted—or shut down. As of 2006, more than two dozen commercially operated plants had built above-ground, air-cooled concrete and steel casks to expand their on-site storage. Many others have committed to do the same, due to uncertainty about the future of underground storage (**Figure 12.10**).

(a) On-site storage casks at the Prairie Island nuclear power plant in Minnesota. Each cask holds 40 spent fuel assemblies (17.6 tons). (Peter Essick/Aurora Photos)

(b) Details of a storage cask. Each cask, designed to last at least 40 years, is monitored and will be replaced if leakage occurs.

Figure 12.10   Storage casks for spent fuel.

Radioactive Wastes with Relatively Short Half-Lives

Some radioactive wastes are produced directly from the fission reaction. Uranium-235, the reactor fuel, may split in several different ways, forming smaller atoms, many of which are radioactive. Most of these, including krypton-85 (half-life, 10.4 years), strontium-90 (half-life, 28 years), and cesium-137 (half-life, 30 years), have relatively short-term radioactivity. In 300 to 600 years they will have decayed to the point where they are safe.

The safe storage of fission products with relatively short half-lives is of concern because fission produces larger amounts of these materials than of the materials with extremely long half-lives. Health concerns exist because many of the shorter-lived fission products mimic essential nutrients and tend to concentrate in the body, where they continue to decay, with harmful effects (see discussion of bioaccumulation in Chapter 7). One of the common fission products, strontium-90, is chemically similar to calcium. If strontium-90 were accidentally released into the environment from improperly stored radioactive waste, it could be incorporated into human and animal bones and teeth in place of calcium. In like manner, cesium-137 replaces potassium in the body and accumulates in muscle tissue, and iodine-131 concentrates in the thyroid gland.

CASE IN POINT

Yucca Mountain

In 1982, passage of the **Nuclear Waste Policy Act** put the burden of developing permanent sites for civilian and military radioactive wastes on the federal government and required the first site to be operational by 1998. It also legally required the U.S. Government to take ownership of nuclear wastes. Since 1999, radioactive wastes from the manufacture of nuclear weapons have been stored permanently in deep underground salt beds near Carlsbad, New Mexico. However, the deadline for completion of an operational high-level radioactive civilian waste repository has been postponed from 1998 to 2010 to 2017, and is likely to be postponed again.

In a 1987 amendment to the Nuclear Waste Policy Act, Congress identified **Yucca Mountain** in Nevada as the only candidate for a permanent underground storage site for 70,000 tons of high-level radioactive wastes from commercially

Figure 12.11   Yucca Mountain.

(a) Yucca Mountain is in an arid, sparsely populated area of Nevada. (Howells David/Gamma-Presse, Inc.)

(b) If and when spent fuel is stored at Yucca Mountain, it will be in a huge complex of interconnected tunnels located in dense volcanic rock 300 m (1,000 ft) beneath the mountain crest. Canisters containing high-level radioactive waste may be stored in the tunnels.

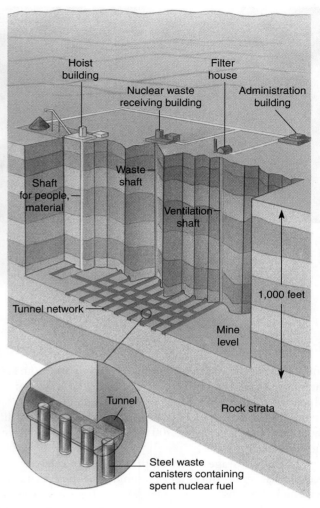

operated power plants (**Figure 12.11**). At the time, such a political decision was relatively easy, since Nevada was a sparsely populated and politically weak state. Yucca Mountain would be able to take the 42,000-plus tons of spent fuel that have been produced in the United States to date plus spent fuel that will be produced until about 2025. At that time, Yucca Mountain will be full, and a new geologic depository will be needed.

Since 1983 the U.S. Department of Energy has spent billions of dollars conducting feasibility studies on Yucca Mountain's geology. Results indicate that the site is safe, at least from volcanic eruptions and earthquakes. However, the suitability of Yucca Mountain has been mired in scientific, management, cost, public opposition, and scheduling controversies, and Nevadans oppose the selection of their state for a radioactive waste site. In 2002 Congress finally approved the choice of Yucca Mountain as the U.S. nuclear-waste repository, despite the state of Nevada's opposition. Nonetheless, it remains unclear whether and when the site will be licensed and opened.

The Yucca Mountain site, some 145 km (90 mi) northwest of Las Vegas, is controversial in part because it is near a volcano (its last eruption may have occurred 20,000 years ago) and active earthquake fault lines. The possibility of a volcanic eruption at Yucca Mountain is currently considered remote (1 chance in 10,000 during the next 10,000 years). Concerns that earthquakes might disturb the site and raise the water table, resulting in radioactive contamination of air and groundwater, were examined when a magnitude 5.6 earthquake occurred in 1992 about 20 km (about 12 mi) from Yucca Mountain. Scientists were already monitoring water table elevation, and they measured a 1-meter change caused by the earthquake. The water table is some 800 m (2625 ft) beneath the mountain crest, so water elevation changes due to earthquake activity are not considered a serious problem by most experts, but remain a concern to others.

Transporting high-level wastes from nuclear reactors and weapons sites by truck or rail is a major concern of opponents of the Yucca Mountain site. The typical shipment would travel an average of 2300 miles, and 43 states would have these dangerous materials passing through on their way to Yucca Mountain. Eight states—Illinois, Indiana, Iowa, Kansas, Missouri, Nebraska, Utah, and Wyoming—would contain major transportation corridors to Yucca Mountain.

Around the world, more than two dozen countries plan to store their spent uranium fuel in deep underground deposits. Many of these countries are having similar challenges of public opposition to proposed sites. France, which has long been considered a nuclear power success story, has been unable to select a permanent site. In Sweden, agreement on a site came only after the country agreed to discontinue nuclear power—although that decision may be revisited as energy demand and climate change concerns increase. ■

High-Level Radioactive Liquid Waste

High-level liquid wastes are dangerously unstable and difficult to monitor. For that reason, they must be converted to solid form before they are stored at Yucca Mountain. The U.S. government wants to store high-level liquid wastes in enormous glass or ceramic logs that are contained in stainless steel canisters. The glass or ceramic contains boron, which absorbs neutrons and so will prevent the logs from exploding. Solidifying liquid waste into solid glass or ceramic logs, known as **vitrification,** is an established practice in Europe. The United States began producing glass logs in 1996, although it does not yet have a permanent disposal site for these hazardous logs. High-level liquid wastes that have not been solidified are temporarily stored in large underground tanks in New York, Washington State, Idaho, and South Carolina.

Decommissioning Nuclear Power Plants

Nuclear power plants are licensed to operate for a maximum of 40 years, although some nuclear power plants with good operating records have asked the NRC to extend their operating licenses past the 40-year cutoff. Several of these reactors have received 20-year extensions.

As nuclear power plants age, certain critical sections, such as the reactor vessel, become brittle or corroded. At the end of their operational usefulness, nuclear power plants are not simply abandoned or demolished because many parts have become contaminated with radioactivity.

Three options exist when a nuclear power plant is closed: storage, entombment, and decommissioning. If an old plant is put into **storage,** the utility company guards it for 50 to 100 years, while some of the radioactive materials decay. This decrease in radioactivity makes it safer to dismantle the plant later, although accidental leaks during the storage period are still a concern.

Most experts do not consider **entombment,** permanently encasing the entire power plant in concrete, a viable option because the tomb would have to remain intact for at least 1000 years. Accidental leaks would likely occur during that time, and we cannot guarantee that future generations would inspect and maintain the "tomb."

The third option for the retirement of a nuclear power plant is to **decommission** the plant immediately after it closes. The workers who dismantle the plant must wear protective clothing and masks. Some portions of the plant are too "hot" (radioactive) for workers to safely dismantle, although advances in robotics may make it feasible to tear down these sections. As the plant is torn down, small sections of it are transported to a permanent storage site.

According to the International Atomic Energy Agency, worldwide, 107 nuclear power plants were permanently retired as of 2004 (23 of these in the United States), and many nuclear power plants are nearing retirement age. In 2004, 143 operational

decommission: To dismantle an old nuclear power plant after it closes.

ENVIRONEWS

Nuclear Power, Climate Change, and the Systems Perspective

In early 2006, one of the founders of Greenpeace published an essay in the *Washington Post* explaining that he had begun to support nuclear power. He argued that the problems with nuclear power are outweighed by the threat of climate change. This idea has been echoed in a variety of settings, from talk radio to the editorial page of *Science* magazine.

However, this narrow conception of *either* nuclear power *or* climate change fails to take a systems perspective, and so may lead us to miss better alternatives. A systems thinker would ask "what are the tradeoffs we make between energy generation, environmental damage, and human health." Systems thinking would probably lead to some increase in nuclear energy generation, but would also assess other options to reduce energy use, increase energy efficiency, explore alternative energy sources, and reduce carbon emissions from fossil fuels.

plants around the world were 25 years old or older (the oldest two were 39 years old). Since 1989, several U.S. nuclear power plants have been decommissioned.

During the 21st century, we may find that we are paying more in our utility bills to decommission old plants than we are to have new plants constructed. For example, decommissioning the Maine Yankee, one of the first large commercial nuclear reactors to be built, is currently underway at a cost of about $635 million; the plant's construction costs in the 1960s and 1970s were $231 million (in today's dollars). For the foreseeable future, however, high level wastes will remain at the site, since no permanent disposal site has been established. Decommissioning other nuclear power plants will probably cost at least as much as for the Maine Yankee.

REVIEW

1. What is low-level radioactive waste, and how is it disposed? What is high-level radioactive waste, and how is it currently stored?

2. What are some of the advantages of storing high-level radioactive waste at Yucca Mountain? What are some of the disadvantages?

3. Why is decommissioning nuclear power plants such a major task?

Fusion: Nuclear Energy for the Future?

LEARNING OBJECTIVE

• Describe some of the technological hurdles that must be overcome before nuclear fusion becomes a reliable energy source.

A common joke among nuclear scientists is that commercially viable fusion will always be 30 years away. In the 1950s when it first became a plausible concept, it was thought to be about 30 years away. In the 1970s, commercial fusion was expected before 2010, and current estimates are for commercial fusion power in the 2030s or 2040s. The atomic reaction that powers the stars, including our sun, is fusion. In fusion, two lighter atomic nuclei are brought together under conditions of high heat and pressure so that they combine, producing a larger nucleus. The energy produced by fusion is considerable; it makes the energy produced by the burning of fossil fuels seem trifling in comparison: 30 mL (1 oz) of fusion fuel has the energy equivalent of 266,000 L (70,000 gal) of gasoline.

Isotopes of hydrogen are the fuel for fusion. In one type of fusion reaction, the nuclei of deuterium and tritium combine to form helium (**Figure 12.12**), releasing huge amounts of thermal energy (heat) in the process; in a fusion power plant, the heat would be converted to electricity. Deuterium, or heavy hydrogen, is present in water and is relatively easy to separate from normal hydrogen. Tritium, which is radioactive, is not found in nature; this human-made hydrogen isotope is formed during the fusion reaction by bombarding another element, lithium, with neutrons. The required isotope of lithium, Li-6, is found in seawater and certain types of surface rocks.

Supporters of nuclear energy view fusion as the best possible form of energy, both because its fuel, hydrogen, is available in virtually limitless supply and because fusion will produce no high-level radioactive waste. (Fusion will produce low-level waste, however.) Unfortunately, many technological difficulties have been encountered in efforts to stage a controlled fusion reaction. It takes phenomenally high temperatures (millions of degrees) to make atoms fuse. To date, the best fusion experiments generate about one-third of the energy used to heat the fuel.

Another challenge is confining the fuel. At extremely high temperatures, a gas separates into negative electrons and positive nuclei. This superheated, ionized gas, called

Figure 12.12 Fusion. In one possible fusion reaction, a deuterium nucleus fuses with a tritium nucleus, producing a helium nucleus, a neutron, and considerable energy. Neutrons formed during the reaction would bombard lithium, splitting it into helium and tritium. Thus, additional tritium fuel would be generated.

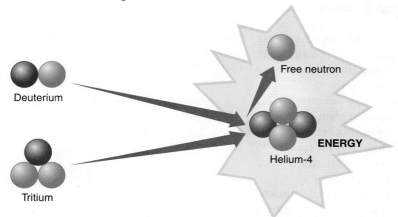

plasma, has a tendency to expand. Confinement of the plasma is necessary so that the nuclei are close enough to one another to fuse, but a regular container does not work because as soon as the nuclei hit the container walls, they lose so much energy they cannot fuse. Several technological approaches to confine plasma are under investigation. One approach studied for decades involves using magnetic confinement to circulate the plasma in a doughnut-shaped chamber called a *tokamak*. Research-sized tokamak reactors are relatively small and fit inside a large room.

Research into fusion power has been marred by several controversies. In the 1990s, researchers prematurely announced that they had discovered "cold fusion," or fusion at temperatures much lower than expected. More recently, researchers published papers on "bubble fusion," which some reviewers think was in fact not fusion at all. The "cold fusion" controversy ruined several scientific careers, while the "bubble fusion" controversy has yet to be resolved. The most promising opportunity for fusion research is the International Thermonuclear Experimental Reactor (ITER), sponsored by a number of countries including the United States. It is scheduled to be constructed over the next several years, and to operate for at least two decades.

REVIEW

1. What are two major hurdles that must be overcome before fusion becomes a practical source of energy?

 ## The Future of Nuclear Power

LEARNING OBJECTIVE

- Briefly summarize the issues that must be addressed if nuclear power is to become a major energy source in the future.

In an effort to promote nuclear energy, nuclear and utility executives have developed a plan that addresses the safety and economic issues associated with nuclear power. They envision building a series of "new generation" nuclear reactors designed to be 10 times safer than current reactors. Costs could be held in line by standardizing nuclear power plants rather than custom-building each one. When France standardized its plant designs, the costs were lowered considerably. The plan calls for improving building schedules and streamlining the regulatory process.

The NRC has already certified several new designs, such as the *pebble-bed modular reactor* (PBMR), which uses small ceramic-encased balls of uranium instead of fuel rods. Nuclear experts consider the new fission reactors much safer than the nuclear reactors that are currently operating in the United States. The PBMR and other new reactors cannot have a meltdown even in a worst-case scenario. Because the fuel cannot melt, radiation cannot be released into the environment during the plant's operation. The PBMR uses helium rather than water to turn the turbines and cool the system. Helium, a gas much less corrosive than steam, is less likely to corrode pipes and cause leaks. The new generation of nuclear power plants, though smaller, simpler in design, less expensive to build, and safer to operate, will still produce high-level radioactive wastes and have a potential link to nuclear weapons.

In early 2007, President George W. Bush called for 30 new commercial nuclear power plants to be built in the United States between 2015 and 2025. While this would increase U.S. nuclear capacity by about one-third, it world not have much impact on the total U.S. energy supply.

Globally, 16 of the 25 nuclear power plants currently under construction are in Asia. To meet their rapidly expanding need for electric power, Asian countries have plans to build dozens more nuclear power plants. Nuclear power appears to be making a comeback in other countries as well. In 2004 France announced its intention to replace its aging nuclear reactors with new ones. Unfortunately, lack of research and training means that designs have not improved as much as they could have, and not enough nuclear engineers are qualified to design, build, and operate nuclear power plants. While we are likely to have more nuclear power over the next few decades, the safety and quality of the plants remains in doubt.

REVIEW

1. If nuclear power is to become a major energy player in the 21st century, what major problems must be addressed?

REVIEW OF LEARNING OBJECTIVES WITH KEY TERMS

• **Distinguish between nuclear energy and chemical energy.**

In ordinary chemical reactions, the atoms of one element do not change into the atoms of another element, nor does any of their mass (matter) change into energy. In contrast, **nuclear energy** is the energy released by nuclear fission or fusion. In nuclear energy small amounts of matter from atomic nuclei are converted into large amounts of energy.

• **Contrast fission and fusion.**

Fission is the splitting of an atomic nucleus into two smaller fragments, accompanied by the release of a large amount of energy. **Fusion** is the joining of two lightweight atomic nuclei into a single, heavier nucleus, accompanied by the release of a large amount of energy. Both fission and fusion result in a significant release of energy, in comparison to the chemical combustion of fossil fuels.

• **Define *radioactive decay.***

Radioactive decay is the emission of energetic particles or rays from unstable atomic nuclei; it includes positively charged alpha particles, negatively charged beta particles, and high-energy electromagnetic gamma rays.

• **Describe the nuclear fuel cycle, including the process of enrichment.**

The **nuclear fuel cycle** is all the processes involved in producing the fuel used in nuclear reactors and in disposing of radioactive wastes (or nuclear wastes). **Enrichment**, which is part of the nuclear fuel cycle, is the process by which uranium ore is refined after mining to increase the concentration of fissionable U-235.

• **Define *nuclear reactor* and describe a typical nuclear power reactor.**

A **nuclear reactor** is a device that initiates and maintains a controlled nuclear fission chain reaction to produce energy for electricity. A typical reactor contains a **reactor core**, where fission occurs; a **steam generator**; a **turbine**; and a **condenser**. The reactor core contains about 250 **fuel assemblies**, each consisting of 200 **fuel rods**. Fuel rods contain pellets of uranium dioxide. Above each fuel assembly is a **control rod** that is moved into or out of the fuel assembly, thereby producing the amount of fission required. The fission of U-235 releases heat that converts water to steam, used to generate electricity. Safety features include a steel **reactor** vessel and a **containment building** built of steel-reinforced concrete.

• **Contrast conventional nuclear fission, breeder nuclear fission, and mixed oxide fuel fission and describe spent fuel.**

Conventional nuclear fission uses U-235, which makes up about 3–5% of uranium after enrichment. **Breeder nuclear fission** is a type of nuclear fission in which nonfissionable U-238 is converted into fissionable Pu-239. Thus, breeder reactors make more fuel than they use. **Mixed Oxide Fuel (MOX)** fission blends plutonium oxide reprocessed from spent fuel or other sources with uranium oxide. **Spent fuel** is the used fuel elements that were irradiated in a nuclear reactor.

• **Discuss the pros and cons of electric power produced by nuclear energy versus coal.**

One reason proponents of nuclear energy argue for the widespread adoption of nuclear energy is it has less of an environmental impact than fossil fuels, particularly coal. The combustion of coal releases carbon dioxide, a greenhouse gas that traps solar heat in our atmosphere and may cause global warming. In comparison, nuclear energy emits few pollutants into the atmosphere. In particular, nuclear energy provides power without producing carbon dioxide. However, it generates highly radioactive waste such as spent fuel, and permanent waste disposal sites are urgently needed. There are also safety concerns about nuclear power plants.

• **Describe the relationship between nuclear power and greenhouse gas reduction.**

Nuclear power can serve as an alternative to electricity generation from coal and natural gas, both of which contribute greenhouse gases to the atmosphere. However, the capacity of nuclear power to make a difference over the next several decades is limited, since many reactors would be needed, and nuclear energy cannot currently substitute for oil—a major greenhouse gas contributor.

• **Explain why nuclear power does not have much effect on U.S. oil needs.**

We use oil to generate only about 3% of the electricity in the United States; we rely on oil primarily for automobiles and for heating buildings. Thus, replacing electricity generated by oil with electricity generated by nuclear energy would do little to lessen our dependence on foreign oil because we would still need oil for heating buildings and driving automotive vehicles. A transition to primarily electric powered or hydrogen fueled vehicles might make nuclear energy more plausible.

• **Describe the nuclear power plant accidents at Three Mile Island and Chornobyl.**

The most serious nuclear reactor accident in the United States occurred in 1979 at the Three Mile Island power plant in Pennsylvania. A partial meltdown of the reactor core took place, although the containment building kept almost all the radioactivity from escaping. In 1986 the world's worst nuclear reactor accident occurred in Chornobyl in the former Soviet Union (now Ukraine). One or two explosions ripped apart a nuclear reactor and expelled large quantities of radioactive material into the atmosphere, resulting in widespread environmental pollution as well as serious local contamination.

• **Discuss the link between nuclear energy and nuclear weapons.**

The increase in global supplies of weapons-grade plutonium and plutonium wastes from commercial nuclear reactors threatens international security because it increases the chance certain nations and terrorist groups could use them to make nuclear weapons.

• **Distinguish between low-level and high-level radioactive wastes.**

Low-level radioactive wastes are radioactive solids, liquids, or gases that give off small amounts of ionizing radiation. **High-level radioactive wastes** are radioactive solids, liquids, or gases that initially give off large amounts of ionizing radiation.

• **Relate the pros and cons of permanent storage of high-level radioactive wastes at Yucca Mountain.**

The United States has selected **Yucca Mountain** in Nevada as a permanent storage site for high-level radioactive wastes from commercially operated nuclear power plants. Since 1983 the U.S. Department of Energy has conducted feasibility studies on Yucca Mountain's geology, and results suggest the site is safe from volcanic eruptions and earthquakes. Transporting high-level wastes from nuclear reactors and weapons sites by truck, rail, or air is a major concern. However, leaving nuclear waste at multiple sites spread across the United States poses a greater risk of theft and, possibly, human health problems.

• **Explain what happens to nuclear power plants after they are closed.**

Nuclear power plants at the end of their operational usefulness are contaminated with radioactivity and must be put into **storage**, **entombed**, or **decommissioned**. Decommissioning is the dismantling of an old nuclear power plant after it closes. Sections of the dismantled plant are transported to a permanent storage site.

• **Describe some of the technological hurdles that must be overcome before nuclear fusion becomes a reliable energy source.**

Commercial fusion as a source of energy is many years from becoming a reality. It takes extremely high temperatures to make atoms fuse, and the best fusion experiments generate only about one-third of the energy used to heat the fuel. Another challenge is confining the fuel, because at extremely high temperatures, a gas separates into negative electrons and positive nuclei. This superheated, ionized gas, called **plasma**, has a tendency to expand. Confinement of the plasma is necessary so that the nuclei are close enough to one another to fuse, but a regular container does not work because as soon as the nuclei hit the container walls, they lose so much energy they cannot fuse.

• **Briefly summarize the issues that must be addressed if nuclear power is to become a major energy source in the future.**

Questions involving safety and management of radioactive waste must be addressed before the public will support a new expansion of nuclear power. It also must prove cost-competitive with other sources, including fossil fuels, hydroelectric, and solar.

Thinking About the Environment

1. What is nuclear fission? How does fission differ from fusion?

2. What are the main steps in the nuclear fuel cycle? Is it a true cycle? Explain your answer.

3. What are the safety features of a pressurized water reactor?

4. What might resolve the conflict between the public's view of the nuclear power industry and the industry's view of the public?

5. How is breeder nuclear fission different from conventional nuclear fission?

6. Breeder reactors produce more fuel than they consume. Does this mean that if we use breeder reactors, we will have a perpetual supply of plutonium for breeder fission? Why or why not?

7. What is spent fuel?

8. Can we prevent catastrophic accidents at nuclear power plants in the future? Why or why not?

9. How does the disposal of radioactive wastes pose technical problems? political problems?

10. What happens to the radioactive waste when a nuclear power plant is decommissioned?

11. What is plasma? How is it related to the development of fusion as a practical source of energy?

12. What are the main arguments for and against the United States developing additional nuclear power plants to provide us with electricity over the next several decades? Which perspective do you find most convincing?

13. Why is it important to think about energy from a systems perspective when deciding the role of nuclear power in reducing climate change?

Quantitative questions relating to this chapter are on our Web site.

Take a Stand

Visit our Web site at http://www.wiley.com/college/raven (select Chapter 12 from the Table of Contents) for links to more information about the controversy involving the transport of high-level radioactive waste to Yucca Mountain. Consider the opposing viewpoints of those who think it is safer to use Yucca Mountain for the permanent storage of wastes from around the country and those who think the transport of these wastes makes a single site at Yucca Mountain too dangerous. You will find tools to help you organize your research, analyze the data, think critically about the issues, and construct a well-considered argument. *Take a Stand* activities can be done individually or as a team, as oral presentations, written exercises, or Web-based (e-mail) assignments.

Additional online materials relating to this chapter, including a Student Testing Section with study aids and self-tests, Environments News, Activity Links, Environmental Investigations, and more, are also on our Web site.

Renewable Energy and Conservation

Array of solar panels in rural Kenya.
Silicon photovoltaic panels can be made of crystalline or amorphous silicon. The former is more efficient—that is, it can convert a larger fraction of solar energy to electricity. Amorphous silicon, while less efficient, costs far less to produce. The panels shown here convert solar energy into electricity, which can then be stored in a battery and used to run a small appliance. (© Liba Taylor/Corbis)

For decades, the standard energy policy for governments, industries, and individuals has been to ask "where can we get more cheap energy?" The world's major sources of energy—fossil fuels, and to a lesser extent wood, nuclear, and hydropower—are relatively inexpensive in the short term. Unfortunately, they can have substantial hidden costs, many of which are transferred to future generations. In addition, traditional energy sources can be habit-forming. We use them even when other energy options, including conservation, would cost less and improve our standard of living.

From a systems perspective, it makes sense to ask "what services do we want from energy, and what combination of characteristics—good and bad—should we be concerned about?" This systems perspective has led researchers to think about appropriate energy sources and uses at different locations.

In the University of California at Berkeley's Renewable and Appropriate Energy Lab (RAEL),

Professor **Daniel M. Kammen** and his students research this topic, and develop and promote energy solutions. They have found, for example, a thriving market for small-scale solar panels in Kenya (see photograph). Several vendors sell systems that include an inexpensive solar panel and a rechargeable battery. The battery charges during the day, and can be used to run small appliances—radio, television, lights. Such systems do not require purchasing gas or diesel to run a generator. They also do not require extending the electrical power lines to remote locations. From a systems perspective, small-scale solar meets the needs of rural Kenyans.

Another RAEL project focuses on using waste biomass from sawmills in Zimbabwe to produce electrical power. *Biomass* is the term used for plants—including wood, corn, grasses, and even plant fibers in animal dung—that are burned for their energy. Biomass can be a controversial energy source. Cutting down trees to burn as wood or charcoal is not necessarily renewable, if it is used more quickly than it grows back. Biomass can be as intensive in terms of energy, fertilizer, and pesticide inputs as any other agricultural product—and can displace important food crops. From a systems perspective, biomass can have many drawbacks.

In the Zimbabwe case—and many other countries—sawdust and other wood processing wastes are often burned, wasting the energy and creating air pollution that includes carbon dioxide. Some of this biomass waste could instead be burned to heat water for steam turbines, generating electricity and

most likely replacing coal. Material that is usually considered waste becomes a resource from a systems perspective.

Like fossil fuels, which trapped solar energy millions of years ago, many renewable sources originate as solar energy. Winds are driven largely by differential heating of Earth's surface. Hydropower relies on the hydrologic cycle, in which evaporation at low altitudes leads to precipitation at higher altitudes. Biomass comes from growing plants. Geothermal power, on the other hand, is more closely related to nuclear energy: Heating in Earth's interior comes close enough to the surface for us to use it to generate steam.

Conservation and efficiency can also be thought of as energy sources, since every unit of energy saved in one place is available for use in another. Conservation has the unfortunate reputation as "living with less," but this is not always, nor even usually, the case. Driving a smaller car, bicycling, or living closer to work or school can save money and time—two things most students would like more of. Technologies that use less energy also pollute less, can use fewer resources (compare the steel in a large SUV to that in a compact hybrid), and can even cost less to produce.

This chapter explores a variety of energy alternatives, including renewable resources, efficient technologies, and conservation practices, as well as their advantages and disadvantages relative to fossil fuels and nuclear power.

World View and Closer to You...
Real news footage relating to renewable energy and conservation around the world and in your own geographic region is available at www.wiley.com/college/raven by clicking on "World View and Closer to You."

Direct Solar Energy

- Distinguish between active and passive solar heating and describe how each is used.
- Contrast the advantages and disadvantages of solar thermal electric generation and photovoltaic solar cells in converting solar energy into electricity.

The sun produces a tremendous amount of energy, and most of it dissipates into space. Only a small portion is radiated to Earth. Solar energy is different from fossil and nuclear fuels because it is perpetually available; we will run out of solar energy only when the sun's nuclear fire burns out. Solar energy is dispersed over Earth's entire surface rather than concentrated in highly localized areas, as are coal, oil, and uranium deposits. To make solar energy useful, we must collect it.

Solar radiation varies in intensity depending on the latitude, season of the year, time of day, and cloud cover. Areas at lower latitudes—closer to the equator—receive more solar radiation annually than do latitudes closer to the North and South Poles. More solar radiation is received during summer than during winter because the sun is directly overhead in the summer and lower on the horizon in winter. Solar radiation is more intense when the sun is high in the sky (noon) than when it is low in the sky (dawn or dusk). Clouds both scatter incident light and absorb some of the sun's energy, thereby reducing its intensity. The southwestern United States, with its lack of cloud cover and lower latitude, receives the greatest amount of solar radiation annually, whereas the Northeast receives the least (**Figure 13.1**).

Although the technology exists to use solar energy directly, it is not widely used, largely because the initial costs associated with converting to solar power are high. However, the long-term energy savings of solar power may offset the high start-up costs. Trapping the sun's energy using current technology is inefficient, meaning relatively little of the sun's energy that hits the solar panels (collecting devices) is actually used. With new technological developments, the efficiency of solar energy collection is increasing, making it an increasingly cost-effective source of energy. The use of solar energy is projected to increase in the future, in both highly developed and developing countries.

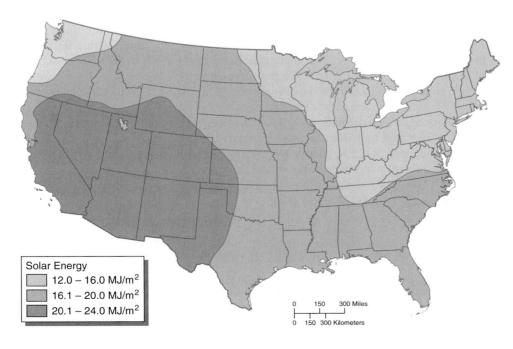

Solar Energy
- 12.0 – 16.0 MJ/m^2
- 16.1 – 20.0 MJ/m^2
- 20.1 – 24.0 MJ/m^2

0   150   300 Miles

0   150 300 Kilometers

Figure 13.1   Solar energy distribution over the United States. This map shows the average daily total of solar energy (on an annual basis) received on a solar collector that tilts to compensate for latitude. The units are in megajoules per square meter. The Southwest is the best area in the United States for year-round solar energy collection.

An estimated 0.5 million people in the rural areas of Africa, Central America, India, and China are currently using solar cookers. Recent designs for solar cookers transmit solar light into the cooker, and the glass cover does not transmit the infrared wavelengths (heat) that would normally escape out of the cooker. Pots containing food are placed inside the box on a black metal plate. The solar cooker can reach a temperature of 177°C (350°F) and can boil, bake, simmer, and sauté foods. In average sunlight, a person can cook a full meal in two to four hours.

Heating Buildings and Water

You have probably noticed the air inside a car sitting in the sun with its windows rolled up becomes much hotter than the surrounding air. Similarly, the air inside a greenhouse remains warmer than the outside air during cold months. (Greenhouses usually require additional heating in cold climates but far less than you might expect.) This kind of warming occurs partly because the material—such as glass—that envelops the air inside the enclosure is transparent to visible light but impenetrable to heat. Visible light from the sun penetrates the glass and warms the surfaces of objects inside, which in turn give off **infrared radiation**—invisible waves of heat. Heat does not escape because infrared radiation cannot penetrate glass, and the air within the glass grows continuously warmer.

passive solar heating: A system of putting the sun's energy to use without requiring mechanical devices to distribute the collected heat.

In **passive solar heating**, solar energy heats buildings without the need for pumps or fans to distribute the heat. Certain design features are incorporated into a passive solar heating system to warm buildings in winter and help them remain cool in summer (**Figure 13.2**). In the Northern Hemisphere large south-facing windows receive more total sunlight during the day than windows facing other directions. The sunlight entering through the windows provides heat that is then stored in floors and walls made of concrete or stone, or in containers of water. This stored heat is transmitted throughout the building naturally by *convection*, the circulation that occurs because warm air rises and cooler air sinks. Buildings with passive solar heating systems must be well insulated so that accumulated heat does not escape. Depending on the building's design and location, passive heating saves as much as 50% of heating costs.[1] Currently, about 7% of new homes built in the United States have passive solar features. The main reasons there are not more such homes is that they cost a bit more initially than do traditional designs. That savings from heating and cooling bills can quickly recover the cost is not always considered.

active solar heating: A system of putting the sun's energy to use in which a series of collectors absorbs the solar energy, and pumps or fans distribute the collected heat.

In **active solar heating**, a series of collection devices mounted on a roof or in a field is used to gather solar energy. The most common collection device is a panel or plate of black metal, which absorbs the sun's energy (**Figure 13.3**). Active solar heating is used primarily for heating water, either for household use or for swimming pools. The heat absorbed by the solar collector is transferred to a liquid inside the panel, which is then pumped to the heat exchanger, where the heat is transferred to water that will be stored in the hot water tank. Because approximately 8% of the energy consumed in the United States goes toward heating water, active solar heating has the potential to supply a significant amount of the nation's energy demand.

The use of active solar energy for space heating, which currently is not as common as for heating water, may become more important when diminishing supplies of fossil fuels force their prices higher. Active solar heating of buildings, now costlier than more conventional forms of space heating, would then become more competitive.

Solar Thermal Electric Generation

solar thermal electric generation: A means of producing electricity in which the sun's energy is concentrated by mirrors or lenses to either heat a fluid-filled pipe or drive a Stirling engine.

Systems that concentrate solar energy to heat fluids have long been used for buildings and industrial processes. In **solar thermal electric generation** electricity is produced by several different systems. One approach is to collect incident sunlight and concentrate

[1] Unless otherwise noted, all energy facts cited in this chapter were obtained from the Energy Information Administration (EIA), the statistical agency of the U.S. Department of Energy (DOE).

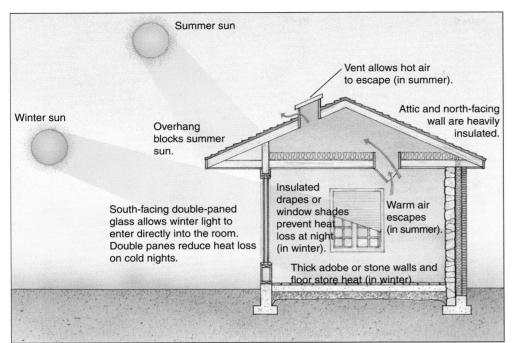

Summer sun

Winter sun

Vent allows hot air
to escape (in summer).

Attic and north-facing
wall are heavily
insulated.

Overhang
blocks summer
sun.

Insulated
drapes or
window shades
prevent heat
loss at night
(in winter).

Warm air
escapes
(in summer).

South-facing double-paned
glass allows winter light to
enter directly into the room.
Double panes reduce heat loss
on cold nights.

Thick adobe or stone walls and
floor store heat (in winter).

(a) Several passive designs are incorporated into this home.

it, using mirrors or lenses, to heat a working fluid to high temperatures. In one such system, trough-shaped mirrors, guided by computers, track the sun for optimum efficiency, center sunlight on oil-filled pipes, and heat the oil to 390°C (735°F) (**Figure 13.4**). The hot oil is circulated to a water storage system and used to change water into superheated steam, which turns a turbine to generate electricity. Alternatively, the heat is used to power a *Stirling engine*. A fluid in a cylinder expands, driving a piston that turns a shaft, providing mechanical energy or producing electricity.

Most electricity in the United States is fed into a grid, a network of cables that carry electricity where it is needed. Since demand is highest during the day, and particularly when air conditioning demand is high, the fact that solar systems only work during the day is not a serious disadvantage. The world's largest solar thermal system is currently operating in the Mojave Desert in Southern California.

The solar power tower is a solar thermal system with a tall tower surrounded by hundreds of mirrors. The computer-controlled mirrors move to follow the sun, focusing solar radiation on a central receiver at the top of the tower. There the concentrated sunlight heats a circulating liquid (molten salt), and the heat is used to produce steam for generating electricity. Molten salt retains heat, and some of the heat is stored for electricity generation during the night, when solar energy is unavailable. Solar power towers have been tested in the United States, several European countries, and Japan; the United States is no longer working on this technology.

Solar thermal energy systems are inherently more efficient than other solar technologies because they concentrate the sun's energy. With improved engineering, manufacturing, and construction methods, solar thermal energy may become cost-competitive with fossil fuels. A newly designed 500 MW Stirling engine system

Figure 13.2 **Passive solar heating designs.**

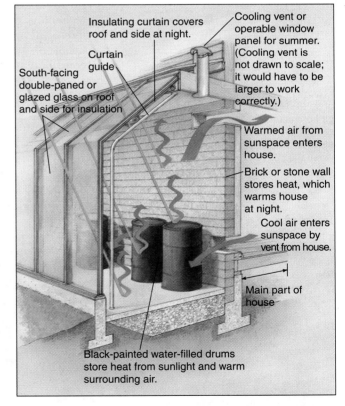

Insulating curtain covers
roof and side at night.

Cooling vent or
operable window
panel for summer.
(Cooling vent is
not drawn to scale;
it would have to be
larger to work
correctly.)

Curtain
guide

South-facing
double-paned or
glazed glass on roof
and side for insulation

Warmed air from
sunspace enters
house.

Brick or stone wall
stores heat, which
warms house
at night.

Cool air enters
sunspace by
vent from house.

Main part of
house

Black-painted water-filled drums
store heat from sunlight and warm
surrounding air.

(b) A solar sunspace can be added to existing homes.

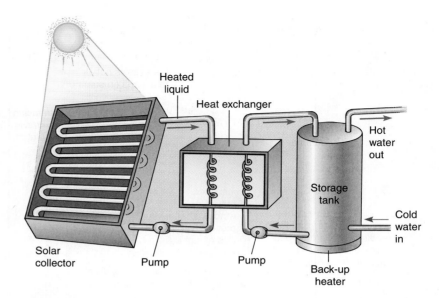

Figure 13.3 Active solar water heating.
Solar collectors are mounted on the roof of a building. Each solar panel is a box with a black metal base and glass covering. Sunlight enters the glass and warms the pipes and the liquid flowing through them. The hot liquid (red arrows) heats water, which is further heated to required temperatures by a backup heater that uses electricity or natural gas. Solar domestic water heating can provide a family's hot water needs year-round.

scheduled to come on line in the southwestern United States in 2009 is expected to produce electricity at \$.05 to \$.13 per kilowatt-hour (kWh); the lower end of this range is competitive with coal-fired plants. This cost should drop over the next decade as the technology matures. (**Table 13.1** compares the generating costs of electricity using different energy sources, including solar thermal.) In addition, the environmental benefits of solar thermal plants are significant because they do not produce air pollution or contribute to acid rain or global climate change.

Photovoltaic Solar Cells

photovoltaic solar cell: A wafer or thin film of solid state materials, such as silicon or gallium arsenide, that is treated with certain metals so that they generate electricity—that is, a flow of electrons—when they absorb solar energy.

Photovoltaic (PV) solar cells currently provide more than 6000 MW of electricity worldwide. This is about the same as six nuclear power plants, but accounts for only about 0.18% of global electricity. PV cells convert sunlight directly into electricity

Figure 13.4 Solar thermal electric generation.

(a) A solar thermal plant in California uses troughs to focus sunlight on a fluid-filled tube. (L. Lefkowitz/Taxi/Getty Images)

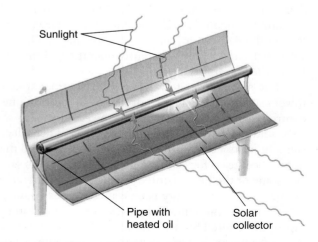

(b) The heated oil is pumped to a water tank where it generates steam used to produce electricity. For simplicity, arrows show sunlight converging on several points; sunlight actually converges on the pipe throughout its length.

E N V I R O N E W S

Campus Sustainability: Energy Alternatives at the University of Nevada, Las Vegas

At the University of Nevada, Las Vegas, where cactus survives only if it is watered, sustainability is a major challenge. But with extreme summer heat and year-round sunlight, it is ideally situated to host solar and alternative energy research. Several projects are currently underway, including:

- Several experimental Stirling engine-based solar electric generators and banks of experimental photovoltaic panels

- Green design for new buildings that, among other things, minimizes southern exposure, incorporates shade structures, puts heat-producing computer labs below ground level, and incorporates rooftop photovoltaics

- An experimental "Zero Energy House," built with substantial insulation, highly efficient air conditioning, rooftop photovoltaics, and other technologies. In its first year, this building returned 1/3 more energy to the electrical grid than it drew

Advances in sustainability in extreme climates remind us how much can be accomplished in more moderate settings.

Table 13.1 2006 Generating Costs of Electric Power Plants

Energy Source	Generating Costs (cents per kilowatt hour)*
Hydropower	4–10¢
Biomass	6–9
Geothermal	3–8
Wind	4–7
Solar thermal	5–13
Photovoltaics	20–25
Natural gas	5–7
Coal	4–6
Nuclear power	2–12

*Electricity production and consumption are measured in kilowatt-hours (kWh). As an example, one 50-watt light bulb that is on for 20 hours uses one kilowatt-hour of electricity ($50 \times 20 = 1000$ watt-hours = 1 kWh).
Source: Daniel M. Kammen, The Rise of Renewable Energy. *Scientific American* (September 2006)

(**Figure 13.5**), and are usually arranged on large panels that absorb sunlight, even on cloudy or rainy days.

PVs generate electricity with no pollution and minimal maintenance. They are used on any scale, from small, portable modules to large, multimegawatt power plants. Our current PV solar cell technology, though used to power satellites, uncrewed airplanes, highway signals, watches, and calculators, has a few limitations that prevent the cells' widespread use to generate electricity. Photovoltaic solar cells are only about 15 to 18% efficient at converting solar energy to electricity (although experimental cells reach 37%), and the number of solar panels for large-scale use requires a great deal of land. At current efficiencies several thousand acres of solar panels would be required to absorb enough solar energy to produce the electricity generated by a single large conventional power plant.

One benefit of PV devices for utility companies is that they can purchase them as small modular units that become operational in a short period. A utility company can purchase PV elements to increase its generating capacity in small increments, rather than committing a billion dollars or more and a decade or more of construction for a massive conventional power plant. Used in this supplementary way, the PV units provide the additional energy, for example, to power irrigation pumps on hot, sunny days.

In remote areas that are not served by electric power plants, such as rural areas of developing countries, it is more economical to use PV solar cells for electricity than to extend power lines. Photovoltaics are the energy choice to pump water, refrigerate vaccines, grind grain, charge batteries, and supply rural homes with lighting. According to the Institute for Sustainable Power, more than one million households in the developing countries of Asia, Latin America, and Africa have installed PV solar cells on the roofs of their homes. A PV panel the size of two pizza boxes supplies a rural household with enough electricity for five lights, a radio, and a television.

The cost of manufacturing PV modules has steadily declined over the past 25 years, from an average factory price of almost $90 per watt in 1975 to about $4.80 per watt in 2007. The cost of producing electricity from PVs has steadily declined from 1970 to the present. Despite this progress, in 2006 the cost was still about $.20 to $.25 per kilowatt-hour.

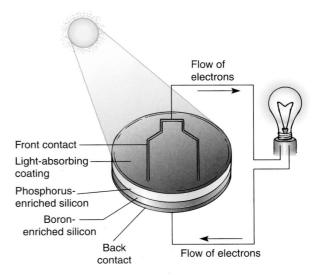

Figure 13.5 **Photovoltaic cells.** Photovoltaic cells contain silicon and other materials. Sunlight excites electrons, which are ejected from silicon atoms. Useful electricity is generated when the ejected electrons flow out of the PV cell through a wire.

Figure 13.6   Solar shingles. These thin-film solar cells look much like conventional roofing materials. (Courtesy of Uni-Solar. Energy Conservation Devices, Discover Magazine; inset, United Solar Systems, Discover Magazine.)

Future technological progress may make PVs economically competitive with electricity produced by conventional energy sources. The production of "thin-film" solar cells, which are much cheaper to manufacture, has decreased costs. Thin-films can be produced as flexible sheets that are incorporated into building materials, such as roofing shingles, tiles, and window glass (**Figure 13.6**). More than 120,000 Japanese homes have installed PV solar-energy roofing in the past few years, and California has committed to a "Million Solar Roofs" by 2018. Other promising technological advances are dye-sensitized solar cells and nano-scale PV technologies.

REVIEW

1. What is active solar energy? passive solar energy?
2. What are the advantages of producing electricity by solar thermal energy? by photovoltaic (PV) solar cells?

Indirect Solar Energy

LEARNING OBJECTIVES

- Define *biomass*, explain why it is an example of indirect solar energy, and outline how it is used as a source of energy.
- Describe the locations that can make optimum use of wind energy and of hydropower, and compare the potential of wind energy and hydropower.

biomass: Plant material, including undigested fiber in animal waste, used as fuel.

wind energy: Electric or mechanical energy obtained from surface air currents caused by solar warming of air.

hydropower: A form of renewable energy that relies on flowing or falling water to generate mechanical energy or electricity.

Combustion of **biomass**—wood and other organic matter—is an example of indirect solar energy because green plants, which use solar energy for photosynthesis, store the energy in biomass. Windmills, or wind turbines, extract **wind energy** to generate mechanical energy or electricity. The damming of rivers and streams to generate electricity is a type of **hydropower**—the energy of flowing water. Hydropower exists because solar energy drives the hydrologic cycle (see Chapter 5).

Biomass Energy

Biomass, one of the oldest fuels known to humans, consists of such materials as wood, fast-growing plant and algal crops, crop wastes, sawdust and wood chips, and animal wastes (**Figure 13.7**). Biomass contains chemical energy that comes from the sun's radiant energy, which photosynthetic organisms use to form organic molecules. Biomass is a renewable form of energy when used no faster than it is produced; deforestation and desertification can result when biomass is overused. Biomass cannot replace fossil fuels. The entire photosynthesis production of the continental United States amounts to only half of our current energy use—and that would mean no other uses, including food, paper, and construction materials.

Figure 13.7   Biomass. Firewood is the major energy source for most of the developing world. (Robert E. Ford/Terraphotographics/BPS)

Biomass fuel, which can be a solid, liquid, or gas, is burned to release its energy. Solid biomass such as wood is burned directly to obtain energy. Biomass—particularly firewood, charcoal (wood that is heated in an oxygen-free environment to concentrate its energy and drive off water), animal dung (primarily undigested plant fiber), and peat (partly decayed plant matter found in bogs and swamps)—supplies a substantial portion of the world's energy. At least half of the human population relies on biomass as their main source of energy. In developing countries, wood is the primary fuel for cooking. In the United States, biomass accounts for about 3% of total U.S. energy production. Biomass in the form of low-cost residues from sawmills, paper mills, and agricultural industries is burned in power plants to generate about 7.6 gigawatts (GWe) of electricity.

It is possible to convert biomass, particularly animal wastes, into **biogas.** Biogas, which is usually composed of a mixture of gases (mostly methane), is stored and transported like natural gas. It is a clean fuel—its combustion produces fewer pollutants than either coal or biomass. In India and China, several million family-sized **biogas digesters** use microbial decomposition of household and agricultural wastes to produce biogas for cooking and lighting. When biogas conversion is complete, the solid remains are removed from the digester and used as fertilizer. Although the technology for biogas digesters is relatively simple, the conditions inside the digester, such as the moisture level and pH, must be carefully monitored if the bacteria are to produce biogas at an optimum level.

Biomass can be converted to liquid fuels, especially **methanol** (methyl alcohol) and **ethanol** (ethyl alcohol), which can be used in internal combustion engines. In many parts of the world, automotive fuels must contain 10% or more ethanol, and the Indy Car racing series now runs entirely on ethanol. *Biodiesel*, made from plant or animal oils, is becoming more popular as an alternative fuel for diesel engines in trucks, farm equipment, and boats. The oil is often refined from waste oil produced at restaurants (for example, the oil used to make french fries); biodiesel burns much cleaner than diesel fuel. Country singer Willy Nelson has been a major biodiesel advocate.

Although some U.S. energy companies convert sugar cane, corn, or wood crops to alcohol, others are interested in the commercial conversion of agricultural and municipal wastes into ethanol. Costs are high, and few companies have successfully invested in waste-to-ethanol. Several companies are currently building plants that convert biomass (using cornstalks, rice straw, the fibrous residues from process sugar cane, and sewage sludge) to ethanol. Currently, the profitability of ethanol is possible only because of government **subsidies** that reduce ethanol's cost. Companies planning waste-to-ethanol processes acknowledge they need government subsidies initially but think they eventually will compete with gasoline without government help.

A major challenge for liquid biomass fuels is that it can take large amounts of energy to produce them. In addition, they are not always "climate neutral," since more CO_2 may be produced than captured. However, recent research suggests that using switch grass for ethanol in the Great Plains may actually store more carbon in the ground (as roots and rhizomes) each year than is released from the ethanol.

Advantages of Biomass Use Biomass is an attractive source of energy because it reduces dependence on fossil fuels and often uses wastes, thereby reducing our waste disposal problem. Biomass is usually burned to produce energy, so the pollution problems of fossil fuel combustion, particularly carbon dioxide emissions, are not completely absent in biomass combustion. However, the low levels of sulfur and ash produced by biomass combustion compare favorably with the levels produced when bituminous coal is burned. It is possible to offset the CO_2 released into the atmosphere from biomass combustion by increasing tree planting. As trees photosynthesize, they absorb atmospheric CO_2 and lock it up in organic molecules that make up the body of the tree, thereby providing a carbon "sink." Thus, if biomass is regenerated to replace the biomass used, there is no net contribution of CO_2 to the atmosphere and to global climate change.

Disadvantages of Biomass Use Use of biomass, especially from plants, poses several problems. For one thing, biomass production requires land, water, and energy. Because use of agricultural land for energy crops competes with the growing of food crops, shifting the balance toward energy production might decrease food production, contributing to higher food prices. In addition, the energy used to produce biomass often requires some use of fossil fuels.

At least half of the world's population relies on biomass as its main source of energy. Unfortunately, in many areas people burn wood faster than they replant trees. Intensive use of wood for energy has resulted in severe damage to the environment, including soil erosion, deforestation and desertification, air pollution (especially when burned indoors), and degradation of water supplies.

Crop residues, a category of biomass that includes cornstalks, wheat stalks, and wood wastes at paper mills and sawmills, are increasingly being used for energy. At first glance, it may seem that crop residues, which normally remain in the soil after harvest, would be a good source of energy if they were collected and burned. After all, they are waste materials that will eventually decompose. However, the systems perspective reminds us that crop residues left in and on the ground prevent erosion by holding the soil in place, and their decomposition enriches the soil by making the minerals originally in plant residues available for new plant growth. If all crop residues were removed from the ground, the soil would eventually be depleted of minerals, and its future productivity would decline. Forest residues, which remain in the soil after trees are harvested, fill similar ecological roles.

Wind Energy

During the past ten years, wind energy capacity has increased at 25% each year—it is the world's fastest growing source of energy. Wind results from the sun warming the atmosphere. Wind is an indirect form of solar energy in which the radiant energy of the sun is transformed into mechanical energy—the movement of air molecules. Wind is sporadic over much of Earth's surface, varying in direction and magnitude. Like direct solar energy, wind power is a highly dispersed form of energy. Harnessing wind energy to generate electricity has great potential, and wind is increasingly important in supplying our energy needs.

New wind turbines can be huge—100 meters tall—and have long blades designed to harness wind energy efficiently (**Figure 13.8**). As turbines have become larger and more efficient, costs for wind power have declined rapidly—from $.40 per kilowatt-hour in 1980 to $.04 to $.07 per kilowatt-hour in 2006. Wind power is cost-competitive with many forms of conventional energy. Advances, such as turbines that use variable-speed operation, may make wind energy an important global source of electricity during the first half of the 21st century. Denmark, one of the world leaders in wind power, currently generates 21% of its electricity using wind energy; much of this power is generated offshore (because ocean winds are strong).

Harnessing wind energy is most profitable in rural areas that receive fairly continual winds, such as islands, coastal areas, mountain passes, and grasslands. Germany currently leads the world as the top producer of wind energy, with 18 GWe of installed

Figure 13.8   Harvesting wind energy.

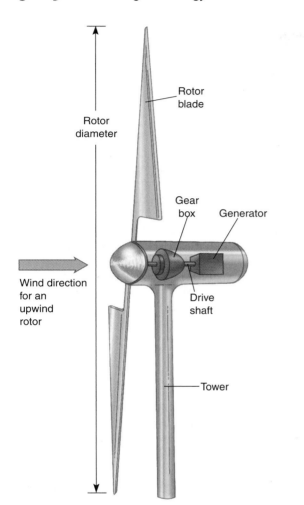

(b) Each of these wind turbines atop Buffalo Ridge in Minnesota generates enough electricity for 250 homes. Buffalo Ridge, which extends from South Dakota across Minnesota into Iowa, is an ideal site for wind farms. (Clean Water Action Alliance of Minnesota, Mark Frederickson/National Renewable Energy Laboratory)

(a) This basic wind turbine design has a horizontal axis (horizontal refers to the orientation of the drive shaft). Airflow causes the turbine's blades to turn 15 to 60 revolutions per minute (rpm). As the blades turn, gears within the turbine spin the drive shaft. This spinning powers the generator, which sends electricity through underground cables to a nearby utility. Wind turbine technology is advancing rapidly, and many changes in design are anticipated. (The tower is not drawn to scale and is much taller than depicted.)

capacity—almost half of the total European production of about 40 GWe. In contrast, the United States—with several times the area and population of Germany, has about half of Germany's wind energy production.

The world's largest concentration of wind turbines is currently located in the Tehachapi Pass at the southern end of the Sierra Nevada mountain range in California. In the continental United States, some of the best locations for large-scale electricity generation from wind energy are on the Great Plains. The 10 states with the greatest wind energy potential, according to the American Wind Energy Association, are North Dakota, Texas, Kansas, South Dakota, Montana, Nebraska, Wyoming, Oklahoma, Minnesota, and Iowa. In fact, if we developed the wind energy in North Dakota, Texas, and Kansas to their full potential, we could supply more than enough electricity to meet the current needs of the entire United States! U.S. wind power projects are under way in these and many other states. Currently, wind energy is captured and placed into regional electricity grids, but deploying wind energy on a national scale (for example, wind energy produced in Texas and used in New York City) requires the development of new technologies for storing and distributing energy.

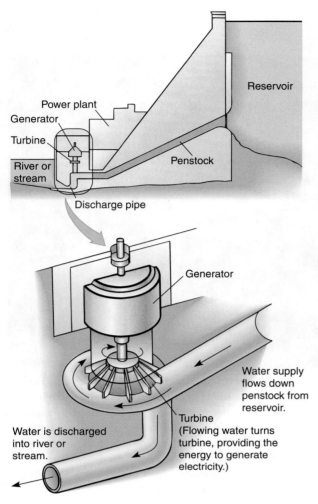

Figure 13.9   Hydroelectric power. A controlled flow of water released down the penstock turns a turbine, which generates electricity.

Table 13.2   Advantages and Disadvantages of Dams

Reasons to Build Dams	*Problems with Dams*
Electrical Power	Ecological disruption
Mechanical Power	downstream
Irrigation	• Sediment stopped in dam
Navigation	• Water source diverted
Flood Control	• Fish migration halted at dam
Commercial Fishing	Ecological disruption in reservoir
Recreation	• Habitat flooded
• Fishing	• Sediment buildup
• Swimming	• Pollution if toxic materials are
• Boating	submerged
	Displacement of people
	Loss of cultural resources
	Catastrophic failure
	Diseases
	Seismisity
	Evaporation

The use of wind power does not cause major environmental problems, although reported bird and bat kills represent one concern. The California Energy Commission estimated that several hundred birds, many of them raptors (birds of prey), turned up dead in the vicinity of the 7000 turbines at Altamont Pass in California during a two-year study; most had collided with the turbines. Studies later determined that Altamont Pass is a major bird migration pathway. Technical "fixes," such as painted blades and antiperching devices to discourage raptors from roosting on the towers, were implemented at Altamont Pass. Other sites have been required to shut down operations during peak migratory periods. Developers of future wind farm sites currently conduct voluntary wildlife studies and try to locate sites away from bird and bat routes.

Wind produces no waste and is a clean source of energy. It produces no emissions of sulfur dioxide, carbon dioxide, or nitrogen oxides. Every kilowatt-hour of electricity generated by wind power rather than fossil fuels prevents 1 to 2 lb of the greenhouse gas CO_2 from entering the atmosphere. The biggest constraints on wind are cost and public resistance. Wind research and production have not been subsidized to the same extent as nuclear energy and biomass; nonetheless, as costs of other energy sources increase, wind is becoming increasingly competitive. The NIMBY effect (Chapter 12) plays a mixed role: Some people consider wind farms to be attractive, while others think they are blights on the landscape.

Hydropower

Hydropower is the world's main renewable source of electrical generation, producing about the same amount of electricity as do the world's nuclear power plants. The sun's energy drives the hydrologic cycle, which encompasses precipitation, evaporation from land and water, transpiration from plants, and drainage and runoff (see Figure 5.6). As water flows from higher elevations back to sea level, we can harness its energy. Unlike the sun's energy, which is highly dispersed, hydropower is a more concentrated energy. The potential energy of water held back by a dam is converted to kinetic energy as the water falls down a penstock, where it turns turbines to generate electricity (**Figure 13.9**). Hydropower is more efficient than any other energy source in producing electricity—about 90% of available hydropower energy is converted into electricity. **Table 13.2** summarizes the reasons for and problems with building dams.

Hydropower generates approximately 19% of the world's electricity, making it the most widely used form of solar energy. The 10 countries with the greatest hydroelectric production are, in decreasing order, Canada, the United States, Brazil, China, Russia, Norway, Japan, India, Sweden, and France. In the United States, approximately 2200 hydropower plants produce between 8 and 12% of its electricity, making it the country's leading renewable energy source. Highly developed countries have already built dams at most of their potential sites, but this is not the case in many developing nations. Particularly in undeveloped, unexploited parts of Africa and South America, hydropower represents a great potential source of electricity.

Although most sites for traditional hydropower plants are already in use in the United States, new technological innovations show promise for expanding our hydropower capacity. About 97% of existing U.S. dams currently do not generate electricity, because traditional hydropower technology is suited only for large dams with rapidly flowing water and large flow capacities. The new technologies can produce electricity at smaller dams. Several companies now manufacture turbines that harness electricity from large, slow-moving rivers or from streams with small flow capacities. As these new technologies improve, they have the potential to increase the amount of electricity generated by hydropower without the construction of a single new dam.

Impacts of Dams   One problem associated with hydropower is that building a dam changes the natural flow of a river. A dam causes water to back up, flooding large areas of land and forming a reservoir, which destroys plant and animal habitats. Native fishes are particularly susceptible to dams because the original river ecosystem is so altered. The migration of spawning fish is also disrupted (see discussion of the Columbia River in Chapter 14). Below the dam, the once powerful river is reduced to a relative trickle. The natural beauty of the countryside is affected, and certain forms of wilderness recreation are made impossible or less enjoyable, although the dams permit water sports in the reservoir.

At least 200 large dams around the world are associated with *reservoir-induced seismicity*—earthquakes that occur during and after the filling of a large reservoir behind a dam. The larger the reservoir and the faster it is filled, the greater the intensity of seismic activity. An area does not have to be seismically active to have earthquakes induced by reservoirs.

In arid regions, the creation of a reservoir results in greater evaporation of water because the reservoir has a larger surface area in contact with the air than the stream or river did. As a result, serious water loss and increased salinity of the remaining water may occur.

If a dam breaks, people and property downstream may be endangered. In addition, waterborne diseases such as schistosomiasis may spread throughout the local population. **Schistosomiasis,** a tropical disease caused by a parasitic worm, damages the liver, urinary tract, nervous system, and lungs. As much as half the population of Egypt suffers from this disease, largely as a result of the Aswan Dam, built on the Nile River in 1902 to control flooding but used since 1960 to provide electric power. (The large reservoir behind the dam provides habitat for the worm, which spends part of its life cycle in the water. The worms infect humans during bathing, swimming, or walking barefoot along water banks, or by their drinking infected water.)

The environmental and social impacts of a dam may not be acceptable to the people living in a particular area. Laws prevent or restrict the building of dams in certain locations. In the United States, the **Wild and Scenic Rivers Act** prevents the hydroelectric development of certain rivers, although the number of rivers protected by this law is less than 1% of the nation's total river systems. Other countries, such as Norway and Sweden, have similar laws.

Dams cost a great deal to build but are relatively inexpensive to operate. A dam has a limited life span, usually 50 to 200 years, because over time the reservoir fills in with silt until it cannot hold enough water to generate electricity. This trapped silt, which is rich in nutrients, is prevented from enriching agricultural lands downstream. The gradual depletion of agricultural productivity downstream from the Aswan Dam in Egypt is well documented. Egypt now relies on heavy applications of chemical fertilizers to maintain the fertility of the Nile River Valley and its delta.

CASE IN POINT

The Three Gorges Dam

For thousands of years, China has wanted to dam the Yangze River. Historically, people living in the river basin have faced drought and flood years, often severe enough to kill many through famine or drowning. In the last century, the additional advantage of hydropower made damming the Yangze even more compelling. Consequently, in the 1990s China began work on the *Three Gorges Dam (TGD)*, so named because the 632-km (412-mi)-long reservoir will flood three upstream gorges. In 2003, the reservoir began to fill, although the dam will not be completed until 2009.

The TGD will meet several goals. First, it is designed to produce 18 GWe of electrical power—equivalent to 18 nuclear power plants or large coal power plants. Given the severe air quality problems facing China (see chapters 20 and 21), and its current reliance on imported energy, this is a great advantage. Second, agricultural productivity downstream will be optimized. (In 1998, a tenth of China's grain supply was destroyed by flooding that would have been controlled by the dam.) In addition, the new reservoir will be used for transportation—large ships will be able to go far upstream—as well as commercial fishing and recreation.

However, the TGD also represents the range of problems associated with building dams. At least 1.5 million people have been displaced, often with inadequate compensation (government officials have already been given death sentences for corruption in handling disbursement of funds intended for displaced people). The dam endangers several species, including the rare Yangze river dolphin, although this animal may already have been driven to extinction by overfishing and pollution. The reservoir may become highly polluted from industrial areas upstream and contaminated sites in the flooded area. Historical and cultural treasures, including temples, ancient hanging coffins, and massive canyon wall writings will be submerged. Also, even as agriculture is improved downstream, thousands of hectares of arable lands upstream will be flooded and thereby removed from production.

China initially had difficulty finding investors, since the uncertainty about design, construction, and effectiveness was so great. Water-borne diseases, including malaria and schistosomiasis, are likely to increase. As people have moved away from the reservoir site, entire cities have been razed and removed. It is likely that toxic materials and human and animals wastes remain, which could contaminate the newly created reservoir. The rate of sedimentation, or build up of silt behind the dam, is not well understood. Too much sedimentation will undermine all of the major reasons for the dam: irrigation, flood control, and hydropower production. ■

Other Indirect Solar Energy

A few other forms of indirect solar energy may become important in the future. Ocean waves are produced by winds, which are caused by the sun; wave energy is therefore considered an indirect form of solar energy. Like other types of flowing water, wave power has the potential to turn a turbine, thereby generating electricity. Norway, Great Britain, Japan, and several other countries are investigating the production of electricity from ocean waves. Only one commercial wave power station is currently operational, on the coast of Islay, a Scottish island.

In the future we may generate power using **ocean temperature gradients,** the differences in temperature at various ocean depths. As much as a 24°C difference exists between warm surface water and cold, deeper ocean water. Ocean temperature gradients, which are greatest in the tropics, are the result of solar energy warming the surface of the ocean. **Ocean Thermal Energy Conversion (OTEC)** would take advantage of this temperature difference to produce electricity or to cool buildings. The first commercial OTEC plant is under construction at the Natural Energy Laboratory of Hawaii Authority on the island of Hawaii. As technology improves and costs of other sources go up, ocean waves and OETC may become more viable.

REVIEW

1. What is biomass?

2. What are the advantages and disadvantages of using wind to produce electricity? of using hydropower to produce electricity?

Other Renewable Energy Sources

LEARNING OBJECTIVE

• Describe tidal energy and geothermal energy, the two forms of renewable energy that are not direct or indirect results of solar energy.

Geothermal and tidal energy are renewable energy sources that are not direct or indirect results of solar energy. **Geothermal energy** is the naturally occurring heat within Earth. **Tidal energy**, caused by changes in water level between high and low tides, is exploited to generate electricity on a limited scale.

Geothermal Energy

Geothermal energy, the natural heat within Earth, arises from the ancient heat within Earth's core, from friction where continental plates slide over one another, and from the decay of radioactive elements. The amount of geothermal energy is enormous. Scientists estimate that just 1% of the heat contained in the uppermost 10 km of Earth's crust is equivalent to 500 times the energy contained in all of Earth's oil and natural gas resources.

Geothermal energy is typically associated with volcanism. Large underground reservoirs of heat exist in areas of geologically recent volcanism. As groundwater in these areas travels downward and is heated, it becomes buoyant and rises until it is trapped by an impermeable layer in Earth's crust, forming a **hydrothermal reservoir.** Hydrothermal reservoirs contain hot water and possibly steam, depending on the temperature and pressure of the fluid. Some of the hot water or steam may escape to the surface, creating hot springs or geysers. Hot springs have been used for thousands of years for bathing, cooking, and heating buildings.

Hydrothermal reservoirs are tapped by drilling wells similar to those used for extracting oil and natural gas. The hot fluid is brought to the surface and used to supply heat directly or to generate electricity. Hydrothermal reservoirs can also generate electricity (**Figure 13.10**). The fluid is brought up a well to the surface, and the resulting steam is expanded through a turbine to spin a generator, creating electricity. The electricity these power stations generate is inexpensive and reliable.

The United States is the world's largest producer of geothermal electricity. Electric power is currently produced at 17 geothermal fields in California, Nevada, Utah, and Hawaii. The world's largest geothermal power plant is The Geysers, a geothermal field in northern California that provides electricity for 1.7 million homes. Other important producers of geothermal energy include the Philippines, Italy, Japan, Mexico, Indonesia, and Iceland. Total geothermal electric capacity is currently about 8 GWe.

Iceland, a country with minimal energy resources other than geothermal energy and hydropower, is geographically situated to optimize the use of geothermal energy. Located on the mid-Atlantic ridge, a boundary between two continental plates, Iceland is an island of intense volcanic activity and consequently has considerable geothermal resources. Iceland uses geothermal energy to generate electricity and to heat two-thirds of its homes. In addition, most of the fruits and vegetables required by the people of Iceland are grown in geothermally heated greenhouses.

Some argument exists as to whether geothermal energy is renewable. As a source of heat for geothermal energy, the planet is inexhaustible on a human time scale. However, the water used to transfer the heat to the surface is not inexhaustible. Some geothermal applications recirculate all the water back into the underground reservoir, ensuring many decades of heat extraction from a given reservoir. Other geothermal applications consume a portion of the water in the process, leading to the eventual depletion of the water in the underground reservoir. In these cases, better fluid

geothermal energy: The use of energy from Earth's interior for either space heating or generation of electricity.

tidal energy: A form of renewable energy that relies of the ebb and flow of the tides to generate electricity.

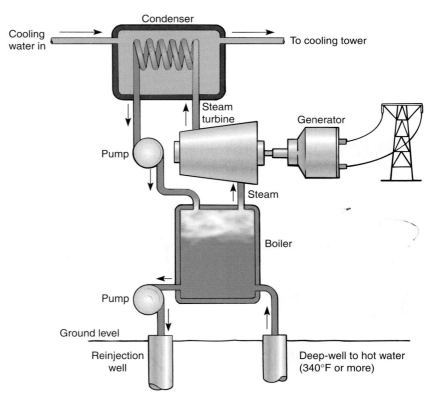

Figure 13.10 Geothermal energy. Shown is one design for a geothermal power plant. Steam separated from hot water pumped from underground turns a turbine and generates electricity. After its use, the steam is condensed and pumped back into the ground. By reinjecting spent water into the ground, geothermal energy remains a renewable energy source because the cooler, reinjected water can be reheated and used again.

management practices are helping to ensure longevity for these reservoirs. A good example is The Geysers geothermal field, where after nearly 40 years of production, experts estimate that about half the water is depleted but about 95% of the heat remains in the rock. Improved fluid management will probably help the field to remain productive for several decades more.

Geothermal energy is considered environmentally benign compared to conventional fossil fuel-based energy technologies because it emits only a fraction of the air pollutants. The most common environmental hazard associated with geothermal energy is the emission of hydrogen sulfide (H_2S) gas, which comes from the large amounts of dissolved minerals and salts found in the steam or hot water. Some geothermal reservoirs contain H_2S in quantities that require mitigation to meet air quality standards. Air pollution control methods are highly effective but do increase the energy cost. A lesser concern associated with geothermal energy is that the surrounding land may subside, or sink, as the water from hot springs and their connecting underground reservoirs is removed. Although experience has shown this is not a problem at most geothermal fields, it has occurred at a few of them.

Geothermal Energy from Hot, Dry Rock Conventional use of geothermal energy relies on hydrothermal reservoirs—that is, on groundwater—to bring the heat to the surface. These geothermal resources are limited geographically and represent only a small fraction of total geothermal energy. Scientists in Australia, Europe, Japan, and the United States are studying how to extract some of the vast amount of geothermal energy stored in hot, dry rock. Researchers at the Los Alamos National Laboratory in New Mexico pioneered the concept of geothermal energy from hot, dry rock and succeeded in demonstrating its feasibility. They drilled a well into the hot, dry rock, used hydraulic pressure to fracture the rock, and then circulated water into the fractured area to make an artificial underground reservoir. When the pressurized water returned to the surface in a second well, it turned to steam, which drove an electricity-generating turbine. Current technology is available to create such systems, but it is expensive. If technology is improved to make hot, dry geothermal energy economically attractive, it could greatly expand the extent and use of geothermal resources.

Heating and Cooling Buildings with Geothermal Energy Increasingly, geothermal energy is employed to heat and cool commercial and residential buildings. **Geothermal heat pumps (GHPs)** take advantage of the difference in temperature between Earth's surface and subsurface (at depths from 1 m to about 100 m). Underground temperatures fluctuate only slightly and are much cooler in summer and warmer in winter than air temperatures. GHPs have an underground arrangement of pipes containing circulating fluids to extract natural heat in winter, when Earth acts as a heat source, and to transfer excess heat underground in summer, when Earth acts as a heat sink (**Figure 13.11**). The hundreds of feet of pipe form a ground loop that feeds into a heat pump, which directs the flow of heated or cooled air. Geothermal heating systems can be modified to provide supplemental hot water.

Although they have been available for many years, GHPs are not widely used because their installation can be expensive. However, with the growth of *green architecture* and rising fuel costs, commercial and residential use of the systems is on the rise. The system's benefits include low operating costs, which may be half those of conventional systems, and high efficiency. The Environmental Protection Agency (EPA) estimates that GHPs are the most efficient heating system available, two to three times more efficient than other heating methods, and produce the lowest carbon dioxide emissions.

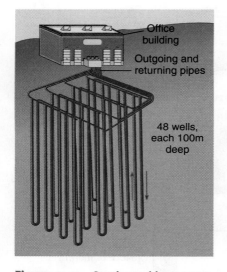

Office building

Outgoing and returning pipes

48 wells, each 100m deep

Figure 13.11 Geothermal heat pump. This diagram is based on the geothermal heat pump used at the Philip Merrill Environmental Center in Annapolis, Maryland. The pipes running through the building and underground are a closed loop. In summer, which is shown in the figure, outgoing warm water from the building (red pipes) returns from the ground cooled (blue pipes). In winter, the opposite happens: Outgoing cool water is warmer when it returns to the building. The ground's temperature is a constant 12.2°C (54°F) year-round.

Tidal Energy

Tides, the alternate rising and falling of the surface waters of the ocean and seas that generally occur twice each day, are the result of the gravitational pull of the moon and the sun. Normally, the difference in water level between high and low tides is about 0.5 m (1 or 2 ft). Certain coastal regions with narrow bays have extremely large differences in water level between high and low tides. The Bay of Fundy in Nova Scotia has the largest tides in the world, with up to 16 m (53 ft) difference between high and low tides.

Water at high tide contains enormous amounts of potential energy as compared to low tide. This energy can be captured (with a dam across a bay or a turbine that works much like a wind turbine) and converted into electricity. Tidal power plants currently operate in France, Russia, China, and Canada. However, total global production is only a few MWe, and is not expected to increase much in the near future.

REVIEW

1. What are the pros and cons of using geothermal energy to produce electricity?
2. What is tidal power?

High and Low Technology Energy Solutions

LEARNING OBJECTIVES

- Describe how a fuel cell works.
- Explain how hydrogen can be generated from any other energy source.
- Distinguish between energy conservation and energy efficiency and give examples of each.
- Define *cogeneration* and give an example of a large-scale cogeneration system.

Human requirements for energy will continue to increase, if only because the human population is growing. In addition, energy consumption continues to increase as developing countries raise their standard of living. We must therefore place a high priority on developing alternative energy sources, technologies that require less energy, and energy conservation. As an example of the difference between energy conservation and energy efficiency, consider gasoline consumption by automobiles. **Energy conservation** measures to reduce gasoline consumption would include carpooling and lowering driving speeds, whereas **energy efficiency** measures would include designing and manufacturing more fuel-efficient automobiles. Both conservation and efficiency accomplish the same goal—saving energy.

Many energy experts consider energy conservation and energy efficiency the most promising energy "sources" available because they save energy for future use and buy us time to explore new energy alternatives. Energy conservation and efficiency can cost less than development of new sources or supplies of energy, and they improve the economy's productivity. The adoption of energy-efficient technologies generates new business opportunities, including the research, development, manufacture, and marketing of those technologies. Many technologies and practices are already known, but are slow to be adopted due to both habit and relatively low energy prices.

In addition to economic benefits and energy resource savings, there are important environmental benefits from greater energy efficiency and conservation. Using more energy-efficient appliances could cut our CO_2 emissions by millions of tons each year, slowing global climate change. Energy conservation and energy efficiency reduce air pollution, acid precipitation, and other environmental damage related to energy production and consumption.

energy conservation: Using less energy, as for example, by reducing energy use and waste.

energy efficiency: Using less energy to accomplish a given task, as for example, with new technology.

Hydrogen and Fuel Cells

Increasingly, people think of hydrogen as the fuel of the future, although it is not likely to see widespread use for at least a decade. Hydrogen is a common element—water molecules contain two hydrogen atoms and one oxygen atom. While water contains little available chemical energy, a hydrogen molecule with two hydrogen atoms (H_2) contains large amounts of available energy. H_2, which is a gas at room temperature, will explode when combined with the plentiful O_2 in the atmosphere, releasing energy and forming water. When chilled to -253°C (-423°F), H_2 becomes a liquid and thus takes up much less space than H_2 gas.

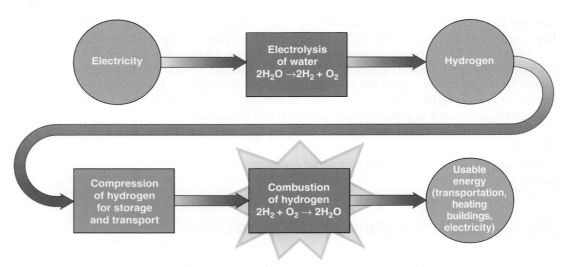

Figure 13.12 Electrolysis. Electricity can split water in a process called electrolysis. This produces hydrogen gas, which represents a chemical form of energy. Following gas compression, pipelines can transport hydrogen to users. When burned in the presence of oxygen, hydrogen produces usable energy and water.

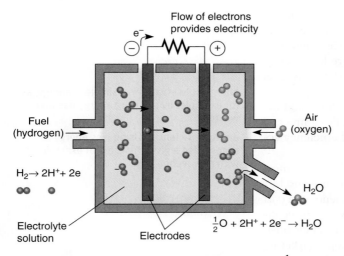

Figure 13.13 Cross-section of a hydrogen fuel cell. Hydrogen is usually produced through electrolysis. If the source of the electrolysis is renewable, the chemical energy in the hydrogen can be considered renewable.

fuel cell: A device that directly converts chemical energy into electricity without needing to produce steam and use a turbine and generator; the fuel cell requires hydrogen from a tank or other source and oxygen from the air.

Hydrogen has both advantages and disadvantages as an energy source. One advantage is that it has very high energy density, comparable to that of gasoline or LNG (see Chapter 11). Thus, unlike coal and nuclear energy, hydrogen could substitute for gasoline in automobiles and other forms of transportation. Another advantage is that hydrogen can be produced from any electrical source. **Electrolysis** is the process of using electricity to separate water into O_2 and H_2, which can be separately captured and stored (**Figure 13.12**). Finally, when H_2 is burned with O_2, the two products are available energy and water—no greenhouse gases, and no other pollutants except for relatively small amounts of oxides of nitrogen (Chapter 20).

Unfortunately, H_2 has several disadvantages. Its extreme volatility means that it has to be stored, handled, and transported very carefully. (Of course, the same is true of gasoline. . .and H_2 has the advantage of being lighter than air, while spilled gasoline spreads on the ground.) Second, the process of converting water into hydrogen is very inefficient, so only a fraction of the energy from electricity can be captured when the hydrogen is burned. Since our main sources of electricity continue to be coal, natural gas, hydropower, and nuclear energy, producing hydrogen from this electricity will still create all the environmental problems associated with those sources.

The most promising way to use hydrogen is in a **fuel cell**. A fuel cell is an electrochemical cell similar to a battery (**Figure 13.13**). Fuel cells differ from batteries because fuel cells produce power as long as they are supplied with fuel, whereas batteries store a fixed amount of energy. Fuel cell reactants (hydrogen and oxygen) are supplied from external reservoirs, whereas battery reactants are contained within the battery. When hydrogen and oxygen react in a fuel cell, water forms and energy is produced as an electric current.

The technology to use hydrogen as a substitute for gasoline, while improving, is not yet commonly available. A few hydrogen fuel cell vehicles are now being used in industrial settings. Experimental hydrogen vehicles for use on public roads currently can travel no more than about 350 km (220 miles) on a single tank of fuel—500 km (310 miles) is the typical minimum distance for commercial vehicles. The scooter

Figure 13.14   Hydrogen fuel cell application. A small hydrogen-powered fuel cell can power a cell phone or laptop, or quickly recharge a battery. The scooter pictured here is powered by hydrogen fuel. (Yoshikazu Tsuno/AFP/Getty Images)

pictured in **Figure 13.14** can travel about 100 km (60 miles) on a single load of hydrogen. Several manufacturers, including General Motors, Honda, and Toyota, expect to have commercial vehicles available sometime after 2010.

Having such vehicles, however, is not particularly useful if H_2 is not readily available. A national hydrogen infrastructure resembling the current infrastructure for gasoline and diesel distribution will have be developed. This means production points (like oil refineries), transportation (pipelines, trains, trucks), and distribution (like the gas stations currently found throughout the United States). Standards for production, delivery, storage, and use are not yet established.

Fuel cells can also provide energy for a range of other uses. Banks of fuel cells provide electricity for buildings or factories. Some companies are looking at smaller applications, producing small fuel cells that would replace batteries in cell phones and laptop computers.

Iceland, which has no domestic fossil fuel resources, plans to build the world's first fleet of fuel cell buses, obtaining its hydrogen fuel by using existing geothermal and hydroelectric resources. This suggests another major advantage of fuel cells—energy independence. This independence could come at a cost, however, if we significantly increase coal and nuclear electrical generation to produce hydrogen.

Energy Consumption Trends and Economics

A country's total energy consumption per unit of its gross domestic product in purchasing power parity (GDP PPP) gives one measure of its **energy intensity.** Lower energy intensity implies that the economy is more energy efficient as it generates wealth. According to the U.S. Department of Energy (DOE), if U.S. energy intensity

Table 13.3 Comparison of 1980 and 2004 Energy Intensities for Selected Countries

Country	Energy Intensity*	
	1980	2004
Japan	7744	6532
Germany	NA	7175
France	8757	7209
China	23,538	9080
United States	15,172	9336
Canada	18,708	13,530

* In Btu per 2000 U.S. Dollars of GDP.
Source: Energy Information Administration

had remained at its 1970 level, which was high, the United States would have spent an additional $150 billion to $200 billion for energy each year. Despite gains in efficiency, the energy intensities of the United States and Canada are still considerably higher than those of Japan and Europe (**Table 13.3**). Although the U.S. economy has become more energy efficient, total energy consumption has still increased in recent years. Importantly, this table shows that reducing energy intensity does not necessarily mean a lower standard of living.

Energy Trends in Developing Countries Per capita consumption of energy in developing nations is substantially less than it is in industrialized countries (see Figure 11.1), although the greatest increase in energy consumption today is occurring in the developing nations. As these countries boost their economic development, their energy demands increase. This is partly because the "new" industrial and agricultural processes being adopted in developing countries are often older, less expensive technologies that are less energy efficient. Also, the burgeoning populations in developing countries contribute to rising energy demands.

Developing countries are faced with the need for economic development and the need to control environmental degradation. At first glance, these two goals appear mutually exclusive. However, both are realized by adopting the new technologies now being developed in industrialized nations to achieve greater energy efficiency. For example, it would cost Brazil $44 billion to build power plants to meet its projected electricity needs for the near future; this cost could be avoided by investing $10 billion in more efficient refrigerators, lighting, and electric motors. The energy efficiency approach in Brazil would cause fewer environmental problems and foster and expand the growth of manufacturing industries devoted to energy-efficient products.

Energy-Efficient Technologies

The development of more efficient appliances, automobiles, buildings, and industrial processes has helped reduce energy consumption in highly developed countries. Compact fluorescent light bulbs produce light of comparable quality but require 25% of the energy used by regular incandescent bulbs and last up to 15 times longer. The energy-efficient bulbs are relatively expensive, but they more than pay for themselves in energy savings. Standard long-tube fluorescent bulbs have become more efficient. New condensing furnaces require approximately 30% less fuel than conventional gas furnaces. "Superinsulated" buildings use 70 to 90% less heat than do buildings insulated by standard methods (**Figure 13.15**).

The **National Appliance Energy Conservation Act (NAECA)** sets national efficiency standards for refrigerators, freezers, washing machines, clothes dryers, dishwashers, room air conditioners, and ranges/ovens (including microwaves). For example, refrigerators built in 2001 consumed 75% less energy than comparable models built in the mid-1970s. This translates to an average saving to the consumer of $135 per year. When 150 million refrigerators are operating at the 2001 standard, the annual electric energy saved will be equivalent to the output of about 32 nuclear power plants!

The NAECA requires appliance manufacturers to provide Energy Guide labels on all new appliances. These yellow labels provide estimates of annual operating costs and efficiency levels. Consumers who use this information to buy energy-efficient appliances save hundreds of dollars on utility bills.

Automobile efficiency has improved dramatically since the mid-1970s as a result of the use of lighter materials and designs that reduce air drag. The U.S. average fuel efficiency of new passenger cars doubled between the mid-1970s and the mid-1980s, although it has declined since then. A trend that has reduced energy efficiency is the popularity of minivans, sport utility vehicles, and light trucks, all of which have higher average gas mileages than sedan-type automobiles (**Figure 13.16**). Despite the reduction

Figure 13.15   Superinsulated buildings.

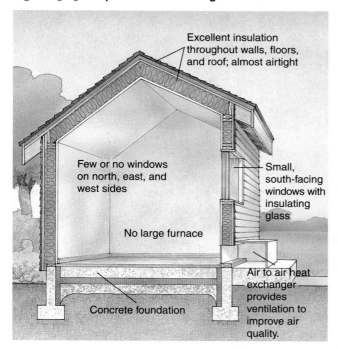

(a) Some of the characteristics of a superinsulated home, which is so well insulated and airtight it does not require a furnace in winter. Heat from the inhabitants, light bulbs, stove, and other appliances provides almost all the necessary heat.

(b) A superinsulated office building in Toronto, Canada, has south-facing windows with insulating glass. The building is so well insulated it uses no furnace. (Courtesy Ontario Hydro)

in energy efficiency caused by recent consumer preferences, significant gains could easily be implemented. Using current technology, automobiles with fuel efficiencies of 60 to 65 mpg could be routinely manufactured within the next decade or so. The main factors that will determine whether these vehicles are available are consumer preference and energy cost.

During the past few decades, many industries have improved their energy efficiencies. New aircraft are much more fuel-efficient than older models, and technological improvements in the papermaking industry make it possible to use less energy to

Figure 13.16   What does this cartoon imply about the relevance of energy efficiency in automobile purchasing decisions? Do you think attitudes will change? Why / why not? (FOXTROT © 2004 Bill Amend/Reprinted with permission of Universal Press Syndicate. All rights reserved.)

Figure 13.17 Cogeneration. In this example of a cogeneration system, fuel combustion generates electricity in a generator. The electricity produced is used in-house or sold to a local utility. The waste heat (leftover hot gases or steam) is recovered for useful purposes, such as industrial processes, heating buildings, hot water heating, and generating additional electricity.

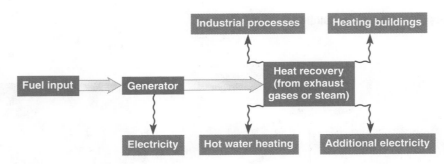

manufacture paper today than was used just a few years ago. The energy savings from such improvements in efficiency translate into greater profits for the companies employing them.

Cogeneration One energy technology with a bright future is **cogeneration,** the production of two useful forms of energy from the same fuel. Cogeneration, or **combined heat and power (CHP),** makes perfect sense from a systems perspective. It involves the generation of electricity through some thermal process (often natural gas); the low-temperature steam produced during this process is used for building or industrial heating. In CHP, the overall conversion efficiency (that is, the ratio of useful energy produced to fuel energy used) is high because some of what is usually waste heat is used.

Cogeneration can be cost effective on a small scale. Modular CHP systems enable hospitals, factories, college and university campuses, and other businesses to harness steam that is otherwise wasted. In a typical CHP system, electricity is produced in a traditional manner—that is, some type of fuel provides heat to form steam from water. Normally, the steam used to turn the electricity-generating turbine is cooled before being pumped back to the boiler for reheating. In cogeneration, after the steam is used to turn the turbine, it supplies energy to heat buildings, cook food, or operate machinery before it is cooled and pumped back to the boiler as water (**Figure 13.17**).

Cogeneration is also done on a large scale. One of the largest CHP systems in the United States is a "combined cycle" natural gas plant in Oswego, New York, that produces electricity for the local utility. It uses natural gas turbines to generate electricity. The exhaust gases, which are at least 1000°C (1832°F), are then used to produce high-pressure steam for a nearby industry. The overall efficiency of the Oswego cogeneration system is 54%, as compared to 33% efficiency in a typical fossil fuel power plant.

Energy Savings in Commercial Buildings Energy costs often account for 30% of a company's operating budget. Unlike cars, which are traded in every few years, buildings are usually used for 50 or 100 years; thus, a company housed in an older building normally does not have the benefits of new energy-saving technologies. It makes good economic sense for these businesses to invest in energy improvements, which often pay for themselves in a few years (**Table 13.4**). More than 20,000 schools, hospitals, commercial office buildings, retail outlets, multifamily residential buildings, and other facilities across the United States have had energy-efficient upgrades that provide energy savings of $.20 to $2.00 per square foot.

To get businesses to install new energy-efficient technologies, energy-services companies, which specialize in designing and installing energy-efficient technologies, offer their assistance so that the business makes little or no financial outlay. Here is how it works. An energy-services company makes a detailed assessment of how a business can improve its energy efficiency. In developing its proposal, the energy-services company guarantees a certain amount of energy savings. It provides the funding to accomplish the improvements, which may be as simple as fine-tuning existing heating,

Table 13.4 Energy Efficiency Upgrades in Selected Commercial Buildings

Project	Energy Payback Time*	Unexpected Benefits Attributed to Project**
Energy-efficient lighting (post office in Nevada)	6 years	6% increase in mail sorting productivity
Energy-efficient (metal-halide) lighting (aircraft assembly plant in Washington)	2 years	Up to 20% better quality control
Energy-efficient lighting (drafting area of utility company in Pennsylvania)	About 4 years	25% lower absenteeism; 12% increase in drawing productivity
Energy-efficient lighting and air conditioning (office building in Wisconsin)	0 years (paid for by utility rebates); energy savings estimated at 40%	16% increase in worker productivity
Energy-saving daylighting, passive solar heating, heat recovery system (bank in Amsterdam)	3 months	15% lower absenteeism

*How long it takes for energy savings to cover the cost of the project.
**Lighting quality as well as lighting efficiency is improved, resulting in greater worker comfort.
Source: Rocky Mountain Institute

ventilation, and air conditioning systems or as major as replacing all existing windows and lights. The reduction in utility costs is used to pay the energy-services company, but once the bill is paid, the business benefits from all additional energy savings. (See "You Can Make a Difference: Saving Energy at Home" for energy-saving suggestions at home.)

Electric Power Companies and Energy Efficiency   Changes in the regulations governing electric utilities have let utilities make more money by generating less electricity. Such programs provide incentives to save energy and thereby reduce power plant emissions that contribute to environmental problems.

Traditionally, to meet future power needs, electric utilities planned to build new power plants or purchase additional power from alternative sources. Now they often avoid these massive expenses by **demand-side management,** in which they help electricity consumers save energy. Some utilities support energy conservation and efficiency by offering cash awards to consumers who install energy-efficient technologies. When voters in California decided to close the nuclear power plant Rancho Seco, the Sacramento Municipal Utility District paid customers to buy more efficient refrigerators and plant trees to shade their houses, thereby lowering air conditioning costs. These efforts helped the utility company to reduce the demand for electricity.

Some utilities give customers energy-efficient compact fluorescent light bulbs, air conditioners, or other appliances. They then charge slightly higher rates or a small leasing fee, but the greater efficiency results in savings for both the utility company and the consumer. The utility company makes more money from selling less electricity because it does not have to invest in additional power generation to meet increased demand. The consumer saves because the efficient light bulbs or appliances use less energy, which more than offsets the higher rates.

According to the American Council for an Energy-Efficient Economy, U.S. electric power plants are themselves an important target for improved energy efficiency. Much heat is lost during the generation of electricity. If all this wasted energy were harnessed, for example, by cogeneration, it could be used productively, thereby conserving energy. Another way to increase energy efficiency would be to improve our electric grids because about 10% of electricity is lost during transmission. To accomplish this, some energy experts envision that future electricity will be generated far from population centers, converted to supercooled hydrogen, and transported through underground superconducting pipelines. The technology to build such conduits has not yet been developed.

REVIEW

1. What is the difference between energy conservation and energy efficiency?

2. What is cogeneration?

YOU CAN MAKE A DIFFERENCE

Saving Energy at Home

The average household spends $1500 each year on utility bills. This cost could be reduced considerably by investments in energy-efficient technologies. When buying a new home, a smart consumer should demand energy efficiency. Although a more energy-efficient house might cost more, depending on the technologies employed, the improvements usually pay for themselves in two or three years. Any time spent in the home after the payback period means substantial energy savings. Energy efficiency has become an essential element of design codes nationwide and will almost certainly be an important part of future home designs.

Some energy-saving improvements, such as thicker wall insulation, are easier to install while the home is being built. Other improvements can be made in older homes to enhance energy efficiency and, as a result, reduce the cost of heating the homes. Examples include installing thicker attic insulation, installing storm windows and doors, caulking cracks around windows and doors, replacing ineffi-

cient furnaces and refrigerators, and adding heat pumps.

Many of the same improvements provide energy savings when a home is air-conditioned. Additional cooling efficiency is achieved by insulating the air conditioner ducts, especially in the attic; buying an energy-efficient air conditioner; and shading the south and west sides of a house with deciduous trees. Window shades and awnings on south- and west-facing windows reduce the heat a building gains from its environment. Ceiling fans can supplement air conditioners by making a room feel comfortable at a higher thermostat setting. Make sure your ceiling fan is set to draw warm air toward the ceiling in the summer, and reverse this setting in the winter.

Other energy savings in the home include:

- Replacing incandescent bulbs with energy-efficient compact fluorescent light bulbs (see figure).
- Installing a programmable thermostat, which cuts heating and air conditioning costs up to 33%.
- Lowering the temperature setting on water heaters to 140°F (with a dishwasher) or 120°F (without).

- Installing low-flow shower heads and faucet aerators to reduce the amount of hot water used.
- Eliminate energy "vampires," or appliances that draw electricity even when they're not in use. (A study by researchers at Cornell University suggests that American households waste $3 billion *each year* on household vampires).

How does a homeowner learn which improvements will result in the most substantial energy savings? In addition to reading the many articles on energy efficiency in newspapers and magazines, a good way to learn about your home is to have a comprehensive energy audit done. Most local utility companies will send an energy expert to your home to perform an audit for little or no charge. The audit will determine the total energy consumed and where thermal losses are occurring (through the ceiling, floors, walls, or windows). On the basis of this assessment, the energy expert will then make recommendations about how to reduce your heating and cooling bills.

Some energy-saving measures for the home.

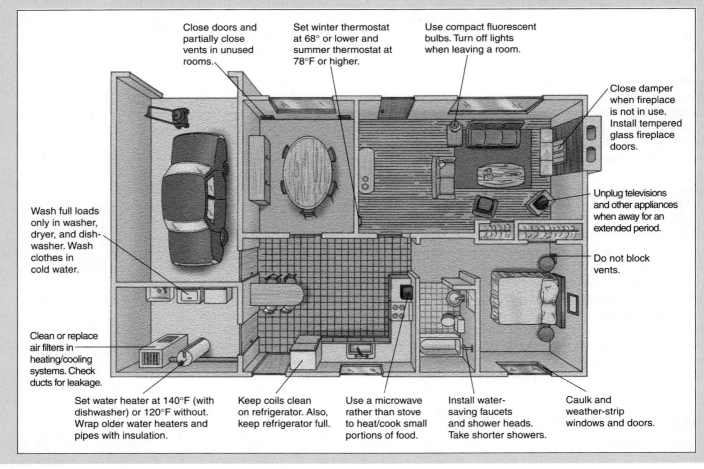

Close doors and partially close vents in unused rooms.

Set winter thermostat at 68° or lower and summer thermostat at 78°F or higher.

Use compact fluorescent bulbs. Turn off lights when leaving a room.

Close damper when fireplace is not in use. Install tempered glass fireplace doors.

Unplug televisions and other appliances when away for an extended period.

Do not block vents.

Wash full loads only in washer, dryer, and dishwasher. Wash clothes in cold water.

Clean or replace air filters in heating/cooling systems. Check ducts for leakage.

Set water heater at 140°F (with dishwasher) or 120°F without. Wrap older water heaters and pipes with insulation.

Keep coils clean on refrigerator. Also, keep refrigerator full.

Use a microwave rather than stove to heat/cook small portions of food.

Install water-saving faucets and shower heads. Take shorter showers.

Caulk and weather-strip windows and doors.

Appropriate Energy: a Systems Perspective

LEARNING OBJECTIVE

• Explain why a systems perspective is likely to lead to appropriate energy solutions

Often, energy sources and types are selected based on short-term needs and consumer habits. A systems perspective that considers the lifestyle, health, financial and environmental implications of energy sources can lead to substantially different decisions.

In the introduction to Chapter 7, we saw that driving children to school affects environmental and public health in several ways. This link is one of many that become clear when we think about energy, environment, and health from a systems perspective. For example, we often have a choice between escalators, elevators, and stairs. Research has shown that such modest lifestyle choices as choosing stairs can contribute substantially to an individual's health. At the same time, it takes far more energy, most of it from fossil fuels, to use an elevator or escalator. Deciding where to live and work, choosing whether to bicycle or ride a motorcycle as recreation, and other such choices simultaneously impact health and the environment.

In developing countries, where energy-intensive habits have not yet developed, selecting appropriate energy technologies may help avoid some of the downfalls of energy intensive lifestyles typical in highly developed countries. Rather than building large coal plants, Kenyans can adopt small-scale solar electric generators. Elsewhere small-scale windmills, microhydro (small-scale hydroelectric power generation), and other technologies may avoid the need for building unwieldy, inefficient energy infrastructure. And the lesson can be learned in reverse as well: The technology that has made the solar option affordable in Kenya can be repackaged for use in the United States and other highly developed countries. Most of us rarely think about our energy decisions, yet small choices can save money and energy while improving our quality of life.

REVIEW

1. How are human health and energy use tightly linked?
2. Why is small-scale energy generation often more appropriate for developing countries?

REVIEW OF LEARNING OBJECTIVES WITH KEY TERMS

• **Distinguish between active and passive solar heating and describe how each is used.**

Passive solar heating is a system of putting the sun's energy to use without requiring mechanical devices (pumps or fans) to distribute the collected heat. Currently, about 7% of new homes built in the United States have passive solar features. **Active solar heating** is a system of putting the sun's energy to use in which a series of collectors absorb the solar energy, and pumps or fans distribute the collected heat. Active solar heating is used for heating water and, to a lesser extent, space heating.

• **Contrast the advantages and disadvantages of solar thermal electric generation and photovoltaic solar cells in converting solar energy into electricity.**

Solar thermal electric generation is a means of producing electricity in which the sun's energy is concentrated by mirrors or lenses onto a fluid-filled pipe; the heated fluid is used to generate electricity. Solar thermal plants are not yet cost-competitive with traditional fuels, but they are more efficient than other direct solar technologies, and they do not produce air pollution or contribute to acid rain or global climate change. A **photovoltaic solar cell** is a wafer or thin film of solid-state materials, such as silicon or gallium arsenide, that are treated with certain metals so that they generate electricity—that is, a flow of electrons—when solar energy is absorbed. Photovoltaics generate electricity with no pollution and minimal maintenance but are only about 10 to 15% efficient at converting solar energy to electricity.

• **Define *biomass*, explain why it is an example of indirect solar energy, and outline how it is used as a source of energy.**

Biomass consists of plant material (including undigested fiber in animal dung) that is used as fuel. Biomass is an example of indirect solar energy because it includes organic materials produced by photosynthesis. Biomass is burned directly to produce heat or electricity or converted to solid (**charcoal**), gas (**biogas**), or liquid (**methanol** and **ethanol**) fuels. Biomass is already being used for energy on a large scale, particularly in developing nations. India and China have several million **biogas digesters** that produce biogas from household and agricultural wastes.

• **Describe the locations that can make optimum use of wind energy and of hydropower, and compare the potential of wind energy and hydropower.**

Wind energy is electric energy obtained from surface air currents caused by the solar warming of air. Harvesting wind energy to generate electricity has great potential because it is currently the most cost-competitive of all forms of solar energy. Harnessing wind energy is most profitable in areas with fairly continual winds, such as islands, coastal areas, mountain passes, and grasslands. **Hydropower** is a form of renewable energy that relies on flowing or falling water to generate electricity. The damming of rivers and streams to generate electricity is the major form of hydropower. Currently, hydropower produces about 19% of the world's electricity. Environmental and social problems associated with hydropower include ecological destruction upstream and downstream, increased evaporation of water, disease and pollution, displacement of people, and inundation of farmland.

• **Describe tidal energy and geothermal energy, the two forms of renewable energy that are not direct or indirect results of solar energy.**

Tidal energy is a form of renewable energy that relies on the ebb and flow of the tides to generate electricity, and is currently used on a very limited scale. **Geothermal energy** is the use of energy from Earth's interior for either space heating or generation of electricity. Geothermal energy can be obtained from **hydrothermal reservoirs** of heated water near Earth's surface. The established technology for extracting geothermal energy from heated areas of Earth's crust involves drilling wells and bringing the steam or hot water to the surface.

• **Describe how a fuel cell works.**

A **fuel cell** is a device that functions much like a battery. It directly converts chemical energy, usually from hydrogen, into electricity without the intermediate step of needing to produce steam and use a turbine and generator. The fuel cell requires oxygen from the air to combine with the hydrogen. Motor vehicles powered by hydrogen fuel cells are currently in use in industrial settings, and are likely to be more widely available sometime in the next decade.

• **Explain how hydrogen can be generated from any other energy source.**

Hydrogen is typically created through **electrolysis**. An electrical current is applied to water, and the energy is absorbed by splitting relatively low-energy water molecules into higher-energy H_2 and O_2 molecules. Any energy source that can be used to generate electricity, including fossil fuels, nuclear, PV cells, and hydropower, can be used to generate hydrogen. Thus the environmental impacts of hydrogen production depend on the energy source used for the electricity.

• **Distinguish between energy conservation and energy efficiency and give examples of each.**

Energy conservation involves using less energy, as, for example, by reducing energy use and waste. **Energy efficiency** involves using less energy to accomplish a given task, as, for example, with new technology. Consider gasoline consumption by automobiles. Energy conservation measures to reduce gasoline consumption would include carpooling and lowering driving speeds, whereas energy efficiency measures would include designing and manufacturing more fuel-efficient automobiles.

• **Define *cogeneration* and give an example of a large-scale cogeneration system.**

Cogeneration is an energy technology that involves recycling "waste" heat. Also called **combined heat and power (CHP)**, cogeneration involves the generation of electricity, and then the steam produced during this process is used rather than wasted. One of the largest CHP systems in the United States is in Oswego, New York. It uses natural gas turbines to generate electricity for the local utility. The hot exhaust gases are then used to produce high-pressure steam for a nearby industry.

• **Explain why a systems perspective is likely to lead to appropriate energy solutions.**

Often, energy sources and types are selected based on short-term needs and consumer habits. A systems perspective that considers the lifestyle, health, financial, and environmental implications of energy sources can lead to substantially different decisions. Riding an elevator or driving a car may be more convenient in the short term, while using stairs, bicycling, or walking is more healthful, less expensive, and less energy intensive.

Thinking About the Environment

1. Explain the following statement: Unlike fossil fuels, solar energy is not resource-limited but is technology-limited.

2. Biomass is considered an example of indirect solar energy because it is the result of photosynthesis. Given that plants are the organisms that photosynthesize, why are animal wastes considered biomass?

3. One advantage of the various forms of renewable energy, such as solar, thermal, and wind energy, is that they cause no net increase in atmospheric carbon dioxide. Is this true for biomass? Why or why not?

4. Some energy experts refer to the Great Plains states as "the Saudi Arabia of wind power." Explain what the reference means.

5. Why is it easier to obtain energy from a small river with a steep grade than from the vast ocean currents?

6. Japan wishes to make use of solar power, but it does not have extensive tracts of land for building large solar power plants. Which solar technology do you think is best suited to Japan's needs? Why?

7. When is hydrogen a "climate-neutral" fuel? Is hydrogen produced using the energy of natural gas or coal climate-neutral? Why or why not?

8. What are some of the challenges we will face if we decide to transition from a carbon-based economy to a hydrogen economy?

9. Explain how energy conservation and efficiency can be considered "sources" of energy.

10. What has allowed energy intensity to drop for most countries since 1980, even as GDPs have gone up?

11. What is cogeneration, and what are its advantages?

12. Evaluate the forms of energy, other than fossil fuels and nuclear power, that have the greatest potential where you live.

13. Give an example of how one or more of the alternative energy sources discussed in this chapter could have a negative effect on each of the following aspects of ecosystems:

    a. Soil preservation

    b. Natural water flow

    c. Foods used by wild plant and animal populations

    d. Preservation of the diversity of organisms found in an area

14. List energy conservation measures you could adopt for each of the following aspects of your life: washing laundry, lighting, bathing, cooking, buying a car, and driving a car.

15. How would the list above differ if you were to take a systems perspective, rather than a short-term perspective? Would it be easy to implement these changes? Why or why not?

Quantitative questions relating to this chapter are on our Web site.

Take a Stand

Visit our Web site at http://www.wiley.com/college/raven (select Chapter 13 from the Table of Contents) for links to more information about the controversy surrounding subsidies for alternative energy development. Consider the views of proponents and opponents, and debate the issue with your classmates. You will find tools to help you organize your research, analyze the data, think critically about the issues, and construct a well-considered argument. *Take a Stand* activities can be done individually or as a team, as oral presentations, written exercises, or Web-based (e-mail) assignments.

Additional online materials relating to this chapter, including a Student Testing Section with study aids and self-tests, Environmental News, Activity Links, Environmental Investigations, and more, are also on our Web site.

Water: A Limited Resource

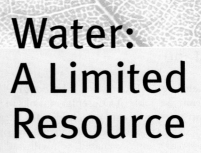

Domestic water in Lagos, Nigeria. In many less-developed countries, access to safe, clean water for household use is limited and expensive. One woman in this photo from Lagos, Nigeria, is filling a bucket with water from a communal tank. Poor residents of large towns often pay more for water than do the wealthy, spend a much larger fraction of their time acquiring it, and use part of their similarly limited fuel supply boiling it. (Stuart Franklin/Magnum Photos, Inc.)

Around the world, about 1.1 billion people live without adequate access to water—many have fewer than 10 L (about 2.6 gallons) of clean water per day. Worldwide, water is often limited in quality, quantity, or both. This is difficult to imagine for most people in the United States, Canada, and other highly developed nations, for whom water is both ample and clean. The typical American uses about 150 L (40 gallons) each day, and has the luxury of watering plants and washing cars with water that is safe enough to drink.

The 2006 Human Development Report, produced by the U.N. Development Program (UNDP), indicates that inhabitants of the slums in Lagos, Nigeria (see photograph), pay 5 to 10 times as much for water as do those in richer neighborhoods. In some places, the poor spend as much as 20% of their income

on water. Contrast this with the 0.15% spent by one of the authors of this book.

Taking a systems perspective is useful in understanding why water is both unsafe and unavailable in much of the world. Consider a typical small Canadian town. A public or private utility company purchases water, often from a single source and with a multi-year contract. The utility company transports and cleans the water, which it then distributes through an infrastructure of pumps and pipes. This complex system requires employees, computerized information processing, coordination with other groups (government, other utilities, consumers), energy to pump the water, and both energy and chemicals to clean it. The total cost might be large, but the coordination makes the cost per person relatively low. Consumers have a reliable source of water at a predictable cost. Different parts of the system are managed by experts, who can be held accountable.

Contrast this with the slums of Lagos. All of the same system components exist: Water must be cleaned, transported, bought, and sold. People, energy, and organization are required to make it happen. However, there is little money to develop and maintain the infrastructure, so water from pipes is sporadic, if available at all. When the water does run, it might be contaminated from holes in the pipes in contact with human waste. An alternative is to buy water from a vendor who comes by with tanks or jugs of water. But these freelance wa-

ter dealers are unpredictable, the quality of the water unknown, and a different price might be negotiated each day. A consumer might boil the water to kill off biological contaminants, but this in turn requires energy, which is also in limited supply and expensive.

As part of its 2006 report, the UNDP proposed that access to enough safe water—at least 20 L (5.2 gals) per day—should be considered a basic human right. Proposals the UNDP advocates that might alleviate some of the problems in Lagos include:

- Having water available at a very low cost to poor residents, with a price that goes up with higher usage. (Right now, water price often stays the same or goes down with increased use, or is available at a flat cost regardless of how much is used).

- Providing public financing for water infrastructure development.

- Including water access as part of broader poverty-reduction programs (the systems approach).

- Holding water providers accountable for consistency and safety.

While there is enough water in the world for all people, problems of distribution and quality assurance, coupled with increasing global population, make universal access to water an issue that will face us for decades to come.

World View and Closer to You...
Real news footage relating to Water around the world and in your own geographic region is available at www.wiley.com/college/raven by clicking on "World View and Closer to You."

The Importance of Water

LEARNING OBJECTIVES

- Describe the structure of a water molecule and explain how hydrogen bonds form between adjacent water molecules.
- Describe surface water and groundwater, using the following terms in your descriptions: wetland, runoff, drainage basin, unconfined and confined aquifer, and water table.

The view of planet Earth from outer space reveals that it is different from other planets in the solar system. Earth is a predominantly blue planet because of the water that covers three-fourths of its surface. Water has a tremendous effect on our planet: It helps shape the continents, it moderates our climate, and it allows organisms to survive.

Life on Earth would be impossible without water. All life forms, from unicellular bacteria to multicellular plants and animals, contain water. Humans are composed of approximately 70% water by body weight. We depend on water for our survival as well as for our convenience: We drink it, cook with it, wash with it (**Figure 14.1**), travel on it, and use an enormous amount of it for agriculture, manufacturing, mining, energy production, and waste disposal.

Although Earth has plenty of water, about 97% of it is salty and not available for use by most terrestrial organisms. Fresh water is distributed unevenly, resulting in serious regional water supply problems. Conflicts often arise over water use because one application decreases the amount available for others. Even regions with readily available fresh water have problems maintaining the quality and quantity of water.

Worldwide, freshwater use is increasing. This is in part because the human population is expanding and in part because, on the average, each person is using more water. Humans now use well over 50% of Earth's accessible, renewable fresh water. An increasing number of countries are experiencing water shortages as population growth and human activities place increasing demands on a limited water supply.

Figure 14.1   **A grandmother washes her grandson, using a bowl to pour water over his head.**
(Nareerat Lertwassana/UNEP/Still Pictures/Peter Arnold, Inc.)

Figure 14.2   Chemical Properties of Water

(a) Each water molecule consists of two hydrogen atoms and one oxygen atom. Water molecules are polar, with positively and negatively charged areas.

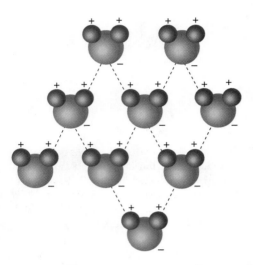

(b) The polarity causes hydrogen bonds (represented by dashed lines) to form between the positive areas of one water molecule and the negative areas of others. Each water molecule forms up to four hydrogen bonds with other water molecules.

Properties of Water

Water is composed of molecules of H_2O, each consisting of two atoms of hydrogen and one atom of oxygen. Water exists in any of three forms: solid (ice), liquid, and vapor (water vapor or steam). Water molecules are **polar**—that is, one end of the molecule has a positive electrical charge, and the other end has a negative charge (**Figure 14.2**). The negative (oxygen) end of one water molecule is attracted to the positive (hydrogen) end of another water molecule, forming a **hydrogen bond** between the two molecules. Hydrogen bonds are the basis for many of water's physical properties, including its high melting/freezing point (0°C, 32°F) and high boiling point (100°C, 212°F). Because most of Earth has a temperature between 0°C and 100°C, most water exists in the liquid form organisms need.

Water absorbs a great deal of solar heat without its temperature rising substantially. This high heat capacity allows the ocean to moderate climate, particularly along coastal areas, and it does not experience the wide temperature fluctuations common on land.

Water must absorb a lot of heat before it **vaporizes,** or changes from a liquid to a vapor. When it does evaporate, it carries the heat, called *heat of vaporization*, with it into the atmosphere. Thus, evaporating water has a cooling effect. That is why your body is cooled when perspiration evaporates from your skin.

Water is sometimes called the "universal solvent." While this is an exaggeration, many materials do dissolve in water. In nature, water is never completely pure, because it contains dissolved gases from the atmosphere and dissolved mineral salts from the land. Seawater contains a variety of dissolved salts, including sodium chloride, magnesium chloride, magnesium sulfate, calcium sulfate, and potassium chloride (**Figure 14.3**). Water's dissolving ability has a major drawback: Many of the substances that dissolve in water cause water pollution.

In general, water expands when heated and contracts when cold. As water cools, it contracts and becomes denser until it reaches 4°C (39°F), the temperature at which it

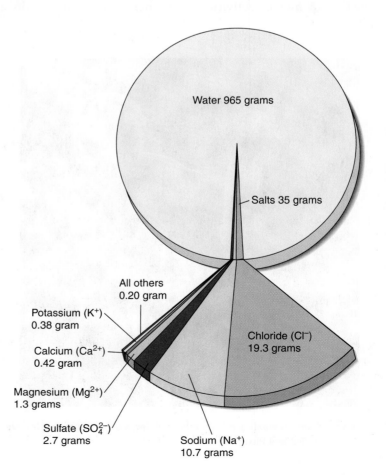

Figure 14.3   Chemical composition of 1 kg (2.2 lb) of seawater. Seawater contains a variety of dissolved salts present as ions.

Water 965 grams

Salts 35 grams

All others
0.20 gram

Potassium (K$^+$)
0.38 gram

Calcium (Ca^{2+})
0.42 gram

Magnesium (Mg^{2+})
1.3 grams

Sulfate (SO$_4^{2-}$)
2.7 grams

Sodium (Na$^+$)
10.7 grams

Chloride (Cl$^-$)
19.3 grams

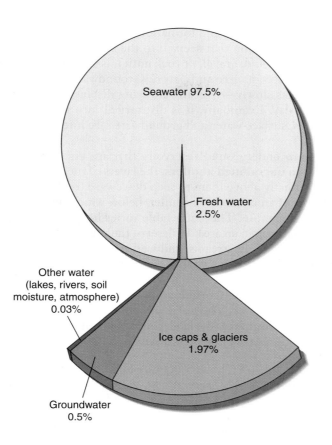

Seawater 97.5%

Fresh water
2.5%

Other water
(lakes, rivers, soil
moisture, atmosphere)
0.03%

Ice caps & glaciers
1.97%

Groundwater
0.5%

Figure 14.4 **Distribution of Water.**
Although three-fourths of Earth's surface is covered with water, substantially less than 1% is available for humans. Most water is salty, frozen, or inaccessible in the soil and atmosphere.

is the densest. When the temperature of water falls below 4°C, it becomes less dense. Ice (at 0°C) floats on the denser, slightly warmer, liquid water. Water freezes from the top down rather than from the bottom up, and aquatic organisms survive beneath the frozen surface.

The Hydrologic Cycle and Our Supply of Fresh Water

In the **hydrologic cycle**, water continuously circulates through the environment, from the ocean to the atmosphere to the land and back to the ocean (see Figure 5.6). The result is a balance among water in the ocean, on the land, and in the atmosphere. The hydrologic cycle continually renews the supply of fresh water on land, which is essential to terrestrial organisms.

Approximately 98% of Earth's water is in the ocean and contains a high amount of dissolved salts (**Figure 14.4**). Seawater is too salty for human consumption and for most other uses. For example, if you watered your garden with seawater, your plants would die. Most fresh water is unavailable for easy consumption because it is frozen as polar or glacial ice or is in the atmosphere or soil. Lakes, creeks, streams, rivers, and groundwater account for only a small portion—about 0.03%—of Earth's fresh water.

Surface water is water found on Earth's surface in streams and rivers, lakes, ponds, reservoirs, and **wetlands**, areas of land covered with water for at least part of the year. The **runoff** of precipitation from the land replenishes surface waters and is considered a renewable, though finite, resource. A **drainage basin** or **watershed** is the area of land drained by a single river or stream. Watersheds range in size from less than 1 km² for a small stream to a huge portion of the continent for a major river system such as the Mississippi River. **Table 14.1** lists the world's 10 largest watersheds.

surface water: Precipitation that remains on the surface of the land and does not seep down through the soil.

runoff: The movement of fresh water from precipitation (including snowmelt) to rivers, lakes, wetlands, and, ultimately, the ocean.

drainage basin: A land area that delivers water into a stream or river system.

Table 14.1 The World's 10 Largest Watersheds

Watershed	Region	Area of Watershed (thousand km²)
Amazon	South America	6145
Congo	Africa	3731
Nile	Africa	3255
Mississippi	North America	3202
Ob	Asia	2972
Paraná	South America	2583
Yenisey	Asia	2554
Lena	Asia	2307
Niger	Africa	2262
Yangtze	Asia	1722

Source: Water Resources eAtlas (World Conservation Union)

Earth contains underground formations that collect and store water. This water originates as precipitation that seeps into the soil and finds its way down through cracks and spaces in sand, gravel, or rock until it is stopped by an impenetrable layer; there it accumulates as **groundwater**. Groundwater flows through permeable sediments or rocks slowly—typically covering distances of several millimeters to a few meters per day. Eventually, it is discharged into rivers, wetlands, springs, or the ocean. Thus, surface water and groundwater are interrelated parts of the hydrologic cycle.

Aquifers are underground reservoirs that are either unconfined or confined (**Figure 14.5**). In **unconfined aquifers**, the layers of rock above are porous and allow surface water directly above them to seep downward, replacing the aquifer contents. The upper limit of an unconfined aquifer, below which the ground is saturated with water, is the **water table**. The water table varies in depth depending on amount of precipitation occurring in an area. In deserts, thewater table is generally far below the surface. In contrast, lakes, streams, and wetlands occur where the water table intersects with the surface. When a well goes dry, the water table has dropped below the depth of the well.

A **confined aquifer,** or **artesian aquifer,** is a groundwater storage area between impermeable layers of rock. The water in a confined aquifer is trapped and often under pressure. Its recharge area (the land from which water percolates to replace groundwater) may be hundreds of kilometers away.

groundwater: The supply of fresh water under Earth's surface that is stored in underground aquifers.

aquifers: Underground caverns and porous layers of sand, gravel, or rock in which groundwater is stored.

water table: The upper surface of the saturated zone of groundwater.

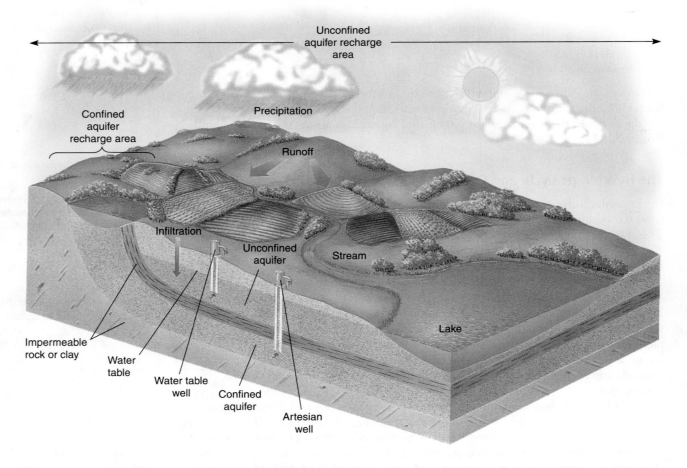

Figure 14.5 Groundwater. Excess surface water seeps downward through soil and porous rock layers until it reaches impermeable rock or clay. An unconfined aquifer has groundwater recharged by surface water directly above it. In a confined aquifer, groundwater is stored between two impermeable layers and is often under pressure. Artesian wells, which produce water from confined aquifers, often do not require pumping because of this pressure.

Most groundwater is considered a nonrenewable resource because it has taken hundreds or even thousands of years to accumulate, and usually only a small portion of it is replaced each year by percolation of precipitation. The recharge of confined aquifers is particularly slow.

REVIEW

1. How do hydrogen bonds form between adjacent water molecules?

2. What is surface water? groundwater?

3. How is runoff related to a drainage basin?

4. How is an aquifer related to the water table?

Water Use and Resource Problems

LEARNING OBJECTIVES

- Describe the role of irrigation in world water consumption.
- Define *flood plain* and explain how humans exacerbate flood damage, using the upper Mississippi River Basin as an example.
- Relate some of the problems caused by overdrawing surface water and aquifer depletion (including saltwater intrusion).

Water consumption varies among countries, ranging from several gallons per person per day in areas of acute shortage to several hundred gallons per person per day in some highly developed nations. This encompasses agricultural and industrial uses as well as direct individual consumption. The greatest user of water worldwide is agriculture. Irrigation accounts for 71% of the world's total water consumption, industry for 20%, and domestic and municipal use for 9%.

Water resource problems fall into three categories: too much, too little, and poor quality/contamination. (Chapter 22 addresses the third category.) We cannot prevent floods and droughts because they are part of natural climate variations. Human activities sometimes exacerbate their seriousness, however. Humans often court disaster when they make environmentally unsound decisions, such as building in an area prone to flooding.

Too Much Water

Many early civilizations—ancient Egypt, for example—developed near rivers that periodically spilled over, inundating the surrounding land with water. When the water receded, a thin layer of sediment rich in organic matter remained and enriched the soil. These civilizations flourished, partly because of their agricultural productivity, which was the result of floods replenishing nutrients in the soil.

Flooding results from system interactions among human activities and natural phenomena. Modern floods are more disastrous in terms of property loss than those of the past because humans often remove water-absorbing plant cover from the soil and construct buildings on **flood plains**. These activities increase the likelihood of both floods and flood damage.

Forests, particularly on hillsides and mountains, trap and absorb precipitation to provide nearby lowlands with some protection from floods. When woodlands are cut down, particularly if they are clear-cut, the area cannot hold water nearly as well. Heavy rainfall then results in rapid runoff from the exposed, barren hillsides. This not only causes soil erosion but also puts lowland areas at extreme risk of flooding.

When a natural area—that is, an area undisturbed by humans—is inundated with heavy precipitation, the plant-protected soil absorbs much of the excess water. What the soil cannot absorb runs off into the river, which may then spill over its banks onto the flood plain. Because rivers meander, the flow is slowed, and the swollen waters rarely cause significant damage to the surrounding area. (See Figure 6.13 for a diagram of a typical river, including its flood plain.)

flood plain: The area bordering a river channel that has the potential to flood.

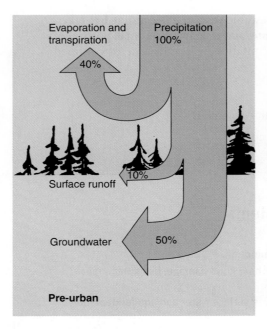

Figure 14.6 **How development changes the natural flow of water.**

(a) The fate of precipitation in Ontario, Canada, before urbanization.

(b) After Ontario was developed, surface runoff increased substantially, from 10 to 43%.

When an area is developed for human use, much of the water-absorbing plant cover is removed. Buildings and paved roads do not absorb water, so runoff, usually in the form of storm sewer runoff, is significantly greater (**Figure 14.6**). People who build homes or businesses on the flood plain of a river will most likely experience flooding at some point.

Increasingly, local governments around the world put zoning restrictions on flood plains to curtail development. Consider the January 1997 floods in California, which killed eight people, destroyed more than 16,000 homes, and caused $1.6 billion in damage. During similar floods in 2006, many expensive levees failed (**Figure 14.7a**). Rather than rebuild levees adjacent to the rivers to try to prevent floods (**Figure 14.7b**), river experts recommended that California try a different approach in which rivers are allowed to occupy part of the flood plain during a flood (**Figure 14.7c**). Smaller levees are built some distance from the river's edge. The new approach is less expensive, results in less damage during floods, and provides some of the natural benefits of floods, such as improved habitat for waterfowl and other wildlife and replenishment of the soil in the flood plain.

CASE IN POINT

The Floods of 1993

The Mississippi River, one of the world's largest rivers, spans 31 states and two Canadian provinces. While most people are well aware of the devastating impacts of Mississippi River flooding caused by Hurricane Katrina (Chapter 23), it was not the first time that flooding has caused widespread damage. In 1993, the Mississippi and its tributaries flooded, spreading over 9.3 million hectares (23 million acres) of flood plains and engulfing farms and towns in nine Midwestern states (**Figure 14.8**). Fifty people were killed, and damage to property was estimated in excess of $12 billion. More than 70,000 people lost their homes, and 8.7 million acres of farmland were damaged. Pesticides and other agricultural chemicals washed off the fields were carried to the Gulf of Mexico. Floodwaters carried the zebra mussel, a highly intrusive exotic species, to new habitats.

The flood, which many at the time considered the worst in U.S. history, was caused by above-average precipitation during the first half of 1993, followed by prolonged

Figure 14.7   Flood Management.

(a) A levee broke and water surged onto the flood plain in Merced, California in 2006. (Jack Bland/The Merced SunStar/©AP/Wide World Photos)

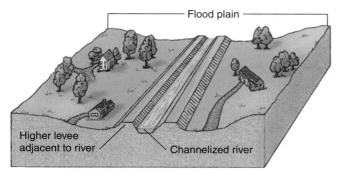

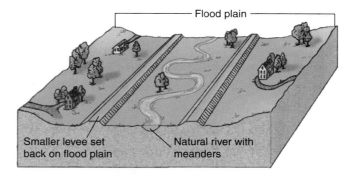

(b) To control flooding, many rivers are channelized (straightened and deepened), with high levees adjacent to the river. This flood control method is expensive and may or may not prevent floods.

(c) Scientists now recommend letting rivers meander naturally through much of their flood plains, with smaller levees set back some distance from the rivers. The flood plains absorb much of the river's water, forming a buffer between the river and developed areas. When a flooding river spills over its banks, it creates a wetland that is an important wildlife habitat.

summer rain. Draining wetlands, building on flood plains, and constructing levees to hold back floodwaters exacerbated the damage.

For the past hundred years or so, people in the Midwest drained wetlands to produce farmland or land on which to build homes. The ability of wetlands to moderate floods was simply unrecognized. It was probably no coincidence that Missouri, Illinois, and Iowa, the three states damaged the most by the floods, had each drained and developed more than 85% of their original wetlands.

Hundreds of levees along the Mississippi and its tributaries hold floodwaters back from the flood plain. Although levees may save lives and property where they are built, they cause floodwaters to surge, damaging less protected farms and towns downstream.

The 1993 floods rekindled an old debate: Should the government rebuild damaged levees, or should it help relocate people away from the flood plain and restore the land to a more natural state? Valmeyer and Hartsburg, two towns that were flooded, represent the two different strategies. The 900 residents of Valmeyer, Illinois, used federal funds to relocate to a nearby hill. The town's old site on the flood plain became a park and wetland. In contrast, the 131 residents of Hartsburg, Missouri, rebuilt and repaired homes in their original location on the flood plain. Hartsburg remains vulnerable to flooding should its levees fail again.

The U.S. government reassessed the national flood policy following the 1993 floods. The Flood Plain Management Task Force issued a 1994 report concluding that people should use flood plains more wisely and rely less on levees to control flooding. Some flood plains should be restored to their natural

Figure 14.8   Floods of 1993. Nine states experienced flooding (green area) during the summer of 1993.

Figure 14.9   Agricultural use of water.
Center-pivot irrigation produces massive green circles. Each circle is the result of a long irrigation pipe that extends along the radius from the circle's center to its edge and slowly rotates, spraying the crop. Photographed in Nebraska. (Mark Wagner, London)

aquifer depletion: The removal of groundwater more rapidly than it can be recharged by precipitation or melting snow.

condition, and the towns built on them moved to higher ground. The report found that the presence of more wetlands would not have prevented flooding of this magnitude because the wetlands would have become saturated with water. The presence of more wetlands in the affected area would have indirectly reduced property damage by keeping people and their property off the flood plain. This observation is significant because the nation's flood plains are being developed in many places.

Meanwhile, flooding continues along the Mississippi River and its tributaries as a result of seasonal snowmelt and rainfall. In 2001, upper Mississippi River communities recorded the second highest water levels ever in floods in parts of Wisconsin, Iowa, and Minnesota, and lower Mississippi River communities in Texas and Louisiana suffered enormous losses from hurricane-related damage in 2005. ■

Too Little Water

Arid lands, or deserts, are fragile ecosystems in which plant growth is limited by lack of precipitation. **Semiarid lands** receive more precipitation than deserts but are subject to frequent and prolonged droughts. For example, a multi-year drought that began in 2000 was still affecting the arid and semiarid lands of western United States as we went to press. No one knows how long the drought will last, but researchers have documented several Arizona droughts in the past few hundred years that each lasted 18 years or longer.

Irrigation increases the agricultural productivity of arid and semiarid lands, and has become increasingly important worldwide in efforts to produce enough food for burgeoning populations (**Figure 14.9**). Since 1955 the amount of irrigated land has more than tripled; Asia has more agricultural land under irrigation than do other continents, with China, India, and Pakistan accounting for most of it. Water use for irrigation will probably continue to increase in the 21st century, particularly in Asia, but at a slower rate than in the last half of the 20th century.

Population growth in arid and semiarid regions intensifies water shortage. More people need more food, so additional water resources are diverted for irrigation. Also, the immediate need for food prompts people to remove natural plant cover to grow crops on marginal lands subject to frequent drought and subsequent crop losses. Livestock overgraze the small amount of plant cover in natural pastures. As a result of the lack of plants, the soil cannot absorb the water as well when the rains do come, and runoff is greater. Because the precipitation does not replenish the soil, crop productivity is poor and the people are forced to cultivate food crops on additional marginal land.

Removing too much fresh water from a river or lake can have disastrous consequences in local ecosystems. Humans can remove perhaps 30% of a river's flow without greatly affecting the natural ecosystem. In some places considerably more is withdrawn for human use. In the arid American Southwest, it is not unusual to remove 70% or more of surface water.

When surface water is overdrawn, wetlands dry up. Natural wetlands play many roles, such as serving as a breeding ground for many species of birds and other animals. Estuaries, where rivers empty into seawater, become saltier when surface waters are overdrawn, and this change in salinity reduces the productivity associated with estuaries.

Aquifer depletion from excessive removal of groundwater lowers the water table. Prolonged aquifer depletion drains an aquifer dry, effectively eliminating it as a water resource. In addition, aquifer depletion from porous sediments causes **subsidence,** or sinking, of the land above it. Some areas of the San Joaquin Valley in California have sunk almost 10 m (33 ft) in the past 50 years.

The limestone bedrock of Florida erodes as groundwater moves through it, sometimes causing a **sinkhole**, a large surface cavity or depression where an underground

Figure 14.10   Saltwater intrusion.

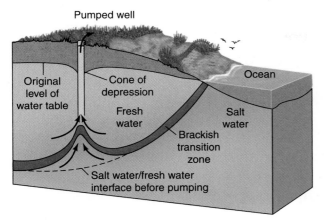

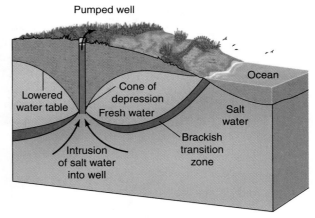

(a) Normally, fresh groundwater overlies salty groundwater. The well supplies fresh water as long as pumping is not excessive.

(b) However, the removal of large amounts of fresh groundwater causes the brackish transition zone to migrate. The well draws up salty groundwater unfit to drink.

cave roof has collapsed. Sinkholes occur more frequently when droughts or excessive pumping of water cause a lowering of the water table.

Saltwater intrusion occurs along coastal areas when groundwater is depleted faster than it recharges (**Figure 14.10**). Well water in such areas eventually becomes too salty for human consumption or other uses. Once it occurs, saltwater intrusion is difficult to reverse.

saltwater intrusion: The movement of seawater into a freshwater aquifer located near the coast; caused by aquifer depletion. Saltwater intrusion is also occurring in low-lying parts of the world due to sea level rise.

REVIEW

1. Which human activity is responsible for almost 70% of global water consumption?

2. How has development along the upper Mississippi River exacerbated property damage during periods of flooding?

3. What are some of the problems associated with overdrawing surface water? with aquifer depletion?

 ## Water Problems in the United States and Canada

LEARNING OBJECTIVES

- Relate the background for each of the following U.S. water problems: Mono Lake, the Colorado River Basin, Delaware, and the Ogallala Aquifer.
- Define *reclaimed water*.

Compared with many countries, the United States has a plentiful supply of fresh water. Despite the overall abundance of fresh water in the United States and Canada, many areas have severe water shortages because of geographic and seasonal variations. (**Figure 14.11** shows the average annual precipitation in North America.)

Surface Water

The increased use of U.S. surface water for agriculture, industry, and personal consumption since the 1960s has caused many water problems. Some U.S. regions that have grown in population during this period include the Southwest, Delaware, and Florida. If water consumption in these and other areas continues to increase, the availability of surface waters could become a serious regional problem, even in places that have never experienced water shortages.

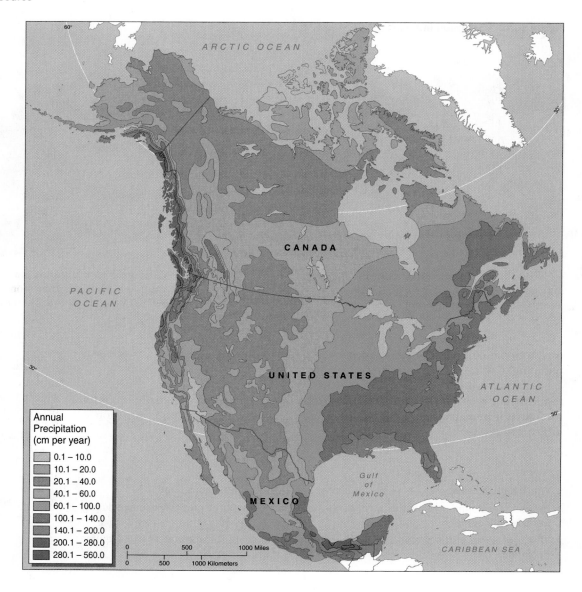

Figure 14.11 Average annual precipitation in North America. Note the arid and semiarid regions of the western United States.

Annual Precipitation (cm per year)

	0.1 – 10.0
	10.1 – 20.0
	20.1 – 40.0
	40.1 – 60.0
	60.1 – 100.0
	100.1 – 140.0
	140.1 – 200.0
	200.1 – 280.0
	280.1 – 560.0

Water problems are particularly severe in the American West and Southwest. Much of this large region is arid or semiarid and receives less than 50 cm (about 20 in.) of precipitation annually. Historically, water in the West was used primarily for irrigation, but municipal, commercial, and industrial uses now compete heavily for available water. Much of the water used in the West and Southwest originates as snow in the Rocky Mountains and the Sierra Nevada Range; climate change (see Chapter 21) is expected to lead to reduced snowfall, and thus less total water available for a growing population.

The development of new sources of water continues to meet expanding water needs in the West and Southwest. Water is diverted from distant sources and transported via **aqueducts** (large conduits) (**Figure 14.12**). As long ago as 1913, Los Angeles started bringing in water from the Owens Valley, an area of California 400 km (250 mi) north, along the east side of the Sierra Nevada. Dams and water-holding basins were established to ensure year-round supply. Now, however, the closest, most practical water sources are already tapped, and experiences such as that at Mono Lake suggest that removing water can dramatically damage existing ecosystems.

Mono Lake Removing too much surface water has serious environmental repercussions. Mono Lake, a salty lake in eastern California, is a striking example of this practice. Rivers and streams that are largely formed from snowmelt in the Sierra Nevada

Range replenish Mono Lake. Evaporation provides the only natural outflow from the lake. Over time, Mono Lake is becoming saltier as rivers deposit dissolved salts (recall fresh water contains some salt) and as water, but not salt, is removed by evaporation.

Beginning in 1941, much of the surface water that fed Mono Lake was diverted to Los Angeles, 442 km (275 mi) away. This change in water flow led to changes throughout the Mono Lake ecosystem. As the water level dropped (about 14 m or 46 ft), increased salinity adversely affected brine shrimp and alkali fly populations. This affected more than 80 species of water birds that feed on the shrimp and flies. Dust storms from the exposed lakebed began to pose a health hazard and violations of federal air pollution standards.

A court order halted water diversions from Mono Lake in 1989, and in 1994 the state of California worked out an agreement on Mono Lake water rights between the Los Angeles water authority and environmental groups. Less water will be diverted to Los Angeles, and Mono Lake will be allowed to return to about 72% of its original volume (by around 2015). The National Audubon Society expects hundreds of thousands of migratory and nesting birds to return to the lake's shores to nest.

The city of Los Angeles is using state funds to develop water conservation and **reclaimed water** projects to replace water supplies from Mono Lake. By 2015 California's reclaimed water projects will produce enough water to make up for the water lost from Mono Lake.

The Colorado River Basin   One of the most serious water supply problems in the United States is in the Colorado River Basin. The river's headwaters are formed from snowmelt in Colorado, Utah, and Wyoming, and major tributaries—collectively called the upper Colorado—extend throughout these states. The lower Colorado River runs through part of Arizona and then along the border between Arizona and both Nevada and California before crossing into Mexico and emptying into the Gulf of California.

The Colorado River provides water for 27 million people, including the cities of Denver, Las Vegas, Salt Lake City, Albuquerque, Phoenix, Los Angeles, and San Diego. It irrigates 3.5 million acres of fruit, vegetable, and field crops worth $1.5 billion per year. The Colorado River has 49 dams, 11 of which produce electricity by hydropower. More than 30 Native American tribes live along the Colorado River and claim rights to some of its water. The river provides $1.25 billion per year in revenues from almost 30 million people who use it for recreation.

The most important of all the treaties regulating use of Colorado River water is the 1922 **Colorado River Compact.** It stipulates an annual allotment of 7.5 million **acre-feet** of water to the lower Colorado (California, Nevada, Arizona, and New Mexico) and the remainder to the upper Colorado (Colorado, Utah, and Wyoming). However, the Colorado River Compact overestimated the average annual flow of the Colorado River, which at the time was thought to be 15 million acre-feet. This over-allocation was enshrined in the multistate agreement.

Traditionally, states in the upper Colorado region and parts of the lower region appropriated little of their water entitlement because they had few people and little development. California used more than its allotment because the water was available. As cities like Denver and Las Vegas have grown, conflicts between the states have increased.

Mexico receives a share of the Colorado stipulated by a 1944 treaty. Consequently, the Colorado River water is often completely consumed before it can reach the Pacific Ocean, causing serious problems for the ecosystem and inhabitants of the Colorado River Delta (**Figure 14.13**). To compound the problem, as more and more water is used, the lower Colorado becomes increasingly salty as it flows toward Mexico; in places, the Colorado River is saltier than the ocean.

Figure 14.12   An aqueduct near Fresno, California. (Dembinsky Photo Associates)

reclaimed water: Treated wastewater that is reused in some way, such as for irrigation, manufacturing processes that require water for cooling, wetland restoration, or groundwater recharge.

acre-foot: The amount of water needed to cover an acre of land one foot deep. An acre-foot is equal to 326,000 gallons, and is enough to supply eight people for one year.

Figure 14.13   Colorado River bed in San Luis Rio Colorado, Mexico. As a result of diversion for irrigation and other uses in the United States, the Colorado River usually dries up before reaching the Gulf of California in Mexico. (Dan Lamont, Seattle, Washington)

In 2003 California agreed to withdraw no more water from the Colorado River than the Colorado River Compact permits. To make up the difference, farmers in California's Imperial Valley agreed to sell some water they would normally use for irrigation to thirsty cities such as San Diego and Los Angeles. Formerly a desert, Imperial Valley is now 200,000 hectares (about 500,000 acres) of irrigated agricultural land. Farmers can use the money earned from water sales to update their water systems to use water more efficiently.

Delaware: A State Without Water?   While the water problems in the West and Southwest are severe, most of the United States and Canadian Provinces face water supply limitations of some sort. Delaware is particularly vulnerable, as there are no water sources that originate within its boundaries and no major lakes to serve almost 800,000 inhabitants. For example, Delaware shares the Christina River Basin, which provides water for both Newark and Wilmington, with parts of Pennsylvania and Maryland. Furthermore, to prevent adverse impacts on local ecosystems, there are restrictions on how much water can be withdrawn from the Christina River Basin. Thus, Delaware water utilities and government must carefully balance development, access for existing residents and businesses, and ecological protection. Recent research suggests that conservation-based pricing, in which consumers are rewarded for using less, can help water managers meet this complex set of needs.

conservation-based pricing: Water supply pricing structures that reward consumers for using less water. These often come in the form of low prices for water use up to some level, and stepped up prices as use increases.

Groundwater

Roughly half the population of the United States uses groundwater for drinking. Many large cities, including Tucson, Miami, San Antonio, and Memphis, have municipal well fields and depend entirely or almost entirely on groundwater for their drinking water. In addition, many rural homes have private wells for their water supply. Groundwater is also used for industry and agriculture. Approximately 40% of the water used for irrigation in the United States comes from groundwater. Increased groundwater consumption since the 1950s has diminished groundwater levels in many areas of heavy use across the United States. **Figure 14.14** shows areas where aquifer depletion is particularly critical. Groundwater has been overdrawn for irrigation in these arid and semiarid areas.

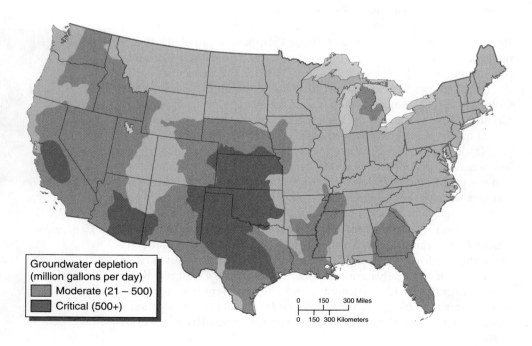

Figure 14.14   Aquifer depletion. Aquifer depletion is a widespread problem in the United States, particularly in the High Plains, California, and southern Arizona.

Groundwater depletion
(million gallons per day)
Moderate (21 – 500)
Critical (500+)

0   150   300 Miles

0   150   300 Kilometers

In certain coastal areas of Louisiana and Texas, the removal of too much groundwater has resulted in the intrusion of salt water from the Gulf of Mexico. Saltwater intrusion from the Pacific Ocean has occurred along parts of the California coast, along coastal areas of Puget Sound in Washington State, and in certain areas of Hawaii. Florida and many coastal regions in the Northeast and Mid-Atlantic states also have saltwater intrusion.

The Ogallala Aquifer   The High Plains cover 6% of U.S. land but produce more than 15% of its wheat, corn, sorghum, and cotton and almost 40% of its livestock. To achieve this productivity, it requires approximately 30% of the irrigation water used in the United States. Farmers on the High Plains rely on water from the **Ogallala Aquifer,** the largest groundwater deposit in the world (**Figure 14.15**).

In some areas farmers are drawing water from the Ogallala Aquifer as much as 40 times faster than nature replaces it, which has lowered the water table more than 30 m (100 ft) in places. Where this has occurred, higher pumping costs have made it too expensive to irrigate. The amount of irrigated Texas farmland has declined 11% in recent years. When farmers revert to dry-land farming in these semiarid regions, they risk economic and ecological ruin during droughts (see discussion of the Dust Bowl in Chapter 15). Areas where the Ogallala is most shallow have experienced recent population declines as farms fail during dry spells. Those areas where the Ogallala is deepest may not experience water shortages during the 21st century. Hydrologists (scientists who deal with water supplies) predict that groundwater will eventually drop in all areas of the Ogallala to a level uneconomical to pump. Their goal is to postpone that day through water conservation, including the use of water-saving irrigation systems.

REVIEW

1. How would you describe the water scarcity problem in the Colorado River Basin?

2. What is aquifer depletion? Why is the Ogallala Aquifer an excellent example of aquifer depletion?

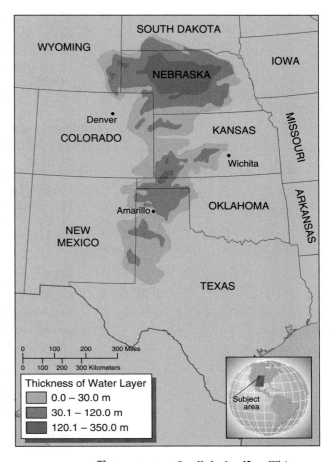

Figure 14.15   Ogallala Aquifer. This massive deposit of groundwater lies under eight midwestern states, with extensive portions in Texas, Kansas, and Nebraska. Water in the Ogallala Aquifer takes hundreds or even thousands of years to renew after it is withdrawn to grow crops and raise cattle.

Global Water Problems

LEARNING OBJECTIVES

- Define *stable runoff.*
- Explain the relationship between climate change and available water.
- Briefly describe each of the following international water problems: drinking water problems, population growth and water problems, the Rhine River Basin, the Aral Sea, and potential international conflicts over water rights.

Data on global water availability and use indicate that the amount of fresh water on the planet is adequate to meet human needs, even taking population growth into account. These data do not consider the *distribution* of water resources in relation to human populations. For example, citizens of Bahrain, a tiny island nation in the Persian Gulf, have no freshwater supply and must rely on ocean water desalinization.

Per capita water use varies greatly from country to country and from continent to continent, depending on the size of the human population and the available water supply. South America and Asia receive more than one-half of the world's renewable fresh water (by precipitation). Although South America has more available water per person than Asia does, it does not have the potential to support as many people as its water

stable runoff: The share of runoff from precipitation that can be depended on every month.

supply would suggest. Most of South America's precipitation falls in the Amazon River Basin, which has poor soil and is unsuitable for large-scale agriculture. In contrast, because most of the precipitation in Asia falls on land suitable for agriculture, the water supply supports more people.

Humans need an adequate supply of water year-round. In some places **stable runoff**, the portion of runoff from precipitation available throughout the year, is low even though total runoff is quite high. India has a wet season—June to September—during which 90% of its annual precipitation occurs. Most of the water that falls during India's monsoon quickly drains away into rivers and is unavailable during the rest of the year. Thus, India's stable runoff is low.

Variation in annual water supply is an important factor in certain areas of the world. The African Sahel region (see Figure 5.21) has wet years and dry years, and the lack of water during the dry years limits human endeavors during the wet years. Since the late 1960s, the Sahel has experienced an ongoing drought that has had a devastating impact on the people and wildlife living in the region.

Water and Climate Change

Climate change is expected to play an important role in future freshwater availability (see Chapter 21). Climate change driven by increases in carbon dioxide and other greenhouse gases impacts more than just global temperatures. It has broad, systemic impacts, including the amount, type, and distribution of precipitation. Precipitation is expected to increase in some areas while it drops in others.

Changes in rainfall may lead to abrupt changes in available surface water, since runoff is influenced by geologic factors such as soil permeability and biological factors such as amount of vegetation. One recent study suggests that a 10% decrease in rainfall in one part of Africa will lead to 17% reduction in drainage, while the same decrease in another area will lead to a 50% reduction in drainage. The study concludes that predicted variations in rainfall due to climate change will affect available surface water for one-fourth of the African continent by 2100.

Climate change will affect available fresh water in other ways as well. The type of precipitation is important: Earlier, we noted how reduced snowfall in the Rocky Mountains and Sierra Nevada will impact stable runoff in the American West and Southwest. Sea level rise—caused by thermal expansion and surface ice melt—has already caused saltwater intrusion into drinking water sources for certain low-lying island nations.

Drinking-Water Problems

Many inhabitants of developing countries have insufficient water to meet the most basic drinking and household needs. The water exists—only about 1% of the Earth's water would suffice for the entire human population. However, this water is not available to many people who have to spend large amounts of money or travel great distances to secure the water they need. Individual governments, the United Nations, the World Bank[1], Non-Governmental Organizations (NGOs), and civic organizations all sponsor water projects in developing countries.

The World Health Organization (WHO) estimates that 1.1 billion people lack access to safe drinking water and about 2.6 billion are without access to a satisfactory means of domestic wastewater and fecal waste disposal. These people risk disease because sewage or industrial wastes contaminate the water they consume. WHO estimates that 80% of human illness results from insufficient water supplies and poor water quality caused by lack of sanitation. Although many developing countries have installed or are installing public water systems, population increases tend to overwhelm efforts to improve the water supply.

[1] The World Bank makes loans to developing countries for projects it thinks will lessen poverty and encourage development.

Population Growth and Water Problems

As the world's population continues to increase, global water problems will become more serious. Asia has the world's largest available water resources—36% of the Earth's total. However, it also houses 60% of the world's people, and the people and the water are not always in the same places. In India, 20% of the world's population has access to only 4% of the world's fresh water, and approximately 8000 Indian villages have no local water. The water supply to some Indian cities—Madras, for example—is so severely depleted that water is rationed from a public tap. Extracting groundwater faster than its rate of recharge has caused water tables to fall 1 to 3 m (3 to 10 ft) per year in parts of Punjab and Haryana.

Water supplies are precarious in much of China, owing to population pressures. The water table across much of the North China Plain, with a population more than twice that of the United States, is falling 2 to 3 m (6 to 10 ft) per year. One-third of the wells in Beijing have gone dry. Much of the water in the Yellow River is diverted for irrigation, leaving downstream areas with little or no water. During the last 15 years of the 20th century, the Yellow River ran dry hundreds of kilometers inland before it reached the Yellow Sea.

Iraq faces a challenge similar to that of Delaware: Headwaters of both the Tigris and the Euphrates Rivers originate outside the country's borders. While current conflicts in Iraq overshadow the water issue, shortages in both quality and quantity of water will be an internal challenge. Water supply will continue to influence Iraq's relations with neighboring countries, especially those upstream.

Pakistan faced an extended drought during the 1990s and early 2000s. In 2001 the water shortages resulted in protests and riots between certain provinces over water use, particularly of the Indus River, the lifeline in Pakistan. The inability to produce enough food in agricultural regions has resulted in an increase in poverty.

Mexico is facing the most serious water shortages of any country in the Western Hemisphere. The main aquifer supplying Mexico City is dropping as much as 3.5 m (about 11 ft) per year. The water table is falling as much as 3 m per year in Guanajuato, an agricultural state in Mexico.

Sharing Water Resources among Countries

Surface water is often an international resource. Around 260 of the world's major watersheds are shared between at least two nations. International cooperation is required to manage rivers that cross international borders.

The Rhine River Basin   The drainage basin for the Rhine River in Europe is in five highly developed and densely populated countries—Switzerland, Germany, France, Luxembourg, and the Netherlands (**Figure 14.16**). Traditionally, Switzerland, Germany, and France used water from the Rhine for industrial purposes and then discharged polluted water back into the river. The Dutch then had to clean up the water so they could drink it. Today, these countries recognize that international cooperation is essential to conserve and protect the supply and quality of the Rhine River.

In 1950 the five countries formed the International Commission for Protection of the Rhine (ICPR) to deal with water issues relating to the Rhine River, but little was accomplished for several decades. River quality began to improve in the mid-1970s, largely in response to international reports on the river's poor condition. In 1986 a severe chemical spill in Switzerland dumped 30 tons of dyes, herbicides, fungicides, insecticides, and mercury into the river. The spill galvanized the ICPR, which initiated a 15-year Rhine Action Plan. It eliminated some major pollution sources, and the water in the Rhine River today is almost as pure as drinking water. Long-absent fishes have returned to the river, including Atlantic salmon, which returned in 1990 after a 30-year absence. The ICPR is currently working on bank restoration, flood control, and cleaning up remaining pollutants.

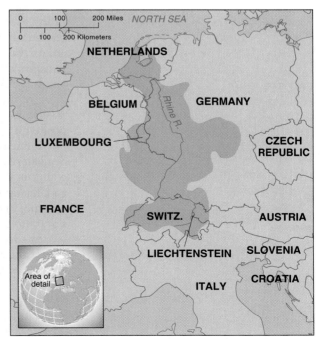

Figure 14.16   Rhine River Basin. The Rhine River drains five European countries—Switzerland, Germany, France, Luxembourg, and the Netherlands. (The green area represents the drainage basin.) Water management of such a river requires international cooperation.

(a)   (b)

Figure 14.17   **Aral Sea.** The satellite images show the Aral Sea in (a) 1976 and (b) 1997. As water was diverted for irrigation, the sea level subsided. It has since recovered considerably. (Courtesy of Worldsat International, Inc.)

The Aral Sea   The Aral Sea, which straddles Kazakhstan and Uzbekistan (both parts of the former Soviet Union), suffers from the same problem as Mono Lake in California. In the 1950s, the Soviet Union began diverting water from the Amu Darya and the Syr Darya, the two rivers that feed into the Aral Sea, to irrigate desert areas surrounding the lake. By the early 1980s, irrigation for growing cotton had diverted more than 95% of the Aral Sea's inflow.

Since 1960 the Aral Sea, once the world's fourth largest freshwater lake, has declined in area more than 50%. Its total volume is down 80%. Much of its biological diversity has disappeared—all 24 fish species originally found there are gone. The satellite photos in **Figure 14.17** demonstrate the shrinking of the Aral Sea.

About 35 million people live in the Aral Sea's watershed. Millions have developed health problems ranging from tuberculosis to severe anemia, and their death rate from respiratory illnesses is among the world's highest. Kidney disease and various cancers are on the rise. International health experts have begun to assess which medical problems are due to the Aral Sea's environmental problems. Toxic salt storms caused by winds that whip the salt on the receding shoreline into the air may be responsible for many of these chronic conditions. Since the 1950s, such storms have increased 60-fold. The wind carries salt hundreds of kilometers from the Aral Sea, and where it is deposited it reduces the productivity of the land.

Immediately following the breakup of the Soviet Union in 1991, plans to save the Aral Sea faltered as responsibility for its rescue shifted from Moscow to the five central Asian countries that share the Aral Basin: Uzbekistan, Kazakhstan, Kyrgyzstan, Turkmenistan, and Tajikistan. In 1994 the five nations established a fund to prevent the complete disappearance of the Aral Sea. The World Bank and the U.N. Environment Program approved a grant to the five countries to help address the environmental problems of the area.

At this time, it appears that the Aral Sea is recovering. The World Bank is sponsoring the **Syr Darya Control and Northern Aral Sea Project,** with goals of ecological restoration and commercial fishery recovery. The Northern Aral Sea experienced a 13% increase in surface area between 2003 and 2006. By 2007, salinity levels were reduced to half of their 1991 values. While recovery of the Southern Aral Sea is proceeding more slowly, salinity has decreased and water volume has increased during each of the past few years.

Potentially Volatile International Water Situations   Because so many watersheds and aquifers are shared between two or more countries, the 21st century may well see countries facing one another in armed conflict over water rights (**Figure 14.18**). Humans remove so much water from the Amu Darya, Ganges, Indus, Nile, Yellow, and Colorado Rivers that their channels run dry at least some parts of the year. Tensions are high along the Mekong River Basin, shared by Laos, Thailand, and Vietnam, and the Indus River Basin is shared between Pakistan and India. India and Bangladesh quarrel over the Ganges River. Slovakia and Hungary both depend on the Danube River. Developing cooperative international agreements on shared water resources is an urgent global issue.

An example of a troublesome international spot is the Jordan River, which supplies water to Israel, Jordan, the West Bank, and Gaza Strip. Water use is increasing because of both population growth and agricultural and economic activity in the region. According to a collaborative study among the U.S. National Academy of Sciences, the Israel Academy of Sciences and Humanities, the Palestine Academy for Sciences and Technology, and the Royal Scientific Society of Jordan, the future outlook for water supplies in the region is one of significant water stress. Participants urged their respective governments to cooperate in developing conservation measures such as reusing wastewater. Nonetheless, differential access between Israeli settlers on the West Bank, who use four to five times as much water as do neighboring Palestinians, may be a significant source of conflict.

Northeastern Africa has a serious water-use situation: the Nile River. Egypt uses most of the Nile's water (and has for millennia), even though 10 nations share the Nile River Basin. Ethiopia and Sudan are expanding their use of the Nile River's flow to meet the demands of their rapidly growing populations; these actions could imperil Egypt's freshwater supply at a time when its population is increasing. The United Nations engineered an international water-use agreement among the Nile River countries to help diffuse this potentially dangerous water situation, and the 10 nations in the region have formed the Nile Basin Association to review past agreements and develop future ones.

IN THE FUTURE, WARS WILL BE FOUGHT OVER WATER

Figure 14.18   Water as a major source of conflict in the future. Do you find this cartoon believable? Has it already begun? Explain your answer. (Chris Madden/Inkline Press)

REVIEW

1. Why is management of the Rhine River so complex?

2. Where is the Aral Sea? Why did it shrink in the 1950s to 1980s? What is its status now?

 ## Water Management

LEARNING OBJECTIVES

- Define *sustainable water use.*
- Contrast the benefits and drawbacks of dams and reservoirs, using the Columbia River to provide specific examples.
- Briefly describe two methods of desalinization.

People have always considered water to be different from other resources. Coal and gold are owned privately and sold as free-market goods, but people variously view water as a public resource or as a private resource. Historically, both in much of the United States and in many other countries, water rights were bound with land ownership. As more and more users compete for the same water, state or provincial governments increasingly make allocation decisions. Some countries have separated land and water ownership so that water rights are sold separately.

Because rivers usually flow through more than one governmental jurisdiction, all affected parties must develop agreements about the management of a river or other shared water resource. Such interstate or transboundary cooperation permits comprehensive rather than piecemeal management. In addition, these arrangements divide the water fairly among the jurisdictions, which then apportion their respective shares to individual users according to an established set of priorities.

Figure 14.19   Grand Coulee Dam on the Columbia River. Shown are the dam and part of its reservoir, the Franklin D. Roosevelt Lake. Dams help to regulate water supply, storing water produced in times when precipitation is plentiful for use during dry periods. The many beneficial uses of dams include electricity generation and flood control, but they destroy the natural river habitat and are expensive to build. (Courtesy U. S. Dept. of Energy)

sustainable water use: The use of water resources in a fashion that does not harm the essential functions of the hydrologic cycle or the ecosystems on which present and future humans depend.

Groundwater management is more complicated, in part because the extent of local groundwater supplies is often not known. Groundwater management includes issuing permits to drill wells, limiting the number of wells in a given area, and restricting the amount of water pumped from each well.

The price of water varies, depending on how it is used. Historically, domestic use is most expensive and agricultural use is least expensive. Consumers rarely pay directly for the entire cost of water, which includes its transportation, storage, and treatment. State and federal governments heavily subsidize water costs, so that we pay for some of the cost of water indirectly, through taxes. Increasingly, state and local governments adjust the price of water as a mechanism to help ensure an adequate supply of water. Raising the price of water to reflect the actual cost generally promotes a more efficient use of water.

Providing a Sustainable Water Supply

The main goal of water management is to provide a sustainable supply of high-quality water. Sustainable water use means humans use water resources carefully, so that water is available for future generations and for existing non-human needs.

Dams and Reservoirs   Dams ensure a year-round supply of water in areas with seasonal precipitation or snowmelt. Dams confine water in reservoirs, from which the flow is regulated (**Figure 14.19**). Dams have other benefits, particularly the generation of electricity (recall the discussion of dams and hydroelectric power in Chapter 13). Many people, however, feel that the drawbacks of dams, including the environmental costs, far outweigh any benefits they provide.

In recent years scientists have come to understand how dams alter river ecosystems, both above and below the dam. Heavy deposition of sediment occurs in the reservoir behind the dam, and the water that passes over the dam does not have its normal sediment load. As a result, the river floor downstream of the dam is scoured, producing a deep-cut channel that is a poor habitat for aquatic organisms.

The Glen Canyon Dam, built in 1963, has profoundly affected the Colorado River in the Grand Canyon National Park. Prior to the dam's construction, powerful spring floods carried sediment that formed beaches and sandbars, providing nesting sites for birds and shallow waters for breeding fishes. The regulated flow of water since the Glen Canyon Dam was constructed changed the ecosystem, to the detriment of some of the Grand Canyon's wildlife. To rectify some of the changes to the river, the Bureau of Reclamation has flooded the Grand Canyon several times, beginning in 1996. Although these floods are small in comparison to some of the natural floods of the past, the sediment-laden floodwater rebuilds beaches and sandbars that are continually eroding (**Figure 14.20**).

The Columbia River.   The Columbia River, the fourth largest river in North America, illustrates the impact of dams on natural fish communities. Its watershed, which covers an area the size of France, includes seven states and two Canadian provinces. Such a large river system has multiple uses. There are more than 100 dams within the Columbia River system, 19 of which are major generators of inexpensive hydroelectric power. The Columbia River system supplies municipal and industrial water to several major urban areas, including Boise, Portland, Seattle, and Spokane. More than 1.2 million hectares (3 million acres) of agricultural land are irrigated with the Columbia's waters, and commercial ships navigate 800 km (500 mi) of the river.

As is often the case in natural resource management, a particular use of the Columbia River system may have a negative impact on other uses. The dam impoundments along the Columbia River that generate electricity and control floods have adversely affected fish populations, particularly salmon. Salmon are migratory fish that spawn in the upper reaches of freshwater rivers and streams. The young offspring, called smolts, migrate to the ocean, where they spend most of their adult lives before returning to their place of birth to reproduce and die.

Before dam construction, natural floods carried sediment that built and maintained sandbars

(a)

During regulated flood, river water is turbid with sediment

Canyon floor flooded with sediment-laden water

(c)

After dam construction, sandbars eroded and sand accumulated on river bottom

(b)

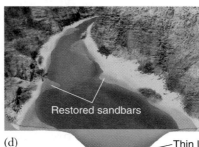

After regulated flood, sandbars are partially restored

Restored sandbars

(d)

Figure 14.20   How periodic flooding of the Grand Canyon helps restore the riverbanks.

The salmon population in the Columbia River system is only a fraction of what it was before the watershed was developed. The many dams that impede salmon migrations are widely considered the most significant cause of this change. Logging around salmon spawning streams contributes sediment pollution that degrades the salmon's habitat. Because streams in logged areas are no longer shaded, the water temperature becomes too warm for the developing salmon eggs.

Several projects to rebuild salmon populations were implemented, but none has been particularly effective. Many of the dams had already installed fish ladders to allow some of the adult salmon to bypass the dams and continue their upstream migration (**Figure 14.21**).

To increase the number of salmon, several hatcheries were built upstream of the dams, in tributaries of the Columbia. Young fish produced at these hatcheries are released to imprint the "smell" of the streams, enabling them to return there as adults to reproduce. Unfortunately, this effort has not reestablished natural spawning, in part because hatchery fish appear genetically incompatible with wild populations.

To protect some of the remaining natural salmon habitats, several streams in the Columbia River system are off-limits for dam development. Underwater screens and passages are being installed at dams to steer smolts away from turbine blades. Trucks and barges transport some of the young fish around dams, while others swim safely over the dam because the electrical generators are periodically turned off to allow passage.

Conservationists prefer a "water-budget" approach to the trucking and barging done for almost 20 years without much success. (Studies show that less than 0.5% of barged salmon return to their spawning grounds.) Extra water is "budgeted" for fish, and released from the dams to simulate spring snowmelt and help wash smolts downstream. However, other interest groups are opposed to increasing water flow for salmon. Farmers want to save the plentiful snowmelt water for irrigation during summer months. The hydropower industry wants to save the water to generate electricity during the winter months, their time of peak demand.

In 1999 the National Marine Fisheries Service (NMFS) extended the protection of the Endangered Species Act (ESA) to all species of salmon and steelhead trout found in Northwest rivers from the Canadian border to northern California, and eastward to Montana. This action, the largest implementation of the ESA ever, affects public and private lands, and includes both rural areas and major cities such as Portland, Oregon.

Figure 14.21   Fish ladder. This ladder is located at the Bonneville Dam along the Oregon side of the Columbia River. Fish ladders help migratory fishes to bypass dams in their migration upstream. (Raymond G. Barnes)

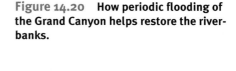

The most controversial plan of the ESA as it applies to salmon and steelhead trout is to tear down four dams on the lower Snake River, a tributary of the Columbia River (see the discussion of the Snake River in Chapter 7). Biologists see this restoration of natural flow as the single most effective way to restore salmon populations. Wild salmon from the Snake River have declined almost 90% since the dams were installed during the 1960s and 1970s. Many biologists, environmentalists, and Native American tribes endorse the proposal, whereas hydroelectric companies and area farmers oppose it.

Clouding the dam removal issue further is the fact that dam removal can have mixed effects. When a dam is dismantled, the water in the reservoir is released, carrying sediment that had been trapped behind the dam. The sediment can replenish stream banks, but it can also contain pollutants that suddenly head downstream. This is another example of how complex environmental systems are, and how making a change in one part of the system can affect many other parts.

The Missouri River. The Missouri River flows from Montana to St. Louis, Missouri, where it joins the Mississippi River and flows on to the Gulf of Mexico (see Figure 14.8). The Army Corps of Engineers has constructed six dams on the Missouri River; these dams provide both benefits and problems for people living along the river. Since 1987 the Corps has increased water flow over the northern dams to protect downstream navigation, including the shipping of two million tons of cargo each year. People who live downstream count on the river water for irrigation, electrical power, and individual water consumption.

The claims on and preferences for Missouri River water are many, complex, and often contested. The area along the northern Missouri River depends on the river for its multimillion-dollar fishing and tourism industry. Farmers want additional dikes and levees to protect their crops on the flood plains, whereas environmentalists want the river restored to its natural state as much as possible. Native Americans with claims to water rights along portions of the river want to use the water in a variety of ways, from generating hydroelectric power to irrigating cropland.

The Missouri River Basin Association, a coalition of eight river-basin states and two dozen Native American tribes, has the unenviable job of working with the Corps to meet the demands of the various competing interest groups as they decide the river's future. The association recognizes that the river does not belong to any single group, and that it must prioritize uses of the river in a way that will at least partially satisfy the environmental groups, farmers, hydroelectric producers, Native Americans, and fishing and tourism interests.

Water Diversion Projects   One way to increase the natural supply of water to a particular area is to divert water from areas where it is in plentiful supply by pumping water through a system of aqueducts. Much of Southern California receives its water supply via aqueducts from Northern California (**Figure 14.22**). Water from the Colorado River is also diverted into Southern California by aqueducts.

Large-scale water diversion projects are controversial and expensive. The Central Arizona Project, which pumps water 540 km (336 mi) from the Colorado River to Phoenix and Tucson, was completed at a cost of almost $4 billion. As you learned earlier, a river or other body of water is damaged when a major portion of its water is diverted. Pollutants that would have been diluted in the normal river flow reach higher concentrations when much of the flow is removed. Fishes and other organisms may decline in number and diversity. Although no one denies people must have water, opponents of water diversion projects contend that serious water conservation efforts would eliminate the need for additional large-scale water diversion.

Figure 14.22   Water Diversion in Southern California. Largely desert, Southern California relies on water diversion for the water needs of its millions of inhabitants. The California Water Project includes 1042 km (648 mi) of aqueducts to transfer large quantities of water to Southern California. This map also shows some of the main reservoirs of the California Water Project.

Desalinization Seawater and salty groundwater are made fit to drink through **desalinization** (or **desalination**). There are two major approaches to desalinization: distillation systems and membrane/filtration systems. In **distillation,** salt water is heated until the water evaporates, leaving behind a crust of salt. The water vapor is then condensed to produce fresh water. **Reverse osmosis,** the most common type of membrane/filtration system, involves forcing salt water through a membrane permeable to water but not to salt. Reverse osmosis removes about 97% of the salt from water.

Desalinization is expensive because it requires a large energy input, although recent advances in reverse osmosis technology have increased its efficiency so that it requires much less energy than distillation. Other expenses involved in desalinization projects include the cost of transporting the desalinized water from the site of production to where it is used. During the early 2000s, the cost of desalinization, excluding transport costs, varied from $0.25 to $3.00 per m³, depending on what technology was used and how salty the water was to begin with. Removing salt from seawater costs three to five times more than removing salt from brackish water. The disposal of salt produced by desalinization is also a concern, since dumping it back into the ocean near productive coastal areas could harm marine organisms. In addition, ocean water desalinization requires the intake of huge amounts of water. This water can contain large numbers of microorganisms and fish, leading to severe localized disruption.

In 2006, desalinization was being done in 130 countries, with a total global capacity of 40 million m³ per day—14 million more than in 2000. Because other freshwater sources are scarce, desalinization is a huge industry in North Africa and the Middle East. In Saudi Arabia, it accounts for 70% of drinking water.

REVIEW
1. What salmon restoration projects are underway in the Columbia River?
2. What is reverse osmosis?

desalinization: The removal of salt from ocean or brackish (somewhat salty) water.

microirrigation: A type of irrigation that conserves water by piping it to crops through sealed systems.

Water Conservation

LEARNING OBJECTIVE

• Give examples of water conservation by agriculture (including microirrigation), industry, and individual homes and buildings (including gray water).

You have seen how population and economic growth have placed an increased demand on our water supply. Today there is more competition than ever among water users with different priorities, and water conservation measures are necessary to guarantee sufficient water supplies. Most water users use more water than they really need, whether it is for agricultural, industrial, or direct personal consumption. With incentives, these users will lower their rates of water consumption. Many studies have shown that programs combining increased prices for water, improved technology, and effective educational tools motivate consumers to conserve water.

Reducing Agricultural Water Waste

Irrigation generally makes inefficient use of water. Traditional irrigation methods practiced for more than 5000 years involve flooding the land or diverting water to fields through open channels. Plants absorb about 40% of the water applied to the soil by flood irrigation; the rest usually evaporates into the atmosphere or seeps into the ground.

One of the most important innovations in agricultural water conservation is **microirrigation**, also called **drip** or **trickle irrigation,** in which pipes with tiny holes bored in them convey water directly to individual plants (**Figure 14.23**). Microirrigation substantially reduces the water needed to irrigate crops, usually by 40 to 60% compared to center pivot irrigation or flood irrigation, and it also reduces the amount of salt left in the soil by irrigation water.

Figure 14.23 Microirrigation. Cutaway view of soil shows a small tube at the root line. Tiny holes in the tube deliver a precise amount of water directly to roots, eliminating much of the waste associated with traditional methods of irrigation. Photographed in Fresno, California. (Courtesy U.S. Dept. of Agriculture)

Another important water-saving measure in irrigation is the use of lasers to level fields, allowing a more even water distribution. As a laser beam sweeps across a field, a field grader receives the beam and scrapes the soil, leveling it. Because farmers must use extra water to ensure that plants growing at higher elevations of a field receive enough, laser leveling of a field reduces the water required for irrigation.

The use of sound water management principles in agriculture reduces water consumption. Traditionally, western farmers were allotted specific amounts of water at specific times, with a "use it or lose it" philosophy. This approach encourages waste. Instead, a field's water needs should be carefully monitored (by measuring rainfall and soil moisture) to determine when to irrigate and how much water to apply. These water management strategies effectively reduce overall water consumption.

Although advances in irrigation technology are improving the efficiency of water use, many challenges remain. For one thing, sophisticated irrigation techniques are expensive. Few farmers in highly developed countries, let alone subsistence farmers in developing nations, can afford to install them. Another challenge is that irrigation needs to make much greater use of recycled wastewater instead of fresh water that could be used for direct human consumption.

Reducing Water Waste in Industry

gray water: Water that has already been used for a relatively non-polluting purpose, such as showers, dishwashing, and laundry; gray water is not potable, but can be reused for toilets, plants, or car washing.

Electric power generators and many industries require water (recall from Chapters 11 to 13 that power plants heat water to form steam, which turns the turbines). In the United States, five major industries—chemical products, paper and pulp, petroleum and coal, primary metals, and food processing—consume almost 90% of industrial water. Water use by these industries does not include water used for cooling purposes.

Stricter pollution control laws provide some incentive for industries to conserve water. Industries usually recapture, purify, and reuse water to reduce their water use and their water treatment costs. The National Steel Corporation plant in Granite City, Illinois, for example, recycles approximately two-thirds of the water it uses daily. The Ghirardelli Chocolate Company in San Leandro, California, installed a recycling system to cool large tanks of its chocolate. IBM plants and laboratories worldwide have adopted many water efficiency projects, resulting in substantial water savings.

Many other companies, especially those located in water-scarce states such as California, are adopting techniques to recycle water. International companies also have to consider water issues in the countries where they establish plants. In 2004, for example, city officials in the drought-plagued state of Kerala, India, ordered PepsiCo Inc. to close its bottling plant for almost a month because water was so scarce. Pepsi officials worked with the local government to dig a deeper well to provide water to the area, including the plant.

It is likely that water scarcity, in addition to more stringent pollution control requirements, will encourage further industrial recycling. The potential for industries to conserve water by recycling is enormous.

Reducing Municipal Water Waste

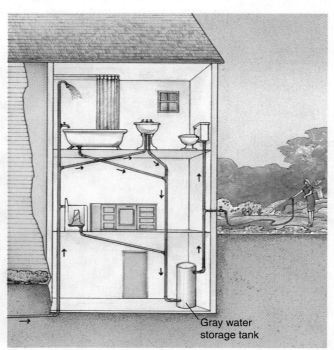

Figure 14.24 Recycling water. Individual homes and buildings can be modified to collect and store "gray water"—water already used in sinks, showers, washing machines, and dishwashers. This "gray water" is used when clean water is not required—for example, in flushing toilets, washing the car, and sprinkling the lawn.

Gray water storage tank

Like industries, regions and cities recycle or reuse water to reduce consumption. Homes and buildings can be modified to match water quality to water use, for example, by collecting and storing gray water[2]. Gray water is water that was already used in sinks, showers, washing machines, and dishwashers. Gray water is recycled to flush toilets, wash the car, or sprinkle the lawn (**Figure 14.24**).

In contrast to water recycling, wastewater reuse occurs when water is collected and treated before being redistributed for use. Israel probably has the world's most highly de-

[2] Permits to install gray water systems vary from state to state. Arizona and other states with severe water shortages are more flexible about allowing gray water systems.

veloped system of treating and reusing municipal wastewater. Israel does this out of necessity because all of its possible freshwater sources are already tapped. Reclaimed water is used for irrigation, leaving higher-quality fresh water for cities. Used water contains pollutants, but most of these are nutrients from treated sewage and are beneficial to crops.

In addition to recycling and reuse, cities decrease water consumption through other conservation measures. These include consumer education of both children and adults, use of water-saving household fixtures, and development of economic incentives to save water (see "You Can Make a Difference: Conserving Water at Home"). These measures successfully pull cities through dry spells; they are effective because individuals are willing to conserve for the common good during water crisis periods.

Increasingly, cities are examining ways to encourage individual water conservation methods all the time. The installation of water meters in residences in Boulder and New York City reduced water consumption by about one-third. Before the installation, homeowners were charged a flat fee, regardless of their water use. For many apartment dwellers, water use is included in the rent; charging each apartment for its water use provides the incentive to use water more efficiently. In addition to installing water meters, a city might encourage water conservation by offering a rebate to any homeowner who installs a conserving device such as a water-saving toilet. Building codes that specify installation of water conservation fixtures (low-flush toilets and water-saving faucets and showerheads) help reduce municipal water consumption.

Some cities are investing in systems to collect and store rainwater. A collection system from roofs of buildings, for example, can provide a substantial amount of water that would normally drain into a city's sewage system. Sun Valley Park, California, is currently developing a project to collect rainwater that floods certain streets during heavy downpours, clean it, and inject it into the Los Angeles aquifer for use at a later time.

Increasing the price of water to reflect its true cost promotes water conservation. As water prices rise, people quickly learn to conserve water. For example, charging more for water during dry periods encourages individuals to conserve water. Although the average cost of water to consumers rose during the 1990s and early 2000s, many U.S. cities still did not charge consumers what the water actually cost them.

The water supply systems (pipes and water mains) in many urban areas are old and leaky. In fact, the average U.S. city loses about one-fourth of its piped water to leaks. Repairing water mains and pipes would improve water accountability, the efficiency of water use. As part of its aggressive water efficiency program, which began in the late 1980s, the Massachusetts Water Resources Authority actively detects and repairs leaks for the greater Boston area. The resulting water conservation has saved taxpayers millions of dollars.

REVIEW

1. What is microirrigation? How does it conserve water?

2. How do cities promote water conservation? (Give at least three examples.)

E N V I R O N E W S

Water and Campus Sustainability

The Sustainability Office Web site at the University of British Columbia (UBC) maintains two sets of running sums: resource usage and resource savings. UBC's Vancouver campus was the first Canadian university to adopt a sustainability program, and despite being located in the rainy Pacific Northwest region of North America, takes sustainable water use seriously. One recent major innovation is a closed-loop storm drain project in a housing unit in part of the planned campus area. Water that drains from this part of campus is used in irrigation and creeks, then diverted to a cistern (a large underground vessel), from which it can be reused. Another innovation is the installation of permeable surfaced parking lots, which allow water to seep into the ground rather than running off and causing erosion. These projects reduce both the campus' demand for municipal water and erosion associated with surface runoff from campus. They also fit UBC's broader, systems-based sustainability initiatives to reduce energy and water use while improving the quality of campus life.

YOU CAN MAKE A DIFFERENCE

Conserving Water at Home

The average U.S. citizen uses 295 L (78 gal) of water per day at home on indoor uses (see figure). Many appliances, such as dishwashers, garbage disposals, and washing machines, need water. The growth of the suburbs, with their expansive landscaping that requires watering, is also responsible for increased water use.

As a water user, you have a responsibility to use water carefully and wisely. The cumulative effect of many people practicing personal water conservation measures has a significant impact on overall water consumption. You can adopt these measures yourself. The bathroom is a good place to start because most of the water used in an average home is for showers, baths, and flushing toilets.

1. Install water-saving showerheads and faucets to cut down significantly on water flow. Low-flow showerheads, for example, reduce water flow from 5 to 9 gal per minute to 2.5 gal per minute. Replacing one old showerhead brings a home $30 to $50 each year in water and energy savings. You can also save water by replacing washers on leaky faucets.

2. Install a low-flush toilet or use a water displacement device in the tank of a conventional toilet. Low-flush toilets require only 2 gal or less per flush, compared with 5 to 9 gal for conventional toilets. To save water with a conventional toilet, fill an empty plastic laundry bottle with water and place it in the tank to displace some of the water. Don't put the bottle where it will interfere with the flushing mechanism; don't add bricks to the tank, because they dissolve over time and can cause costly plumbing repairs.

3. An important way to conserve water at home is to fix leaky fixtures. For example, a toilet with a silent leak could waste 30 to 50 gal of water each day.

4. If you are in the market for a washing machine, high-efficiency washing machines require less water than traditional models, and at the same time require less energy and less detergent. Wash full loads of clothes, or adjust the water level to match the size of the load.

5. Modify your personal habits to conserve water. Avoid leaving the faucet running. Allowing the faucet to run while shaving consumes an average of 20 gal of water; you will use only 1 gal if you simply fill the basin with water or run the water only to rinse your razor. You may save as much as 10 gal of water a day by wetting your toothbrush and then turning off the tap while you brush your teeth, as opposed to running the water during the entire process. Also, most of us take longer showers than are needed. Time yourself the next time you take a shower, and if it is 10 minutes or longer, work on reducing your shower time.

6. Surprisingly, you will save water by using a dishwasher, which typically consumes about 12 gal per run, instead of washing dishes by hand with the tap running—but only if you run the dishwasher with a full load of dishes. That 12 gal of water is used regardless of whether the dishwasher is full or half-empty.

Remember that wasting water costs you money. Conserving water at home reduces your water bill and heating bill: If you are using less hot water, you are using less energy to heat that water.

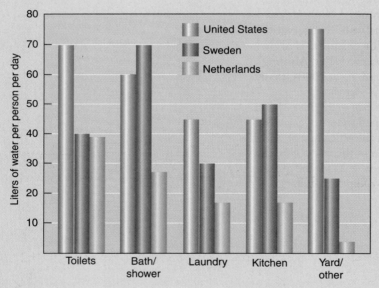

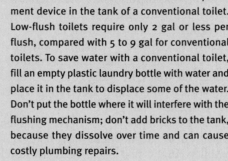

Residential water use in three highly developed countries. Data, in liters of water per person per day, are for the United States, Sweden, and the Netherlands.

REVIEW OF LEARNING OBJECTIVES WITH KEY TERMS

• **Describe the structure of a water molecule and explain how hydrogen bonds form between adjacent water molecules.**

Many of the properties of water, such as its high heat capacity and high dissolving ability, are the result of its **polarity**—one end of the molecule has a positive charge and the other end has a negative charge. The negative end of one molecule is attracted to the positive end of another, forming a **hydrogen bond** between the molecules.

• **Describe surface water and groundwater, using the following terms in your descriptions: wetland, runoff, drainage basin, unconfined and confined aquifer, and water table.**

Surface water comes from precipitation that remains on the surface of the land and does not seep down through the soil. **Wetlands** are areas of land covered with water for at least part of the year. **Runoff** is the movement of fresh water from precipitation and snowmelt to rivers, lakes, wetlands, and, ultimately, the ocean. A **drainage basin** is a land area that delivers water into a stream or river system. **Groundwater** is the supply of fresh water under Earth's surface that is stored in underground aquifers. **Aquifers** are underground caverns and porous layers of sand, gravel, or rock in which groundwater is stored. **Unconfined aquifers** have porous layers of rock above them, whereas **confined aquifers** have impermeable layers of rock above them. The **water table** is the upper surface of the saturated zone of groundwater.

- **Describe the role of irrigation in world water consumption.**

The greatest user of water worldwide is agriculture, for irrigation. Irrigation accounts for about 70% of the world's total water consumption.

- **Define *flood plain* and explain how humans exacerbate flood damage, using the upper Mississippi River Basin as an example.**

A **flood plain** is the area bordering a river channel that has the potential to flood. Flood damage is exacerbated by the deforestation of hillsides and mountains and by the development of flood plains. Many experts consider the 1993 flood in nine midwestern states to be the worst flood in U.S. history. It was exacerbated by the development of the flood plain in the upper Mississippi River Basin.

- **Relate some of the problems caused by overdrawing surface water and aquifer depletion (including saltwater intrusion).**

When surface water is overdrawn, the organisms in freshwater ecosystems suffer; also, natural wetlands dry up and estuaries become saltier. **Aquifer depletion** is the removal of groundwater faster than it can be recharged by precipitation or melting snow. Aquifer depletion from porous rocks can cause **subsidence**, or sinking land. In some areas, **sinkholes** occur when droughts or excessive pumping of water cause a lowering of the water table. Aquifer depletion can cause **saltwater intrusion**, the movement of seawater into a freshwater aquifer located near the coast.

- **Relate the background for each of the following U.S. water problems: Mono Lake, the Colorado River Basin, Delaware, and the Ogallala Aquifer.**

Mono Lake had surface water diverted to Los Angeles, lowering its water level and increasing its salinity. In 1994 California decided that Mono Lake would be returned almost to its original level. The Colorado River Basin is overdiverted for human use. The lower Colorado supplies water to Tucson, Phoenix, San Diego, and Los Angeles. Population growth in Colorado, Utah, and Wyoming threatens the lower Colorado's water supply. Also, the water available for Mexico is insufficient. Delaware faces the awkward situation of having no water sources that originate within its borders. Aquifer depletion of the Ogallala Aquifer on the High Plains has lowered the water table in some places by more than 30 m. Water experts predict that groundwater in the Ogallala will eventually drop to a level uneconomical to pump. These cases demonstrate how balancing limited water supply with demands from existing users, new users, and ecosystems can be a challenge.

- **Define *reclaimed water*.**

Reclaimed water is treated wastewater that is reused in some way, such as for irrigation, manufacturing processes that require water for cooling, wetland restoration, or groundwater recharge.

- **Define *stable runoff*.**

Stable runoff is the share of runoff from precipitation that can be depended on every month.

- **Explain the relationship between climate change and available water.**

Climate change is expected to lead to changes in precipitation which, due to biological and geological factors, may have severe reductions in some areas. Other effects of climate change on fresh water include reduced snow pack and saltwater intrusion from sea level rise.

- **Briefly describe each of the following international water problems: drinking water problems, population growth and water problems, the Rhine River basin, the Aral Sea, and potential international conflicts over water rights.**

People in many developing countries lack access to safe drinking water and wastewater disposal. Population growth is outstripping water supplies in countries such as India, China, and Mexico. The river basin for the Rhine River is in five European countries, which cooperate to conserve and protect the supply and quantity of the Rhine River. For decades, water flowing into the Aral Sea was overdiverted for irrigation of farmland, and airborne salt from the dry lakebed likely harmed the health of people living nearby. The condition of the Aral Sea has improved significantly over the past decade. International tensions over water rights in such rivers as the Mekong, Indus, Ganges, Tigris-Euphrates, Jordan, and Nile could contribute to or result in armed conflicts.

- **Define *sustainable water use*.**

Sustainable water use is the wise use of water resources, without harming the essential functioning of the hydrologic cycle or the ecosystems on which humans depend.

- **Contrast the benefits and drawbacks of dams and reservoirs, using the Columbia River to provide specific examples.**

Dams ensure a year-round supply of water in areas that have seasonal precipitation or snowmelt. Many people think the drawbacks of dams outweigh any benefits they provide. The Columbia River, which has more than 100 dams, is used for shipping, hydroelectric power, and municipal and industrial water. Dams along the Columbia River have adversely affected salmon populations.

- **Briefly describe two methods of desalinization.**

Desalinization is the removal of salt from ocean or brackish (somewhat salty) water. One method of desalinization is **distillation**, heating salt water until water evaporates, leaving behind salt; the water vapor is then condensed. Membrane/filtration methods are more energy-efficient than distillation. The main method is **reverse osmosis**, in which salt water is forced through a membrane permeable to water but not to salt.

- **Give examples of water conservation by agriculture (including microirrigation), industry, and individual homes and buildings (including gray water).**

Certain agricultural techniques can significantly cut agricultural water consumption. **Microirrigation** is a type of irrigation that conserves water by piping it to crops through sealed systems. Water conservation, including recycling and reuse, can reduce both industrial and municipal water consumption. **Gray water,** which is water that has already been used for light household applications, can substitute for fresh water where drinkability is not required.

Thinking About the Environment

1. Explain why the poor in many countries have to pay more for their water than do the wealthy.

2. Diagram hydrogen bonding between water molecules and explain how it affects the properties of water.

3. Discuss the dissolving ability of water as it relates to ocean salinity and to water pollution.

4. Are our water supply problems largely the result of too many people? Give reasons why you support or refute this idea.

5. Explain the problem that was created by overdrawing surface water that flows into Mono Lake. How is this problem being addressed?

6. Briefly describe the complexity of international water use, using the Rhine River or the Aral Sea as an example.

7. Explain the relationship between global climate change and local availability of fresh water.

8. How is water used in agriculture? Discuss two ways to use agricultural water more sustainably.

9. Which industries consume the most water? Discuss one way to use industrial water more sustainably.

10. Explain how water resource problems might contribute to economic or political instability.

11. Should we allow housing on the flood plain of a river? Should taxpayers provide federal disaster assistance for those who choose to live on flood plains? Explain your answers.

12. Imagine you are a water manager for a Southwestern metropolitan district with a severe water shortage. What strategies would you use to develop a sustainable water supply? Why would a systems perspective help you achieve this?

13. Taking a systems perspective, develop a brief water conservation plan for your own personal daily use.

Quantitative questions relating to this chapter are on our Web site.

Take a Stand

Visit our Web site at http://www.wiley.com/college/raven (select Chapter 14 from the Table of Contents) for links to more information about the controversy surrounding dam removal on the Snake River to help salmon species recover. Consider the views of proponents and opponents, and debate the issue with your classmates. You will find tools to help you organize your research, analyze the data, think critically about the issues, and construct a well-considered argument. *Take a Stand* activities can be done individually or as a team, as oral presentations, written exercises, or Web-based (e-mail) assignments.

Additional online materials relating to this chapter, including a Student Testing Section with study aids and self-tests, Environmental News, Activity Links, Environmental Investigations, and more, are also on our Web site.

Soil Resources

Section of a shelter-forest near Dunhuang, China. Millions of trees defend China against raging dust storms from the Gobi Desert. Shown are pines and poplars developed by Chinese scientists to grow rapidly in poor soil. Note the agricultural land protected by the Great Green Wall. (George Steinmetz)

During its long history, China has had to contend with both invaders and devastating dust storms sweeping across its northern border. The Great Wall of China, a centuries-old earth and stone structure some 2400 km (1500 mi) long, was built across northern China to defend it against Mongolian raids. Although China's Great Wall was only partially successful in keeping out invaders, a more recently constructed "wall" has been effective in controlling the dust storms from the expanding Gobi Desert. The new wall, often called the Great Green Wall, consists of some 300 million trees planted beginning in the 1950s. This wall of shelter-forest extends for some 4800 km (3000 mi) and is 800 km (500 mi) wide in places (see photograph). The Great Green Wall and other shelter-forests in China's deserts now exceed 100,000 km² in area.

An average dust storm in northern China transports as much as 100 million tons of soil particles for hundreds or even thousands of kilometers. The dust damages crops, grounds air traffic, and causes many other inconveniences. The Great Green Wall has reduced both the frequency and severity of dust storms because the trees slow the wind's velocity, and the moist forest floor discourages additional soil from being picked up and carried by the wind. During the 1950s Beijing, located 480 km (300 mi) downwind of the Gobi Desert, experienced 10 to 20 major dust storms per year. By the 1970s, the annual number had declined to fewer than five, and during the 1990s and early 2000s, even a single dust storm per year was unusual.

Although climate was responsible for forming China's deserts, human activities probably caused them to enlarge. Removal of forests, overcollection of firewood, and overgrazing by sheep and goats on semiarid steppes (shortgrass prairies) loosened the fertile soil and exacerbated the dust storms coming out of the Gobi Desert before the Great Green Wall was planted. At least 8% of China's land area is desert, and *desertification,* the progressive degradation of grassland and other productive lands into unproductive desert, is an ongoing problem in China. (See Chapter 18 for a more detailed discussion of land degradation and desertification.) Planting trees to slow or stop the encroachment of desert gradually improves soil quality and reduces the severity of dust storms caused by wind erosion. Students and other volunteers continue to plant green walls of thousands of trees in Chinese deserts.

In this chapter you will learn about soil as a valuable natural resource on which humans and many other organisms depend. Many human activities cause or accentuate soil problems, such as erosion and nutrient mineral depletion. The goals of soil conservation are to minimize soil erosion and maintain soil fertility so that this resource can be used in a sustainable fashion. Conserving our soil resources is critical to human survival because more than 99% of our food comes from the land.

World View and Closer to You...
Real news footage relating to soil preservation around the world and in your own geographic region is available at www.wiley.com/college/raven by clicking on "World View and Closer to You."

What Is Soil?

LEARNING OBJECTIVES

- Define *soil* and identify the factors involved in formation of the soil system.
- List the four components of the soil system and give the ecological significance of each.
- Describe the various soil horizons.
- Relate at least two ecosystem services performed by soil organisms, and briefly discuss nutrient cycling.

soil: The uppermost layer of Earth's crust, which supports terrestrial plants, animals, and microorganisms.

Soil is the relatively thin surface layer of Earth's crust consisting of mineral and organic matter modified by the natural actions of weather, wind, water, and organisms. It is easy to take soil for granted. We walk on and over it throughout our lives but rarely stop to think about how important it is to our survival.

Vast numbers and kinds of organisms, mainly microorganisms, inhabit soil and depend on it for shelter, food, and water. Plants anchor themselves in soil, and from it they receive essential nutrient minerals and water. Terrestrial plants could not survive without soil, and because we depend on plants for our food, humans could not exist without soil either.

Soil represents a system in which a range of organisms and physical processes interact. The following sections describe portions of the soil system.

Soil-Forming Factors

Soil is formed from *parent material*, rock that is slowly broken down, or fragmented, into smaller and smaller particles by biological, chemical, and physical **weathering processes** in nature. It takes a long time, sometimes thousands of years, for rock to disintegrate into finer and finer mineral particles. To form 2.5 cm (1 in.) of topsoil may require between 200 and 1000 years. Time is also required for organic material to accumulate in the soil. Soil formation is a continuous process that involves interactions between Earth's solid crust and the biosphere. The weathering of parent material beneath already-formed soil continues to add new soil. The thickness of soil varies from a thin film on young lands, near the North and South Poles and on the slopes near the tops of mountains, to more than 3 m (10 ft) on old lands, such as certain forests.

Organisms and climate both play essential roles in weathering, sometimes working together. When plant roots and other soil organisms respire, they produce carbon dioxide, CO_2, which diffuses into the soil and reacts with soil water to form carbonic acid, H_2CO_3. Organisms such as lichens produce other kinds of acids. These acids etch tiny cracks in the rock; water then seeps into these cracks. If the parent material is located in a temperate climate, the alternate freezing and thawing of the water during the winter causes the cracks to enlarge, breaking off small pieces of rocks. Small plants then become established and send their roots into the larger cracks, fracturing the rock further.

Topography, a region's surface features, such as the presence or absence of mountains and valleys, is also involved in soil formation. Steep slopes often have little or no soil on them because soil and rock are continually transported down the slopes by gravity; runoff from precipitation tends to amplify erosion on steep slopes. Moderate slopes and valleys, on the other hand, may encourage the formation of deep soils.

Soil Composition

The soil system is composed of four distinct parts: mineral particles, which make up about 45% of a soil; organic matter (about 5%); water (about 25%); and air (about 25%). Soil occurs in layers, each of which has a certain composition and special properties. The plants, animals, fungi, and microorganisms that inhabit soil interact with it, and nutrient minerals are continually cycled from the soil to organisms, which use them in their biological processes. When the organisms die, bacteria and other soil organisms decompose the dead material, returning the nutrient minerals to the soil.

The mineral portion, which comes from weathered rock, constitutes most of soil. It provides anchorage and essential nutrient minerals for plants, as well as pore space for water and air. Because different rocks are composed of different minerals, soils vary in mineral composition and chemical properties. Rocks rich in aluminum form acidic soils, whereas rocks with silicates of magnesium and iron form soils that may be deficient in calcium, nitrogen, and phosphorus. Also, soils formed from the same kind of parent material may not develop in the same way because other factors such as weather, topography, and organisms differ.

The age of a soil affects its mineral composition. In general, older soils are more weathered and lower in certain essential nutrient minerals. Large portions of Australia, South America, and India have old, infertile soils. In contrast, in geologically recent time, glaciers passed across much of the Northern Hemisphere, pulverizing bedrock and forming fertile soils.[1] Essential nutrient minerals are readily available in these geologically young soils and in young soils formed in areas of volcanic activity.

Litter (dead leaves and branches on the soil's surface), animal dung, and dead remains of plants, animals, and microorganisms in various stages of decomposition constitute *soil organic material*. Microorganisms, particularly bacteria and fungi, gradually decompose this material. During decomposition, essential nutrient mineral ions are released into the soil, where they may be bound to soil particles, absorbed by plant roots, or leached through the soil. Organic matter increases the soil's water-holding capacity by acting much like a sponge. For these reasons gardeners often add organic matter to soils, especially sandy soils, which are naturally low in organic matter.

The black or dark brown organic material that remains after much decomposition has occurred is **humus (Figure 15.1)**. Humus, which is not a single chemical compound but a mix of many organic compounds, binds to nutrient mineral ions and holds water. On average, humus persists in agricultural soil for about 20 years. Certain components of humus, however, may persist in the soil for hundreds of years. Although humus is somewhat resistant to further decay, a succession of microorganisms gradually reduces it to carbon dioxide, water, and nutrient minerals. Detritus-feeding animals such as earthworms, termites, and ants also help break down humus.

Soil has numerous pore spaces around and among the soil particles. The pore spaces occupy roughly 50% of a soil's volume and are filled with varying proportions of water (called **soil water**) and air (called **soil air**) (**Figure 15.2**); both are necessary to produce a moist but aerated soil that sustains plants and other soil-dwelling organisms. Generally speaking, water is held in the smaller pores (less than 0.05 mm in diameter), whereas air is found in the larger pores. After a prolonged rain, almost all of the pore spaces may be filled with water, but water drains rapidly from the larger pore spaces, drawing air from the atmosphere into those spaces.

Soil water originates as precipitation, which drains downward, or as groundwater, which rises upward from the water table (see Chapter 14). Soil water contains low concentrations of dissolved nutrient mineral salts that enter the roots of plants as they absorb the water. Water not bound to soil particles or absorbed by roots **leaches,** or percolates (moves down) through the various layers of the soil, carrying dissolved nutrient minerals with it. The deposition of leached material in the lower layers of soil is known as **illuviation.** Iron and aluminum compounds, humus, and clay are some illuvial materials that gather in the subsurface portion of the soil. Some substances completely leach out of the soil because they are so soluble they

[1] The Pleistocene epoch, which began approximately two million years ago, was marked by four periods of glaciation. At their greatest extent these ice sheets covered nearly 10.4 million square kilometers (4 million square miles) of North America, extending south as far as the Ohio and Missouri Rivers.

Figure 15.1   Soil rich in humus. Humus is partially decomposed organic material, primarily from plant and animal remains. Soil rich in humus has a loose, somewhat spongy structure with several properties, such as increased water-holding capacity, that are beneficial for plants and other organisms living in it. (Courtesy U. S. Department of Agriculture)

Figure 15.2   Pore space.

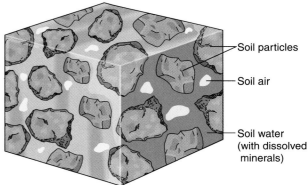

(a) In a wet soil, most of the pore space is filled with water.

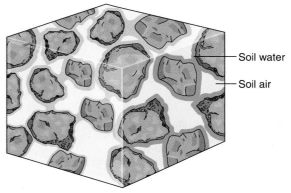

(b) In a dry soil, a thin film of water is tightly bound to soil particles, and soil air occupies most of the pore space.

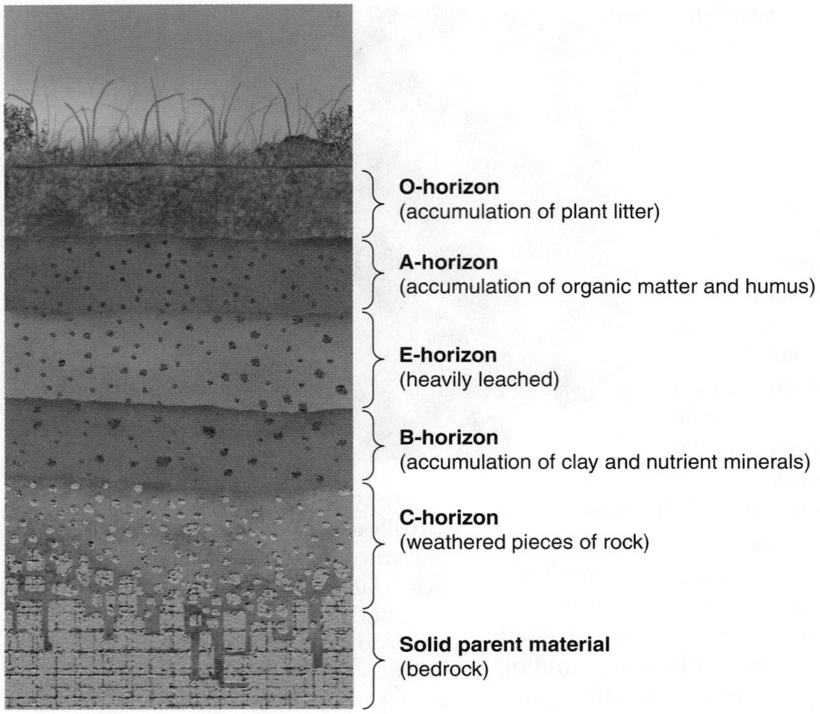

O-horizon
(accumulation of plant litter)

A-horizon
(accumulation of organic matter and humus)

E-horizon
(heavily leached)

B-horizon
(accumulation of clay and nutrient minerals)

C-horizon
(weathered pieces of rock)

Solid parent material
(bedrock)

Figure 15.3   Soil profile. Each horizon has its own chemical and physical properties.

soil horizons: The horizontal layers into which many soils are organized, from the surface to the underlying parent material.

ecosystem services: Important environmental benefits that ecosystems provide to people; include clear air to breathe, clean water to drink, and fertile soil in which to grow crops.

migrate down into the groundwater. It is also possible for water to move upward in the soil, transporting dissolved materials with it, such as when the water table rises.

Soil air contains the same gases as atmospheric air, although they are usually present in different proportions. Generally, as a result of cellular respiration by soil organisms, there is more carbon dioxide and less oxygen in soil air than in atmospheric air. (Recall from Chapter 3 that cellular respiration uses oxygen and produces carbon dioxide.) Among the important gases in soil air are oxygen, required by soil organisms for cellular respiration; nitrogen, used by nitrogen-fixing bacteria (see Chapter 5); and carbon dioxide, involved in soil weathering. As mentioned earlier in the chapter, carbon dioxide dissolves in water to form carbonic acid, a weak acid that accelerates the weathering process during soil formation.

Soil Horizons

A deep vertical cut through many soils reveals that they are organized into distinctive horizontal layers called **soil horizons**. A **soil profile** is a vertical section from surface to parent material, showing the soil horizons (**Figure 15.3**).

The uppermost layer of soil, the **O-horizon,** is rich in organic material. Plant litter accumulates in the O-horizon and gradually decays. In desert soils the O-horizon is often completely absent, but in certain organically rich soils it may be the dominant layer.

Just beneath the O-horizon is the topsoil, or **A-horizon,** which is dark and rich in accumulated organic matter and humus. The A-horizon is somewhat nutrient-poor owing to the gradual loss of many nutrient minerals to deeper layers by leaching. In some soils, a heavily leached **E-horizon** develops between the A- and B-horizons.

The **B-horizon,** the lighter-colored subsoil beneath the A-horizon, is often a zone of accumulation in which nutrient minerals that leached out of the topsoil and litter accumulate. It is typically rich in iron and aluminum compounds and clay.

Beneath the B-horizon is the **C-horizon,** which contains weathered pieces of rock and borders the unweathered solid parent material. The C-horizon is below the extent of most roots and is often saturated with groundwater.

Soil Organisms

Although soil organisms are usually hidden underground, their numbers are huge. Millions of microorganisms, including bacteria, fungi, algae, microscopic worms, and protozoa, may inhabit just one teaspoon of fertile agricultural soil. Many other organisms colonize the soil ecosystem, including plant roots, insects such as termites and ants, earthworms, moles, snakes, and groundhogs (**Figure 15.4**). Most numerous in soil are bacteria, which number in the hundreds of millions per gram of soil. Scientists have identified about 170,000 species of soil organisms, but thousands remain to be identified. Moreover, scientists know little about the roles of most soil organisms, in part because it is hard to study their activities under natural conditions.

Soil organisms provide several essential **ecosystem services**, such as maintaining soil fertility by decaying and cycling organic material, preventing soil erosion, breaking down toxic materials, cleansing water, and affecting the composition of the atmosphere. Many of these services are discussed in greater detail in this chapter.

Worms are important organisms living in soil. Earthworms, probably one of the most familiar soil inhabitants, ingest soil and obtain energy and raw materials by digesting some of the compounds that make up humus. **Castings,** bits of soil that have

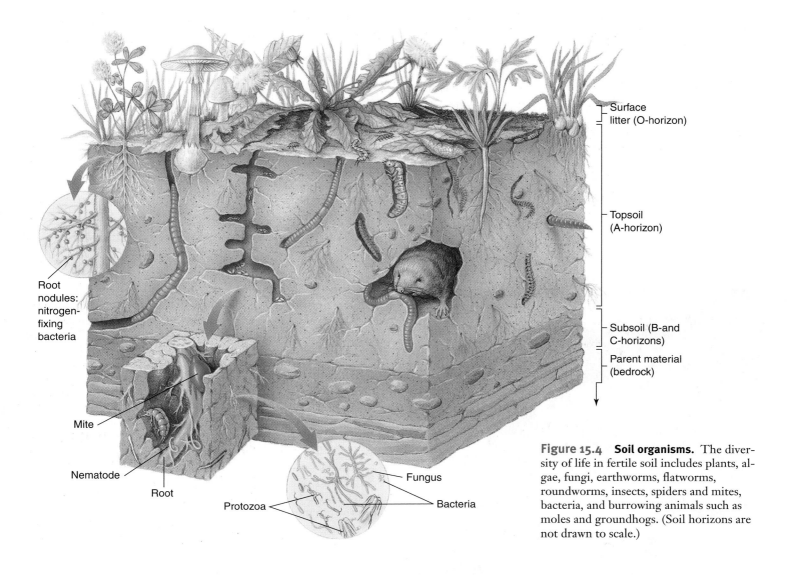

Figure 15.4 Soil organisms. The diversity of life in fertile soil includes plants, algae, fungi, earthworms, flatworms, roundworms, insects, spiders and mites, bacteria, and burrowing animals such as moles and groundhogs. (Soil horizons are not drawn to scale.)

passed through the gut of an earthworm, are deposited on the soil surface. In this way, nutrient minerals from deeper layers in the soil are brought to upper layers. Earthworm tunnels serve to aerate the soil, and the worms' waste products and corpses add organic material to the soil.

Ants live in the soil in enormous numbers, constructing tunnels and chambers that aerate it. Members of soil-dwelling ant colonies forage on the surface for bits of food, which they carry back to their nests. Not all of this food is eaten, and its eventual decomposition helps increase the organic matter in the soil. Many ants are also indispensable in plant reproduction because they bury seeds in the soil.

The properties of soil affect plant growth, although most plants tolerate a wide range of soil types. In turn, the types of plants that grow in soil affect it. As a result of the interactions among plants, climate, and soil, it is hard to specify cause and effect in their relationships. Are the plants growing in a certain locality because of the soil found there, or is the soil's type determined by the plants?

One important symbiotic relationship in the soil occurs between fungi and the roots of vascular plants. These associations, called **mycorrhizae,** help plants absorb adequate amounts of essential nutrient minerals from the soil. The threadlike body of the fungal partner, its **mycelium,** extends into the soil well beyond the roots. Nutrient minerals absorbed from the soil by the fungus are transferred to the plant, whereas food produced by photosynthesis in the plant is delivered to the fungus. Mycorrhizal fungi

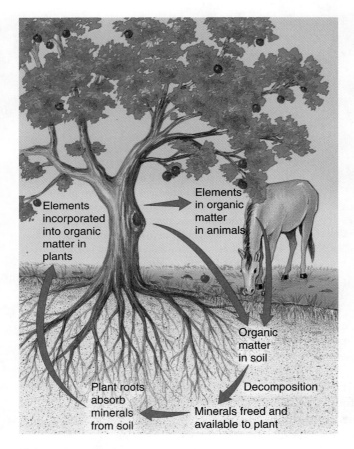

Figure 15.5 **Nutrient cycling.** In a balanced ecosystem, nutrient minerals cycle from the soil to organisms and then back to the soil.

nutrient cycling: The pathway of various nutrient minerals or elements from the environment through organisms and back to the environment.

enhance the growth of plants (see Figure 4.7). When mycorrhizal fungi are absent from the soil following a natural or human-induced disturbance, the reestablishment of certain tree species is retarded.

Nutrient Cycling

In a balanced ecosystem, the relationships among soil and the organisms that live in and on it ensure soil fertility. Essential nutrient minerals such as nitrogen and phosphorus are cycled from the soil to organisms and back again to the soil (**Figure 15.5**; also see Chapter 5). Decomposition is part of **nutrient cycling**. Bacteria and fungi decompose plant and animal detritus and wastes, transforming large organic molecules to small inorganic molecules, including carbon dioxide, water, and nutrient minerals; the nutrient minerals are released into the soil to be used again.

Abiotic (nonliving) processes are involved in nutrient cycling. Although leaching causes some nutrient minerals to be lost from the soil ecosystem to groundwater, the weathering of the parent material replaces much or all of them. In addition, dusts carried in the atmosphere for hundreds or thousands of kilometers help replace nutrient minerals in certain soils. Hawaiian rainforest soils, for example, receive dust inputs from central Asia, a distance more than 6000 km (3730 mi) away.

REVIEW

1. How do weathering processes affect soil formation?
2. What are the four components of soil, and how is each important?
3. Name two ecosystem services that soil organisms perform.

Soil Properties and Major Soil Types

LEARNING OBJECTIVES

• Briefly describe soil texture and soil acidity.
• Distinguish among spodosols, alfisols, mollisols, aridosols, and oxisols.

Texture and acidity are two parameters that characterize soils. Soil **texture** refers to the relative proportions of different-sized inorganic mineral particles of sand, silt, and clay. The size assignments for sand, silt, and clay give soil scientists a way to classify soil

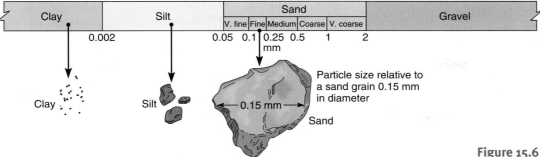

Figure 15.6   **Relative sizes of soil particles.**

texture. Particles larger than 2 mm in diameter, called gravel or stones, are not considered soil particles because they do not have any direct value to plants. The largest soil particles (0.05 to 2 mm in diameter) are **sand,** medium-sized particles (0.002 to 0.05 mm in diameter) are **silt,** and small particles (less than 0.002 mm in diameter) are **clay** (**Figure 15.6**). Sand particles are large enough to be seen easily with the eye; silt particles are about the size of flour particles and are barely visible with the eye; and clay particles are only seen under an electron microscope. A soil's texture affects many of a soil's properties, which in turn influence plant growth. Clay is particularly important in determining many soil characteristics because clay particles have the greatest surface area of all soil particles. If the surface areas of about 450 g (1 lb) of clay particles were laid out side by side, they would occupy 1 hectare (2.5 acres).

Soil minerals are often present in charged forms, or **ions.** Mineral ions may be positively charged (K^+, for example) or negatively charged (NO_3^-, for example). Each clay particle has predominantly negative electrical charges on its outer surface that attract and reversibly bind positively charged mineral ions (**Figure 15.7**). Many of these mineral ions, such as potassium (K^+) and magnesium (Mg^{2+}), are essential for plant growth and are "held" in the soil for plant use by their interactions with clay particles. In contrast, negatively charged mineral ions are usually not held as tightly in the soil and are often washed out of the root zone.

Soil always contains a mixture of different-sized particles, but the proportions vary from soil to soil. A **loam,** which is an ideal agricultural soil, has an optimum combination

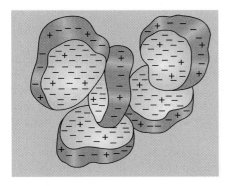

(a) The outer surfaces of clay particles are predominantly negatively charged.

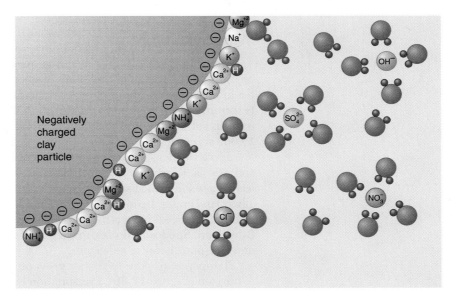

Figure 15.7   Availability of nutrient minerals.

(b) In this close-up of part of a clay particle and the thin film of water around it, note the large number of positively charged ions attracted to the surface of the clay particle. Also note how the positive (blue) ends of water molecules surround the individual negatively charged ions dissolved in solution.

Table 15.1 Soil Properties Affected by Soil Texture

| Soil Property | Soil Texture Type | | |
	Sandy Soil	Loam	Clay Soil
Aeration	Excellent	Good	Poor
Drainage	Excellent	Good	Poor
Nutrient mineral-holding capacity	Low	Medium	High
Water-holding capacity	Low	Medium	High
Workability (tillage)	Easy	Moderate	Difficult

of different soil particle sizes. It contains approximately 40% each of sand and silt, and about 20% of clay. Generally, the larger particles provide structural support, aeration, and permeability to the soil, whereas the smaller particles bind into aggregates, or clumps, and hold nutrient minerals and water (**Table 15.1**). Soils with larger proportions of sand are not as desirable for most plants because they do not hold mineral ions or water as well. Plants grown in such soils are more susceptible to mineral deficiencies and drought. Soils with larger proportions of clay are not as desirable for most plants because they provide poor drainage and often do not contain enough oxygen. Clay soils used for agriculture tend to get compacted, which reduces the number of pore spaces that water and air can fill.

Soil Acidity

Soil acidity is measured using the pH scale, which extends from 0 (extremely acidic) through 7 (neutral) to 14 (extremely alkaline). (See Appendix I for a discussion of pH.) The pH of most soils ranges from 4 to 8, but some soils are outside this range. The soil of the Pygmy Forest in Mendocino County, California, is extremely acidic, with a pH of 2.8 to 3.9. At the other extreme, certain soils in Death Valley, California, have a pH of 10.5.

Soil pH affects plants partly because the solubility of certain nutrient minerals varies with differences in pH. Plants absorb soluble mineral elements, but cannot absorb insoluble forms. At a low pH, the aluminum and manganese in soil water are more soluble, and the roots sometimes absorb them in toxic concentrations. Certain mineral salts essential for plant growth, such as calcium phosphate, become less soluble and less available to plants at a higher pH.

Soil pH greatly affects the leaching of nutrient minerals. An acidic soil has a reduced ability to bind positively charged ions to it. As a result, certain nutrient mineral ions essential for plant growth, such as potassium (K^+), are leached more readily from an acidic soil. The optimum soil pH for most plant growth is 6.0 to 7.0 because most nutrient minerals needed by plants are available in that pH range.

Soil pH affects plants and, in turn, is influenced by plants and other soil organisms. Litter composed of the needles of conifers contains acids that leach into the soil, lowering its pH. The decomposition of humus and the cellular respiration by soil organisms also decrease the pH of soil. (The effects of acid rain on soil are discussed in Chapter 20.)

Major Soil Groups

Variations in climate, local vegetation, parent material, underlying geology, topography, and soil age result in thousands of soil types throughout the world. These soils differ in color, depth, mineral content, acidity, pore space, and other properties. **Soil taxonomy** is the method of classification of these soils into 12 distinctive *orders*; each order is subdivided into many different *series*. In the United States alone, as many as 19,000 soil series are known.

Here we focus on five common soil orders: spodosols, alfisols, mollisols, aridisols, and oxisols. Regions with colder climates, ample precipitation, and good drainage typically have soils called **spodosols,** with distinct layers (**Figure 15.8a**). A spodosol usually forms under a coniferous forest and has an O-horizon of acidic litter composed primarily of needles; an ash-gray, acidic, leached E-horizon; and a dark brown, illuvial B-horizon. Spodosols do not make good farmland because they are too acidic and are nutrient-poor because of leaching.

Temperate deciduous forests grow on **alfisols,** soils with a brown to gray-brown A-horizon (**Figure 15.8b**). Precipitation is great enough to wash much of the clay and soluble nutrient minerals out of the A- and E-horizons and into the B-horizon. When the deciduous forest is intact, soil fertility is maintained by a continual supply of plant litter such as leaves and twigs. When the soil is cleared for farmland, fertilizers (which contain nutrient minerals such as nitrogen, potassium, and phosphorus) must be used to maintain fertility.

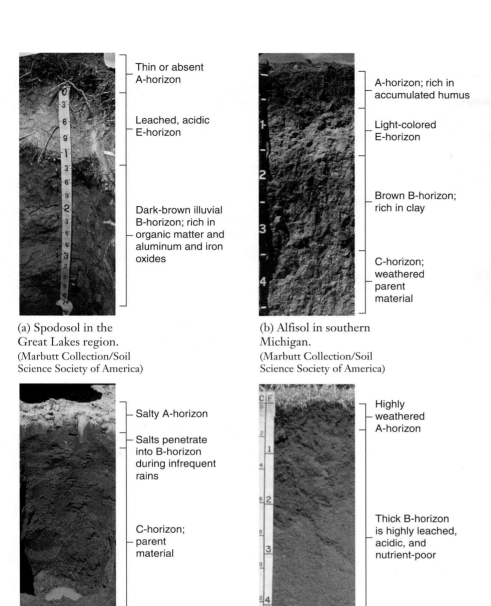

(a) Spodosol in the
Great Lakes region.
(Marbutt Collection/Soil
Science Society of America)

Thin or absent
A-horizon

Leached, acidic
E-horizon

Dark-brown illuvial
B-horizon; rich in
organic matter and
aluminum and iron
oxides

(b) Alfisol in southern
Michigan.
(Marbutt Collection/Soil
Science Society of America)

A-horizon; rich in
accumulated humus

Light-colored
E-horizon

Brown B-horizon;
rich in clay

C-horizon;
weathered
parent
material

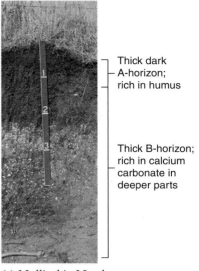

(c) Mollisol in North
Dakota.
(Marbutt Collection/Soil
Science Society of America)

Thick dark
A-horizon;
rich in humus

Thick B-horizon;
rich in calcium
carbonate in
deeper parts

(d) Aridosol in Nevada.
(Marbutt Collection/Soil
Science Society of America)

Salty A-horizon

Salts penetrate
into B-horizon
during infrequent
rains

C-horizon;
parent
material

(e) Oxisol in Hawaii.
(Soil Science Society of America)

Highly
weathered
A-horizon

Thick B-horizon
is highly leached,
acidic, and
nutrient-poor

Figure 15.8 **Five representative soil orders.**

Mollisols, found primarily in temperate, semiarid grasslands, are fertile soils (**Figure 15.8c**). They possess a thick, dark brown to black A-horizon rich in humus. Some soluble nutrient minerals remain in the upper layers because precipitation is not great enough to leach them into lower layers. Most of the world's grain crops are grown on mollisols.

Aridisols are found in arid regions of all continents. The lack of precipitation in these deserts precludes much leaching, and the lack of lush vegetation precludes the accumulation of much organic matter. As a result, aridisols do not usually have distinct layers of leaching and illuviation. Some aridisols have a salic (salty) horizon (**Figure 15.8d**). Some aridisols provide rangeland for grazing animals, and crops can be grown on aridisols if water is supplied by irrigation.

Oxisols, which are low in nutrient minerals, exist in tropical and subtropical areas with ample precipitation (**Figure 15.8e**). Little organic material accumulates on the forest floor (O-horizon) because leaves and twigs are rapidly decomposed. The A-horizon is enriched with humus derived from the rapidly decaying plant parts. The B-horizon, which is quite thick, is highly leached, acidic, and nutrient-poor. Oddly enough, tropical rain forests, with their lush vegetation, grow on oxisols. Most of the nutrient minerals in tropical rain forests are locked up in the vegetation rather than in the soil. As soon as plant and animal remains touch the forest floor, they promptly begin to decay, and plant roots quickly reabsorb the nutrient minerals. Even wood, which may take years to decompose in temperate soils, is decomposed in a matter of months in tropical rain forests, largely by subterranean termites.

REVIEW

1. What is soil texture?

2. What is the pH of most soils? What happens if the soil pH falls outside this range?

3. Which of the five soil orders (spodosols, alfisols, mollisols, aridosols, and oxisols) is associated with deserts? with tropical rain forests? with semiarid grasslands?

 ## Soil Problems

LEARNING OBJECTIVES

- Define *sustainable soil use.*
- Explain the impacts of soil erosion, mineral depletion, soil salinization, and desertification on plant growth and on other resources such as water.
- Describe the American Dust Bowl and explain how a combination of natural and human-induced factors caused this disaster.

Humans disrupt soil systems that would be balanced in nature. Human activities often cause or exacerbate soil problems, including erosion, mineral depletion of the soil, soil salinization, and desertification, all of which occur worldwide. Understanding how soil systems work is essential to mitigating these disruptive effects and promoting **sustainable soil use**. Soil used in a sustainable way renews itself by natural processes year after year.

sustainable soil use: The wise use of soil resources, without a reduction in the amount or fertility of soil, so that it is productive for future generations.

soil erosion: The wearing away or removal of soil from the land.

Soil Erosion

Water, wind, ice, and other agents promote **soil erosion**. Water and wind are particularly effective in moving soil from one place to another. Rainfall loosens soil particles, which are then transported away by moving water (**Figure 15.9**). Wind loosens soil and blows it away, particularly if the soil is barren and dry. Soil erosion is a natural process accelerated by human activities.

Erosion reduces the amount of soil in an area and therefore limits the growth of plants. Erosion causes a loss of soil fertility because essential nutrient minerals and organic matter in the soil are removed. As a result of these losses, the productivity of eroded agricultural soil drops, and more fertilizer must be used to replace the nutrient minerals lost to erosion.

Humans often accelerate soil erosion with poor soil management practices. Although soil erosion is often caused by poor agricultural practices, agriculture is not the only culprit. Removal of natural plant communities, such as during the construction of roads and buildings, and unsound logging practices, such as clearcutting large forested areas, accelerate soil erosion.

Soil erosion has an impact on other natural resources as well. Sediment that gets into streams, rivers, and lakes affects water quality and fish habitats. If the sediment contains pesticide and fertilizer residues, they further pollute the water. When forests are removed within the watershed of a hydroelectric power facility, accelerated soil erosion causes the reservoir behind the dam to fill with sediment much faster than usual. This process results in a reduction of electricity production at that facility.

Sufficient plant cover limits the amount of soil erosion. Leaves and stems cushion the impact of rainfall, and roots help to hold the soil in place. Although soil erosion is a natural process, abundant plant cover makes it negligible in many natural ecosystems.

Figure 15.9   Soil erosion caused by water. The branching gullies shown here are the most serious form of erosion and will continue to advance unless checked by some type of erosion control. Photographed in Colorado. (Grant Heilman/Grant Heilman Photography)

Soil Erosion in the United States and the World   Every five years the Natural Resources Conservation Service (NRCS), formerly called the Soil Conservation Service, measures the rate of soil erosion at thousands of sites across the United States. It also uses satellite data and models to estimate annual soil erosion. These measurements and estimates indicate that erosion remains a serious threat to cultivated soils in many regions throughout the United States, particularly in parts of southern Iowa, northern Missouri, western and southern Texas, and eastern Tennessee. The good news is that soil erosion on all U.S. croplands declined about 38% between 1982 and 1997, as calculated in the early 2000s, although significant erosion still occurs.

Water erosion is particularly severe in the midwestern grain belt along the Mississippi and Missouri Rivers, as well as in the Central Valley of California and in the hilly Palouse River region of the Pacific Northwest. The NRCS estimates that about 25% of U.S. agricultural lands are losing topsoil faster than natural soil-forming processes regenerate it. This loss is often so gradual that even farmers fail to notice it. A severe rainstorm may wash away 1 mm (0.04 in.) of soil, which seems insignificant until the cumulative effects of many storms are taken into account. Twenty years of soil erosion amounts to the loss of about 2.5 cm (1 in.) of soil, an amount that could take hundreds of years for natural soil-forming processes to replace.

Soil erosion is a significant problem worldwide. Although estimates vary widely depending on what assumptions are made, soil erosion results in an annual loss of as much as 75 billion metric tons (83 billion tons) of topsoil around the world. Soil erosion is greatest in certain parts of Asia, Africa, and Central and South America. In India and China, soil experts estimate that erosion causes an annual loss of as much as 6.6 billion metric tons and 5.5 billion metric tons of soil, respectively. These two countries have 13% of the world's total land area from which they must feed 2.4 billion people—more than 37% of the world's human population.

The American Dust Bowl

Semiarid lands, such as the Great Plains of North America, have low annual precipitation and are subject to periodic droughts. Prairie grasses, the plants that grow best in semiarid lands, are adapted to survive droughts. Although the above-ground portions of the plant may die, the root systems survive several years of drought. When the rains return, the root systems send up new leaves. Soil erosion is minimal because the dormant but living root systems hold the soil in place and resist the assault by wind and water.

The soils of semiarid lands are often of high quality, owing largely to the accumulation over many centuries of a thick, rich humus. These lands are excellent for grazing and for growing crops on a small scale. Problems arise when large areas of land are cleared for crops or when animals overgraze the land. The removal of the natural plant cover opens the way for climate conditions to "attack" the soil, and it gradually deteriorates from the onslaught of hot summer sun, occasional violent rainstorms, and wind. If a prolonged drought occurs under such conditions, disaster can strike.

The effects of wind on soil erosion were vividly experienced over a wide region of the central United States during the 1930s **(Figure 15.10)**. Throughout the late 19th and early 20th centuries, much of the native grasses were removed to plant wheat. Then, between 1930 and 1937, the semiarid lands stretching from Oklahoma and Texas into Canada received 65% less annual precipitation than was normal. The rugged prairie grasses that were replaced by crops could have survived these conditions, but not the wheat. The prolonged drought caused crop failures, which left fields barren and particularly vulnerable to wind erosion.

Winds from the west swept across the barren, exposed soil, causing dust storms of incredible magnitude **(Figure 15.11)**. Topsoil from Colorado, Texas, Oklahoma, and other prairie states was blown eastward for hundreds of kilometers. Women hanging out clean laundry in Georgia went outside later to find it dust-covered. Bakers in New York City and Washington, D.C., had to keep freshly baked bread away from open windows so that it would not get dirty. The dust even discolored the Atlantic Ocean several hundred kilometers off the coast. On April 14, 1937, known as Black Sunday, the most severe dust storm in U.S. history darkened the sky and blotted out the sun.

The Dust Bowl years occurred during the Great Depression, and ranchers and farmers quickly went bankrupt. Many abandoned their dust-choked land and dead livestock and migrated west to the promise of California. The plight of these dispossessed farmers is movingly portrayed in the novel *The Grapes of Wrath*, written in 1939 by John Steinbeck, for which he won the Pulitzer Prize in 1940.

When the rains finally arrived in the Great Plains, many areas were too eroded to support agriculture. In the years following the Dust Bowl, the U.S. Soil Conservation Service planted fence rows of shrubs and trees (to slow the winds) and seeded many badly eroded areas with native prairie grasses. Agriculture returned, largely because irrigation protected against crop loss from droughts. However, the Ogallala Aquifer, which provides irrigation water for this region, is being drained faster than its rate of replacement (see Chapter 14). When farmers are forced to abandon irrigation and revert to dryland farming, only careful management practices will prevent Dust Bowl conditions from returning during the next drought.

Although the United States no longer has a dust bowl, the Great Plains are still subject to droughts and soil erosion. For example, a severe five-year drought in the early 2000s ruined crops in parts of Montana, Wyoming, Colorado, Kansas, Texas, and New Mexico. Dust storms (brownouts) reoccurred on the High Plains, reminding people of the "Dirty Thirties." ■

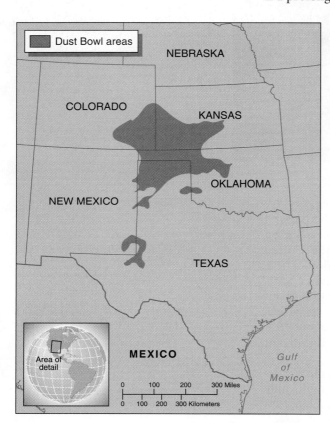

Figure 15.10   Location of the greatest damage from the American Dust Bowl. More than 30 million hectares (74 million acres) of land in the Great Plains were damaged during the Dust Bowl years. Shaded parts of Colorado, Kansas, Oklahoma, Texas, and New Mexico suffered the most extensive damage.

Nutrient Mineral Depletion

In a natural ecosystem, essential nutrient minerals cycle from the soil to organisms, particularly plants, which absorb them through their roots. When those organisms die and microorganisms decompose them, the essential nutrient minerals are released into the soil, where they become available for use by organisms again. An agricultural system disrupts this pattern of nutrient cycling when the crops are harvested. Much of the plant material, containing nutrient minerals, is removed from the cycle, so it fails to decay and release its nutrient minerals back to the soil. Over time, soil that is farmed inevitably loses its fertility (**Figure 15.12**).

Worldwide, more than one billion people depend on agricultural soils that are not productive enough to adequately support them. A combination of factors has caused this situation, including unsound farming methods, extensive soil erosion, and desertification; all of these factors contribute to nutrient mineral depletion. Along with these factors, the needs of a rapidly expanding population exacerbate soil problems worldwide.

In 2001 the International Food Policy Research Institute issued the results of a study of world agricultural lands using satellite maps. The institute found widespread damage to soil quality. Only 16% of the world's farmland does not have soil fertility problems such as poor drainage, aluminum toxicity, acidity, salinity, low soil nutrients, or depletion of organic matter. North America, with 29% of its soil free of problems, has the largest share of good soil. In contrast, parts of Asia have as little as 6% of farmland without soil quality problems.

Figure 15.11   The Dust Bowl years. This historic photo shows an abandoned Oklahoma farm in 1937. Total devastation was often the aftermath of dust storms. (Courtesy U. S. Department of Agriculture)

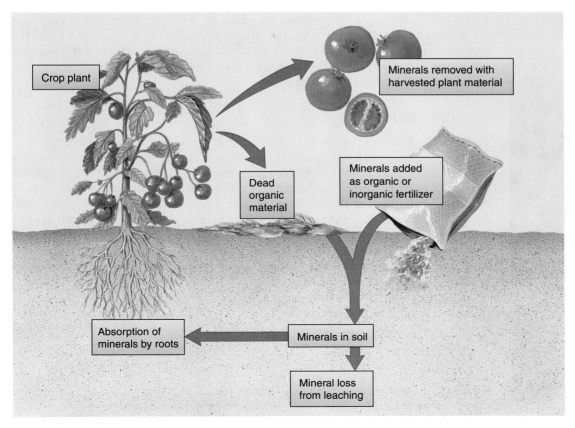

Crop plant

Minerals removed with harvested plant material

Dead organic material

Minerals added as organic or inorganic fertilizer

Absorption of minerals by roots

Minerals in soil

Mineral loss from leaching

Figure 15.12   Mineral depletion in agricultural soil. As plant and animal detritus decomposes in natural ecosystems, nutrient minerals are cycled back to the soil for reuse. In agriculture, much of the plant material is harvested. Because the nutrient minerals in the harvested portions are unavailable to the soil, the nutrient cycle is broken, and fertilizer must be added periodically to the soil.

Figure 15.13   Salinization of irrigated soil. As water evaporates from the soil's surface, mineral salts in irrigation water are left behind and gradually accumulate as a visible white powder, making the soil unfit for agriculture. (Courtesy U.S. Department of Agriculture)

salinization The gradual accumulation of salt in a soil, often as a result of improper irrigation methods.

desertification: Degradation of once-fertile rangeland, agricultural land, or tropical dry forest into nonproductive desert.

Mineral Depletion in Tropical Rainforest Soils   In tropical rain forests, the climate, the typical soil type, and the removal by humans of the natural forest community result in a particularly severe type of mineral depletion. Soils found in tropical rain forests are somewhat nutrient-poor because the nutrient minerals are stored primarily in the vegetation. Any nutrient minerals released as dead organisms decay in the soil and are promptly reabsorbed by plant roots and their mutualistic fungi. If this did not occur, the heavy rainfall would quickly leach the nutrient minerals away. Nutrient reabsorption by vegetation is so effective that tropical rainforest soils support luxuriant plant growth despite the relative infertility of the soil, as long as the forest remains intact.

When the forest is cleared, whether to sell the wood or to make way for crops or rangeland, its efficient nutrient cycling is disrupted. Removal of the vegetation that so effectively stores the forest's nutrient minerals allows them to leach out of the system. Crops can be grown on these soils for only a few years before the small mineral reserves in the soil are depleted. When cultivation is abandoned, the forest may eventually return to its original state, provided there is a nearby forest to serve as a source of seeds. The regrowth of forest is a slow process, however, and not all forest ecosystems return after being destroyed. (Chapter 18 discusses aspects of deforestation other than soil degradation.)

Soil Salinization

Soils found in arid and semiarid regions often contain high natural concentrations of inorganic compounds as mineral salts (see Figure 15.8d). In these areas, the amount of water that drains into lower soil area is minimal, because the little precipitation that falls quickly evaporates, leaving behind the salt. In contrast, humid climates have enough precipitation to leach salts out of soils and into waterways and groundwater.

Irrigation of agricultural fields often results in their becoming increasingly saline (salty), an occurrence known as **salinization** (**Figure 15.13**). Irrigation water contains small amounts of dissolved salts (fresh water always contains some salts). The continued application of such water, season after season, year after year, leads to the gradual accumulation of salt in the soil. When the water evaporates, the salts are left behind, particularly in the upper layers of the soil, which are the layers that are most important for agriculture. Given enough time, the salt concentration can rise to such a high level that plants are poisoned or their roots dehydrated. Also, when irrigated soil becomes waterlogged, capillary movement may carry salts from groundwater to the soil surface, where they are deposited as a crust of salt.

Desertification

Asia and Africa have the largest land areas with extensive soil damage, and in both places rapid population growth compounds the problem. Consider the Sahel, a broad band of semiarid land that stretches across Africa just south of the Sahara Desert and includes all or parts of many countries (see Figure 5.21). The Sahel normally experiences periodic droughts, but for the past 30 years there has been a sustained rainfall deficit—that is, substantially less precipitation than normal. During droughts, the soil cannot support as many crops or grazing animals. Despite the drought, the Sahelians must use their land to grow crops and animals for food or they will starve. The soil is so overexploited under these circumstances that it supports fewer and fewer people; the day is approaching when the Sahel could become unproductive desert (**Figure 15.14**).

As noted in the chapter introduction, the degradation of once-fertile rangeland or forest into nonproductive desert caused partly by soil erosion, forest removal, overcultivation, and overgrazing is called **desertification**. To reclaim the land would require

Figure 15.14   Onset of human-induced desertification. Cattle in Burkina Faso have eaten all the ground cover; the trees that remain will probably be stripped of branches to feed the hungry cattle. Overexploitation of the Sahel, a semiarid region south of the Sahara Desert, is increasing the amount of unproductive desert area. (Robert E. Ford/Terraphotographics/BPS)

restricting its use for many years so it could recover. If these measures were taken, the Sahelians would have no means of obtaining food. Desertification is discussed in greater detail in Chapter 18.

1. What is sustainable soil use?
2. How would you distinguish between soil erosion and nutrient mineral depletion?
3. What was the American Dust Bowl?

Soil Conservation and Regeneration

LEARNING OBJECTIVES

- Summarize how conservation tillage, crop rotation, contour plowing, strip cropping, terracing, shelterbelts, and agroforestry minimize erosion and mineral depletion of the soil.
- Briefly describe the Conservation Reserve Program.

Although agriculture may cause or accelerate soil degradation, good soil conservation practices promote sustainable soil use. Conservation tillage, crop rotation, contour plowing, strip cropping, terracing, and shelterbelts help to minimize erosion and mineral depletion of the soil. Land badly damaged by soil erosion and mineral depletion can be successfully restored, but it is a costly, time-consuming process.

conservation tillage: A method of cultivation in which residues from previous crops are left in the soil, partially covering it and helping to hold it in place until the newly planted seeds are established.

crop rotation: The planting of a series of different crops in the same field over a period of years.

Conservation Tillage

Conventional methods of tillage, or working the land, include spring plowing, in which the soil is cut and turned in preparation for planting seeds; and harrowing, in which the plowed soil is leveled, seeds are covered, and weeds are removed. Conventional tillage prepares the land for crops, but in removing all plant cover, it greatly increases the likelihood of soil erosion. Conventionally tilled fields contain less organic material and generally hold less water than undisturbed soil.

An increasing number of farmers have adopted **conservation tillage** (**Figure 15.15**). Several types of conservation tillage have been developed to fit different areas of the country and different crops. One of these, **no-tillage,** leaves the soil undisturbed over the winter. During planting, special machines cut a narrow furrow in the soil for seeds. Conservation tillage is one of the fastest growing trends in U.S. agriculture. During the early 2000s, almost 40% of U.S. farmland was planted using conservation tillage; the remainder was planted using conventional tillage.

In addition to reducing soil erosion, conservation tillage increases the organic material in the soil, which in turn improves the soil's water-holding capacity. Decomposing organic matter releases nutrient minerals more gradually than when conventional tillage methods are employed. Farmers who adopt no-tillage save on fuel costs, machinery wear and tear, and labor time when they do not plow their land. However, use of conservation tillage requires new equipment, new techniques, and greater use of herbicides to control weeds. Research to develop alternative methods of weed control for use with conservation tillage is underway. (Chapter 19 discusses *sustainable agriculture*, which includes conservation tillage and the other soil conservation practices presented in this chapter.)

Figure 15.15   Conservation tillage.
Decaying residues from the previous year's crop (rye) surround young soybean plants in a field in Iowa. Conservation tillage reduces soil erosion as much as 70% because plant residues from the previous season's crops are left in the soil. (Courtesy U.S. Department of Agriculture)

Crop Rotation

Farmers who practice effective soil conservation measures often use a combination of conservation tillage and **crop rotation**. When the same crop is grown continuously, pests for that crop tend to accumulate to destructive levels, so crop rotation lessens insect damage and disease. Many studies have shown that continuously growing the same crop for many years depletes the soil of certain essential nutrient minerals faster and makes the soil more prone to erosion. Crop rotation is therefore effective in maintaining soil fertility and in reducing soil erosion.

A typical crop rotation would be corn → soybeans → oats → alfalfa. Soybeans and alfalfa, both members of the legume family, increase soil fertility through their association with bacteria that fix atmospheric nitrogen into the soil. Thus, soybeans and alfalfa provide nutrients for the grain crops they alternate with in crop rotation.

Contour Plowing, Strip Cropping, and Terracing

Hilly terrain must be cultivated with care because it is more prone to soil erosion than flatland. Contour plowing, strip cropping, and terracing help control erosion of farmland with variable topography.

contour plowing: Plowing that matches the natural contour of the land.

In **contour plowing,** furrows run around hills rather than in straight rows. **Strip cropping,** a special type of contour plowing, produces alternating strips of different crops along natural contours (**Figure 15.16a**). For example, alternating a row crop such as corn with a closely sown crop such as wheat reduces soil erosion. Even more effective control of soil erosion is achieved when strip cropping is done in conjunction with conservation tillage.

Farming is undesirable on steep slopes, but if it must be done, **terracing** produces level areas and thereby reduces soil erosion (**Figure 15.16b**). Nutrient minerals and soil are retained on the horizontal platforms instead of being washed away. Soils are preserved in a somewhat similar manner in low-lying areas that are diked to make rice paddies. The water forms a shallow pool, retaining sediments and nutrient minerals.

Preserving Soil Fertility

The two main types of fertilizer are organic and commercial inorganic. *Organic fertilizers* include such natural materials as animal manure, crop residues, bone meal, and compost. Organic fertilizers are chemically complex, and their exact compositions vary. The nutrient minerals in organic fertilizers become available to plants only as the organic material decomposes. For that reason, organic fertilizers are slow-acting and long-

Figure 15.16 **Strip cropping and terracing.**

(a) Strip cropping is evident in this well-managed farm. Quite often crop rotations in such strips include a legume, which reduces the need for nitrogen fertilizers. Photographed in Wisconsin. (Courtesy U. S. Department of Agriculture)

(b) Terracing hilly or mountainous areas, such as the Luzon rice fields, Philippines, reduces the amount of soil erosion. However, some slopes are so steep that they should be left covered by natural vegetation to prevent extensive erosion. (David Cavagnaro)

lasting. (For two discussions of compost, see "You Can Make a Difference: Practicing Environmental Principles." Also see "Meeting the Challenge: Municipal Solid Waste Composting" in Chapter 24.)

Commercial inorganic fertilizers are manufactured from chemical compounds, and their exact compositions are known. Because they are soluble, they are immediately available to plants. However, commercial inorganic fertilizers are available in the soil for only a short period because they quickly leach away.

It is environmentally sound to avoid or limit the use of manufactured fertilizers, for several reasons. First, because of their high solubility, commercial inorganic fertilizers are mobile and often leach into groundwater or surface runoff, polluting the water. Second, manufactured fertilizers do not improve the water-holding capacity of the soil as organic fertilizers do. Another advantage of organic fertilizers is that, in ways not completely understood, they change the types of organisms that live in the soil, sometimes suppressing microorganisms that cause certain plant diseases. Commercial inorganic fertilizers are a source of nitrogen-containing gases (nitrous and nitric oxides) that are air pollutants. Finally, the production of commercial inorganic fertilizers requires a great deal of energy, which is largely obtained from our declining reserves of fossil fuels.

Soil Reclamation

It is possible to reclaim land badly damaged from erosion. The United States has largely reversed the effects of the 1930s Dust Bowl, and China has reclaimed badly eroded land in Inner Mongolia (northern China). Soil reclamation involves two steps: (1) stabilizing the land to prevent further erosion, and (2) restoring the soil to its former fertility. To stabilize the land, the bare ground is seeded with plants that eventually

YOU CAN MAKE A DIFFERENCE

Practicing Environmental Principles

Gardeners often dispose of grass clippings, leaves, and other plant refuse by bagging it for garbage collection or burning it. But these materials are a valuable resource for making **compost,** a natural soil and humus mixture that improves both soil fertility and soil structure. Grass clippings, leaves, weeds, sawdust, coffee grounds, ashes from the fireplace or grill, shredded newspapers, potato peels, and eggshells are just some of the materials that can be composted, or transformed by microbial action, to compost.

To make compost, spread a 6- to 12-inch layer of grass clippings, leaves, or other plant material in a shady area, sprinkle it with an organic garden fertilizer or a thin layer of farm animal manure, and cover it with several inches of soil. Add layers as you collect more organic debris. Water the mixture thoroughly, and turn it over with a pitchfork each month to aerate it. Although it is possible to make compost on the open ground, it is more efficient to construct an enclosure, which allows temperatures to build from the heat generated by microbial action. (Organic material decomposing

efficiently in a compost heap has a really hot core.) An enclosed compost heap is also less likely to attract animals.

When the compost is uniformly dark in color, is crumbly, and has a pleasant, "woodsy" odor, it is ready to use. The time it takes for decomposition will vary from one to six months, depending on the climate, the materials you are using, and how often you turn it and water it.

Whereas compost is mixed into soil to improve the soil's fertility, **mulch** is placed on the surface of soil, around the bases of plants (see figure). Mulch helps control weeds and increases the amount of water in the upper levels of the soil by reducing evaporation. It lowers the soil temperature in the summer and extends the growing season slightly by providing protection against cold in the fall. Mulch decreases erosion by lessening the amount of precipitation runoff.

Although mulches can consist of inorganic materials such as plastic sheets or gravel, organic mulches of compost, grass clippings, straw, chopped corncobs, or shredded bark have the added benefit of increasing the organic content of the soil. Grass clippings are an effective mulch when placed around the bases of garden plants because they mat together,

making it difficult for weeds to grow. You must replace grass mulches often because they decay rapidly. Some gardeners prefer mulches of more expensive materials, such as shredded bark, because they take longer to decompose and are more attractive.

Mulch. Mulch discourages the growth of weeds and helps keep the soil damp. Organic mulches such as this shredded bark have the added benefit of gradually decaying, thereby increasing soil fertility. (Courtesy U. S. Department of Agriculture.)

grow to cover the soil, holding it in place. After the Dust Bowl, land in Oklahoma and Texas was seeded with drought-resistant native grasses. The plants start to improve the quality of the soil almost immediately, as dead portions are converted to humus. The humus holds nutrient minerals in place and releases them a little at a time; it also improves the water-holding capacity of the soil.

One of the best ways to reduce the effects of wind on soil erosion is to plant **shelterbelts** to lessen the impact of wind (**Figure 15.17**; also recall the Great Green Wall of China discussed in the chapter introduction).

Restoration of soil fertility to its original level is a slow process. During the soil's recovery, use of the land must be restricted: It cannot be farmed or grazed. Disaster is likely if the land is put back to use before the soil has completely recovered. But restriction of land use for a period of several to many years is sometimes difficult to accomplish. How can a government tell landowners they may not use their own land? How can land use be restricted when people's livelihoods and perhaps even their lives depend on it?

shelterbelt: A row of trees planted as a windbreak to reduce soil erosion of agricultural land.

agroforestry: Concurrent use of forestry and agricultural techniques on the same land area to improve degraded soil and offer economic benefits.

Agroforestry

Organizations such as the International Centre for Research in Agroforestry, established in Nairobi, Kenya, are developing techniques to lessen the environmental degradation of the Sahel and other tropical areas. One of their research aims is to use **agroforestry**—land-use practices in which trees and crops are planted together to improve soil fertility in degraded soils.

For example, nitrogen-fixing acacias and other trees might be intercropped with traditional crops such as millet and sorghum. The trees grow for many years and provide several environmental benefits, such as reducing soil erosion, regulating the release of rainwater into groundwater and surface waters, and providing habitat for the natural enemies of crop pests. Acacia trees fix nitrogen, thereby improving soil fertility. When the leaves fall off the trees, they gradually decompose, returning mineral nutrients to the soil. The leaf layer also improves the soil's ability to hold moisture (less moisture evaporates from leaf-covered soil). Over time, the degraded land slowly improves. The result is higher crop yields. When the trees are so tall that they shade out the crops, the forest provides the farmers with food (such as fruits and nuts), fuelwood, lumber, and other forestry products.

Soil Conservation Policies in the United States

During the late 1920s and early 1930s, **Hugh H. Bennett,** a soil scientist in the U.S. Department of Agriculture, spoke out about the dangers of soil erosion. A report he published in 1928 was largely ignored until the disastrous effects of the Dust Bowl years focused attention on soil as a valuable natural resource. **The Soil Conservation Act** of 1935 authorized the formation of the Soil Conservation Service (now called the Natural Resources Conservation Service); its mission is to work with U.S. citizens to conserve natural resources on private lands. To that end, the NRCS assesses soil damage and develops policies to improve and sustain soil resources.

Historically, farmers are more likely to practice soil conservation during hard financial times and periods of agricultural surpluses, both of which translate into lower prices for agricultural products. When prices are high, with a good market for agricultural products, farmers have more incentive to put every parcel of land into production, including marginal, highly erodible lands.

The **Food Security Act (Farm Bill)** of 1985 contained provisions for two main soil conservation programs, a conservation compliance program and the Conservation Reserve Program. The conservation compliance program requires farmers with highly

Figure 15.17 Shelterbelts surrounding kiwi orchards. Trees protect the delicate fruits from the wind and reduce the ability of the wind to pick up soil from farmland. Photographed in North Island, New Zealand. (© Kevin Flaming/Corbis Images)

erodible land to develop and adopt a five-year conservation plan for their farms that includes erosion-control measures. If they don't comply, they lose federal agricultural subsidies such as price supports.

The **Conservation Reserve Program (CRP)** is a voluntary subsidy program that pays farmers to stop producing crops on highly erodible farmland. It requires planting native grasses or trees on such land and then "retiring" it from further use for 10 to 15 years. During that period the land may not be grazed, nor may the grass be harvested for hay. The CRP has benefited the environment. Annual loss of soil on CRP lands planted with grasses or trees has been reduced from an average of 7.7 metric tons of soil per hectare (8.5 tons per acre) to 0.6 metric ton per hectare (0.7 ton per acre). Because the vegetation is not disturbed once it is established, it provides biological habitat. Small and large mammals, birds of prey, and ground-nesting birds such as ducks have increased in number and kind on CRP lands. The reduction in soil erosion has improved water quality and enhanced fish populations in surrounding rivers and streams.

REVIEW

1. What is crop rotation? strip cropping?
2. How can degraded soils be reclaimed?
3. What is the Conservation Reserve Program?

 ## REVIEW OF LEARNING OBJECTIVES WITH KEY TERMS

• **Define *soil* and identify the factors involved in formation of the soil system.**

Soil is the uppermost layer of Earth's crust, which supports terrestrial plants, animals, and microorganisms. Soil formation involves interactions among parent material, climate, organisms, time, and topography. Biological, chemical, and physical weathering processes slowly break parent material into smaller and smaller particles.

• **List the four components of the soil system and give the ecological significance of each.**

The soil system is composed of inorganic nutrient minerals, organic materials, soil air, and soil water. The inorganic portion, which comes from weathered parent material, provides anchorage and essential nutrient minerals for plants, as well as pore space for water and air. Organic material decomposes, releasing essential nutrient mineral ions into the soil, where they may be absorbed by plant roots. Soil air and soil water produce a moist but aerated soil that sustains plants and other soil-dwelling organisms.

• **Describe the various soil horizons.**

Soil horizons are the horizontal layers into which many soils are organized, from the surface to the underlying parent material. The uppermost layer of soil, the **O-horizon**, contains plant litter that gradually decays. Just beneath the O-horizon is the topsoil, or **A-horizon**. In some soils, a heavily leached **E-horizon** develops between the A- and B-horizons. The **B-horizon** beneath the A-horizon is often a zone of illuviation in which nutrient minerals that leached out of the topsoil and litter accumulate. Beneath the B-horizon is the **C-horizon**, which contains weathered pieces of rock and borders the unweathered solid parent material.

• **Relate at least two ecosystem services performed by soil organisms, and briefly discuss nutrient cycling.**

Ecosystem services, important environmental benefits that ecosystems provide to people, include clear air to breathe, clean water to drink, and fertile soil in which to grow crops. Soil organisms are important in forming soil and in cycling nutrient minerals. **Nutrient cycling** is the pathway of various nutrient minerals or elements from the environment through organisms and back to the environment. In nutrient cycling the nutrient minerals removed from the soil by plants are returned when plants or the animals that eat plants die and are decomposed by soil microorganisms.

• **Briefly describe soil texture and soil acidity.**

The **texture** of a soil refers to the relative proportions of **sand, silt,** and **clay**. The pH of most soils ranges from 4 to 8, but some soils are outside this range. The soil properties of texture and acidity affect a soil's water-holding capacity and nutrient availability, which in turn determine how well plants grow.

• **Distinguish among spodosols, alfisols, mollisols, aridosols, and oxisols.**

Regions with colder climates, ample precipitation, and good drainage typically have **spodosols** with distinct layers. Temperate deciduous forests grow on **alfisols**, soils with a brown to gray-brown A-horizon. **Mollisols**, found primarily in temperate, semiarid grasslands, are fertile. **Aridisols** are found in arid regions of all continents. **Oxisols** are low in nutrient minerals and exist in tropical and subtropical areas with ample precipitation.

• **Define *sustainable soil use.***

Sustainable soil use is the wise use of soil resources, without a reduction in the amount or fertility of soil, so that it is productive for future generations.

• **Explain the impacts of soil erosion, mineral depletion, soil salinization, and desertification on plant growth and on other resources such as water.**

Soil erosion is the wearing away or removal of soil from the land. Mineral depletion occurs in all soils that are farmed. **Salinization** is the gradual accumulation of salt in a soil, often as a result of improper irrigation methods. **Desertification** is degradation of once-fertile rangeland, agricultural land, or tropical dry forest into nonproductive desert. Soil erosion, mineral depletion, salinization, and desertification limit the growth of plants. Sediment from erosion affects water quality and fish habitats when it gets into streams.

• **Describe the American Dust Bowl and explain how a combination of natural and human-induced factors caused this disaster.**

The 1930s Dust Bowl in the western United States is an example of accelerated wind erosion caused by use of marginal land for agriculture. Much of the native grasses were removed to plant wheat. Then, for several years the semiarid lands received less annual precipitation than was normal. The prolonged drought caused crop failures. Winds from the west swept across the barren, exposed soil, causing dust storms.

• **Summarize how conservation tillage, crop rotation, contour plowing, strip cropping, terracing, shelterbelts, and agroforestry minimize erosion and mineral depletion of the soil.**

Conservation tillage is a method of cultivation in which residues from previous crops are left in the soil, partially covering it and helping to hold it in place until the newly planted seeds are established. **Crop rotation** is the planting of a series of different crops in the same field over a period of years, thereby maintaining soil fertility. **Contour plowing** reduces soil erosion because it matches the natural contour of the land. **Strip cropping** is a type of contour plowing that produces alternating strips of different crops along the contours. Farming is undesirable on steep slopes, but if it must be done, **terracing** produces level areas that erode less. A **shelterbelt** is a row of trees planted as a windbreak to reduce soil erosion of agricultural lands. **Agroforestry** is the use of both forestry and agricultural techniques to improve degraded areas and offer economic benefits.

• **Briefly describe the Conservation Reserve Program.**

The Conservation Reserve Program, which is voluntary, pays farmers to stop producing crops on highly erodible farmland.

Thinking About the Environment

1. Explain the roles of weathering, organisms, climate, and topography in soil formation.

2. How does the presence of various-sized particles (sand, silt, and clay) affect soil characteristics?

3. Give an example of how plants affect soil pH. Give an example of how soil pH affects plants.

4. Charles Darwin once wrote that the land is plowed by earthworms. Explain.

5. Which two of these soil orders are best suited for agriculture: spodosol, alfisoil, mollisol, aridosol, and oxisol? Why?

6. Describe two ways in which nutrient minerals are lost from the soil.

7. It could be said that unlike other communities, tropical rain forests live *on* the soil rather than *in* it. What does this statement imply about tropical soils?

8. The American Dust Bowl is sometimes portrayed as a "natural" disaster brought on by drought and high winds. Present a case for the view that this disaster was not caused by nature as much as by humans.

9. Which soil horizons are most prone to erosion? What is the significance of your answer?

10. Where does eroded soil go after it is transported by water, wind, or ice?

11. Distinguish among conservation tillage, crop rotation, contour plowing, strip cropping, terracing, shelterbelts, and agroforestry as methods of sustainable soil use.

12. Conservation tillage has many benefits, including reduction of soil erosion. However, certain pests that cause plant disease reside in the plant residues left on the ground with conservation tillage. Knowing that disease-causing organisms are often quite specific for the plants they attack, recommend a way to control such disease organisms. Base your answer on the soil conservation methods discussed in this chapter.

13. How is degraded soil reclaimed?

14. How does the Conservation Reserve Program help preserve highly erodible farmland?

15. How is human overpopulation related to world soil problems?

16. President Franklin D. Roosevelt once sent a letter to the state governors in which he said, "A nation that destroys its soils, destroys itself." Explain his reasoning.

17. Why is soil considered a system? How does sustainable soil use protect the soil system?

Quantitative questions relating to this chapter are on our Web site.

Take a Stand

Visit our Web site at http://www.wiley.com/college/raven (select Chapter 15 from the Table of Contents) for links to more information about the environmental implications of the 2002 Farm Bill. Debate with your classmates whether the Grasslands Reserve Program is an effective deterrent against sod busting. You will find tools to help you organize your research, analyze the data, think critically about the issues, and construct a well-considered argument. *Take a Stand* activities can be done individually or as a team, as oral presentations, written exercises, or Web-based (e-mail) assignments.

Additional online materials relating to this chapter, including a Student Testing Section with study aids and self-tests, Environmental News, Activity Links, Environmental Investigations, and more, are also on our Web site.

Acid mine drainage contaminating a stream. Shown is the characteristic orange-red acid runoff that contains sulfuric acid contaminated with lead, arsenic, cadmium, silver, and zinc. (David Hiser/Stone/Getty Images)

16

Minerals: A Nonrenewable Resource

T he **General Mining Law of 1872** was established to encourage settlement in the sparsely populated western states. It allows companies or individuals, regardless of whether they are U.S. citizens or foreigners, to stake mining claims on federal land. They then purchase the land for $2.50 to $5 an acre, extract the valuable hardrock minerals such as gold, silver, copper, lead, or zinc, and keep all the profits. In contrast, 12.5% of profits on lumber, coal, oil, and natural gas obtained from federal land is paid to the government.

In 1995, for example, the General Mining Law permitted a company (Asarco) to obtain federal land in Arizona that contains copper and silver reserves worth an estimated $2.9 billion; the company paid $1745 for this land. The Congressional Budget Office determined that the federal treasury loses an estimated $150 million in annual revenue from the law's inequitable provisions. Although Congress put a hold on such sales, several hundred pending applications worth almost $16 billion in minerals were filed before the 1996 deadline.

The General Mining Law contains no provisions for environmental protection such as the replacement of topsoil and vegetation, or the reestablishment of biological habitat. As a result, hardrock mining has left a legacy of ravaged land, poisoned water, and lifeless ecosystems throughout the West. Acid draining from loose rocks produced during the mining process has made many streams and rivers totally lifeless (see photograph; also see Appendix I for a discussion of acids and pH).

More than 50 of the estimated 100,000 to 500,000 abandoned mines in the United States are designated Superfund sites. The federal government—that is, U.S. taxpayers—will finance the cleanup of these sites (see Chapter 24 for a discussion of Superfund sites). For example, after a mining company extracted $105 million of gold from a mine in Summitville, Colorado, it declared bankruptcy, leaving behind an environmental disaster. Now a Superfund site, the Summitville mine will cost more than $140 million to clean up. Cleanup of all Superfund mining sites will cost an estimated $12.5 billion to $17.5 billion.

In 1872, the same year that President Ulysses S. Grant signed the General Mining Law, Yellowstone was designated the first U.S. national park. During the 1990s a proposed 81-hectare (200-acre) mine site located in Montana less than 5 km (3 mi) from the Yellowstone border threatened its pristine condition. A Canadian company (Noranda, Inc.) had the rights to this land, which was surrounded by federal wilderness lands, and the company planned to establish the Crown Butte gold, silver, and copper mine there. Opponents of the mine, including the superintendent of Yellowstone, worried that some of the pollution from the mine would contaminate Clark's Fork of the Yellowstone River and Yellowstone Park. Because Yellowstone is considered a national treasure, President Clinton stopped the mine by presidential order and negotiated with the Canadian company. In 1997 Noranda agreed to sell the property to the U.S. government for about $65 million.

In 2000 Congress enacted new mining laws to protect taxpayers and the environment from the worst abuses of the General Mining Law, but when George W. Bush became president, he moved to weaken the new rules in response to pressure from the mining industry. Ultimately, the issue should be decided by examining the tradeoffs among interconnected systems—economics, ethics, resources, and the environment.

In this chapter you will consider the distribution and abundance of minerals as well as the environmental damage from obtaining and processing them. You will also examine options for the future, when mineral deposits become depleted.

 World View and Closer to You...
Real news footage relating to mineral conservation around the world and in your own geographic region is available at www.wiley.com/college/raven by clicking on "World View and Closer to You."

Introduction to Minerals

LEARNING OBJECTIVES

- Define *minerals* and explain the difference between high-grade ores and low-grade ores, and between metallic and nonmetallic minerals.
- Distinguish between surface mining and subsurface mining.
- Describe several natural processes that concentrate minerals in Earth's crust.
- Briefly describe how mineral deposits are discovered, extracted, and processed; make sure you use the terms *overburden, spoil bank,* and *smelting* in your description.

minerals: Elements or compounds of elements that occur naturally in Earth's crust.

Minerals are such an integral part of our lives that we often take them for granted (**Figure 16.1**). Steel, an essential building material, is a blend of iron and other metals. Beverage cans, aircraft, automobiles, and buildings all contain aluminum. Copper, which readily conducts electricity, is used for electrical and communications wiring. The concrete used in buildings and roads is made from sand and gravel, as well as cement, which contains crushed limestone. Sulfur, a component of sulfuric acid, is an indispensable industrial mineral with many applications in the chemical industry. It is used to make plastics and fertilizers and to refine oil. Other important minerals include platinum, mercury, manganese, and titanium.

Human need and desire for minerals have influenced the course of history. Phoenicians and Romans explored Britain in a search for tin. One of the first metals that humans used, tin came into its own during the Bronze Age (3500 to 1000 BCE), when tin and copper were combined to produce the tougher and more durable alloy, bronze. The desire for gold and silver was directly responsible for the Spanish conquest of the New World. In 1849, a gold rush in California led to the arrival of a large number of settlers from the eastern United States. More recently, the lure of gold in Amazonian and Indonesian rain forests has contributed to the destruction of indigenous homelands and ecosystems.

Earth's minerals are elements or (usually) compounds of elements and have precise chemical compositions. For example, **sulfides** are mineral compounds in which certain elements are combined chemically with sulfur, and **oxides** are mineral compounds in which elements are combined chemically with oxygen.

Rocks are naturally formed aggregates, or mixtures, of minerals and have varied chemical compositions. An **ore** is rock that contains a large enough concentration of a particular mineral to be profitably mined and extracted. **High-grade ores** contain relatively large amounts of particular minerals, whereas **low-grade ores** contain lesser amounts.

Figure 16.1 Examples of minerals.
(a) Concrete highways are made of sand, gravel, and crushed limestone. (b) Table salt is a nonmetallic mineral. (c) Copper, a metallic mineral, is often shaped into wire for electrical equipment or sheets for roofing, gutters, and downspouts. (Dennis Drenner/George Semple/Charles D. Winters)

(a)

(b)

(c)

Minerals are metallic or nonmetallic (**Table 16.1**). **Metals** are minerals such as iron, aluminum, and copper, which are malleable, lustrous, and good conductors of heat and electricity. **Nonmetallic minerals,** such as sand, stone, salt, and phosphates, lack these characteristics.

Mineral Distribution and Formation

Certain minerals, such as aluminum and iron, are relatively abundant in Earth's crust. Others, including copper, chromium, and molybdenum, are relatively scarce. Abundance does not necessarily mean that the mineral is easily accessible or profitable to extract. It is possible that you have gold and other expensive minerals in your own backyard. However, unless the concentrations are large enough to make them profitable to mine, they will remain there.

Like other natural resources, mineral deposits in Earth's crust are distributed unevenly. Some countries have extremely rich mineral deposits, whereas others have few or none. Although iron is widely distributed, Africa has less iron than the other continents. Many copper deposits are concentrated in North and South America, particularly in Chile and the United States, whereas most of Asia (other than Indonesia) has a relatively small amount of copper. Much of the world's tin is in China and Indonesia, and most chromium reserves are in South Africa. We discuss the international implications of the unequal distribution of important minerals later in the chapter.

Table 16.1 Some Important Minerals and Their Uses

Mineral	Type	Some Uses
Aluminum (Al)	Metal element	Structural materials (airplanes, automobiles), packaging (beverage cans, toothpaste tubes), fireworks
Borax ($Na_2B_4O_7$)	Nonmetal	Diverse manufacturing uses—glass, enamel, artificial gems, soaps, antiseptics
Chromium (Cr)	Metal element	Chrome plate, pigments, steel alloys (tools, jet engines, bearings)
Cobalt (Co)	Metal element	Pigments, alloys (jet engines, tool bits), medicine, varnishes
Copper (Cu)	Metal element	Alloy ingredient in gold jewelry, silverware, brass, and bronze; electrical wiring, pipes, cooking utensils
Gold (Au)	Metal element	Jewelry, money, dentistry, alloys
Gravel	Nonmetal	Concrete (buildings, roads)
Gypsum ($CaSO_4-2H_2O$)	Nonmetal	Plaster of Paris, soil treatments, wallboard
Iron (Fe)	Metal element	Basic ingredient of steel (buildings, machinery)
Lead (Pb)	Metal element	Lead pipes, solder, battery electrodes, pigments
Magnesium (Mg)	Metal element	Alloys (aircraft), firecrackers, bombs
Manganese (Mn)	Metal element	Steel, alloys (steamship propellers, gears), batteries, chemicals
Mercury (Hg)	Liquid metal element	Thermometers, barometers, dental inlays, electric switches, streetlights, medicine
Molybdenum (Mo)	Metal element	High-temperature applications, lamp filaments, boiler plates, rifle barrels
Nickel (Ni)	Metal element	Money, alloys, metal plating
Phosphorus (P)	Nonmetal element	Medicine, fertilizers, detergents
Platinum (Pt)	Metal element	Jewelry, delicate instruments, electrical equipment, cancer chemotherapy, industrial catalyst
Potassium (K)*	Metal element	Salts used in fertilizers, soaps, glass, photography, medicine, explosives, matches, gunpowder
Common salt (NaCl)	Nonmetal	Food additive, raw material for synthetics
Sand (largely SiO_2)	Nonmetal	Glass, concrete (buildings, roads)
Silicon (Si)	Nonmetal element	Electronics, solar batteries, ceramics, silicones
Silver (Ag)	Metal element	Jewelry, silverware, photography, alloys
Sulfur (S)	Nonmetal element	Insecticides, rubber tires, paint, matches, papermaking, photography, rayon, medicine, explosives
Tin (Sn)	Metal element	Cans and containers, alloys, solder, utensils
Titanium (Ti)	Metal element	Paints; manufacture of aircraft, satellites, and chemical equipment
Tungsten (W)	Metal element	High-temperature applications, light bulb filaments, dentistry
Uranium (U)	Metal element	Electricity generation in nuclear power plants; nuclear weapons
Zinc (Zn)	Metal element	Brass, metal coatings, electrodes in batteries, medicine (zinc salts)

* Potassium, which is very reactive chemically, is never found free in nature; it is always combined with other elements.

Formation of Mineral Deposits   Concentrations of minerals within Earth's crust are the result of several natural processes, including magmatic concentration, hydrothermal processes, sedimentation, and evaporation. As magma (molten rock) cools and solidifies deep in Earth's crust, it often separates into layers, with the heavier iron- and magnesium-containing rock[1] settling on the bottom and the lighter silicates (rocks containing silicon) rising to the top. Varying concentrations of minerals are often found in the different rock layers. This layering, called **magmatic concentration,** is responsible for some deposits of iron, copper, nickel, chromium, and other metals.

Hydrothermal processes involve water that was heated deep in Earth's crust. This water seeps through cracks and fissures and dissolves certain minerals in the rocks. The minerals are then carried along in the hot water solution. The dissolving ability of the water is greater if chlorine or fluorine is present because these elements react with many metals (such as copper) to form salts (copper chloride, for example) that are soluble in water. When the hot solution encounters sulfur, a common element in Earth's crust, a chemical reaction between the metal salts and the sulfur produces metal sulfides. Because metal sulfides are not soluble in water, they form deposits by settling out of the solution. Hydrothermal processes are responsible for deposits of minerals such as gold, silver, copper, lead, and zinc.

The chemical and physical weathering processes that break rock into finer and finer particles are important not only in soil formation (as you saw in Chapter 15) but in the production of mineral deposits. Weathered particles are transported by water and deposited as sediment on riverbanks, deltas, and the sea floor in a process called **sedimentation.** During their transport, certain minerals in the weathered particles dissolve in the water. They later settle out of solution. When the warm water of a river meets the cold water of the ocean, settling occurs because less material dissolves in cold water than in warm water. Sedimentation has formed important deposits of iron, manganese, phosphorus, sulfur, copper, and other minerals.

Significant amounts of dissolved material accumulate in inland lakes and in seas that have no outlet or only a small outlet to the ocean. If these bodies of water dry up by **evaporation,** a large amount of salt is left behind. Over time, it may be covered with sediment and incorporated into rock layers. Evaporation has formed significant deposits of common table salt, borax, potassium salts, and gypsum.

How Minerals Are Found, Extracted, and Processed

The process of making mineral deposits available for human consumption occurs in several steps. First, a particular mineral deposit is located. Second, mining extracts the mineral from the ground. Third, the mineral is processed, or refined, by concentrating it and removing impurities. During the fourth and final step, the purified mineral is used to make a product. As you will see, each of these steps has environmental implications.

Discovering Mineral Deposits   Geologists employ a variety of instruments and measurements to help locate valuable mineral deposits. Aerial or satellite photography sometimes discloses geologic formations associated with certain types of mineral deposits. Aircraft and satellite instruments that measure Earth's magnetic field and gravity reveal certain deposits. Seismographs, used to detect earthquakes, also provide clues about mineral deposits. Geologists analyze these data, along with their knowledge of Earth's crust and how minerals are formed, to estimate locations of possible mineral deposits. Once these sites are identified, mining companies drill or tunnel for mineral samples and analyze their composition.

Deposits on the ocean floor are estimated after detailed three-dimensional maps of the sea floor are produced, usually with the aid of depth-measuring devices. Sophisticated computer analysis evaluates the data recorded by such instruments.

---

[1] Pure magnesium is a relatively lightweight element. Rock usually contains magnesium in the form of magnesium oxide, which is heavier.

Extracting Minerals   The depth of a particular deposit determines whether surface or subsurface mining will be used. In **surface mining**, minerals are extracted near the surface, whereas in **subsurface mining**, minerals too deep to be removed by surface mining are extracted. Surface mining is more common because it is less expensive than subsurface mining. Because even surface mineral deposits occur in rock layers beneath Earth's surface, the overlying soil and rock layers, called **overburden**, must first be removed, along with the vegetation growing in the soil. Then giant power shovels scoop the minerals out.

There are two kinds of surface mining, open-pit surface mining and strip mining. Iron, copper, stone, and gravel are usually extracted by **open-pit surface mining**, in which a giant hole is dug (**Figure 16.2**). Large holes formed by open-pit surface mining are called quarries. In **strip mining**, a trench is dug to extract the minerals. Then a new trench is dug parallel to the old one; the overburden from the new trench is put into the old trench, creating a hill of loose rock called a **spoil bank**.

Subsurface mining, which is underground, may be done with a shaft mine or a slope mine. A **shaft mine** is a direct vertical shaft to the vein of ore. The ore is broken up underground and then hoisted through the shaft to the surface in buckets. A **slope mine** has a slanting passage that makes it possible to haul the broken ore out of the mine in cars rather than hoisting it up in buckets. Sump pumps keep the subsurface mine dry, and a second shaft is usually installed for ventilation.

Subsurface mining disturbs the land less than surface mining, but it is more expensive and more hazardous for miners. There is always a risk of death or injury from explosions or collapsing walls, and prolonged breathing of dust in subsurface mines can result in lung disease. (Chapter 11 discusses coal mining and its hazards.)

Processing Minerals   Processing metallic minerals often involves **smelting**. Purified copper, tin, lead, iron, manganese, cobalt, or nickel smelting is done in a chimneylike *blast furnace*. **Figure 16.3** shows a blast furnace used to smelt iron. Iron ore, limestone rock, and coke (modified coal used as an industrial fuel) are added at the top

Figure 16.2   Open-pit surface mining. Shown is an open-pit copper mine near Tucson, Arizona. (Bruce F. Molnia/Terraphotographics/BPS)

surface mining: The extraction of mineral and energy resources near Earth's surface by first removing the soil, subsoil, and overlying rock strata.

subsurface mining: The extraction of mineral and energy resources from deep underground deposits.

overburden: Soil and rock overlying a useful mineral deposit.

spoil bank: A hill of loose rock created when the overburden from a new trench is put into the already excavated trench during strip mining.

smelting: The process in which ore is melted at high temperatures to separate impurities from the molten metal.

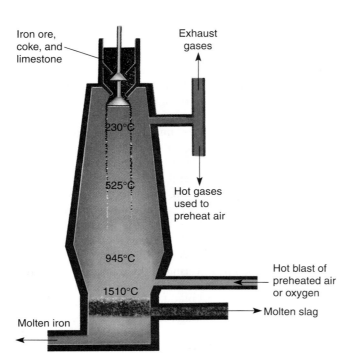

Figure 16.3   Blast furnace. Such towerlike furnaces are used to separate metal from impurities in the ore. The energy for smelting comes from a blast of heated air.

of the furnace, while heated air or oxygen is added at the bottom. Chemical reactions take place throughout the furnace as the ore moves downward: The iron ore reacts with coke to form molten iron and carbon dioxide, whereas the limestone reacts with impurities in the ore to form a molten mixture called **slag.** Both molten iron and slag collect at the bottom, but slag floats on molten iron because it is less dense than iron. The slag is cooled and then disposed of. Note the vent for exhaust gases near the top of the iron smelter (see Figure 16.3). If air pollution-control devices are not installed, dangerous gases such as sulfur oxides are emitted during smelting.

REVIEW

1. What is the difference between high-grade and low-grade ores?
2. How does magmatic concentration form mineral deposits?
3. How are mineral deposits discovered?

Environmental Impacts of Minerals

LEARNING OBJECTIVES

- Relate the environmental impacts of mining and refining minerals, including a brief description of acid mine drainage.
- Explain how mining lands can be restored.

There is no question that the extraction, processing, and disposal of minerals harm the environment. Mining disturbs and damages the land, and processing and disposal of minerals pollute the air, soil, and water. As noted in the discussion of coal in Chapter 11, pollution can be controlled and damaged lands can be fully or partially restored, but these remedies are costly. Historically, the environmental cost of extracting, processing, and disposal of minerals has not been incorporated into the actual price of mineral products to consumers.

Most highly developed countries have regulatory mechanisms in place to minimize environmental damage from mineral consumption, and many developing nations are in the process of putting them in place. Such mechanisms include policies to prevent or reduce pollution, restore mining sites, and exclude certain recreational and wilderness sites from mineral development.

Mining and the Environment

Mining, particularly surface mining, disturbs large areas of land. In the United States, current and abandoned metal and coal mines occupy an estimated 9 million hectares (22 million acres). Because mining destroys existing vegetation, this land is particularly prone to erosion, with wind erosion causing air pollution and water erosion polluting nearby waterways and damaging aquatic habitats.

Open-pit mining of gold and other minerals uses huge quantities of water. As miners dig deeper into the ground to obtain the ore, they eventually hit the water table and have to pump out the water to keep the pit dry. In northern Nevada, scientists from the U.S. Geological Survey surveyed several wells and measured a drop in the water table of as much as 305 m (1000 ft). This drop, which took place during the 1990s, was linked to gold mining in the region. (Nevada has numerous gold mines that provide the United States with half of its gold.) At the same time, the Western Shoshone tribe living in the area noticed that springs spiritually important to their culture began to dry up. The region's farmers and ranchers are concerned that gold mining is depleting the groundwater they need for irrigation. The lowering of the water table in the desert ecosystem can also affect fish and other organisms that rely on the rare pools of water where this ground water reaches the surface. Ranchers, farmers, environmentalists, and others would like the mines to reinject the water into the ground after they have pumped it out.

Mining affects water quality. According to the Worldwatch Institute, mining has contributed to the contamination of at least 19,000 km (11,800 mi) of streams and rivers in the United States. Rocks rich in minerals often contain high concentrations of heavy metals such as arsenic and lead. When rainwater seeps through sulfide minerals exposed in mine wastes, sulfuric acid is produced that in turn dissolves other toxic substances, such as lead, arsenic, and cadmium, in the spoil banks of coal and metal ore mines. These acids and highly toxic substances, called acid mine drainage, are washed into soil and water, including groundwater, by precipitation runoff (see chapter opening photograph). When such acids and toxic compounds make their way into nearby lakes and streams, they adversely affect the numbers and kinds of aquatic life. Rapid drainage during thunderstorms or a spring snowmelt produces "toxic pulses" of poisonous water that are particularly harmful to waterfowl, fish, and other wildlife in the watershed. Two examples of the many acid mine drainage sites in North America are the Berkeley Pit superfund site near Butte, Montana, and Britannia Beach in British Columbia, Canada.

acid mine drainage: Pollution caused when sulfuric acid and dangerous dissolved materials such as lead, arsenic, and cadmium wash from mines into nearby lakes and streams.

Cost-Benefit Analysis of Mine Development Environmental economists suggest that before a decision is made to develop a mine, a cost-benefit analysis should be performed that includes the benefits of the mine in dollar terms versus the benefits in dollar terms of preserving the land intact for wildlife habitat, ranchers, farmers, indigenous people, watershed protection, and recreation. This economic analysis should include the cost of damage to the environment caused by extracting and processing mineral resources. The evaluation should take into account that, over time, the benefits of the mine will decline as the mineral ore is exhausted, whereas the benefits of the natural environment will likely increase, in part because natural areas are becoming rarer as the number of developed areas increases. When a cost-benefit analysis of this type is performed, it may indicate that present and future benefits of preserving the land are greater than present and future benefits of developing a mine.

Environmental Impacts of Refining Minerals

On average, approximately 80% or more of mined ore consists of impurities that become wastes after processing (**Table 16.2**). These wastes, called **tailings,** are usually left in giant piles on the ground or in ponds near the processing plants (**Figure 16.4**). The tailings contain toxic materials such as cyanide, mercury, and sulfuric acid. Left exposed in this way, these toxic substances contaminate the air, soil, and water. Heavy metals in mine tailings at the Bunker Hill Superfund site in northern Idaho, for example, have leached (washed) into the south fork of the Coeur d'Alene River, killing fishes and waterfowl.

Smelting plants have the potential to emit large quantities of air pollutants during mineral processing. One impurity in many mineral ores is sulfur. Unless expensive pollution-control devices are added to smelters, the sulfur escapes into the atmosphere, where it forms sulfuric acid. The environmental implications of the

Figure 16.4 Aerial view of gold tailings. These tailings are stacked in long, parallel mounds. The truck in the lower right gives a sense of scale. Photographed near Yankee Fork of the Salmon River, Idaho. (©Natalie Fobes/Corbis)

Table 16.2 Ore and Waste Production for Selected Minerals

Mineral	Amount of Mined Ore (Million Tons)	Percentage of Ore That Becomes Waste During Refining*
Iron ore	2958	60
Copper	1663	99
Gold	745	99.99
Lead	267	97.5
Aluminum	128	81

*Data do not include the overburden of rock and soil that originally covered the ore deposits.
Source: Adapted form Gardner, G. et al. *State of the World,* 2003.

resulting acid precipitation are discussed in Chapter 20. Pollution-control devices for smelters are the same as those used when sulfur-containing coal is burned—scrubbers and electrostatic precipitators (see Figure 20.11).

Other contaminants found in many ores include the heavy metals lead, cadmium, arsenic, and zinc. These elements have the potential to pollute the atmosphere during the smelting process. Cadmium, for example, is found in zinc ores, and emissions from zinc smelters are a major source of environmental cadmium contamination. In humans, cadmium is linked to high blood pressure; diseases of the liver, kidneys, and heart; and certain types of cancer. In addition to airborne pollutants, smelters emit hazardous liquid and solid wastes that can cause soil and water pollution. Pollution-control devices prevent such hazardous emissions, although the toxic materials captured must be safely disposed of or they will still cause environmental pollution.

One of the most significant environmental impacts in mineral production is the large amount of energy required to mine and refine minerals, particularly if they are being refined from low-grade ore. Gold is currently being extracted from low-grade ores in Nevada. For every 0.9 metric ton (2000 pounds) of rock that is dug and crushed, as little as 0.7 g (0.025 oz) of gold is refined. Huge amounts of energy are required to dig and crush the tons of rock. Most of this energy is obtained by burning fossil fuels, which depletes energy reserves and produces carbon dioxide and other air pollutants.

CASE IN POINT

Copper Basin, Tennessee

Copper Basin, Tennessee, in the southeast corner of Tennessee, near its borders with Georgia and North Carolina, is an historic example of environmental degradation caused by smelting. Until relatively recently, the Copper Basin area progressed from lush forests to a panorama of red, barren hills baking in the sun (**Figure 16.5**). Few plant or animal species could be found—just 130 km² (50 mi²) of hills with deep ruts gouged into them. The ruined land had a stark, otherworldly appearance. How did this situation develop?

During the middle of the 19th century, copper ore was discovered near Ducktown in southeastern Tennessee. Copper mining companies extracted the ore from the ground and dug vast pits to serve as open-air smelters. They cut down the surrounding

Figure 16.5 Environmental devastation near Ducktown, Tennessee. Air pollution from a copper smelter in Tennessee killed the vegetation, and then water erosion carved gullies into the hillsides. This unrestored section of Copper Basin was photographed in the early 1980s. (The telephone pole at right provides a sense of scale.) (Pat Armstrong/Visuals Unlimited)

trees and burned them in the smelters to produce the high temperatures needed for the separation of copper metal from other contaminants in the ore. The ore contained great quantities of sulfur, which reacted with oxygen in the air to form sulfur dioxide. As sulfur dioxide from the open-air smelters billowed into the atmosphere, it reacted with water, forming sulfuric acid that fell as acid precipitation.

As a result of deforestation and acid precipitation, ecological ruin of the area occurred in a few short years. Acid precipitation quickly killed any plants attempting a comeback after removal of the forests. Because plants no longer covered the soil and held it in place, soil erosion cut massive gullies in the gently rolling hills. Of course, the forest animals disappeared with the plants, which had provided their shelter and food. The damage did not stop here. Soil eroding from the Copper Basin, along with acid precipitation, ended up in the Ocoee River, killing its entire aquatic community.

Beginning in the 1920s and 1930s, several government agencies, including the Tennessee Valley Authority and the U.S. Soil Conservation Service, tried to replant a portion of the area. They planted millions of loblolly pine and black locust trees as well as shorter ground-cover grasses and legume plants that tolerate acid conditions, but most of the plants died. The success of such efforts was marginal until the 1970s, when land reclamation specialists began using new techniques such as application of seed and time-released fertilizer by helicopter. These plants had a greater survival rate, and as they became established, their roots held the soil in place. Leaves dropping to the ground contributed organic material to the soil. The plants provided shade and food for animals such as birds and field mice, which slowly began to return.

Today reclamation of Copper Basin continues under a 2001 agreement among the state of Tennessee, the U.S. Environmental Protection Agency, and OXY USA, Inc., with the goal of having the entire area under plant cover early in the 21st century. Of course, the return of the forest ecosystem that originally covered the land before the 1850s will take at least a century or two. ◼

Restoration of Mining Lands

When a mine is no longer profitable to operate, the land can be reclaimed, or restored to a seminatural condition, as has been done to most of the Copper Basin in Tennessee. The goals of reclamation include preventing further degradation and erosion of the land, eliminating or neutralizing local sources of toxic pollutants, and making the land productive for purposes other than mining (**Figure 16.6**). Restoration also makes such areas visually attractive.

A great deal of research is available on techniques of restoring lands degraded by mining, called **derelict lands.** Restoration involves filling in and grading the derelict land to its natural contours, then planting vegetation to hold the soil in place. The establishment of plant cover is not as simple as throwing a few seeds on the ground. Often the topsoil is completely gone or contains toxic levels of metals, so special types of plants that tolerate such a challenging environment must be used. According to experts, the main limitation on the restoration of derelict lands is not a lack of knowledge but the lack of funding.

The **Surface Mining Control and Reclamation Act** of 1977 requires reclamation of areas that were surface mined for coal (see Chapter 11). However, no federal law is in place to require restoration of derelict lands produced by other kinds of mines. Recall from the chapter introduction that the General Mining Law makes no provision for reclamation.

Creative Approaches to Cleaning Up Mining Areas

Although wetlands are widely known to provide beneficial wildlife habitats, few people realize the potential

Figure 16.6 Restoration of mining lands. Part of a phosphate mine near Fort Meade, Florida, was reclaimed and is currently used as a pasture (background). The unrestored area that remains is in the foreground. Restoration of mining lands makes them usable once again, or at least stabilizes them so that further degradation does not occur. (William Felger/Grant Heilman Photography)

of wetlands to help clean up former mining lands. Wetlands tend to trap sediments and pollutants that enter them from upstream areas, so the quality of water resources located downstream from wetlands is improved. Although a single wetland provides these benefits, a series of wetlands constructed in the affected drainage basin is much more effective.

Consider the area around Butte, Montana, where copper was mined for 100 years. This area comprises the largest Superfund site in the United States. Its soil and water are contaminated with copper, zinc, nickel, cadmium, and arsenic. Many cleanup technologies are being developed and tested in Butte, including the design and construction of artificial wetlands. As contaminated water seeps into the wetland, bacteria consume the sulfur draining from the mines, making the water less acidic. As the water becomes less acidic, zinc and copper precipitate (settle out of solution) and enter the sediments. Constructed wetlands typically take 50 to 100 years to neutralize the acid enough for aquatic life to return to rivers and streams downstream from acid mine drainage. This time estimate is based on observations of more than 800 wetland systems constructed at coal mining sites in Appalachia, the region in the eastern United States that encompasses the central and southern Appalachian Mountains.

Creating and maintaining wetlands is expensive, although it is cost effective when compared to using lime to reduce the water's acidity. Recently, scientists at Butte have tried an inexpensive approach that involves cow manure, a "resource" readily available in Montana. Dumping cow manure onto mining lands causes the pH of the mine drainage water to increase as bacteria consume the manure. Toxic materials precipitate out of the basic water, just as they do in the artificial wetlands, improving the water quality.

Scientists are also using plants to remove heavy metals from former mining lands. **Phytoremediation** is the use of specific plants to absorb and accumulate toxic materials such as nickel from the soil. Although most plants do not tolerate soils rich in nickel, some plants, such as twist flower (*Streptanthus polygaloides*), thrive on it. This species is a *hyperaccumulator*, a plant that absorbs high quantities of a metal and stores it in its cells. The plants can be grown on nickel-contaminated land, harvested, and hauled to a hazardous waste site for disposal. Alternatively, the plants are burned, and nickel is obtained from the ashes. Phytoremediation has great potential to decontaminate mining and other hazardous waste sites and to extract valuable metals from soil in an environmentally benign way. (See Chapter 24 for further discussion of phytoremediation.)

REVIEW

1. What are three harmful environmental effects of mining and processing minerals?

2. Are mining lands usually restored when the mine is no longed profitable to operate? Why or why not?

Minerals: An International Perspective

LEARNING OBJECTIVES

• Contrast the consumption of minerals by developing countries and by industrialized nations such as the United States and Canada.

• Distinguish between mineral reserves and mineral resources.

The economies of industrialized countries require the extraction and processing of large amounts of minerals to make products. Most of these highly developed countries rely on the mineral deposits in developing countries, having long since exhausted their own supplies. As developing countries become more industrialized, their own mineral requirements increase correspondingly, adding further pressure to a nonrenewable resource. In fact, humans have consumed more minerals since World War II than were consumed in the previous 5000 years, from the beginning of the Bronze Age to the middle of the 20th century.

You have seen that mining in the United States has caused many serious environmental problems. The problems in developing countries that rely on mining for a significant part of their economies are as great or greater than the ones faced by highly developed countries. The governments in developing nations lack the financial resources and political will to deal with acid mine drainage and other serious environmental problems caused by hardrock mining. To complicate the issue, foreign companies often have significant mining interests in developing countries. For example, France, Germany, Great Britain, Japan, Russia, Spain, and the United States have been involved at various times during the past two centuries in mining (some would say exploiting) ores containing tin, zinc, copper, and lead in Bolivia. The mining district of Bolivia currently faces a catastrophic environmental nightmare from decades of mining abuse. Yet the Bolivian government does not address the issue because mining is the predominant industry in Bolivia. (Recall from the chapter introduction that the federal government has been reluctant to deal with mining issues, and hardrock mining is a relatively minor part of the U.S. economy.)

United States and World Use

At one time, most of the highly developed nations had rich resource bases, including abundant mineral deposits that enabled them to industrialize. In the process of industrialization, they largely depleted their domestic reserves of minerals so that they must increasingly turn to developing countries. This is particularly true for Europe, Japan, and, to a lesser extent, the United States.

As with the consumption of other natural resources, there is a large difference in consumption of minerals between highly developed and developing countries. The United States and Canada, which have about 5.1% of the world's population, consume about 25% of many of the world's metals (**Figure 16.7**). It is too simplistic, however, to divide the world into two groups, the mineral consumers (highly developed countries) and the mineral producers (developing countries). Four of the world's top five mineral producers are highly developed countries: the United States, Canada, Australia, and the Russian Federation. South Africa, a moderately developed, middle-income country, is the other mineral producer in the top five. Many developing countries lack any significant mineral deposits.

Mineral production in China is increasing dramatically, as is its mineral consumption as China industrializes. For example, in 2004 China smelted more than 20% of the world's primary aluminum (aluminum obtained from ores and not from recycling). China also consumed almost all of this aluminum, making it the world's largest producer and largest consumer of primary aluminum.

Because industrialization increases the demand for minerals, developing countries that at one time met their mineral needs with domestic supplies become increasingly reliant on foreign supplies as economic development occurs. South Korea is one such nation. During the 1950s it exported iron, copper, and other minerals. South Korea experienced dramatic economic growth from the 1960s to the present and, as a result, must now import iron and copper to meet its needs.

Distribution Versus Consumption

The metallic element chromium provides a useful example of global versus national distribution and consumption. Chromium is used to make vivid red, orange, yellow, and green pigments for paints; chrome plate; and, combined with other metals, certain types of hard steel. There is no substitute for chromium in many of its important applications, including jet engine parts. Industrialized nations that lack significant chromium deposits, such as the United States, must import essentially all of their chromium. South Africa is one of only a few countries with significant deposits of chromium. Zimbabwe and Turkey also export chromium. Although world reserves of chromium are adequate for the immediate future, the United States and several other industrialized countries are utterly dependent on a few countries for their chromium supplies.

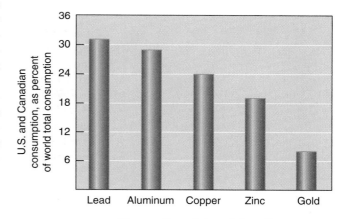

Figure 16.7 U.S. and Canadian consumption of selected metals. The heavily industrialized United States and Canada, which have about 5% of the world's population, consume a disproportionate share of many of the world's metals. (CRU International)

Many industrialized nations have stockpiled strategically important minerals to reduce their dependence on potentially unstable suppliers. The United States and others have stockpiles of *strategic minerals* such as titanium, tin, manganese, chromium, platinum, and cobalt, mainly because these metals are critically important to industry and defense. These stockpiles are large enough to provide strategic metals for approximately three years.

Will We Run out of Important Minerals?

To address this question, we must first examine how large the global supply of various mineral reserves and mineral resources is. **Mineral reserves** are currently profitable to extract, whereas **mineral resources** are potential resources that may be profitable to extract in the future. The combination of a mineral's reserves and resources is its **total resources** or **world reserve base.**

Estimates of mineral reserves and resources fluctuate with economic, technological, and political changes. If the price of a mineral on the world market falls, certain borderline mineral reserves may slip into the mineral resource category; increasing prices may restore them to the mineral reserve category. When new technological methods decrease the cost of extracting ores, deposits ranked in the mineral resource category are reclassified as mineral reserves. If the political situation in a country becomes so unstable that mineral reserves cannot be mined, they are reclassified as mineral resources; another change in the political situation at a later time may cause the minerals to be placed in the mineral reserve list again.

It is extremely difficult to forecast future mineral supplies. In the 1970s, projections of escalating demand and impending shortages of many important minerals were commonplace. There are three reasons that none of these shortages actually materialized. One, new discoveries of major deposits occurred in recent decades—iron and aluminum deposits in Brazil and Australia, for example. Two, plastics, synthetic polymers, ceramics, and other materials replaced metals in many products. Three, a global economic slump resulted in a lower consumption of minerals. Today on the world market there is even a glut of some minerals, which has caused their value to spiral downward. However, there is always the possibility that changes in the world economic situation will contribute to future mineral shortages.

Economic factors aside, the prediction of future mineral needs is difficult because it is impossible to know when or if there will be new discoveries of mineral reserves or replacements for minerals (such as the development of plastics). It is impossible to know when or if new technological developments will make it economically feasible to extract minerals from low-grade ores.

With these reservations in mind, most experts currently think mineral supplies, both metallic and nonmetallic, will be adequate during the 21st century. However, several important minerals—mercury, tungsten, and tin, for example—may become increasingly scarce during that period. Another reasonable projection is that the prices of even relatively plentiful minerals, such as iron and aluminum, will increase during your lifetime. The eventual depletion of large, rich, and easily accessible deposits of these metals means that we will have to mine and refine low-grade ores, which will be more expensive.

mineral reserves: Mineral deposits that have been identified and are currently profitable to extract.

mineral resources: Any undiscovered mineral deposits or known deposits of low grade ore that are currently unprofitable to extract.

REVIEW

1. How does mineral consumption differ between industrialized and developing countries?

Increasing the Supply of Minerals

LEARNING OBJECTIVE

• Briefly discuss efforts to discover new mineral supplies.

As a resource becomes scarce, efforts intensify to discover new supplies, to conserve existing supplies of that resource, and to develop new substitutes for it. Although many reserves have been discovered and exploited, unknown deposits may be found. In addition, the development of advanced mining technologies may make it possible to exploit known resources that are too expensive to develop using existing techniques.

Locating and Mining New Deposits

Many known mineral reserves have not yet been exploited. Although Indonesia is known to have many rich mineral deposits, its thick forests and mosquitoes that carry the malaria parasite have made accessibility to these deposits difficult. Both northern and southern polar regions have had little mineral development. This is due in part to a lack of technology for mining in frigid environments. Normal offshore drilling rigs cannot be used in Antarctic waters because the shifting ice formed during its harsh winter would tear the rigs apart. As new technologies become available, increasing pressure will be exerted to mine in northern Canada, Siberia, and Antarctica.

Exploitation of the rich mineral deposits in Siberia is planned, although new technologies must be developed to make this feasible. Some of the ore deposits in Siberia have unusual combinations of minerals (for example, potassium combined with aluminum) that cannot be separated using existing technology.

Is there a possibility that currently unknown mineral deposits will be discovered at some future time? The U.S. Geological Survey thinks that undiscovered mineral deposits may exist, particularly in developing countries where detailed geologic surveys are not yet available. It is likely that a detailed survey of the western portion of South America, along the Andes Mountains, will reveal significant mineral deposits. Geologists presume that minerals will be found in the Amazon Basin, although in many ways the rain forest and thick overlying alluvial layers of the river basin make these deposits as inaccessible as those in Antarctica. Logistic problems hamper examination of certain areas deep in the rain forest to assess the likelihood of deposits being present. As in other regions, mining in the Amazon Basin would pose a grave environmental threat.

Geologists consider it likely that deep deposits buried 10 km (6.2 mi) or more in Earth's crust will someday be discovered and exploited. The special technology required to mine deposits that deep is not yet available.

Minerals in Antarctica

To date, no substantial mineral deposits have been found in Antarctica, although smaller amounts of valuable minerals have been discovered. Geologists think it likely that major deposits of valuable metals and oil are present and that they will be discovered in the future. Nobody owns Antarctica, and many nations are involved in negotiations on the future of this continent and its possible mineral wealth.

The **Antarctic Treaty,** an international agreement in effect since 1961, limits activity in Antarctica to peaceful uses such as scientific studies. Twenty-six nations are voting members to the Antarctic Treaty. During the 1980s, nearly a decade of delicate negotiations resulted in a pact that would have permitted exploitation of Antarctica's minerals. The pact, the *Convention on the Regulation of Antarctic Mineral Resource Activities*, required unanimous agreement for ratification. In 1989 several countries refused to support the pact because of concerns that any mineral exploitation would damage Antarctica's environment. As a result of these concerns, an international agreement, the **Environmental Protection Protocol to the Antarctic Treaty,** or the **Madrid Protocol,** was established. The protocol, which went into effect in 1990, includes a moratorium on mineral exploration and development for a minimum of 50 years. It designates Antarctica and its marine ecosystem as a "natural reserve dedicated to peace and society."

Why be concerned about Antarctica's environment? For one thing, polar regions are extremely vulnerable to human activities. Even scientific investigations and tourists, with their trash, pollution, and noise, have negatively affected the wildlife, such as emperor penguins, leopard seals, and blue whales, along Antarctica's coastline. No one doubts that large-scale mining operations would wreak havoc on such a fragile environment. Maintaining Antarctica in a pristine state is important because this continent plays a pivotal role in regulating many aspects of the global environment, such as global changes in sea level. Studying the natural environment of

Antarctica helps scientists gain valuable insights into such important environmental issues as global climate change and stratospheric ozone depletion.

Minerals from the Ocean

The mineral reserves of the ocean may provide us with future supplies. Minerals could be extracted from seawater. Alternatively, the sea floor may be mined where minerals have accumulated in the loose ocean sediments or near underwater volcanoes. Mining the seabed has environmental implications for marine ecosystems where minerals are extracted, as well as for land, where the minerals must be processed.

Seawater, which covers approximately three-fourths of our planet, contains many dissolved minerals. The total amount of minerals available in seawater is staggeringly high, but their concentrations are low. Currently, sodium chloride (common table salt), bromine, and magnesium are profitably extracted from seawater. It may be possible in the future to profitably extract other minerals from seawater and concentrate them, but current mineral prices and technology make this impossible now.

Large deposits of minerals lie on the ocean floor. **Manganese nodules**—small rocks the size of potatoes that contain manganese and other minerals, such as copper, cobalt, and nickel—are widespread on the ocean floor, particularly in the Pacific (**Figure 16.8**). According to the Marine Policy Center at the Woods Hole Oceanographic Institute, the estimates of these reserves are quite large. The Pacific Ocean may contain as much as 1.4 million metric tons (1.5 million tons) of these minerals. Dredging manganese nodules from the ocean floor would adversely affect sea life, and the current market value for these minerals would not cover the expense of obtaining them using existing technology. Furthermore, it is not clear which country has the legal right to minerals in international waters. Despite these concerns, many experts think that deep-sea mining will be feasible in a few decades, and several industrialized nations such as the United States have staked out claims in a region of the Pacific known for its large number of nodules. To date, none has been mined.

Such potential exploitations of the ocean floor are controversial. Many people think it is inevitable that minerals will be mined from the floor of the deep sea, but others think the seabed should be declared off limits because of the potential ecological havoc that mining could cause on the diverse life forms inhabiting the ocean floor. Sea urchins, sea cucumbers, sea stars, acorn worms, sea squirts, sea lilies, and lamp shells are but a few of the animals known to inhabit the seabed environment.

These problems have been considered since the 1960s, when industrialized countries first expressed an interest in removing manganese nodules from the ocean floor. Their interest triggered the formation of an international treaty, the **U.N. Convention on the Law of the Sea (UNCLOS).** UNCLOS, which became effective in 1994, is generally considered a "constitution for the ocean" that protects its resources. As of 2007, 153 countries had ratified this treaty. (The United States has not yet ratified UNCLOS but voluntarily observes its provisions.)

It seems likely that ocean mining will become technologically feasible and profitable sometime during the 21st century. The provisions of UNCLOS are not binding for territorial waters, only for international waters, so there is no legal reason to block seabed mining in territorial waters.

Advanced Mining and Processing Technologies

We have already mentioned that special technologies will be needed to mine minerals in inaccessible areas such as polar regions and deposits deep in the ground. Making use of large, low-grade mineral deposits throughout the world will also require the development of special techniques. As minerals grow scarcer, economic and political pressure to exploit low-grade ores will increase. Obtaining high-grade metals from low-grade ores is an expensive proposition, in part because a great deal of energy must be expended to obtain enough ore. Future technology may make such exploitation more energy-efficient, thereby reducing costs.

E N V I R O N E W S

Diamonds in North America

Diamonds had never been found in large quantities in the Western Hemisphere until 1989, when Canadian geologists pinpointed the site of a whole cluster of "pipes," the veins that carry diamonds to the surface from great depths. The region became a prime target for diamond discovery when seismologists determined that the rocks were old: Diamonds are found in ancient deposits. At least a few of the pipes are extremely rich in diamonds, producing 3 carats per metric ton of mined material, and estimates of the potential deposits run as high as 1000 pipes. The deposits, worth billions of dollars, are in Canada's Northwest Territories, in a sub-Arctic region called the Barren Lands. Now the area bustles with the activity of several hundred companies that have staked out millions of acres for exploration.

Several diamond mines are currently operating in Canada, and they now produce 15% of the world's diamonds. Environmentalists are understandably concerned about the possible impact mining will have on the isolated area, which supports a caribou herd of 325,000. But jobs in the region are scarce and potential profits are huge (Canada's diamond industry is now worth more than $2.0 billion per year), so drilling for diamonds is under way.

Even if advanced technology makes obtaining minerals from low-grade ores feasible, other factors may limit exploitation of this potential source. In arid regions, the vast amounts of water required during the extraction and processing of minerals may be the limiting factor. The environmental costs may be too high, because obtaining minerals from low-grade ores causes greater land disruption and produces far more pollution than does the development of high-grade ores.

Biomining   In some cases, microorganisms are used to extract minerals from low-grade ores. Microorganisms have proved efficient for copper mining, allowing the U.S. copper industry to become more competitive internationally. When mixed with sulfuric acid, a bacterium (*Thiobacillus ferrooxidans*) promotes a chemical reaction that leaches copper into an acidic solution, releasing larger quantities of the metal more efficiently than traditional methods.

Other important applications of **biomining** are emerging. Although it is still in the development stage, treating low-grade gold ores with bacteria such as *Thiobacillus* allows a 90% recovery of gold, compared to 75% recovery for the more expensive and energy-intensive conventional methods. Phosphates, used primarily for fertilizers and additives in some manufactured goods, are traditionally extracted by inefficient burning at high temperatures or by wasteful acid treatment processes. New biological processes extract phosphates at room temperature.

Figure 16.8   **Manganese nodules on the ocean floor.**

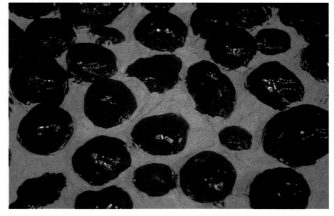

(a) These potato-sized nodules have enticed miners, but it is not yet commercially feasible to obtain them. Photographed in the Pacific Ocean. (Science VU/Visuals Unlimited)

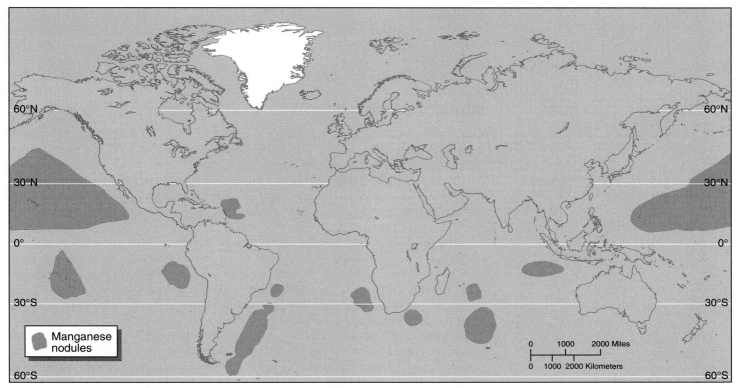

(b) Location of known manganese nodules in the ocean. (Adapted from P.A. Rona)

REVIEW

1. What is a problem that limits mining new deposits in Indonesia? in Siberia?

2. Have mineral deposits been mined in Antarctica? Why or why not?

Using Substitution and Conservation to Expand Mineral Supplies

LEARNING OBJECTIVES

• Summarize the conservation of minerals by reuse, recycling, and changing our mineral requirements.

• Explain how sustainable manufacturing and dematerialization contribute to mineral conservation.

Because much of our civilization's technology depends on minerals, and because certain minerals may be unavailable or quite limited in the future, we should extend existing mineral supplies as far as possible through substitution and conservation.

Finding Mineral Substitutes

The substitution of more abundant materials for scarce minerals is an important goal of manufacturing. Economics partly drives the search for substitutes; one effective way to cut production costs is to substitute an inexpensive or abundant material for an expensive or scarce one. In recent years, plastics, ceramic composites, and high-strength glass fibers have been substituted for scarcer materials in many industries.

Earlier in the 20th century, tin was a critical metal for can making and packaging industries; since then, other materials have been substituted for tin, including plastic, glass, and aluminum. The amounts of lead and steel used in telecommunications cables have decreased dramatically during the past 35 years, and the amount of plastics has had a corresponding increase. In addition, glass fibers have replaced copper wiring in telephone cables.

Although substitution extends mineral supplies, it is not a cure-all for dwindling resources. Certain minerals have no known substitutes. For example, platinum catalyzes many chemical reactions important in industry. So far, no other substance has been found with the catalyzing abilities of platinum.

Mineral Conservation

Conservation, including both reuse and recycling, extends mineral supplies. The **reuse** of items such as beverage bottles, which are collected, washed, and refilled, is one way to extend mineral resources. In **recycling**, used items such as beverage cans and scrap iron are collected, remelted, and reprocessed into new products. In addition to the introduction of specific conservation techniques such as reuse and recycling, public awareness and attitudes about resource conservation can be modified to encourage low waste.

Reuse When the same product is used over and over again, both mineral consumption and pollution are reduced. The benefits of reuse are greater than those of recycling (see Chapter 24). To recycle a glass bottle requires crushing it, melting the glass, and forming a new bottle. Reuse of a glass bottle simply requires washing it, which obviously expends less energy than recycling. Reuse is a national policy in Denmark, where nonreusable beverage containers are prohibited.

Several countries and states have adopted beverage container deposit laws, which require consumers to pay a deposit, usually a nickel or dime, for each beverage bottle or can they purchase. The deposit is refunded when the container is returned to the retailer or to special redemption centers. Unredeemed deposits are generally used to provide revenue for environmental programs such as hazardous waste cleanups. In addition to encouraging reuse and recycling, thereby reducing mineral resource con-

reuse: Conservation of the resources in used items by using them over and over again.

recycling: Conservation of the resources in used items by converting them into new products.

sumption, beverage container deposit laws save tax money by reducing litter and solid waste. Countries that have adopted beverage container deposit laws include the Netherlands, Germany, Norway, Sweden, and Switzerland. Parts of Canada and the United States have deposit laws.

Recycling   A large percentage of the products made from minerals—such as cans, bottles, chemical products, electronic devices, and batteries—is typically discarded after use. The minerals in some of these products—batteries and electronic devices, for instance—are difficult to recycle. Minerals in other products, such as paints containing lead, zinc, or chromium, are lost through normal use. However, the technology exists to recycle many other mineral products. Recycling of certain minerals is already a common practice throughout the industrialized world. Significant amounts of gold, lead, nickel, steel, copper, silver, zinc, and aluminum are recycled.

Recycling has several advantages in addition to extending mineral resources. It saves unspoiled land from the disruption of mining, reduces the amount of solid waste that must be disposed, and decreases energy consumption and pollution. Recycling an aluminum beverage can saves the energy equivalent of about 180 mL (6 oz) of gasoline. Recycling aluminum reduces the emission of aluminum fluoride, a toxic air pollutant produced during the processing of aluminum ore.

About 44% of the aluminum cans in the United States are currently recycled. The aluminum industry, local governments, and private groups have established thousands of recycling centers across the country. It takes approximately six weeks for a used can to be melted, re-formed, filled, and put back on a supermarket shelf. Clearly, even more recycling is possible. It may be that today's sanitary landfills will become tomorrow's mines, as valuable minerals and other materials are extracted from them.

Changing our Mineral Requirements   We can reduce mineral consumption by becoming a low-waste society. U.S. citizens have developed a "throwaway" mentality in which damaged or unneeded articles are discarded (**Figure 16.9**). Industries looking for short-term economic profits encourage this attitude, even though the long-term economic and environmental costs of such an attitude are high. We consume fewer resources if products are durable and repairable. Laws such as those requiring a deposit on beverage containers reduce consumption by encouraging reuse and recycling.

The throwaway mentality is also evident in manufacturing industries. Traditionally, industries consume raw materials and produce goods *and* a large amount of waste that is simply discarded (**Figure 16.10a**). Increasingly, manufacturers are finding that the waste products from one manufacturing process can become the raw materials for another industry. By selling these "wastes," industries gain additional profits and lessen the amounts of materials that must be disposed.

The chemical and petrochemical industries are among the first businesses that pioneered the minimization of wastes by converting such wastes into useful products. For example, some chemical companies buy used aluminum wastes from other

Figure 16.9   The throwaway mentality of our industrialized society. Many of these discarded materials could be recycled, and some could easily have been repaired and reused. (Courtesy Institute of Scrap Recycling Industries, Inc.)

Figure 16.10　**Mineral flow in an industrial society.**

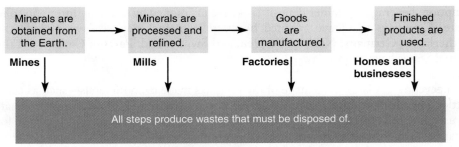

(a) Massive amounts of solid waste are produced at all steps in the traditional flow of minerals, from mining the mineral to discarding the used-up product.

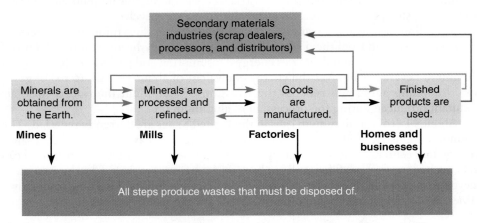

Key:　——— Sustainable manufacturing
　　　 ——— Consumer reuse
　　　 ——— Consumer recycling

(b) The flow of minerals in a low-waste society is more complex, with sustainable manufacturing, consumer reuse, and consumer recycling practiced at intermediate steps.

sustainable manufacturing: A manufacturing system based on industrial waste minimization.

companies and convert the aluminum in the wastes to aluminum sulfate, a chemical used to treat municipal water supplies. Such minimization of waste is known as **sustainable manufacturing** (see **Figure 16.10b** and "Meeting the Challenge: Industrial Ecosystems"). Sustainable manufacturing requires that companies provide information about their waste products to other industries. However, many companies are reluctant to reveal the kinds of wastes they produce because their competitors may deduce trade secrets from the nature of their wastes. This difficulty will have to be overcome before sustainable manufacturing is fully implemented.

Dematerialization　As products evolve, they tend to become lighter in weight and often smaller. Washing machines manufactured in the 1960s were much heavier than comparable machines manufactured today. The same is true of other household appliances, automobiles, and electronic items. This decrease in the weight of products over time is known as **dematerialization.** Ideally, dematerialization is beneficial to the environment because it reduces the quantity of waste during both production and consumption.

Although dematerialization gives the appearance of reducing consumption of minerals and other materials, it sometimes has the opposite effect. Smaller and lighter products may be of lower quality. Because repairing broken, lightweight items is difficult and may cost more than the original products, retailers and manufacturers encourage consumers to replace rather than repair the items. Although the weight of materials being used to make each item has decreased, the number of such items being used in a given period may actually have increased.

REVIEW

1. What is the difference between reuse and recycling?
2. What is sustainable manufacturing?

MEETING THE CHALLENGE

Industrial Ecosystems

Traditional industries operate in a one-way, linear fashion: natural resources from the environment → products → wastes dumped back into environment. However, natural resources such as minerals and fossil fuels are present in finite amounts, and the environment has a limited capacity to absorb waste. The field of **industrial ecology** has emerged to address these issues. An extension of the concept of sustainable manufacturing, industrial ecology seeks to use resources efficiently and regards "wastes" as potential products. Industrial ecology tries to create **industrial ecosystems** that compare in many ways to natural ecosystems.

Consider the industrial ecosystem in Kalundborg, Denmark, that consists of an electric power plant, an oil refinery, a pharmaceutical plant, a wallboard factory, a sulfuric acid producer, a cement manufacturer, fish farming, horticulture (greenhouses), and area homes and farms. At first glance, these entities seem to have little in common, but they are linked to one another in ways that resemble a food web

in a natural ecosystem (see figure). In this industrial ecosystem, the wastes produced by one company are sold to another company as raw materials for their processes, in a manner analogous to nutrient cycling in nature.

The coal-fired electric power plant originally cooled its waste steam and released it into the local fjord. The steam is now supplied to the oil refinery and the pharmaceutical plant, and additional surplus heat produced by the power plant warms greenhouses, the fish farm, and area homes. The need for 3500 oil-burning home heating systems was eliminated as a result.

Surplus natural gas from the oil refinery is sold to the power plant and the wallboard factory. The power plant now saves tons of coal each year by burning the less expensive natural gas. Before selling the natural gas, the oil refinery removes excess sulfur from it, as required by air pollution-control laws. This sulfur is sold to a company that uses it to manufacture sulfuric acid.

To meet environmental regulations, the power plant installed pollution-control equipment to remove sulfur from its coal smoke. This

sulfur, in the form of calcium sulfate, is sold to the wallboard plant and used as a substitute for gypsum, which is naturally occurring calcium sulfate. The fly ash produced by the power plant goes to the cement manufacturer for use in road building.

Local farmers use the sludge from the fish farm as a fertilizer for their fields. The fermentation vats at the pharmaceutical plant also generate a high-nutrient sludge used by local farmers. Most pharmaceutical companies discard this sludge because it contains living microorganisms, but the Kalundborg plant heats the sludge to kill the microorganisms, converting a waste material into a commodity.

These interactions did not spring into existence at the same time; each represents a separately negotiated deal. It took a decade to develop the entire industrial ecosystem. Although these examples of industrial cooperation were initiated for economic reasons, each has distinct environmental benefits, from energy conservation to a reduction of pollution.

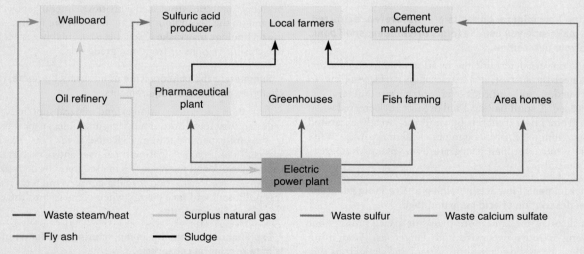

Industrial ecosystem in Kalundborg, Denmark. During the production of energy, food, and other products, resource recovery is maximized and waste production is minimized.

REVIEW OF LEARNING OBJECTIVES WITH KEY TERMS

• **Define** *minerals* **and explain the difference between high-grade ores and low-grade ores, and between metallic and nonmetallic minerals.**

Minerals are elements or compounds of elements that occur naturally in Earth's crust. **High-grade ores** contain relatively large amounts of particular minerals, whereas **low-grade ores** contain lesser amounts. Minerals may be **metals** such as iron, aluminum, and copper, or **nonmetals** such as phosphates, salt, and sand.

• **Distinguish between surface mining and subsurface mining.**

Surface mining is the extraction of mineral and energy resources near Earth's surface by first removing the soil, subsoil, and overlying rock strata. In **strip mining**, a type of surface mining, a trench is dug to extract the minerals. **Subsurface mining** is the extraction of mineral and energy resources from deep underground deposits.

• **Describe several natural processes that concentrate minerals in Earth's crust.**

Magmatic concentration is the formation of mineral deposits as liquid magma separates into layers, cools, and solidifies. **Hydrothermal processes** form mineral deposits as hot water dissolves minerals from rock, and the solution seeps through cracks until the minerals encounter sulfur, react to produce metal sulfides, cool, and settle out of solution. In **sedimentation,** certain minerals dissolve in water and later settle out of solution on riverbanks, deltas, and the sea floor. In **evaporation,** dissolved materials in lakes with no outlet to the ocean form mineral deposits when the water evaporates and the materials are left behind.

• **Briefly describe how mineral deposits are discovered, extracted, and processed; make sure you use the terms** *overburden, spoil bank,* **and** *smelting* **in your description.**

Detailed geologic surveys determine the location of mineral deposits. If the deposit is near the surface, it is extracted by surface mining, in which the **overburden,** soil and rock overlying a useful mineral deposit, is first removed. A **spoil bank** is a hill of loose rock created when the overburden from a new trench is put into the already excavated trench during strip mining. Processing minerals often involves **smelting,** in which ore is melted at high temperatures to separate impurities from the molten metal.

• **Relate the environmental impacts of mining and refining minerals, including a brief description of acid mine drainage.**

Surface mining disturbs the land more than subsurface mining, but subsurface mining is more expensive and dangerous. **Acid mine drainage** is pollution caused when sulfuric acid and dangerous dissolved materials such as lead, arsenic, and cadmium wash from mines into nearby lakes and streams. **Tailings,** the impurities that make up about 80% of mined ore, are often left in giant piles near mineral processing plants. Mercury, cyanide, and sulfuric acid leach into the soil and water from tailings. Unless pollution-control devices are used, smelting plants may emit large amounts of air pollutants during mineral processing.

• **Explain how mining lands can be restored.**

Derelict lands are extensively damaged due to mining but can be restored to prevent further degradation and to make the land productive for other purposes. Land reclamation is expensive, and no federal law exists to require restoration of derelict lands other than those produced by coal mines.

• **Contrast the consumption of minerals by developing countries and by industrialized nations such as the United States and Canada.**

Highly developed nations consume a disproportionate share of the world's minerals, but as developing countries industrialize, their need for minerals increases. The richest concentrations of minerals in highly industrialized countries have been largely exploited. As a result, these nations have increasingly turned to developing countries for the minerals they require. Sometimes highly developed nations must rely on potentially unstable developing nations for strategically important minerals.

• **Distinguish between mineral reserves and mineral resources.**

Mineral reserves are mineral deposits that have been identified and are currently profitable to extract. **Mineral resources** are any undiscovered mineral deposits or known deposits of low-grade ore that are currently unprofitable to extract. It is difficult to estimate a mineral's reserves and resources accurately because mineral consumption changes over time, and economic factors influence all aspects of mineral consumption.

• **Briefly discuss efforts to discover new mineral supplies.**

Mineral deposits will probably be discovered in some developing countries as they are surveyed geologically. Minerals that possibly exist in Antarctica may be mined in the future, but current international law protects Antarctica from such activities. Mineral deposits in ocean water and on the ocean floor may be mined in the future. Advanced mining technology may make it possible to profitably extract minerals from inaccessible regions or from low-grade ores.

• **Summarize the conservation of minerals by reuse, recycling, and changing our mineral requirements.**

Substitution and conservation extend mineral supplies. Manufacturing industries try to substitute more common, less expensive minerals for scarce and expensive minerals. **Reuse** is the conservation of the resources in used items by using them over and over again. **Recycling** is the conservation of the resources in used items by converting them into new products. Reuse and recycling conserve mineral resources, cause less pollution, and save energy when compared to the extraction and processing of virgin ores.

• **Explain how sustainable manufacturing and dematerialization contribute to mineral conservation.**

If mineral supplies are to last and if our standard of living is to remain high, consumers must decrease consumption. To accomplish this goal, manufacturers must make high-quality, durable, repairable products. **Sustainable manufacturing** is a manufacturing system based on industrial waste minimization. **Industrial ecology** is an extension of the concept of sustainable manufacturing in which resources are used efficiently and "wastes" are regarded as potential products. **Dematerialization** is the decrease in size and weight of a product as a result of technological improvements over time. Dematerialization reduces consumption only if products are durable and easily and inexpensively repaired.

Thinking About the Environment

1. What is the difference between rocks and minerals? between metals and nonmetallic minerals?

2. Distinguish among the following ways in which mineral deposits may form: magmatic concentration, hydrothermal processes, sedimentation, and evaporation.

3. Explain why it is difficult to obtain an accurate appraisal of total mineral resources.

4. Distinguish between surface and subsurface mining, between open-pit and strip mines, between shaft and slope mines.

5. What is overburden? What is a spoil bank?

6. What is smelting? What are tailings?

7. Explain why it is more environmentally damaging to obtain minerals from low-grade ores than to extract them from high-grade ores.

8. How did Copper Basin, Tennessee, become an environmental disaster?

9. What are manganese nodules? Where are they located?

10. Historically, the cost of environmental damage arising from mining and processing minerals was not included in the price of consumer products. Do you think it should be? Why or why not?

11. Outline the benefits of beverage container deposit laws.

12. Sketch mineral flow in a traditional industrialized society. Make sure you include mines, mills, factories, homes, and businesses. Now sketch mineral flow in a low-waste industrialized society. Which diagram is more complex, and why?

13. What is dematerialization?

14. Some people in industry argue that the planned obsolescence of products, which means they must be replaced often, creates jobs. Others think that the production of smaller quantities of durable, repairable products would generate jobs and stimulate the economy. Explain each viewpoint.

15. How does the industrial ecosystem at Kalundborg resemble a natural ecosystem? What are some of the environmental benefits of Kalundborg's industrial ecosystem?

Quantitative questions relating to this chapter are on our Web site.

Take a Stand

Visit our Web site at http://www.wiley.com/college/raven (select Chapter 16 from the Table of Contents) for links to more information about the controversy surrounding mining the ocean floor. Consider the views of proponents and opponents, and debate the issue with your classmates. You will find tools to help you organize your research, analyze the data, think critically about the issues, and construct a well-considered argument. *Take a Stand* activities can be done individually or as a team, as oral presentations, written exercises, or Web-based (e-mail) assignments.

Additional online materials relating to this chapter, including a Student Testing Section with study aids and self-tests, Environmental News, Activity Links, Environmental Investigations, and more, are also on our Web site.

Preserving Earth's Biological Diversity

Nesting pair of bald eagles. (Matthew Beck, Inverness, Florida)

The American bald eagle—the symbol of the United States and an emblem of strength—was a common sight throughout colonial North America. More recently, the bald eagle fell on hard times. Its numbers dropped precipitously to only 417 nesting pairs in the lower 48 states in 1963, and it was in danger of extinction.

Several factors contributed to its decline. As European settlers pushed across North America, they cleared many forests near lakes and rivers, destroying the bald eagle's habitat. Eagles were hunted for sport and because it was thought they had a significant impact on commercially important fishes. In fact, bounties were offered for dead bald eagles as recently as 1952. In addition, the eagles' numbers dwindled because they could not reproduce at high enough levels to ensure their population growth or their survival. Their reproductive failure was the direct result of ingesting food contaminated with the pesticide DDT (dichlorodiphenyltrichloroethane), which made the eagles' eggs so thin-shelled that they cracked open before the embryos could mature and hatch (see Figure 7.5). Mercury, lead, and selenium were other environmental pollutants that harmed bald eagles.

Banning the use of DDT in the United States in 1972 started the recovery efforts for the bald eagle, which was listed as an endangered species following enactment of the *Endangered Species Act* (ESA) in 1973. Conservation efforts involved the U.S. Fish and Wildlife Service (FWS), other federal agencies, state and local governments, Native American tribes, conservation organizations, universities, corporations, and individuals. In addition to raising birds in captive-breeding programs, biologists removed eagle eggs from their nests in nature, raised the baby eagles in wildlife refuges, and returned them to nature. Removal of eggs helped increase the number of eagles because nesting eagles commonly lay additional eggs to replace those that were removed.

As a result of continuing efforts, the number of nesting pairs in the continental United States increased to more than 7000 pairs in 2007. (Nesting pairs are counted because bald eagles mate for life.) In 1994 the bald eagle was removed from the endangered list and transferred to the less critical threatened list; in 2007, bald eagles had recovered enough to be removed from the list of endangered and threatened species. Although the ESA no longer applies to bald eagles, these birds will still have federal protection under the *Bald and Gold Eagle Protection Act* and the *Migratory Bird Treaty Act*.

Today the bald eagle symbolizes more than a country, for the bald eagle demonstrates that we can preserve our biological heritage if enough people care to do something about it. In this chapter we first examine the importance of all forms of life, and then we consider extinction, which is an increasing threat to so many organisms. Finally, we explore how to preserve biological resources and save at least some endangered species from disappearing forever.

World View and Closer to You...
Real news footage relating to biodiversity preservation around the world and in your own geographic region is available at www.wiley.com/college/raven by clicking on "World View and Closer to You."

Biological Diversity

LEARNING OBJECTIVES

- Define *biological diversity* and distinguish among genetic diversity, species richness, and ecosystem diversity.
- Relate several important ecosystem services provided by biological diversity.

A **species** is a group of more or less distinct organisms that are capable of interbreeding with one another in nature to produce fertile offspring but do not interbreed with other organisms. We do not know exactly how many species exist. In fact, biologists now realize how little we know about Earth's diverse organisms. Scientists estimate there may be as few as 5 million to 10 million or as many as 100 million species. To date, about 1.8 million species have been scientifically named and described, including more than 330,000 plant species, 45,000 vertebrate animal species, and some 950,000 insect species. About 10,000 new species are identified each year.

The variation among organisms is referred to as biological diversity or **biodiversity,** but the concept includes much more than simply the number of species, called **species richness.** Biological diversity occurs at all levels of biological organization, from populations to ecosystems. It takes into account **genetic diversity,** the genetic variety *within* all populations of that species (**Figure 17.1**). Biological diversity also includes **ecosystem diversity,** the variety of ecosystems found on Earth: the forests, prairies, deserts, coral reefs, lakes, coastal estuaries, and other ecosystems of our planet. Ecosystem diversity also encompasses the variety of interactions among organisms in natural communities. For example, a forest community with its trees, shrubs, vines, herbs, insects, worms, vertebrate animals, fungi, bacteria, and other microorganisms has greater ecosystem diversity than a wheat field.

biological diversity: The number, variety, and variability of Earth's organisms; consists of three components: genetic diversity, species richness, and ecosystem diversity.

Why We Need Organisms

Humans depend on the contributions of thousands of species for their survival. In primitive societies, these contributions are direct: Plants, animals, and other organisms are the sources of food, clothing, and shelter. In industrialized societies most people do not hunt for their morning breakfasts or cut down trees for their shelter and firewood. Nevertheless, we still depend on organisms.

Although all societies make use of many kinds of plants, animals, fungi, and microorganisms, most species have not been evaluated for their potential usefulness. There are more than 330,000 known plant species, but at least 250,000 of them have not been assessed for their industrial, medicinal, or agricultural potential. The same is true of most of the millions of microorganisms, fungi, and animals. Most people do not think of insects as an important biological resource, but insects are instrumental in several important ecological and agricultural processes, including pollination of crops, weed control, and insect pest control. In addition, many insects produce unique chemicals that may have important applications for human society. Bacteria and fungi provide us with foods, antibiotics, and other medicines, as well as important biological processes such as nitrogen fixation (see Chapter 5). Biological diversity represents a rich, untapped resource for future uses and benefits, and many as-yet-unknown species may someday provide us with products. A reduction in biological diversity decreases this treasure prematurely and permanently.

Ecosystem Services and Species Richness   The living world is a complex system. Each ecosystem is composed of many separate parts, the functions of which are organized and integrated to maintain the ecosystem's overall performance. The activities of all organisms are interrelated; we are linked and dependent on one another and on the physical environment, often in subtle ways. When one species declines, other species linked to it may decline or increase in number.

Consider, for example, the role of alligators in the environment (**Figure 17.2**). The American alligator helps maintain populations of smaller fishes by eating the gar, a fish

Figure 17.1   Genetic diversity in corn.
The variation in corn kernels and ears is evidence of the genetic diversity in the species *Zea mays*. (David Cavagnaro)

Figure 17.2   Role of alligators in the environment. The American alligator *(Alligator mississippiensis)* plays an integral, but often subtle, role in its natural ecosystem. Photographed in the Florida Everglades. (Ed Darack/Visuals Unlimited)

that preys on them. Alligators dig underwater holes that other aquatic organisms use during droughts when the water level is low. The nest mounds they build are enlarged each year and eventually form small islands colonized by trees and other plants. In turn, the trees on these islands support heron and egret populations. The alligator habitat is maintained in part by underwater "gator trails," which help clear out aquatic vegetation that might eventually form a marsh.

Plants, animals, fungi, and microorganisms are instrumental in many environmental processes essential to human existence. Forests are not just a potential source of lumber; they provide watersheds from which we obtain fresh water, they control the number and severity of local floods, and they reduce soil erosion. Many flowering plant species depend on insects to transfer pollen for reproduction. Animals, fungi, and microorganisms help keep the populations of various species in check so that the numbers of one species do not increase enough to damage the stability of the entire ecosystem. Soil dwellers, from earthworms to bacteria, develop and maintain soil fertility for plants. Bacteria and fungi perform the crucial task of decomposition, which allows nutrients to cycle in the ecosystem. All these processes are **ecosystem services,** important environmental benefits that ecosystems provide to people, such as clean air to breathe, clean water to drink, and fertile soil in which to grow crops. Ecosystem services maintain the living world, including human societies, and we are completely dependent on these basic support services. (Table 4.1 summarizes some important ecosystem services.)

You might think that the loss of some species from an ecosystem would not endanger the rest of the organisms, but this is far from true. Imagine trying to assemble an automobile if some of the parts were missing. You might piece it all together so that it resembled a car, but it probably would not run as well. Similarly, the removal of species from a community makes an ecosystem run less smoothly. If enough species are removed, the entire ecosystem will change. Species richness within an ecosystem provides the ecosystem with resilience, that is, the ability to recover from environmental changes or disasters (see the discussion of species richness and community stability in Chapter 4).

Genetic Reserves   The maintenance of a broad genetic base is critical for each species' long-term health and survival. Consider economically important crop plants. During the 20th century, plant scientists developed genetically uniform, high-yielding varieties of important food crops such as wheat. It quickly became apparent, however, that genetic uniformity resulted in increased susceptibility to pests and disease.

By crossing the "super strains" with more genetically diverse relatives, disease and pest resistance can be reintroduced into such plants. A corn blight fungus that ruined the corn crop in the United States in 1970 was brought under control by crossing the cultivated, highly uniform U.S. corn varieties with genetically diverse ancestral varieties from Mexico. When some of the genes from Mexican corn were incorporated into the U.S. varieties, the U.S. strains became resistant to the corn blight fungus. (The global decline in domesticated plant and animal varieties is discussed in Chapter 19.)

Scientific Importance of Genetic Diversity   *Genetic engineering*, the incorporation of genes from one organism into a different species (see Figure 19.9), makes it possible to use the genetic resources of organisms on a wide scale. The gene for human insulin, for example, was engineered into bacteria. These bacteria subsequently become tiny chemical factories, manufacturing at a relatively low cost the insulin required in large amounts by diabetics. Genetic engineering, available since the mid-1970s, has provided new vaccines, more productive farm animals, and agricultural plants with desirable characteristics such as disease resistance.

Although we have the skills to transfer genes from one organism to another, we do not have the ability to *make* genes that encode for specific traits. Genetic engineering depends on a broad base of genetic diversity from which it obtains genes. It has taken hundreds of millions of years for **evolution** to produce the genetic diversity found in organisms living on our planet today. This diversity may hold solutions to today's problems and to future problems we have not begun to imagine. It would be unwise to allow such an important part of our heritage to disappear.

Medicinal, Agricultural, and Industrial Importance of Organisms   The genetic resources of organisms are vitally important to the pharmaceutical industry, which incorporates into its medicines many hundreds of chemicals derived from organisms. From extracts of cherry and horehound for cough medicines to certain ingredients of periwinkle and mayapple for cancer therapy, derivatives of plants play important roles in the treatment of illness and disease (**Figure 17.3**). Many of the natural products taken directly from marine organisms, such as tunicates, red algae, mollusks, corals, and sponges, are promising anticancer or antiviral drugs. The AIDS (acquired immune deficiency syndrome) drug AZT (azidothymidine), for example, is a synthetic derivative of a compound from a sponge. The 20 best-selling prescription drugs in the United States are either natural products, natural products that are slightly modified chemically, or synthetic drugs whose chemical structures were originally obtained from organisms.

The agricultural importance of plants and animals is indisputable, because we must eat to survive. However, the kinds of foods we eat are limited compared with the total number of edible species. There are probably many species that are nutritionally superior to our common foods. Quinoa, a plant long cultivated as food in the Andes Mountains in South America, looks and tastes somewhat like rice but has a much higher concentration of protein and is more nutritionally balanced. Winged beans are a tropical legume from Southeast Asia and Papua New Guinea. Because the seeds of the winged bean contain large quantities of protein and oil, they may be the tropical equivalent of soybeans. Almost all parts of the plant are edible, from the young, green fruits to the starchy storage roots.

Modern industrial technology depends on a broad range of products from organisms. Plants supply oils and lubricants, perfumes and fragrances, dyes, paper, lumber, waxes, rubber and other elastic latexes, resins, poisons, cork, and fibers. Animals provide wool, silk, fur, leather, lubricants, waxes, and transportation, and they are important in medical research. The armadillo, for example, is used for research in Hansen's disease (leprosy) because it is one of only two species known to be susceptible to that disease (the other species is humans).

Insects secrete a large assortment of chemicals that represent a wealth of potential products. Certain beetles produce steroids with birth-control potential, and fireflies produce a compound that may be useful in treating viral infections. Centipedes secrete a fungicide over the eggs of their young that could help control the fungi that attack crops. Because biologists estimate that perhaps 90% of all insects have not yet been identified, insects represent an important potential biological resource.

Aesthetic, Ethical, and Spiritual Value of Organisms   Organisms not only contribute to human survival and physical comfort, but they also provide recreation, inspiration, and spiritual solace. Our natural world is a thing of beauty largely because of the diversity of living forms found in it. Artists have attempted to capture this beauty in drawings, paintings, sculpture, and photography, and poets, writers, architects, and musicians have created works reflecting and celebrating the natural world.

The strongest ethical consideration involving the value of organisms is how humans perceive themselves in relation to other species. Traditionally, many human cultures view themselves as superior beings, subduing and exploiting other forms of life for their benefit. An alternative view is that organisms have intrinsic value in and of themselves and that as stewards of the life forms on Earth, humans should watch over and protect their existence (see Chapter 2 for a discussion of *environmental ethics*).

Figure 17.3   Medicinal value of the rosy periwinkle. The rosy periwinkle *(Catharanthus roseus)* produces chemicals effective against certain cancers. Drugs from the rosy periwinkle have increased the chance of surviving childhood leukemia from about 5% to more than 95%. (Doug Wechsler)

REVIEW

1. What is biological diversity?

2. What are ecosystem services?

Endangered and Extinct Species

LEARNING OBJECTIVES

- Define *extinction* and distinguish between background extinction and mass extinction.
- Contrast threatened and endangered species, and list four characteristics common to many endangered species.
- Define *biodiversity hotspots* and explain where most of the world's biodiversity hotspots are located.
- Describe four human causes of species endangerment and extinction and tell which cause is the most important.
- Explain how invasive species endanger native species.

extinction: The elimination of a species from Earth.

Extinction, the death of a life form, occurs when the last individual member of a species dies. Extinction is an irreversible loss: Once a species is extinct it will never reappear. Biological extinction appears to be the eventual fate of all species, much as death is the eventual fate of all individuals. Biologists estimate that for every 2000 species that have ever lived, 1999 of them are extinct today.

During the time span in which organisms have occupied Earth, a continuous, low-level extinction of species, or **background extinction,** has occurred. At certain periods in Earth's history, maybe five or six times, there has been a second kind of extinction, **mass extinction,** in which numerous species disappeared during a relatively short period of geologic time. The course of a mass extinction episode may have taken millions of years, but that is a short time compared with Earth's age, which is estimated at 4.6 billion years.

The causes of past mass extinctions are not well understood, but biological and environmental factors were probably involved. A major climate change could have triggered the mass extinction of species. Marine organisms are particularly vulnerable to temperature changes; if Earth's temperature changed just a few degrees, it is likely that many marine species would have become extinct. It is possible that mass extinctions of the past were triggered by catastrophes, such as the collision of Earth and a large asteroid or comet. The impact could have forced massive quantities of dust into the atmosphere, blocking the sun's rays and cooling the planet.

Although extinction is a natural biological process, it is greatly accelerated by human activities. The burgeoning human population has spread into almost all areas of Earth. Whenever humans invade an area, the habitats of many organisms are disrupted or destroyed, which contributes to their extinction. For example, the dusky seaside sparrow, a small bird that was found only in the marshes of St. Johns River in Florida, became extinct in 1987, largely due to human destruction of its habitat.

Currently, Earth's biological diversity is disappearing at an unprecedented rate (**Figure 17.4**). Conservation biologists estimate that species are presently becoming extinct at a rate at 100 to 1000 times the natural rate of background extinctions. For example, the first World Conservation Union *Red List of Threatened Plants*, issued in 1997 and based on 20 years of data collection and analysis around the world, lists about 34,000 species of plants currently threatened with extinction (see You Can Make a Difference: Declining Biological Diversity).

Endangered and Threatened Species

endangered species: A species that faces threats that may cause it to become extinct within a short period.

The legal definition of an **endangered species**, as stipulated in the Endangered Species Act, is a species in imminent danger of extinction throughout all or a significant portion of its range. (The area in which a particular species is found is its **range.**) A species is endangered when its numbers are so severely reduced that it is in danger of becoming extinct without human intervention.

threatened species: A species whose population has declined to the point that it may be at risk of extinction.

A species is defined as **threatened** when extinction is less imminent but its population is quite low. The legal definition of a threatened species is a species likely to become endangered in the foreseeable future, throughout all or a significant portion of its range.

Figure 17.4 **Representative endangered or extinct species.** Officials at the U.S. Fish and Wildlife Service estimate that more than 500 U.S. species have gone extinct during the past 200 years. Of these, roughly 250 have become extinct since 1980.

Endangered and threatened species represent a decline in biological diversity because as their numbers decrease, their genetic variability is severely diminished. Long-term survival and evolution depend on genetic diversity, so a decline in genetic diversity adds to the risk of extinction for endangered and threatened species, as compared to species that have greater genetic variability.

Characteristics of Endangered Species Many endangered species share certain characteristics that seem to have made them more vulnerable to extinction. Some of these characteristics are: having an extremely small (localized) range; requiring a large territory; living on islands; having a low reproductive success, often the result of a small population size or low reproductive rates; needing specialized breeding areas; and having specialized feeding habits.

Many endangered species have a limited natural range, which makes them particularly prone to extinction if their habitat is altered. The Tiburon mariposa lily consists of a single population growing on a hilltop near San Francisco. Development of that area would almost certainly cause the extinction of this species.

YOU CAN MAKE A DIFFERENCE

Declining Biological Diversity

Your children and grandchildren are faced with inheriting a biologically impoverished world, but people dedicated to preserving our biological heritage can reverse the trend toward extinction. You do not have to be a biologist to make a contribution; some of the most important contributions come from outside the biological arena. Following is a partial list of actions that would help maintain the biological diversity that is our heritage.

1. A political commitment to protect organisms is necessary because no immediate or short-term economic benefit is obtained from conserving species. This commitment must take place at all political levels, from local to international. Lawmaking will not ensure the protection of organisms without strong public support. Increasing public awareness of the importance of biological diversity is critical.

2. Providing publicity on species conservation issues costs money. Private funds raised by organizations such as the Sierra Club, the Nature Conservancy, and the World Wildlife Fund support such endeavors, but clearly more money is needed. As an individual, you can join and actively support conservation organizations.

3. Before an endangered species can be saved, its numbers, range, ecology, biological nature, and vulnerability to changes in its environment must be determined. Basic research provides this information. We cannot preserve a given species effectively until we know how large to make a protected habitat and what features are essential in its design. As an individual, you can inform state and national politicians of your desire to have conservation research funded with tax dollars.

4. A worldwide system of protected parks and reserves that includes every major ecosystem must be established. The protected land would provide other benefits in addition to the preservation of biological diversity. It would safeguard the watersheds that supply water, and it would serve as a renewable source of important biological products in areas with multiple uses. It would provide people with unspoiled lands for aesthetic and recreational enjoyment. As an individual, you can help establish parks by writing to national lawmakers.

5. The establishment of parks and refuges will not prevent biological impoverishment if we continue to pollute the planet, because it is impossible to protect parks and refuges from threats such as acid rain, stratospheric ozone depletion, and climate change. Strong steps must be taken to curb the practice of dumping toxins into the air, soil, and water—for human health and well-being as well as for the well-being of the species so important to ecosystem stability. Specific recommendations on how you as an individual can reduce pollution are discussed in Chapters 20 through 24.

6. Developing nations in the tropics, the repositories of most of Earth's genetic diversity, do not have much money to spend on conservation. Their governments are consumed with human problems such as overpopulation, disease, and foreign debts. One way to help such countries appreciate the importance of the biological resources they possess is to allow them to charge fees for the use of genetic material. Much of the money earned could be used to help alleviate human problems. And some of the money from genetic resources could be used to provide protection for organisms, thus preserving biological diversity for continued, sustained exploitation. Promoting ecotourism is a second way that people benefit financially by protecting their biological resources. In ecotourism people pay to visit natural environments and view native species. When done correctly, ecotourism conserves natural areas and improves the well-being of local people. A third way of providing economic incentives to developing nations is for highly developed countries to forgive or reduce debts owed by such nations. In exchange, the developing countries would agree to support local conservation efforts, including the protection of endangered species. You can help in the formulation of such policies. Let your lawmakers know where you stand. Join and support conservation groups. Campaign to preserve our biological heritage for future generations. All of these actions require thinking about biological diversity from a systems perspective. The choices you make and the policies you support at all levels of government can impact ecosystems and species around the world.

Species that require extremely large territories to survive—often because they are tertiary consumers at the top of the food web—may be threatened with extinction when all or part of their territory is modified by human activity. The California condor, a scavenger bird that lives off of carrion and requires a large, undisturbed territory—hundreds of square kilometers—to find adequate food, is slowly recovering from the brink of extinction. In 1983 the California condor population reached a low of 22 birds, and from 1987 to 1992, it was no longer found in nature (**Figure 17.5**). A program to reintroduce zoo-bred California condors into the wild began in 1992. Currently, there are about 200 condors flying free in California and adjacent areas in Mexico and Arizona. The condor story is not an unqualified success, because they have yet to become a self-sustaining population that replaces its numbers without augmentation by captive breeding.

Many species **endemic** to certain islands (that is, they are not found anywhere else in the world) are endangered. These organisms often have small populations that cannot be replaced by immigration if their numbers are destroyed. Because they evolved in isolation from competitors, predators, and disease organisms, island species have few defenses when such organisms are introduced, usually by humans. It is not surprising that of the 171 bird species that have become extinct in the past few centuries, 155 of them lived on islands.

In ecological terms, *island* refers not only to any landmass surrounded by water but also to any isolated habitat surrounded by an expanse of unsuitable territory. Accordingly, a small patch of forest surrounded by agricultural and suburban lands is considered an island. **Habitat fragmentation,** the breakup of large areas of habitat into small, isolated patches (that is, islands), is a major threat to the long-term survival of many species. (National parks are discussed as islands in Chapter 18.)

For a species to survive, its members must be present within their range in large enough numbers for males and females to mate. The minimum population density and size that ensure reproductive success vary from one type of organism to another. For all organisms, if the population density and size fall below a critical minimum level, the population declines, becoming susceptible to extinction.

Endangered species often share other characteristics. Some have low reproductive rates. The female blue whale produces a single calf every other year, and no more than 6% of swamp pinks, an endangered species of small flowering plant, produce flowers in a given year (**Figure 17.6**). Some endangered species breed only in specialized areas; the Kemp's Ridley sea turtle, for example, lays its eggs on a single beach in Mexico.

Highly specialized feeding habits may endanger a species. In nature, the giant panda eats only bamboo. Periodically, all the bamboo plants in a given area flower and die; when this occurs, panda populations face starvation. Like many other endangered species, giant pandas are also endangered because their habitat has been fragmented into small islands, and there are few intact habitats where they can survive. China's 1600 wild giant pandas live in isolated habitats that occupy a small fraction of their historic range. China, in partnership with the World Wildlife Fund (WWF)[1], has recently increased the number of panda reserves to 40 and added five forested **wildlife corridors.** These wildlife corridors allow pandas in isolated habitats to interbreed.

Where Is Declining Biological Diversity the Greatest Problem?

Declining biological diversity is a concern throughout the United States, but it is most serious in the states of Hawaii (where 63% of species are at risk) and California (where about 29% of species are at risk). Hawaii has lost hundreds of species and has more species listed as endangered than any other state. At least two-thirds of Hawaii's native forests are gone.

As serious as declining biological diversity is in the United States, it is even more serious abroad, particularly in tropical rain forests. Tropical rain forests are found in South and Central America, central Africa, and Southeast Asia. Although tropical rain forests cover only 7% of Earth's surface, as many as 50% of all species inhabit them.

Ecosystem loss and degradation are occurring in many places around the world, but tropical rain forests are being destroyed faster than almost all other ecosystems. Using remote sensing surveys, scientists have determined that approximately 1% of tropical rain forests are being cleared or severely degraded each year. The forests are making way for human settlements, banana plantations, oil and mineral explorations, and other human activities.

Many species in tropical rain forests are endemic, and the clearing of tropical rain forests contributes to their extinction. Tropical rain forests provide important ecosystem services that help maintain their ecosystem. The forest itself generates much of the rainfall in tropical rain forests. If half of the existing rain forest in the Amazon region of

Figure 17.5   California condor. California condors *(Gymnogyps californianus)*, which have up to a 3-meter (10-foot) wingspan, are critically endangered largely because development has reduced the size of their wilderness habitat. Photographed in the San Diego Wild Animal Park. (Tom Mc Hugh/ Photo Researchers, Inc.)

Figure 17.6   The swamp pink. This endangered species lives in the boggy areas of the eastern United States. Photographed in Killens Pond State Park, Delaware. (Jeffrey Lepore/Photo Researchers, Inc.)

[1] Outside of the United States, Canada, and Australia, the World Wildlife Fund is known as the World Wide Fund for Nature.

South America were destroyed, precipitation in the remaining forest would decrease. As the land became drier, organisms adapted to moister conditions would be replaced by organisms that tolerate the drier conditions. Many of the original species, being endemic and unable to tolerate the drier conditions, would become extinct.

Perhaps the most unsettling outcome of tropical **deforestation** is its disruptive effect on evolution. In Earth's past, mass extinctions were followed during the next several million years by the evolution of new species to replace those that died out. After the dinosaurs became extinct, ancestral mammals evolved into the variety of running, swimming, flying, and burrowing mammals that exist today. The evolution of a large number of related species from an ancestral organism is known as **adaptive radiation.** In the past, tropical rain forests may have supplied ancestral organisms from which adaptive radiations could occur. Destroying tropical rain forests may be reducing nature's ability to replace its species through adaptive radiation. (Of course, even if adaptive radiation occurs following the current mass extinction episode, it will require millions of years—a timescale that is useless to humans.)

A few countries, primarily developing nations, hold most of the biological diversity that is so ecologically and economically important to the entire world. The situation is complicated because these countries often cannot afford protective measures to maintain biological diversity. (Tropical rain forests are also discussed in Chapters 6 and 18.)

Earth's Biodiversity Hotspots   In the 1980s ecologist **Norman Myers** of Oxford University coined the term biodiversity hotspots. In 2000, using plants as their criteria, Myers and ecologists at Conservation International identified 25 biological hotspots around the world (**Figure 17.7**). As many as 44% of all species of vascular plants live within the hotspots. Interestingly, these 25 hotspots for plants contain 29% of the world's endemic bird species, 27% of endemic mammal species, 38% of endemic reptile species, and 53% of endemic amphibian species. Many humans—nearly 20% of the world's population—live in the hotspots. Fifteen of the 25 hotspots are tropical, and nine are mostly or solely islands.

Many biologists recommend that conservation planners focus on preserving land in these hotspots to reduce the mass extinction of species currently underway. Not all biologists agree. Some critics think that concentrating most of our efforts on the 25 biodiversity hotspots causes us to neglect species living in other habitats, such as deserts, grasslands, tundra, and temperate forests, all of which are also at risk.

biodiversity hotspots: Relatively small areas of land that contain an exceptional number of endemic species and are at high risk from human activities.

Figure 17.7   Biological hotspots. These 25 hotspots, which are rich in endemic species, are at great risk from human activities. (Conservation International)

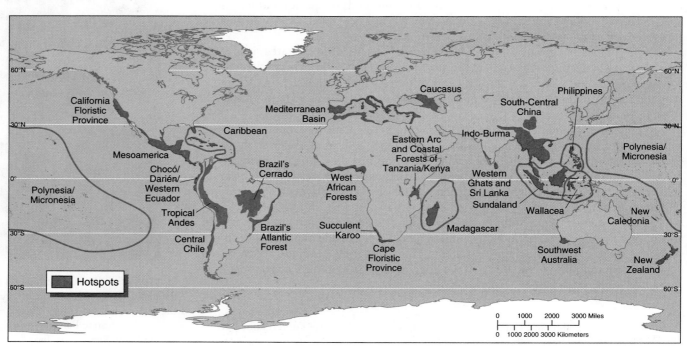

Human Causes of Species Endangerment

In 2001 the United Nations requested a **Millennium Ecosystem Assessment** to gather scientific information about ecosystem changes and the effects of these changes on human well-being. Several reports were published in 2005 based on the work of more than 1300 scientists from 95 countries. One of these reports, the *Biodiversity Synthesis Report*, examined the important links between ecosystem health and biological diversity. The report stated that biological diversity is declining rapidly due to several factors.

Scientists generally agree that the single greatest threat to biological diversity is land use change, which causes loss of habitat. The spread of invasive species, overexploitation, and pollution (including climate change from carbon dioxide pollution) are also important. Underlying these direct causes of declining biological diversity are human population increase, increasing economic activity, increased use of technology, and social, political, and cultural factors (**Figure 17.8**). All of these direct and indirect factors interact in complex ways, and so it is most effective to deal with the problem of declining biological diversity using a system perspective. Addressing a single factor, such as overexploitation, without considering the other factors that may amplify declining biological diversity is probably doomed to failure.

Land Use Change   Most species facing extinction today are endangered because of the destruction, fragmentation, or degradation of habitats by human activities (**Figure 17.9**; also see Figure 18.11). We demolish or alter habitats when we build roads, parking lots, bridges, and buildings; clear forests to grow crops or graze domestic animals; and log forests for timber. We drain marshes to build on aquatic habitats, thus converting them to terrestrial ones, and we flood terrestrial habitats when we build dams with their reservoirs. Exploration and mining of minerals, including fossil fuels, disrupt the land and destroy habitats. Habitats are altered by outdoor recreation, including off-road vehicles, hiking off-trail, golfing, skiing, and camping. Because most organisms are utterly dependent on a particular type of environment, habitat destruction reduces their biological range and ability to survive.

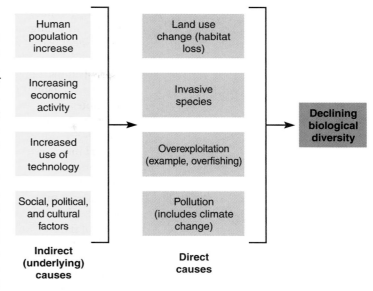

Figure 17.8   Indirect and direct causes of declining biological diversity. These causes interact with and amplify the effects of each other.

Figure 17.9   Destruction of the world's wildlife habitats. What is one of the main causes of habitat loss in terrestrial ecosystems? (© Steve Greenberg)

As the human population has grown, the need for increased amounts of food has resulted in a huge conversion of forests and other natural lands into croplands and permanent pastures. According to the U.N. Food and Agriculture Organization, total agricultural lands currently occupy 38% of Earth's land area (see Figure 18.1). Agriculture has a major impact on aquatic ecosystems because of the diversion of water for irrigation.

Little habitat remains for many endangered species. The grizzly bear, for example, occupies about 2% of its original habitat in the lower 48 states of the United States. Human population growth and the extraction of resources have destroyed most of the grizzly's wilderness habitat.

Habitat destruction, fragmentation, and degradation are happening around the world. As entire habitats are transformed for human purposes, many species are becoming extinct, and the genetic diversity within many surviving species is being reduced.

Africa provides a vivid example of a systems issue—the conflict between humans and species such as elephants over land use. African elephants are nomads that require a lot of natural landscape in which to forage for the hundreds of kilograms of food that each consumes daily. In Africa people are increasingly pushing into the elephants' territory to grow crops and graze farm animals. The elephants often trample and devour crops, ruining a year's growth of crops in a single night; they have even killed people. Farmers cannot shoot at or kill elephants because they are protected. (Before elephants were listed as a protected species, their numbers had declined precipitously because of overhunting by ivory hunters.) A study of 25 African regions with both wild areas and human settlements found that when human density increases to a certain level, the elephants migrate out of the area. The problem is that the wild areas to which elephants can move are steadily shrinking. One of the great challenges is finding a way to allow people and elephants to coexist in an increasingly crowded world.

Invasive Species **Biotic pollution,** the introduction of a foreign species into an ecosystem in which it did not evolve, often upsets the balance among the organisms living in that area and interferes with the ecosystem's normal functioning. Unlike other forms of pollution, which may be cleaned up, biotic pollution is usually permanent. The foreign species may compete with native species for food or habitat or may prey on them. Generally, an introduced competitor or predator has a greater negative effect on local organisms than do native competitors or predators. Foreign species whose introduction causes economic or environmental harm are called invasive species (**Figure 17.10**).

Although invasive species may be introduced into new areas by natural means, humans are usually responsible for such introductions, either knowingly or unknowingly. The water hyacinth was deliberately brought from South America to the United States because it has lovely flowers. Today it has become a nuisance in Florida waterways, clogging them so that boats cannot easily move and crowding out native species.

Islands are particularly susceptible to the introduction of invasive species. Less than 50 years ago, the brown tree snake was accidentally introduced in Guam, an island in the West Pacific. Thought to have arrived from the Solomon Islands on a U.S. Navy ship shortly after the end of World War II, the brown tree snake thrived and is now estimated to number about three million. It started consuming rainforest birds in large numbers, and as a result, 9 of the 12 native species of forest birds are extinct in nature. The Guam rail, a flightless bird endemic to Guam, numbered about 80,000 in 1968 but was extinct in nature by 1986; today it exists only in small captive populations safely isolated from the snakes. The snakes have also decimated Guam's small reptiles and mammals.

In the mid-1980s an aggressive aquarium-bred strain of alga (*Caulerpa*) was accidentally released into the Mediterranean Sea when a seaside aquarium cleaned out its tanks. The alga spread like a dense carpet over large parts of the Mediterranean seabed, crowding out biologically diverse seafloor communities of native sea grasses, sponges, corals, sea fans, anemones, sea stars, and lobsters. *Caulerpa* is toxic to many Mediterranean species. Other regions far from the Mediterranean are concerned that *Caulerpa* could cause havoc to their marine ecosystems. It is now well established in southeast Australia. The United States banned the importation of *Caulerpa*

invasive species: Foreign species that spread rapidly in a new area where they are free of predators, parasites, or resource limitations that may have controlled their populations in their native habitat.

Figure 17.10 **Invasive species.** Selected examples of the more than 6500 established foreign species accidentally or deliberately introduced into the United States.

under the **Federal Noxious Weed Act.** However, in 2000 *Caulerpa* was discovered off the coast of California. A successful eradication effort that involved covering *Caulerpa* patches with tarps and pumping in poisonous chlorine gas took six years and cost $7 million.

Overexploitation Sometimes species become endangered or extinct as a result of deliberate efforts to eradicate or control their numbers. Many of these species prey on game animals or livestock. Ranchers, hunters, and government agents have reduced populations of large predators such as the wolf and grizzly bear. Predators of game animals and livestock are not the only animals vulnerable to human control efforts. Some animals are killed because their lifestyles cause problems for humans. The Carolina parakeet, a beautiful green, red, and yellow bird endemic to the southern United States, was extinct by 1920, exterminated by farmers because it ate fruit and grain crops.

Unregulated hunting, or overhunting, was a factor contributing to the extinction of certain species in the past but is now strictly controlled in most countries. The passenger pigeon was one of the most common birds in North America in the early 1800s, but a century of overhunting, coupled with habitat loss, resulted in its extinction in the early 1900s. Unregulated hunting was one of several factors that caused the near extinction of the American bison. Bison were decimated by the U.S. Army, which killed bison to disrupt the food supply of the Plains Indians, and by commercial hunters, who killed bison for their hides and tongues (considered a choice food) as well as meat for work crews of railroad companies.

Illegal commercial hunting, or *poaching*, endangers many larger animals, such as the tiger, cheetah, and snow leopard, whose beautiful furs are quite valuable (**Figure 17.11**). Rhinoceroses are slaughtered primarily for their horns, used for ceremonial dagger handles in the Middle East and for purported medicinal purposes in Asian medicine. Bears are killed for their gallbladders, used in Asian medicine to treat ailments ranging from indigestion to heart disease. Endangered American turtles are captured and exported illegally to China, where they are killed for food. Caimans (reptiles similar to crocodiles) are killed for their skins and made into shoes and handbags. Although all of these animals are legally protected, the demand for their products on the black market has caused them to be hunted illegally.

In West Africa, poaching has contributed to the decline in lowland gorilla and chimpanzee populations. The meat (called *bushmeat*) of these rare primates and other protected species such as anteaters, elephants, and mandrill baboons provides an important source of protein for indigenous people. Bushmeat is also sold to urban restaurants.

Commercial harvest is the collection of a live organism from nature. Commercially harvested organisms end up in zoos, aquaria, biomedical research laboratories, circuses, and pet stores. Several million birds are commercially harvested each year for the pet trade, but unfortunately many of them die in transit, and many more die from improper treatment after they are in their owners' homes. Although it is illegal to capture endangered animals from nature, there is a thriving black market, mainly because collectors in the United States, Canada, Europe, and Japan are willing to pay large amounts to obtain a variety of species, particularly rare tropical birds (**Figure 17.12**). The United States passed the **Wild Bird Conservation Act** of 1992 that imposed a moratorium on importing rare bird species. Poaching data collected before and after 1992 indicated a drop in poaching rates after the law went into effect.

Animals are not the only organisms threatened by excessive commercial harvest. Many unique and rare plants have been collected from nature to the point that they are endangered. These include carnivorous plants, wildflower bulbs, certain cacti, and orchids. On the other hand, carefully monitored and regulated commercial use of animal and plant resources can create an economic incentive to ensure that these resources do not disappear.

Figure 17.11 Illegal trade in products made from endangered species. (Steve Hillebrand/U.S. Fish & Wildlife Service)

Figure 17.12 Illegal animal trade. These hyacinth macaws (*Anodorhynchus hyacinthus*) were seized in French Guiana in South America as part of the illegal animal trade there. (Jany Sauvanet/Natural History Photographic Agency)

Pollution Human-produced acid rain, stratospheric ozone depletion, and climate change degrade even wilderness habitats that are "totally" natural and undisturbed. Acid rain is thought to have contributed to the decline of large stands of forest trees and the biological death of many freshwater lakes. Because ozone in the upper atmosphere shields the ground from a large proportion of the sun's harmful ultraviolet (UV) radiation, ozone depletion in the upper atmosphere represents a threat to all terrestrial life.

Climate warming, caused in part by an increase in atmospheric carbon dioxide released when fossil fuels are burned, is another threat. Overwhelming evidence indicates that recent climate changes have already affected biological diversity. Further climate change is expected to increase the rate of extinction, particularly in certain regions such as polar areas. Such habitat modifications particularly reduce the biological diversity of species with extremely narrow and rigid environmental requirements (discussed further in Chapter 21).

Excessive fertilizer use has contributed to high levels of nutrients in soil and aquatic ecosystems. Other types of pollutants that affect organisms include industrial chemicals, agricultural pesticides, organic pollutants from sewage, antibiotics and hormones from agriculture and human prescriptions, acid mine drainage seeping from mines, thermal pollution from the heated wastewater of industrial plants, and plastics (**Figure 17.13**). The effects of various forms of pollution on biological diversity are discussed throughout the text.

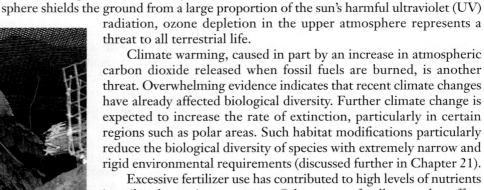

Disappearing Frogs

Of all this chapter's examples of species in trouble, frogs and other amphibians deserve special notice. Many scientists think amphibians are **indicator species** that provide an early warning of environmental damage with the potential to affect other species. Amphibians also merit attention because precipitous declines in amphibian populations are occurring around the world. The discovery in some areas of frogs and other amphibians with deformities adds another layer of complexity to the amphibian crisis.

Amphibians, represented by about 5700 species of frogs, toads, and salamanders, are survivors. These tough little animals, which typically spend part of their life in the water and part on land, have existed as a group for more than 350 million years. Despite their evolutionary resilience, amphibians are remarkably sensitive environmental indicators in both aquatic and terrestrial ecosystems. Most frogs lay gelatinous and unprotected eggs in ponds and other pools of standing water. The water is where tadpoles (immature frogs) undergo metamorphosis, maturing into adult frogs that live on land. As adults, frogs breathe primarily through their permeable skin. This moist, absorptive skin makes frogs susceptible to environmental contaminants.

Since the 1970s, many of the world's frog populations have dwindled or disappeared. According to the *Global Amphibian Assessment*, 168 amphibian species have gone extinct around the world in the last two decades, and about one-third of all amphibian species are in decline. Habitat loss is the greatest threat to amphibians, but researchers have observed that the declines are not limited to areas with obvious habitat destruction. Some remote, pristine locations also show dramatic declines in amphibians. Biologists are not certain what is causing these declines, and it appears no single factor is responsible. Potential factors for which there is strong evidence include pollutants, infectious diseases, and global climate change.

Agricultural chemicals are implicated in amphibian declines in California's Sierra Nevada Mountains. Frog populations on the eastern slopes are relatively healthy, but about eight species are declining on the western slopes, where prevailing winds carry residues of pesticides from the Central Valley, a huge agricultural region. Agricultural chemicals may also be contributing to amphibian declines in Maryland and in Ontario, Canada.

In South America and Australia, data suggest that a *chytrid* (a fungus) is responsible for massive die-offs. Climate change may be exacerbating chytrid-induced amphibian deaths. Researchers have found that amphibian populations decline the most during warmer years. Chytrids thrive in warmer temperatures. Thus, warmer temperatures cause higher rates of chytrid transmission to susceptible amphibians.

Amphibian Deformities   In 1995 Minnesota schoolchildren found that almost half of the leopard frogs they caught at a local pond were deformed—with extra legs, extra toes, eyes located on the shoulder or back, deformed jaws, bent spines, missing legs, missing toes, and missing eyes (**Figure 17.14**). (Less than 1% of frogs exhibit deformities in healthy frog populations.) Deformed frogs usually die before they reproduce. Predators easily catch frogs with extra or missing legs. Since the discovery, most states have reported abnormally large numbers of deformities in many amphibian species. Frog deformities have now been found on four continents.

Several factors may produce amphibian deformities. Several pesticides affect normal development in frog embryos. Also, infecting tadpoles with a *trematode* (a parasitic flatworm) causes the adults that develop from the tadpoles to exhibit limb deformities. Multiple environmental stressors, such as habitat loss, disease, and air and water pollution, may interact synergistically with one another to cause deformities. For example, an amphibian stressed by pesticide residues or drought may be more susceptible to a parasite. ■

Figure 17.13   Plastic pollution harms wildlife. The plastic from a broken fishing net has ensnared a stellar sea lion (*Eumetopias jubatus*), and part of the net is cutting into the seal's neck. Photographed in Alaska. (Ron Sanford/Photo Researchers, Inc)

Figure 17.14   Frog deformities. Pollution and parasites are implicated in developmental abnormalities in amphibians. (Frans Lanting/Minden Pictures, Inc.)

E N V I R O N E W S

Pollinators in Decline

Two-thirds of flowering plants depend on insects for successful pollination, and nearly a third of human food crops are pollinated by bees. Scientists in recent years have documented sharp declines in pollinator populations worldwide. In North America, the number of domesticated honeybees declined about 50% during the past several decades. The declines in bees and other pollinators are attributed to several likely threats—disease, habitat alteration through such activities as agriculture and grazing, introduction of competing pollinator species, and insect mortality from the use of pesticides.

Threats to pollinator species—and to the plants they pollinate—are serious because of the nature of many plant-pollinator relationships. This mutualistic relationship can be highly specific, meaning that the extinction of one pollinating insect species may lead to the extinction of its dependent plant species. Pollination requirements are not known for most wild plant species, or even for many crop plants. Scientists and natural resource managers increasingly stress the importance of managing and protecting wild pollinators. The future success of many wild plants and food crops depends on our learning more about the biology of important pollinators and guarding these species from additional environmental threats.

REVIEW

1. What are endangered species? threatened species?
2. What are biodiversity hotspots? Where are most biodiversity hotspots located?
3. What is the most significant cause of species endangerment and extinction?

Conservation Biology

LEARNING OBJECTIVES

- Define *conservation biology* and compare in situ and ex situ conservation.
- Describe restoration ecology.

conservation biology: The scientific study of how humans impact organisms and of the development of ways to protect biological diversity.

What strategies should we develop to cope with declining biological diversity? The broad field of **conservation biology** addresses these concerns. Conservation biology ranges from studying the processes that influence a decline in biological diversity to protecting and restoring populations of endangered species, to preserving entire ecosystems and landscapes.

Conservation biologists have demonstrated that a single large area of habitat, which has the potential to support large populations, is generally more effective at safeguarding species than several habitat fragments, each with the potential to support small populations. A large area of habitat also has the potential to support greater species richness than several habitat fragments.

It is better if areas of habitat for a given species are located close together rather than far apart. If an area of habitat is isolated from other areas, individuals of a species may not effectively disperse from one habitat to another. Because the presence of humans adversely affects many species, areas that lack roads or are inaccessible to humans are better habitats than human-accessible areas.

Virtually all conservation biologists think it is more effective and, ultimately, more economical to preserve intact ecosystems in which many species live than to work on preserving individual species one at a time. Conservation biologists typically assign a higher priority to preserving areas that are more biologically diverse than other areas (recall the earlier discussion of biodiversity hotspots).

Conservation biology includes two problem-solving techniques to save organisms from extinction: in situ and ex situ conservation. **In situ conservation** (on-site conservation), which includes the establishment of parks and reserves, concentrates on preserving biological diversity in nature. With increasing demands on land, in situ conservation cannot guarantee the preservation of all types of biological diversity.

Sometimes only ex situ conservation can save a species. **Ex situ conservation** (off-site conservation) involves conserving biological diversity in human-controlled settings. The breeding of captive species in zoos (such as the condors discussed earlier in the chapter), and the seed storage of genetically diverse plant crops are examples of ex situ conservation.

Protecting Habitats

Protecting animal and plant habitats—that is, conserving and managing the ecosystem as a whole—is the single best way to preserve biological diversity. Because human activities have adversely affected the sustainability of many of Earth's ecosystems, direct conservation management of protected areas is often required (**Figure 17.15**).

Many nations appreciate the need to protect their biological heritage and have set aside areas for biological habitats. Ecuador, Venezuela, Denmark, and the Dominican Republic have established protected areas totaling more than 30% of their land. Austria, Germany, New Zealand, Slovakia, Bhutan, and Belize have more than 20% of their land areas protected.

Currently, more than 3000 national parks, sanctuaries, refuges, forests, and other protected areas exist worldwide. These encompass some one billion hectares, an area almost as large as Canada. Some of these areas were set aside to protect specific endangered species. The world's first such refuge was established in 1903 at Pelican Island, Florida, to protect the brown pelican. Today the U.S. National Wildlife Refuge System has land set aside in more than 535 refuges. Although the bulk of the protected land is in Alaska, refuges exist in all 50 states.

Many protected areas have multiple uses that sometimes conflict with the goal of preserving species. National forests may be open for logging, grazing, and mineral extraction. The mineral rights to many refuges are privately owned, and some refuges have had military exercises conducted on them.

Protected areas are not always effective in preserving biological diversity, particularly in developing countries where biological diversity is greatest, because there is little money or expertise to manage them. According to the Worldwatch Institute, each guard in Brazil's national parks is responsible for an average of 6000 km^2, an area larger than the state of Delaware. Such "paper parks," where governments are unable or unwilling to enforce conservation laws, are even more vulnerable to logging, farming, mining, and poaching than are protected areas in the United States, Canada, and other highly developed countries.

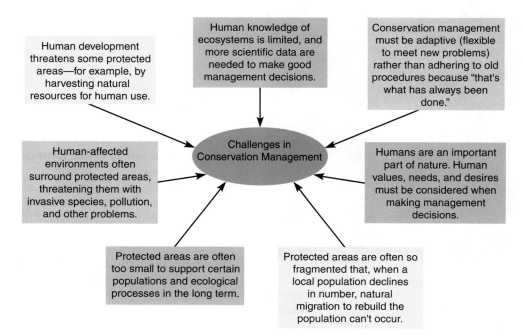

Figure 17.15   Some challenges in conservation management.

restoration ecology: The study of the historical condition of a human-damaged ecosystem, with the goal of returning it as close as possible to its former state.

Figure 17.16 **Prairie restoration.** The University of Wisconsin-Madison Arboretum pioneered restoration ecology.

(a) The restoration of the prairie was at an early stage in November 1935. (University of Wisconsin Madison Arboretum)

(b) The prairie as it looks today. This picture was taken at approximately the same location as the 1935 photograph. (Virginia Kline/University of Wisconsin, Madison Arboretum)

Another shortcoming of the world's protected areas is that many are in lightly populated mountain areas, tundra, and the driest deserts, places that often have spectacular scenery but relatively few kinds of species. In contrast, ecosystems in which biological diversity is greatest often receive little protection. A high concentration of endemic species occurs in the Philippines and in Madagascar, yet these island nations have each set aside less than 2% of their land. Protected areas are urgently needed in tropical rain forests, the tropical grasslands and savannas of Brazil and Australia, and dry forests widely scattered around the world. Desert organisms are underprotected in northern Africa and Argentina, and many islands and temperate river basins need protection.

Worldwide, about 11.5% of Earth's land area—nearly 19 million km^2 (7.3 million mi^2)—has been set aside to protect biological diversity. However, many existing protected areas are too small or too isolated from other protected areas to effectively conserve species. Also, the habitats for more than 700 highly endangered mammals, birds, reptiles, and amphibians are not included in the global network of protected areas.

Restoring Damaged or Destroyed Habitats

Although preserving habitats is an important part of conservation biology, the realities of our world, including the fact that the land-hungry human population continues to increase, dictate a variety of other conservation measures. Scientists can reclaim disturbed lands and convert them into areas with high biological diversity. **Restoration ecology**, in which the principles of ecology are used to help return a degraded environment to a more functional and sustainable one, is an important part of in situ conservation.

Since 1934 the University of Wisconsin-Madison Arboretum has carried out one of the world's most famous examples of restoration ecology (**Figure 17.16**). During that time, several distinct natural communities of Wisconsin were carefully developed on damaged agricultural land. These communities include a tallgrass prairie, a dry prairie, and several types of pine and maple forests.

Restoration of disturbed lands creates biological habitats and has additional benefits such as the regeneration of soil damaged by agriculture or mining. The disadvantages of restoration include the expense and the amount of time it requires to restore an area. Even so, restoration is an important aspect of conservation biology, because it is thought that restoration will reduce extinction.

Zoos, Aquaria, Botanical Gardens, and Seed Banks

Zoos, aquaria, and botanical gardens often play a critical role in saving individual species on the brink of extinction. Eggs may be collected from nature, or the remaining few wild animals may be captured and bred in zoos, aquaria, and other research environments.

Special techniques, such as artificial insemination and embryo transfer, are used to increase the number of wild animal offspring. In **artificial insemination,** sperm collected from a suitable male of a rare species is used to artificially impregnate a female, perhaps located in another zoo in a different city or even in another country. In **embryo transfer,** a female of a rare species is treated with fertility drugs, which cause her to produce multiple eggs. Some of these eggs are collected, fertilized with sperm, and surgically implanted into a female of a related but less rare species, who later gives birth to offspring of the rare species (**Figure 17.17**). For example, frozen embryos from the endangered African wildcat have been successfully transferred to female house cats. Another technique involves hormone patches to stimulate reproduction in endangered birds; the patch is attached under the female bird's wing.

Scientists have learned to successfully breed the endangered whooping crane in captivity should the species become extinct in the wild. The captive population now totals 145 cranes. Three wild whooping crane populations currently exist—in Florida (nonmigratory), Texas-Canada (migratory), and Wisconsin-Florida (migratory). In late 2006 the total number of wild cranes was 373. Scientists are trying to enhance existing flocks and establish additional flocks in the wild by soft-releasing cranes (for a nonmigratory flock) or flying them behind an ultralight (for a migratory flock). (In *soft release*, the birds are initially placed in a specially constructed pen to help them adjust before being released into the wild.) These efforts have not yet led to additional self-sustaining flocks.

Attempting to save a species on the brink of extinction is expensive, and only a small proportion of endangered species can be saved. Because zoos, aquaria, and botanical gardens do not have the space or money to save all endangered species, conservation biologists must prioritize which species they will attempt to save. Traditionally, the public has supported efforts to save large, charismatic species such as pandas, bald eagles, and black-footed ferrets. Smaller, less "attractive" endangered species, many of which are important providers of ecosystem services, have largely been ignored. Clearly, it is more cost-effective to maintain existing natural habitats so that species will never become endangered in the first place.

Reintroducing Endangered Species to Nature   The ultimate goal of captive-breeding programs is to produce offspring in captivity and then release them into nature so that wild populations are restored. However, only one of every 10 reintroductions using animals raised in captivity is successful. What guarantees that a reintroduced population will survive?

Whether such reintroductions actually succeed has been evaluated only in recent years. The Hawaiian goose, or nene (pronounced nay-nay), was down to 30 individuals in the 1960s. It was reintroduced to the islands of Hawaii (the Big Island) and Maui beginning in the 1970s, but although hundreds of birds have been released, a self-sustaining nene population has not developed on either island. Apparently, some of the same factors that originally caused the nene's extinction in nature, including habitat destruction and foreign predators such as the Indian mongoose, are responsible for the failure of the reintroduced birds. (On Kauai a group of captive nenes escaped during a hurricane and may be successfully reestablishing themselves. Kauai is less developed than Maui or the Big Island and does not have the Indian mongoose.)

Before attempting a reintroduction, conservation biologists now make a feasibility study. This study includes determining (1) what factors originally caused the species to become extinct in nature, (2) whether these factors still exist, and (3) whether any suitable habitat still remains.

If the animal to be reintroduced is a social animal, a small herd is usually released together. This is accomplished by first placing the herd in a large, semi-wild enclosure that is somewhat protected from predators but that requires the herd to obtain its own food. When the herd's behavior begins to resemble the behavior of wild herds, it is released.

Sometimes it is impossible to teach critical survival skills to animals raised in captivity. The effort to reintroduce captive-raised thick-billed parrots to the Chiricahua Mountains of Arizona was canceled in 1993 because all 88 birds released between 1986 and 1993 died or disappeared. Wild thick-billed parrots are loud, sociable birds whose flocking instinct contributes to their survival because individual birds loudly announce

Figure 17.17   Bongo calf and its surrogate mother. This young bongo *(Tragelaphus eurycerus)* calf was transferred as an embryo to the uterus of a female eland *(Taurotragus oryx)*, where it completed development. The bongo, a shy, elusive species inhabiting dense bamboo forests and deep jungle in equatorial Africa, is an endangered species, primarily because of habitat fragmentation and overhunting. The larger and more common eland, a different but related species, inhabits plains and open brush from Kenya to South Africa and Angola.

Figure 17.18 Seeds from a seed bank.
Shown are small vials and packets of seeds from the seed bank in Svalbard, Norway. (Courtesy Nordiska Genbanken, Alnarp, Sverige)

the presence of danger, such as hawks or other birds of prey, to the group. Those parrots raised in captivity lacked such social behavior, and despite efforts to teach them to stay together, they separated from the flock after they were released. Future releases will have to await the availability of wild-caught, thick-billed parrots from Mexico.

Once animals are released, they must be monitored. If any animals die, their cause of death is determined to search for ways to prevent unnecessary deaths in future reintroductions.

Seed Banks More than 100 seed collections called **seed banks,** or *gene banks*, exist around the world and collectively hold more than three million samples at low temperatures (**Figure 17.18**). They offer the advantage of storing a large amount of plant genetic material in a small space. Seeds stored in seed banks are safe from habitat destruction, climate change, and general neglect. There have even been some instances of using seeds from seed banks to reintroduce to nature a plant species that had become extinct.

Some disadvantages to seed banks exist. Many types of plants, such as avocados and coconuts, cannot be stored as seeds. The seeds of these plants do not tolerate being dried out, a necessary step before the seeds are sealed in moisture-proof containers for storage at -18°C (-0.4°F).

Seeds do not remain alive indefinitely and must be germinated periodically so that new seeds can be collected. Growing, harvesting, and returning seeds to storage is the most expensive aspect of storing plant material in seed banks. (Cryopreservation in liquid nitrogen at -160°C, or -256°F, is being developed for certain kinds of seeds. Seeds stored at this temperature survive for longer periods than seeds stored at warmer temperatures.) Because accidents such as fires or power failures can result in the permanent loss of the genetic diversity in the seeds, biologists typically subdivide seed samples and store them in several seed banks.

Perhaps the most important disadvantage of seed banks is that plants stored in this manner remain stagnant in an evolutionary sense. They do not evolve in response to changes in their natural environments. As a result, they may be less fit for survival when they are reintroduced into nature. Despite their shortcomings, seed banks are increasingly viewed as an important method of safeguarding seeds for future generations.

Conservation Organizations

Conservation organizations are an essential part of the effort to maintain biological diversity. These groups help educate policymakers and the public about the importance of biological diversity. In certain instances, they serve as catalysts and galvanize public support for important biodiversity preservation efforts. They provide financial support for conservation projects, from basic research to the purchase of land that is a critical habitat for a particular species or group of species.

The World Conservation Union (IUCN)[2] assists countries with hundreds of conservation biology projects. The IUCN and other conservation organizations are currently assessing how effective established wildlife refuges are in maintaining biological diversity (**Figure 17.19**). In addition, the IUCN and the WWF have identified major conservation priorities by determining which biomes and ecosystems do not have protected areas. The IUCN maintains a data bank on the status of the world's species; its material is published in the *IUCN Red Data Books* about organisms and their habitats.

REVIEW

1. What is conservation biology?

2. What is the difference between in situ and ex situ conservation?

[2] Formerly called the International Union for Conservation of Nature and Natural Resources, the World Conservation Union still goes by the acronym IUCN.

Conservation Policies and Laws

LEARNING OBJECTIVES

- Briefly describe the benefits and shortcomings of the U.S. Endangered Species Act.
- Relate the purpose of the World Conservation Strategy.

In 1973 the **Endangered Species Act (ESA)** was passed in the United States, authorizing the FWS to protect endangered and threatened species in the United States and abroad. Many other countries now have similar legislation. The ESA conducts a detailed study of a species to determine if it should be listed as endangered or threatened. Currently, more than 1300 species in the United States are listed as endangered or threatened. The ESA provides legal protection to listed species to help reduce their danger of extinction. This act makes it illegal to sell or buy any product made from an endangered or threatened species.

The ESA requires the FWS to select critical habitats and design a detailed recovery plan for each species listed. The recovery plan includes an estimate of the current population size, an analysis of what factors contributed to its endangerment, and a list of activities to help the population recover. Currently, 1075 U.S. species have approved recovery plans.

The ESA was updated in 1982, 1985, and 1988. It is considered one of the strongest pieces of U.S. environmental legislation, in part because species are designated as endangered or threatened entirely on biological grounds. Currently, economic considerations cannot influence the designation of endangered or threatened species. Biologists generally agree that as a result of passage of the ESA in 1973, fewer species became extinct than had the law never been passed.

The ESA is one of the most controversial pieces of environmental legislation. It does not provide compensation for private property owners who suffer financial losses because they cannot develop their land if a threatened or endangered species lives there. The ESA has also interfered with some federally funded development projects.

The ESA was scheduled for congressional reauthorization in 1992 but has been entangled since then in political wrangling between conservation advocates and those who support private property rights. Conservation advocates think the ESA does not do enough to save endangered species, whereas those who own land on which rare species live think the law goes too far and infringes on property rights. Another contentious issue is over the financial cost of the law. Critics say that federal and state governments spend too much, given the little bit of environmental gain that the ESA accomplishes. Some critics—notably business interests and private property owners—view the ESA as an impediment to economic progress. To protect the habitat of the northern spotted owl, for example, the timber industry was blocked from logging old-growth forests in certain parts of the Pacific Northwest (see Chapter 2 introduction).

Those who defend the ESA point out that of 34,000 past cases of endangered species versus economic development, only 21 cases were not resolved by some sort of compromise. Compromise is crucial to the success of saving endangered species because, according to the U.S. General Accounting Office, more than 90% of endangered species live on at least some privately owned lands. Some critics of the ESA think the law should be changed so that private landowners are given economic incentives to help save endangered species living on their lands. For example, tax cuts for property owners who are good land stewards could make the presence of endangered species on their properties an asset instead of a liability.

Defenders of the ESA agree that the law is not perfect. Relatively few endangered species have recovered enough to be delisted—that is, removed from protection of the ESA. However, the FWS reports that hundreds of listed species are stable or improving; they expect as many as several dozen additional species to be delisted in the next decade or so.

Figure 17.19   Biological Dynamics of Forest Fragment Project. When a protected area is set aside, it is important to know what the minimum size of that area must be so that it is not affected by encroaching species from surrounding areas. Shown are 1-hectare and 10-hectare plots of a long-term study on the effects of habitat fragmentation on Amazonian rain forest by the World Wildlife Fund and Brazil's National Institute for Amazon Research. Plots with an area of 100 hectares are also under study, along with identically sized sections of intact forest, used as controls. Preliminary data indicate that the smaller forest fragments do not maintain their ecological integrity. For example, large trees often die or are damaged from exposure to wind and weather. (Richard O. Bierregaard, Jr.)

The ESA is geared more to saving a few popular or unique endangered species rather than the much larger number of less glamorous species, such as fungi and insects, that perform valuable ecosystem services. Yet it is the less glamorous organisms that play central roles in ecosystems and contribute the most to their functioning.

Conservationists would like the ESA strengthened in such a way as to manage whole ecosystems and maintain complete biological diversity rather than attempt to save endangered species as isolated entities. This approach offers collective protection to many declining species rather than to single species.

Habitat Conservation Plans

The 1982 amendment of the ESA provided a way to resolve conflicts between protection of endangered species and development interests on private property: **habitat conservation plans (HCPs).** HCPs vary greatly, from small projects to regional conservation and development plans.

Habitat conservation plans allow a landowner to "take" (injure, kill, or modify the habitat of) a rare species if the "taking" doesn't threaten the survival or recovery of the threatened or endangered species on that property. If a landowner sets aside land as habitat for the rare species, he or she then has the right to develop part of the property without threat of legal action by the FWS. Conservationists are wary of HCPs because HCPs do not provide any promise of recovery of rare species. Conservationists are concerned that HCPs may actually contribute to a species' extinction.

International Conservation Policies and Laws

The **World Conservation Strategy,** a plan designed to conserve biological diversity worldwide, was formulated in 1980 by the IUCN, the World Wildlife Fund, and the U.N. Environment Program. In addition to providing guidelines for conserving biological diversity, the World Conservation Strategy seeks to preserve the vital ecosystem services on which all life depends for survival and to develop sustainable uses of organisms and the ecosystems that they comprise.

The **Convention on Biological Diversity** was produced by the 1992 Earth Summit to decrease the rate of extinction of the world's endangered species. This treaty requires that each signatory nation inventory its own biodiversity and develop a **national conservation strategy,** a detailed plan for managing and preserving the biological diversity of that specific country.

The exploitation of endangered species is somewhat controlled at the international level by the **Convention on International Trade in Endangered Species of Wild Flora and Fauna (CITES),** which went into effect in 1975. Originally drawn up to protect endangered animals and plants considered valuable in the highly lucrative international wildlife trade, CITES bans the hunting, capturing, and selling of endangered or threatened species and regulates the trade of organisms listed as potentially threatened. Unfortunately, enforcement of this treaty varies from country to country. Even where enforcement exists, the penalties aren't severe. As a result, illegal trade in rare, commercially valuable species continues.

The goals of CITES often stir up controversy over such issues as who actually owns the world's wildlife and whether global conservation concerns take precedence over competing local interests. These conflicts often highlight socioeconomic differences between wealthy consumers of CITES products and poor people who trade the endangered organisms.

The case of the African elephant, discussed earlier in the chapter, bears out these controversies. Listed as an endangered species since 1989 to halt the slaughter of elephants driven by the ivory trade, the species seems to have recovered in southern Africa. Organizations such as the Humane Society in the United States are developing a birth control vaccine to reduce the number of elephant births. However, the African people living near the elephants want to cull the herd periodically and sell elephant meat, hides, and ivory for profit. In the late 1990s and early 2000s, CITES transferred elephant populations in Namibia, Botswana, and South Africa to a less restrictive listing to allow a one-time trade of legally obtained stockpiled ivory (from animals that died of natural causes). The money earned from the sale of ivory is funding elephant conservation programs and community development projects for people living near the elephants.

REVIEW

1. What is the Endangered Species Act?
2. What is the World Conservation Strategy?

 ## Wildlife Management

LEARNING OBJECTIVE

• Distinguish between conservation biology and wildlife management.

Wildlife management is an applied field of conservation biology that focuses on the continued productivity of plants and animals. Wildlife management programs often have different priorities than those of conservation biology. In contrast to conservation biology, which often focuses on threatened or endangered species, most attention in wildlife management is focused on common organisms. Wildlife management includes the regulation of hunting and fishing and the management of food, water, and habitat.

The natural predators of many game animals have been largely eliminated in the United States. As a result of the near-disappearance of predators such as wolves, the populations of animals such as squirrels, ducks, and deer sometimes exceed the carrying capacity of their environment (see Chapter 8). When this occurs, the habitat deteriorates, and many animals starve to death. Sport hunting effectively controls the overpopulation of game animals, provided restrictions are observed to prevent overhunting. Laws in the United States determine the time of year and length of hunting seasons for various species, as well as the number, sex, and physical size of each species that may be harvested.

Wildlife managers manipulate the plant cover, food, and water supplies of a specific animal's habitat. Because different animals predominate in different stages of *ecological succession*, controlling the stage of ecological succession of an area's vegetation encourages the presence of certain animals and discourages others (see Chapter 4). Quail and ring-necked pheasant are found in grassy, open areas characteristic of early-succession stages. Moose, deer, and elk predominate in partially open forest, such as an abandoned field or meadow adjacent to a forest; the field provides food, and the forest provides protective cover. Other animals, such as grizzly bear and bighorn sheep, require undisturbed vegetation. Wildlife managers control the stage of succession with techniques such as planting certain types of vegetation, burning the undergrowth with controlled fires, and building artificial ponds.

Management of Migratory Animals

International agreements are established to protect migratory animals. Ducks, geese, and shorebirds spend their summers in Canada and their winters in the United States and Central America. During the course of their annual migrations, which usually follow established routes, or **flyways,** they must have areas in which to rest and feed. Wetlands, the habitat of these animals, must be protected in both their winter and summer homes.

Arctic Snow Geese   The Arctic snow goose has become a major challenge for wildlife managers because its population expanded rapidly during the last two decades of the 20th century. This goose breeds in large colonies along coastal salt marshes of the Arctic during the short Arctic summer. The population migrates south during autumn and traditionally wintered in salt marshes along the Texas and Louisiana coasts. The snow goose has successfully expanded its winter range into Arkansas, Mississippi, Oklahoma, New Mexico, and northern Mexico, largely because the geese obtain seeds

wildlife management: The application of conservation principles to manage wild species and their habitats for human benefit or for the welfare of other species.

and other food from agricultural lands. Because snow geese have expanded their winter range so successfully, more of the adults survive to return to the Arctic. Their adaptability to human-induced changes in the environment has enabled them to avoid *density-dependent factors* (i.e., lack of food during winter months) that would normally keep the population in check. The huge population of snow geese has damaged much of the Arctic's fragile coastal ecosystem as the geese forage there for a variety of plants and insects (**Figure 17.20**).

Wildlife managers want to avoid a massive die-off of geese in the Arctic, a catastrophe that is unavoidable if the goose population is not brought under control. To reduce population numbers, U.S. and Canadian wildlife managers have increased the "taking" of snow geese by sport hunters. Animal rights groups oppose hunting and suggest that farmers in the winter range of the geese should modify their agricultural methods to reduce the amount of food available for the geese. However, this approach would be difficult to implement because of the impact it would have on so many farmers. Wildlife managers are looking at other options if increased sport hunting is not sufficient to reduce the number of snow geese. For example, the geese could be commercially harvested for human consumption.

Management of Aquatic Organisms

Fishes with commercial or sport value must be managed to ensure they are not over-exploited to the point of extinction. Freshwater fishes such as trout and salmon are managed in several ways. Fishing laws regulate the time of year, size of fish, and maximum allowable catch. Natural habitats are maintained to maximize population size. Ponds, lakes, and streams may be restocked with young hatchlings from hatcheries.

Traditionally, the ocean's resources have been considered common property, available to the first people to exploit them. As a result, commercial fishing is severely reducing the number of marine fishes. (Chapter 19 discusses this dwindling resource.)

During the 19th and 20th centuries, many whale species were harvested to the point of **commercial extinction,** meaning that so few remained that it was unprofitable to hunt them. Although commercially extinct species still have living representatives, their numbers are so reduced that they are endangered. In 1946 the International Whaling Commission set an annual limit on killed whales for each whale species in an attempt to secure sustainable whale populations. Unfortunately, these limits were set too high, resulting in further population declines during the next 20 years. Conservationists began to call for a global ban on commercial whaling; such a moratorium went into effect in 1986.

Figure 17.20 Damage caused by snow geese. The only part of this once-lush salt marsh that survived the onslaught of snow geese was this small patch enclosed in protective fencing to prevent the geese from foraging. The rest of the marsh was reduced to a wasteland. Photographed in northern Manitoba, Canada, along the Hudson Bay. (Courtesy of Robert L. Jefferies, University of Toronto)

Scientists have since monitored whale populations and concluded that overall the ban is working. The populations of most whales, such as humpbacks and bowheads, seem to be growing. One species, the Pacific gray whale, has recovered sufficiently to be removed from the endangered and threatened species lists. The North Atlantic right whale and southern blue whales, however, are still endangered. In 1994 the International Whaling Commission established the Southern Ocean Whale Sanctuary in Antarctic waters, where many of the world's great whales feed and reproduce. This vast sanctuary, which bars commercial hunting, would continue to exist should the current ban on whaling ever be lifted.

Despite international pressure, Japan does not honor either the global ban on commercial whaling or the Southern Ocean Whale Sanctuary. Japan has justified its continuing whale harvests by saying that the whales are killed for "scientific purposes," although the whale meat from these harvests is sold in Japanese markets and restaurants.

REVIEW

1. What is wildlife management?
2. How do the goals of wildlife management and conservation biology differ?

REVIEW OF LEARNING OBJECTIVES WITH KEY TERMS

• **Define *biological diversity* and distinguish among genetic diversity, species richness, and ecosystem diversity.**

Biological diversity is the number and variety of Earth's organisms; it consists of three components: **genetic diversity** (the variety within a species), **species richness** (the number of species), and **ecosystem diversity** (variety within and among ecosystems).

• **Relate several important ecosystem services provided by biological diversity.**

Ecosystem services are important environmental benefits that ecosystems provide to people; examples include clean air to breathe, clean water to drink, and fertile soil in which to grow crops. Bacteria and fungi perform the important ecosystem service of decomposition. Forests provide watersheds from which we obtain fresh water. Insects transfer plant pollen for reproduction. Soil organisms maintain soil fertility. Plant roots anchor in the soil and reduce soil erosion.

• **Define *extinction* and distinguish between background extinction and mass extinction.**

Extinction is the elimination of a species from Earth. **Background extinction** is the continuous, low-level extinction of species. **Mass extinctions** are episodes in Earth's history in which numerous species became extinct in a relatively short period.

• **Contrast threatened and endangered species, and list four characteristics common to many endangered species.**

A **threatened species** is a species whose population has declined to the point that it may be at risk of extinction. An **endangered species** is a species that faces threats that may cause it to become extinct within a short period. Endangered and threatened species often have limited natural ranges and low population densities. Endangered and threatened species may have low reproductive rates or specialized food or reproduction requirements. Many island species are endangered.

• **Define *biodiversity hotspots* and explain where most of the world's biodiversity hotspots are located.**

Biodiversity hotspots are relatively small areas of land that contain an exceptional number of endemic species and are at high risk from

human activities. **Norman Myers** and Conservation International have identified 25 biodiversity hotspots around the world; 15 hotspots are tropical, and nine are mostly or entirely islands.

• **Describe four human causes of species endangerment and extinction and tell which cause is the most important.**

Habitat destruction is the most significant cause of declining biological diversity because it reduces a species' biological range. **Habitat fragmentation** is a type of habitat destruction in which large areas of habitat are broken into small, isolated patches. Other causes of declining biological diversity are the introduction of foreign species, overexploitation, and pollution.

• **Explain how invasive species endanger native species.**

Invasive species are foreign species that spread rapidly in a new area where they are free of predators, parasites, or resource limitations that may have controlled their population in their native habitat. Invasive species often upset the balance among the organisms living in that area and interfere with an ecosystem's normal functioning. The foreign species may compete with native species for food or habitat or may prey on them.

• **Define *conservation biology* and compare in situ and ex situ conservation.**

Conservation biology is the scientific study of how humans impact organisms and of the development of ways to protect biological diversity. Efforts to preserve biological diversity in nature, called **in situ conservation**, include establishing parks, wildlife sanctuaries and refuges, and other protected areas. **Ex situ conservation**, which includes captive breeding and storing genetic material, occurs in human-controlled settings. Zoos, aquaria, botanical gardens, and seed banks are examples of ex situ conservation.

• **Describe restoration ecology.**

Restoration ecology is the study of the historical condition of a human-damaged ecosystem, with the goal of returning it as close as possible to its former state. Restoration ecology is an important part of in situ conservation.

• **Briefly describe the benefits and shortcomings of the U.S. Endangered Species Act.**

The **Endangered Species Act (ESA)** authorizes the U.S. Fish and Wildlife Service to protect from extinction endangered and threatened species, both in the United States and abroad. The FWS selects critical habitats and designs detailed recovery plans for each species listed. The ESA does not include economic considerations, such as compensation for private property owners who suffer financial losses as a result of the law. **Habitat conservation plans** help resolve ESA conflicts between conservation and development interests on private lands.

• **Relate the purpose of the World Conservation Strategy.**

The World Conservation Union (IUCN), World Wildlife Fund, and U.N. Environment Program developed the **World Conservation Strategy** to conserve biological diversity and vital ecosystem services worldwide.

• **Distinguish between conservation biology and wildlife management.**

Wildlife management is the application of conservation principles to manage wild species and their habitats for human benefit or for the welfare of other species. In contrast to conservation biology, which often focuses on threatened or endangered species, most attention in wildlife management is focused on common organisms.

Thinking About the Environment

1. Is biological diversity a renewable or nonrenewable resource? Why could it be seen both ways?

2. Give at least five important ecosystem services provided by living organisms.

3. If we preserve species solely on the basis of their potential economic value—such as a source of a novel drug—does this mean that they lose their value after we have capitalized on a newly discovered chemical? Why or why not?

4. Do you think the alligator (see Figure 17.2) is a keystone species (discussed in Chapter 4)? Explain your answer.

5. Give four characteristics common to many endangered species.

6. What are the four main causes of species endangerment and extinction? Give examples of specific organisms harmed by each cause. Which cause do biologists consider most important?

7. What invasive species are problems in your area? How do they affect ecosystems?

8. Why are frogs and other amphibians considered indicator species?

9. Give examples of in situ and ex situ conservation.

10. Why is the Arctic snow goose such a challenge for U.S. wildlife managers? for Canadian wildlife managers?

11. Explain why being pro-nature does not imply also being anti-development?

12. If you had the assets and authority to take any measure to protect and preserve biological diversity, but could take only one, what would it be?

13. The most recent version of the World Conservation Strategy includes stabilizing the human population. How would stabilizing the human population affect biological diversity?

14. In *A Sand County Almanac*, Aldo Leopold wrote, "To keep every cog and wheel is the first precaution of intelligent tinkering." How does his statement relate to biological systems?

Quantitative questions relating to this chapter are on our Web site.

Take a Stand

Visit our Web site at http://www.wiley.com/college/raven (select Chapter 17 from the Table of Contents) for links to more information about the elephant versus people issue in Africa. Consider the opposing views of conservationists and farmers, and debate the issue with your classmates. You will find tools to help you organize your research, analyze the data, think critically about the issues, and construct a well-considered argument. *Take a Stand* activities can be done individually or as a team, as oral presentations, written exercises, or Web-based (e-mail) assignments.

Additional online materials relating to this chapter, including a Student Testing Section with study aids and self-tests, Environmental News, Activity Links, Environmental Investigations, and more, are also on our Web site.

Land Resources

Canoeing in Congaree National Park.
(QT Luong/Terra Galleria Photography)

In 2003 and 2004, two new U.S. national parks were created, Congaree National Park and Great Sand Dunes National Park. These parks bring the total number of national parks in the United States to 58.

Congaree National Park in South Carolina incorporated Congaree Swamp National Monument in 2003. (National monuments are similar to national parks except that they are given less funding and therefore less protection.) One of the smallest national parks, Congaree is significant because it preserves the largest remaining intact hardwood bottomland forest in the United States (see figure). Water oaks, tupelos, and swamp magnolias abound in Congaree Swamp. Although it is small, Congaree provides crucial wildlife habitat for many species, including black bears, bobcats, coyotes, deer, and numerous water birds. The Congaree River that flows through the swamp provides a habitat for alligators and various fishes, amphibians, snakes, and turtles. The park offers visitors such recreational activities as hiking, camping, and canoeing.

Designating a national park does not happen overnight. The Sierra Club began campaigning to preserve the Congaree Swamp in 1969. In 1976, Congress declared the area a national monument. Congaree Swamp National Monument was increasingly recognized as a unique and important land resource over the next 25 years. In 1983, it became an International Biosphere Reserve, one of about 500 international areas of significant ecological importance; this designation by the United Nations Educational, Scientific, and Cultural Organization means that the country in which the International Biosphere Reserve is located should make strong efforts to preserve it. In 2001, Congaree Swamp National Monument was declared an Important Bird Area; this designation by the Audubon Society recognizes the area as a uniquely vital habitat for birds and other wildlife.

Great Sand Dunes National Park and Preserve in Colorado was formerly a national monument that was redesignated a national park in 2004. The tallest sand dunes in North America are located here, on the eastern side of the San Luis Valley. Some dunes are as tall as 230 meters (750 feet) above the valley floor. They formed as rain and wind eroded the surrounding mountains and prevailing winds pushed the eroded sand into dunes. As the winds blow, the dunes continually change shape but they remain more or less in the same position.

As with Congaree National Park, it took years to obtain the land and get approval for Great Sand Dunes National Park. The Nature Conservancy, an important nongovernmental conservation organization, helped purchase nearby ranches that ultimately contributed land to the national park. Federal, state, and private donors also assisted. For recreation, the park offers nature walks, short tours, hiking, and camping, including limited camping on the dunes themselves.

National parks represent much more than wildlife sanctuaries. Parks preserve the land as much as possible so that future generations of people can appreciate and enjoy the beauty of fast-disappearing natural areas. If you get a chance, you should visit these latest additions to the National Park System.

 World View and Closer to You...
Real news footage relating to land resources around the world and in your own geogra-phic region is available at www.wiley.com/college/raven by clicking on "World View and Closer to You."

Land Use

LEARNING OBJECTIVES

- Relate at least four ecosystem services provided by natural areas.
- Describe world land use and summarize current land ownership in the United States.
- Contrast the views of the wise-use movement and the environmental movement regarding the use of federal lands.

Most of Earth's land area has a low density of humans. These sparsely populated **nonurban** or **rural** lands include forests, grasslands, deserts, and wetlands. Most people living in rural areas have jobs directly connected with natural resources—such as farming or logging. The many **ecosystem services** performed by rural lands enable the majority of humans to live in concentrated urban environments. Maintaining parcels of undisturbed land adjacent to agricultural and urban areas provides vital ecosystem services such as wildlife habitat, flood and erosion control, and groundwater recharge. Undisturbed land breaks down pollutants and recycles wastes. Natural environments provide homes for organisms. One of the best ways to maintain biological diversity and to protect endangered and threatened species is by preserving or restoring the natural areas to which these organisms are adapted. (Table 4.1 summarizes some important ecosystem services.)

Scientists use undisturbed rural lands as a benchmark, or point of reference, to determine the impacts of human activity. Geologists, zoologists, botanists, ecologists, and soil scientists are some of the scientists who use undisturbed rural lands for scientific inquiry. These areas provide perfect settings for educational experiences in science as well as history (because they demonstrate the way the land was when humans originally settled here).

Unspoiled natural areas are important for their recreational value, providing places for hiking, swimming, boating, rafting, sport hunting, and fishing. Wild areas are important to the human spirit. Forest-covered mountains, rolling prairies, barren deserts, and other undeveloped areas are aesthetically pleasing and also help us recover from the stresses of urban and suburban living. We can escape the tensions of the civilized world by retreating, even temporarily, to the solitude of natural areas.

World Land Use

The extent of human land use is increasingly having global impacts. Currently, humans use an estimated 3% of the world's total land area for cities and 38% for agriculture—that is, for raising crops and livestock (**Figure 18.1**). Another 30% of the land surface consists of rock, ice, tundra, and desert—areas considered unsuitable for long-term human use. This leaves 29% of the land surface as natural ecosystems, such as forests, that could potentially be developed for human purposes. You have just seen, however, that these natural ecosystems provide many valuable ecosystem services important to human survival. Yet urban areas, croplands, and pastures have expanded worldwide in recent decades in response to the need to support more than six billion humans.

Land Use in the United States

About 55% of the land in the United States is privately owned by citizens, corporations, and nonprofit organizations, and about 3% by Native American tribes. The rest is owned by the federal government (about 35% of U.S. land) and by state and local governments (about 7% of U.S. land). Government-owned land encompasses all types of ecosystems, from tundra to desert, and includes land that contains important resources such as minerals and fossil fuels, land that possesses historical or cultural significance, and land that provides critical biological habitat. Most federally owned land is in Alaska and 11 western states (**Figure 18.2**). It is managed primarily by four agencies, three in the U.S. Department

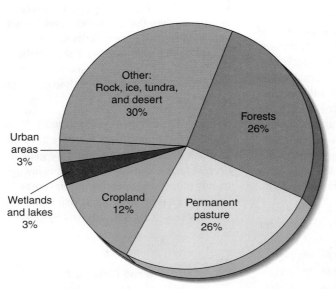

Figure 18.1 World land use. Note that 3% of the world's total land area is used for cities and 38% for agriculture (cropland and pastureland). (World Resources Institute and Earth Institute at Columbia University)

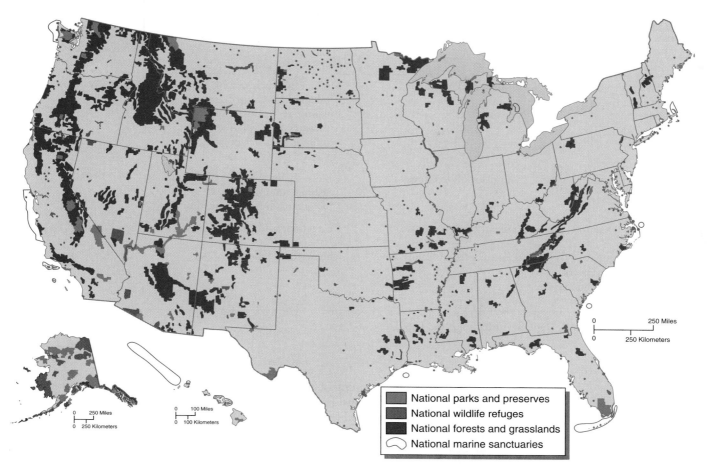

National parks and preserves
National wildlife refuges
National forests and grasslands
National marine sanctuaries

Figure 18.2   Selected federal lands.
Shown are national parks and preserves, national wildlife refuges, national forests and grasslands, and national marine sanctuaries in the United States. Note the preponderance of federal lands in western states and Alaska. Other federal lands, such as military installations and research facilities, are not included. (National Geographic Society)

of the Interior—the Bureau of Land Management (BLM), the Fish and Wildlife Service (FWS), and the National Park Service (NPS)—and one in the U.S. Department of Agriculture—the U.S. Forest Service (USFS) (**Table 18.1**).

Managing Public and Private Land   Many environmental concerns converge in the matter of land use. Pollution, population issues, preservation of our biological resources, mineral and energy requirements, and production of food are all tied to land use. Overriding these concerns are economic factors, such as the way privately owned land is taxed. Sometimes forest or agricultural land located near urban and suburban areas is taxed as potential urban land. Because of the higher taxes on this land, its owners fall under greater pressure to sell it, which ultimately hastens its development. However, if such land is taxed as forest or farmland, the lower taxes are an incentive for owners to hold onto the land and maintain it in its undeveloped condition. Thus, economic factors largely control land use.

Public Planning of Land Use.   Examine the use of land where you live. You may be surrounded by high rises and factories or by tree-lined streets interspersed with open parkland. Regardless of your surroundings, they probably have a land use plan that includes zoning. However, land use plans seldom take into account all aspects of land as a resource both before and after development. The philosophy of most land use plans is that development is good because it increases the tax base, even though the revenue from these taxes is usually consumed providing services to the developed area.

Land use decisions are complex because they have multiple effects. If a tract of land is developed for housing, then roads, sewage lines, hospitals, and schools must be built nearby to accommodate the influx of people. This also requires services, such as new restaurants and shopping areas, which take up more land.

Table 18.1 Administration of Federal Lands

Agency	Land Held	Primary Uses	Area in Millions of Hectares (Acres)
Bureau of Land Management (Dept. of Interior)	National resource lands	Mining, livestock grazing, oil and natural gas extraction	109 (270)
U.S. Forest Service (Dept. of Agriculture)	National forests	Logging, recreation, conservation of watersheds, wildlife habitat, mining, livestock grazing, oil and natural gas extraction	77 (191)
U.S. Fish and Wildlife Service (Dept. of Interior)	National wildlife refuges	Wildlife habitat; also logging, hunting, fishing, mining, livestock grazing, oil and natural gas extraction	38 (95)
National Park Service (Dept. of Interior)	National Park System	Recreation, wildlife habitat	34 (84)
Other—includes Department of Defense, Corps of Engineers (Dept. of the Army), and Bureau of Reclamation (Dept. of Interior)	Remaining federal lands	Military uses, wildlife habitat	29 (72)

Source: U.S. Dept. of Interior, U.S. Dept. of Agriculture, and U.S. Dept. of Defense.

Ideally, public planning of land use would take into account all repercussions of the proposed land use, not just its immediate effects. It helps to begin with an inventory of the land, including its soil type, topography, types of organisms, endangered or threatened species, and historic or archaeological sites. By doing this, a public planning commission attempts to understand the value of the land as it currently exists, as well as its potential value after any proposed change.

In addition to providing people with open space for recreation and mental health, undeveloped land provides ecosystem services that must be recognized. These benefits should be compared with the possible economic benefits of development. In the long term, the best use of land may not be the use that provides immediate economic gain.

If the land will ultimately be developed, a well-designed development plan will be comprehensive. It will indicate which areas will remain open space, which will remain agricultural, and which will be zoned for high-, medium-, and low-density housing. (Chapter 10 contains additional discussion of land use patterns and zoning.)

Management of Federal Lands. How do we best manage the legacy of federal lands? Should federal lands be managed under multiple uses, or should they be preserved so that they benefit U.S. citizens for generations to come? These questions have divided many Americans into two groups: those who wish to exploit resources on federal lands now (a coalition of several hundred corporate and grassroots organizations known collectively as the *wise-use movement*) and those who wish to preserve the resources on federally owned lands (a coalition of several hundred grassroots organizations known collectively as the *environmental movement*). The descriptions of wise-use and environmental movements that follow are mainstream; certain groups associated with one or the other of these movements may not support all of the listed goals.

People who support the **wise-use movement** think that the government has too many regulations protecting the environment and that property owners should have more flexibility to use natural resources. They believe that a primary purpose of federal lands is to enhance economic growth. Some of their goals include the following:

1. Put all national forests under timber management, including old-growth forests.
2. Permit mining and commercial development of wilderness areas, wildlife refuges, and national parks, where appropriate.

3. Allow unrestricted development of wetlands.

4. Sell parts of resource-rich federal lands to private interests, such as mining, oil, coal, ranching, and timber groups, for sustainable resource extraction.

Many organizations that embrace the wise-use movement have environmentally friendly names. The National Wetlands Coalition, for example, consists primarily of real estate developers and energy companies who wish to drain and develop wetlands. Similarly, logging companies support the American Forest Resource Alliance.

In contrast to the wise-use movement, the **environmental movement** views federal lands as a legacy of U.S. citizens. They think that:

1. The primary purpose of public lands is to protect biological diversity and ecosystem integrity.

2. Those who extract resources from public lands should pay U.S. citizens compensation equal to the fair market value of the resource and not be subsidized by taxpayers.

3. Those who use public lands should be held accountable for any environmental damage they cause.

REVIEW

1. What are ecosystem services?

2. What percentage of land in the United States is privately owned? What percentage is public land owned by the federal government?

3. How do the wise-use and environmental movements differ regarding their views on the use of public lands?

 ## Wilderness, Parks, and Wildlife Refuges

LEARNING OBJECTIVES

- Describe the following federal lands, stating which government agency administers them and current issues of concern: wilderness, national parks, and national wildlife refuges.
- Define *natural regulation*.

Wilderness encompasses regions where the land and its community of organisms are not greatly disturbed by human activities and where humans visit but do not permanently inhabit. The 88th Congress recognized that increased human population and expansion into undeveloped areas might result in a future in which no lands occurred in their natural condition. Accordingly, the **Wilderness Act** of 1964 authorized the U.S. government to set aside federally owned land that retains its primeval character and lacks permanent improvements or human habitation as part of the **National Wilderness Preservation System.** These federal lands range in size from tiny islands to portions of national parks (42% of wilderness areas are in national parks), national forests (33% of wilderness areas), and national wildlife refuges (22% of wilderness areas). The Big Gum Swamp wilderness in Florida is only 5528 hectares (13,660 acres), whereas the Selway-Bitterroot Wilderness in Idaho is more than 530,000 hectares (1.3 million acres). Areas designated as wilderness are given the highest protection of any public lands. These areas are to remain natural and unchanged so they will be unimpaired for future generations to enjoy (**Figure 18.3**). Wilderness areas are located within other types of federal lands, and the agencies that manage those lands also manage the wilderness areas. Thus, four government agencies—the NPS, USFS, FWS, and BLM—oversee the 630 wilderness areas that encompass 40.3 million hectares (102 million acres) of land.

Although mountains are the most common land safeguarded by this system, representative examples of other ecosystems have been set aside, including tundra,

wilderness: A protected area of land in which no human development is permitted.

Figure 18.3   Wilderness areas. These hikers are in the Selway-Bitterroot Wilderness in Idaho. (David Hiser/Stone/Getty Images)

desert, and wetlands. More than one-half of the lands in the National Wilderness Preservation System lie in Alaska, and western states contain much of the remaining lands. Because few sites untouched by humans exist in the eastern states, requirements were modified in 1975 so that the wilderness designation could be applied to certain federally owned lands where forests are recovering from logging.

Millions of people visit U.S. wilderness areas each year, and some areas are overwhelmed by this traffic: Eroded trails, soil and water pollution, litter and trash, and human congestion predominate over quiet, unspoiled land. Government agencies now restrict the number of people allowed into each wilderness area at one time, so that human use does not seriously affect the wilderness. Some of the most popular wilderness areas may require more intensive future management, such as building trails, outhouses, cabins, and campsites. These amenities are not encountered in true wilderness, posing a dilemma between wilderness preservation and human use and enjoyment of wild lands.

Limiting the number of human guests in a wilderness area does not control all factors that threaten wilderness. **Invasive species** that become established in wilderness have the potential to upset the balance among native species. The white pine blister rust, a foreign fungus that kills white pine trees, has invaded the wilderness in the northern Rocky Mountains. Wilderness managers are concerned that declining white pine populations could harm the population of grizzly bears in the region. (Pine seeds are a major part of the grizzlies' diet.) The Wilderness Act specifies both the preservation of natural conditions and the avoidance of intentional ecological management. In this example, the only way to preserve as much as possible of the original wilderness may be to intentionally manipulate the white pine population by breeding and planting fungus-resistant trees.

Large tracts of wilderness, most of it in Alaska, have been added to the National Wilderness Preservation System since passage of the Wilderness Act in 1964. People who view wilderness as a nonrenewable resource support the designation of wilderness areas. They think it is particularly important to preserve additional land in the lower 48 states, where currently less than 2% of the total land area is specified as wilderness. Increasing the amount of federal land in the National Wilderness Preservation System is opposed by groups who operate businesses on public lands (such as timber, mining, ranching, and energy companies) and by their political representatives.

National Parks

In 1872 Congress established the world's first national park, Yellowstone National Park, in federal lands in the territories of Montana and Wyoming. The purpose of the park was to protect this land of great scenic beauty and biological diversity in an unimpaired condition for present and future generations. The **National Park System** was originally composed of such large, scenic areas in the West as Yellowstone, Grand Canyon, and Yosemite Valley (**Figure 18.4**). Today the National Park System has more cultural and historical sites—battlefields and historically important buildings and towns—than places of scenic wilderness. Additions to the National Park System are made through acts of Congress, although the president has the authority to establish national monuments on federally owned lands.

The NPS was created in 1916 as a new federal bureau in the Department of Interior and given the responsibility to administer the national parks and monuments. The NPS currently administers 388 sites, 58 of which are national parks. The total acreage administered by the NPS is 34.1 million hectares (84.3 million acres).

Because the NPS believes that knowledge and understanding increase enjoyment, one of its primary roles is to teach people about the natural environment, management of natural resources, and history of a site by providing nature walks and guided tours of its parks. Exhibits along roads and trails, evening campfire programs, museum displays, and lectures are other educational tools.

Figure 18.4   Yosemite National Park in California. This winter view shows the Merced River flowing past the rock formation, El Capitan. (Larry Ulrich/©Larry Ulrich)

The popularity and success of U.S. national parks have led many other nations to follow the example of the United States in establishing national parks. Today about 1200 national parks exist in more than 100 countries. As in the United States, these parks usually have multiple roles, from providing biological habitat to facilitating human recreation.

Threats to U.S Parks   All the problems plaguing urban areas are found in popular national parks during peak seasonal use, including crime, vandalism, litter, traffic jams, and pollution of the soil, water, and air. In addition, thousands of resource violations, from cutting live trees and collecting plants, minerals, and fossils to defacing historical structures with graffiti and setting fires, are investigated in national parks each year. Park managers have had to reduce visitor access to environmentally fragile park areas degraded from overuse.

Many people think more funding is needed to maintain and repair existing parks. Facilities at some of the largest, most popular parks, such as Yosemite, the Grand Canyon, and Yellowstone, were last upgraded some 30 years ago. Entrance fees account for about $132 million of the $2.2 billion a year the Park Service spends. Although steps were taken to make parks more self-sufficient, they still depend on general tax revenues to pay for their operations.

Some national parks have imbalances in wildlife populations that involve declining populations of many species of mammals, including bears, white-tailed jackrabbits, and red foxes. Grizzly bears in national parks in the western United States are threatened. Grizzly bears require large areas of wilderness, and the human presence in national parks may adversely affect them. More important, the parks may be too small to support grizzlies. Fortunately, so far grizzly bears have survived in sustainable numbers in Alaska and Canada.

Other mammal populations, notably elk, have proliferated. Elk in Yellowstone National Park's northern range increased from a population of 3100 in 1968 to a record high of 19,000 in 1994. Ecologists have documented that the elk have reduced the abundance of native vegetation, such as willow and aspen, and seriously eroded stream banks. The reintroduction of wolves to Yellowstone, which began in 1995 and 1996, is helping to reduce and redistribute the elk population (see Chapter 4 opener).

Human activities beyond park borders affect national parks (recall the Chapter 16 introductory discussion of a proposed gold mine near Yellowstone National Park). Pollution does not respect park boundaries, and parks are increasingly becoming *islands* of natural habitat surrounded by human development. Development on the borders of national parks limits the areas in which wild animals may range, forcing them into isolated populations. Ecologists have found that when environmental stressors occur, several small "island" populations are more likely to become threatened than a single large population occupying a sizable range (see Chapter 17).

Natural Regulation   A park management policy called natural regulation was introduced in Yellowstone National Park and many other U.S. national parks in 1968. Under natural regulation, the population of the elk herd in Yellowstone is allowed to fluctuate naturally because of varying weather conditions and predator populations such as grizzlies and wolves. Park managers do not try to maintain the herd at a consistent level by culling or artificially propagating elk. Because fires are an integral part of the Yellowstone ecosystem, wildfires in the park are not suppressed unless they threaten people or buildings. Park managers may intervene in other situations, such as to control invasive species or to restore native species (such as wolves).

natural regulation: A park management policy that involves letting nature take its course most of the time, with corrective actions undertaken as needed to adjust for changes caused by pervasive human activities.

Wildlife Refuges

The **National Wildlife Refuge System,** established in 1903 by President Theodore Roosevelt, is the most extensive network of lands and waters committed to wildlife habitat in the world. The National Wildlife Refuge System contains more than 535 refuges, with at least one in each of the 50 states, and encompasses 38.4 million hectares (95 million acres) of land. The various refuges represent all major ecosystems found in the United States, from tundra to temperate rain forest to desert, and are home to some of North America's most endangered species, such as the whooping crane. The mission of the National Wildlife Refuge System, which the FWS administers, is to preserve lands and waters for the conservation of fishes, wildlife, and plants of the United States. Wildlife-dependent activities, such as hunting, fishing, wildlife observation, photography, and environmental education, are permitted on parts of some wildlife refuges as long as they are compatible with scientific principles of fish and wildlife management.

REVIEW

1. What is the U.S. National Wilderness Preservation System? Which state has the most wilderness land?
2. Which government agency administers the National Park System?
3. What is natural regulation, and why is it controversial?
4. What is the purpose of the National Wildlife Refuge System?

Forests

LEARNING OBJECTIVES

- Define *sustainable forestry* and explain how monocultures and wildlife corridors are related to forestry.
- Describe national forests, stating which government agencies administer them and current issues of concern.
- Define *deforestation,* including clearcutting, and relate the main causes of tropical deforestation.
- Relate how conservation easements help private landowners protect forests from development.

Figure 18.5 Role of forests in the hydrologic cycle. Forests return most of the water that falls as precipitation to the atmosphere by transpiration. In contrast, when an area is deforested, almost all precipitation is lost as runoff.

Up to 75% water recycled by transpiration and evaporation

25% or more water seeps into ground or runs off to rivers, streams, and lakes

Forests, important ecosystems that provide many goods and services to support human society, occupy about one-fourth of Earth's total land area. Timber harvested from forests is used for fuel, construction materials, and paper products. Forests supply nuts, mushrooms, fruits, and medicines and provide employment for millions of people worldwide. They also offer recreation and spiritual sustenance to an increasingly crowded world.

Forests provide a variety of beneficial ecosystem services. Forests influence local and regional climate conditions. If you walk into a forest on a hot summer day, you will notice that the air is cooler and moister than it is outside the forest. This is the result of a biological cooling process called *transpiration*, in which water from the soil is absorbed by roots, transported through plants, and then evaporated from their leaves and stems. Transpiration provides moisture for clouds, eventually resulting in precipitation (**Figure 18.5**). Thus, forests help maintain local and regional precipitation.

Forests play an essential role in regulating global biogeochemical cycles, such as those for carbon and nitrogen. Photosynthesis by trees removes large quantities of heat-trapping carbon dioxide from the atmosphere and fixes it into carbon compounds. Forests thus act as carbon "sinks" that help mitigate global climate change. At the same time, oxygen that almost all organisms require for cellular respiration is released into the atmosphere.

Tree roots hold vast tracts of soil in place, reducing erosion and mudslides. Forests protect watersheds because they absorb, hold, and slowly release water; this moderation of water flow provides a more regulated flow of water downstream, even during dry periods, and helps control floods and droughts. Forest soils remove impurities from water, improving its quality. In addition, forests provide a variety of essential habitats for many organisms, such as mammals, reptiles, amphibians, fishes, insects, lichens and fungi, mosses, ferns, conifers, and numerous kinds of flowering plants.

Forest Management

When forests are managed for timber production, their species composition and other characteristics are altered from their natural condition. Specific varieties of commercially important trees are planted, and those trees not as commercially desirable are thinned out or removed. *Traditional forest management* often results in low-diversity forests. In the southeastern United States, many tree plantations of young pine that are grown for timber and paper production are all the same age and are planted in rows a fixed distance apart (**Figure 18.6**). These "forests" are essentially monocultures—areas uniformly covered by one crop, like a field of corn. Herbicides are sprayed to kill shrubs and herbaceous plants between the rows. One of the disadvantages of monocultures is that they are more prone to damage from insect pests and disease-causing microorganisms. Consequently, pests and diseases must be controlled in managed forests, usually by applying insecticides and fungicides. Because managed forests contain few kinds of food, they cannot support the variety of organisms typically found in natural forests. Tree plantations have the potential to benefit remaining natural forests, provided that remaining forests are conserved and protected and that the plantations themselves do not replace natural forests.

In recognition of the many ecosystem services performed by natural forests, a newer method of forest management, known as **ecologically sustainable forest management** or, simply, sustainable forestry, is evolving. Sustainable forestry maintains a mix of forest trees, by age and species, rather than a monoculture. This broader approach seeks to conserve forests for the long-term commercial harvest of timber and nontimber forest products. Sustainable forestry also attempts to sustain biological diversity by providing improved habitats for a variety of species; to prevent soil erosion and improve soil conditions; and to preserve watersheds that produce clean water. Effective sustainable forest management involves cooperation among environmentalists, loggers, farmers, indigenous people, and local, state, and federal governments.

When logging occurs using sustainable forestry principles, unlogged areas are set aside as sanctuaries for organisms, along with **wildlife corridors**. The purpose of wildlife corridors is to provide escape routes should they be needed and to allow animals to migrate so that they can interbreed. (Small, isolated, inbred populations may have a higher risk of extinction.) Wildlife corridors may allow large animals such as the Florida panther to maintain large territories. Some scientists question the effectiveness of wildlife corridors, although recent research on wildlife corridors in fragmented landscapes suggests that wildlife corridors help certain wildlife populations persist. Additional research is needed to resolve the question for all threatened and endangered species.

The actual methods of ecologically sustainable forest management that distinguish it from traditional forest management are gradually being developed. These vary from one forest ecosystem to another, in response to different ecological, cultural, and economic conditions. In Mexico many sustainable forestry projects involve communities that are economically dependent on forests. Because trees have such long life spans, scientists and forest managers of the future will judge the results of today's efforts.

Harvesting Trees According to the U.N. Food and Agriculture Organization (FAO), 3.4 million cubic meters of wood (for fuelwood, timber, and other products) were harvested in 2004. The five countries with the greatest tree harvests are the United States, Canada, Russia, Brazil, and China; these countries currently produce more than half the world's timber. About 50% of harvested wood is burned directly as fuelwood or used to make charcoal. (Partially burning wood in a large kiln from which air is excluded converts the wood into charcoal.) Most fuelwood and charcoal are used in developing countries (discussed shortly). Highly developed countries consume more than three-fourths of the remaining 50% of harvested wood for paper and wood products.

Figure 18.6   Tree plantation. This intensively managed pine plantation is a monoculture, with trees of uniform size and age. Such plantations supplement the harvesting of trees in wild forests to provide the United States with the timber it requires. According to a Forest Service report, the United States annually consumes nearly 20% more wood than it produces. Photographed in the southern U.S. pulpwood region (from Alabama to Georgia). (Kirtley-Perkins/Visuals Unlimited)

monoculture: Ecological simplification in which only one type of plant is cultivated over a large area.

sustainable forestry: The use and management of forest ecosystems in a way that meets the needs of the present generation without compromising the ability of future generations to use the forests.

wildlife corridor: A protected zone that connects isolated unlogged or undeveloped areas.

Figure 18.7 Harvesting trees.

(a) In selective cutting, the older, mature trees are selectively harvested from time to time, and the forest regenerates itself naturally.

(b) In shelterwood cutting, less desirable and dead trees are harvested. As younger trees mature, they produce seedlings, which continue to grow as the now-mature trees are harvested.

(c) Seed tree cutting involves the removal of all but a few trees, which are allowed to remain, providing seeds for natural regeneration.

(d) In clearcutting, all trees are removed from a particular site. Clearcut areas may be reseeded or allowed to regenerate naturally.

(e) Aerial view of a large patch of clear-cut forest in Washington state. Clearcutting is the most common but most controversial type of logging. The lines are roads built to haul away the logs. (Calvin Larsen/Photo Researchers, Inc.)

clearcutting: A logging practice in which all the trees in a stand of forest are cut, leaving just the stumps.

Loggers harvest trees in several ways—by selective cutting, shelterwood cutting, seed tree cutting, and clearcutting (**Figure 18.7**). **Selective cutting,** in which mature trees are cut individually or in small clusters while the rest of the forest remains intact, allows the forest to regenerate naturally. The trees left by selective cutting produce seeds that germinate to fill the void. Selective cutting has fewer negative effects on the forest environment than other methods of tree harvest, but it is not as profitable in the short term because timber is not removed in great enough quantities.

The removal of all mature trees in an area over an extended period is **shelterwood cutting.** In the first year of harvest, undesirable tree species and dead or diseased trees are removed. The forest is then left alone for perhaps a decade, during which the remaining trees continue to grow, and new seedlings become established. During the second harvest, many mature trees are removed, but some of the largest trees are left to shelter the young trees. The forest is then allowed to regenerate on its own for perhaps another decade. A third harvest removes the remaining mature trees, but by this time a healthy stand of younger trees is replacing the mature ones. Little soil erosion occurs with this method of tree removal, even though more trees are removed than in selective cutting.

In **seed tree cutting,** almost all trees are harvested from an area; a scattering of desirable trees is left behind to provide seeds for the regeneration of the forest. **Clearcutting** is harvesting timber by removing all trees from an area. After the trees are removed by clearcutting, the area is either allowed to reseed and regenerate itself naturally or is planted with one or more specific varieties of trees. Timber companies prefer clearcutting because it is the most cost-effective way to harvest trees. Clearcutting in small patches actually benefits some wildlife species, such as deer and certain songbirds. These species thrive in the regrowth of trees and shrubs that follows removal of the overhead canopy. However, clearcutting over wide areas is ecologically unsound. It destroys biological habitats and increases soil erosion, particularly on sloping land. In 1996 hundreds of mudslides from steep hillsides that had been clearcut occurred in Oregon following heavy rains; properties and roads were damaged, and several people were killed. Sometimes the land is so degraded from clearcutting that reforestation does not take place; whereas lower elevations are usually regenerated successfully, higher elevations are often difficult to regenerate. Obviously, the recreational benefits of forests are lost when clearcutting occurs.

Deforestation

The most serious problem facing the world's forests is deforestation. According to the FAO, forests shrank more than 36 million hectares (89 million acres) between 2000 and 2005. This amounts to a net annual loss of 0.2%. This estimate of forest loss does not take into account remaining forests that have been thinned or degraded by overharvesting, declining biological diversity, and reduced soil fertility.

Causes of deforestation include fires caused by drought and land-clearing practices, expansion of agriculture, construction of roads in forests, tree harvests, and insects and diseases. When forests are converted to other land uses, they no longer make valuable contributions to the environment or to the people who depend on them. Forest destruction, particularly in the tropics, threatens indigenous people whose cultural and physical survival depends on the forests.

Deforestation results in decreased soil fertility through rapid leaching of the essential mineral nutrients found in most forest soils. Uncontrolled soil erosion, particularly on steep deforested slopes, affects the production of hydroelectric power as silt builds up behind dams. Increased sedimentation of waterways caused by soil erosion harms downstream fisheries. In drier areas, deforestation contributes to the formation of deserts (discussed shortly). When a forest is removed, the total amount of surface water that flows into rivers and streams actually increases. Because the forest no longer regulates this water flow, the affected region experiences alternating periods of flood and drought.

Deforestation contributes to the extinction of many species. (The importance of tropical forests, particularly tropical rain forests, as the repositories of much of the world's biological diversity was discussed in Chapter 17.) Many tropical species, in particular, have limited ranges within a forest, so they are especially vulnerable to habitat modification and destruction. Migratory species, including birds and butterflies, suffer from deforestation.

Deforestation contributes to regional and global climate changes. Trees release substantial amounts of moisture into the air; about 97% of the water that roots absorb from the soil is evaporated directly into the atmosphere. This moisture falls back to Earth in the hydrologic cycle (see Chapter 5). When a large forest is removed, rainfall may decline and droughts may become more common in that region. Studies suggest that the local climate has become warmer and drier in parts of Brazil where tracts of the rain forest were burned. Where deforestation has occurred in Central America, the nearby cloud forests have lost much of their moisture-providing clouds. Temperatures may rise slightly in a deforested area because there is less evaporative cooling from the trees.

Deforestation also contributes to an increase in global temperature, ocean acidification, and other global changes by releasing carbon originally stored in the trees into the atmosphere as carbon dioxide. Carbon dioxide enables the air to retain heat. The carbon in forests is released immediately if the trees are burned, or more slowly when unburned parts decay. If trees are harvested and logs are removed, roughly one-half of the forest carbon remains as dead materials (branches, twigs, roots, and leaves) that decompose, releasing carbon dioxide. When an old-growth forest is harvested, researchers estimate that it takes about 200 years for the replacement forest to accumulate the amount of carbon that was stored in the original forest.

deforestation: The temporary or permanent clearance of large expanses of forest for agriculture or other uses.

Forest Trends in the United States

Figure 18.8 **Forest ownership in the United States.** Most forests are privately owned. (From Marsh and Grossa)

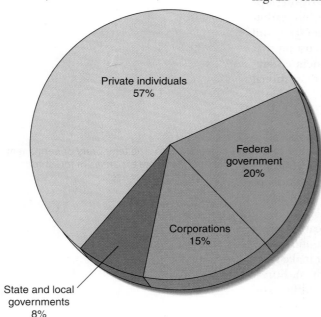

Figure 18.8 **Forest ownership in the United States.** Most forests are privately owned. (From Marsh and Grossa)

Private individuals 57%

Federal government 20%

Corporations 15%

State and local governments 8%

conservation easement: A legal agreement that protects privately owned forest or other property from development for a specified number of years.

In recent years, most temperate forests in the Rocky Mountains, Great Lakes region, and New England and other eastern states have been holding steady or even expanding. In Vermont the amount of land covered by forests has increased from 35% in 1850 to 80% today. Expanding forests are the result of *secondary succession* on abandoned farms (see Chapter 4), commercial planting (tree plantations) on both private and public lands, and government protection. Although the returning forests generally do not have the biological diversity of virgin stands, many forest organisms have successfully become reestablished in regenerated forests. The good news about these returning forests must be tempered by the fact that we are contributing to deforestation elsewhere by importing more timber to meet the increased demand for lumber, paper, and other wood products; we also import beef raised on former forest lands.

Slightly more than one-half of U.S. forests are privately owned (**Figure 18.8**). Many private owners are under economic pressure, such as high property taxes, to subdivide the land and develop tracts of it for housing or shopping malls (see Meeting the Challenge: Preserving Forests in the Eastern United States). Projected conversion of forests to agricultural, urban, and suburban lands over the next 40 years will probably have the greatest impact in the South, where more than 85% of forest is privately owned and logging is largely unregulated.

The **Forest Legacy Program** is a provision of the 1990 **Farm Bill** that helps private landowners protect environmentally important forestlands from development. Here's how the program, managed by the U.S. Department of Agriculture, works. A willing landowner sells some or all ownership rights, such as the right to develop the land, to the U.S. government, which then holds a **conservation easement**. The landowner usually continues to live and/or work on the property and can sell those rights not already sold to the government to other private individuals. All future property owners must honor the provisions of the conservation easement.

MEETING THE CHALLENGE

Preserving Forests in the Eastern States

Timber companies own most of the large tracts of privately owned forests in the eastern United States. Many of these forests are becoming available for purchase, because developers are willing to pay huge amounts of money for them. The timber companies find that selling the land is worth far more than they can earn from logging it. After purchasing the forest property, the developers cut down the trees and build residential communities, golf courses, and hunting clubs. Once development has occurred, there is no chance for the forest to ever regenerate.

Conservation organizations are buying some of these properties to preserve the forest. The

mission of the Nature Conservancy, which was founded in 1951, is "to preserve the plants, animals, and natural communities that represent the diversity of life on Earth by protecting the lands and waters they need to survive." The Nature Conservancy has about one million members who contribute funds to purchase land and aquatic ecosystems. This organization typically collaborates with individual landowners, corporations (such as timber companies), state and federal governments, and indigenous people in a collaborative effort to preserve natural areas. (Recall the role of the Nature Conservancy in obtaining land for Great Sand Dunes National Park, discussed in the chapter introduction.)

The Nature Conservancy and other conservation organizations have only enough

money to purchase a small fraction (less than 2%) of the eastern forests that will probably be sold in the next few years. However, Nature Conservancy scientists prioritize their "wish list" to select the ecosystems that are most uncommon and that may provide habitat for endangered or threatened species. The lands encompassing those ecosystems are the ones that the Nature Conservancy works to obtain. Frequently, a deal is brokered with the timber companies that allows them to continue logging the ecologically valuable forests for five years or so. But at the end of the logging cycle, the land is allowed to undergo secondary succession and revert to forest. More important, the land is no longer available for economic development.

U.S. National Forests   According to the USFS, the United States has 155 national forests encompassing 77 million hectares (191 million acres) of land, mostly in Alaska and western states. The USFS manages most national forests, with the remainder overseen by the BLM. National forests were established for multiple uses—to provide U.S. citizens with the maximum benefits of natural resources such as fish, wildlife, and timber. Multiple uses include timber harvest; livestock forage; water resources and watershed protection; mining; hunting, fishing, and other forms of outdoor recreation; and habitat for fishes and wildlife. Recreation, which increased dramatically in national forests during the 1990s and early 2000s, ranges from camping at designated campsites to backpacking in the wilderness. Visitors swim, boat, picnic, and observe nature in national forests. With so many possible uses of national forests, conflicts inevitably arise, particularly between timber interests and those who wish to preserve the trees for other purposes.

Road building is a contentious issue in national forests, in part because the USFS builds taxpayer-funded roads that allow private logging companies access to the forest to remove timber. The money the USFS charges for timber concessions does not cover the cost for taxpayer-funded roads. The BLM also subsidizes logging operations on lands it manages. About 697,000 km (433,000 mi) of logging roads have been built throughout U.S. national forests. Road building in national forests is environmentally destructive if improper construction accelerates soil erosion and mudslides (particularly on steep terrain) and causes water pollution in streams. Biologists are concerned that so many roads fragment wildlife habitat and provide entries for disease organisms and invasive species.

Another issue in national forests is clearcutting. We now examine the clearcutting debate in the Tongass National Forest, which is located along Alaska's southeastern coast.

CASE IN POINT

Tongass National Forest

Despite its northern location, the Tongass National Forest is one of the world's few temperate rain forests (**Figure 18.9**; also see Chapter 6 for a description of temperate rain forests). It is one of the wettest places in the United States. This moisture supports old-growth forest of giant Sitka spruce, yellow cedar, and western hemlock, some of which are 700 years old. Stream banks are lined with bent willows, ferns, mosses, and other vegetation. This forest, the largest in the National Forest System, provides habitat for a wealth of wildlife, such as grizzly bears and bald eagles.

The Tongass is a prime logging area because a single large Sitka spruce may yield as much as 10,000 board feet of high-quality timber. The logging industry forms the basis of much of the local economy. Regeneration of mature forest after it was clearcut is slow, on the order of several centuries. On Vancouver Island to the south of the Tongass, the trees in temperate rain forest that was clearcut in 1911 have regrown to only 20% of their original size.

As in most national forests, it is expensive to log in the Tongass. To cover the high costs of operating, timber interests such as pulp mills rely on obtaining the timber from the federal government at below-market prices. This right was granted in 1954 by a 50-year contract that expired in the 1990s. In 1990 some members of Congress tried to pass the Tongass Timber Reform Act to force timber interests to pay market prices, but the legislation was bitterly opposed by other members of Congress. The compromise agreement, reached in 1997, provided timber to the mills at market prices. As a result of this legislation, clearcut logging continued in the Tongass but at lower rates than in the past.

In 1999 the Tongass Land Management Plan of 1997 was modified after several dozen appeals were filed against the plan. The 1999 decision, called the *Modified 1997 Forest Plan*, protects an additional 100,000 acres of old-growth forest from logging.

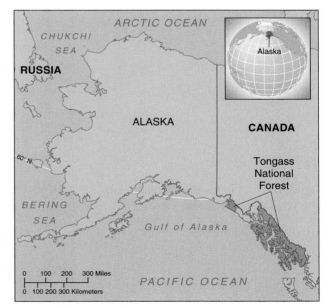

Figure 18.9   Alaska's Tongass National Forest. This temperate rain forest (light green area) is in southeastern Alaska along the Pacific Ocean.

This brings the total protected area in the Tongass to 234,000 acres. (The total area of the Tongass National Forest is 17 million acres.) The modified plan increases timber harvest rotations from 100 years to 200 years in specially designated wildlife areas. This change reduces the impact of forest fragmentation and protects the Sitka black-tailed deer population, used for food by native tribes. The 1999 decision specifies that road density will be reduced to protect the habitat of wolves, bears, and other wildlife. It reduces the maximum quantity of timber harvested on a sustainable basis from 267 million board feet (the 1997 value) to 187 million board feet. This quantity is considered sufficient to meet the requirements of timber operators in the region.

The USFS officially adopted the *Roadless Area Conservation Rule* in 2000. A series of lawsuits challenging the rule ensued, initiated mainly by logging, mining, and gas and oil interests. In 2001 a U.S. district judge blocked the roadless rule, and plans to log several areas of Tongass National Forest that lack roads proceeded. In 2005 the Bush administration issued new rules making it easier to build roads in pristine parts of national forests for logging, mining, and other development; these rules put the decision to log in the hands of individual states.

The take-home message from all of this legal wrangling is that the USFS, like many other government agencies, is a political organization that takes its lead from current presidential policies. Changes in administrations often leave the USFS and other government agencies floundering as they strive to implement rulings from a previous administration that are no longer supported by the current administration. ■

Trends in Tropical Forests

There are two main types of tropical forests: tropical rain forests and tropical dry forests. In places where the climate is warm and moist throughout the year—with about 200 or more cm (at least 79 in.) of precipitation annually—**tropical rain forests** prevail. Tropical rain forests are found in Central and South America, Africa, and Southeast Asia, but almost half of them are in just three countries: Brazil, Democratic Republic of the Congo, and Indonesia (**Figure 18.10**).

In other tropical areas where annual precipitation is less but is still enough to support trees, including regions subjected to a wet season and a prolonged dry season, **tropical dry forests** occur. During the dry season, tropical trees shed their leaves and remain dormant, much as temperate trees do during the winter. India, Kenya, Zimbabwe, Egypt, and Brazil are a few of the countries that have tropical dry forests.

Most of the remaining undisturbed tropical forests, which lie in the Amazon and Congo River basins of South America and Africa, are being cleared and burned at a rate unprecedented in human history. Tropical forests are also being destroyed at an extremely rapid rate in southern Asia, Indonesia, Central America, and the Philippines.

Figure 18.10 Distribution of tropical rain forests. Rain forests (green areas) are located in Central and South America, Africa, and Southeast Asia. Much of the remaining tropical rain forests are highly fragmented. (Based on Goode Base map)

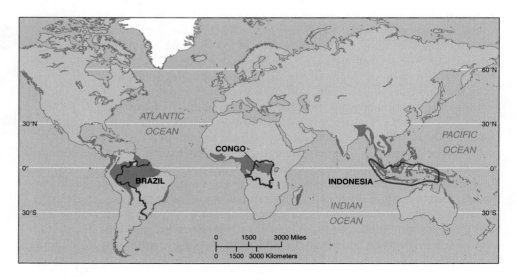

Why Are Tropical Rain Forests Disappearing? Several studies show a strong correlation between population growth and deforestation. More people need more food, and so the forests are cleared for agricultural expansion. Tropical deforestation is a problem that cannot be attributed simply to population pressures, however. The main causes of deforestation vary from place to place, and a variety of economic, social, and government factors interact to cause deforestation. Government policies sometimes provide incentives that favor the removal of forests. The Brazilian government opened the Amazonian frontier, beginning in the late 1950s, by constructing the Belem-Brasilia Highway, which cut through the Amazon Basin. Such roads open the forest for settlement (**Figure 18.11**). Sometimes economic conditions encourage deforestation. The farmer who converts more forest to pasture can maintain a larger herd cattle, which is a good hedge against inflation.

Keeping in mind that tropical deforestation is complex, three agents—subsistence agriculture, commercial logging, and cattle ranching—are considered the most immediate causes of deforestation. Other reasons for the destruction of tropical rain forests include mining, particularly when ore smelters burn charcoal produced from rainforest trees, and dam construction for hydroelectric power, which inundates large areas of forest.

Figure 18.11 **Satellite photograph of human settlements along a portion of road in Brazil's tropical rain forest.** Numerous smaller roads extend perpendicularly from the main roads. As the farmers settle along the roads, they clear out more and more forest (dark green) for their croplands and pastures (tans and pinks). (NRSC/Photo Researchers, Inc.)

Subsistence Agriculture. Subsistence agriculture, in which a family produces just enough food to feed itself, accounts for more than half of tropical deforestation. In many developing countries where tropical rain forests occur, the majority of people do not own the land they live and work on. In Brazil, 5% of the farmers own 70% of the land. Most subsistence farmers have no place to go except into the forest, which they clear to grow food. Land reform in Brazil, Madagascar, Mexico, the Philippines, Thailand, and many other countries would make the land owned by a few available to everyone, thereby easing the pressure of subsistence farmers on tropical forests. This scenario is unlikely, because wealthy landowners have more economic and political clout than impoverished, landless peasants.

Subsistence farmers often follow loggers' access roads until they find a suitable spot. They first cut down the trees and allow them to dry, then they burn the area and plant crops immediately after burning; this is known as **slash-and-burn agriculture** (discussed further in Chapter 19). The yield from the first crop is often quite high because the nutrients that were in the burned trees are now available in the soil. However, soil productivity declines at a rapid rate, and subsequent crops are poor. In a short time, the people farming the land must move to a new part of the forest and repeat the process. Cattle ranchers often claim the abandoned land for grazing, because land not rich enough to support crops can still support livestock.

Slash-and-burn agriculture done on a small scale, with plenty of forest to shift around in so that there are periods of 20 to 100 years between cycles, is sustainable. The forest regrows rapidly after a few years of farming. But when millions of people try to obtain a living in this way, the land is not allowed to lie uncultivated long enough to recover. Globally, at least 200 million subsistence farmers obtain a living from slash-and-burn agriculture, and the number is growing rapidly. Moreover, there is only half as much forest available today as there was 50 years ago.

Commercial Logging. About 20% of tropical deforestation is the result of commercial logging. Vast tracts of tropical rain forests, particularly in Southeast Asia, are harvested for export abroad. Most tropical countries allow commercial logging to proceed at a much faster rate than is sustainable. Unmanaged tropical deforestation does not contribute to economic development; rather, it depletes a valuable natural resource faster than it can regenerate for sustainable use.

Cattle Ranching and Agriculture for Export. Approximately 10% of tropical deforestation is carried out to provide open rangeland for cattle. Cattle ranching,

Figure 18.12 **Deforestation for fuel.** Indian women gather firewood in the Ranthambore National Park buffer zone. Note the branches trimmed off the trees in the background. About 94% of wood removed from Indian forests is burned as fuel. (Marin Hervey/Natural History Photographic Agency)

Figure 18.13 **Logging in Canada's boreal forest.** About 80% of Canada's forest products are exported to the United States. Photographed in Ontario. (Medford Taylor/National Geographic Society)

which employs relatively few local people (ranching does not require much labor), is particularly important in Latin America. Some of the beef raised on these ranches, often owned by foreign companies, is exported to highly developed countries, although much is consumed locally. After the forests are cleared, cattle graze on the land for perhaps 20 years, after which time the soil fertility is depleted. When this occurs, shrubby plants, known as **scrub savanna,** take over the range.

A considerable portion of forestland is cleared for plantation-style agriculture, which produces export crops such as citrus fruits, bananas, and soy. (Soy is exported to use as animal feed.) Plantation-style agriculture, often practiced by foreign corporations, is sustainable on forest soil as long as fertilizers and other treatments are applied.

Why Are Tropical Dry Forests Disappearing? Tropical dry forests are being destroyed at a rapid rate, primarily for fuelwood (**Figure 18.12**). Wood—about half of the wood consumed worldwide—is used as heating and cooking fuel by much of the developing world. The unsustainable use of wood has led to a fuelwood crisis in many developing countries. The two billion or so people that the FAO says cannot get enough fuelwood to meet basic needs such as boiling water and cooking are at risk of waterborne infectious diseases (see Table 22.1 for a list of some human diseases transmitted by polluted water). Often the wood cut for fuel is converted to charcoal, which is then used to power steel, brick, and cement factories. Charcoal production is extremely wasteful: 3.6 metric tons (4 tons) of wood produce enough charcoal to fuel an average-sized iron smelter for only five minutes.

Boreal Forests and Deforestation

Although tropical forests have been depleted extensively, they are not the only forests at risk. Extensive deforestation due to the logging of certain boreal forests began in the late 1980s and early 1990s and continues today. **Boreal forests** occur in Alaska, Canada, Scandinavia, and northern Russia. Coniferous evergreen trees such as spruce, fir, cedar, and hemlock dominate these northern forests. Boreal forests comprise the world's largest biome, covering about 11% of Earth's land area.

Boreal forests, harvested primarily by clearcut logging, are currently the primary source of the world's industrial wood and wood fiber. The annual loss of boreal forests is estimated to encompass an area twice as large as the Amazonian rain forests of Brazil.

About one million hectares (2.5 million acres) of Canadian forests are logged annually, and most of Canada's forests are under logging tenures. (*Tenures* are agreements between provinces and companies that give companies the right to cut timber.) Canada is the world's biggest timber exporter (**Figure 18.13**). On the basis of current harvest quotas, logging in Canada is unsustainable. According to the World Resources Institute, the Canadian government promotes but does not always implement sustainable forest policies.

Extensive tracts of Siberian forests in Russia are also harvested, although exact estimates are unavailable. Alaska's boreal forests are at risk because the U.S. government may increase logging on public lands (recall the earlier discussion of logging in Alaska's temperate rain forest).

E N V I R O N E W S

Reforestation in Kenya

The East African country of Kenya is predominantly arid or semiarid. By the 1970s, most of its tropical dry forests had been destroyed, and the little that remained was disappearing rapidly. A young woman, **Wangari Maathai**, was concerned about the destruction of her country's natural resources. Dr. Maathai, who earned a Ph.D. from the University of Nairobi, founded a grassroots, nongovernmental tree-planting organization called the Green Belt Movement (GBM). The GBM mobilized local groups of women to work for the continued improvement of Kenya's environment. Since 1977, these groups have planted about 30 million trees across eastern and southern Africa. The trees provide many ecosystem services, including cover that protects Kenya from encroaching desert. In 2002 Dr. Maathai was elected to Kenya's Parliament and appointed an Assistant Minister for the Environment. In 2004, she was awarded the Nobel Peace Prize for her environmental work. The Nobel Committee chair selected Ms. Maathai because "peace depends on our ability to secure our environment."

REVIEW

1. What is sustainable forestry?

2. Why is the fact that U.S. national forests were created for multiple uses often a contentious issue?

3. What is deforestation?

4. What are four important causes of tropical deforestation?

Rangelands and Agricultural Lands

LEARNING OBJECTIVES

- Describe public rangelands, stating which government agencies administer them and current issues of concern.
- Define *desertification* and explain its relationship to overgrazing.
- Discuss trends in U.S. agricultural land, such as encroachment of suburban sprawl.

rangeland: Land that is not intensively managed and is used for grazing livestock.

Figure 18.14   Rangeland. When the carrying capacity of rangeland is not exceeded, it is a renewable resource. Photographed along the Salmon River in Idaho. (Steve Smith/Superstock)

Rangelands are grasslands, in both temperate and tropical climates, that serve as important areas of food production for humans by providing fodder for livestock such as cattle, sheep, and goats (**Figure 18.14**). Rangelands may be mined for minerals and energy resources, used for recreation, and preserved for biological habitat and for soil and water resources. The predominant vegetation of rangelands includes grasses, forbs (small herbaceous plants other than grasses), and shrubs.

Rangeland Degradation and Desertification

Grasses, the predominant vegetation of rangelands, have a *fibrous root system*, in which many roots form a diffuse network in the soil to anchor the plant. Plants with fibrous roots hold the soil in place quite well, thereby reducing soil erosion. Grazing animals eat the leafy shoots of the grass, and the fibrous roots continue to develop, allowing the plants to recover and regrow to their original size.

Carefully managed grazing is beneficial for grasslands. Because the vegetation is adapted to grazing, the removal of mature vegetation by grazing animals stimulates rapid regrowth. At the same time, the hooves of grazing animals disturb the soil surface enough to

overgrazing: Destruction of vegetation caused by too many grazing animals consuming the plants in a particular area so they cannot recover.

land degradation: The natural or human-induced process that decreases the future ability of the land to support crops or livestock.

desertification: Degradation of once-fertile rangeland, agricultural land, or tropical dry forest into nonproductive desert.

allow rainfall to more effectively reach the root systems of grazing plants. Several studies around the world have reported that moderate levels of grazing encourage greater plant diversity.

The **carrying capacity** of a rangeland is the maximum number of animals the rangeland plants can sustain over an indefinite period without deterioration of the rangeland. When the carrying capacity of a rangeland is exceeded, grasses and other plants are overgrazed. When plants die, the ground is left barren, and the exposed soil is susceptible to erosion. Sometimes other plants that can tolerate the depleted soil invade an overgrazed area. In parts of Texas Hill Country that were overgrazed, junipers (which are not good forage food) replaced the lush grasses.

Most of the world's rangelands lie in semiarid areas that have natural extended droughts. During dry periods, the carrying capacity of the rangeland is considerably lower, because the lack of precipitation reduces plant productivity. Native grasses in these drylands can survive a severe drought: The aerial portion of the plant dies back, but the extensive root system remains alive and holds the soil in place. When the rains return, the roots develop new shoots.

When overgrazing occurs in combination with an extended drought, once-fertile rangeland can be converted to desert. The reduced grass cover caused by overgrazing allows winds to erode the soil. Even when the rains return, land degradation is so extensive that the rangeland may not recover. Water erosion removes the little bit of remaining topsoil and the sand left behind forms dunes. This progressive degradation, which induces unproductive desertlike conditions on formerly productive rangeland (or tropical dry forest), is called desertification. It reduces the agricultural productivity of economically valuable land, forces many organisms out, and threatens endangered species. Worldwide, desertification seems to be on the increase. Each year since the mid-1990s, the United Nations estimates that 3560 km^2 (1374 mi^2)—an area about the size of Rhode Island—has turned into desert.

There is still much that scientists do not understand about desertification and the processes related to it. We do not know if the declines in productivity caused by land degradation and desertification are permanent. Nor do we know the extent to which desertification results from natural fluctuations in climate versus population pressures and human activities. Currently, many ecologists who study desertification do not think human activities are the main cause of desert encroachment. For example, several years of satellite data support the hypothesis that natural climate variation is the primary factor responsible for the shifting boundary between the southern edge of the Sahara Desert and the African Sahel (see Figure 5.21). Ecologists acknowledge that human activities and overpopulation contribute to degradation in the Sahel but say that fluctuations in climate are causing the desert to advance southward into the Sahel (during drier years) and then retreat (during wetter years). Ecologists prefer to use the term *degradation* rather than desertification when referring to the effects of human activities in arid grasslands, whereas international policy makers prefer the term *desertification*.

About 30% of the total human population lives in the Sahel or other rangelands that border deserts. One of the consequences of desertification/degradation is a decline in agricultural productivity, and studies have shown that when an area cannot feed its people, they emigrate. For example, there are more Senegalese in certain regions of France than in their native villages in Africa. According to the U.N. Environment Program, about 135 million people worldwide are in danger of displacement as a result of desertification. The World Bank, which estimates that desertification costs $42.3 billion per year in economic losses, is beginning to emphasize programs with sustainable agricultural systems for dryland areas.

Rangeland Trends in the United States

Rangelands make up approximately 30% of the total land area in the United States and occur mostly in the western states. Of this, approximately one-third is publicly owned and two-thirds is privately owned. Much of the private rangeland is under increasing pressure from developers, who want to subdivide the land into lots for homes and condominiums. To preserve the open land, conservation groups often pay ranchers for conservation easements that prevent future owners from developing the land. An estimated 405,000 hectares (one million acres) of private rangeland are now protected by conservation easements.

Excluding Alaska, there are at least 89 million hectares (220 million acres) of public rangelands. The BLM, guided by the **Taylor Grazing Act** of 1934, the **Federal Land Policy and Management Act** of 1976, and the **Public Rangelands Improvement Act** of 1978, manages most of the approximately 69 million hectares (170 million acres) of public rangelands. The USFS manages an additional 20 million hectares (50 million acres).

Overall, the condition of public rangelands in the United States has slowly improved since the low point of the Dust Bowl in the 1930s (see Chapter 15). Much of this improvement is attributed to fewer livestock being permitted to graze the rangelands after passage of the Taylor Grazing Act in 1934. Better livestock management practices, such as controlling the distribution of animals on the range through fencing or herding, as well as scientific monitoring, have contributed to rangeland recovery. But restoration is slow and costly, and more is needed. Rangeland management includes seeding in places where plant cover is sparse or absent, conducting controlled burns to suppress shrubby plants, constructing fences to allow rotational grazing, controlling invasive weeds, and protecting habitats of endangered species. Most livestock operators today use public rangelands in a way that results in their overall improvement.

Grazing Fees on Public Rangelands   The federal government distributes permits to private livestock operators that allow them to use public rangelands for grazing in exchange for a fee. (In 2006 the monthly grazing fee was $1.56 per cow on lands managed by the BLM and USFS and $1.73 per month on National Grasslands. The comparable monthly cost on private land is more than $13.) The permits are held for many years and are not open to free-market bidding by the general public—that is, only ranchers who live in the local area are allowed to obtain grazing permits.

Some environmental groups are concerned about the ecological damage caused by overgrazing of public rangelands and want to reduce the number of livestock animals allowed to graze. They want public rangelands managed for other uses, such as biological habitat, recreation, and scenic value, rather than exclusively for livestock grazing. To accomplish this goal, they would like to purchase grazing permits and then set aside the land.

Conservative economists have joined environmentalists in criticizing the management of federal rangelands. According to policy analysts at Taxpayers for Common Sense, in 2003 taxpayers contributed at least $67 million more than the grazing fees collected on public rangelands. This money is used to manage and maintain the rangelands, including installing water tanks and fences, and to fix the damage from overgrazing. Taxpayers for Common Sense and other free-market groups want grazing fees to cover all costs of maintaining herds on publicly owned rangelands.

Agricultural Lands

The United States has more than 121 million hectares (300 million acres) of **prime farmland,** land that has the soil type, growing conditions, and available water to produce food, forage, fiber, and oilseed crops. Certain areas of the country have large amounts of prime farmland. For example, 90% of the *Corn Belt*, a corn-growing region in the Midwest encompassing parts of six states, is considered prime farmland. Not all prime farmland is used to grow crops; approximately one-third contains roads, pastures, rangelands, forests, feedlots, and farm buildings.

Traditionally, farming was a family business. However, larger agribusiness conglomerates that operate more efficiently are rapidly replacing the family farm. As of 2003, there were 2.1 million farms in the United States, as compared to 6.4 million in 1920. In the same period, the average farm size increased from 60 hectares (148 acres) to 178 hectares (441 acres).

Figure 18.15 **Suburban spread onto agricultural land.** Homes and businesses occupy land that was once cornfields in York County, Pennsylvania. Loss of farmland to urban and suburban development is a problem in certain areas of the United States. (© AP/Wide World Photos)

There is considerable concern that much of our prime agricultural land is falling victim to urbanization and suburban sprawl by being converted to parking lots, housing developments, and shopping malls. Both natural ecosystems and agricultural lands adjacent to urban areas are being developed (**Figure 18.15**). In certain areas of the United States, loss of rural land is a significant problem. According to the American Farmland Trust, the top five U.S. farm areas threatened by population growth and urban/suburban spread are California's Central Valley, South Florida, California's coastal region, mid-Atlantic Chesapeake region (Maryland to New Jersey), and North Carolina Piedmont. More than 162,000 hectares (400,000 acres) of prime U.S. farmland are lost each year.

The 1996 Farm Bill included funding for the establishment of a national *Farmland Protection Program*. (Twenty-five states and several local jurisdictions also have farmland protection programs.) This voluntary program helps farmers keep their land in agriculture. The farmers sell conservation easements that prevent them from converting their land to nonagricultural uses. The easements are in effect from a minimum of 30 years to forever. As with other conservation easements, the farmers retain full rights to use their property, in this case, for agricultural purposes. Funding of the Farmland Protection Program is subject to annual appropriations from Congress.

REVIEW

1. What are rangelands? What is the carrying capacity of a rangeland?
2. What is desertification?
3. What human activity is threatening many acres of prime farmland?

 ## Wetlands and Coastal Areas

LEARNING OBJECTIVE

- Describe the current threats to freshwater and coastal wetlands, and explain why the definition of *wetlands* is controversial.

wetlands: Lands that are usually covered by shallow water for at least part of the year and that have characteristic soils and water-tolerant vegetation.

Wetlands are lands transitional between aquatic and terrestrial ecosystems (see Figure 6.16). People used to think the only benefit of wetlands was to provide habitat for migratory waterfowl and other wildlife. In recent years the many ecosystem services that wetlands perform have been recognized, and estimates of their economic value have increased accordingly. Wetlands recharge groundwater and reduce damage from flooding because they hold excess water when rivers flood their banks. Wetlands improve water quality by trapping and holding nitrates and phosphates from fertilizers, and they help cleanse water that contains sewage, pesticides, and other pollutants. Wetlands provide habitat for many species listed as endangered or threatened; these include up to one-half of all fish species, one-third of all bird species, and one-sixth of all mammal species listed on the U.S. endangered or threatened species lists. Freshwater wetlands produce many commercially important products, including wild rice, blackberries, cranberries, blueberries, and peat moss. They are sites for fishing, hunting, boating, bird watching, photography, and nature study.

Many human activities threaten wetlands, including drainage for agriculture or mosquito control and dredging for navigation. Other threats include channelization and construction of dams, dykes, or seawalls for flood control; filling in for solid waste disposal, road building, and residential and industrial development; conversion for aquaculture (fish farming); mining for gravel, phosphate, and fossil fuels; and logging.

In the United States, wetlands have been shrinking an estimated 23,675 hectares (58,500 acres) per year since 1985. This represents a slower rate of loss than in previous years. In the contiguous 48 states, more than half of the more than 89.4 million hectares (221 million acres) of wetlands that originally existed during colonial times are gone; only 42 million hectares (104 million acres) remain. Most of the loss since the 1950s is from farmers converting wetlands to cropland. The greatest percentage declines in wetlands have occurred in the agricultural states of Ohio, Indiana, Iowa, and California. The greatest total acreage declines in wetlands have occurred in Florida, Louisiana, and Texas.

The loss of wetlands is legislatively controlled by a section of the 1972 **Clean Water Act.** This legislation, up for renewal since 1997, does a reasonably good job of protecting coastal wetlands but a poor job of protecting inland wetlands, which is where most wetlands are. The **Emergency Wetlands Resources Act** of 1986 authorized the FWS to inventory and map U.S. wetlands. These maps have many uses, such as planning for drinking water supply protection, siting of development projects, floodplain planning, and making endangered species recovery plans.

Since President George H.W. Bush became president in 1989, the United States has attempted to prevent any new net loss of wetlands by conserving existing wetlands and restoring some that were lost. Development of wetlands is allowed only if a corresponding amount of previously converted wetlands is restored. However, not all wetland restorations are successful, and there is no routine tracking of compliance. Furthermore, there are many things we do not understand about wetland dynamics.

For example, the Sweetwater Marsh was constructed along San Diego Bay in 1984 by the California Department of Transportation, which was legally required to do so when it destroyed a similar marsh during road construction (**Figure 18.16**). One of the main purposes of the reconstructed marsh was to provide habitat for the light-footed clapper rail, an endangered species. The clapper rail never became established in the reconstructed marsh. Ecologists determined that the marsh grass in Sweetwater Marsh was too short for the bird to use for nesting. That species of marsh grass grows significantly taller in natural marshes, where the sediments retain enough nitrogen to fertilize it. The reconstructed marsh's sediments, which were dredged from San Diego Bay's shipping channel and obtained from an old urban dump, are too sandy to retain nitrogen.

Thus, the policy of no net loss of wetlands is only partially successful, and wetlands loss continues, though at a slower rate. Two factors complicate the wetlands policy: (1) confusion and dissent about the definition of wetlands, which was not spelled out in the original Clean Water Act, and (2) the question of who owns wetlands. In 1989 a team of government scientists developed a comprehensive, scientifically correct definition of wetlands. (The definition is technical and beyond the scope of this text.) It provoked an outcry from farmers and real estate developers, who perceived it as an economic threat to their property values. Largely in response to their criticisms, politicians attempted to narrow the definition of wetlands several times during the 1990s, removing marginal wetlands that were not as wet as swamps or marshes. This narrower definition, which relaxed rules on private land use, ignored decades of wetlands research and excluded about one-half of existing U.S. wetlands from protection.

In 1992 Congress asked the National Research Council to help settle the issue because it had become so controversial. The council published its wetlands study in 1995, urging Congress to put the wetlands debate back on a more scientific footing. The council recommended that because shallow wetlands and intermittently wet wetlands perform the same ecosystem services as swamps and marshes, they should be regulated by the same principles. The council study generated an even greater debate in Congress, illustrating that more complete scientific information does not necessarily solve

Figure 18.16 The Sweetwater Marsh.

(a) The reconstructed wetland is in the upper left part of the photograph. (Courtesy State of California, Department of Transportation)

(b) The light-footed clapper rail is an endangered species found in certain California salt and brackish marshes. (Richard R. Hansen/Photo Researchers, Inc.)

problems when stakes are high and values diverge. During the late 1990s and early 2000s, a series of federal court cases involving wetland protection resulted in conflicting decisions. To further complicate the matter, in 2006, the Supreme Court asked the U.S. Army Corp of Engineers to determine which wetlands deserve federal protection under the Clean Water Act.

The federal government owns less than 25% of wetlands in the lower 48 states; the remaining 75% is privately owned. This means that private citizens control whether wetlands are protected and preserved or developed and destroyed. Because of the traditional rights of private land ownership in the United States, landowners resent the federal government's telling them what they may or may not do with their properties. Property-rights advocates in Congress side with landowners in thinking the government has overprotected wetlands. It is therefore important that private landowners become informed of the environmental importance of wetlands and the critical need to maintain wetlands where feasible. Although some private owners recognize the value of wetlands and voluntarily protect them, others are constrained by economic realities that dictate what they must do with their land.

Congress authorized the establishment of the *Wetlands Reserve Program* (WRP) under the **Food Security Act** of 1985 and its amendments. The WRP is a voluntary program that seeks to restore and protect privately owned freshwater wetlands that were previously drained, such as those drained for conversion to cropland. Participants are offered financial incentives to restore the wetlands, and they can establish conservation easements to protect the wetlands. If a property owner establishes a permanent easement, the WRP pays all restoration costs and an amount equal to the agricultural value of the land. As with other conservation easements, the landowner continues to control access to the land. The Natural Resources Conservation Service administers the WRP, which is funded annually by Congress and is subject to budget cuts.

Coastlines

Coastal wetlands, also called saltwater wetlands, provide food and protective habitats for many aquatic animals. They could be considered the ocean's nurseries because so many marine fishes and shellfish spend the first parts of their lives there. In addition to being highly productive, coastal wetlands protect coastlines from erosion and reduce damage from hurricanes.

Historically, tidal marshes and other coastal wetlands were regarded as wasteland, good only for breeding large populations of mosquitoes. Coastal wetlands throughout the world have been drained, filled in, or dredged out to turn them into "productive" endeavors such as industrial parks, housing developments, and marinas. Agriculture contributes to the demise of coastal wetlands, which are drained for their rich soil. Other human endeavors that cause the destruction of coastal wetlands include timber harvesting and fish farming (see Figure 19.15).

In the United States, people have belatedly recognized the importance of coastal wetlands and have passed some legislation to slow their destruction. Intact coastal wetlands moderate the effects of tides more inexpensively than engineering structures such as retaining walls. Sandy beaches along intact coastal wetlands are alternatively washed away and replenished by wave action.

When retaining seawalls are built parallel to the beach, the beach rapidly erodes away, including beach on properties adjacent to the wall (**Figure 18.17**). Most legislation protects beaches at the shoreline from seawalls by *rolling easements*, which prioritize public access to the shore over property owners' rights to build walls. Few laws protect shoreline along bays and sounds, and walls are rapidly

Figure 18.17   Seawalls and beach erosion. To protect their property from rising water levels and storm surges, many property owners along coasts build seawalls, which erode beaches between the wall and the water. Such walls accelerate beach erosion on adjacent properties where no seawall exists. Seawalls also prevent public access to beaches.

Figure 18.18 **Coastal erosion.** How do humans affect coastal erosion? (Bizarro © Dan Piraro/Reprinted with special permission of King Features Syndicate)

replacing these tidal shorelines in many areas of the United States. Because sea levels have begun to rise due to global climate change and are expected to rise even more rapidly as the 21st century progresses, these walls will become even more commonplace unless states enact legislation against their construction.

Coastal Demographics Many coastal areas are overdeveloped, highly polluted, and overfished. Although more than 50 countries have coastal management strategies, their focus is narrow and usually deals only with the economic development of a thin strip of land that directly borders the ocean. Coastal management plans generally do not integrate the management of both land and offshore waters, nor do they take into account the main reason for coastal degradation—human numbers. Perhaps 3.8 billion people—about two-thirds of the world's population—live within 150 km (93 mi) of a coastline. Demographers project that three-fourths of all humans—perhaps as many as 6.4 billion—will live in that area by 2025. Many of the world's largest cities are situated in coastal areas, and these cities are currently growing more rapidly than noncoastal cities.

The United States is not immune to environmental destruction along its coasts (**Figure 18.18**). A 2002 report from the Pew Oceans Commission estimated that more than 27 million additional people will settle along U.S. coasts over the next 15 years. Already, 14 of the 20 largest U.S. cities and 19 of the country's 20 most densely populated counties lie along coasts. These population statistics do not take into account the additional effects of seasonal visitors to coastal resorts. The report says that 14% of U.S. coasts were developed in 1997, and the percentage is expected to rise to 25% by 2025.

If the world's natural coastal areas are not to become urban sprawl or continuous strips of tourist resorts during the 21st century, coastal management strategies must be developed that take into account projections of human population growth and distribution. Such comprehensive management plans will be difficult to formulate and execute because they must regulate coastal development and prevent resource degradation, both on land and in offshore waters. The key to successful planning is local community involvement. If the public understands the importance of natural coastal areas, people may become committed to the sustainable development of coastlines.

REVIEW

1. What is a wetland? What are three ecosystem services that wetlands provide?

✦ Conservation of Land Resources

LEARNING OBJECTIVE

• Name at least three of the most endangered ecosystems in the United States.

Our ancestors considered natural areas as an unlimited resource to exploit. They appreciated prairies as valuable agricultural land and forests as immediate sources of lumber and eventual farmland. This outlook was practical as long as there was more land than people needed. But as the population increased and the amount of available land decreased, it became necessary to consider land as a limited resource. Increasingly, the emphasis has shifted from exploitation to preservation and restoration of the remaining natural areas.

Although all types of ecosystems must be conserved, several are in particular need of protection. The National Biological Service (now the Biological Resources Discipline) and the Defenders of Wildlife commissioned studies that developed a numerical ranking of the most endangered ecosystems in the United States. They used four criteria:

1. The area lost or degraded since Europeans colonized North America.
2. The number of present examples of a particular ecosystem, or the total area.
3. An estimate of the likelihood that a given ecosystem will lose a significant area or be degraded during the next 10 years.
4. The number of threatened and endangered species living in that ecosystem.

Table 18.2 The Top 10 Most Endangered Ecosystems in the United States

Ecosystems (in order of priority)

South Florida landscape
Southern Appalachian spruce-fir forests
Longleaf pine forests and savannas
Eastern grasslands, savannas, and barrens
Northwestern grasslands and savannas
California native grasslands
Coastal communities in lower 48 states and Hawaii
Southwestern riparian communities
Southern California coastal sage scrub
Hawaiian dry forest

Source: From Noss, R.F., et al.

Table 18.2 lists the 10 most endangered U.S. ecosystems based on these criteria. Examples include the South Florida landscape, Southern Appalachian spruce-fir forests, and longleaf pine forests and savannas. As these ecosystems are lost and degraded, the organisms that compose them decline in number and in genetic diversity. Researchers have also found that some rare types of soils are threatened. Soils can take hundreds of years to develop but can be eroded very quickly when disrupted, threatening individual species that rely on these specialized soils. Conservation strategies that set aside ecosystems are the best way to preserve an area's biodiversity (as well as its soil).

As you have seen in this chapter, government agencies, private conservation groups, and private citizens have begun to set aside natural areas for permanent preservation. Such activities ensure that our children and grandchildren will inherit a world with wild places and other natural ecosystems.

REVIEW

1. What are three U.S. ecosystems that need protection?

✦ REVIEW OF LEARNING OBJECTIVES WITH KEY TERMS

• **Relate at least four ecosystem services provided by natural areas.**

Natural areas provide many **ecosystem services**, including watershed management, soil erosion protection, climate regulation, and wildlife habitat.

• **Describe world land use and summarize current land ownership in the United States.**

Globally, about 41% of the world's land areas are for human use (urban areas and agriculture); 30% are rock, ice, tundra, and desert; and 29% are natural ecosystems such as forests. Private citizens,

corporations, and nonprofit organizations own about 55% of land in the United States. The federal government owns about 35%, state and local governments about 7%, and Native American tribes about 3%.

• **Contrast the views of the wise-use movement and the environmental movement regarding the use of federal lands.**

The **wise-use movement** thinks a primary purpose of federal lands is to enhance economic growth. The **environmental movement** considers federal lands a legacy for U.S. citizens that should be preserved to protect biological diversity and ecosystem integrity.

- **Describe the following federal lands, stating which government agencies administer them and current issues of concern: wilderness, national parks, and national wildlife refuges.**

Wilderness is a protected area of land in which no human development is permitted. The National Park Service (NPS), Forest Service (USFS), Fish and Wildlife Service (FWS), and Bureau of Land Management (BLM) administer the National Wilderness Preservation System; problems include overuse of some areas and the introduction of **invasive species.** The NPS administers national parks, which have multiple roles, including recreation and ecosystem preservation; problems include overuse, high operating costs, imbalances in wildlife populations, and development of lands adjoining park boundaries. The National Wildlife Refuge System, administered by the FWS, contains habitat for the conservation of fishes, wildlife, and plants of the United States.

- **Define *natural regulation.***

Natural regulation is a park management policy that involves letting nature take its course most of the time, with corrective actions undertaken as needed to adjust for changes caused by pervasive human activities.

- **Define *sustainable forestry* and explain how monocultures and wildlife corridors are related to forestry.**

Sustainable forestry is the use and management of forest ecosystems in an environmentally balanced and enduring way. Sustainable forestry seeks to conserve forests for timber harvest, sustain biological diversity, prevent soil erosion, and preserve watersheds. A **monoculture** represents ecological simplification in which only one type of plant is cultivated over a large area; tree plantations that are monocultures have little species richness and are not a desirable practice for sustainable forestry. When logging occurs, sustainable forestry dictates that unlogged areas are set aside as wildlife sanctuaries; a **wildlife corridor** is a protected zone that connects isolated unlogged or undeveloped areas.

- **Describe national forests, stating which government agencies administer them and current issues of concern.**

The federal government owns 20% of U.S. forests. The USFS manages most national forests, with the rest overseen by the BLM. National forests have multiple uses, including timber harvest, livestock forage, water resources and watershed protection, mining, recreation, and habitat for fishes and wildlife. Issues in national forests include confrontations over multiple uses, building logging roads with general tax revenues, and cutting old-growth forest.

- **Define *deforestation,* including clearcutting, and relate the main causes of tropical deforestation.**

Deforestation is the temporary or permanent clearance of large expanses of forest for agriculture or other uses. Deforestation causes decreased soil fertility and increased soil erosion; impairs watershed functioning; contributes to the extinction of species; and contributes to regional and global climate changes. **Clearcutting** is a logging practice in which all the trees in a stand of forest are cut, leaving just the stumps. Tropical forests are cut to provide temporary agricultural land, obtain timber, obtain open rangeland for cattle, and supply firewood and charcoal.

- **Relate how conservation easements help private landowners protect forests from development.**

A **conservation easement** is a legal agreement that protects privately owned forest or other property from development for a specified number of years. In the Forest Legacy Program the landowner grants a conservation easement to the U.S. government.

- **Describe public rangelands, stating which government agencies administer them and current issues of concern.**

Rangeland is land that is not intensively managed and is used for grazing livestock. Provided the number of grazing animals is balanced with the rangeland's **carrying capacity,** the rangeland remains a renewable resource. About two-thirds of U.S. rangelands are privately owned. The federal government owns much of the remainder, and the BLM manages most of these lands, with the rest overseen by the USFS. The federal government allows private livestock operators to use public rangelands for grazing; environmental groups want public rangelands managed for other uses, such as wildlife habitat, recreation, and scenic value.

- **Define *desertification* and explain its relationship to overgrazing.**

Desertification is degradation of once-fertile rangeland or tropical dry forest into nonproductive desert. **Overgrazing** is the destruction of vegetation caused by too many grazing animals consuming the plants in a particular area so that they cannot recover. Overgrazing results in barren, exposed soil susceptible to erosion; if this **land degradation** continues, it contributes to desertification.

- **Discuss trends in U.S. agricultural land, such as encroachment of suburban sprawl.**

Agricultural lands are former forests or grasslands that were plowed for cultivation. Increasingly, agribusiness conglomerates farm large tracts of land and operate more cost-effectively than do family farmers on small pieces of land. In certain areas, agricultural lands are threatened by expanding urban and suburban areas.

- **Describe the current threats to freshwater and coastal wetlands, and explain why the definition of *wetlands* is controversial.**

Wetlands are lands that are usually covered by shallow water for at least part of the year and that have characteristic soils and water-tolerant vegetation. Freshwater wetlands provide habitat for many organisms, purify natural bodies of water, and recharge groundwater. Despite their many ecosystem services, wetlands are often drained or dredged for other purposes, such as conversion to agricultural land. Coastlines include tidal marshes and other coastal wetlands. Ecosystem services provided by coastal wetlands include food and habitat for many wildlife species and protection from coastal erosion and storm damage. Failing to recognize the importance of coastal wetlands has led to their destruction and development. The U.S. policy of preventing any net loss of wetlands is complicated by confusion and dissent about the definition of wetlands and by property-rights advocates, who view the scientific definition of wetlands as an economic threat.

- **Name at least three of the most endangered ecosystems in the United States.**

According to the Biological Resources Discipline and the Defenders of Wildlife, the top three most endangered ecosystems in the United States are the South Florida landscape, Southern Appalachian spruce-fir forests, and longleaf pine forests and savannas.

Thinking About the Environment

1. Give at least five ecosystem services provided by nonurban lands. Why is it difficult to assign economic values to many of these benefits?

2. What are the main types of federally owned land in the United States? What uses are permitted on each type of land? What are current issues of concern for each type?

3. Do you think additional federal lands should be added to the wilderness system? Why or why not?

4. Suppose a valley contains a small city surrounded by agricultural land. The valley is encircled by mountain wilderness. Explain why the preservation of the mountain ecosystem would support both urban and agricultural land in the valley.

5. How would a park manager of a national park that adheres to natural regulation probably respond to a lightning-induced forest fire? to the establishment of a noxious invasive weed species?

6. Why is deforestation a serious global environmental problem?

7. What are the environmental effects of clearcutting on steep mountain slopes? on tropical rainforest land?

8. Some analysts think an effective way to combat the influx of unwanted immigrants to such countries as the United States is to curb deforestation. Explain the connection.

9. Distinguish between rangeland degradation and desertification.

10. What is a wetland? Why is the scientific definition of wetlands controversial?

11. List at least five causes of wetland destruction.

12. Describe current wetland protection policies in the United States and give their strengths and weaknesses.

13. Explain how certain tax and zoning laws increase the conversion of prime farmland to urban and suburban development.

14. Should private landowners have control over what they wish to do to their land? How would you as a landowner handle land use decisions that may affect the public? Present arguments for both sides of this issue.

15. How are economic growth and sustainable use of natural resources compatible goals?

16. Why does natural regulation in national parks require paying careful attention to ecosystem processes before deciding to intervene?

Quantitative questions relating to this chapter are on our Web site.

Take a Stand

Visit our Web site at http://www.wiley.com/college/raven (select Chapter 18 from the Table of Contents) for links to more information about the management of federal lands. Consider the opposing views of the wise-use and environment movements, and debate the issue with your classmates. You will find tools to help you organize your research, analyze the data, think critically about the issues, and construct a well-considered argument. *Take a Stand* activities can be done individually or as a team, as oral presentations, written exercises, or Web-based (e-mail) assignments.

Additional online materials relating to this chapter, including a Student Testing Section with study aids and self-tests, Environmental News, Activity Links, Environmental Investigations, and more, are also on our Web site.

Food Resources: A Challenge for Agriculture

Kwashiorkor. Millions of children suffer from this disease, caused by severe protein deficiency. Note the characteristic swollen belly, which results from fluid retention. Photographed in Sudan in 2005. (Antony Njuguna/Reuters/NewsCom)

The U.N. Food and Agriculture Organization (FAO) reported in its annual *State of Food Insecurity* that 852 million people lack access to the food needed for healthy, productive lives. Most of these people, which represent about 13% of the total human population, live in rural areas of the poorest developing countries.

People who receive fewer calories than needed are *undernourished*. Worldwide, an estimated 182 million children under the age of 5 are seriously underweight for their age, according to the World Health Organization (WHO).

People can receive enough calories in their diets but still be *malnourished* because they are not receiving enough of specific, essential nutrients such as proteins or vitamin A. For ex-

ample, a diet of rice lacks sufficient amounts of proteins, lipids, minerals, and vitamins to maintain normal body functions. Adults suffering from malnutrition are more susceptible to disease than those who are well fed. In addition to poor physical development and increased disease susceptibility, children who are malnourished do not grow normally. Because malnutrition affects cognitive development, malnourished children typically do not perform well in school. The WHO estimates that more than three billion people worldwide—the greatest number in history—are malnourished.

The two most common diseases of malnutrition are marasmus and kwashiorkor. **Marasmus** (from the Greek word *marasmos,* meaning "a wasting away") is progressive emaciation caused by a diet low in both total calories and protein. Marasmus is most common among children in their first year of life—particularly children of poor families in developing nations. Symptoms include a pronounced slowing of growth and extreme atrophy (wasting) of muscles. It is possible to reverse the effects of marasmus with an adequate diet.

Kwashiorkor (a native word in Ghana meaning "displaced child") is malnutrition resulting from protein deficiency. It is common among children in all poor areas of the world. The main symptoms include edema (fluid retention and swelling); dry, brittle hair; apathy; stunted growth; and sometimes mental retardation. One of the most typical features of kwashiorkor is a pronounced swelling of the abdomen (see photograph). Kwashiorkor is treated by gradually restoring a balanced diet.

People who eat food in excess of that required are *overnourished*. Generally, a person suffering from overnutrition has a diet high in saturated (animal) fats, sugar, and salt. Overnutrition results in obesity, high blood pressure, and an increased likelihood of such disorders as diabetes and heart disease. Many human studies show a correlation between diets high in animal fat and red meat and certain kinds of cancer (colon and prostate). Overnutrition is most common among people in highly developed nations, such as the United States, where the Pan American Health Organization estimates that two out of three adults are overweight, and nearly one in three is obese. Overnutrition is also emerging in some developing countries, particularly in urban areas.

The production of food in an environmentally sustainable way is one of the principal challenges facing humanity today. In this chapter you will examine the nature and magnitude of the world's food problems, ranging from the farmers in developing countries who produce barely enough food to feed their families to the energy-intensive but questionably sustainable agricultural methods of highly developed countries.

 World View and Closer to You...
Real news footage relating to food resources around the world and in your own geographic region is available at www.wiley.com/college/raven by clicking on "World View and Closer to You."

Food and Nutrition

LEARNING OBJECTIVE

• Identify the main components of human nutritional requirements.

The foods humans eat are composed of several major types of biological molecules necessary to maintain health: carbohydrates, proteins, and lipids. **Carbohydrates,** such as sugars and starches, are important primarily because they are metabolized readily by the body in **cellular respiration.** In this process, the energy of these biological molecules is transferred to the molecule *adenosine triphosphate (ATP).* The body uses the energy in ATP to contract muscles, produce heat, repair damaged tissues, grow, fight off infections, and reproduce.

Proteins are large, complex molecules composed of repeating subunits called **amino acids.** Proteins perform several critical roles in the body. When plant and animal proteins in foods are digested, or broken down into amino acids, the body absorbs these amino acids, which may be reassembled in different orders to form human proteins. A substantial part of the human body, from hair and nails to muscles, is made up of protein. Proteins are also metabolized in cellular respiration to release energy.

Twenty amino acids are required for human nutrition. The human body manufactures about half of these, using starting materials such as carbohydrates. The remaining amino acids, called **essential amino acids,** must be obtained from food.

Lipids are a diverse group of biological molecules that includes fats and oils. Like carbohydrates and proteins, lipids are metabolized by cellular respiration to provide the body with a high level of energy. Lipids have several other important roles in the body. Some lipids are hormones, and others are essential components of cell membranes.

In addition to carbohydrates, proteins, and lipids, humans require minerals, vitamins, and water in our diets. **Minerals,** such as iron, iodine, and calcium, are inorganic elements essential for the normal functioning of the human body. Minerals are ingested in the form of salts dissolved in food and water. **Vitamins** are complex biological molecules that living cells require in small quantities. Vitamins help regulate metabolism and the normal functioning of the human body. Whereas plants synthesize most vitamins, humans and other animals must obtain vitamins from food.

Human Foods

Biologists estimate that there are more than 330,000 species of plants. Of these, slightly more than 100 provide about 90% of the food that humans consume, either directly or indirectly. (Humans consume foods indirectly when cereal grains are used to feed livestock that humans eat as meat.) Just 15 species of plants provide the bulk of food for humans (**Table 19.1**). Of these plants, three cereal grains—rice, wheat, and corn—provide about half of the calories that people consume. Our dependence on so few species of plants for the bulk of our food puts us in an extremely vulnerable position. Should disease or some other factor wipe out one of the important food crops, severe food shortages would threaten us. There are tens of thousands of kinds of plants that have been used as sources of food at one time or another. Many of these doubtless could be developed into important sources of food. We must identify them, find out how to use them, and study their cultivation requirements.

Animals provide us with foods that are particularly rich in protein. These foods include fishes, shellfish, meat, eggs, milk, and cheese. Cows, sheep, pigs, chickens, turkeys, geese, ducks, goats, and water buffalo are the most important types of about 80 species of livestock. Though nutritious, livestock is an expensive source of food because animals are inefficient converters of plant food. Of every 100 calories of plant material a cow consumes, it burns off approximately 86 in its normal metabolic functioning. That means humans consume only 14 calories out of 100 (14%) stored in the cow. Meat consumption is high in affluent societies, so large portions of the crops grown in highly developed countries are used to produce livestock animals for human consumption. Almost half of the cereal grains grown in highly developed countries are used to feed livestock (see "You Can Make a Difference: Vegetarian Diets" on page 449).

REVIEW

1. What three main types of biological molecules are metabolized by cellular respiration to provide the body with energy?

2. What are essential amino acids?

World Food Problems

LEARNING OBJECTIVES

- Explain how famines differ from chronic hunger.
- Define *world grain carryover stocks* and explain how they are a measure of world food security.

Table 19.1 The 15 Most Important Food Crops in Terms of Production

Plant Crop	Type of Crop	2005 World Production* (1000 metric tons)
Sugarcane	Sugar plant (stem)	1,168,773
Corn (maize)	Cereal grain	615,758
Rice, paddy	Cereal grain	586,743
Wheat	Cereal grain	537,978
White potato	Ground crop (tuber)	255,681
Soybean	Legume	207,834
Cassava (manioc)	Ground crop (root)	187,862
Sweet potato	Ground crop (root)	125,626
Barley	Cereal grain	113,499
Sorghum	Cereal grain	53,721
Peanuts (ground nuts)	Legume	33,395
Oats	Cereal grain	21,369
Beans, dry	Legume	15,967
Rye	Cereal grain	15,132
Peas, dry	Legume	10,639

* Based on the 20 highest producing countries for a specific agricultural commodity.
Source: U.N. Food and Agricultural Organization.

Producing enough food to feed the world's people is the largest challenge in agriculture today, and the challenge grows more difficult each year because the human population is continually expanding. As shown in **Figure 19.1**, annual grain production increased from 1970 to 2005. However, the world population increased more than two billion during that period, so the amount of grain *per person* has not changed appreciably.

During the 1990s and early 2000s, world increases in food production barely kept pace with population growth. (We will ignore for the moment the fact that chronic hunger still persists in many places in the world.) Grain production increased from 247 kg per person in 1950 to its highest level, 342 kg per person, in 1984. Since then, it has declined to a 2005 level of 312 kg per person. Global food production can be increased in the short term, although whether this increase is sustainable is questionable.

The long-term solution to the food supply problem is stabilization of the human population. Most promising in this regard are observations of the human *demographic transition*—that reductions in birth rates seem to follow reductions in death rates by a generation or two, for reasons that are not entirely clear (see Figure 8-14). Therefore,

Figure 19.1   Total world grain production and grain production per person, 1970–2005.

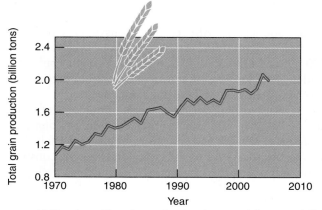

(a) Total world grain production increased from 1.1 billion tons in 1970 to 2.0 billion tons in 2005.

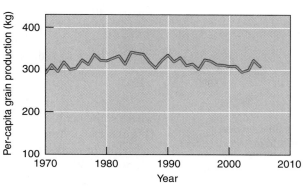

(b) The amount of grain produced per person has remained more or less stable, at about 300 kg per person.

helping people in developing countries get enough to eat, live in safe environments, and have access to good medical care may be the most promising way to stabilize the human population.

Famines

Crop failures caused by drought, war, flood, or some other catastrophic event may result in **famine,** a temporary but severe food shortage. Throughout human history, famine has struck one or more regions of the world every few years. The developing nations of Africa, Asia, and Latin America are most at risk. The worst African famine in history, caused in part by widespread drought, occurred from 1983 to 1985. Hardest hit were Ethiopia and Sudan, in which 1.5 million people died of starvation. The people living in this region lacked sufficient money to purchase food and did not have stored food reserves to protect them against several years of crop failures. In the early 1990s, drought and civil unrest in Somalia resulted in famine for an estimated two million Somalis. More than two million North Koreans died from starvation or hunger-related illnesses during a famine in the late 1990s. The North Korean famine was caused by several years of floods and drought that destroyed its economy and collective farming system. In 2005, drought and an invasion of locusts threatened 3.6 million people in Niger with famine.

Famines get a great deal of media attention because of the huge and obvious amount of human suffering they cause. However, many more people die from undernutrition and malnutrition than from the starvation associated with famine.

Maintaining Grain Stockpiles

world grain carryover stocks: The amounts of rice, wheat, corn, and other grains remaining from previous harvests, as estimated at the start of a new harvest.

World grain carryover stocks provide a measure of world *food security*, a goal in which all people have access at all times to adequate amounts and kinds of food needed for healthy, active lives. Stockpiles of grain have decreased each year since their all-time high in 1987 of 422 million metric tons (466 million) tons. This amount would have been enough, assuming it was evenly distributed, to feed the world's people for 104 days. The amount of grain stockpiled in 2002 would feed the world's people for only 60 days (**Figure 19.2**). According to the United Nations, world grain carryover stocks should not fall below the minimum amount of 70 days' supply in a given year.

When food is scarce and prices increase, the risk of political instability is a real concern in poor nations. Some experts think it will take at least three or four bountiful harvests to rebuild grain stocks. Others are more pessimistic about the low grain stocks and think they indicate the beginning of a period of agricultural scarcity and higher food prices. Supporting this view is the fact that world grain carryover stocks have reached 70 days or higher only two times since 1990 (in 1991 and 1993).

World grain stocks have dropped in the past few years for several reasons. Environmental conditions such as rising temperatures, falling water tables, and droughts have caused poor harvests. Also, crop yields have slowed because of declining investments in agricultural research to improve crops and livestock. As the United States and other countries search for gasoline substitutions to reduce dependency on foreign oil, corn yields will be increasingly diverted to ethanol production instead of food and animal feed.

World grain stocks have also dropped because consumption of beef, pork, poultry, and eggs has increased in China and other developing countries, where some people are becoming more affluent and can afford to diversify their diets (**Figure 19.3**). Animal products account for 40% of the kilocalories people consume in highly developed countries and only 5% of the kilocalories consumed in developing countries. **Table 19.2** compares meat consumption in India, China, Italy, and the United States.

The increased consumption of meat and meat products has prompted a surge in grain used to feed the world's 45 billion livestock animals. The U.S. Office of Technology calculates that every kilogram of chicken meat requires an average input of 2.7

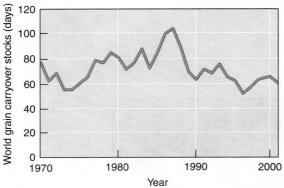

Figure 19.2   World grain carryover stocks, 1970–2002. World grain carryover stocks, expressed as the number of days' supply on hand, should be at least 70 days' worth to allow a cushion against poor harvests and to help provide stability in world grain prices. (USDA)

kg of livestock feed grain. Similarly, 1 kg of pork meat requires 6 kg of grain, and 1 kg of beef meat requires 7 kg of grain.

Poverty and Food: Making Food Affordable for the Poor

Despite the fact that enough food is currently produced to feed the world's people with an adequate but not generous diet, the harsh reality is that food is not shared equally by all people. The main cause of undernutrition and malnutrition is poverty. The world's poorest people—those living in developing countries in Asia, Africa, and Latin America—do not own land on which to grow food and do not have sufficient money to purchase food. Over 1.3 billion people in developing countries have incomes equivalent to less than $1 a day and are so poor that they cannot afford to eat enough food or enough of the right kinds of food.

Worldwide, chronic hunger is more common in rural areas than in urban areas. The lack of rural infrastructure, such as roads, contributes to hunger. Poor rural people living in areas without roads have few employment opportunities to improve their incomes. Studies in China and India have shown that when the government builds roads into rural areas, poverty and chronic hunger are reduced because people use the roads to find jobs with higher wages.

Infants, children, and the elderly are more susceptible to poverty and chronic hunger. The greatest proportion of chronically hungry people is in Africa. The largest number of hungry people is in Asia. Poverty and hunger are not restricted to developing nations, however. Poor hungry people are found in the United States, Canada, Europe, and Australia.

Economic and Political Effects on Human Nutrition

Despite the fact that the major food-producing countries currently produce enough grain to feed the rest of the world, there is still an enormous economic problem. It costs money to produce, store, transport, and distribute food. Asian, African, and Latin American countries, which have the greatest need for imported food, are least able to pay for it. On the other hand, the food-producing nations cannot afford simply to give food away and absorb the costs indefinitely. (Developing countries, however, rarely if ever need food aid indefinitely. They usually need it to get through one or several difficult seasons.)

The observation that led to Indian economist **Amartya Sen**'s Nobel Prize in Economics in 1998 is that the leading cause of hunger and famine in the world is neither lack of total food nor inefficient distribution. Instead, the leading cause is the type of government: Democratic governments are more likely to get their people fed in difficult times than are totalitarian regimes. In addition, government inefficiency and bureaucratic red tape add to food problems, sometimes making it difficult to distribute the food to the hungriest people and to ensure those who need it get it instead of those who do not need it. (Sometimes dishonest government officials, military personnel, or civilians sell food intended for hungry people for personal profit.) Thus, getting food to the people who need it is largely a political problem.

One solution to food problems that is often suggested is for developing countries to shift to more local food production and consumption—that is, their food should be produced in the area where it is consumed. If food production matches market demands in developing countries, people are fed and economic growth occurs as agriculture

Figure 19.3   Meat market in Guangzhou (Canton), China. As their incomes rise, people in China and many other developing countries are eating more meat and meat products. Because livestock animals eat grains, the increased demand for meat has contributed to the decline in world grain carryover stocks. According to Norman Myers, a scientist at Oxford University, if each of China's people were to eat just one extra chicken per year, the grain required to raise those chickens would equal all the grain exports of Canada. (Canada is the world's second largest exporter of grain.) (H. Donnezan/Photo Researchers, Inc.)

Table 19.2 Annual Per Capita Consumption of Meat in Selected Countries

Country	Beef (kg)	Pork (kg)	Poultry (kg)	Mutton (kg)
India	1	0.4	1	1
China	4	30	6	2
Italy	26	33	19	2
United States	45	31	46	1

Source: Brown, L. R. *Tough Choices.*

ENVIRONEWS

Food Safety

In 2007, U.S. pet owners panicked when many pet foods were recalled. The pet foods contained wheat gluten that had been tainted with melamine, a chemical used to make plastics. An underappreciated fact associated with the tainted pet food is that the wheat gluten originated in China. In 2007, the U.S. Food and Drug Administration also blocked the import of five kinds of seafood from China because they found traces of antimicrobials, some of which can cause cancer in laboratory animals when they are fed the chemicals over an extended period.

Globalization of farming has made food safety—for both pets and humans—more of a challenge. Our food travels longer distances, and many raw foods—or the ingredients in processed foods—are imported. More people come into contact with the food as it journeys around the world, making it a challenge to trace an outbreak to its source. Also, imported food is distributed over large areas, making it difficult to contain an outbreak when it does occur.

According to the U.S. Department of Agriculture, the five most common food-borne illnesses cost the U.S. economy almost $7 billion per year in medical expenses and lost productivity. To address the challenge of food safety, food processors are developing new methods to screen foods for contamination. For example, meat inspectors who visually examined meat for problems are slowly being replaced by light systems that scan the meat and detect traces of *E. coli* and other food-borne pathogens.

generates incomes and employment. This approach, however, is not easy to implement in today's global market. An alternative that seem more plausible is for developing countries to earn money to purchase food by growing high-value crops for export to the United States and other highly developed countries.

World food problems are many, as are their solutions. We must increase the sustainable production of food, assist overall economic development, and improve food distribution. Highly developed nations can help developing countries become agriculturally self-sufficient by providing economic assistance and technical aid. But the ultimate solution to chronic hunger is tied to achieving a stable population in each nation at a level that its environment can support.

REVIEW

1. What are world grain carryover stocks?

The Principal Types of Agriculture

LEARNING OBJECTIVE

• Contrast industrialized agriculture with subsistence agriculture and describe three kinds of subsistence agriculture.

industrialized agriculture: Modern agricultural methods, which require a large capital input and less land and labor than traditional methods.

Agriculture can be roughly divided into two types: industrialized agriculture and subsistence agriculture. Most farmers in highly developed countries and some in developing countries practice **industrialized agriculture** or **high-input agriculture.** Industrialized agriculture relies on large inputs of capital and energy, in the form of fossil fuels, to produce and run machinery, irrigate crops, and produce agrochemicals, such as commercial inorganic fertilizers and pesticides (**Figure 19.4**). Industrialized agriculture produces high **yields** (the amount of a food crop produced per unit of land), enabling forests and other natural areas to remain wild instead of being converted to agricultural land. The productivity of industrialized agriculture has not been without costs, however. Industrialized agriculture causes several problems, such as soil degradation and increases in pesticide resistance in agricultural pests. We discuss these and other problems later in the chapter and in Chapter 23.

subsistence agriculture: Traditional agricultural methods, which are dependent on labor and a large amount of land to produce enough food to feed oneself and one's family.

Most farmers in developing countries practice **subsistence agriculture**, the production of enough food to feed oneself and one's family, with little left over to sell or

Figure 19.4 Energy inputs in industrialized agriculture. Fossil fuel inputs occur at virtually all stages in agricultural production. (Adapted from G. H. Heichel)

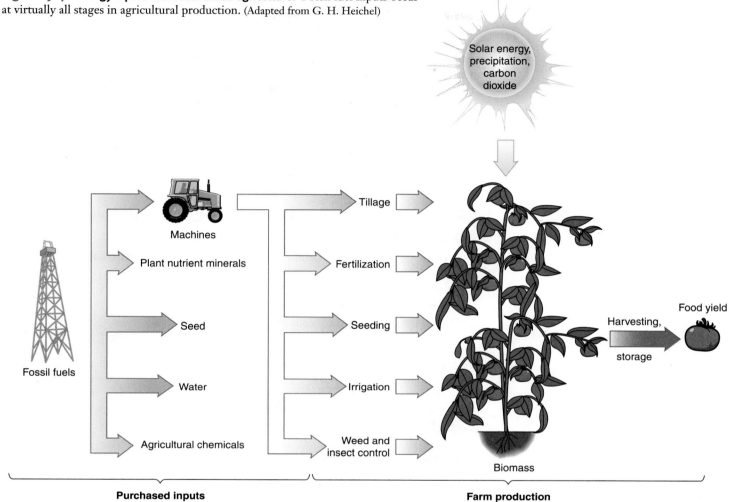

reserve for hard times. Subsistence agriculture, too, requires a large input of energy, but from humans and draft animals rather than from fossil fuels.

Some types of subsistence agriculture require large tracts of land. **Shifting cultivation** is a form of subsistence agriculture in which short periods of cultivation are followed by longer periods of fallow (land left uncultivated) in which the land reverts to forest. Shifting cultivation supports relatively small populations. **Slash-and-burn agriculture** is one of several distinct types of shifting cultivation that involves clearing small patches of tropical forest to plant crops (see Chapter 18). Because tropical soils lose their productivity quickly when they are cultivated, farmers using slash-and-burn agriculture must move from one area of forest to another every three years or so, so slash-and-burn agriculture is land-intensive. **Nomadic herding,** in which livestock is supported by land too arid for successful crop growth, is another type of land-intensive subsistence agriculture. Nomadic herders must continually move their livestock to find adequate food for them.

Intercropping is a form of intensive subsistence agriculture that involves growing a variety of plants simultaneously on the same field. When certain crops are grown together, they produce higher yields than when they are grown as **monocultures.** One reason for higher yields is that different pests are found on each crop, and intercropping discourages the buildup of any single pest species to economically destructive levels. Native Americans practiced intercropping when they planted corn, bean, and squash seeds in the same mound of soil. Because the root systems of these plants grow to different depths, they do not compete with one another for water and essential minerals. In addition, the protein-rich bean crop fixes nitrogen that fertilizes the corn and squash plants naturally. **Polyculture** is a type of intercropping in which several kinds of plants that mature at different times are planted together. In polyculture practiced in the tropics, fast- and slow-maturing crops are often planted together so that crops are harvested throughout the year. Vegetable crops and cereal grains, which mature first, might be planted with papayas and bananas, which mature later.

REVIEW

1. What are some differences between industrialized agriculture and subsistence agriculture?
2. What are shifting cultivation, nomadic herding, and intercropping?

Challenges of Producing More Crops and Livestock

LEARNING OBJECTIVES

- Describe the beneficial and harmful effects of domestication on crop plants and livestock, defining *germplasm* in the process.
- Relate the benefits and problems associated with the green revolution.
- Explain the roles of hormones and antibiotics in industrialized agriculture.
- Identify the potential benefits and problems with genetic engineering.

Having a general grasp of world food problems, let's consider some of the challenges facing agriculture today. If you were to travel around the world, you would find many kinds of agriculture and types of food. Despite this diversity, an overall trend toward greater uniformity in the plants and animals we eat has occurred as we have come to rely on fewer and fewer types of plants and animals for the bulk of food production.

The Effect of Domestication on Genetic Diversity

Wild plant and animal populations usually have a lot of genetic diversity—that is, variation in their genes, units of hereditary information that specify certain traits (see Figure 17.1 for a visual demonstration of genetic diversity in corn). **Genetic diversity** contributes to a species' long-term survival by providing the variation that enables each population to adapt to changing environmental conditions. During the **domestication** of plants and animals, much of this genetic diversity is lost because the farmer selects for propagation only those plants and animals with the most desirable agricultural characteristics. At the same time, other traits not of obvious value to humans are selected against. Hence, many of the high-yielding crops produced by modern agriculture are genetically uniform. Most of the vegetable crops grown in the United States are of only a few varieties. Similarly, dairy cattle and poultry in the United States have low genetic diversity. The familiar black-and-white Holsteins comprise 91% of U.S. dairy cattle, and white leghorn chickens produce almost all the white eggs consumed in the United States.

The loss of genetic diversity that accompanies modern agriculture usually does not prove disastrous to crop plants, because they do not have to survive in the wild under natural conditions. Under cultivation, they are watered, fertilized, and protected as much as possible from pests, including weeds, insects, and disease organisms. Domesticated animals are also protected from the challenges of nature.

The lower genetic diversity of domesticated plants and animals increases the likelihood that they will succumb to new strains of disease-causing organisms. Bacteria, fungi, viruses, and similar organisms evolve quite rapidly because of their high reproductive rates. When a disease breaks out in a domesticated plant or animal population, the entire uniform population is susceptible. The loss is greater than it would be in a natural, varied population, in which at least some individuals would contain genes to resist the disease-causing organism.

domestication: The process of taming wild animals or adapting wild plants to serve humans; domestication markedly alters the characteristics of the domesticated organism.

The Global Decline in Domesticated Plant and Animal Varieties Although domestication contributes in general to less genetic diversity than is found in wild relatives, the many farmer-breeders around the world who have selected for specific traits have developed many local varieties of each domesticated plant and animal. French farmers, for example, developed 200 breeds of cattle during the 18th and 19th centuries. Each traditional variety represents the legacy of the hundreds of farmers who developed it over the centuries. A traditional variety is adapted to the climate where it was bred and contains a unique combination of traits conferred by its unique combination of genes.

A global trend is currently under way to replace the many local varieties of a particular crop or domesticated farm animal with just a few kinds. According to a Worldwatch Institute report, Mexican farmers currently grow only about 20% of the corn varieties that they grew in the 1930s. The modern varieties, which are bred for uniformity and maximum production, are generally more susceptible to insect pests and disease and less able to adapt to environmental changes, including climate change. When farmers abandon their traditional varieties in favor of more modern ones, the former varieties frequently face extinction (**Figure 19.5**). This represents a great loss in genetic diversity, because each variety's characteristic combination of genes gives it distinctive nutritional value, size, color, flavor, resistance to disease, and adaptability to different climates and soil types.

The gene combinations of local varieties are potentially valuable to agricultural breeders because they can be transferred to other varieties by traditional breeding methods or by genetic engineering (discussed later in this chapter). For example, in the late 1980s Russian aphids, tiny insects that suck out the juices of plant leaves and stems, infested U.S. wheat and barley crops. Over several years, aphid-resistant varieties of wheat and barley were developed using genes of wheat and barley varieties from the Middle East.

To preserve older, more diverse varieties of plants and livestock, many countries are collecting **germplasm**. It includes seeds, plants, and plant tissues of traditional crop varieties, and the sperm and eggs of traditional livestock breeds. The International Plant Genetics Resources Institute in Rome, Italy, is the organization that oversees plant germplasm collections worldwide. National germplasm collections range in size from the U.S. National Plant Germplasm System in Colorado, which holds almost half a million varieties, to the national gene bank in the African country Malawi, which holds about 8000 varieties of native crops and fruits. (See the section on seed banks in Chapter 17 for more information about plant germplasm collections.)

Who actually owns germplasm, particularly crop diversity, has been a contentious international issue for about 20 years. Developing nations, which possess much of the world's crop genetic diversity, have maintained that they own the germplasm found in their countries. However, plant breeders want free access to that germplasm, and agricultural firms want the right to patent any crop improvements they make using germplasm. In 2004 the **International Treaty on Plant Genetic Resources for Food and Agriculture** went into effect. (The United States has signed but not ratified it.) The treaty limits the genetic materials that agricultural companies are allowed to patent and affirms the right of farmers to save, use, exchange, and sell farm-saved seeds. Germplasm will be made available to plant breeders and agricultural companies in exchange for royalty payments that will be used for conservation programs.

Increasing Crop Yields

Until the 1940s, agricultural yields among various countries, both highly developed and developing, were generally equal. Advances by research scientists since then have dramatically increased food production in highly developed countries (**Figure 19.6**). Greater knowledge of plant nutrition has resulted in fertilizers that promote high yields.

Figure 19.5 Dutch Belted cow. The USDA recognizes Dutch Belts as a viable dairy breed that produces high-quality, flavorful milk. The breed is endangered in the United States, with a population of about 200 purebred cows, and rare in Canada. The Netherlands was the original source of this variety, known in the 1600s, but the American Dutch Belt is genetically closer to the original breed than the few Dutch Belts still remaining in the Netherlands. Photographed in Wisconsin. (Photo by Agri-Graphics, Ltd, Courtesy Winifred Hoffman)

germplasm: Any plant or animal material that may be used in breeding.

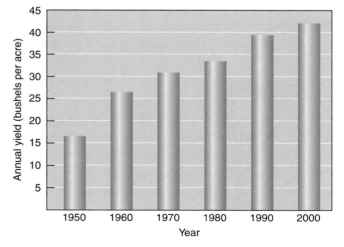

Figure 19.6 Average U.S. wheat yields, 1950–2000. Each year shown is actually an average of three years to minimize the effects that poor weather conditions might have in a given year. Similar increases in yield occurred in other grain crops. (USDA)

Figure 19.7 Development of high-yielding rice varieties.
(Courtesy International Rice Research Institute)

(a) Traditional rice plants are taller and do not yield as much grain (clusters at top of plant) as the more modern varieties.

(b) This rice plant was developed during the 1960s by crossing a high-yielding, disease-resistant variety with a dwarf variety to prevent the grain-heavy plants from lodging, or falling over.

(c) Improvements since the green revolution have been modest, as this rice variety, developed during the 1990s, shows. Some researchers think rice and certain other genetically improved crops are near their physical limits of productivity.

The use of pesticides to control insects, weeds, and disease-causing organisms has improved crop yields. Selective breeding programs have resulted in agricultural plants with more desirable features. Breeders developed wheat plants with larger, heavier grain heads (for higher yield). Because of the weight of the heads, other traits were gradually incorporated into wheat, such as shorter, thicker stalks, which prevent the plants from falling over during storms.

CASE IN POINT

The Green Revolution

By the middle of the 20th century, serious food shortages occurred in many developing countries, and it was widely recognized that additional food supplies were needed to feed their growing populations. The development and introduction during the 1960s of high-yielding varieties of wheat and rice to Asian and Latin American countries gave these nations the chance to provide their people with adequate supplies of food. But the high-yielding varieties required intensive cultivation methods, including the use of commercial inorganic fertilizers, pesticides, and mechanized machinery, to realize their potential. These agricultural technologies were passed from highly developed nations to developing nations.

Using modern cultivation methods and the high-yielding varieties of certain staple crops to produce more food per acre of cropland is known as the **green revolution. Norman Borlaug,** a U.S. scientist who worked on wheat in Mexico during the 1940s and 1950s, is credited with beginning the green revolution. Borlaug's introduction of a short-stemmed, hybrid strain of "miracle wheat" to Mexico and countries in South Asia was followed by the development by other plant breeders of high-yielding varieties of rice and other grain crops (**Figure 19.7**). For his contributions, Borlaug was awarded the Nobel Peace Prize in 1970.

Some of the success stories of the green revolution are remarkable. During the 1920s, Mexico produced less than 700 kg (0.77 ton) of wheat per hectare annually. During the green revolution years that began in 1965, Mexico's annual wheat production rose more than 2400 kg (2.65 tons) per hectare. Indonesia used to import more rice than any other country in the world. Today Indonesia produces enough rice to both feed its people and export some.

Critics of the green revolution argue that it has made developing countries dependent on imported technologies, such as agrochemicals and tractors, at the expense of traditional agriculture. The two most important problems associated with higher crop production are the high energy costs built into this form of agriculture and the serious environment problems caused by the intensive use of commercial inorganic fertilizers and pesticides. The environmental impacts of intensive agriculture, including the green revolution, are discussed later in the chapter.

Rice and wheat are not the only crops improved by the green revolution. High-yielding varieties of crops such as potatoes, barley, and corn have been developed. Nonetheless, many important food crops have not been improved yet. People in Africa eat sorghum, millet, cassava, and sweet potatoes, but none of these has been greatly improved by green revolution technology. Overall, the green revolution has benefited large landowners but not subsistence farmers, who represent a substantial segment of the agricultural community in most developing nations. Subsistence farmers need improved crops that respond to labor-intensive agriculture (human and animal labor) and do not require large outlays of energy and capital. ■

Increasing Crop Yields in the Post-Green Revolution Era The International Food Policy Research Institute has projected that the world demand for rice, wheat, and corn will increase 40% between 2000 and 2020. This increase in grain production will be needed to feed the increasing human population and to satisfy the appetites of increasing numbers of affluent people who can afford to buy meat.

This challenge cannot be met by increasing the amount of land under cultivation, as the best arable lands are already being cultivated. Projected freshwater shortages, increasing costs of agricultural chemicals, and deteriorating soil quality caused by intensive agricultural techniques may further constrain productivity. As Figure 19.7 demonstrates, recent progress in coaxing more grain out of crops that were genetically improved during the green revolution has resulted in diminishing returns. Grain yields have continued to rise since the 1960s, but in recent years, the rates of increase have not been as great as in earlier years. Many scientists think that genetic engineering, discussed shortly, will assist in the endeavor to breed more productive varieties. In addition to developing improved crop varieties, modern agricultural methods, such as water-efficient irrigation, will have to be introduced to developing countries that do not currently have them if we are going to continue increasing crop yields.

Efforts are under way to increase food security in low-income, food-deficient countries. During the 1990s the FAO initiated a special program for farmers in 19 nations, most in Africa. Participating farmers are given genetically improved seeds, commercial inorganic fertilizers, and pesticides and are trained in improved agricultural techniques. These farmers then provide demonstrations to neighboring farmers on how to increase and diversify food production, reduce water use, control pests, and protect the soil and other natural resources. In Ethiopia this program doubled grain yields at demonstration sites and resulted in substantial profits after the costs of the seeds and agricultural chemicals were subtracted.

Increasing Livestock Yields

The use of hormones and antibiotics, though controversial, increases animal production. **Hormones,** usually administered by ear implants, regulate livestock bodily functions and promote faster growth. Although U.S. and Canadian farmers use hormones, the European Union (EU) currently bans all imports of hormone-treated beef because of health concerns for human consumers. They cite a few studies that suggest that these hormones or their breakdown products, both found in trace amounts in meat and meat products, could cause cancer or affect the growth of young children. In 1999 an international scientific committee organized by the FAO and WHO examined the hormone issue in detail. They concluded that the traces of hormones found in beef are safe because they are so low compared to normal hormone concentrations in the human body. However, they did not consider the question of whether excess hormones excreted in cattle feces could end up in other foods or drinking water. Critics of the EU ban contend that it is a move to reduce competition and protect the EU beef industry. These critics note that many foods, such as ice cream, peas, butter, wheat germ, and soybean oil, have natural hormone levels higher than hormone-implanted beef.

Modern agriculture has embraced the routine addition of low doses of **antibiotics** to the feed for pigs, chickens, and cattle. These animals typically gain 4 to 5% more weight than animals that do not receive antibiotics, presumably because they must expend less energy to fight infections. According to the *New England Journal of Medicine*, 40% of the 25,000 tons of antibiotics produced annually in the United States is used in livestock operations, particularly those in which large numbers of animals are confined in small areas. Most of these antibiotics are administered continuously to healthy animals.

Several studies link the indiscriminate use of antibiotics in humans and livestock to the increasing resistance of bacteria to antibiotics. The development of bacterial resistance to antibiotics is an example of **evolution.** Bacteria are continually evolving, even inside the bodies of human and animal hosts. When an antibiotic is used to treat a bacterial infection, a few bacteria may survive because they are genetically resistant to the antibiotic, and they pass these genes to future generations. As a result, the bacterial population contains a larger percentage of antibiotic-resistant bacteria than before (**Figure 19.8**). The Worldwatch Institute reports that antibiotic resistance has been documented in more than 20 kinds of potentially harmful bacteria (the tuberculosis bacterium is one example), and some bacterial strains are resistant to every antibiotic known, a total of more than 100 drugs.

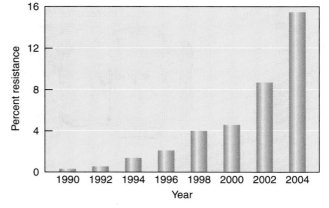

Figure 19.8   Evolution of antibiotic resistance. Shown is the increasing resistance of *E.coli* in blood and cerebrospinal infections to the antibiotic ciprofloxacin. (Adapted from D. Livermore)

Because there is increasing evidence that the use of antibiotics in agriculture reduces their medical effectiveness for humans, in 2003 WHO recommended that routine use of antibiotics in livestock be eliminated. Many European countries have stopped administering low doses of antibiotics for growth promotion in livestock, but the United States and many other countries continue the practice.

Genetic Engineering

genetic engineering: The manipulation of genes, for example, by taking a specific gene from a cell of one species and placing it into a cell of an unrelated species, where it is expressed.

Genetic engineering is a controversial technology that has begun to revolutionize medicine and has the potential to improve agriculture as well. The agricultural goals of genetic engineering are not new. Using traditional breeding methods, farmers and scientists have developed desirable characteristics in crop plants and agricultural animals for centuries. It takes time to develop such genetically improved organisms. Through use of traditional breeding methods, it might take 15 years or more to incorporate genes for disease resistance into a particular crop plant. Genetic engineering has the potential to accomplish the same goal in a fraction of that time.

Genetic engineering differs from traditional breeding methods in that desirable genes from any organism can be used, not just those from the species of the plant or animal being improved. If a gene for disease resistance found in soybeans would be beneficial in tomatoes, the genetic engineer can splice the soybean gene into the tomato plant (**Figure 19.9**). Traditional breeding methods could not do this because soybeans and tomatoes belong to separate groups of plants and do not interbreed.

Genetic engineering may produce food plants that will be more nutritious because they will contain all the essential amino acids. (Currently, no single food crop has this

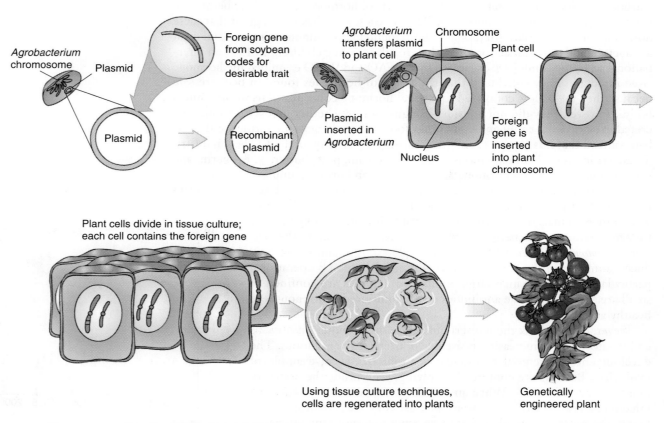

Figure 19.9 Genetic engineering. This example of genetic engineering uses a plasmid, a small circular molecule of DNA (genetic material) found in many bacteria. The plasmid of the bacterium *Agrobacterium* introduces desirable genes from another organism into a plant. After the foreign DNA is spliced into the plasmid, the plasmid is inserted into *Agrobacterium*, which then infects plant cells in culture. The foreign gene is inserted in the plant's chromosome, and genetically modified plants are then produced from the cultured plant cells.

trait.) Genetic engineering may produce crops rich in beta-carotene, which the body uses to make vitamin A. According to WHO, 250 million of the world's children are at risk of vitamin A deficiency. A deficiency in vitamin A causes poor vision, protein deficiency (vitamin A helps the body absorb and use amino acids), and an impaired immune system. In 2000 an international team of scientists reported that they had successfully engineered rice grains to produce beta-carotene. The so-called golden rice has great potential for improved world health because about half the world's people eat rice as their staple food, and rice is a poor source of many vitamins, including vitamin A. (Golden rice is still not available commercially.)

Crop plants resistant to insect pests, viral diseases, drought, heat, cold, herbicides, or salty or acidic soils are being developed. Scientists have inserted a gene from a virus into yellow squash and zucchini, thereby making them resistant to viral diseases that yellow their leaves and reduce crop yields. As another example, researchers at the U.S. Department of Agriculture (USDA) identified a gene in rye that codes for a protein that prevents plant roots from absorbing aluminum, a metal more soluble in acidic soils. (Aluminum is a natural component of the inorganic minerals of soils, but it is normally insoluble and is not absorbed by plant roots.) Crops grown in acidic soils often contain toxic levels of aluminum. Acidic soils are widespread in the tropics; in Latin America, 51% of all soils are acidified. Incorporation of an anti-aluminum gene into crop plants such as wheat would allow them to grow in such soils.

Genetic engineering has been used to develop more productive farm animals, including rapidly growing hogs and fishes. Perhaps the greatest potential contribution of animal genetic engineering is the production of vaccines against disease organisms that harm agricultural animals. For example, genetically engineered vaccines have been developed to protect cattle against the deadly viral disease rinderpest, which has reached epidemic proportions in parts of Asia and Africa.

Several hundred private genetic engineering firms, as well as thousands of scientists in colleges, universities, and government research labs around the world, are involved in agricultural genetic engineering. Although a great deal of research must be done before most of the envisioned benefits from genetic engineering are fully realized, genetic engineering has already transformed agriculture.

The first **genetically modified (GM)** crops were approved for commercial planting in the United States in the early 1990s. The United States is the world's top producer of GM crops, with 80% of its soybean acres, 70% of its cotton acres, and 38% of its corn acres planted with GM crops in 2003. The U.S. Food and Drug Administration (FDA) regulates most of these crops. The FDA applies the same policies and rules on health risk assessment to genetically modified crops as it does to new crops developed by traditional breeding methods.

The Safety of Genetic Engineering   A growing body of evidence has determined that current GM crop plants are as safe for human consumption as crops grown by conventional or organic agriculture. According to the FAO, GM crops widely planted in North America and elsewhere appear to have posed little threat to the environment.

However, more research on the environmental impact of GM crops is required. To that end, strict guidelines exist in areas of genetic engineering research in which there are unanswered questions about possible effects on the environment. Much research is currently being conducted to assess the effects of introducing GM crops whose seeds or pollen might spread in an uncontrolled manner. It is important to assess the biology of each genetically modified organism to determine if it has characteristics that might cause an environmental hazard under certain conditions.

The **Cartagena Protocol on Biosafety,** an outgrowth of the 1992 U.N. Convention on Biological Diversity, lessens the threat of gene transfer from GM organisms to their wild relatives by providing appropriate procedures in the handling and use of GM organisms. The protocol went into effect in 2003. (Although the United States has neither signed nor ratified the protocol, it has participated as an active observer. Other major GM crop exporters, such as Canada, Australia, and Argentina, also have not ratified the protocol.)

The Backlash Against Genetically Modified Foods   During the late 1990s and early 2000s, opposition to genetically engineered crops increased in many countries in Europe and Africa. In 1999 the EU placed a five-year moratorium on virtually all approvals of GM crops (the moratorium is now lifted). The EU refused to buy U.S. corn because it might be genetically modified. Some opposition may have been the result of economic considerations, such as protecting the market for homegrown foods by banning imports, and some opposition was based on legitimate scientific concerns. One concern is that the inserted genes could spread in an uncontrolled manner from GM crops to weeds or wild relatives of crop plants and possibly harm natural ecosystems in the process. Scientists recognize this concern as legitimate and must take special precautions to avoid this possibility. Critics also worry that some consumers might develop food allergies to GM foods, although scientists routinely screen new GM crops for allergenicity.

Should Foods from Genetically Modified Crops and Livestock Be Labeled?   The question of whether or not to label foods containing GM organisms is controversial. Some consumers want such labels because they view genetic engineering as "unnatural." The position of the FDA and most scientists is that such labeling would be counterproductive, in the sense that it would increase public anxiety over a technology that is essentially the same as conventional breeding methods. Moreover, labeling would be expensive because the GM food would have to be kept separate from other foods during planting, harvesting, processing, and distribution; thus, production costs would be higher.

The scientific consensus is that the risks associated with consuming food derived from GM varieties are the same as those associated with consuming food derived from new varieties produced by traditional genetic techniques. Dozens of new plant varieties produced by traditional breeding methods enter the marketplace every year, from corn and rice to pumpkins and tomatoes; all are safe, and none are labeled. In 1996 the U.S. Court of Appeals upheld the FDA view that labeling should not be required just because some consumers want it.

REVIEW

1. Why does decreased genetic diversity in farm plants and animals increase the likelihood of economic disaster from disease?

2. What is the green revolution? What are some benefits and problems associated with the green revolution?

3. What is one pro and one con of genetically engineering crops?

The Environmental Impacts of Agriculture

LEARNING OBJECTIVE

- Describe the environmental impacts of industrialized agriculture, including land degradation and habitat fragmentation.

The practices of industrialized agriculture have resulted in several environmental problems that impair the ability of nonagricultural terrestrial and aquatic ecosystems to provide essential **ecosystem services.** This raises questions about the sustainability of intensive agriculture.

The agricultural use of fossil fuels and pesticides produces air pollution. Untreated animal wastes and agricultural chemicals such as fertilizers and pesticides cause water pollution that reduces biological diversity, harms fisheries, and leads to outbreaks of nuisance species (see EnviroNews on harmful algal blooms in Chapter 22). According to the Environmental Protection Agency, agricultural practices are the single largest cause of surface-water pollution in the United States. Water pollution from agriculture is particularly significant in midwestern states such as Iowa, Wisconsin, and Illinois. Some of these contaminants flow into the Mississippi River and, from there, into the Gulf of Mexico (see Chapter 22). Some agricultural chemicals have been detected in water deep underground, as well as in surface waters. Nitrates from animal wastes and commercial inorganic fertilizers are probably the most widespread groundwater contaminant in agricultural areas.

Industrialized agriculture has favored the replacement of traditional family farms by large agribusiness conglomerates. In the United States, most cattle, hogs, and poultry are now grown in feedlots and livestock factories (**Figure 19.10**). In livestock factories, thousands of animals are confined to small pens in buildings the size of football fields. Such large concentrations of animals create many environmental problems, including air and water pollution. The quantity of manure produced by several hundred thousand pigs in one livestock factory causes a severe waste disposal problem. At hog factories, the manure is often stored in deep lagoons that have the potential to pollute the soil, surface water, and groundwater. This happened during Hurricane Fran in 1996, when manure from 22 large animal waste lagoons in North Carolina spilled onto the floodplain and into streams, causing major fish kills. People living near livestock factories dislike the odor, which often exceeds federal and state guidelines for emissions and causes their property values to decline.

Many insects, weeds, and disease-causing organisms have developed or are developing resistance to pesticides. Pesticide resistance forces farmers to apply progressively larger quantities of pesticides (**Figure 19.11**). Residues of pesticides contaminate our food supply and reduce the number and diversity of beneficial microorganisms in the soil. Fishes and other aquatic organisms are sometimes killed by pesticide runoff into lakes, rivers, and estuaries.

Land degradation is a reduction in the potential productivity of land. Soil erosion, which is exacerbated by large-scale mechanized operations, causes a decline in soil fertility, and the sediments lost by erosion damage water quality. The USDA estimates that about one-fifth of U.S. cropland is vulnerable to soil erosion damage, and soil erosion is an even greater problem in some developing nations. Examples of other types of degradation are compaction of soil by heavy farm machinery and waterlogging and salinization of soil from improper irrigation methods.

Crop production requires enormous amounts of water. According to the Worldwatch Institute, it takes 1000 tons of water to produce 1 ton of grain. Worldwide, irrigation consumes almost 70% of the total fresh water that humans withdraw from aquifers and surface waters. Some agricultural regions remove water from aquifers faster than it is recharged by precipitation, lowering water tables. The huge Ogallala Aquifer under Nebraska, Kansas, Texas, and other states is one of the best known examples of overdrawing an aquifer for agriculture (see Figure 14.15). Most of the water in the Ogallala is ancient, left by melting glaciers at the end of the last Ice Age. Thus, the Ogallala Aquifer is largely a nonrenewable resource. Lack of water increasingly affects agricultural yields in parts of Africa, Australia, China, India, and the former Soviet Union. As a result of mismanagement of water resources, vast areas of irrigated land have become too waterlogged or too salty to grow crops.

Clearing grasslands and forests and draining wetlands to grow crops have resulted in **habitat fragmentation** that reduces biological diversity (see Chapter 17). Many species have become endangered or threatened as a result of habitat loss caused by agriculture. The most dramatic example of habitat loss in North America is tallgrass prairie, more than 90% of which was converted to agriculture.

Figure 19.10   Hog factory in Iowa. The hogs remain indoors and are fed and watered by machine at timed intervals throughout the day. Although livestock factories are efficient and produce meat relatively inexpensively, they cause environmental problems such as sewage disposal. Some livestock factories produce as much sewage as a small city, yet the wastes are currently not covered by water quality laws and do not go to a sewage treatment plant. (© Macduff Everton/Corbis Images)

land degradation: The natural or human-induced process that decreases the future ability of the land to support crops or livestock.

habitat fragmentation: The breakup of large areas of habitat into small, isolated patches.

Figure 19.11   Colorado potato beetles on potato leaves. As a result of being exposed to heavy applications of pesticides over the years, Colorado potato beetles are resistant to most insecticides registered for use on potatoes. Photographed in Virginia. (Grant Heilman/Grant Heilman Photography)

Cultivating Marginal Lands

The United States had agricultural surpluses during the 1980s in part because farmers brought large amounts of previously unused land into production. Unfortunately, much of this land was marginal as agricultural land, because it was prone to soil erosion caused by intermittent floods or frequent droughts (and therefore wind erosion when the ground cover was removed). Harvesting crops from highly erodible land is ecologically unsound and cannot be done indefinitely. Some of the marginal farmlands in the United States are now retired from use (see discussion of the Conservation Reserve Program in Chapter 15).

Other countries have paid a high price for cultivating land highly prone to erosion. The former Soviet Union began to cultivate large areas of marginal land during the 1950s. Although the initial production of cereal crops was high, much of this land had to be abandoned by the 1980s. The annual per capita food production in Haiti, a poor Caribbean nation, is half what it was in 1950 as a result of both population growth and lower production on eroded soils.

From the 1960s to the 21st century, the amount of dry land being irrigated for agriculture was greatly increased. About 70% of the world's total irrigated land is found in Asia, and the amount of irrigated land there continues to expand each year. Since 1995, however, the amount of irrigated land in the rest of the world has remained steady. In Europe and Oceania, the amount of irrigated land has actually decreased. This change is due to the increasing cost of irrigation, the depletion of aquifers, the abandonment of salty soil, and the diversion of irrigation water to residential and industrial uses.

REVIEW

1. What are the major environmental problems associated with industrialized agriculture?
2. How does agriculture contribute to habitat fragmentation?

Solutions to Agricultural Problems

LEARNING OBJECTIVE

- Define *sustainable agriculture* and contrast it with industrialized agriculture.

Food production poses an environmental quandary. The green revolution and industrialized agriculture have unquestionably met the food requirements of most of the human population even as that population has more than doubled since 1960. But we have had to pay for food gains with serious environmental problems, and we do not know if industrialized agriculture is sustainable for more than a few decades. To compound the issue, we must continue to increase food production to feed the growing human population, but the resulting damage to the environment may lessen our chances of increasing food production in the future.

Fortunately, the dilemma is not as hopeless as it seems. Farming practices and techniques exist that ensure a sustainable output at yields comparable to industrialized agriculture. Farmers who practice industrialized agriculture can adopt these alternative agricultural methods, which cost less and are less damaging to the environment. Advances are also being made in sustainable subsistence agriculture.

Sustainable Agriculture

sustainable agriculture: Agricultural methods that maintain soil productivity and a healthy ecological balance while having minimal long-term impacts.

In **sustainable agriculture**, also called **alternative** or **low-input agriculture,** certain modern agricultural techniques are carefully combined with traditional farming methods from agriculture's past. Sustainable agriculture is modeled after natural ecosystems, with their high biological diversity, biodegradation of materials, and maintenance of soil fertility. To this end, sustainable agriculture relies on beneficial biological processes and environmentally friendly chemicals that disintegrate quickly and do not persist as residues in the environment. The sustainable farm consists of

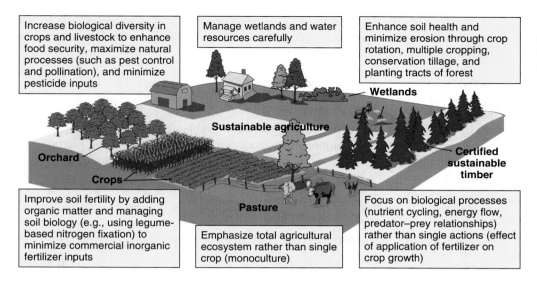

Figure 19.12 **Some goals of sustainable agriculture.** Natural ecosystems provide a model for sustainable agriculture.

Increase biological diversity in crops and livestock to enhance food security, maximize natural processes (such as pest control and pollination), and minimize pesticide inputs

Manage wetlands and water resources carefully

Enhance soil health and minimize erosion through crop rotation, multiple cropping, conservation tillage, and planting tracts of forest

Wetlands

Sustainable agriculture

Orchard

Certified sustainable timber

Crops

Improve soil fertility by adding organic matter and managing soil biology (e.g., using legume-based nitrogen fixation) to minimize commercial inorganic fertilizer inputs

Pasture

Emphasize total agricultural ecosystem rather than single crop (monoculture)

Focus on biological processes (nutrient cycling, energy flow, predator–prey relationships) rather than single actions (effect of application of fertilizer on crop growth)

field crops, trees that bear fruits and nuts, small herds of livestock, and even tracts of forest (**Figure 19.12**). Such diversification protects the farmer against unexpected changes in the marketplace. The breeding of disease-resistant crop plants and the maintenance of animal health rather than the continual use of antibiotics to prevent disease are important parts of sustainable agriculture. Water and energy conservation are practiced in sustainable agriculture.

Instead of using large quantities of chemical pesticides, sustainable agriculture controls pests by enhancing natural predator-prey relationships. For example, apple growers in Maryland monitor and encourage the presence of ladybird beetles in their orchards because these insects feed voraciously on European red mites, a major pest of apples. As a general rule, sustainable agriculture tries to maintain biological diversity on farms as a way to minimize pest problems. Providing hedgerows (rows of shrubs) between fields provides a habitat for birds and other insect predators.

Crop selection helps control pests without heavy pesticide use. In parts of Oregon, apples are grown without major pest problems, but insects often infest peaches, whereas in western Colorado, apples have major pest problems but peaches do well. Therefore, apples would be the preferred crop for sustainable agriculture in Oregon, as would peaches in Colorado.

An important goal of sustainable agriculture is to preserve the quality of agricultural soil. Crop rotation, conservation tillage, and contour plowing help control erosion and maintain soil fertility (see Figures 15.15 and 15.16a). Sloping hills converted to mixed-grass pastures erode less than do hills planted with field crops, thereby conserving the soil and supporting livestock.

A combination of manure and crop rotations with legumes is environmentally superior to intensive agricultural methods that use commercial inorganic fertilizers to supply nitrogen. Animal manure added to soil decreases the need for high levels of commercial inorganic fertilizers and cuts costs. Using biological nitrogen fixation (by planting legumes) to convert atmospheric nitrogen into a form that plants can use lessens the need for nitrogen fertilizers.

As you see from this discussion, sustainable agriculture is not a single program but a series of programs adapted for specific soils, climates, and farming requirements. Some sustainable farmers—those who use a system of **integrated pest management (IPM)**—incorporate a limited use of pesticides with such practices as crop rotation, continual monitoring for potential pest problems, use of disease-resistant varieties, and biological pest controls (see Chapter 23).

Other sustainable farmers practice **organic agriculture** and use no commercial inorganic fertilizers or pesticides. According to guidelines established by the **Organic Food Production Act** in 1990, *organic foods* are crops grown in soil free of commercial inorganic fertilizers and pesticides for at least three years. If the land that the crops are grown on has been inspected, private or state agencies label it *certified organic*, which specifies the highest standards. Cattle and other livestock labeled *USDA organic* are not treated with antibiotics or hormones and are fed organic feed grown without commercial inorganic fertilizers or pesticides. Animals certified *human raised and handled* get fresh air, water, exercise, and are not raised in cramped pens. Federal standards for certification of organically grown food went into effect in 2002, replacing standards that varied from state to state.

In growing recognition of the environmental problems associated with industrialized agriculture, more and more mainstream farmers are trying some methods of sustainable agriculture. These methods cause fewer environmental problems to the agricultural ecosystem, or **agroecosystem**, than industrialized agriculture. The trend from intensive techniques that produce high yields to methods that focus on long-term sustainability of the soil has sometimes been referred to as the **second green revolution.**

Making Subsistence Agriculture Sustainable and More Productive

Because there are many kinds of agriculture around the world, there is no single solution to increase yields while minimizing environmental damage. Traditional slash-and-burn agriculture is sustainable as long as there are few farmers and large areas of rain forest. Because relatively small patches of forest are cleared for raising crops, the trees quickly return when the land is abandoned and the farmer has moved on to clear another plot of forest. If the abandoned land lies fallow for a period of 20 to 100 years, the soil recovers to the point where subsistence farmers can again clear the forest for planting. Burning the trees releases nutrients into the soil so that crops can again be grown. However, too many people practice slash-and-burn agriculture, and as a result, more and more tropical forests are being destroyed (see Chapter 18). Because so many people are trying to grow crops on rainforest land, the soil does not lie uncultivated between farming cycles long enough for it to recover.

Some researchers are studying ways to make former rainforest land retain its productivity for longer periods than is usual in shifting cultivation. Consider Papua New Guinea, a small island nation in which approximately 80% of the people are subsistence farmers. Research scientists in this country have developed methods to deal with some of the most troublesome problems associated with shifting cultivation: soil erosion, declining fertility, and attacks by insects and diseases. Their research has helped forest plots remain productive for longer periods. Heavy mulching with organic material, such as weed and grass clippings, has lessened soil infertility and erosion. The composted mulch is then piled into rows that follow the contours of the land, further reducing erosion. Several crops are planted together, reducing insect damage. One of the crops is always a legume (such as beans), which helps restore nitrogen fertility to the soil. An extension program demonstrates these methods to rural farmers.

REVIEW

1. What is sustainable agriculture?
2. What are some features of a sustainable farm?

Fisheries of the World

LEARNING OBJECTIVES

- Define *bycatch*.
- Contrast fishing and aquaculture and relate the environmental challenges of each activity.

The ocean contains valuable food resources. About 90% of the world's total marine catch is fishes, with clams, oysters, squid, octopus, and other mollusks representing an additional 6% of the total catch. Crustaceans, including lobsters, shrimp, and crabs, make up about 3%, and marine algae constitute the remaining 1%.

Fishes and other seafood are highly nutritious because they contain easily digestible, high-quality protein (protein with a good balance of essential amino acids). Humans obtain approximately 5% of the total protein in their diet from fishes and other seafood; the rest is obtained from milk, eggs, meat, and plants. In certain countries, particularly in developing nations that border the ocean, seafood makes a much larger contribution to the total protein in the human diet.

Fleets of fishing vessels obtain most of the world's marine catch. In addition, numerous fishes are captured in shallow coastal waters and inland waters. According to the FAO, the world annual fish harvest increased substantially from 1950 (19.3 million tons) to 2003, when the world fish catch was 132.5 million tons.

Problems and Challenges for the Fishing Industry

No nation lays legal claim to the open ocean. Consequently, resources in the ocean are more susceptible to overuse and degradation than are resources on the land, which individual nations own and for which they feel responsible (see section on the Tragedy of the Commons in Chapter 1).

The most serious problem for marine fisheries is that many marine species have been overharvested to the point that their numbers are severely depleted (**Figure 19.13**). Large predatory fish such as tuna, marlin, and swordfish have declined by 90% since the 1950s, according to Canadian researchers who analyzed data from ocean and coastal regions around the world. Each fish species has a maximum sustainable harvest level; if a particular species is overharvested, its numbers drop, and harvest is no longer economically feasible.

Figure 19.13 Declining cod fisheries. Declining fisheries, which have been observed worldwide, are a global environmental problem.

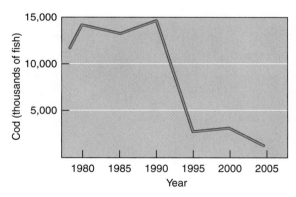

(a) Collapse of the cod fishery on Georges Bank. In late 1994, the U.S. Commerce Department closed two large portions of the Georges Bank off the coast of Massachusetts. Since then, haddock and some other commercially important fish populations have begun to recover but cod has not. (O'Brien et al.)

(b) Cod fishing off the coast of Lofoten, Norway. (Terje Rakke/The Image Bank/Getty Images)

According to the FAO, 62% of the world's fish stocks are in urgent need of management action. Fisheries have experienced such pressure for two reasons: One, the growing human population requires protein in its diets, leading to a greater demand; and two, technological advances in fishing gear have made it possible to fish so efficiently that every single fish is often removed from an area.

Sophisticated fishing equipment includes sonar, radar, computers, airplanes, and even satellites to locate fish schools (**Figure 19.14**). Some boats set out **longlines,** fishing lines with thousands of baited hooks; each longline is up to 128 km (80 mi) long. **Purse-seine nets** are huge nets, as long as 2000 m (1.25 mi), set out by small powerboats to encircle large schools of tuna and other fishes; after the fishes are completely surrounded, the bottom of the net is closed to trap them. A **trawl net** is a weighted, funnel-shaped net pulled along the bottom of the ocean to catch bottom-feeding fishes and shrimp; as much as 27 metric tons (30 tons) of fishes, shrimp, and other seafood are caught in a single net. Trawl nets, some of which are large enough to hold 12 Boeing 747s, destroy the ocean floor habitat. **Drift nets** are plastic nets up to 64 km (40 mi) long that entangle thousands of fishes and other marine organisms. Although most countries have banned drift nets, they are still used illegally.

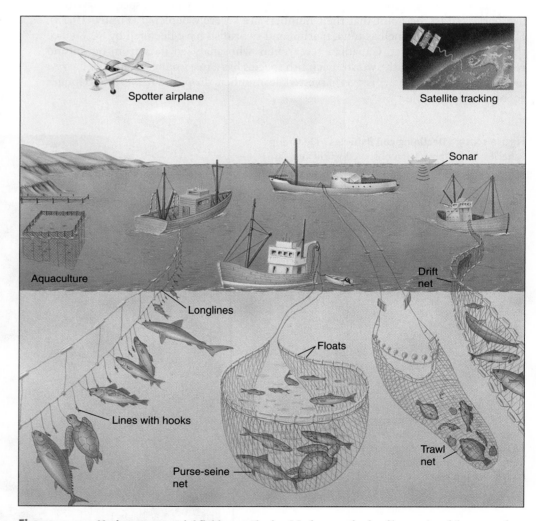

Figure 19.14   **Modern commercial fishing methods.** Modern methods of harvesting fish are so effective that many fish species have become rare. Sea turtles, dolphins, seals, whales, and other aquatic organisms are accidentally caught and killed in addition to the target fish. The depth of longlines is adjusted to catch open-water fishes, such as sharks and tuna, or bottom fishes, such as cod and halibut. Purse seines catch anchovies, herring, mackerel, tuna, and other fishes that swim near the water's surface. Trawls catch cod, flounder, red snapper, scallops, shrimp, and other fishes and shellfish that live on or near the ocean floor. Drift nets catch salmon, tuna, and other fishes that swim in open waters.

Fishermen tend to concentrate on a few fish species with high commercial value, such as menhaden, salmon, tuna, and flounder, while other species, collectively called **bycatch**, are unintentionally caught and then discarded. The FAO reports that about 25% of all marine organisms caught—some 27 million metric tons (30 million tons)—are dumped back into the ocean. Most of these unwanted animals are dead or soon die because they are crushed by the fishing gear or are out of the water too long. The United States and other countries are trying to significantly reduce the amount of bycatch or develop uses for it.

In response to overharvesting, many nations have extended their limits of jurisdiction to 320 km (200 mi) offshore. This action removed most fisheries from international use, because more than 90% of the world's fisheries are harvested in relatively shallow waters close to land. This policy was supposed to prevent overharvesting by allowing nations to regulate the amounts of fishes and other seafood harvested from their waters. However, many countries have a policy of **open management,** in which all fishing boats of that country are given unrestricted access to fishes in national waters.

The **Magnuson Fishery Conservation Act,** which went into effect in 1977, regulates marine fisheries in the United States. This law established eight regional fishery councils, each of which developed a management plan for its region. Until 1996, the act was not particularly successful because managers were often pressured to set quotas too high, and the National Marine Fisheries Service estimated that more than one-third of U.S. fish stocks were being fished at higher levels than could be sustained. In 1996 the act was reauthorized as the **Magnuson-Stevens Fishery Conservation and Management Act.** It required the regional councils and the National Marine Fisheries Service to protect "essential fish habitat" for more than 600 fish species, reduce overfishing, rebuild the populations of overfished species, and minimize bycatch. Fishing quotas, restrictions of certain types of fishing gear, limits on the number of fishing boats, and closure of fisheries during spawning periods are some of the management tools used to reduce overfishing. The Magnuson-Stevens Act, which was reauthorized in 2007, strengthened controls on illegal and unreported fishing in U.S. waters.

bycatch: The fishes, marine mammals, sea turtles, seabirds, and other animals caught unintentionally in a commercial fishing catch.

Ocean Pollution and Deteriorating Habitat

One of the great paradoxes of human civilization is that the same ocean used to provide food to a hungry world is used as a dumping ground. Pollution increasingly threatens the world's fisheries. Everything from accidental oil spills to the deliberate dumping of litter pollutes the water. Heavy metals such as lead, mercury, and cadmium are widely used in industry but enter the environment and find their way into aquatic food webs, where they are highly toxic to both fishes and humans who eat fish.

Between 60 and 80% of all commercially important ocean fishes spend at least part of their lives in coastal areas. Tidal marshes, mangrove swamps, estuaries, and the like serve as spawning areas, nurseries, and feeding grounds. Coastal areas are in high demand for recreational and residential development, and many coastal waters are polluted. The World Resources Institute estimates that about 80% of global ocean pollution comes from human activities on land. Stormwater runoff from cities, agricultural areas, and roads is the single largest source of ocean pollution. As development and pollution degrade coastal ecosystems, the habitats of young marine animals are undermined, contributing to further depletion of fish stocks already suffering from overfishing. (Chapter 6 discusses the impact of human activities on the ocean, and Chapter 18 discusses coastlines.)

Aquaculture: Fish Farming

Aquaculture is more closely related to agriculture on land than it is to the fishing industry just described. Aquaculture is carried out in both fresh water and marine water; the cultivation of marine organisms is sometimes called **mariculture.** To optimize the quality and productivity of their "crops," aquaculture farmers control the diets, breeding cycles, and environmental conditions of their ponds or enclosures.

aquaculture: The growing of aquatic organisms (fishes, shellfish, and seaweeds) for human consumption.

Aquaculturists try to reduce pollutants that might harm the organisms they are growing, and they keep them safe from potential predators.

Although aquaculture is an ancient practice that probably originated in China several thousand years ago, its enormous potential to provide food has been appreciated only recently. Aquaculture contributes variety to the diets of people in highly developed countries. Inhabitants of developing nations benefit even more from aquaculture: It may provide them with much-needed protein and even serve as a source of foreign exchange when they export such delicacies as aquaculture-grown shrimp.

According to the FAO, world aquaculture production of finfish and shellfish reached 454.8 million tons in 2003. Other important aquaculture crops include seaweeds, oysters, mussels, clams, lobsters, and crabs. Currently, aquaculture is the fastest growing type of food production in the world, and one out of every three fish destined for human consumption comes from fish farms. The nation with the largest aquaculture harvest is China, which accounts for about 68% of total world production.

Aquaculture is a $900 million-a-year industry in the United States, where it accounts for 6% of all fish and seafood consumed. All striped bass and rainbow trout available at U.S. retail markets are produced by aquaculture, as are more than half the fresh salmon served in the United States. Catfish, tilapia, salmon, shrimp, and oysters are the most important types of seafood grown by aquaculture in the United States.

Aquaculture differs from fishing in several respects. For one thing, although highly developed nations harvest more fishes from the ocean, developing nations produce much more seafood by aquaculture. One reason for this is that developing nations have an abundant supply of cheap labor, a requirement of aquaculture because it, like land-based agriculture, is labor-intensive. Another difference between fishing and aquaculture is that the limit on the size of a catch in fishing is the size of the natural population, whereas the limit on aquacultural production is largely the size of the area in which they are grown.

In addition to being done inland, aquaculture is practiced in estuaries and in the ocean, both near the shore and offshore. Other uses of coastlines compete with aquaculture for available space. Developing countries that grow shrimp by aquaculture cut down the coastal mangroves that provide so many important environmental benefits (**Figure 19.15**; also see Chapter 6). Many marine fishes breed in the tangled roots of mangroves, and there are concerns that an expansion of shrimp farming could contribute to a decline in marine fish populations.

Offshore aquaculture facilities, sometimes called "ocean ranches," are becoming increasingly common. Ocean ranches, which increasingly use cutting edge technologies such as submersible cages with robotic surveillance, may avoid damaging coastlines but often lack any pollution-restricting legal oversight that other aquaculture facilities are subject to.

Figure 19.15 Aquaculture in a coastal mangrove forest. The dykes enclose shrimp ponds. Aquaculture of shrimp is the single largest factor responsible for mangrove habitat losses worldwide. Photographed in Borneo. (Frans Lanting/Minden Pictures, Inc.)

Because so many fishes are concentrated in a relatively small area, aquaculture produces wastes that pollute the adjacent water and harm other organisms. Aquaculture causes a net loss of wild fish because many of the fishes farmed are carnivorous. Sea bass and salmon eat up to 5 kg of wild fish to gain 1 kg of weight.

Although interest in aquaculture is increasing worldwide, several factors are slowing its expansion. Setting up and running an aquaculture facility is expensive. Also, research is needed to make aquaculture of certain organisms profitable. An organism's requirements for breeding must be established, and ways to control excessive breeding must be available so that the population does not overbreed and produce many stunted individuals rather than fewer large ones. The population must be continually monitored for diseases, which have a tendency to spread rapidly in the crowded conditions characteristic of aquaculture.

YOU CAN MAKE A DIFFERENCE

Vegetarian Diets

A vegetarian is a person who does not eat the flesh of any animal, including that of fish and poultry. People embrace vegetarian diets for many reasons. Balanced vegetarian diets provide good nutrition without high levels of saturated fats or cholesterol, both of which cause health problems such as heart disease and obesity. Some studies in the United States indicate that people who are vegetarians live longer, healthier lives than nonvegetarians.

Some people become vegetarians because they are morally or philosophically opposed to killing animals, even for food. Certain religious groups, notably Hindus and Seventh Day Adventists, exclude animal products from their diets. Other people convert to vegetarianism out of their sense of responsibility for land use and its wide repercussions. In general, fewer plants are required to support vegetarians than to support meat eaters.*

The amount of usable energy in the food chain is decreased approximately 90% by adding an additional level—that is, the animals we eat—to the chain (see discussion of pyramids of energy in Chapter 3). The actual percentage varies because not all animals are alike in their efficiency at turning food into meat. Simply stated, if everyone were to become a vegetarian, much more food would be available for human consumption. (You do not have to be a strict vegetarian to have a positive environmental effect. Try adopting a meatless day once or twice a week.)

Vegetarians are divided into four groups—lacto-ovo vegetarians, lacto vegetarians, ovo vegetarians, and vegans—based on whether they eat milk and/or egg products. *Lacto-ovo vegetarians* eat milk, eggs, and foods made from milk and eggs. *Lacto vegetarians* do not eat eggs, but they eat milk and milk products such as cheese, yogurt, and butter. *Ovo vegetarians* do not eat dairy, but eggs are allowed. *Vegans* exclude milk and eggs in all forms from their diets.

Some people are reluctant to switch to a vegetarian diet because they fear they will not get enough protein. (Plant foods generally have a lower percentage of protein than do animal foods.) Meat eaters in highly developed countries usually consume much more protein than they need, and the problem with a vegetarian diet is usually not a lack of protein but obtaining the proper balance of essential amino acids. It is relatively easy for lacto-ovo vegetarians, lacto vegetarians, and ovo vegetarians to plan a healthy diet because milk and eggs contain all the essential amino acids. Milk is also rich in calcium, which is important for strong bones and teeth. Both milk and eggs contain vitamin B_{12}, which helps form red blood cells. Vegans, however, must plan their diets carefully to obtain a proper balance of amino acids, as well as enough calcium and vitamin B_{12}. Sesame seeds, broccoli, spinach, and other leafy green vegetables are rich in calcium, so these foods must be included in the vegetarian diet every day. Because vitamin B_{12} is found almost exclusively in animal products, most vegans take vitamin B_{12} supplements.

A nutritious vegetarian diet includes a combination of foods that contains all the essential amino acids. A meal of rice and beans or corn and beans provides the proper complement of essential amino acids. Cookbooks and other references contain menus and recipes that give the vegetarian diet adequate amounts of high-quality protein. The following list provides an overview of combinations of foods that offer a proper balance of essential amino acids. At least one food from each group should be consumed at every meal.

Group I

Grains
Barley, corn, oats, rice, rye, wheat

Nuts and seeds
Almonds, beechnuts, Brazil nuts, cashews, filberts, pecans, pumpkin seeds, sunflower seeds, walnuts

Group II

Legumes
Peas, black-eyed peas, chick peas, black beans, fava beans, kidney beans, lima beans, pinto beans, mung beans, navy beans, soybeans (usually eaten as bean curd, or tofu)

Dairy products
Cheese, cottage cheese, eggs, milk, yogurt

*For people living in marginal lands, such as semi-arid grasslands, the soil will not support extensive crop production but will support livestock that consume native plants for forage. In these areas, it is more efficient to grow livestock animals for food than to grow crop plants.

One of the most important limits on aquaculture's potential is the receptivity of animals to the domestication process itself. Land animals such as cows, pigs, and sheep were domesticated over thousands of years. During this time, there were undoubtedly failed attempts to domesticate other animals, which for one reason or another could not be domesticated. The same is true of aquaculture. It is not simply a matter of observing a need for more tuna, for instance, and opening an aquaculture facility that produces tuna. The organisms to be produced profitably by aquaculture must have certain traits that make their domestication possible. Aquatic organisms that are social and do not exhibit territoriality or aggressive behavior are possible candidates for domestication.

REVIEW

1. What is aquaculture?
2. What are some of the harmful environmental effects associated with aquaculture?

REVIEW OF LEARNING OBJECTIVES WITH KEY TERMS

• **Identify the main components of human nutritional requirements.**

Humans require a balanced diet that includes **carbohydrates, proteins,** and **lipids** in addition to **vitamins, minerals,** and water.

• **Explain how famines differ from chronic hunger.**

People who are chronically hungry lack access to the food they need to have healthy, productive lives. The two regions of the world with the greatest chronic hunger are South Asia and sub-Saharan Africa. Crop failures caused by drought, war, flood, or some other catastrophe may result in temporary but severe food shortages called **famines**.

• **Define** *world grain carryover stocks* **and explain how they are a measure of world food security.**

World grain carryover stocks are the amounts of rice, wheat, corn, and other grains remaining from previous harvests, as estimated at the start of a new harvest. World grain carryover stocks provide a measure of food security, a goal in which all people have access at all times to adequate amounts and kinds of food needed to lead healthy, active lives. According to the United Nations, world grain carryover stocks should not fall below the minimum amount of 70 days' supply.

• **Contrast industrialized agriculture with subsistence agriculture and describe three kinds of subsistence agriculture.**

Industrialized agriculture includes modern agricultural methods, which require a large capital input and less land and labor than traditional methods. **Subsistence agriculture** includes traditional agricultural methods, which depend on labor and a large amount of land to produce enough food to feed oneself and one's family. In subsistence agriculture, human and animal energy are used instead of fossil fuels. **Shifting cultivation**, in which short periods of cultivation are followed by long periods of the land lying fallow, and **nomadic herding**, in which herders wander freely over rangelands, are examples of subsistence agriculture. **Polyculture** is a type of intensive subsistence agriculture in which several crops that mature at different times are grown together.

• **Describe the beneficial and harmful effects of domestication on crop plants and livestock, defining** *germplasm* **in the process.**

Domestication is the process of taming wild animals or adapting wild plants to serve humans; domestication markedly alters the characteristics of the domesticated organism. **Germplasm** consists of any plant or animal material that may be used in breeding. When plants and animals are domesticated, much of the genetic diversity in wild germplasm is lost. Globally, a few agricultural varieties are replacing the hundreds of varieties developed by farmer-breeders over the centuries.

• **Relate the benefits and problems associated with the green revolution.**

The **green revolution**, in which new, high-yielding varieties are intensively cultivated with mechanized machinery, commercial inorganic fertilizers, and pesticides, began in the 1960s. The green revolution gave Latin American and Asian countries the chance to produce adequate supplies of food; Africa has not benefited much from the green revolution because its common crops have not been greatly improved by green revolution technology. The two most important problems associated with the green revolution are the energy costs built into this form of agriculture and the serious environmental problems caused by the intensive use of commercial inorganic fertilizers and pesticides.

• **Explain the roles of hormones and antibiotics in industrialized agriculture.**

The use of hormones and antibiotics has increased animal production. **Hormones** regulate livestock bodily functions and promote faster growth. The routine addition of low doses of **antibiotics** to the livestock feed causes these animals to gain more weight than animals that do not receive antibiotics, presumably because the antibiotic-fed animals expend less energy to fight infections.

• **Identify the potential benefits and problems with genetic engineering.**

Genetic engineering is the manipulation of genes, for example, by taking a specific gene from a cell of one species and placing it into a cell of an unrelated species, where it is expressed. Genetic engineering may produce food plants that are more nutritious, resistant to insect pests and viral diseases, or tolerant of drought, heat, cold, herbicides, or salty soil. Research is currently being conducted to assess the negative effects of introducing GM organisms into natural environments—for example, agricultural strains of plants whose seeds or pollen might spread in an uncontrolled manner.

• **Describe the environmental impacts of industrialized agriculture, including land degradation and habitat fragmentation.**

Soil erosion causes a decline in soil fertility as well as downstream sediment pollution. Agricultural chemicals such as pesticides and commercial inorganic fertilizers cause air, water, and soil pollution. **Land degradation** is the natural or human-induced process that decreases the future ability of the land to support crops or livestock. Many insects, weeds, and disease-causing organisms have developed resistance to pesticides, forcing farmers to apply larger quantities. Irrigation consumes huge quantities of fresh water. **Habitat fragmentation** is the breakup of large areas of habitat into small, isolated patches. Expanding the amount of agricultural land has resulted in habitat fragmentation that reduces biological diversity.

• **Define** *sustainable agriculture* **and contrast it with industrialized agriculture.**

Sustainable agriculture consists of agricultural methods that maintain soil productivity and a healthy ecological balance while having minimal long-term impacts. Unlike the **monocultures** of industrialized agriculture, the sustainable farm consists of field crops, trees that bear fruits and nuts, small herds of livestock, and tracts of forest. Sustainable agriculture avoids the continual use of antibiotics, large quantities of chemical pesticides, and high levels of commercial inorganic fertilizers.

• **Define** *bycatch*.

Bycatch consists of the fishes, marine mammals, sea turtles, seabirds, and other animals caught unintentionally in a commercial fishing catch.

• **Contrast fishing and aquaculture and relate the environmental challenges of each activity.**

Fishing is like hunting, whereas aquaculture is like agriculture. Currently, 62% of the world's fish stocks are either fully exploited, overexploited, or depleted because of the growing human population and technological advances in fishing gear. **Aquaculture** is the growing of aquatic organisms (fishes, shellfish, and seaweeds) for human consumption. Currently, aquaculture is the fastest growing type of food production in the world, and one out of every three fish destined for human consumption comes from fish farms. Although aquaculture has an enormous potential to provide food, it causes environmental problems, such as loss of coastlines and water pollution.

Thinking About the Environment

1. What age group in humans is usually most affected by undernutrition, malnutrition, and famine? Why?

2. Distinguish among carbohydrates, proteins, and lipids.

3. Why is population control the most fundamental solution to world food problems?

4. How is poverty related to chronic hunger?

5. Distinguish between shifting cultivation and slash-and-burn agriculture.

6. Distinguish between intercropping and polyculture.

7. What are two environmental problems associated with industrialized agriculture?

8. What was the green revolution?

9. Describe the environmental problems associated with farming each of these areas: tropical rain forests; hillsides; arid regions.

10. Give at least three examples of ways that industrialized agriculture could be made more sustainable.

11. Current research has not demonstrated conclusively that organic foods are healthier to eat than foods grown by conventional agriculture. If that is the case, then what is the benefit of organic farming? (*Hint*: Is organic agriculture better for the environment? If so, how?)

12. Read the cartoon to the right. What point is the cartoonist trying to make?

13. What is the problem with open management?

14. Explain why aquaculture is more like agriculture than it is like traditional fishing.

15. How does a sustainable agricultural system resemble a natural ecosystem?

Quantitative questions relating to this chapter are on our Web site.

(Bizarro © Dan Piraro/Reprinted with special permission of King Features Syndicate)

Take a Stand

Visit our Web site at http://www.wiley.com/college/raven (select Chapter 19 from the Table of Contents) for links to more information about issues involving genetically modified (GM) foods. Consider the opposing views of those who support and those who oppose the development and use of GM foods and debate the issue with your classmates. You will find tools to help you organize your research, analyze the data, think critically about the issues, and construct a well-considered argument. *Take a Stand* activities can be done individually or as a team, as oral presentations, written exercises, or Web-based (e-mail) assignments.

Additional online materials relating to this chapter, including a Student Testing Section with study aids and self-tests, Environmental News, Activity Links, Environmental Investigations, and more, are also on our Web site.

Air Pollution

20

Air pollution concerns in Jakarta, Indonesia. The caption on this poster—one of several around Jakarta—reads "Welcome to Pollution City." In addition to severe smoke pollution during agricultural burns, Jakarta suffers from substantial industrial, transportation, and other urban air pollution sources. (Jewel Samad/AFP/Getty Images)

For the past several years, large sections of Indonesia, Malaysia, Thailand, and other areas in Southeast Asia have suffered from periods of severe air pollution. While air pollution in many places is associated with industry and transportation, in Southeast Asia, agriculture is often the culprit. Farmers set fires to clear forested areas for use as plantations. These fires can burn uncontrolled for days and weeks, creating a smoky haze that covers entire countries and causing extremely hazardous conditions.

1997 was a particularly dry year throughout Southeast Asia, with reduced rainfall due to the El Niño effect (Chapter 5). Intentionally set fires lasted from June to December, burning over 160,000 hectares (600 mi²). The resulting haze was directly responsible for hundreds of deaths in Indonesia. These direct deaths, however, were only a small part of the problem. Other illnesses included hundreds of thousands of cases of asthma and bronchitis, and millions of acute respiratory infections. Schools were closed; business, traffic, and tourism were impacted; and economic damages were estimated to be close to $9 billion. It is unclear whether the resulting agricultural benefits exceeded these costs, a reminder of the need to think about impacts on systems, not just one small part of the system.

After the 1997 fires, several laws and restrictions were placed on agricultural burning, but with limited effect. While fires have been set every year since 1997, 2006 was another exceptionally bad year. In October, papers around the world carried news of extremely poor air quality in Singapore. Residents flooded hospitals, airplane flights were cancelled, and some people wore air masks even while indoors. In one case, two cargo ships collided due to poor visibility.

Smoke from burning biomass—whether from fireplaces or forests—contains a range of hazardous chemicals. Small particles settle deep in the lungs, causing irritation that leads to infection, as well as reduced ability to breathe. These particles can contain toxic organic or metallic materials. Incomplete combustion of carbon-containing materials releases volatile organic chemicals and carbon monoxide, which interferes with oxygen uptake. Other chemicals containing phosphorus, potassium, nitrogen, and sulfur are released. Most of these chemicals are essential to life as nutrients in the soil, but can be toxic when inhaled.

Smoke is even a problem around San Francisco, California, which has some of the strictest air quality regulations in the world. On still winter nights, thousands of fireplaces provide comfort to people in their homes, while smoke and carbon monoxide levels rise outside, often to dangerous levels. And in the late summer, legal agriculture-related fires are set throughout California's Central Valley.

One of the earliest recorded environmental laws can be traced to 13th-century London, in which the burning of soft coal that resulted in dense, acidic smoke was banned. Yet London air quality continued to be notoriously bad for the ensuing seven centuries. Smoke is a major pollutant, from indoor wood smoke in Kenya to fluorine-laden coal smoke in central China. As long as fire remains a tool for energy and agriculture, smoke will remain a problem for humans.

World View and Closer to You...
Real news footage relating to air pollution around the world and in your own geographic region is available at www.wiley.com/college/raven by clicking on "World View and Closer to You."

The Atmosphere as a Resource

LEARNING OBJECTIVE

- Describe the composition and function of the atmosphere.

The atmosphere is a gaseous envelope surrounding Earth (see Figure 5.10). Excluding water vapor, four gases comprise most of the atmosphere: nitrogen (N_2, 78.08%), oxygen (O_2, 20.95%), argon (Ar, 0.93%), and carbon dioxide (CO_2, 0.04%). Other gases and particles, including those we call **pollutants,** occur in much smaller concentrations. The two atmospheric gases most important to humans and other organisms are carbon dioxide and oxygen. During *photosynthesis*, plants, algae, and certain bacteria use carbon dioxide to manufacture sugars and other organic molecules; this process produces oxygen. During *cellular respiration*, most organisms use oxygen to break down food molecules and supply themselves with chemical energy; this process produces carbon dioxide. Nitrogen gas is an important component of the nitrogen cycle. The atmosphere performs additional **ecosystem services,** namely, blocking Earth's surface from much of the ultraviolet (UV) radiation coming from the sun, moderating the climate, and redistributing water in the hydrologic cycle.

Living at the interface of Earth's surface and atmosphere, humans think of the atmosphere as an unlimited resource, but we should reconsider. **Ulf Merbold,** a German space shuttle astronaut, felt differently about the atmosphere after viewing it in space (**Figure 20.1**). "For the first time in my life, I saw the horizon as a curved line. It was accentuated by the thin seam of dark blue light—our atmosphere. Obviously, this wasn't the 'ocean' of air I had been told it was so many times in my life. I was terrified by its fragile appearance."

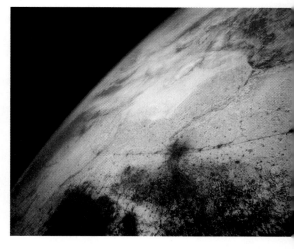

Figure 20.1   The atmosphere. The "ocean of air" is an extremely thin layer compared to the size of Earth. In this photo, shot in a low-Earth orbit from the space shuttle Endeavor, the atmosphere is a thin blue layer that separates the planet from the blackness of space. (The Trans-Siberian Railway, the world's longest, appears as a dark line against a background of snow.) (© NASA/Corbis Images)

REVIEW

1. What are two ecosystem services performed by the atmosphere?

Types and Sources of Air Pollution

LEARNING OBJECTIVES

- Define *air pollution* and distinguish between primary and secondary air pollutants.
- List the seven major classes of air pollutants, including ozone and hazardous air pollutants, and describe their characteristics and effects.
- Describe industrial smog, photochemical smog, and temperature inversions.

Air pollution consists of gases, liquids, or solids present in the atmosphere in high enough levels to harm humans, other organisms, or materials. Although air pollutants can come from natural sources—as when lightning causes a forest fire or a volcano erupts—human activities release many kinds of substances into the atmosphere and make a major contribution to air pollution. Some of these substances are harmful when they precipitate (form a solid) and settle on land and surface waters, whereas other substances are harmful because they alter the chemistry of the atmosphere. From the standpoint of human health, probably more significant than the overall human contribution of air pollution is the fact that much of the air pollution released by humans is concentrated in densely populated urban areas.

Although many different air pollutants exist, we will focus on the seven most important types from a regulatory perspective: particulate matter, nitrogen oxides, sulfur oxides, carbon oxides, hydrocarbons, ozone, and air toxics (**Table 20.1**). Air pollutants are often divided into two categories, primary and secondary (**Figure 20.2**). **Primary air pollutants** are harmful chemicals that enter directly into the atmosphere. The major ones are carbon oxides, nitrogen oxides, sulfur dioxide, particulate matter, and hydrocarbons. **Secondary air pollutants** are harmful chemicals that form from other substances released into the atmosphere. Ozone and sulfur trioxide are secondary

air pollution: Various chemicals added to the atmosphere by natural events or human activities in high enough concentrations to be harmful.

primary air pollutant: A harmful substance, such as soot or carbon monoxide, that is emitted directly into the atmosphere.

secondary air pollutant: A harmful substance formed in the atmosphere when a primary air pollutant reacts with substances normally found in the atmosphere or with other air pollutants.

Table 20.1 Major Air Pollutants

Pollutant	Composition	Primary or Secondary	Characteristics
Particulate matter			
Dust	Variable	Primary	Solid particles
Lead	Pb	Primary	Solid particles
Sulfuric acid	H_2SO_4	Secondary	Liquid droplets
Nitrogen oxides			
Nitrogen dioxide	NO_2	Primary	Reddish-brown gas
Sulfur oxides			
Sulfur dioxide	SO_2	Primary	Colorless gas with strong odor
Carbon oxides			
Carbon monoxide	CO	Primary	Colorless, odorless gas
Carbon dioxide*	CO_2	Primary	Colorless, odorless gas
Hydrocarbons			
Methane	CH_4	Primary	Colorless, odorless gas
Benzene	C_6H_6	Primary	Liquid with sweet smell
Ozone	O_3	Secondary	Pale blue gas with acrid odor
Air toxics			
Chlorine	Cl_2	Primary	Yellow-green gas

* Discussed in Chapter 21.
Source: Environmental Protection Agency.

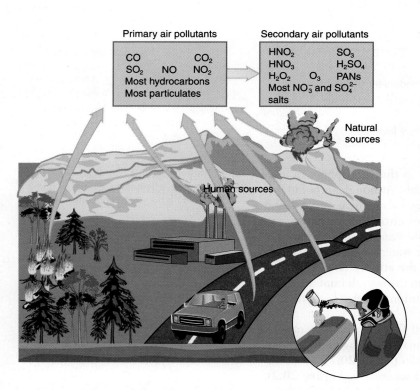

Figure 20.2 Primary and secondary air pollutants. Primary air pollutants are emitted, unchanged, from a source directly into the atmosphere, whereas secondary air pollutants are produced in the atmosphere from chemical reactions involving primary air pollutants.

air pollutants because both are formed by chemical reactions that take place in the atmosphere.

Major Classes of Air Pollutants

Particulate matter consists of thousands of different solid and liquid particles suspended in the atmosphere. **Solid particulate matter** is generally referred to as dust, whereas liquid suspensions are commonly called mists. Particulate matter includes a variety of pollutants, such as soil particles, soot, lead, asbestos, sea salt, and sulfuric acid droplets. Particulate matter reduces visibility by scattering and absorbing sunlight. Urban areas receive less sunlight than rural areas, partly as a result of greater quantities of particulate matter in the air. Particulate matter corrodes metals, erodes buildings and works of sculpture when the air is humid, and soils clothing and draperies.

Particulate matter can be dangerous for one of two reasons. First, it may contain materials—such as heavy metals, asbestos, or organic chemicals—that have toxic or carcinogenic effects. These toxins, upon contacting or being absorbed into the body, have a range of effects. Second, extremely small particles, even if not toxic, can get lodged deep in the lungs. Microscopic particles may be classified as PM-10 (particular matter less than 10 μm [10 micrometers] in diameter) or PM-2.5 (particular matter less than 2.5 μm in diameter). The EPA samples microscopic particulate matter at 1000 locations around the

United States so that they can better understand its composition, which varies with location and season.

Nitrogen oxides are gases produced by the chemical interactions between atmospheric nitrogen and oxygen when a source of energy, such as combustion of fuels, produces high temperatures. Collectively known as NO_x, nitrogen oxides consist mainly of nitric oxide (NO), nitrogen dioxide (NO_2), and nitrous oxide (N_2O). Nitrogen oxides inhibit plant growth and, when breathed, aggravate health problems such as asthma, a disease in which breathing is wheezy and labored because of airway constriction. They are involved in the production of photochemical smog and acid deposition (when nitrogen dioxide reacts with water to form nitric acid and nitrous acid). Nitrous oxide is associated with global warming (nitrous oxide traps heat in the atmosphere and is therefore a **greenhouse gas**) and depletes ozone in the stratosphere. Nitrogen oxides cause metals to corrode and textiles to fade and deteriorate.

Sulfur oxides are gases produced by the chemical interactions between sulfur and oxygen. Sulfur dioxide (SO_2), a colorless, nonflammable gas with a strong, irritating odor, is a major sulfur oxide emitted as a primary air pollutant. Another major sulfur oxide is sulfur trioxide (SO_3), a secondary air pollutant that forms when sulfur dioxide reacts with oxygen in the air. Sulfur trioxide, in turn, reacts with water to form another secondary air pollutant, sulfuric acid. Sulfur oxides cause acid deposition, and they corrode metals and damage stone and other materials. Sulfuric acid and sulfate salts produced in the atmosphere from sulfur oxides damage plants and irritate the respiratory tracts of humans and other animals.

Carbon oxides are the gases carbon monoxide (CO) and carbon dioxide (CO_2). Carbon monoxide, a colorless, odorless, and tasteless gas produced in the largest quantities of any atmospheric pollutant except carbon dioxide, is poisonous and interferes with the blood's ability to transport oxygen. Carbon dioxide, also colorless, odorless, and tasteless, is a greenhouse gas; its buildup in the atmosphere is associated with global climate change.

Hydrocarbons are a diverse group of organic compounds that contain only hydrogen and carbon; the simplest hydrocarbon is methane (CH_4). Small hydrocarbon molecules are gaseous at room temperature. Methane is a colorless, odorless gas that is the principal component of natural gas. (The odor of natural gas comes from sulfur compounds deliberately added so that humans can indirectly detect the presence of the explosive methane gas by smelling the sulfur-containing compounds.) Medium-sized hydrocarbons such as benzene (C_6H_6) are liquids at room temperature, although many are volatile and evaporate readily. The largest hydrocarbons, such as the waxy substance paraffin, are solids at room temperature. The many different hydrocarbons have a variety of effects on human and animal health. Some cause no adverse effects, others injure the respiratory tract, and still others cause cancer. All except methane are important in the production of photochemical smog. Methane is a potent greenhouse gas linked to global climate change.

Ozone (O_3) is a form of oxygen considered a pollutant in one part of the atmosphere but an essential component in another. In the **stratosphere,** which extends from 10 to 45 km (6.2 to 28 mi) above Earth's surface, oxygen reacts with UV radiation coming from the sun to form ozone. Stratospheric ozone prevents much of the solar UV radiation from penetrating to Earth's surface. Unfortunately, certain human-made pollutants (chlorofluorocarbons, or CFCs) react with stratospheric ozone, breaking it down into molecular oxygen, O_2.

Unlike stratospheric ozone, ozone in the **troposphere**—the layer of atmosphere closest to Earth's surface—is a human-made air pollutant. (Ground-level ozone does not replenish the ozone depleted from the stratosphere because ground-level ozone breaks down to form oxygen long before it drifts up to the stratosphere.) Ozone in the troposphere is a secondary air pollutant that forms when sunlight catalyzes reactions between nitrogen oxides and volatile hydrocarbons. The most harmful component of photochemical smog, ozone reduces air visibility and causes health problems. Ozone stresses plants and reduces their vigor, and chronic (of long duration) ozone exposure lowers crop yields (**Figure 20.3**). Chronic exposure to ozone is a possible contributor to forest decline, and ground-level ozone is a greenhouse gas associated with global climate change.

ozone: A pale blue gas that is both a pollutant in the lower atmosphere (troposphere) and an essential component that screens out UV radiation in the upper atmosphere (stratosphere).

Figure 20.3   Ozone damage. Compare the soybean leaf grown in clean air (right) with the one damaged by ozone (left). Note the darkened, dead tissue along the leaf's margin and the buckling of leaf tissue between the veins. The ozone-damaged leaf is lighter green, indicating that it does not have as much chlorophyll. Plants exposed to ozone exhibit other symptoms, including reduced root growth and a lowered productivity. On average, soybeans exposed to ozone pollution exhibit a 15% reduction in yield. (Runk/Schoenberger/Grant Heilman Photography)

hazardous air pollutants: Air pollutants that are potentially harmful and may pose long-term health risks to people who live and work around chemical factories, incinerators, or other facilities that produce or use them.

Most of the hundreds of other air pollutants—such as chlorine, lead, hydrochloric acid, formaldehyde, radioactive substances, and fluorides—are present in low concentrations, although it is possible to have high local concentrations of specific pollutants. Some of these air pollutants, known as **hazardous air pollutants (HAPS)**, or **air toxics,** are potentially harmful. To limit the release of more than 180 hazardous air pollutants, the Clean Air Act Amendments of 1990 (discussed later in the chapter) regulate emissions of both large and small businesses, such as bakeries, distilleries, dry cleaners, furniture makers, gasoline service stations, hospitals, auto paint shops, and print shops.

Sources of Outdoor Air Pollution

The two main human sources of primary air pollutants are transportation (**mobile sources**) and industries (**stationary sources**) (**Figure 20.4**), although intentional fires can also contribute significantly. Automobiles and trucks, known as mobile sources, generate significant quantities of nitrogen oxides, carbon oxides, particulate matter, and hydrocarbons as a result of the combustion of gasoline. Although diesel engines in trucks, buses, trains, and ships consume less fuel than other types of combustion engines, they produce more air pollution. One heavy-duty truck emits as much particulate matter as 150 automobiles, whereas one diesel train engine produces, on average, 10 times the particulate matter of a diesel truck. Pollutants from engines used in outboard motorboats, jet skis, and other mobile sources cause both air and water pollution.

Electric power plants and other industrial facilities, known as stationary sources, emit most of the particulate matter and sulfur oxides released in the United States; they emit sizable amounts of nitrogen oxides, hydrocarbons, and carbon oxides. The combustion of fossil fuels, especially coal, is responsible for most of these emissions. The top three industrial sources of toxic air pollutants—that is, chemicals released into the air that are fatal to humans at specified concentrations—are the chemical industry, the metals industry, and the paper industry.

Not all air pollution is generated by human activities. On a hot summer day in the Blue Ridge Mountains, which are part of the Appalachians, a blue haze hangs over the forested hills. This haze is caused by hydrocarbon emissions from tree leaves. Many plants produce a variety of hydrocarbons in response to heat. Isoprene and other hydrocarbons are volatile and evaporate into the atmosphere, where they affect

Figure 20.4   Air pollutants in the United States. (Environmental Protection Agency)

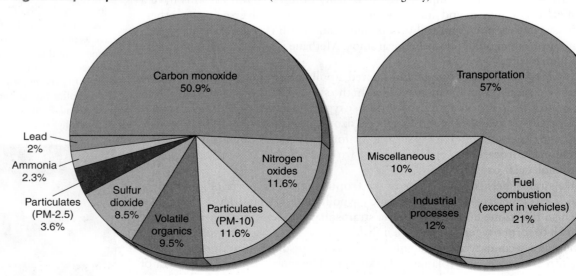

(a) The concentrations of nitrogen oxides and volatile organics are an indirect measure of ozone, which is a secondary air pollutant formed in their presence.

(b) Transportation and industrial fuel combustion (such as electric power plants) are major contributors of pollutants.

atmospheric chemistry. These hydrocarbons are highly reactive and contribute to ozone formation, a key ingredient in photochemical smog (discussed shortly). Carbon monoxide is one of the breakdown products of isoprene. The contribution of biologically generated hydrocarbon emissions in southeastern cities such as Atlanta, which is heavily wooded, is substantial. Elsewhere in the country, however, the "natural" contribution is a much smaller percentage.

Urban Air Pollution

Air pollution localized in urban areas, where it reduces visibility, is often called **smog.** The word "smog" was coined at the beginning of the 20th century for the smoky fog that was so prevalent in London because of coal combustion. Today there are several different types of smog. Traditional London-type smog—that is, smoke pollution—is sometimes called **industrial smog.** The principal pollutants in industrial smog are sulfur oxides and particulate matter. The worst episodes of industrial smog typically occur during winter months, when combustion of household fuel such as heating oil or coal is high. In December 1952, 4000 Londoners died in the world's worst industrial smog incident. An additional 8000 people died within the next two months, possibly due to the lingering effects of the smog, although the exact causes of death for these people have never been explained. Because of air quality laws and pollution-control devices, industrial smog is generally not a significant problem in highly developed countries today, but it is often serious in many communities and industrial regions of developing countries.

Another important type of smog is photochemical smog. This brownish-orange smog is called photochemical because light—that is, sunlight—initiates several chemical reactions that collectively form the ingredients in photochemical smog. First noted in Los Angeles in the 1940s, photochemical smog is generally worst during the summer months. Both nitrogen oxides and hydrocarbons are involved in its formation. One of the photochemical reactions occurs among nitrogen oxides (largely from automobile exhaust), volatile hydrocarbons, and oxygen in the atmosphere to produce ground-level ozone; this reaction requires solar energy (**Figure 20.5**). The ozone formed in this way then reacts with other air pollutants, including hydrocarbons, to form more than 100 different secondary air pollutants (peroxyacyl nitrates, or PANs, for example) which injure plant tissues, irritate eyes, and aggravate respiratory illnesses in humans.

photochemical smog: A brownish-orange haze formed by chemical reactions involving sunlight, nitrogen oxide, and hydrocarbons.

CASE IN POINT

Efforts to Reduce Ozone in Southern California

Many human sources contribute to the ingredients of photochemical smog. Automobiles, as well as emissions from gasoline stations and oil refineries, are major significant contributors. But any process that releases **volatile organic compounds (VOCs)** contributes to smog production. For this reason, in many places where ozone is a problem there are restrictions on paints, cleaning products, dry cleaners, and even bakeries (when bread is baked, yeast byproducts that are VOCs are released to the atmosphere).

Both weather and topography affect air pollution. Variation in temperature during the day usually results in air circulation patterns that help dilute and disperse air pollutants. As the sun increases surface temperatures, the air near the ground is warmed. This heated air expands and rises to higher levels in the atmosphere (warm air is lighter and more buoyant than cool air), causing a low-pressure area near the ground. The surrounding air then moves into the low-pressure area. Thus, under normal conditions, air circulation patterns prevent toxic pollutants from increasing to dangerous levels near the ground.

During periods of temperature inversion, also called *thermal inversion*, polluting gases and particulate matter remain trapped in high concentrations close to the ground, where people live and breathe. Temperature inversions usually persist for only a few hours before being broken up by solar heating that warms the air near the ground. Sometimes, however, atmospheric stagnation caused by a stalled high-pressure air mass allows a temperature inversion to persist for several days.

temperature inversion: A deviation from the normal temperature distribution in the atmosphere, resulting in a layer of cold air temporarily trapped near the ground by a warmer, upper layer.

Figure 20.5 Photochemical smog.

(a) Smog in Los Angeles is sometimes so bad it blocks out the sun. (Jodi Cobb/National Geographic Society)

Certain types of topography (surface features) increase the likelihood of temperature inversions. Cities located in valleys, near the coast, or on the leeward side of mountains (the side toward which the wind blows) are prime candidates for temperature inversions. The Los Angeles Basin is a plain that lies between the Pacific Ocean and mountains to the north and east. During the summer, the sunny climate produces a layer of warm, dry air at upper elevations. However, a region of upwelling occurs just off the Pacific coast, bringing cold ocean water to the surface and cooling the ocean air. As this cool air blows inland over the basin, the mountains block its movement further. Thus, a layer of warm, dry air overlies cool air at the surface, producing a temperature inversion.

The South Coast Air Quality Management District (SCAQMD) is a governmental organization established in 1977 to deal with the historically poor air quality in the area surrounding Los Angeles. Forming the SCAQMD consolidated efforts to curb air pollution in southern California that had begun shortly after WWII. A special agency was needed because air pollutants follow geographic, and not political, boundaries. The borders of the SCAQMD roughly represent an "air basin" in which the air generally moves from the west (the Pacific Ocean) to the east (toward Arizona and Nevada). It covers an area of nearly 18,000 km^2 (7,000 mi^2), including part or all of four southern California counties, and almost 15 million people.

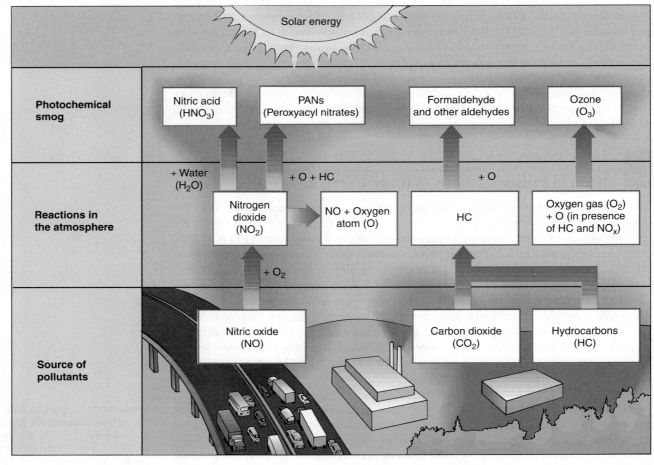

(b) Several chemical reactions result in the complex mixture known as smog: ozone, peroxyacyl nitrates (PANs), nitric acid, and various organic compounds such as formaldehyde.

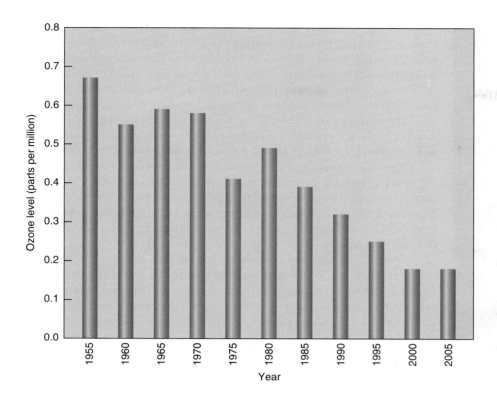

Figure 20.6   **Peak ozone concentrations in Southern California, 1955–2005.** Peak ozone is the highest level of ozone recorded on any single day during the year. Average daily ozone, number of days above federal and state standards, and other measures show similar patterns. Air quality has improved steadily over the past half century, but still presents a health threat.

While the rules and regulations of the SCAQMD comprise several large books; evidence suggests that just a few major regulations have been responsible for the substantial improvements in air quality over the past five decades (**Figure 20.6**). In 1976, ozone levels in the air basin exceeded the federal "safe" standard of 0.12 ppm on 194 days, and the state standard of 0.09 ppm on 237 days. This means that the ozone levels were considered unsafe nearly two-thirds of the year! These numbers had already dropped steadily through the 1960s and 1970s, a trend that has continued. They appeared to reach a low in 1998, when 114 days exceeded the state standard, with the same or more excess ozone days from 1999 to 2003. However, both 2004 and 2005 had fewer excess ozone days than did 2003.

Since ozone is created by a combination of high temperature, sunlight, hydrocarbons, water vapor, and oxides of nitrogen, some of those have to be controlled in order to reduce ozone. Only two of these components, hydrocarbons and oxides of nitrogen, can be controlled. Regulations to limit ozone formation have required restrictions on everything from combustion engines and oil refineries to charcoal lighter fluid and spray paints (**Figure 20.7**). Regulations in southern California, which has a thriving economy, have shown that environmental quality and financial success may be complements, not opposites. ■

CASE IN POINT

Air Pollution in Beijing and Mexico City

Los Angeles, California, was once known for the worst smog in the world. However, while L.A. Smog is still legendary, five decades of regulation have significantly reduced this trend. Now, Beijing, China, and Mexico City complete for having the worst air quality in the world, and seven of the world's worst cities for air pollution can be found in China.

With rapid industrialization and a growing population, urban China faces some of the worst air quality in the world (**Figure 20.8**), with an estimated 400,000 premature deaths each year associated with air pollution in 2004. Automobiles, dust from construction sites, and electricity generated with coal outside of town are the

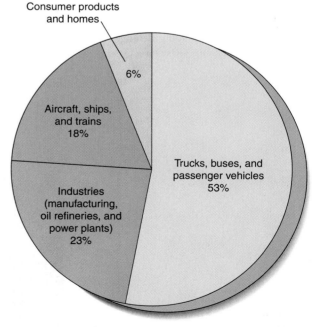

Figure 20.7   **Sources of smog in Los Angeles, CA.** Trucks, buses, and passenger vehicles account for more than half of the emissions that produce smog.

Figure 20.8 **Air pollution in Liaonign Province, China.** Coal smoke pollutes the air above workers' houses in Liaoning Province, China. All forms of pollution are increasing threats as China becomes industrialized. China has many of the worlds' most polluted cities, including Beijing. (Alain Le Garsmeur/Stone/Getty Images)

biggest culprits in Beijing, which has the worst air quality in the world. However, in preparation for hosting the 2008 summer Olympics, Beijing took a number of steps to curb its air pollution. These included increased public transportation (buses, subways, and light rail). Much of the bus fleet has been converted to natural gas, which burns more cleanly than gasoline. Existing power plants have been improved, and have switched to low-sulfur coal. New power plants around the Beijing area will burn natural gas instead of coal. Automobile emission standards, while lagging behind those of Europe and the United States, are steadily improving.

Other big challenges are the population (10.7 million and growing) and the desire of more and more Beijing residents to own cars. Even as Beijing passes some regulations designed to improve air quality, other rules undermine those efforts. Beijing was once known as a town where bicycles ruled; now, bicycles are banned from certain parts of the city to improve the flow of automobile traffic.

Mexico City, the world's second largest city, has the second most polluted air of any major metropolitan area in the world (**Figure 20.9**). Average visibility has dropped from 11 km (7 mi) in the 1940s, when surrounding snow-capped volcanoes were commonly seen, to 1.6 km (1 mi). Mexico City's air pollution is due in part to its large population growth in the past several decades (Mexico City grew from 5.4 million in 1960 to 19.4 million in 2005) and in part to its location. Mexico City is in a bowl-shaped valley ringed on three sides by mountains; winds coming in from the open northern end are trapped in the valley. Air quality is at its worst from October to January, largely as a result of temperature inversions caused by seasonal variations in atmospheric conditions.

The city has more than three million passenger vehicles, 360 gasoline stations, and about 36,000 businesses, which the Mexican government says release 3.94 million metric tons (4.35 million tons) of pollutants into the air each year. Mexican gasoline contains a lot of contaminants, and the average automobile is 10 years old and produces more pollutants than do newer cars. The air contains particles of dried fecal matter from the millions of gallons of sewage dumped onto land near the city. In addition, liquefied petroleum gas—a major source of energy for cooking and heating in Mexico City—escapes unburned into the atmosphere from thousands of leaks, increasing the level of hydrocarbons in the city's air. Simply breathing the air in Mexico City is equivalent to smoking two packs of cigarettes a day.

Figure 20.9 **Smog in Mexico City, Mexico.** The Mexican flag blows in an extremely polluted breeze. Mexico City has the dubious distinction of having some of the worst air quality in the world. (Roberto Velazquez/© AP/Wide World Photos)

During the 1990s, Mexico embarked on an ambitious plan to improve Mexico City's air quality. It spent more than $5 billion to replace old buses, taxis, delivery trucks, and cars with cleaner vehicles, such as those with catalytic converters. Mexico switched to unleaded gasoline and reforested some of the nearby hillsides to reduce particulate matter produced by wind erosion. Driving restrictions apply when the air quality is particularly poor, and exhaust emissions are periodically checked on autos.

In addition, Pemex, Mexico's national oil company, has upgraded its refineries and increased gas imports from the United States, which produces a cleaner fuel. An old, polluting oil refinery within the city limits was closed, and several large industries installed pollution-control devices. All these changes translate into cleaner air. In the early 2000s, Mexico City's air pollution, though still often dangerously high, was gradually improving. ■

REVIEW

1. What are particulates? Give at least three examples of different kinds of particulates.
2. Why is ozone considered an air pollutant even though it is essential in the stratosphere?
3. How is sunlight involved in the production of photochemical smog?

 ## Effects of Air Pollution

LEARNING OBJECTIVES

- Distinguish between emphysema and chronic bronchitis.
- Relate the adverse health effects of specific air pollutants and explain why children are particularly susceptible to air pollution.

Air pollution injures organisms, reduces visibility, and attacks and corrodes materials such as metals, plastics, rubber, and fabrics. The respiratory tracts of animals, including humans, are particularly harmed by air pollutants, which worsen existing medical conditions such as chronic lung disease, pneumonia, and cardiovascular problems (**Figure 20.10**). Most forms of air pollution reduce the overall productivity of crop plants, and when combined with other environmental stressors, such as low winter temperatures or prolonged droughts, air pollution causes plants to decline and die. Air pollution is involved in acid deposition, global climate changes, and stratospheric ozone depletion.

Air Pollution and Human Health

Generally speaking, exposure to low levels of pollutants such as ozone, sulfur oxides, nitrogen oxides, and particulate matter irritates the eyes and causes inflammation of the respiratory tract (**Table 20.2**). Evidence shows that many air pollutants suppress

Figure 20.10 **Caution: Air pollution is hazardous to your health.** How does this cartoon depict a system out of balance? (From *The Beast That Ate The Earth* /Chris Madden/Inkline Press)

Table 20.2 Health Effects of Several Major Air Pollutants

Pollutant	Source	Effects
Particulate	Industries, electric power plants, motor vehicles, construction, agriculture	Aggravates respiratory illnesses; long-term exposure may cause increased incidence of chronic conditions such as bronchitis; linked to heart disease; suppresses immune system; some particles, such as heavy metals and organic chemicals, may cause cancer or other tissue damage
Nitrogen oxides	Motor vehicles, industries, heavily fertilized farmland	Irritate respiratory tract; aggravate respiratory conditions such as asthma and chronic bronchitis
Sulfur oxides	Electric power plants and other industries	Irritate respiratory tract; same effects as particulates
Carbon monoxide	Motor vehicles, industries, fireplaces	Reduces blood's ability to transport oxygen; headache and fatigue at lower levels; mental impairment or death at high levels
Ozone	Formed in atmosphere (secondary air pollutant)	Irritates eyes; irritates respiratory tract; produces chest discomfort; aggravates respiratory conditions such as asthma and chronic bronchitis

emphysema: A disease in which the air sacs (alveoli) in the lungs become irreversibly distended, decreasing the efficiency of respiration and causing breathlessness and wheezy breathing.

chronic bronchitis: A disease in which the air passages (bronchi) of the lungs become permanently inflamed, causing breathlessness and chronic coughing.

the immune system, increasing susceptibility to infection. In addition, evidence continues to accumulate that exposure to air pollution during respiratory illnesses may result in the development later in life of chronic respiratory diseases, such as **emphysema** and **chronic bronchitis**.

Health Effects of Specific Air Pollutants   Both sulfur dioxide and particulate matter irritate the respiratory tract and, because they cause the airways to constrict, actually impair the lungs' ability to exchange gases. People suffering from emphysema and asthma are sensitive to sulfur dioxide and particulate pollution. Nitrogen dioxide also causes airway constriction and, in people suffering from asthma, an increased sensitivity to pollen and dust mites (microscopic animals found in household dust).

One of the largest studies ever conducted on the effects of particulate pollution on human health was published in 2003. The National Institute of Environmental Health financed the landmark Brigham Young University study, which tracked the causes of death for more than 319,000 people in all 50 states. The study compared mortality data with air pollution levels—specifically, PM-2.5—in their communities. (Coal-fired power plants and diesel engines are the main emitters of tiny airborne soot particles.) Because so many people were included in the study, scientists could discount the effects of tobacco, obesity, diet, and other disease factors related to death rates. The study found that people who live and work in the country's most polluted areas are more likely to die prematurely from specific kinds of heart disease than those living in U.S. cities with the cleanest air. The soot is inhaled and lodges in the lung tissue, causing an inflammation reaction that triggers a number of processes that clog arteries leading to the heart.

Carbon monoxide binds irreversibly with iron in the blood's hemoglobin, eliminating its ability to transport oxygen. At medium concentrations, carbon monoxide causes headaches and fatigue. As the concentration of carbon monoxide increases, reflexes slow down and drowsiness occurs; at a certain high level, carbon monoxide causes death. People at greatest risk from carbon monoxide include pregnant women, infants, and those with heart or respiratory diseases. A four-year study in seven U.S. cities—Chicago, Detroit, Houston, Los Angeles, Milwaukee, New York, and Philadelphia—linked carbon monoxide concentrations in the air to increases in hospital admissions for congestive heart failure.

Ozone and the volatile compounds in smog are irritants that cause a variety of health problems, including burning eyes, coughing, and chest discomfort. Ozone brings on asthma attacks and suppresses the immune system. A 2002 study at the University of California, Los Angeles, concluded that pregnant women exposed to high levels of ozone and carbon monoxide are three times more likely to give birth to infants with serious heart defects. In 2004 the results of a 14-year study of 95 urban areas across the United States linked elevated ozone levels to increases in deaths from cardiovascular and respiratory ailments.

The health effects of about 150 hazardous air pollutants produced by motor vehicles, businesses, and industries have not been widely studied, although long-term exposure to certain air toxics has been linked to cancer. The Environmental Defense Fund estimates that 360 people out of every million Americans develop cancer as a result of air toxics, although the cancer rate varies widely from one place to another.

Children and Air Pollution   As is true of essentially all environmental stressors, air pollution is a greater health threat to children than it is to adults. The lungs continue to develop throughout childhood, and air pollution can restrict lung development, making children more vulnerable to health problems later in life. In addition, a child has a higher metabolic rate than an adult and needs more oxygen. To obtain this oxygen, a child breathes more air—about two times as much air per pound of body weight as an adult. This means that a child breathes more air pollutants into the lungs. A 1990 study in which autopsies were performed on 100 Los Angeles children who died for unrelated reasons found that more than 80% had early stage lung disease.

A 10-year study of about 5000 children in 12 communities in southern California examined the effects of chronic exposure to air pollution on children's developing lungs. Results from this study, which ended in 2001, indicate that children who live in high-ozone areas and participate in sports are more likely to develop asthma than children who live there but do not participate in sports. In addition, results indicate that children who breathe the most polluted air (higher concentrations of nitrogen dioxide, particulate matter, and acid vapor) have less lung growth than children who breathe cleaner air. If the children moved to areas with less particulate air pollution, their lung development increased, but if they moved to areas with worse particulate air pollution, their lung development decreased.

REVIEW

1. What is the general effect of air pollution on the body's immune system?
2. What are some of the health effects of exposure to ozone?
3. Why is air pollution a greater threat to children than it is to adults?

Controlling Air Pollution in the United States

LEARNING OBJECTIVES

- Summarize how electrostatic precipitators and scrubbers work.
- Explain how Phase I vapor recovery reduces gasoline-related emissions.
- Summarize the effects of the Clean Air Act on U.S. air pollution.

There is bad and good news about air pollution in the United States. The bad news is that many locations throughout the country still have unacceptably high levels of one or more air pollutants. Moreover, most health experts estimate that air pollution causes the premature deaths of thousands of people in the United States each year. The good news is that, overall air quality has improved since 1970. This improvement in air quality is largely due to the U.S. Clean Air Act and state-level air quality initiatives.

Controlling Air Pollutants

Many measures already discussed for energy efficiency and conservation (see Chapters 11 and 13) also reduce air pollution. Hybrid cars and natural gas-powered buses, for example, produce much lower emissions than do their predecessors. Technologies exist to control all the forms of air pollution discussed in this chapter.

Historically, "command and control" technologies have been used to reduce emissions. Usually, this means equipment that limits the emissions after they have been generated. However, this technological approach can be more expensive than alternatives such as changing industrial processes to reduce emissions.

Smokestacks fitted with *electrostatic precipitators*, fabric filters, *scrubbers*, or other technologies remove particulate matter (**Figure 20.11**). In addition, particulate matter is controlled by careful land-excavating activities, such as sprinkling water on dry soil being moved during road construction.

Several methods exist for removing sulfur oxides from flue (chimney) gases, but it is often less expensive simply to switch to a low-sulfur fuel such as natural gas or even to a non-fossil fuel energy source such as solar energy. Sulfur can be removed from fuels before they are burned, as in coal gasification (see Figure 11.19).

Gasoline is extremely volatile, and gasoline vapors can be a major source of VOCs. In order to reduce these emissions, gasoline sellers in most urban (and many rural) parts of the world require some form of **vapor recovery**. *Phase I vapor recovery* involves underground storage tanks at gas stations (**Figure 20.12**). As one hose from a delivery truck fills the underground tank, another returns the vapors in the tank—which otherwise would be vented to the atmosphere—to the truck. The truck then returns to the gasoline depot, where the vapors are either combusted, or condensed into gasoline. *Phase II vapor recovery*

vapor recovery: The removal of unburned gasoline vapors from gasoline containers, including underground tanks at gas stations and automobile gas tanks. Recovered vapors are either burned or condensed and recovered.

Figure 20.11　Electrostatic precipitator and scrubber.

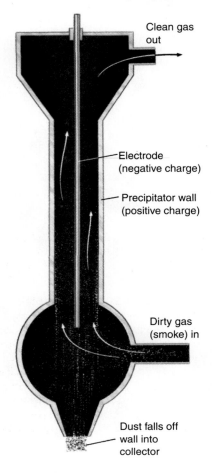

Clean gas out

Electrode (negative charge)

Precipitator wall (positive charge)

Dirty gas (smoke) in

Dust falls off wall into collector

(a) In an electrostatic precipitator, the electrode imparts a negative charge to particulates in the dirty gas. These particles are attracted to the positively charged precipitator wall, and then fall off into the collector.

(b) Emissions from a Delaware Valley steel mill with the electrostatic precipitator turned off. (John D. Cunningham/Visuals Unlimited)

(c) Emissions from the same steel mill when the electrostatic precipitator is turned on. (John D. Cunningham/Visuals Unlimited)

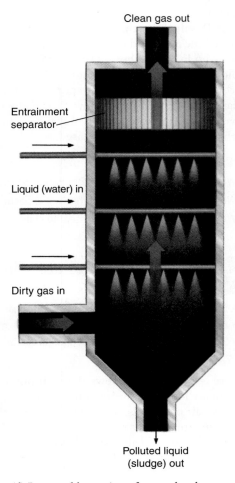

Clean gas out

Entrainment separator

Liquid (water) in

Dirty gas in

Polluted liquid (sludge) out

(d) In a scrubber, mists of water droplets trap particulates in the dirty gas. The toxic dust produced by electrostatic precipitators and the polluted sludge produced by scrubbers must be safely disposed of or they will cause soil and water pollution.

involves removing vapor from the gas tank in cars as the gas is pumped in. These vapors are usually returned to the underground tank for removal in the Phase I process.

Lower combustion temperature in automobile engines reduces the formation of nitrogen oxides. Mass transit reduces automobile use, thereby decreasing nitrogen oxide emissions. Nitrogen oxides produced during high-temperature combustion processes in industry can be removed from smokestack exhausts. The release of nitrogen oxides from cultivated fields to which nitrogen fertilizers were applied is reduced significantly when no-tillage is practiced.

Advanced furnaces and engines burn more cleanly, reducing production of both carbon monoxide and hydrocarbons. Catalytic afterburners, used immediately following combustion, oxidize most unburned gases. The use of properly functioning catalytic converters to treat auto exhaust reduces carbon monoxide and volatile hydrocarbon emissions by about 85% over the life of the car. Careful handling of petroleum and hydrocarbons, such as gasoline, paint thinner, and lighter fluid, reduces air pollution from spills and evaporation.

Figure 20.12　Phase I gasoline vapor recovery. Phase I vapor recovery removes vapors from the empty space above the gasoline in underground gasoline tanks.

Hose B (Vapor Tube)

Hose A (Gasoline Tube)

Vapor

Gasoline

The Clean Air Act

The first air quality legislation in the United States was the Air Pollution Control Act of 1955. However, the **Clean Air Act** of 1970 (with updates in 1977 and 1990) set the standard for modern air quality regulation. This law authorizes the EPA to set limits on the amount of specific air pollutants permitted everywhere in the United States. Individual states are responsible for meeting deadlines for air pollution standards. States may pass stronger pollution controls than the EPA authorizes, but they cannot mandate weaker limits than those stipulated in the Clean Air Act.

The EPA, which oversees the Clean Air Act, has focused on six air pollutants (lead, particulate matter, sulfur dioxide, carbon monoxide, nitrogen oxides, and ozone) and has established maximum acceptable concentrations for each. The most dramatic improvement is in the amount of lead in the atmosphere, which showed a 98% decrease between 1970 and 2006, primarily because of the switch from leaded to unleaded gasoline. Atmospheric levels of the other pollutants are also reduced (**Figure 20.13**). For example, between 1970 and 2006, sulfur dioxide emissions declined 55%. During this same period, the U.S. gross domestic product tripled, vehicle miles travelled nearly tripled, and energy consumption increased 47%.

Although air quality is gradually improving, more than 154 million metric tons (170 million tons) of pollutants are emitted into U.S. air each year. The atmosphere in many urban areas still contains higher levels of pollutants than are recommended on the basis of health standards, and photochemical smog remains a major problem in many metropolitan areas (**Table 20.3**). In EPA language, these cities are classified as nonattainment areas for one or more criteria air pollutants. Nonattainment areas are classified on the basis of how badly polluted they are. This classification ranges from marginal (somewhat polluted and relatively easy to clean up) to extreme (so polluted that it will take many years to clean up the air). The EPA estimates that over 200 million Americans currently live in nonattainment areas.

The Clean Air Act and its amendments required progressively stricter controls of motor vehicle emissions. The provisions of the Clean Air Act Amendments of 1990 include the development of "superclean" cars, which emit lower amounts of nitrogen oxides and hydrocarbons, and the use of cleaner-burning gasoline in the most polluted cities in the United States. These changes were phased in gradually by the year 2000. More recent automobile models do not produce as many pollutants as older models. Yet despite the increasing percentage of newer automobile models on the road, air quality has not improved in some areas of the United States because of the large increase in the number of cars being driven.

The Clean Air Act Amendments of 1990 focus on industrial airborne toxic chemicals in addition to motor vehicle emissions. Between 1970 and 1990, the airborne emissions of only seven toxic chemicals were regulated. In comparison, the Clean Air Act Amendments of 1990 required a 90% reduction in the atmospheric emissions of 189 toxic chemicals by 2003. To comply with this requirement, both small businesses such as dry cleaners and large manufacturers such as chemical companies had to install pollution control equipment if they had not already done so.

Changes to the Clean Air Act in 1997 Evidence that had accumulated since passage of the Clean Air Act Amendments of 1990 indicated that the standards the EPA had originally established for ground-level ozone and particulate matter were not strict enough to protect the health of U.S. citizens. The Clean Air Act Amendments of 1990 limited the emission of PM-10. Because of concern over the potential health effects of microscopic particulate matter, the EPA proposed

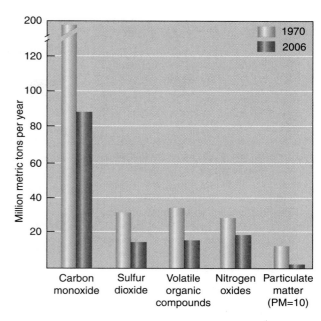

Figure 20.13 Emissions in the United States, 1970 and 2006. Carbon monoxide, sulfur dioxide, volatile organic compounds (many of which are hydrocarbons), particulate matter, and nitrogen oxides showed decreases. PM-10 applies to particles less than or equal to 10 µm (10 micrometers) in size. Since 1990, the EPA has also monitored PM-2.5, which are small particles less than or equal to 2.5 µm. (Air Quality Planning and Standards)

Table 20.3 U.S. Urban Areas with The Worst Air Quality (Ozone Nonattainment Areas), 2002

Extreme

Los Angeles South Coast Air Basin, California

Very Severe

Chicago, Gary and Lake County, Illinois-Indiana
Houston, Galveston, and Brazoria, Texas
Milwaukee and Racine, Wisconsin
New York City, Northern New Jersey, and Long Island, New York-New Jersey-Connecticut
Southeast Desert, California

Severe

Baltimore, Maryland
Philadelphia, Wilmington, Trenton, Pennsylvania-New Jersey-Delaware-Maryland
Sacramento, California
San Joaquin Valley, California
Ventura country (between Santa Barbara and Los Angeles), California

separate standards on the emission of PM-2.5. The EPA also revised ozone standards.

The revised standards provoked an outcry from industries such as the Chemical Manufacturers Association, the American Trucking Association, and other industry and state groups, some of which challenged the new standards in court. The U.S. Supreme Court reaffirmed that the EPA must set these standards on the basis of health considerations, not cost. Counties that do not meet particulate standards by 2010 may lose federal highway funding. (The EPA under the George W. Bush administration has added new regulations to reduce the soot emission of diesel trucks, beginning in 2007, and diesel tractors, bulldozers, locomotives, and other diesel engines by 2012. However, another Bush regulation prolongs the operation of older—and more polluting—coal-fired power plants without requiring them to install the pollution-control equipment mandated by the Clean Air Act.)

Other Ways to Improve Air Quality

Reducing the sulfur content in gasoline from its current average of 330 parts per million (ppm) to 30 ppm or lower would significantly reduce air pollution. (Parts per million is the number of molecules of a particular pollutant found in a million molecules of air, water, or some other material.) Sulfur clogs catalytic converters so that they cannot effectively remove emissions from automobile exhaust. The technology to remove sulfur from gasoline exists, but it would be expensive to implement because many oil refineries would have to be modernized. The American Petroleum Institute estimates that the reduction in sulfur in U.S. gasoline would add $0.03 to $0.06 to the cost of each gallon of gasoline, but the EPA's estimate is much lower, $0.01 to $0.02 per gallon. The EPA says that the overall improvement in the air would be equivalent to removing 50 million cars from U.S. roads each year and would result in a more than $16 billion savings in health benefits. California, a leader in environmental quality, has imposed a sulfur limit in gasoline of 40 ppm, as have Canada and the European Union.

Minivans, sport utility vehicles (SUVs), and light pickup trucks, which account for almost 50% of new passenger vehicles, currently do not have the same federal emissions and efficiency standards as automobiles. Many of these larger vehicles produce more than twice the pollution of a car. In 1999 the federal government issued strict pollution standards for these motor vehicles. The federal regulations called for cutting tailpipe emissions of nitrogen oxides to 0.7 gram per mile in regular automobiles by the 2004 model year; SUVs and minivans had to comply with this requirement by the 2007 model year (see "Meeting the Challenge: Cleaner Cars, Cleaner Fuels").

Many states, such as California, New York, New Jersey, and Connecticut, now require diesel trucks and buses to undergo emissions tests similar to those that these states have required for automobiles for many years. Diesel exhaust, particularly particulate matter, is viewed as a toxic pollutant. Public health advocates maintain that breathing diesel exhaust may contribute to asthma, an increased incidence of lung cancer, and other lung diseases.

REVIEW

1. What air pollutants do electrostatic precipitators and scrubbers remove?
2. What is the U.S. Clean Air Act, and how has it reduced outdoor air pollution?

Ozone Depletion in the Stratosphere

LEARNING OBJECTIVES

- Define *ultraviolet (UV) radiation*.
- Distinguish between tropospheric and stratospheric ozone and describe the importance of the stratospheric ozone layer.
- Define *stratospheric ozone thinning* and relate some of the harmful effects of ozone depletion.
- Define *chlorofluorocarbons* and explain how they and other chemicals attack ozone.

MEETING THE CHALLENGE

Cleaner Cars, Cleaner Fuels

Performance and style have traditionally been higher priorities for automobile manufacturers than reducing environmental pollutants. However, today's carmakers are exploring new ways to reduce pollution and save on fuel use. Carmakers continue to improve traditional gasoline-powered internal combustion engines so that they require less fuel and produce less air pollution. For example, some manufacturers are using cylinder shutdown, in which an eight-cylinder engine runs on four or six cylinders when it does not need the extra power. Other manufacturers are looking at ways to improve fuel efficiency when the vehicle is idling. As good as all of this may sound, we still retain our dependence on CO_2-producing fossil fuels with this technology, so it does not offer a long-term solution to air pollution.

Diesel engines have always been more fuel-efficient than gasoline-powered engines, but they have also produced more air pollution, particularly smog-causing nitrogen oxides and cancer-causing soot. However, carmakers are under pressure to improve the fuel economy of SUVs, minivans, and pickup trucks because the George W. Bush administration requires improvement in fuel mileage every year through 2011. Diesel engines, which offer excellent fuel economy, are a logical choice for these vehicles, provided that they can be cleaned up. Carmakers are now designing advanced diesel engines that produce less nitrogen oxides and soot, but they are not yet available on the market.

Hybrid cars that use a combination of gasoline and electricity are a promising technology that is available today. The hybrid car starts with the gasoline engine but uses the electric motor alone when driving at low speeds or when idling in traffic. At normal driving speeds, both engine and motor contribute power, and when the car is slowing down, the wheels power the electric generator to store electricity in the battery for future use, so the hybrid does not have to be plugged in to be recharged. Hybrids are particularly promising for urban vehicles. In city driving, hybrids get remarkable gas mileage and produce relatively few air pollutants. Hybrid vehicles are relatively expensive, but consumer demand remains large and price is dropping. The federal government and many state governments offer tax deductions for hybrid owners. In 2002, two hybrid models were available in the United States; by 2008, there were twenty-seven models, including SUVs, trucks, and sedans (but not minivans).

Liquid hydrogen is an extremely clean fuel (the only waste product is water). Some car designs have fuel cells that combine stored hydrogen with oxygen from the air to produce electricity. (See Figure 13.13 for a diagram of a hydrogen fuel cell.) Fuel cells do not produce the harmful emissions that conventional gasoline and diesel engines do. While most major auto manufacturers are doing research and development on fuel cells for automobiles, recent prototypes remain expensive, and vehicle range (distance a vehicle can travel on a single load of fuel) remains low.

The widespread adoption of pressurized hydrogen fuel would require an infrastructure of service stations that would allow motorists to fill up with pressurized liquid hydrogen. To get around the need for hydrogen filling stations, some auto manufacturers are experimenting with on-board "reformers" that use gasoline or methanol to produce hydrogen for the fuel cell. Although such "reformer" autos would produce air pollutants, the quantity of emissions would be a fraction of those emitted by today's gasoline- and diesel-powered motor vehicles.

Hydrogen for "nonreformer" fuel cells can be produced from any energy supply; currently, only natural gas- and coal-generated hydrogen are cost-competitive with gasoline. When production of hydrogen from natural gas or coal is taken into account, hydrogen fuel does not produce zero emissions. Other hydrogen production options—including nuclear, wind, and biomass—may become more cost competitive in the future. Honda, General Motors, and Toyota hope to have fuel cell vehicles on the market by around 2010.

Ozone (O_3) is a form of oxygen that is a human-made pollutant in the troposphere but a naturally produced, essential component in the stratosphere, which encircles our planet some 10 to 45 km (6 to 28 mi) above the surface. The relatively high concentrations of ozone in the stratosphere form a layer that shields the surface from much of the **ultraviolet (UV) radiation** coming from the sun (**Figure 20.14**).

Scientists divide UV radiation into three bands: UV-A (with wavelengths of 320 to 400 nm), UV-B (280 to 320 nm), and UV-C (200 to 280 nm). (A nanometer, abbreviated nm, is one billionth of a meter.) The shorter the wavelength, the more energetic and more dangerous UV radiation is. Fortunately, oxygen and ozone in the atmosphere absorb all incoming UV-C (the most lethal wavelengths). The ozone layer absorbs most incoming UV-B radiation. UV-A is not affected by ozone, and most reaches the surface.

The ozone layer over Antarctica thins naturally for a few months each year. In 1985, however, **stratospheric ozone thinning** was first observed to be greater than could be explained by natural causes. This increased thinning, which occurs each September, is commonly referred to as the "ozone hole" (**Figure 20.15**). During the 1990s, the ozone-thinned area continued to grow. By 2000 it had reached the record size of 29.2 million km^2 (11.4 million mi^2). A smaller thinning has also been detected in the stratospheric ozone layer over the Arctic. In addition, world levels of stratospheric ozone have decreased for several decades. According to the National Center for Atmospheric Research, ozone levels over Europe and North America have dropped almost 10% since the 1970s.

ultraviolet radiation: That part of the electromagnetic spectrum with wavelengths just shorter than visible light.

stratospheric ozone thinning: The accelerated destruction of ozone in the stratosphere by human-produced chlorine- and bromine-containing chemicals.

Figure 20.14   Stratospheric ozone layer.

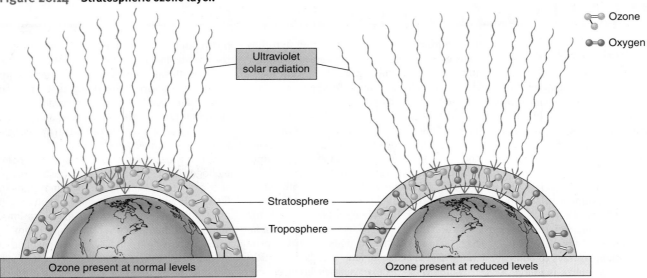

(a) Stratospheric ozone absorbs about 99% of incoming solar ultraviolet (UV) radiation, effectively shielding the surface.

(b) When stratospheric ozone is present at reduced levels, more high-energy UV radiation penetrates the atmosphere to the surface, where its presence harms organisms.

The Causes of Ozone Depletion

Both chlorine- and bromine-containing substances catalyze ozone destruction. The primary chemicals responsible for release of chlorine in the stratosphere, thus causing ozone depletion, are **chlorofluorocarbons (CFCs)**. Chlorofluorocarbons were used as propellants for aerosol cans, as coolants in air conditioners and refrigerators (for example, Freon), as foam-blowing agents for insulation and packaging, and as solvents.

Halons, methyl bromide, methyl chloroform, and carbon tetrachloride also release chlorine or bromine and thus lead to ozone depletion. Halons are used as fire retardants; methyl bromide is a widely used pesticide effective against a variety of pests, such as nematodes (parasitic worms), fungi, and weed seeds. Methyl chloroform and carbon tetrachloride are industrial solvents.

The evidence linking CFCs and other human-made compounds to stratospheric ozone destruction includes laboratory measurements, atmospheric observations, and calculations by computer models. In 1995 the Nobel Prize in chemistry was awarded to **Sherwood Rowland, Mario Molina,** and **Paul Crutzen,** the scientists who first explained the connection between the thinning ozone layer and chemicals such as CFCs. This Nobel Prize was the first one ever given for work in environmental science.

CFCs and other chlorine-containing compounds released at ground level slowly drift up to the stratosphere, where UV radiation breaks them down, releasing chlorine. Similarly, the breakdown of halons and methyl bromide releases bromine. The hole in the ozone layer that was discovered over Antarctica occurs annually between September and November (spring in the southern hemisphere). At this time, two important conditions are present: Sunlight returns to the polar region, and the **circumpolar vortex** develops—a mass of cold air that circulates around the southern polar region and isolates it from the warmer air on the rest of the planet.

The cold air causes polar stratospheric clouds to form; these clouds contain ice crystals to which chlorine and bromine adhere, making them available to destroy ozone. The sunlight catalyzes the chemical reaction in which chlorine or bromine breaks ozone molecules apart, converting them into oxygen molecules. The chemical reaction that destroys ozone does not alter the chlorine or bromine, and one chlorine or bromine atom can break down many thousands of ozone molecules. The chlorine and bromine remain in the stratosphere for many years. When the circumpolar vortex breaks up, the ozone-depleted air spreads northward, diluting ozone levels in the stratosphere over South America, New Zealand, and Australia.

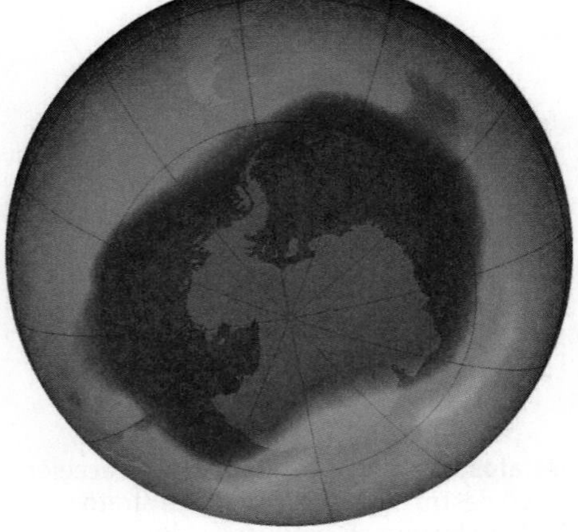

chlorofluorocarbons: Human-made organic compounds of carbon, chlorine, and fluorine that had many industrial and commercial applications but were banned because they attack the stratospheric ozone layer.

Figure 20.15   Ozone thinning.
A computer-generated image of part of the Southern Hemisphere, taken on September 24, 2006, reveals the ozone thinning (bluish purple area over Antarctica). The ozone-thin area is not stationary but moves about as a result of air currents. (Courtesy of NASA)

The Effects of Ozone Depletion

With depletion of the ozone layer, higher levels of UV radiation reach Earth's surface. A study conducted in Toronto from 1989 to 1993 showed that wintertime levels of UV-B increased more than 5% each year as a result of lower ozone levels. A study in New Zealand showed that summertime levels of peak UV-B radiation were 12% higher during the 1998–1999 summer than during similar periods a decade earlier.

Excessive exposure to UV radiation is linked to several health problems in humans, including eye cataracts, skin cancer, and weakened immunity. The lens of the eye contains transparent proteins that are replaced at a slow rate. Exposure to excessive UV radiation damages these proteins, and over time, the damage accumulates so that the lens becomes cloudy, forming a cataract. Cataracts can be cured by surgery, but millions of people in developing nations cannot afford the operation and so remain partially or totally blind.

Excessive, chronic exposure to UV radiation causes most cases of skin cancer. Ultraviolet B radiation causes mutations, or changes, in the deoxyribonucleic acid (DNA) residing in skin cells. Such changes gradually accumulate and may lead to skin cancer. Globally, about 2.2 million cases of skin cancer occur each year. Malignant melanoma, the most dangerous type of skin cancer, is increasing faster than any other type of cancer (**Figure 20.16**). Some forms of malignant melanoma spread rapidly through the body and may cause death a few months after diagnosis.

Scientists are concerned that increased levels of UV radiation may disrupt ecosystems. For example, the productivity of Antarctic phytoplankton, the microscopic drifting algae that are the base of the Antarctic food web, has declined from increased exposure to UV radiation. Research shows that surface UV-B inhibits photosynthesis in these phytoplankton. Direct damage to natural populations of Antarctic fish is documented: Increased DNA mutations in ice-fish eggs and larvae (young fish) were matched to increased levels of UV radiation. Researchers are currently studying whether these mutations lessen the animals' ability to survive. Because organisms live in interdependent systems, a negative effect on one species has ramifications throughout the ecosystem.

High levels of UV radiation may also damage crops and forests. Plants interact with many other species in both natural ecosystems and agricultural ecosystems, and the impacts of UV radiation on each of these organisms affect plants indirectly. Exposure to higher levels of UV-B radiation may increase wheat yields by inhibiting fungi that cause disease in wheat. On the other hand, exposure to higher levels of UV radiation decreases cucumber yields by making them more susceptible to disease.

Facilitating the Recovery of the Ozone Layer

In 1978 the United States, the world's largest user of CFCs, banned the use of CFC propellants in products such as antiperspirants and hair sprays. Although this ban was a step in the right direction, it did not solve the problem. Most nations did not follow suit, and besides, propellants represented only the tip of the iceberg in terms of CFC use.

In 1987 representatives from many countries met in Montreal to sign the **Montreal Protocol,** an agreement that originally stipulated a 50% reduction of CFC production by 1998. Despite this effort, the environmental news about CFCs continually worsened in the early 1990s. After scientists reported that decreases in stratospheric ozone occurred over the heavily populated midlatitudes of the Northern Hemisphere in all seasons, the Montreal Protocol was modified to include stricter measures to limit CFC production.

Industrial companies that manufacture CFCs quickly developed substitutes, such as hydrofluorocarbons (HFCs) and hydrochlorofluorocarbons (HCFCs) (see "Meeting the Challenge: Business Leadership in the Phase-Out of CFCs"). HFCs do not attack ozone, although they are potent greenhouse gases. HCFCs attack ozone but are not as destructive as the chemicals they are replacing. Although production of HFCs and HCFCs has increased rapidly, these chemicals are transitional substances that will be used only until industry develops substitutes.

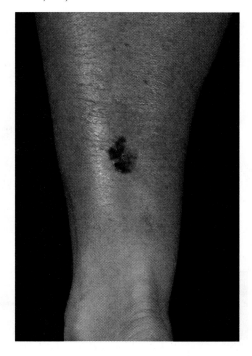

Figure 20.16   Malignant melanoma on the back of the leg. These tumors of pigmented cells sometimes, but not always, arise from preexisting moles. Early diagnosis is important because this form of skin cancer spreads to other parts of the body and can be fatal. (Custom Medical Stock Photo, Inc.)

Production of CFCs, carbon tetrachloride, and methyl chloroform was completely phased out in the United States and other highly developed countries in 1996, except for a relatively small amount exported to developing countries. Developing countries phased out CFC use in 2005. Methyl bromide was supposed to be phased out by 2005 in highly developed countries, which are responsible for 80% of the global use of that chemical, but they have been given extra time to adopt effective substitutes. HCFCs will be phased out in 2030.

Satellite measurements taken in 1997 provided the first evidence that the levels of ozone-depleting chemicals were starting to decline in the stratosphere. In the early 2000s, the first signs of recovery of the ozone layer were evident: Measurement of the rate of stratospheric ozone depletion indicated that it was declining.

Two chemicals—CFC-12 and halon-1211—may have increased and still represent a threat to ozone recovery. Although highly developed countries no longer manufacture CFC-12, it continues to leak into the atmosphere from old refrigerators and vehicle air conditioners discarded in those countries. Moreover, developing countries such as China, India, and Mexico have increased their production of CFC-12. In contrast, halon use has been phased out worldwide since 2006. An international fund, the Montreal Multilateral Fund, is available to help developing countries during their transition from ozone-depleting chemicals to safer alternatives.

Unfortunately, CFCs are extremely stable, and those being used today probably will continue to deplete stratospheric ozone for at least 50 years. Scientists expect human-exacerbated ozone thinning to reappear over Antarctica each year, although the

MEETING THE CHALLENGE

Business Leadership in the Phase-Out of CFCs

As evidence began to accumulate linking chlorofluorocarbons to the thinning of the ozone layer, some business leaders predicted that the economy would suffer and our standard of living would decline when CFCs were banned. Other companies viewed the impending phase-out of CFCs as an opportunity—a chance to achieve a competitive advantage by being one of the first to adopt environmentally friendly products. These companies moved quickly and voluntarily to restrict their use of CFCs. Let us examine how three companies met the challenge: McDonald's, Nortel (formerly Northern Telecom), and York International. We then discuss how the chemical industry worked to develop safe, effective alternatives to CFCs.

Amid growing public concern over the effect of CFCs on the ozone layer, a group of schoolchildren petitioned McDonald's to stop using plastic-foam food containers made with CFCs as the foam-blowing agent. In 1987 McDonald's told its suppliers of disposable foam packaging that they had 18 months to come up with alternatives. McDonald's announcement not only generated a great deal of positive publicity for the company but also galvanized the foam packaging industry, which voluntarily ended all use of CFCs in

1988. Today all foam packaging made in the United States is ozone-friendly.

Nortel is a major electronics firm that used CFCs to clean circuit boards and other electronic components. In 1988 the company's engineers met with government officials to develop a planned phaseout of CFCs. They decided to eliminate all use of CFCs in three years ("free in three"). Nortel tested possible CFC substitutes and tried to reduce CFC consumption at its factories, which competed with one another to be the first to "get to zero." Meanwhile, Nortel worked within the electronics industry to modify the production process so that the need for CFCs and other cleaning solvents was eliminated altogether. Nortel's leadership not only made it one of the first electronics companies to stop using CFCs, but gave it unanticipated benefits because the new manufacturing processes are cheaper and more effective.

Manufacturers of refrigeration and air conditioning systems, including "chillers" for larger buildings, were initially skeptical that they could eliminate CFCs, which were used as refrigerants, without reducing energy efficiency. Nonetheless, the industry began an aggressive search for alternatives to CFCs. York International was the first company in the chiller industry to develop an alternative, HCFC-123. Other companies followed suit and stopped

selling CFC-based chillers in the United States by 1993. Many existing CFC chillers in buildings have been retrofitted or replaced by new chillers, which are more energy-efficient and save owners money in electricity costs. The chiller story is not an unqualified success because most new chillers still rely on hydrofluorocarbons or hydrochlorofluorocarbons. Carrier was the first manufacturer to offer a chlorine-free refrigerant (Puron) in its air conditioners, beginning in 1996.

The international chemical industry worked hard to ensure that deadlines in the global phaseout of CFCs could be met. Seventeen major chemical companies collectively established two programs, the Alternative Fluorocarbons Environmental Acceptability Study (AFEAS) and the Program for Alternative Fluorocarbon Toxicity Testing (PAFT). AFEAS was created to provide data on the potential environmental effects of CFC alternatives, and PAFT, to provide information about potential effects on human health. Both programs relied on international cooperation among scientists from academic institutions, government research programs, and chemical companies. This collaborative effort enabled a more rapid phase-out of CFCs than would usually be the case when new chemicals have to be subjected to environmental and toxicity testing.

area and degree of thinning will gradually decline over time, until full recovery takes place sometime after 2050.

REVIEW

1. What is the stratospheric ozone layer? How does it protect life on Earth?
2. What is stratospheric ozone thinning? How does it occur?
3. What is the Montreal Protocol?

 ## Acid Deposition

LEARNING OBJECTIVES

- Describe the pH scale.
- Explain how acid deposition develops and relate some of the effects of acid deposition.
- Define *forest decline* and relate its possible causes.

What do fishless lakes in the Adirondack Mountains, recently damaged Mayan ruins in southern Mexico, and dead trees in the Czech Republic have in common? The answer is that these damages are the result of acid precipitation or, more properly, **acid deposition**. It includes sulfuric and nitric acids in precipitation (**wet deposition**) as well as dry, sulfuric acid- and nitric acid-containing particles that settle out of the air (**dry deposition**).

Robert Angus Smith, a British chemist, coined the term *acid rain* in 1872 after he noticed that buildings in areas with heavy industrial activity were being worn away by rain. Acid precipitation, including acid rain, sleet, snow, and fog, poses a serious threat to the environment. Until recently, industrialized countries in the Northern Hemisphere had been hurt the most, especially the Scandinavian countries, Central Europe, Russia, and North America. More recently, largely due to high-sulfur coal, acid rain has become a major concern in China, where SO_2 releases in 2010 are expected to be three times those of 1990.

acid deposition: Sulfur dioxide and nitrogen dioxide emissions react with water vapor in the atmosphere to form acids that return to the surface as either dry or wet deposition.

Measuring Acidity

The **pH scale**, which runs from 0 to 14, expresses the relative degree of acidity or basicity of a substance (**Figure 20.17**; also see Appendix I). A pH of 7 is neither acidic nor basic, whereas a pH less than 7 indicates an acidic solution. The pH scale is logarithmic, so a solution with a pH of 6 is 10 times more acidic than a solution with a pH of 7. Similarly, a solution with a pH of 5 is 10 times more acidic than a solution with a pH of 6 and 100 times more acidic than a solution with a pH of 7. A solution with a pH greater than 7 is basic, or alkaline.

For purposes of comparison, distilled water has a pH of 7, tomato juice has a pH of 4, vinegar has a pH of 3, and lemon juice has a pH of 2. Normally, rainfall is slightly acidic (with a pH from 5 to 6) because CO_2 and other naturally occurring compounds in the air dissolve in rainwater, forming dilute acids. However, the pH of precipitation in the northeastern United States averages 4 and is often 3 or even lower.

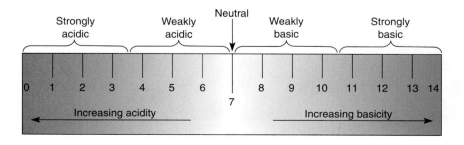

Figure 20.17   The pH scale. This scale, which includes values between 0 and 14, expresses the concentrations of acidic and basic solutions. Pure water, which is neutral, has a pH of 7. Each decrease of one pH unit represents a tenfold increase in acidity.

How Acid Deposition Develops

Acid deposition occurs when sulfur dioxide and nitrogen oxides are released into the atmosphere, combine with moisture to form acids, and then are deposited on land through rain, snow, or condensate (dew) (**Figure 20.18**). Motor vehicles are a major source of nitrogen oxides. Coal-burning power plants, large smelters, and industrial boilers are the main sources of sulfur dioxide emissions and produce substantial amounts of nitrogen oxides as well. Sulfur dioxide and nitrogen oxides, released into the air from tall smokestacks, are carried long distances by winds. Tall smokestacks were an early attempt to control local air pollution—under the premise that "the solution to pollution is dilution." Tall smokestacks allow England to "export" its acid deposition problem to the Scandinavian countries and the midwestern United States to "export" its acid emissions to New England and Canada.

During their stay in the atmosphere, sulfur dioxide and nitrogen oxides react with water to produce dilute solutions of sulfuric acid (H_2SO_4), nitric acid (HNO_3), and nitrous acid (HNO_2). Acid deposition returns these acids to the ground, causing the pH of surface waters and soil to decrease.

The Effects of Acid Deposition

Acid deposition affects living and non-living things. It corrodes metals and building materials, damaging, for example, the Washington Monument in Washington, D.C., historic sites in Venice and Rome, and ancient Mayan ruins in southern Mexico.

The link between acid deposition and declining aquatic animal populations is well established. Field investigations were conducted by the Adirondack Lakes Survey Corporation (ALSC), a nonprofit group that works with various universities, the Environmental Protection Agency (EPA), and various state and local organizations. Of the 1469 Adirondack lakes and ponds examined, 352 had pH values of 5.0 or less, and 346 of those acidified lakes and ponds had no fish populations. Toxic metals such as aluminum dissolve in acidic lakes and streams and enter food webs. This increased concentration of toxic metals may explain how acidic water adversely affects fishes.

Although impacts on fish have received the lion's share of attention regarding acid deposition, other animals are also harmed. Several studies have found that birds living in areas with pronounced acid deposition were much more likely to lay eggs with thin, fragile shells that break or dry out before the chicks hatch. The inability to produce

Figure 20.18 **Acid deposition.** Sulfur dioxide and nitrogen oxide emissions react with water vapor in the atmosphere to form acids that return to the surface as either dry or wet deposition.

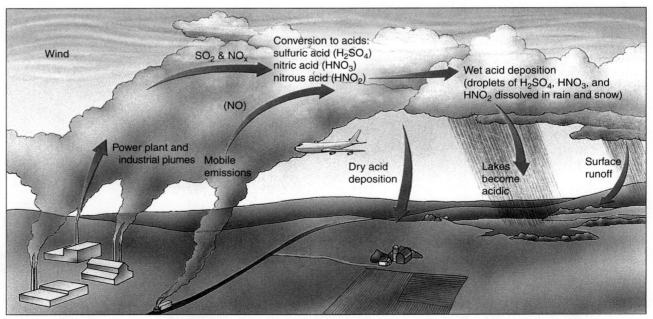

strong eggshells was attributed to reduced calcium in the birds' diets. Calcium is unavailable because in acidic soils it becomes soluble and is washed away, with little left for plant roots to absorb. A smaller amount of calcium in plant tissues means a smaller amount of calcium in the insects and snails that eat the plants. Less calcium is therefore available to the birds that eat these insects and snails.

Acid Deposition and Forest Decline   Forest surveys in the Black Forest of southwestern Germany indicate that up to 50% of trees in the areas surveyed are dead or severely damaged. (Monitoring the loss of leaves and needles from trees determines the damage.) The same is true for trees in many other European forests. More than half of the red spruce trees in the mountains of the northeastern United States have died since the mid-1970s, and sugar maples in eastern Canada and the United States are dying. Beginning in the late 1990s, the Northern Hardwood Damage Survey has observed and mapped tree death at higher elevations of the Appalachian Mountains from Georgia to Maine.

Many living trees exhibit symptoms of **forest decline.** The general symptoms of forest decline are reduced vigor and growth, but some plants exhibit specific symptoms, such as yellowing of needles in conifers. Forest decline is more pronounced at higher elevations, possibly because most trees growing at high elevations are at the limits of their normal range and are less vigorous and more susceptible to stressors of any type.

Many factors interact to decrease the health of trees (**Figure 20.19**), and no single factor accounts for the recent instances of forest decline. Although acid deposition correlates well with areas experiencing tree damage, it is only partly responsible. Several other human-induced air pollutants are implicated, including tropospheric (surface-level) ozone and toxic heavy metals such as lead, cadmium, and copper. Power plants, ore smelters, refineries, and motor vehicles produce these pollutants in addition to the sulfur and nitrogen oxides that interact to form acid deposition. Insects and weather factors such as drought and severe winters (cold and wind can injure susceptible plants) may also be important.

To complicate matters further, the actual causes of forest decline may vary from one tree species to another and from one location to another. Thus, forest decline appears to result from the combination of multiple stressors—acid deposition, tropospheric ozone, UV radiation (which is more intense at higher altitudes), insect attack, drought, and so on. When one or more stressors weaken a tree, then an additional stressor, such as air pollution, may be decisive in causing its death.

One way in which acid deposition harms plants is well established: Acid deposition alters the chemistry of soils, which affects the development of plant roots as well as their uptake of dissolved minerals and water from soil. Essential plant minerals such as calcium and potassium readily wash out of acidic soil, whereas others, such as nitrogen, become available in larger amounts. Heavy metals such as manganese and aluminum dissolve in acidic soil water, becoming available for absorption in toxic amounts. A study completed in 1989 in Central Europe, which has experienced greater forest damage than North America, found a strong correlation between forest damage and soil chemistry altered by acid deposition.

The Politics of Acid Deposition

One reason acid deposition is so hard to combat is that it does not occur only in the locations where the gases that cause it are emitted. Acid deposition does not recognize borders between states or countries; it is entirely possible for sulfur and nitrogen oxides released in one spot to return to the ground hundreds of kilometers from their source.

The United States has wrestled with this issue. Several states in the Midwest and East—Illinois, Indiana, Missouri, Ohio, Pennsylvania, Tennessee, and West Virginia—produce between 50 and 75% of the acid deposition that contaminates New England and southeastern Canada. When legislation was formulated to deal with the problem, arguments ensued about who should pay for the installation of expensive air pollution devices to reduce emissions of sulfur and nitrogen oxides. Should the states emitting

Figure 20.19   Forest decline

(a) These trees, photographed in the Black Forest, Germany, show signs of forest decline. (© Hans Reinhard/Bruce Coleman, Inc.)

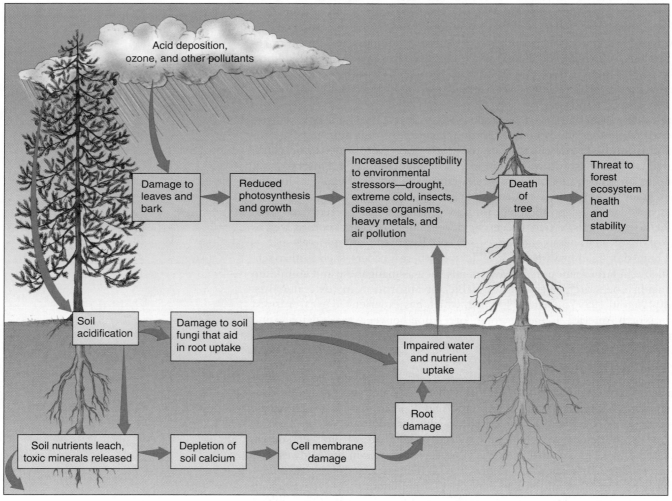

(b) Acid deposition is one of several stressors that interact, contributing to the decline and death of trees. Acid deposition increases soil acidity, causing certain essential mineral ions, such as calcium, to leach out of the soil.

the gases be required to pay all the expenses to clean up the air, or should some of the cost be absorbed by the areas that stand to benefit most from a reduction in pollution?

Pollution abatement issues are complex within one country but are magnified even more in international disputes. England uses its large reserves of coal to generate electricity. Gases from coal-burning power plants in England move eastward with prevailing winds and return to the surface as acid deposition in Sweden and Norway. Similarly, emissions from mainland China produce acid deposition in Japan, Taiwan, and North and South Korea.

Facilitating Recovery from Acid Deposition

The basic concept of acid deposition control is straightforward: reducing emissions of sulfur dioxide and nitrogen oxides curbs acid deposition. Simply stated, if sulfur dioxide and nitrogen oxides are not released into the atmosphere, they cannot come down as acid deposition. Installation of scrubbers in the smokestacks of coal-fired power plants (see Figure 20.11) and use of clean-coal technologies to burn coal without excessive emissions effectively diminish acid deposition (see Chapter 11). In turn, a decrease in acid deposition prevents surface waters and soil from becoming more acidic than they already are.

Despite the fact that the United States, Canada, and many European countries have reduced emissions of sulfur, acid precipitation remains a serious problem. Acidified forests and bodies of water have not recovered as quickly as hoped. Trees in the U.S. Forest Service's Hubbard Brook Experimental Forest in New Hampshire, an area damaged by acid deposition, have grown little since 1987, when emissions started to decline. Many northeastern streams and lakes, such as those in New York's Adirondack Mountains, remain acidic. A primary reason for the slow recovery is probably that the past 30 or more years of acid rain have profoundly altered soil chemistry in many areas. Essential plant minerals such as calcium and magnesium have leached from forest and lake soils. Because soils take hundreds or even thousands of years to develop, it may be decades or centuries before they recover from the effects of acid rain.

Many scientists are convinced that ecosystems will not recover from acid rain damage until substantial reductions in nitrogen oxide emissions occur, a trend which seems to be under way. Nitrogen oxide emissions are harder to control than sulfur dioxide emissions because motor vehicles produce a substantial portion of nitrogen oxides. Engine improvements may reduce nitrogen oxide emissions, but as the population continues to grow, the engineering gains may be offset by an increase in the number of motor vehicles. Dramatic cuts in nitrogen oxide emissions will require a reduction in high-temperature energy generation, especially in gasoline and diesel engines.

REVIEW

1. What is acid deposition?
2. What causes acid deposition?
3. Is forest decline related to acid deposition? Explain your answer.

Air Pollution Around the World

LEARNING OBJECTIVES

- Explain why air pollution is generally worse in developing countries than in highly developed countries.
- Describe the global distillation effect and tell where it commonly occurs.

As developing nations become more industrialized, they produce more air pollution. The leaders of most developing countries believe they must become industrialized rapidly to compete economically with highly developed countries. Environmental quality is usually a low priority in the race to develop. Outdated technologies are often adopted, and air pollution laws, where they exist, are not enforced. Thus, air quality is deteriorating rapidly in many developing nations.

Shenyang and neighboring cities in China have so many smokestacks belching coal smoke (coal is burned to heat many homes) that residents see the sun only a few weeks of the year (see Figure 20.8). The rest of the time residents are choked in a haze of orange-colored coal dust. In other developing countries, such as India and Nepal, biomass (wood or animal dung) is burned indoors, often in stoves with little or no outside ventilation, thereby exposing residents to serious indoor air pollution. Scientists have determined that one of the principal causes of acute respiratory infections, a serious worldwide health threat, is exposure to the pollutants in biomass fuels when they are burned indoors.

The growing number of automobiles in developing countries contributes to air pollution, particularly in urban areas. Many vehicles in these countries are 10 or more years old and have no pollution-control devices. Motor vehicles produce about 60 to 70% of the air pollutants in urban areas of Central America, and 50 to 60% in urban areas of India. Since the mid-1990s the most rapid proliferation of motor vehicles worldwide occurred in Latin America, Asia, and Eastern Europe.

Lead pollution from heavily leaded gasoline is an especially serious problem in developing nations. The gasoline refineries in these countries are generally not equipped to remove lead from gasoline. (The same situation occurred in the United States until federal law mandated that the refineries upgrade their equipment.) In Cairo, children's blood lead levels are more than two times higher than the level considered at-risk in the United States. Lead can retard children's growth and cause brain damage.

According to a World Health Organization study, the five worst cities in the world in terms of exposing children to air pollution are Beijing, China; Mexico City, Mexico; Shanghai, China; Tehran, Iran; and Calcutta, India. The study determined that respiratory disease is now the leading cause of death for children worldwide. More than 80% of these deaths occur in young children (under the age of five) who live in cities in developing countries.

global distillation effect: The process whereby volatile chemicals evaporate from land as far away as the tropics and are carried by air currents to higher latitudes, where they condense and fall to the ground.

Figure 20.20 **Global distillation effect.**
(a) Long-distance atmospheric transport occurs in part because evaporation exceeds deposition onto the land and ocean at low latitudes. (b) The more volatile the chemical, the farther it travels before being deposited onto the land and ocean. (c) Some chemicals move to higher latitudes by repeatedly evaporating and settling ("leapfrogging"), sometimes taking several decades before being permanently deposited.

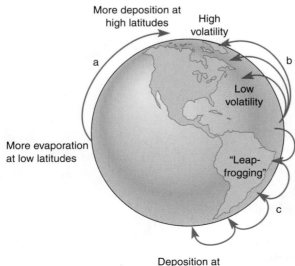

More deposition at high latitudes

High volatility

a

b

Low volatility

More evaporation at low latitudes

"Leap-frogging"

c

Deposition at high latitudes

Long-Distance Transport of Air Pollution

Certain hazardous air pollutants are distributed globally by atmospheric transport in a process known as the **global distillation effect**. The air toxics involved in the global distillation effect are persistent compounds, such as polychlorinated biphenyls (PCBs, industrial compounds) and dichlorodiphenyltrichloroethane (DDT, a pesticide), that do not readily break down and so accumulate in the environment. Many of these persistent compounds are restricted or even banned by many countries. Yet because they are volatile, they move through the atmosphere. The pathway of movement is generally from warmer developing countries, where they are still used, to colder highly developed nations, where they condense and are deposited on land and surface water (**Figure 20.20**).

Many industrialized countries remain highly contaminated by persistent compounds despite their restricted use. The effect is more pronounced where it is colder—that is, at higher latitudes and higher elevations. Dangerous levels of certain persistent toxic compounds have been measured in the Yukon (northwestern Canada) and in other pristine arctic regions. These chemicals enter food webs and become concentrated in the body fat of animals at the top of the food chain (see the discussion of biological magnification in Chapter 7). Fishes, seals, polar bears, and arctic people such as the Inuit are particularly vulnerable. When an Inuit consumes a single bite of raw whale skin, she ingests more PCBs than scientists think should be consumed in a week. The level of PCBs in the breast milk of that Inuit woman is five times higher than that in the milk of women who live in southern Canada.

In 1996 concern over protecting the Arctic led to the formation of the Arctic Council, consisting of Canada, the United States, Russia, Finland, Norway, Sweden, Denmark, and Iceland. Since then scientists from the Arctic Monitoring and Assessment Program, based in Norway, have conducted additional environmental studies that reinforce earlier studies. The Stockholm Convention on Persistent Organic Pollutants was adopted in 2001, and went into effect in 2004. Its goal, according to the U.N. Environment Program, is to phase out the use of at least 12 persistent toxic chemicals, including PCBs, dioxins and furans (chemical contaminants), and DDT and eight other pesticides. (For additional discussion, see the section on the global ban of persistent organic pollutants in Chapter 23.)

Movement of Air Pollution over the Ocean   The Indian Ocean has generally been considered one of the world's cleanest areas because the countries that surround it are not heavily industrialized. During a six-week study in the winter of 1999, scientists from the National Science Foundation and the Scripps Institution of Oceanography reported that a large portion of the Indian Ocean was covered by hazy, polluted air. The pollution extended over 9.5 million km^2 (3.8 million mi^2), an area the size of the United States. It is thought that prevailing winds during the winter monsoon blow the pollution, which includes particulate matter and sulfur droplets, from the Indian subcontinent, China, and Southeast Asia. The hazy area is expected to increase as industry develops in these areas.

Examples of pollution traveling from one continent to another have not been well documented until recently. Certain atmospheric conditions (that is, a low-pressure system over the Aleutians and a high-pressure system near Hawaii) cause a strong wind toward North America that allows air pollution from Asia to cross the Pacific Ocean. In 1997 scientists from the University of Washington detected carbon monoxide, particulate matter, and PANs in the atmosphere over the western United States. Computer models suggested that these pollutants were produced in Asia six days earlier. In 1998 more definitive evidence of pollution from Asia affecting air quality in North America occurred when a major dust storm in China produced a visible cloud of particulate matter that was tracked by satellite across the Pacific Ocean. The polluted air was analyzed when it reached the United States a few days later and found to contain arsenic, copper, lead, and zinc from ore smelters in Manchuria.

Air pollution generated in North America may travel across the Atlantic Ocean to Europe. However, definitive evidence of a North American–European connection has not been demonstrated to date.

REVIEW

1. Is air pollution worse in highly developed nations or in developing countries? Why?

2. What is the global distillation effect? What kinds of air pollutants are involved in the global distillation effect?

Indoor Air Pollution

LEARNING OBJECTIVES

- Summarize the sick building syndrome.
- Explain why tobacco smoke and radon are considered major indoor air pollutants.

The air in enclosed places such as automobiles, homes, schools, and offices may have significantly higher levels of air pollutants than the air outdoors. In congested traffic, levels of harmful pollutants such as carbon monoxide, benzene, and airborne lead may be several times higher inside an automobile than in the air immediately outside. The concentrations of certain indoor air pollutants may be two to five times greater—and sometimes more than 100 times—than outdoors. Indoor pollution is of particular concern to urban residents because they may spend as much as 90% to 95% of their time indoors. The EPA considers indoor air pollution as one of the top five environmental health risks in the United States.

Because illnesses caused by indoor air pollution usually resemble common ailments such as colds, influenza, or upset stomachs, they are often not recognized. The most common contaminants of indoor air are radon (discussed shortly), cigarette smoke, carbon monoxide, nitrogen dioxide (from gas stoves), formaldehyde (from carpeting, fabrics, and furniture), household pesticides, lead, cleaning solvents, ozone (from photocopiers), and asbestos (**Figure 20.21**). In addition, the reaction of indoor ozone, generally present at lower levels than outdoors, with volatile chemicals in air fresheners, aromatherapy candles, and cleaning agents forms secondary air pollutants such as formaldehyde.

Viruses, bacteria, fungi (yeasts, molds, and mildews), dust mites, pollen, and other organisms or their toxic parts are important forms of indoor air pollution often found in heating, air conditioning, and ventilation ducts. Excessive indoor dampness exacerbates indoor microbial growth (particularly fungal growth), dust mite populations, and cockroach and rodent infestations. A 2004 report by the U.S. Institute of Medicine said that the presence of mold in a damp indoor environment is linked to upper respiratory tract (nose and throat) symptoms, including wheezing and coughing, and to asthma symptoms in people who already have asthma. There was also suggestive evidence of an association between molds, a damp indoor environment, and illness in the lower respiratory tracts in otherwise healthy children.

Health officials are paying increasing attention to the **sick building syndrome.** The Labor Department estimates that more than 20 million employees are exposed to

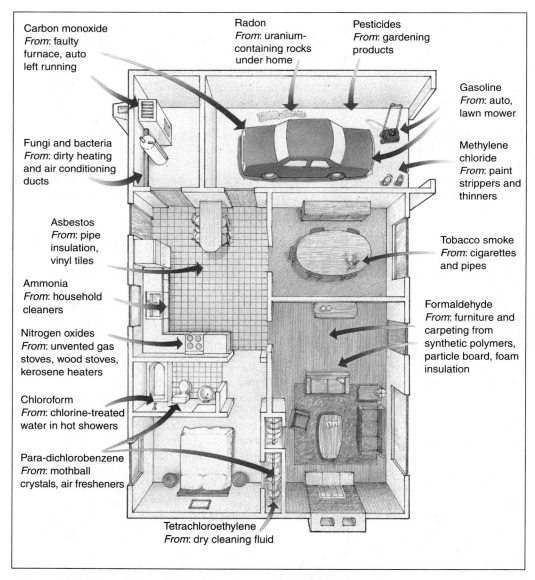

Carbon monoxide
From: faulty furnace, auto left running

Radon
From: uranium-containing rocks under home

Pesticides
From: gardening products

Gasoline
From: auto, lawn mower

Fungi and bacteria
From: dirty heating and air conditioning ducts

Methylene chloride
From: paint strippers and thinners

Asbestos
From: pipe insulation, vinyl tiles

Tobacco smoke
From: cigarettes and pipes

Ammonia
From: household cleaners

Formaldehyde
From: furniture and carpeting from synthetic polymers, particle board, foam insulation

Nitrogen oxides
From: unvented gas stoves, wood stoves, kerosene heaters

Chloroform
From: chlorine-treated water in hot showers

Para-dichlorobenzene
From: mothball crystals, air fresheners

Tetrachloroethylene
From: dry cleaning fluid

Figure 20.21   **Sources of household air pollution.** Homes may contain higher levels of toxic pollutants than outside air, even near polluted industrial sites.

health risks from indoor air pollution. The EPA estimates that the annual medical costs for treating the health effects of indoor air pollution in the United States exceed $1 billion. When lost work time and diminished productivity are added to healthcare costs, the total annual cost to the economy may be as much as $50 billion. Fortunately, many building problems are relatively inexpensive to detect and alleviate. For example, carbon monoxide detectors, which cost about $25 to $50, should be installed in bedrooms as well as rooms with fireplaces, furnaces, and gas appliances.

Indoor Air Pollution and the Asthma Epidemic

Asthma was considered a rare disease until the middle of the 20th century, and it is far more common in industrialized nations than in developing countries. Since 1970 the number of people in the United States who suffer from asthma has doubled, to more than 15 million; nine million asthma sufferers are children. Health officials are worried by this trend, which is due in part to indoor air pollution. Indoor exposure to different air pollutants contributes to the development and exacerbation of asthma. Exactly what pollutant(s) is/are causing the increase in asthma is unknown, although some evidence suggests that exposure to allergens (substances that stimulate an allergic reaction) such as dust mites and cockroach feces is a major cause.

Indoor Tobacco Smoke

Smoking, which causes serious diseases such as lung cancer, emphysema, and heart disease, is responsible for the premature deaths of nearly half a million people in the United States each year. Cigarette smoking annually causes about 120,000 of the 140,000 deaths from lung cancer in the United States. Smoking also contributes to heart attacks, strokes, male impotence, and cancers of the bladder, mouth, throat, pancreas, kidney, stomach, voice box, and esophagus. It also causes substantial property damage through fires, burns, and smoke odor and discoloration.

Cigarette smoke is a mixture of air pollutants that includes hydrocarbons, carbon dioxide, carbon monoxide, particulate matter, cyanide, and a small amount of radioactive materials that come from the fertilizer used to grow the tobacco plants. Smokers exhale tobacco smoke into the air we all breathe. Passive smoking, which is the non-smokers' chronic breathing of smoke from cigarette smokers, also increases the risk of cancer, especially in business settings (bars, casinos, restaurants) and homes. For this reason, it has been banned as a workplace hazard in many locations. Passive smokers suffer more cancer, respiratory infections, allergies, and other chronic respiratory diseases than other nonsmokers. Passive smoking is particularly harmful to infants and young children, pregnant women, the elderly, and people with chronic lung disease. When parents of infants smoke, the infant has double the chance of pneumonia or bronchitis in its first year of life. Smoking during pregnancy adversely affects fetal development, leading to lower birth weights and smaller head circumferences.

A worldwide trend is that more total people in developing countries are smoking, even while fewer people in highly developed nations are smoking. A poll in the mid-1990s found that about 25% of U.S. adults were currently smoking compared with a peak of 41% in the mid-1970s. Smoking has also declined in Japan, and most European countries. However, more people are taking up the habit in Brazil, Pakistan, and many other developing nations. Tobacco companies in the United States promote smoking abroad, and a substantial portion of our tobacco crop is exported. Cigarette sales in developing countries have increased by 80% since 1990.

The World Health Organization (WHO) estimates that, worldwide, over five million people die each year of smoking-related causes, and it wants a global ban on tobacco advertising. To meet this goal, WHO developed the Framework Convention on Tobacco Control, which calls for a ban on cigarette advertising, higher taxes on tobacco products, and restrictions on smoking in public places. This treaty went into effect for signatory nations in 2005. (The United States had not ratified this treaty as we went to press.)

Within the United States, Canada, and other highly developed countries, bans on smoking in many public places—including government buildings, restaurants and airplanes—have substantially reduced both smoking and exposure to smoke. Although fewer U.S. citizens are smoking, certain groups in our society still have high numbers of tobacco addicts, including certain minority groups and those with the least education. A need exists to continue educating these groups, as well as all young people (more than one million U.S. children and teenagers take up smoking each year), about the dangers of smoking before they become addicted.

Radon

Radon is a serious indoor air pollutant in many places in highly developed countries. Radon seeps through the ground and enters buildings, where it sometimes accumulates to dangerous levels (**Figure 20.22**). Although radon is also emitted into the atmosphere, it gets diluted and dispersed and is of little consequence outdoors.

Radon and its decay products emit alpha particles, a form of ionizing radiation that is damaging to tissue but cannot penetrate very far into the body. Consequently, radon harms the body only when it is ingested or inhaled. The radioactive particles lodge in the tiny passages of the lungs and damage surrounding tissue. There is compelling evidence, based primarily on several studies of uranium miners, that inhaling large amounts of radon increases the risk of lung cancer. Other studies suggest that people who are exposed to relatively low levels of radon over an extended time are at risk for lung cancer. In 1998 the National Research Council of the National Academy of Sciences released an extensive evaluation of radon's effects on human health. It estimated that residential exposure to radon causes 12% of all lung cancers—between 15,000 and 22,000 lung cancers annually. Cigarette smoking exacerbates the risk from radon exposure; about 90% of radon-related cancers occur among current or former smokers.

According to the EPA, about 6% of U.S. homes have high enough levels of radon to warrant corrective action—that is, a radon level above 4 picocuries per liter of air. (As a standard of reference, outdoor radon concentrations range from 0.1 to 0.15 picocuries per liter of air worldwide.) The highest radon levels in the United States occur in homes on a geologic formation, the Reading Prong, which runs across southeastern Pennsylvania into northern New Jersey and New York. Iowa has the most pervasive radon problem, where 71% of the homes tested in 1989 had radon levels high enough to warrant corrective action.

Ironically, efforts to make our homes more energy-efficient have increased the hazard of indoor air pollutants, including radon. Drafty homes waste energy but allow radon to escape outdoors so that it does not build up inside. Testing for radon can be inexpensive, and corrective actions are generally reasonably priced. Radon concentrations in homes are minimized by sealing basement concrete floors and by ventilating crawl spaces and basements.

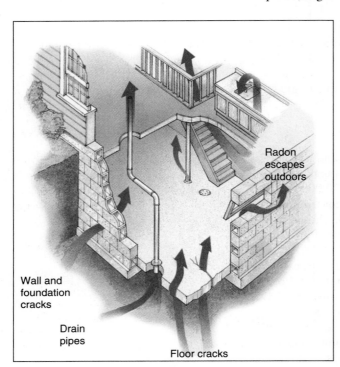

Radon
escapes
outdoors

Wall and
foundation
cracks

Drain
pipes

Floor cracks

Figure 20.22   Household radon infiltration. Cracks in basement walls or floors, openings around pipes, and pores in concrete blocks provide some of the entries for radon.

REVIEW

1. What is the sick building syndrome?

2. How does radon enter buildings?

REVIEW OF LEARNING OBJECTIVES WITH KEY TERMS

- **Describe the composition and function of the atmosphere.**

Excluding water vapor and air pollutants, four gases comprise the atmosphere: nitrogen, oxygen, argon, and carbon dioxide. The two atmospheric gases most important to living organisms are carbon dioxide and oxygen. Nitrogen gas is an important component of the nitrogen cycle. The atmosphere performs ecosystem services, namely, blocking Earth's surface from much of the ultraviolet (UV) radiation coming from the sun, moderating the climate, and redistributing water in the hydrologic cycle.

- **Define** *air pollution* **and distinguish between primary and secondary air pollutants.**

Air pollution consists of various chemicals added to the atmosphere by natural events or human activities in high enough concentrations to be harmful. A **primary air pollutant** is a harmful substance, such as soot or carbon monoxide, that is emitted directly into the atmosphere. A **secondary air pollutant** is a harmful substance formed in the atmosphere when a primary air pollutant reacts with substances normally found in the atmosphere or with other air pollutants.

- **List the seven major classes of air pollutants, including ozone and hazardous air pollutants, and describe their characteristics and effects.**

The main classes of air pollutants produced by human activities are particulate matter, nitrogen oxides, sulfur oxides, carbon oxides, hydrocarbons, ozone, and hazardous air pollutants. **Particulate matter**, solid particles and liquid droplets suspended in the atmosphere, corrodes metals, erodes buildings, and soils fabrics. **Nitrogen oxides**, such as nitric oxide, nitrogen dioxide, and nitrous oxide, are associated with photochemical smog and acid deposition; nitrous oxide is associated with global climate change as well as stratospheric ozone depletion; nitrogen oxides corrode metals and fade textiles. **Sulfur oxides**, such as sulfur dioxide and sulfur trioxide, are associated with acid deposition and corrode metals and damage stone and other materials. **Carbon oxides** include carbon monoxide, which is poisonous, and carbon dioxide, a **greenhouse gas**. **Hydrocarbons** are solids, liquids, or gases associated with photochemical smog; methane is a greenhouse gas, and some hydrocarbons are dangerous to human health. **Ozone** is a pale blue gas that is both a pollutant in the lower atmosphere (the **troposphere**) and an essential component that screens out UV radiation in another (the **stratosphere**). In the troposphere ozone reduces air visibility, causes health problems, stresses plants, and is a greenhouse gas. **Hazardous air pollutants** are air pollutants that are potentially harmful and may pose long-term health risks to people who live and work around chemical factories, incinerators, or other facilities that produce or use them.

- **Describe industrial smog, photochemical smog, and temperature inversions.**

Industrial smog, or smoke pollution, is composed primarily of sulfur oxides and particulate matter. **Photochemical smog** is a brownish-orange haze formed by chemical reactions involving sunlight, nitrogen oxide, and hydrocarbons. A **temperature inversion** is a deviation from the normal temperature distribution in the atmosphere, resulting in a layer of cold air temporarily trapped near the ground by a warmer, upper layer. Temperature inversions that persist in congested urban areas cause air pollutants to accumulate to dangerous levels.

- **Distinguish between emphysema and chronic bronchitis.**

Exposure to air pollution may result in the development of chronic respiratory diseases such as emphysema and chronic bronchitis. **Emphysema** is a disease in which the air sacs (alveoli) in the lungs become irreversibly distended, decreasing the efficiency of respiration and causing breathlessness and wheezy breathing. **Chronic bronchitis** is a disease in which the air passages (bronchi) of the lungs become permanently inflamed, causing breathlessness and chronic coughing.

- **Relate the adverse health effects of specific air pollutants and explain why children are particularly susceptible to air pollution.**

Air pollutants irritate the eyes, inflame the respiratory tract, and suppress the immune system. Sulfur dioxide, particulate matter, and nitrogen dioxide constrict airways, impairing the lungs' ability to exchange gases. Carbon monoxide combines with hemoglobin and reduces its ability to transport oxygen; carbon monoxide poisoning can cause death. Adults at greatest risk from air pollution include those with heart and respiratory diseases. Air pollution is a greater health threat to children than it is to adults, in part because air pollution impedes lung development. Children with weaker lungs are more likely to develop respiratory problems, including chronic respiratory diseases.

- **Summarize how electrostatic precipitators and scrubbers work.**

In an electrostatic precipitator, the electrode imparts a negative charge to particulates in the dirty gas. These particles are attracted to the pos-itively charged precipitator wall and then fall off into the collector. In a scrubber, mists of water droplets trap particulates in the dirty gas. The toxic dust produced by electrostatic precipitators and the polluted sludge produced by scrubbers must be safely disposed of or they will cause soil and water pollution.

- **Explain how Phase I vapor recovery reduces gasoline-related emissions.**

Phase I **vapor recovery** captures gasoline vapors that otherwise would be vented to the atmosphere. When a gasoline fueling truck arrives at a gas station, it hooks up two hoses: one that pours gasoline into an underground tank, and another that returns vapors to the truck. The truck then returns to a central facility, where the vapors are either condensed and recovered or combusted.

- **Summarize the effects of the Clean Air Act on U.S. air pollution.**

Air quality in the United States has slowly improved since passage of the **Clean Air Act** in 1970. This law authorizes the EPA to set limits on how much of specific air pollutants are permitted in the United States. The most dramatic improvement is the decline in lead in the atmosphere, although levels of sulfur oxides, ozone, carbon monoxide, volatile organic compounds (many of which are hydrocarbons), and particulate matter are also reduced.

- **Define *ultraviolet (UV) radiation*.**

Ultraviolet radiation is that part of the electromagnetic spectrum with wavelengths just shorter than visible light. UV radiation is high-energy radiation lethal to organisms at higher levels of exposure.

- **Distinguish between tropospheric and stratospheric ozone and describe the importance of the stratospheric ozone layer.**

Ozone (O_3) is a naturally occurring gas in the atmosphere. Human activities contributing to air pollution can cause its concentration in the lower troposphere to reach health-endangering levels; in the stratosphere, ozone is naturally produced, fortunately reaching levels far above air quality standards because the ozone layer in the stratosphere is vital in shielding Earth's surface from harmful ultraviolet (UV) radiation.

- **Define *stratospheric ozone thinning* and relate some of the harmful effects of ozone depletion.**

Stratospheric ozone thinning is the accelerated destruction of ozone in the stratosphere by human-produced chlorine- and bromine-containing chemicals. With depletion of the ozone layer, higher levels of UV radiation reach Earth's surface. In humans, excessive exposure to UV radiation causes cataracts, weakened immunity, and skin cancer. Increased levels of UV radiation may disrupt ecosystems, such as the Antarctic food web, because the negative effect of UV radiation on one species has ramifications throughout the ecosystem. The effects of increased levels of UV radiation on plants require additional study.

- **Define *chlorofluorocarbons* and explain how they and other chemicals attack ozone.**

Chlorofluorocarbons are human-made organic compounds of carbon, chlorine, and fluorine that had many industrial and commercial applications but were banned because they attack the stratospheric ozone layer. Additional compounds that attack ozone include halons, methyl bromide, methyl chloroform, carbon tetrachloride, and nitrous oxide. Thinning of the ozone layer over Antarctica occurs when the **circumpolar vortex**, a mass of cold air that circulates around the southern polar region, isolates the cold air from the rest of the planet's warmer air. The cold causes ice crystals to form in stratospheric clouds. Chlorine and other chemicals adhere to the ice particles and, from there, attack ozone. Sunlight catalyzes the chemical reaction in which chlorine or bromine breaks ozone molecules apart, converting them into oxygen molecules.

- **Describe the pH scale.**

The **pH scale** is a measure of acidity or basicity of an aqueous solution. A pH of 7 is neither acidic nor basic, whereas a pH less than 7 indicates an acidic solution. The pH scale is logarithmic, so a solution with a pH of 6 is 10 times more acidic than a solution with a pH of 7. A solution with a pH greater than 7 is basic.

- **Explain how acid deposition develops and relate some of the effects of acid deposition.**

Acid deposition is a type of air pollution that includes acid that falls from the atmosphere as precipitation or as dry acidic particles. Acid deposition occurs when sulfur dioxide and nitrogen oxides are released into the atmosphere. These pollutants react with water to produce sulfuric acid, nitric acid, and nitrous acid. Acid deposition kills aquatic organisms and may harm forests. Acid deposition attacks materials such as metals and stone.

- **Define *forest decline* and relate its possible causes.**

Forest decline is a gradual deterioration and often death of many trees in a forest; air pollution and acid deposition contribute to forest decline in many areas. No single factor accounts for forest decline, which appears to result from the combination of multiple stressors—acid deposition, tropospheric ozone, UV radiation (which is more intense at higher altitudes), insect attack, drought, climate change, and so on.

- **Explain why air pollution is generally worse in developing countries than in highly developed countries.**

Air quality is deteriorating in developing nations. Rapid industrialization, a growing number of automobiles in developing countries, and lack of emissions standards are contributing to air pollution, particularly in urban areas.

- **Describe the global distillation effect and tell where it commonly occurs.**

The **global distillation effect** is the process in which volatile chemicals evaporate from land as far away as the tropics and are transported by winds to higher latitudes, where they condense and fall to the ground. Volatile chemicals contaminate some remote polar regions as a result of the global distillation effect.

- **Summarize the sick building syndrome.**

The **sick building syndrome** is the presence of air pollution inside office buildings that can cause eye irritations, nausea, headaches, respiratory infections, depression, and fatigue.

- **Explain why tobacco smoke and radon are considered major indoor air pollutants.**

Tobacco smoke contains many hazardous chemicals, and causes many diseases in smokers and passive smokers. Indoor tobacco smoke can be a significant workplace hazard to non-smokers who spend large amounts of time in smoky areas. **Radon** is a colorless, tasteless radioactive gas produced naturally during the radioactive decay of uranium in Earth's crust. Radon seeps through the ground and enters buildings, where it sometimes accumulates to dangerous levels. Radon is inhaled into the lungs, where it damages surrounding tissue and increases the risk of lung cancer.

Thinking About the Environment

1. The atmosphere of Earth has been compared to the peel covering an apple. Explain the comparison.

2. List the seven main kinds of air pollutants and briefly describe their sources and effects.

3. Distinguish between primary and secondary air pollutants.

4. Why might global warming lead to more photochemical smog, even if emissions of nitrogen oxides and volatile organic chemicals remain constant?

5. Which is a more stable atmospheric condition, cool air layered over warm air or warm air layered over cool air? Explain. Which condition is a temperature inversion?

6. Briefly describe the health effects of exposure to various air pollutants.

7. What does the Environmental Protection Agency mean by "nonattainment areas for one or more criteria air pollutants"?

8. Why might we be more worried about organic or metallic particles than inert particles?

9. What country has the worst air pollution in the world? Is this likely to change? Why or why not?

10. Distinguish between the benefits of the ozone layer in the stratosphere and the harmful effects of ozone at ground level.

11. Discuss at least two harmful effects of stratospheric ozone depletion.

12. Discuss the harmful effects of acid deposition on materials, aquatic organisms, and soils.

13. Discuss some of the possible causes of forest decline. How might these factors interact to speed the rate of decline?

14. Why is the global distillation effect likely to become an increasingly challenging problem in the future?

15. One of the most effective ways to reduce the threat of radon-induced lung cancer is to quit smoking. Explain.

16. Conserving energy by reducing the rate at which indoor air is replaced with outdoor air can improve energy efficiency (less heating and cooling), but can contribute to indoor air pollution and the sick building syndrome. Explain how a systems approach to building design might help solve this problem.

Quantitative questions relating to this chapter are on our Web site.

Take a Stand

Visit our Web site at http://www.wiley.com/college/raven (select Chapter 20 from the Table of Contents) to learn more about the growing number of clean car choices. You will find tools to help you organize your research, analyze the data, think critically about the issues, and construct a well-considered argument. *Take a Stand* activities can be done individually or in a team, as oral presentations, written exercises, or Web-based (e-mail) assignments.

Additional online materials relating to this chapter, including a Student Testing Section with study aids and self-tests, Environmental News, Activity Links, Environmental Investigations, and more, are also on our Web site.

21
Global Climate Change

Reforestation project in Guatemala. Slowing the rate of deforestation to zero, coupled to reforestation, could be a component of climate change mitigation. (© 2001 CARE, photo by Jason Sangster)

Anthropogenic (human-caused) climate change is an established phenomenon. Within the scientific community, the question is no longer whether climate change will occur, but at what rate, with what effects, and what, if anything, can we do about it. The biggest culprit in climate change is an increase in atmospheric carbon dioxide (CO_2), which is generated primarily through burning fossil fuels.

In the September 2006 issue of *Scientific American,* **Robert Socolow** and **Stephen Pacala** at the Princeton University Carbon Mitigation Initiative suggest that somewhere near but below a doubling of atmospheric carbon from preindustrial levels lies the "boundary separating the truly dangerous consequences of emissions from the merely unwise." They propose that to

keep future atmospheric carbon below a doubling of the preindustrial level (in the early 1800's, the atmosphere contained 600 billion tons of carbon) would require seven billion fewer tons generated each year by 2056 than are currently expected.

Many people, seeing the enormity of this challenge, despair of finding a solution. Socolow and Pacala, however, propose that reductions can be thought of in terms of "wedges," each of which would result in a one-billion-ton-per-year reduction by 2056. A combination of any seven wedges would put us on a path to avoid the critical doubling of CO_2. They identify fifteen technologies in five categories, any one of which could serve as one of the seven wedges. Three of these wedges are:

Increase fuel economy of two billion cars from 30 to 60 mpg. Current projections are that there will be two billion cars on the world's roads by 2056, each traveling an average of 10,000 miles per year. If the typical car operates at 60 mpg, there will be one billion fewer tons of carbon generated each year than if they operate at 30 mpg.

Install carbon capture and storage at 800 large coal-fired power plants. Currently, the carbon dioxide produced from burning coal is released to the atmosphere. Instead, this carbon could be captured and stored (as discussed later in the chapter). If 90% of the

carbon released each year is captured and stored, one billion fewer tons will be released to the atmosphere.

Stop all deforestation. Deforestation worldwide currently releases two billion tons of carbon to the atmosphere each year. However, it is expected to slow to one billion tons per year without intervention. Thus, to achieve a one-billion-ton wedge below the expected amount would require a complete cessation of deforestation (or a balance between deforestation and reforestation) (see photo).

In this chapter we examine climate change, which is a truly global challenge. Climate change is perhaps the best example of the systemic nature of environmental problems. The interaction of economics, politics, energy, agriculture, and human values with the natural world has led to climate change. Making changes in economics, politics, energy, agriculture, and human behavior such as those associated with these wedges will be necessary to curb climate change.

Calculations for the individual wedges can be found at www.princeton.edu/~cmi.

World View and Closer to You...
Real news footage relating to climate change around the world and in your own geographic region is available at www.wiley.com./college/raven by clicking on "World View and Closer to You."

Introduction to Climate Change

LEARNING OBJECTIVES

- Describe radiative forcing and the enhanced greenhouse effect, and list the five main greenhouse gases.
- Distinguish between positive and negative feedback and discuss water vapor and the aerosol effect as they relate to global climate change.
- Explain how climate models project future climate conditions.
- Describe the importance of extreme and unpredictable climate change.

Earth's average temperature is based on daily measurements taken at several thousand land-based meteorological stations around the world, as well as data from weather balloons, orbiting satellites, transoceanic ships, and hundreds of sea-surface buoys with temperature sensors. Data show that eleven of the twelve years between 1995 and 2006 were among the twelve warmest years since the mid-1800s. According to the National Oceanic and Atmospheric Administration (NOAA), global temperatures in those years may have been the highest in the last millennium. (Although widespread thermometer records have been assembled only since the mid-19th century, scientists reconstruct earlier temperatures using indirect climate evidence in tree rings, lake and ocean sediments, small air bubbles in ancient ice, and coral reefs.) The last two decades of the 20th century were its warmest (**Figure 21.1**).

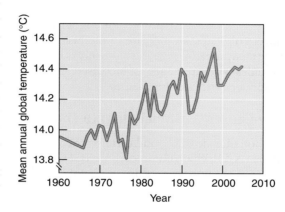

Figure 21.1 **Mean annual global temperature, 1960–2005.** Data are presented as surface temperatures (°C) for 1960, 1965, and every year thereafter. The measurements, which naturally fluctuate, clearly show the warming trend of the last several decades. (The dip in global temperatures in the early 1990s, caused by the eruption of Mount Pinatubo in 1991, is discussed later in the chapter.) (NASA)

Other evidence confirms the increase in global temperature. Several studies have documented that phenological spring in the Northern Hemisphere now comes about six days earlier than it did in 1959, and autumn is delayed five days. (*Phenological spring* is determined by when buds of specific plants open, and autumn by when leaves of specific leaves turn color and fall.) Since 1949, the United States has experienced an increased frequency of extreme heat stress events, which are extremely hot, humid days and nights during summer months; medical records indicate that heat-related deaths among elderly and other vulnerable people increase during these events. In the past few decades the rate of sea level rise has increased. For most of the 20th century, the rate was about 1.5 to 2 cm/decade—it is now at about 3 cm/decade. Glaciers worldwide have retreated, and extreme weather events such as severe rainstorms have occurred with increasing frequency in certain regions.

Scientists around the world have researched global climate change for the past 50 years. As the evidence has accumulated, those most qualified to address the issue have concluded that temperatures have increased over the past century, that it is extremely unlikely that the warming can be explained by natural causes, and that human-produced greenhouse gases are the most plausible explanation for the warming that has occurred. Further, the remainder of this century will experience significant additional climate change, and human activities will be largely responsible for this change.

In response to the growing scientific agreement about both the change and its human cause, governments around the world organized the U.N. Intergovernmental Panel on Climate Change (IPCC) in 1988. With input from and review by hundreds of climate experts, the IPCC provides the most definitive scientific statements about global climate change. The IPCC reviews all the published literature, especially that published over the previous five years, and summarizes the current state of knowledge and uncertainty as it relates to global climate change. The IPCC Fourth Assessment Report, issued in 2007, concluded that human-produced air pollutants have caused

Table 21.1 Increases in Selected Atmospheric Greenhouse Gases Preindustrial to the Present

Gas	Estimated Preindustrial Concentration[1]	2006 Concentration[5]
Carbon dioxide	288 ppm[2]	382 ppm
Methane	848 ppb[3]	1,783 ppb
Nitrous oxide	285 ppb	320 ppb
Chlorofluorocarbon-12	0 ppt[4]	535 ppt
Chlorofluorocarbon-11	0 ppt	249 ppt

[1] The preindustrial value is for the 17th and 18th centuries. There have been significant variations, as, for example, over the course of the ice ages.
[2] ppm = parts per million.
[3] ppb = parts per billion.
[4] ppt = parts per trillion.
[5] 2006 annual average.
Source: Carbon Dioxide Information Analysis Center, Environmental Sciences Division, Oak Ridge National Laboratory.

most of the climate warming observed over the last 50 years. Depending on the assumed **emissions scenario** (a prediction about the amounts, rates, and mix of future greenhouse gases), and on the intensity of the climatic response, the IPCC report projects a 0.2°C (0.4°F) increase in global average temperature in each of the next two decades. By the year 2100, depending on whether and how much we can control greenhouse gas emissions, temperatures are expected to have risen anywhere from 1.8 to 4.0°C (3.2 to 7.2°F). Based on reconstructions of the Earth's past climates, such warming would make the Earth warmer during the 21st century than in several tens of millions of years.

The fourth IPCC report also projected that it would be very likely that higher maximum temperatures and more hot days would be experienced over nearly all land areas. A similar level of confidence exists for projections of higher minimum temperatures, fewer frost days, fewer cold days, an increase in the heat index, and more intense precipitation events over many areas. There is a 66 to 90% chance that the continental interiors in the mid-latitudes will experience an increased risk of drought and that some coastal areas will experience stronger hurricanes.

The Causes of Global Climate Change

greenhouse gas: A gas that absorbs infrared radiation; carbon dioxide, methane, nitrous oxide, chlorofluorocarbons, and tropospheric ozone are all greenhouse gases.

parts per million: The number of molecules of a particular substance found in one million molecules of air, water, or some other material; abbreviated ppm.

Carbon dioxide (CO_2) and certain other trace gases, including methane (CH_4), nitrous oxide (N_2O), chlorofluorocarbons (CFCs), and tropospheric ozone (O_3), are accumulating in the atmosphere as a result of human activities (**Table 21.1**). All of these are **greenhouse gases**, or gases that absorb radiated heat from the sun, thereby increasing the temperature of the atmosphere. Additional, but minor, greenhouse gases include carbon tetrachloride, methyl chloroform, chlorodifluoromethane (HCFC-22), sulfur hexafluoride, trifluoromethyl sulfur pentafluoride, fluoroform (HFC-23), and perfluoroethane.

The concentration of atmospheric carbon dioxide has grown from about 288 **parts per million (ppm)** approximately 200 years ago (before the Industrial Revolution began) to 382 ppm in 2006 (**Figure 21.2**). Burning carbon-containing fossil fuels—coal, oil, and natural gas—accounts for most of the human contribution to total carbon dioxide. Land conversion, such as when tracts of tropical forests are logged or burned, also releases CO_2 and causes an increase in the atmospheric CO_2 concentration.

Not only does the burning of vegetation release CO_2 into the atmosphere, but it also reduces the capacity of the biosphere to remove and store carbon in roots and tree trunks by photosynthesis. Scientists estimate that without aggressive efforts to reduce carbon emissions, during the second half of the 21st century the concentration of atmospheric CO_2 will reach double what it was in the 1700s.

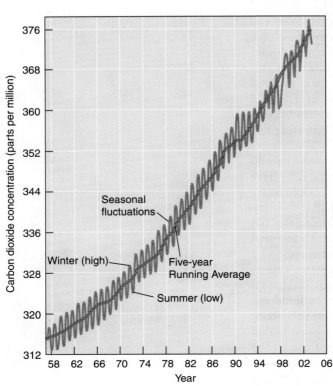

Figure 21.2 Carbon dioxide (CO_2) in the atmosphere, 1958–2003. The annual mean concentration of atmospheric CO_2 has increased steadily since 1958, when measurements began at the Mauna Loa Observatory in Hawaii. This location was selected because it is far from urban areas where factories, power plants, and motor vehicles that emit CO_2 might bias the measurements. The seasonally varying concentrations are highest in late winter when plants are not actively growing and absorbing CO_2, and lowest in late summer as a result of plants growing and absorbing CO_2. (Scripps Institute of Oceanography)

Because these gases absorb **infrared radiation**—that is, radiated heat from the sun—higher concentrations lead to warming and global climate change. This occurs because the absorption of heat slows its eventual re-radiation into space, thereby warming the lower atmosphere.

The capacity of various gases to affect the balance of energy entering and leaving the atmosphere is referred to as **radiative forcing**. A major fraction of the trapped heat is transferred to the ocean and raises its temperature as well, although the large heat capacity of the ocean means that it will take many decades for enough warming to occur to reestablish the disturbed energy balance. This retention of heat in the atmosphere is a natural phenomenon that has made Earth habitable for its millions of species. However, as human activities increase the atmospheric concentration of greenhouse gases, the atmosphere and ocean will continue to warm, and the overall global temperature will rise. Carbon dioxide accounts for 60% of the increased radiative forcing and heat retention caused by greenhouse gases.

Because CO_2 and other gases slow the loss of heat generated by the incoming solar radiation in a manner parallel to how the glass enclosure reduces energy loss in a greenhouse, the natural trapping of heat in the atmosphere is often referred to as the **greenhouse effect,** and the gases that absorb infrared radiation are called greenhouse gases. Greenhouse gases accumulating in the atmosphere as a result of human activities are thus causing an **enhanced greenhouse effect** (**Figure 21.3**).

The levels of the other trace gases associated with global climate change are also rising. Every time you drive your car, the combustion of gasoline in the car's engine releases CO_2 along with other pollution-creating gases (see Chapter 20). Decomposition of carbon-containing organic material by anaerobic bacteria in moist places as varied as rice paddies, sanitary landfills, and the intestinal tracts of cattle and other large animals (humans included) is a major source of methane (CH_4). Various industrial processes, land use conversion, and the use of fertilizers produce nitrous oxide. CFCs are refrigerants released into the atmosphere from old, leaking refrigerators and air conditioners (see Chapter 20). Although CFC emissions are decreasing, the very long lifetime of emissions in the past—from a variety of sources including aerosol spray cans and foam insulation—means that they will continue to contribute to future climate change. Water vapor, which is also a greenhouse gas, exerts a **positive feedback** on the climate that amplifies warming. Warmer temperatures cause greater evaporation from the ocean and a higher concentration of atmospheric water vapor, which in turn causes warmer air and ocean temperatures, causing further evaporation.

Although current rates of fossil fuel combustion and deforestation are high, causing the CO_2 level in the atmosphere to increase markedly, scientists think the warming trend will be slower than the increasing level of CO_2 might indicate. The reason is that it requires more heat to raise the temperature of the ocean than of the air (recall the Chapter 14 discussion of the high heat capacity of water). In addition, the atmosphere is well mixed, while the ocean is stratified, so that the ocean takes longer than the atmosphere to absorb heat. For this reason, climate scientists expect that ocean warming will be more pronounced in the 21st century than it was in the 20th century.

Other Pollutants Cool the Atmosphere

One of the complications that makes the rate and extent of global climate change difficult to predict is that other air pollutants, known as atmospheric aerosols, tend to cool the atmosphere in what is called the **aerosol effect**. Aerosols, which come from both natural and human sources, are tiny particles so small they remain suspended in the troposphere for days, weeks, or months. Because sulfate particles are efficient at scattering radiation, a sulfate-laden haze tends to cool the planet by reflecting some of the incoming sunlight back into space, away from Earth. Temperature observations indicate that sulfur-laden haze has significantly moderated warming in some of the

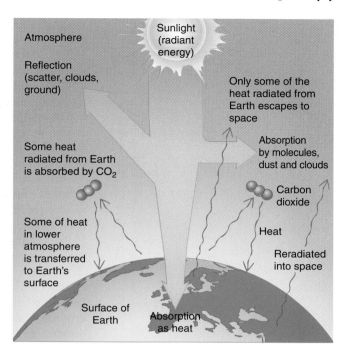

Figure 21.3   Enhanced greenhouse effect. The buildup of carbon dioxide (CO_2) and other greenhouse gases absorbs some of the outgoing infrared (heat) radiation, warming the atmosphere. Some of the heat in the warmed atmosphere is transferred back to Earth's surface, warming the land and ocean. The percent of incoming radiation absorbed is increasing, while percent reflected is decreasing.

infrared radiation: Radiation that has wavelengths longer than that of visible light, but shorter than that of radio waves; most of the energy absorbed by Earth is radiated as infrared radiation, which can be absorbed by greenhouse gases.

radiative forcing: The capacity of a gas to affect the balance of energy that enters and leaves Earth's atmosphere; measured in units of power per unit area, usually watts per square meter (W/m^2).

positive feedback: A situation in which a change in some condition triggers a response that intensifies the changed condition.

aerosol effect: Atmospheric cooling that occurs where and when aerosol pollution is the greatest.

industrialized parts of the world. By contrast, sooty aerosols generally absorb radiation, and so they tend to warm the planet. In the atmosphere, there are complex mixtures of aerosols of various types, making the actual aerosol effect on the climate relatively uncertain, although likely exerting an overall cooling influence.

The sulfur dioxide emissions that produce the sulfur-laden haze come mainly from the stacks of the same power plants that are responsible for much of the CO_2 emissions. In addition, volcanic eruptions inject sulfur-containing particles into the atmosphere. The explosion of Mount Pinatubo in the Philippines in June 1991 was the largest volcanic eruption in the 20th century. The force of this eruption injected massive amounts of sulfur into the stratosphere (the layer of the atmosphere above the troposphere), where the particles stay aloft longer (up to a few years) than for aerosols emitted into the troposphere. Because a sulfur-laden layer in the stratosphere reduces the amount of sunlight that reaches Earth's surface (although not the total amount that reaches the planet), this eruption caused a temporary period of global cooling. Compared to the rest of the 1990s, 1992 and 1993 global temperatures were relatively cool (see Figure 21.1).

Human-produced sulfur emissions should not be viewed as a panacea for human-enhanced global climate change, despite their cooling effect. For one thing, sulfur emissions are produced in heavily populated industrial areas, primarily in the Northern Hemisphere. Because they do not remain in the atmosphere for long, they do not disperse globally. Thus, sulfur pollution likely contributes to regional cooling, somewhat offsetting the greenhouse gas-induced warming.

Overall, the effect of the greenhouse gas increase is considerably more potent than that of the sulfur-laden haze. The increased concentrations of some greenhouse gases will persist in the atmosphere for years to hundreds of years, whereas human-produced sulfur emissions remain for only days, weeks, or months. And carbon dioxide and other greenhouse gases warm the planet 24 hours a day, whereas sulfur haze cools the planet only during the daytime. In addition, because sulfur emissions are a respiratory irritant and cause acid deposition (Chapter 20), most nations are trying to reduce their sulfur emissions, not maintain or increase them.

Developing and Using Climate Models

Many interacting factors, such as winds, clouds, ocean currents, and **albedo** (a measure of reflectivity; ice has a higher albedo than does asphalt), affect the complex climate system (see Chapter 5), each exerting its influence on the climate. Because interactions among the atmosphere, ocean, and land are too complex and too large to construct in a testing laboratory, climate scientists develop simulation models using powerful computers to test how the Earth system works. Such models use well-established physical laws along with approximations for treating small-scale features of the climate to represent the effects of competing processes and thereby to describe Earth's climate system in numerical terms (**Figure 21.4**). Models can be used to explore and analyze past climate events and, in the most advanced models, to project future warming and suggest the consequences ("what-if" scenarios) of warming on the biosphere and its life-support systems.

A climate model is only as good as its representation of the physical laws and processes. Global climate change models have been refined in recent years and are now capable of representing many features of the present climate and the climate of the last few centuries. However, limitations do remain, particularly in the representation of clouds and changes that are likely to occur as the climate changes. If global climate change leads to more low-lying clouds, they would reflect some of the incoming sunlight and decrease the amount of warming, acting as **negative feedback**. On the other hand, if global climate change leads to more high, thin cirrus clouds, they would only reflect a little more solar radiation, but would trap a lot more infrared radiation, intensifying the warming (that is, serving as a positive feedback mechanism). As additional case studies are carried out and as new understanding about these and other uncertainties is becoming available, the predictions of the models are being viewed with greater confidence.

negative feedback: A situation in which a change in some condition triggers a response that counteracts, or reverses, the changed condition.

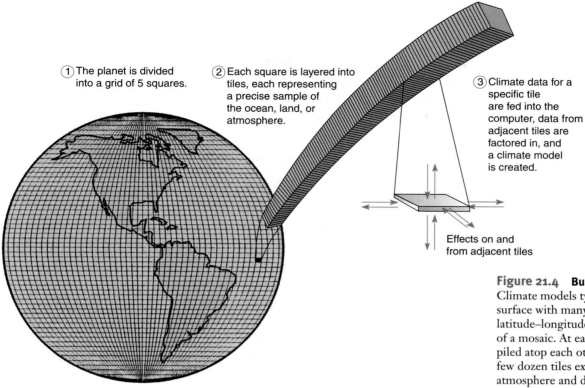

① The planet is divided into a grid of 5 squares.

② Each square is layered into tiles, each representing a precise sample of the ocean, land, or atmosphere.

③ Climate data for a specific tile are fed into the computer, data from adjacent tiles are factored in, and a climate model is created.

Effects on and from adjacent tiles

Figure 21.4   Building a climate model.
Climate models typically cover the planet's surface with many hundreds to thousands of latitude–longitude tiles, much like the tiles of a mosaic. At each location, the tiles are piled atop each other, creating stacks of a few dozen tiles extending up into the upper atmosphere and down into the ocean depths. The model's program considers how hour by hour (or even minute by minute) changes in sunlight, temperature, air pressure, currents or winds, and water vapor in one tile affect conditions in that tile and in each of the adjacent tiles. The computer continually performs calculations, taking into account the fundamental laws involving conservation of mass, momentum, and energy. These calculations can stretch out to centuries, specifying, as appropriate, any changes in the concentrations of greenhouse gases, solar radiation, or some other parameter.

Most climate models are used to project what the climate will be like a few decades or a century from now. One model, developed at Princeton University, has been used to examine the implications of climate warming five centuries from now. The model simulation assumed that limitations on emissions will be implemented in time to stabilize the CO_2 concentration in 2050 at double the **preindustrial CO_2 level.** It presents a dramatic scenario of a warmer climate that future generations will have to live with. Among other changes, the model projects that the rising sea level would inundate the southern end of Florida from Key Largo to Fort Lauderdale. Average summer temperatures in the southeastern and mideastern states (north to Pennsylvania) would increase from the current 27°C (80°F) to 31°C (87°F); but because this warm air would hold more moisture, the average temperature would feel more like 36°C (97°F).

Climate models present humans with a potential ethical dilemma. How do we balance scientific uncertainty about the rate and extent of climate change with the similar uncertainty about the economic impacts of reducing greenhouse gas emissions? The IPCC has established a "business as usual" scenario, which estimates the amount of carbon dioxide that will be released over the next century if economies develop as expected with no intentional attempts to reduce emissions on a large scale. This scenario projects a doubling of atmospheric carbon by around 2050. Knowing this baseline allows us to think about what combination of strategies will most effectively and efficiently avoid this doubling.

Unpredictable and Extreme Climate Change

Our current knowledge of global climate is so incomplete that unanticipated effects from a globally warmed world will undoubtedly occur. Some of these effects are simply unpredictable—that is, there will be complete surprises. Others are conceptually predictable, but there are thresholds (as in the case of dying coral reefs discussed later in the chapter) and tipping points, and we don't know when they might occur. As an example of a tipping point, there could be a disruption of the **ocean conveyor belt,** which transports heat around the globe (see Figure 5.15). The ocean conveyor belt delivers heat from the tropics into the northern part of the North Atlantic Ocean. Some of this heat is transferred to the atmosphere, thereby helping to warm Europe and adjacent lands as much as 10°C (18°F). As the warm North Atlantic water transfers heat to the atmosphere, it cools, sinks, and flows southward. The cooler, sinking water carries some of the CO_2 from the atmosphere deep into the ocean where, through incompletely understood mechanisms, much of the carbon is sequestered (stored).

Models based on the behavior of the ocean conveyor belt during past episodes of climate warming—immediately following the ice ages, for example—suggest that an abrupt climate change could occur. Climate warming, with its associated freshwater melting off the Greenland ice sheet, could weaken or even shut down the ocean conveyor belt in as short a period as a decade. Data reported at a 2002 ocean science conference suggest that such a change may be underway. No one knows for sure, but changes in the ocean conveyor belt could cause major cooling in Europe even as greater global warming occurs elsewhere. In addition, a weakened ocean conveyor belt would not sequester as much carbon in the ocean, leading to a positive feedback loop. Less CO_2 stored in the ocean would mean more CO_2 in the atmosphere, which would cause additional atmospheric warming, which in turn would cause the ocean conveyor belt to weaken even further.

Climate models project expected or most likely outcomes and ranges of possible outcomes. The outcomes usually reported represent a range that the modelers feel is reasonably likely to include the actual outcome. Sometimes these ranges include low ends that might be somewhat troubling, and high ends that might cause serious disruptions. For example a 0.5°C (0.9°F) increase in average summer high temperature at some location might not make much difference, whereas a 4°C (7.2°F) increase certainly would.

In addition, climate model outputs often include possible extreme cases, such as a 6°C (11°F) increase in global annual average temperatures or a 6 m (19 ft) sea level rise. As you read the next section on effects of climate change, keep in mind that there may be effects beyond what is described here, and there are likely to be surprises that we cannot currently predict.

REVIEW

1. What is the enhanced greenhouse effect, and what are the five main greenhouse gases that contribute to it?
2. What are positive and negative feedback?
3. How do water vapor and the aerosol effect relate to global climate change?
4. How do climate models project future climate conditions?
5. Why are unpredictable and extreme climate change important?

The Effects of Global Climate Change

LEARNING OBJECTIVES

- Differentiate between sea level rise due to melting and that due to thermal expansion of water.
- Describe how climate change will impact non-human life on Earth.
- Give examples of effects of climate change on humans.
- Describe how interactions among global climate change, ozone depletion, and acid deposition affect North American lakes.

The effects of global climate change are many and varied. Scientists have determined that the ocean absorbs about half of the CO_2 released into the atmosphere by human activities. This is good news for the planet because it means there is less CO_2 in the atmosphere to cause global warming; if the ocean were not acting as a sink for CO_2, warming would be much more severe now than it is. However, in 2004 scientists reported that the excess CO_2 dissolved in the ocean is beginning to threaten many forms of life there.

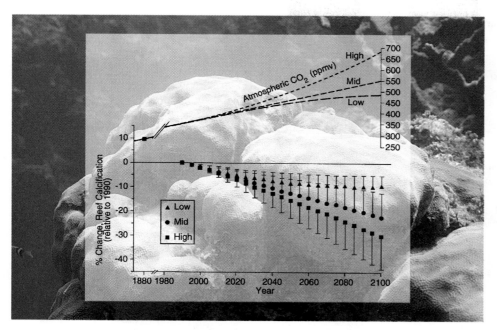

Figure 21.5   Simulated effect of atmospheric CO₂ on coral reef calcification. Increases in atmospheric CO_2 damage coral reefs directly and indirectly. Increasing surface water acidity weakens the coral skeletons, which are made of secreted calcium carbonate. In addition, as water temperatures rise, the symbiotic algae that support the corals begin to die, causing the coral to lose their characteristic color (hence the term "bleaching"). With an extended period of increased temperature and acidity, the coral reefs could die off irrecoverably. (U.S. National Climate Assessment; Graph, Reprinted with permission from *Science*, photo courtesy Paul Grabhorn.)

CO_2 reacts with water to form the weak acid, carbonic acid, H_2CO_3. As more and more CO_2 is absorbed, it has begun to change the chemistry of ocean waters, making them more acidic. The acid attacks calcium carbonate, $CaCO_3$, which is normally abundant in the ocean. Many organisms, such as corals, shellfish, and plankton, incorporate $CaCO_3$ into their protective shells. If the rise in the CO_2 concentration continues, it could be catastrophic to sea life, especially as warming is also occurring (**Figure 21.5**).

Let's now consider some other observed and potential effects of global climate change, including changes in sea level, precipitation patterns (including the frequency and intensity of storms), distributions of organisms, human health, and agriculture. A more complete, but still not exhaustive, list of impacts of climate change would also include forests (and thus the timber industry), tourism and recreation, and coastal infrastructure.

Melting Ice and Rising Sea Levels

The IPCC projects 18 to 59 cm (0.6 to 1.9 ft) sea level rise by 2100, while noting that it could be much more. Sea level rises can be caused in two ways. Water, like other substances, expands as it warms. The IPCC reports that during the 20th century, sea level rose about 0.2 m (8 in.), most of it due to **thermal expansion.** Thermal expansion contributes more than half of sea level rise. The current rate of increase is about 3 mm per year, and that rate is increasing. In addition, sea level rises due to the retreat of glaciers and thawing of ice at the South Pole. Water absorbs more heat than does ice, which is highly reflective. So melting ice has a positive feedback effect on heating: Water absorbs more heat, which causes more ice to melt.

As the overall temperature of Earth has increased, a major thawing of glaciers and the polar ice caps has occurred. During the Antarctic summer in 2002, most of the Larsen B ice shelf, an area roughly the size of Rhode Island, broke off the Antarctic Peninsula. (Antarctic ice shelves are thick sheets of floating ice fed mainly by glaciers that flow off the land.) This loss of ice coincided with a decades-long trend of atmospheric warming in the Antarctic.

The area of ice-covered ocean in the Arctic has decreased significantly over the past several decades. The average latitude of the southern edge of arctic ice (between 71°N and 72°N latitude during the 1970s) has retreated northward, to 75° N. Sonar measurements from naval submarines operating under the ice indicate that the remaining arctic ice pack has thinned rapidly, losing 40% of its volume in less than three decades.

Figure 21.6   **Grinnell Glacier, a monitor of climate change.**

(a) Glacier National Park, Montana, photographed in July, 1957. Grinnell Glacier is in the foreground. (Gregory G. Dimijian, MD/Photo Researchers, Inc.)

(b) A photograph at approximately the same location in July, 1998. Due to dramatic shrinkage, Grinnell Glacier is out of the picture. (Gregory G. Dimijian, MD/Photo Researchers, Inc.)

Mountain glaciers around the world are also melting at accelerating rates, contributing to sea level rise. Qori Kalis Glacier, the largest glacier in the Peruvian Andes, is currently retreating about 30 m (100 ft) a year. The Gangotri Glacier in India is retreating at a similar rate. According to the Worldwatch Institute, 100 of Glacier National Park's 150 glaciers have completely melted since 1850; the remaining 50 are retreating so rapidly they will probably be gone by 2030 (**Figure 21.6**).

A 2006 analysis of the Greenland ice sheet (the world's second largest expanse of landbound ice) determined that Greenland is losing about 249 km^3 (57 mi^3) of ice each year, up from estimates of 44 km^3 (11 mi^3) in 2002, and 8.3 km^3 (2 mi^3) in the period 1993–1998. If half of the Greenland ice sheet were to melt (as it did during the Eemian Interglacial, just over 100,000 years ago), sea level would rise several meters.

CASE IN POINT

Impacts in Fragile Areas

The Eskimo Inuit, the indigenous people of Alaska's and Canada's far north, pursue a way of life dictated by the frigid climate. Effects of global climate change are changing the Inuit traditional existence. Many populations of wildlife, which the Inuit harvest for food, are smaller or displaced. Other changes that threaten subsistence livelihoods include reduced snow cover, shorter river ice seasons, and thawing of permafrost. Warmer temperatures increase the risk of contaminating water supplies, as bacteria move more freely through thawed soil. Larger thawed areas could lead to the collapse of bridges, buildings, roads, and oil pipelines.

Glimpses of these observed and potential changes are gained from the region's natives, who report changes linked to warmer temperatures: Drying tundra, thinner and retreating sea ice, warmer winters, and changes in the numbers, distribution, and migration of some wildlife species. Climate change data support the Inuit's observations. Scientists analyzing lake-sediment cores to explore climate changes for the past 400 years note that the greatest warming trend occurred from 1840 to the late 20th century. With temperatures projected to rise even faster in the 21st century, scientists caution that the relatively undisturbed Arctic may be particularly susceptible to the effects of climate change.

Melting ice that drains into the ocean raises the sea level, but what about land-bound melting ice? Evidence indicates that **permafrost,** the permanently frozen subsoil characteristic of the tundra and boreal forests of Alaska, Canada, Russia, China, and Mongolia, is thawing. Permafrost provides the foundation on which tundra plants and forest trees are anchored and on which houses and roads are built. As the permafrost thaws, this foundation collapses. Near Fairbanks, Alaska, hundreds of homes and telephone poles are sinking at odd angles into the ground (**Figure 21.7**). Thawing permafrost also releases methane and other greenhouse gases—another positive feedback.

While melting ice and thawing permafrost are important in the extreme north, sea level rise has begun to impact small island nations. In 1999 two uninhabited islands (Tebua Tarawa and Abanuea) in the South Pacific were submerged under rising water. In 2001 the 11,000 residents of nearby Tuvalu announced that they must evacuate because the rise in sea level has caused lowland flooding, harming their water supply and food production. New Zealand has offered to allow some Tuvaluans to immigrate each year. Small island nations such as the Maldives, a chain of 1200 islands in the Indian Ocean, are considered highly vulnerable to a rise in sea level. About 80% of the Maldives is lower than 1 m (39 in.) above sea level, and the country's highest point is only 2 m above sea level. As sea levels rise, storm surges could easily sweep over entire islands.

Other countries vulnerable to a rise in sea level—such as Bangladesh, Egypt, Vietnam, and Mozambique—have dense populations living in low-lying river deltas. A rise in sea level could cause Bangladesh to lose as much as 18% of its land, displacing millions of people in this densely populated nation. The rising sea level is not only affecting developing countries. Coastal farmers in England have begun to lose land and crops to sea level increases. ■

Figure 21.7   Twawing permafrost. The telephone poles tilt as the permafrost in which they are anchored thaws. Thawing permafrost causes ground subsidence, erosion, and landslides. (Cliff Riedinger/AlaskaStock)

Changes in Precipitation Patterns

Computer models indicate that, as global climate change occurs, precipitation patterns will change, causing some areas to have more frequent droughts (**Figure 21.8**). At the same time, heavier snow and rainstorms are projected to cause more frequent flooding in other areas.

Changes in precipitation patterns are likely to affect the availability and quality of fresh water in many locations. Arid or semiarid areas, such as the Sahel region just south of the Sahara Desert (see Figure 5.21), may have the most troublesome water shortages

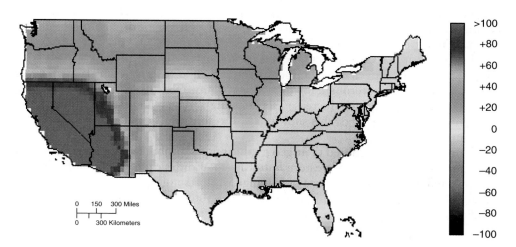

Figure 21.8   Effect of climate warming on precipitation This model shows one scenario (the Canadian Model Scenario) of how warmer global temperatures might alter precipitation in the United States over the next 100 years. Colors indicate the percent change in annual precipitation per century. Locate the state you live in. Will it be wetter, drier, or about the same? (U. S. Climate Change Research Program)

as the climate changes. Closer to home, water experts predict water shortages in the American West because warmer winter temperatures will cause more precipitation to fall as rain rather than snow; melting snow currently provides 70% of stream flows in the West during summer months. The U.N. Security Council began meetings in 2007 to consider the security implications of climate change-related drought.

The frequency and intensity of storms over warm surface waters appear likely to increase. In 1998, NOAA developed a computer model examining how global climate change might affect hurricanes, and since hurricane Katrina devastated New Orleans in 2005, there has been increased interest in the relationship between climate change and the intensity and frequency of hurricanes. Recent research suggests that a sea-surface temperature 2.2°C (4.0°F) warmer than today will result in hurricanes with higher maximum winds speeds and increased total precipitation. Changes in storm intensity are expected because as Earth warms, more water evaporates, which in turn releases more energy into the atmosphere (recall the discussion of water's heat of vaporization in Chapter 14). This energy can generate more powerful storms. While the number of hurricanes each year does not appear to have changed over the past several decades in response to climate change, the mean and peak intensities have increased in correlation to increased surface water temperature.

As discussed in Chapter 5, the El Niño-Southern Oscillation (ENSO), the periodic warming (El Niño) and cooling (La Niña) of the tropical Pacific Ocean, affects precipitation and other aspects of the entire global climate system. Until recently climate scientists could not predict whether human-induced global climate change would affect ENSO. The IPCC's most current analysis of computer models of a globally warming world indicates greater extremes of drying and heavy rainfall during El Niño events. Scientists are still uncertain whether El Niño events will occur more frequently with global climate change.

Effects on Organisms

An increasing number of studies report measurable changes in the biology of plant and animal species as a result of climate warming. Such effects range from earlier flowering times for plant species to migrations of aquatic species. Changes are also evident in many populations, communities, and ecosystems. Other human-induced factors, such as pollution and changes in land use, exacerbate threats posed by climate change. Here we report on the results of several of the hundreds of studies conducted to date.

Researchers determined that populations of zooplankton in parts of the California Current, which flows from Oregon southward along the California coast, had declined 80% since 1951, apparently because the current has warmed slightly. The decline in zooplankton in the California Current has affected the entire food web there, and populations of plankton-eating fishes and seabirds have declined.

As temperatures have risen in the waters around Antarctica during the past two decades, a similar decline in shrimp-like krill has contributed to a reduction in Adélie penguin populations (see "Case in Point: How Humans Have Affected the Antarctic Food Web" in Chapter 3). Because there are fewer krill, the birds do not get enough food. Warmer temperatures in Antarctica—during the past 50 years the average annual temperature on the Antarctic Peninsula has increased 2.6°C (4.5°F)—have contributed to reproductive failure in Adélie penguins. The birds normally lay their eggs in snow-free rocky outcrops. However, there is now more open water close to the nesting ground, resulting in increased air moisture and increased snowfall. When the penguins incubate the eggs, this snow melts into cold pools of slush that kill the developing chick embryos.

Some species have shifted their geographic ranges, probably in response to global climate change. One type of western butterfly (the Edith's checkerspot butterfly) has disappeared in the southern parts of its range while its northern range has expanded by about 160 km (about 100 mi) northward, establishing new colonies there. Similar studies conducted in Europe have found that 22 butterfly species out of the 35 species examined have shifted their ranges northward, anywhere from 32 to 240 km (20 to

Figure 21.9   **Polar bears are encountering difficulties surviving in a world with less contiguous (unbroken) polar ice.** What does this cartoon imply about human obligations to other animals? (Mike Peters, Mother Goose & Grimmm © 2006 Grimmy, Inc. All rights reserved. Used with permission of Grimmy, Inc. /Cartoonist Group)

150 mi). Scientists studying birds in Great Britain also report that the ranges of several dozen species have moved northward an average of 19 km (12 mi).

The ranges of some species are also shifting up in altitude. For over 100 years, researchers at the University of California at Berkeley have been collecting small mammal specimens from Yosemite National Park. This collection clearly shows that small mammals which were previously found only at relatively low elevations are now found at higher elevations; a steady upward shift has been documented during this time.

Biologists have determined that many migrating bird species such as robins are showing up at their summer homes earlier than in past years, as a result of increasingly warmer spring temperatures. The prospect of an early spring may sound inviting until you consider all the ramifications. Species are interdependent; birds, for example, need large amounts of food when raising chicks. If spring comes early, food may not be available when the birds need it. Biologists in the Netherlands have determined that in the past (as recently as 1980), trees leafed out, then winter moth caterpillars hatched, and then great tit eggs hatched, and the parents fed and successfully raised their chicks. However, each species reacts to warming temperatures differently. The interdependent "leaves/caterpillars/birds" system has become uncoupled because the peak number of caterpillars now occurs when the trees leaf out, which is before the chicks are hatched (egg-laying time for this species has not changed), to the detriment of the bird populations.

As warming accelerates in the 21st century, many species will undoubtedly become extinct, particularly those with narrow temperature requirements; those confined to small, specialized habitats; and those living in fragile ecosystems. Other species may survive in greatly reduced numbers and ranges. Ecosystems considered at greatest risk of species loss in the short term are polar seas, coral reefs, mountain ecosystems, coastal wetlands, and tundra (**Figure 21.9**).

Coral reefs are systems that include the corals, symbiotic organisms living within the reefs, and other organisms including fish that live, eat, and reproduce around the reefs (**Figure 2.10a**). Temperature-related coral bleaching occurs when water temperature exceeds a threshold, affecting the coral symbiotes and making them and the corals more susceptible to disease-causing organisms that healthy corals are normally resistant to (**Figure 21.10b**). Bleaching is exacerbated by the increased acidity of the ocean—two parts of the delicately balanced ocean system are changing at once. The IPCC predicts that with a 2°C (3.6°F) increase in global mean annual temperature, most corals worldwide will experience bleaching, with widespread coral mortality occurring above an increase of 3°C (5.4°F). In 1998 scientists documented the most geographically extensive and most severe epidemic of coral bleaching ever observed. About 10% of the world's corals died that year, in many cases from viral, bacterial, or fungal infections. In 1998 tropical waters were the warmest recorded to that time (Coral bleaching is also discussed in Chapter 6).

Figure 21.10 **Effects of climate on coral reefs.**

(a) The corals in this image are normal and healthy. (Digital Vision)

(b) Bleached corals. Coral researchers expect that many of Earth's corals will die from the combined effects of ocean warming and acidification. (Peter Scoones/Photo Researchers, Inc.)

Biologists generally agree that global climate change will have an especially severe impact on plants because they cannot move about when environmental conditions change. Although wind and animals disperse seeds, sometimes over long distances, limits in seed dispersal rates can limit the speed of migration. During past climate warmings, such as during the glacial retreat that took place some 12,000 years ago, analysis of tree pollens indicate that species were able to migrate at rates of only 4 to 200 km (2.5 to 124 mi) per century.

If Earth warms the projected 1.4 to 5.8°C (2.5 to 10°F) during the 21st century, the ideal ranges for some temperate tree species (that is, the environment where they grow the best) may shift northward as much as about 500 km (about 300 mi) (**Figure 21.11**). The U.S. Department of Agriculture reports hardiness zones, based on the average low temperature in an area, to determine which plants will do well around the country. These zones have shifted considerably over the past two decades . Moreover, soil characteristics, water availability, competition with other plant species, and habitat fragmentation will affect the rate at which plants can move into a new area.

Although generally not a benefit to humans, some species will come out of global climate change as winners, with greatly expanded numbers and range. Those organisms considered most likely to prosper include certain weeds, insect pests, and disease-carrying organisms common in a wide range of environments. A 3°C (5.4°F) increase in average surface temperature, for example could allow the Mediterranean fruit fly, an economically important insect pest, to expand its range into northern Europe.

Effects on Human Health

While the precise extent to which climate change contributes to adverse health effects is uncertain, it is clear that climate change can significantly impact human health, and will have an increasing effect in the future. The relationships between human health and climate work at the systemic level, and are both complex and inseparable. Some immediate health impacts are clear—for example, the 2003 heat wave in France resulted in around 15,000 deaths. If this heat wave had occurred in an area better prepared for heat, however, many of those deaths could have been avoided (for example, by going to air-conditioned locations and drinking plenty of fluids). Most of the health effects associated with climate change are indirect and have multiple, interrelated causes.

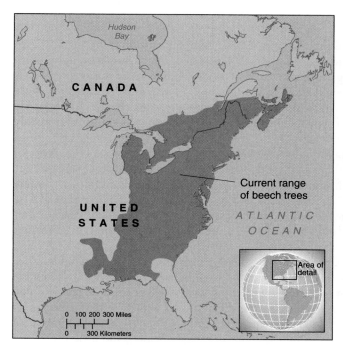

(a) Present geographical range of American beech trees. These large shade trees produce edible beechnuts that are an important food source for squirrel, raccoons, bears, game birds, and other wildlife.

(b) The projected range of beech trees after global climate change occurs, based on a model by the U.S. Geophysical Fluid Dynamics Laboratory (GFDL). According to the GFDL scenario of changed temperature and precipitation patterns, beeches would continue to grow in parts of Maine and Nova Scotia, and their range would expand into Quebec and northern Ontario.

Figure 21.11   Cimate change and beech trees in North America.

Climate warming is also likely to affect human health indirectly. The ranges of mosquitoes and other disease carriers are likely to expand into the newly warm areas and could, in the absence of other limiting factors, spread malaria, dengue fever, schistosomiasis, and yellow fever (**Table 21.2**). As many as 50 million to 80 million additional cases of malaria could occur annually in tropical, subtropical, and temperate areas. According to the World Health Organization, during 1998, the warmest year on record, the incidence of malaria, Rift Valley fever, and cholera surged in developing countries. Highly developed countries are less vulnerable to such disease outbreaks because of better housing (which keeps mosquitoes outside), medical care, pest control, and public health measures such as water treatment plants, but all of these require funds and commitment to maintain.

Table 21.2 Likelihood of Altered Distribution of Infectious Diseases as a Result of Climate Change

Disease	Vector	Present Distribution	Likelihood of Distribution Changes with Climate Warming
Malaria	Mosquito	Tropics, subtropics	Highly likely
Dengue fever	Mosquito	Tropics, subtropics	Very likely
Schistosomiasis	Water snails	Tropics, subtropics	Very likely
Yellow fever	Mosquito	Tropical South America and Africa	Very likely
Onchocerciasis	Blackfly	Africa, Latin America	Very likely
Lymphatic filariasis	Mosquito	Tropics, subtropics	Likely
Leishmaniasis	Phlebotomine sandfly	Asia, southern Europe, Africa, Americas	Likely
American trypanosomiasis	Triatomine bug	Central and South America	Likely
African trypanosomiasis	Tsetse fly	Tropical Africa	Likely
Dracunculiasis	Copepod	Central and West Africa	Unknown

Source: From Table 4.21 in *Health and Environment in Sustainable Development: Five Years After the Summit*, World Health Organization (1997), and based on an unpublished assessment prepared by A.J. McMichael et al.

Effects on Agriculture

Agriculture, as demonstrated in Chapter 19, is a complex and carefully manipulated ecological system. Consequently, the impacts of global climate change on agriculture are difficult to anticipate. Agricultural productivity could increase because higher levels of CO_2 in the atmosphere could allow a higher rate of photosynthesis. However, there are many interacting factors at work. How will growing seasons change? Will certain pests become more or less of a problem? How much of the types of crops traditionally grown in a given area have to change?

The rise in sea level may cause water to inundate river deltas, which are some of the world's best agricultural lands. Certain agricultural pests and disease-causing organisms will probably proliferate and reduce yields. Scientists think global climate change will increase the frequency and duration of droughts, a problem that will be particularly serious for countries with limited water resources. It is likely that warmer temperatures will result in decreases in soil moisture in many agricultural soils (warmer temperatures cause increased evaporation).

Another effect of climate warming on agriculture involves nighttime temperatures, which have generally increased more than daytime temperatures since 1950, when measurements began. Changing nighttime temperatures will have positive effects on some crops, but others like tomatoes, which set their fruit only if nighttime temperatures go below a certain level, will become more difficult to grow. Other crops that require cool summers and/or winter freezes include blueberries, maple, apples, and broccoli.

In 1999 a study at the National Science Foundation's Long-Term Ecological Research Site in Colorado linked warmer nighttime temperatures to changes in the types and distribution of grasses on the prairie. Most notably, weeds and nonnative grasses have largely replaced buffalo grass, an important food for cattle and other livestock. Range scientists point out that buffalo grass can withstand continuous grazing, but the invading plants replacing buffalo grass may be more sensitive to grazing pressures. Ecologists suggest that such changes in the structure and dynamics of rangeland ecosystems worldwide could have a profound effect on livestock production.

On a regional scale, current models of modest warming forecast that agricultural productivity will increase in some areas and decline in others. Models suggest that Canada and Russia may be able to increase their agricultural productivity in a warmer climate, whereas tropical and subtropical regions where many of the world's poorest people live will be hardest hit by declining agricultural productivity. Central America and Southeast Asia may experience some of the greatest declines in agricultural productivity.

In addition to these expected impacts of climate warming on agriculture, modern, energy-intensive agricultural methods may have to be altered so that they are less reliant on CO_2-producing fossil fuels (see Chapter 19). The manufacture of fertilizers, pesticides, and other agricultural chemicals requires a huge input of energy from fossil fuels, as does the production and use of modern farm equipment.

Climate-induced changes to agriculture will require cultural, economic, and infrastructure adaptation. Some farmers will do much better while others can no longer compete. The agriculture-related fraction of GDP will go up in some countries and down in others. The types of crops that do well—and therefore the diets of many people—will change. The extent to which such adaptation can keep pace with climate change is uncertain.

International Implications of Global Climate Change

Dealing with global climate change is complicated by social, economic, and political factors that vary from one country to another. How will the global community deal with the environmental refugees of global climate change, such as those who might be affected by extreme weather events that lead to agricultural failures? Where will they go? Who will help them resettle? It will be difficult for all countries to develop a consensus on dealing with global climate change, partly because global climate change will clearly have greater impacts on some nations than on others. All major nations must cooperate if we are to effectively address global climate change and its impacts.

Highly Developed versus Developing Nations   Although highly developed countries are the primary producers of greenhouse gases, the rate of production by certain developing countries is rapidly increasing (even though their per capita emissions remain well below those for highly developed nations). In 2007, China surpassed the United States as the largest single contributor of CO_2, although the per capita rate in the United States remains about three times that of China. Furthermore, although highly developed nations have huge amounts of coastal infrastructure at risk, many developing nations may experience the greatest impacts of global climate change. Because developing countries have less technical expertise and fewer economic resources, they are the ones likely to be least able to respond to the challenges of global climate change.

Consider Bangladesh. In 2000 the Bangladeshi environment minister estimated that rising sea levels may displace perhaps 20 million people from coastal areas of Bangladesh by the end of the 21st century. Because there is not enough land at higher elevations to accommodate these refugees, she asked highly developed countries to rethink their immigration policies.

Tensions have increased among nations, especially between the highly developed and developing countries, over their differing self-interests. Most developing countries see increased use of fossil fuels as their route to industrial development and resist pressure from highly developed nations to decrease fossil fuel consumption. Developing countries such as India ask why they should have to take actions to curb CO_2 emissions when the rich industrialized nations historically have been the main cause of the problem. Since 1950, the 20% of the world's population living in highly developed nations have produced 74% of the CO_2 emissions. Currently, highly developed countries produce about ten times more CO_2 emissions per person than developing countries (**Figure 21.12**).

In contrast, highly developed countries argue that the booming economic growth and much greater number of people living in developing countries threaten to overwhelm the world with carbon dioxide emissions as the developing countries become industrialized. According to the U.S. Department of Energy, CO_2 emissions from developing countries will surpass those from highly developed countries by 2020, assuming current trends in fossil fuel consumption continue. Developing countries respond that even when they are producing half of the world's CO_2 emissions, it will still be unequal because 80% of the world's population living in developing countries will be producing only half the emissions.

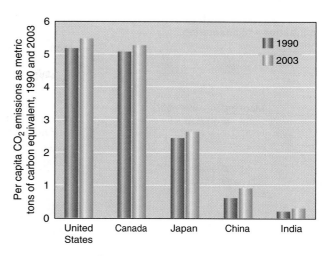

Figure 21.12   **Per capita carbon dioxide (CO_2) emission estimates for selected countries, 1990 and 2003.** For each country, the column on the left is the 1990 value, and the column on the right is the 2003 value. In all cases, per capita CO_2 emissions have grown. Currently, industrialized nations produce a disproportionate share of CO_2 emissions. As developing nations such as China and India industrialize, however, their per capita CO_2 emissions increase. China's per capita use has increased by 50% over a decade and a half. (In conformance with scientific practice, CO_2 emissions are quantified by their weight in carbon. One metric ton of carbon equivalent equals 3.67 metric tons of CO_2.) (World Bank)

Links among Global Climate Change, Ozone Depletion, and Acid Deposition

Often, environmental studies examine a single issue, such as global climate change, acid deposition, or ozone depletion (Chapter 20). Canadian researchers at the Experimental Lakes Area in Ontario, Canada, decided to take a different approach and explore the interactions of all three environmental problems simultaneously. In 1996 the scientists reported that organisms in North American lakes may be more susceptible to damage from UV radiation than the thinning of the ozone layer would indicate. The combined effects of acid deposition and climate warming may increase how far UV radiation penetrates lake water. Some of the possible effects of increased UV penetration are to disrupt photosynthesis in algae and aquatic plants and to cause sunburn damage (skin lesions) in fishes.

How far UV radiation penetrates lake water is related to the presence of dissolved organic compounds. This organic material, which is present even in the clearest, least polluted lakes, comes from the decomposition of dead organisms. Dissolved organic material acts like a sunscreen, absorbing UV radiation so that it penetrates only a few inches into the water.

A warmer climate increases evaporation, which reduces the amount of water flowing into a lake from the surrounding watershed. Because most of a lake's dissolved organic compounds wash into the lake in the stream flow, even a slightly drier climate reduces how much organic material is present in the lake. Therefore, UV radiation penetrates deeper into the lake as a result of climate warming.

Figure 21.13 Relationship between mitigation and adaptation.

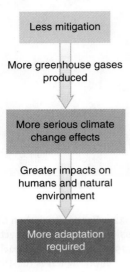

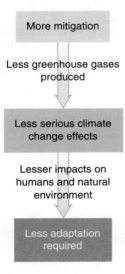

(a) If we make fewer efforts to mitigate climate change, we will have to adapt to more serious problems that impact food and water resources, biological diversity, and human health.

(b) If we take aggressive mitigation measures, the long-range changes in climate will be less and have less serious impacts on humans and the natural environment.

Acid deposition also affects the amount of dissolved organic compounds in a lake. The presence of acid causes organic matter to clump and settle on the lake floor. The removal of organic material from the water allows UV radiation to penetrate farther.

REVIEW

1. How do melting ice and thermal expansion of water contribute to sea level rise?
2. How might global climate change affect each of the following: sea level, precipitation patterns, living organisms, human health, and agriculture?
3. What do the relationships between acid deposition, stratospheric ozone, and climate change tell us about the importance of thinking about the environment as a system?

 ## Dealing with Global Climate Change

LEARNING OBJECTIVES

- Explain why reducing atmospheric carbon will require reductions or changes in global energy use.
- Differentiate between carbon capture and storage (including carbon sequestration) and reduced CO_2 production.
- Discuss the need for international commitment in order for climate change mitigation to succeed.

Our present understanding of the changing global climate and its potential impact on human society and other species gives us many legitimate reasons to develop strategies to deal with this problem. Even if we immediately stopped greenhouse gas emissions (which we cannot do, given the importance to society of the energy generated from them), global temperature would continue increasing for several decades. Sea levels would continue to rise for centuries as the Earth's climate system adjusts to the influences of the greenhouse gases that have accumulated during the past two centuries. For the same reasons that the global climate system is slow to respond to increases in greenhouse gas concentrations, it will be slow to respond even if emissions are turned off quickly.

Although stopping climate change will require that we deal with all greenhouse gases, our focus will be on CO_2 because it is produced in the greatest quantities and has the largest influence (about 60%) of all the greenhouse gases. The increase in the CO_2 level caused by human activities will persist for centuries, so the emissions we are producing today will still be around in the 22nd century and beyond. The amount and severity of global climate change that will be left to our grandchildren are thus directly related to how much additional greenhouse gases we add to the atmosphere during our lifetimes.

To avoid the most dangerous consequences of climate change, many studies indicate that the atmospheric CO_2 concentration needs to be stabilized at 550 ppm. This is roughly twice the concentration of atmospheric CO_2 that scientists estimate existed in the preindustrial world and only about 50% higher than the current CO_2 concentration.

There are two basic ways to attempt to manage global climate change: mitigation and adaptation (**Figure 21.13**). *Mitigation* focuses on limiting greenhouse gas emissions to moderate or postpone global climate change, thus buying time to pursue solutions that stop or reverse the change. *Adaptation* focuses on learning to live with the environmental changes and societal consequences brought about by global climate change. While some people have objected to the development of strategies to adapt to climate change because they feel this implies an acceptance that global climate change is unavoidable, the widespread observations that the climate is changing make clear that adaptation will be unavoidable. The only question is whether we prepare for the changes or simply react to them. On the other hand, sufficient mitigation to halt climate change is a choice that the world (meaning us and everyone else) may or may not make in time to slow the pace of change and limit the damage from climate change that our children and grandchildren will face. How we deal with climate change is a decision with multi-generational consequences.

Mitigation of Global Climate Change

Because global warming is essentially a consequence of the choices that are made in the generation and use of energy, the development and use of alternatives to fossil fuels have the potential to eventually halt the warming caused by CO_2 emissions. Alternatives to petroleum (oil) and natural gas (but not coal) are also likely to become necessary over coming decades because these fuels are present in limited amounts. Some alternatives to fossil fuels—solar energy and nuclear energy—were discussed in Chapters 12 and 13. Energy conservation techniques, discussed in Chapter 13, could also help address a significant part of the climate change problem.

Many studies indicate that energy use and greenhouse gas emissions can be significantly reduced with little cost to society by adopting the best currently available technologies and implementing certain policies to encourage their use. For example, increasing the efficiency of automobiles and appliances would reduce the use of fossil fuels and the output of CO_2. In California, regulatory programs that increase minimum energy efficiency standards for appliances and buildings have cut per capita electricity use to half the U.S. average. The higher price of gasoline has led consumers to want cars with increased gas mileage, which carmakers have proven they can make profitably,

E N V I R O N E W S

California Takes the Lead

Due to its large and diverse economy, California currently emits more greenhouse gases than do most countries. That is about to change. On September 27, 2006, California Governor Arnold Schwarzenegger signed into law Assembly Bill 32, which mandates reductions in the State's emissions of greenhouse gases (see photo). California's plan is to combine regulatory and economic tools to reduce CO_2 emissions by 25% between now and 2020, with a long-term goal having 2050 emissions significantly below those in 1990.

According to the bill's sponsors, two forces are behind California's decision. First, climate change will have substantial impacts on California's population and economy. While some business leaders are concerned about negative impacts of CO_2 reductions, others expect that the long-term economic effects will be positive. Second, the United States lacks a national-level climate management policy. California is not new to taking a lead on air pollution mitigation—many environmental laws that the United States passed in the 1970s were modeled on California's legislation in the 1950s and 1960s. A big difference between this and earlier policies, though, is that previous laws managed emissions and effects that occurred within the State, whereas the causes and effects of climate change are global. In 2007, the U.S. Supreme Court decided the Environmental Protection Agency is required under the **Clean Air Act** to regulate CO_2 and other greenhouse gas emissions. The United States may again be following the California example.

California Governor signs climate change mitigation bill on Treasure Island. Accompanying California Governor Arnold Schwarzenegger are a number of environmental, civic, and world leaders including New York Governor George Pataki. (Office of California Governor Arnold Schwarzenegger.)

carbon sequestration: Placing carbon that has been produced when generating usable energy from fossil fuels into some sort of permanent storage.

as shown by the rising sales of hybrid vehicles. Energy-pricing strategies, such as carbon taxes and the elimination of energy subsidies (see Chapter 11), are other policies that could help lead to reductions in emissions. A carbon tax (See Chapter 2) could be levied against users of fossil fuels, based on the proportion of CO_2 emissions produced per unit of heat released when the fuel is burned. Because coal is carbon-intensive, it would have the highest carbon taxes of all fossil fuels.

Carbon Capture and Storage   In addition to taking steps to curb greenhouse gas emissions, many countries are investigating **carbon management,** ways to capture and **sequester,** or store CO_2. Several power plants currently capture CO_2 emitted in their flue gases, but the technology is new. Technological innovations that efficiently trap the CO_2 being emitted from smokestacks would help to slow the growth in carbon emissions and yet allow continued use of fossil fuels (while they last) for the generation of electricity. Government incentives, such as providing research grants for the development of such technologies, most likely will be necessary to inspire innovations. Several nations have imposed taxes on greenhouse gases; the taxes are intended to motivate industrial emitters to improve efficiency and to develop CO_2-capturing technologies.

Carbon sequestration would require a dramatic shift in where fossil fuels are used. Trying to remove the CO_2 from the engine in a gasoline-powered car or diesel train would be a huge logistical challenge: Capturing CO_2 is difficult, and transporting CO_2 as a gas or as calcium carbonate ($CaCO_3$) to some separate location would require additional energy. Most proposals for carbon sequestration involve generating electricity or hydrogen at a fixed location where the CO_2 can most easily be captured.

Carbon dioxide or some other carbon compound could be sequestered in geologic formations or depleted oil or natural gas wells on the land or injected as a liquid into the ocean depths. Carbon management is a new, unproven technology, and its costs and potential ecological effects will have to be studied fully before it is adopted on a large scale. England, Norway, Japan, the United States, Canada, and Australia are among the countries most interested in developing carbon management.

Sequestering Carbon in Trees   One way to mitigate global climate change involves removing atmospheric carbon dioxide from the air by planting and maintaining forests (recall the chapter introduction). Like other green plants, trees incorporate the carbon into organic matter in leaves, stems, and roots through the process of photosynthesis. Because trees typically live for 100 or more years, the carbon in their roots and stems remains sequestered away from the atmosphere for a relatively long time. Although estimates vary widely, a reasonable estimate is that trees could remove 10 to 15% of the excess CO_2 in the atmosphere, but it would require enormous plantings. Scientists who have studied sequestering carbon in trees think this proposal might provide short-term benefits for the climate, although it is no substitute for cutting emissions of greenhouse gases.

Adaptation to Global Climate Change

Because the overwhelming majority of climate experts think that significant human-induced global climate change is inevitable (the only question being how significant), government planners and social scientists are developing strategies to help various regions and sectors of society adapt to climate warming. One of the most pressing issues is rising sea level. People living in coastal areas could be moved inland, away from the dangers of storm surges. This solution would have high societal and economic costs, especially given the increasing fraction of the population living near the ocean. Another alternative is the construction of dikes and levees to protect coastal land—an expensive option, but perhaps less so than relocating the people protected by such constructions. Rivers and canals that spill into the ocean would have to be channeled to prevent saltwater intrusion into fresh water and agricultural land.

We also must adapt to shifting agricultural zones. Many countries with temperate climates are evaluating semitropical crops to determine the best substitutes for traditional crops as the climate warms. Large lumber companies are currently developing drought-resistant strains of trees. Trees planted today will be harvested in the last half of the 21st century, when global climate change may be well advanced. These adaptations will result in economic shifts, as certain low-yield acreage gains value, and other high-yield acreage loses value.

Adaptation to global climate change is currently under study at locations around the United States. Study groups typically consist of scientists and various representatives of municipal governments, state and federal agencies, local businesses, and community organizations. One of the potential problems identified in the New York City study group involves its sewer system. The waterways for storm runoff normally close during high tides to prevent salt water from the Atlantic Ocean from backing into the system. As the sea level rises in response to global climate change, the waterways will have to be shut during many low tides, which will greatly increase the risk of flooding during storms. (When the waterways are closed, excess water does not drain away.) City planners will have to rebuild the storm runoff system, which is an expensive proposition, or find some other way to prevent flooding. Evaluating such problems and finding and implementing solutions now will ease future stresses of climate change.

International Efforts to Reduce Greenhouse Gas Emissions

Despite all the posturing, the international community recognizes that it must stabilize CO_2 emissions. At least 174 nations, including the United States, have now signed the U.N. Framework Convention on Climate Change (UNFCCC) developed at the 1992 Earth Summit. Its ultimate goal was to stabilize greenhouse gas concentrations in the atmosphere at levels low enough to prevent dangerous human influences on the climate. The details of how that goal was to be accomplished were left to future conferences.

At the 1996 meeting of the parties to the UNFCCC held in Geneva, Switzerland, highly developed countries agreed to establish legally binding timetables to cut emissions of greenhouse gases. These timetables were decided at a meeting of representatives from 160 countries held in Kyoto, Japan, in 1997. By 2005 enough countries had ratified the resulting **Kyoto Protocol** for it to come into force. This international treaty, which is legally binding, provides operational rules on reducing greenhouse gas emissions.

It is noteworthy that the United States and Australia have not ratified the Kyoto Protocol. The United States initialed the Kyoto Protocol in 1998, but the Clinton Administration never forwarded it to the Senate for ratification (they were continuing to negotiate details and implementation). President George W. Bush withdrew the United States commitment in 2001 on the grounds that its implementation via a tax or permit system aimed at reducing greenhouse gases would create an unacceptable economic burden for the nation (such as causing the loss of jobs as energy costs climb). There is, however, a considerable range of estimates of potential economic costs. When the environmental and societal costs of climate change, including damage to coastal infrastructure, ecosystems, and human health, are considered, many climate experts calculate that ratification of the Kyoto Protocol would be economically beneficial.

Most analysts agree that the Kyoto Protocol will prove difficult to implement without the full participation of the United States, since in the short term, a country with strict CO_2 restrictions will be at a competitive disadvantage compared to a country with more relaxed standards. In place of the Kyoto Protocol, President Bush's strategy is to encourage U.S. companies and citizens to voluntarily reduce their emissions by taking steps to use fossil fuels more efficiently and, where possible, to adopt renewable energy in place of fossil fuels. While the long trend of improving efficiency has continued since the President's decision, total U.S. emissions have continued to rise. In response, some U.S. states, cities, and corporations have initiated actions to reduce greenhouse gases; these policies are part of a wider ranging effort to improve their environment and economy, enhance energy security, and promote price stability by increasing the diversity of their energy sources.

Even if all countries, including the United States, were to implement the Kyoto Protocol fully, it would not prevent the continuing buildup of atmospheric greenhouse gases; it would only modestly slow their rate of buildup. The international community recognizes that the Kyoto Protocol is only the first step in addressing climate change to the extent called for in the UNFCCC. However, the protocol provides the legal framework whereby countries can forge agreements to make additional cuts in greenhouse gases in the future. Importantly, ratifying the Kyoto Protocol might signal the start of international commitment and actions to address climate change, and to understand the effectiveness and costs associated with various policy alternatives.

REVIEW

1. Why do we focus on CO_2 when there are several other greenhouse gases?
2. Why is adaptation an important issue when considering climate change?
3. What are two approaches to carbon sequestration?
4. Why are international relations essential to solving problems associated with climate change?

REVIEW OF LEARNING OBJECTIVES WITH KEY TERMS

• **Describe radiative forcing and the enhanced greenhouse effect, and list the five main greenhouse gases.**

Radiative forcing is the capacity of a gas to affect the balance of energy that enters and leaves the atmosphere. The **enhanced greenhouse effect** is the additional warming produced by increased levels of gases that absorb **infrared radiation (heat)**. **Greenhouse gases**, the gases that absorb infrared radiation, include carbon dioxide, methane, nitrous oxide, chlorofluorocarbons, and tropospheric ozone. Based on numerous studies, the IPCC, a United Nations panel of experts, concluded in 2007 that human-produced air pollutants have caused most of the recent climate warming. Climate scientists conclude that the world will almost certainly warm substantially during the 21st century. Scientists are less certain about what regional patterns may emerge.

• **Distinguish between positive and negative feedback and discuss water vapor and the aerosol effect as they relate to global climate change.**

Positive feedback is a situation in which a change in some condition triggers a response that intensifies the changed condition. **Negative feedback** is a situation in which a change in some condition triggers a response that moderates the changed condition. Water vapor, a greenhouse gas, is an example of positive feedback, because warmer air temperatures cause greater evaporation from the ocean, which in turn causes warmer temperatures. The **aerosol effect** is atmospheric cooling or warming that occurs where and when aerosol pollution is greatest (e.g., sulfates tend to reflect radiation and lead to a cooling influence; soot aerosols tend to absorb solar radiation and contribute to a warming influence). The effect of greenhouse gases is substantially more potent than that of aerosols.

• **Explain how climate models project future climate conditions.**

Climate models are computer models that describe the global climate as a system. The models divide the atmosphere and oceans into small, three-dimensional parts, and evaluate the effects of changes in one part on adjacent parts. These models incorporate positive and negative feedbacks that influence such factors as temperature, wind patterns, cloud moisture, and ice cover. Running these models based on different predicted levels of CO_2 leads to projections of possible future climate conditions.

• **Describe the importance of extreme and unpredictable climate change.**

While we often look at expected results of climate models, the extreme or "worst case" scenarios are often of much greater concern. Knowing, for example, that sea level may rise one meter may be more important than knowing it will probably rise one-fifth of that amount. Unpredictable changes are perhaps more problematic. It is almost certain that there will be surprises, or things we cannot predict. But not knowing what the surprises will be makes it impossible to prepare for or mitigate them. The only way to avoid such surprises is to eliminate climate change.

• **Differentiate between sea level rise due to melting and that due to thermal expansion of water.**

More than half of recent and projected sea level rise is associated with the fact that water expands as it heats up; this is known as **thermal expansion**. How quickly the ocean absorbs increased atmospheric heat will significantly influence the rate of sea level rise. Melting of glaciers and land-based ice sheets is the other major contributor to sea level rise, as water previously "stored" on land will shift to the ocean.

• **Describe how climate change will impact non-human life on Earth.**

Climate change will impact non-human life in several ways. The ocean will become more acidic as CO_2 is dissolved. As temperatures get warmer and precipitation patterns shift, the geographic range of different species will shift. Depending on how extreme and rapid these changes occur, some species may not be able to adapt quickly enough. In addition, non-human life interacts as a system, so changes in one or two species in one location can impact many, geographically dispersed species.

• **Give examples of effects of climate change on humans.**

Humans have already been affected by the direct impact of hotter summer extremes. In addition to this, the ranges of many tropical diseases, including malaria, are likely to expand into areas with large human populations. Agriculture will also be affected, which may impact food availability, especially at the local level. Sea level rise has impacted people who live in coastal areas, from inhabitants of low-lying islands to farmers in England. Climate change may even lead to conflicts, especially over access to water.

• **Describe how interactions among global climate change, ozone depletion, and acid deposition affect North American lakes.**

How far UV radiation, which has increased because of the thinning ozone layer, penetrates water is related to the presence of dissolved organic material. Organic material acts like a sunscreen, absorbing UV radiation so that it penetrates only a few inches. A warmer climate can increase evaporation from the watersheds for lakes, thereby reducing the amount of water flowing into a lake if precipitation does not increase sufficiently. Because most of a lake's dissolved organic compounds wash in from the watershed, even a slightly drier climate reduces how much organic material is present in the lake. As a result, UV radiation penetrates deeper into the lake. In addition, acid from acid deposition causes organic matter to clump and settle on the lake floor; the removal of organic material from the water allows UV radiation to penetrate farther, causing greater impacts for life in the lake.

• **Explain why reducing atmospheric carbon will require reductions or changes in global energy use.**

Fossil fuels—coal, oil, and natural gas—are the major sources of both energy use and CO_2 production. One way to reduce CO_2 production is to reduce the use of coal, oil, and natural gas. Reducing CO_2 emissions will require a shift in energy sources, such as more wind, solar, or nuclear energy.

• **Differentiate between carbon capture and storage (including carbon sequestration) and reduced CO_2 production.**

Carbon capture and storage, or **sequestration,** would allow us to continue burning fossil fuels, while finding some way to prevent the resulting CO_2 from being released into the atmosphere. This could involve underground storage, conversion into $CaCO_3$, or other major technological developments. Reducing CO_2 production would require shifting away from carbon-based (fossil) fuels.

• **Discuss the need for international commitment in order for climate change mitigation to succeed.**

Climate change has global impacts. Reducing carbon emissions and adapting to a changed climate cannot succeed when only a few countries participate. One concern is that, in the short term, a country with strict CO_2 restrictions will be at a competitive disadvantage compared to a country without. Thus, having international commitments may be necessary for success in any location.

Thinking About the Environment

1. How does the enhanced greenhouse effect impact global climate?

2. Within the scientific community, as well as major groups such as the IPCC and the U.S. National Academies of Science, there is agreement that human-caused climate change has begun and will be significant in the next few years. Yet, a look at the media, blogs, and political debate suggests that there is scientific uncertainty. What might account for this difference in perspective?

3. Austrian biologists who study plants growing high in the Alps found that plants adapted to cold-mountain conditions migrated up the peaks as fast as 3.7 m (12 ft) every decade during the 20th century, apparently in response to climate warming. Assuming that warming continues during the 21st century, what will happen to the plants if they reach the tops of the mountains?

4. With regard to global climate change, less reduction of greenhouse gases will mean more need for adaptation. Discuss the implications of this on people today and people 50 years from now.

5. Suggest some ways in which seasonal changes in plant growth in Canada might affect songbirds in Mexico.

6. One of Socolow and Pacala's wedges (see introduction) involves reducing the number of miles driven by each car by 50% (they predict two billion cars worldwide by 2056). Considering what you know about how people drive now, how could people reduce their driving by 50%? How would this affect their lives?

7. What aspects of adaptation to global climate change would be easier for highly developed nations, and which for developing nations? Explain.

8. Based on what you've read in this chapter and the information on environmental economics in Chapter 2, explain how a global market for CO_2 permits might function.

9. Insurance companies that provide policies for hurricanes and other natural disasters may shift hundreds of millions of dollars of their investments from fossil fuels to solar energy. On the basis of what you have learned in this chapter, explain why insurance companies consider such an investment in their best interest.

10. Some environmentalists contend that the wisest way to "use" fossil fuels is to leave them in the ground. How would this affect air pollution? global climate change? energy supplies?

11. Explain how acid deposition, climate warming, and ozone depletion interact synergistically to increase the penetration of ultraviolet radiation in North American lakes.

12. Why does mitigating climate change require us to think about problem solving from a systems perspective?

Quantitative questions relating to this chapter are on our Web site.

Take a Stand

Visit our Web site at http://www.wiley.com/college/raven (select Chapter 21 from the Table of Contents) for links to more information about the Kyoto Protocol. Consider the opposing views of proponents and opponents of the protocol, and debate the issue with your classmates. You will find tools to help you organize your research, analyze the data, think critically about the issues, and construct a well-considered argument. *Take a Stand* activities can be done individually or in a team, as oral presentations, written exercises, or Web-based (e-mail) assignments.

Additional online materials relating to this chapter, including a Student Testing Section with study aids and self-tests, Environmental News, Activity Links, Environmental Investigations, and more, are also on our Web site.

Water Pollution

Checking plant life in the wastewater treatment system in Arcata, California. This constructed wetland is a successful way to treat sewage in a small community like Arcata. (© Ted Shreshinsky/Corbis Images)

S eventeen thousand people inhabit the town of Arcata on the coast of northern California. Home to Humboldt State University, this small college town has an international reputation for ecological innovation. Faced with financing a $25 million regional wastewater treatment plant in 1975, Arcata decided in 1978 to pioneer a low-tech, systems-based, natural approach. The city restored and constructed a series of freshwater wetlands in a former industrial area and then routed the wastewater through these wetlands. The wetland wastewater treatment plant was completed in 1986 at a cost of $7 million.

Arcata's treatment of wastewater initially follows the steps used in most municipalities. The solid contaminants are allowed to settle out, and the dissolved organic wastes are biologically degraded and then treated with chlorine to remove disease-causing agents. However, conventional treatment does not remove other pollutants, such as nitrogen and phosphorus, because it is too

expensive. Such pollutants are usually left in the treated wastewater. Unfortunately, when treated wastewater is discharged into rivers, streams, or the ocean, these contaminants sometimes cause problems.

Arcata developed a way to remove such contaminants from treated water for a fraction of the cost of a normal advanced treatment plant and, at the same time, to increase the amount of ecologically important wetlands in the town's vicinity. The town hired biologists, who worked with city engineers to develop a series of six marshes that occupy about 62 hectares (154 acres). Essentially, Arcata uses cattails, bulrushes, and other marsh plants to absorb and assimilate the contaminants, thus cleaning the wastewater (see photograph). Algae, fungi, and bacteria living in the marsh also feed on these contaminants. After water is purified in the series of marshes, it is pumped to a treatment center, where it is chlorinated to kill bacteria, treated with sulfur dioxide to remove any remaining chlorine, and finally released into nearby Humboldt Bay.

The highly productive marsh ecosystem—the Arcata Marsh and Wildlife Sanctuary—provides wildlife habitat for many organisms, such as fishes, muskrats, raccoons, and river otters. Thousands of birds reside in the wetlands permanently or temporarily, and more than 200 species of birds such as ducks, coots, herons, egrets, grebes, and osprey have been observed in the sanctuary. For human recreation the sanctuary has 7.2 km (4.5 mi) of trails that meander through the wetlands.

At least 800 towns and cities around the world have followed Arcata's example and built wetlands to treat wastewater. Although most of the constructed wetlands for wastewater treatment are found in small coastal communities, even large urban cities such as Phoenix and Orlando have made use of this approach. Orlando, Florida, restored a 486-hectare (1200-acre) wetland that was drained and used as a cow pasture since the late 1800s. This wetland now removes phosphorus and nitrogen contaminants from 49 million L (13 million gal) of treated city wastewater each day.

As you saw in Chapter 14, all organisms require water for their survival. However, having water of good quality is just as important as having enough water. Water is used over and over—as, for example, when a downstream town uses river water that an upstream city first used. Therefore, wastewater treatment is an important part of **sustainable water use.**

This chapter discusses some of the pollutants found in fresh water and how we can improve water quality. (Ocean pollution is discussed in Chapters 6 and 19.) Although the United States has made progress in cleaning up its water since the early 1980s, much remains to be done. In some areas, water quality has actually deteriorated, whereas in other areas, strong cleanup efforts have allowed us only to hold our ground, and not make any gains.

World View and Closer to You...
Real news footage relating to water pollution around the world and in your own geographic region is available at www.wiley.com/college/raven by clicking on "World View and Closer to You."

Types of Water Pollution

LEARNING OBJECTIVES

- Define *water pollution*.
- List and briefly describe eight categories of water pollutants.
- Discuss how sewage is related to eutrophication, biochemical oxygen demand (BOD), and dissolved oxygen.
- Distinguish between oligotrophic and eutrophic lakes and explain how humans induce artificial eutrophication.
- State the purpose of the fecal coliform test.

Water pollution is a global problem that varies in magnitude and type of pollutant from one region to another. In many locations, particularly in developing countries, the main water pollution issue is lack of disease-free drinking water. Water pollutants are divided into eight categories: sewage, disease-causing agents, sediment pollution, inorganic plant and algal nutrients, organic compounds, inorganic chemicals, radioactive substances, and thermal pollution. These eight types are not exclusive: For example, sewage can contain disease-causing agents, inorganic plant and algal nutrients, and organic compounds. Let us examine each of these types of water pollution.

water pollution: Any physical or chemical change in water that adversely affects the health of humans and other organisms.

Sewage

The release of **sewage** into water causes several pollution problems. First, because it carries disease-causing agents, water polluted with sewage poses a threat to public health (see the next section on disease-causing agents). Sewage also generates two serious environmental problems in water, enrichment and oxygen demand. **Enrichment,** the fertilization of a body of water, is due to the presence of high levels of plant and algal nutrients such as nitrogen and phosphorus. Microorganisms decompose sewage and other organic materials into carbon dioxide (CO_2), water, and similar inoffensive materials. This degradation process, known as **cellular respiration,** requires the presence of oxygen. Fishes and other organisms in healthy aquatic ecosystems also use oxygen. But oxygen has a limited ability to dissolve in water, and when an aquatic ecosystem contains high levels of sewage or other organic material, the decomposing microorganisms use up most of the dissolved oxygen, leaving little for fishes or other aquatic animals. At extremely low oxygen levels, fishes and other animals leave or die.

Sewage and other organic wastes are measured in terms of their **biochemical oxygen demand (BOD),** or **biological oxygen demand.** BOD is usually expressed as milligrams of dissolved oxygen per liter of water for a specific number of days at a given temperature. A large amount of sewage in water generates a high BOD, which robs the water of dissolved oxygen (**Figure 22.1**). When dissolved oxygen levels are low, anaerobic (without oxygen) microorganisms produce compounds with unpleasant odors, further deteriorating water quality.

sewage: Wastewater from drains or sewers (from toilets, washing machines, and showers); includes human wastes, soaps, and detergents.

biochemical oxygen demand (BOD): The amount of oxygen needed by microorganisms to decompose biological wastes into carbon dioxide, water, and minerals.

Figure 22.1   Effect of sewage on dissolved oxygen and biochemical oxygen demand (BOD). Note the initial oxygen depletion (blue line) and increasing BOD (red line) close to the sewage spill (at distance 0). The stream gradually recovers as the sewage is diluted and degraded. As indicated on the graph, fishes cannot live in water that contains less than 4 mg of dissolved oxygen per liter of water.

Eutrophication: An Enrichment Problem   Lakes, estuaries, and slow-flowing streams that have minimal levels of nutrients are unenriched, or oligotrophic. An **oligotrophic** lake has clear water and supports small populations of aquatic organisms (**Figure 22.2a**). **Eutrophication** is the enrichment of a lake, estuary, or slow-flowing stream by inorganic plant and algal nutrients such as phosphorus; an enriched body of water is said to be **eutrophic.** The enrichment of water results in an increased photosynthetic productivity. The water in a eutrophic lake is cloudy and usually resembles pea soup because of the presence of vast numbers of algae and cyanobacteria (**Figure 22.2b**).

Although eutrophic lakes contain large populations of aquatic animals, these organisms are different from those predominant in oligotrophic lakes. For example, an unenriched lake in the northeastern United States may contain pike, sturgeon, and whitefish (**Figure 22.2c**). All three are found in the deeper,

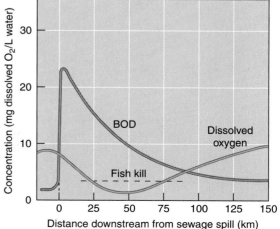

(a) Crater lake, an oligotrophic lake in Oregon.
(Rich Buzzelli/Tom Stack & Associates)

(b) A small eutrophic lake in western New York.
(W.A. Banaszewski/Visuals Unlimited)

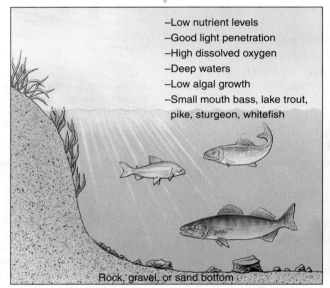

—Low nutrient levels
—Good light penetration
—High dissolved oxygen
—Deep waters
—Low algal growth
—Small mouth bass, lake trout, pike, sturgeon, whitefish

Rock, gravel, or sand bottom

(c) An oligotrophic lake has a low level of inorganic plant and algal nutrients.

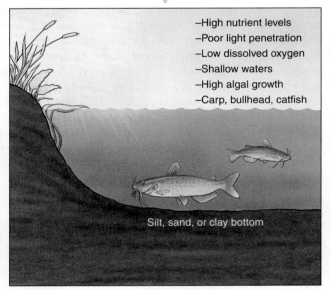

—High nutrient levels
—Poor light penetration
—Low dissolved oxygen
—Shallow waters
—High algal growth
—Carp, bullhead, catfish

Silt, sand, or clay bottom

(d) A eutrophic lake has a high level of these nutrients.

Figure 22.2   Oligotrophic and eutrophic lakes.

colder part of the lake, where there is a higher concentration of dissolved oxygen. In eutrophic lakes, on the other hand, the deeper, colder levels of water are depleted of dissolved oxygen because when the excessive numbers of algae die, they settle to the lake's bottom and stimulate an increased amount of decay. Microorganisms that decompose the dead algae use up much of the lake's dissolved oxygen in the process. There is a high BOD from decomposition on the lake floor, and fishes such as pike, sturgeon, and whitefish die out and are replaced by warm water fishes, such as catfish and carp, that tolerate smaller amounts of dissolved oxygen (**Figure 22.2d**).

Over vast periods, oligotrophic lakes, estuaries, and slow-moving streams become eutrophic naturally. As natural eutrophication occurs, these bodies of water are slowly enriched and grow shallower from the immense number of dead organisms that have settled in the sediments over a long period. Gradually, plants such as water lilies and cat-tails take root in the nutrient-rich sediments and begin to fill in the shallow waters, forming a marsh. Some human activities, however, greatly accelerate eutrophication. This fast, human-induced process is usually called **artificial eutrophication**, or *cultural eutrophication*, to distinguish it from natural eutrophication. Artificial eutrophication results from the enrichment of aquatic ecosystems by nutrients found predominantly in fertilizer runoff and sewage.

artificial eutrophication: Overnourishment of an aquatic ecosystem by nutrients such as nitrates and phosphates; due to human activities such as agriculture and discharge from sewage treatment plants.

Disease-Causing Agents

Disease-causing agents are infectious organisms that cause diseases; they come from the wastes of infected individuals. Municipal wastewater usually contains many bacteria, viruses, protozoa, parasitic worms, and other infectious agents that cause human or animal diseases (**Table 22.1**). Typhoid, cholera, bacterial dysentery, polio, and infectious hepatitis are some of the more common bacterial or viral diseases transmitted through contaminated food and water. However, many human diseases such as acquired immune deficiency syndrome (AIDS) are not transmissible through water.

The vulnerability of our public water supplies to waterborne disease-causing agents was dramatically demonstrated in 1993 when a microorganism (*Cryptosporidium*) contaminated the water supply in the greater Milwaukee area. About 370,000 people developed diarrhea, making it the largest outbreak of a waterborne disease ever recorded in the United States, and several people with weakened immune systems died. Smaller outbreaks of contamination by salmonella and other bacteria have occurred in several cities and towns since the Milwaukee episode. In 2000 the first waterborne outbreak in North America of the deadly strain of *Escherichia coli* (0157:H7) occurred in an Ontario town (Walkerton) in Canada. Several people were killed, and several thousand became sick. Prior to this outbreak, scientists thought this *E. coli* strain was transmitted almost exclusively through contaminated food. These and similar outbreaks have triggered additional concerns about the safety of our drinking water.

Table 22.1 Some Human Diseases Transmitted by Polluted Water

Disease	Infectious Agent	Type of Organism	Symptoms
Cholera	*Vibrio cholerae*	Bacterium	Severe diarrhea, vomiting; fluid loss of as much as 20 quarts per day causes cramps and collapse
Dysentery	*Shigella dysenteriae*	Bacterium	Infection of the colon causes painful diarrhea with mucus and blood in the stools; abdominal pain
Enteritis	*Clostridium perfringens,* other bacteria	Bacterium	Inflammation of the small intestine causes general discomfort, loss of appetite, abdominal cramps, and diarrhea
Typhoid	*Salmonella typhi*	Bacterium	Early symptoms include headache, loss of energy, fever; later, a pink rash appears along with (sometimes) hemorrhaging in the intestines
Infectious hepatitis	Hepatitis virus A	Virus	Inflammation of liver causes jaundice, fever, headache, nausea, vomiting, severe loss of appetite, muscle aches, and general discomfort
Poliomyelitis	Poliovirus	Virus	Early symptoms include sore throat, fever, diarrhea, and aching in limbs and back; when infection spreads to spinal cord, paralysis and atrophy of muscles occur
Crytosporidiosis	*Cryptosoporidium* sp.	Protozoon	Diarrhea and cramps last up to 22 days
Amoebic dysentery	*Entamoeba histolytica*	Protozoon	Infection of the colon causes painful diarrhea with mucus and blood in the stools; abdominal pain
Schistosomiasis	*Schistosoma* sp.	Fluke	Tropical disorder of the liver and bladder causes blood in urine, diarrhea, weakness, lack of energy, repeated attacks of abdominal pain
Ancylostomiasis	*Ancylostoma* sp.	Hookworm	Symptoms are severe anemia and sometimes symptoms of bronchitis

fecal coliform test: A water quality test for the presence of fecal bacteria, which indicates a chance that pathogenic organisms may be present as well.

Figure 22.3 **Fecal coliform test.**

(a) A water sample is first passed through a filtering apparatus. The filter disk is then placed on a medium that supports coliform bacteria for a period of 24 hours. (Courtesy Millipore Corp.)

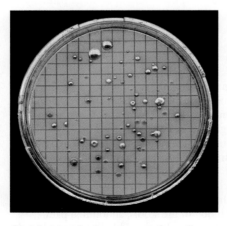

(b) After incubation, the number of bacterial colonies is counted. Each colony of *Escherichia coli* arose from a single coliform bacterium in the original water sample. (Courtesy Millipore Corp.)

Monitoring for Sewage Because sewage-contaminated water is a threat to public health, periodic tests are made for the presence of sewage in our water supplies. Although many different microorganisms thrive in sewage, the common intestinal bacterium, *E. coli*, is typically used as an indication of the amount of sewage present in water and as an indirect measure of the presence of disease-causing agents. *E. coli* is perfect for monitoring sewage because it is not present in the environment except from human and animal feces, where it is found in large numbers. To test for the presence of *E. coli* in water, the fecal coliform test is performed (**Figure 22.3**). A small sample of water is passed through a filter to trap all bacteria. The filter is then transferred to a petri dish that contains nutrients. After an incubation period, the number of greenish colonies present indicates the number of *E. coli*. Safe drinking water should contain no more than one coliform bacterium per 100 mL of water (about 1/2 cup), safe swimming water should have no more than 200 per 100 mL of water, and general recreational water (for boating) should have no more than 2000 per 100 mL. In contrast, raw sewage may contain several million coliform bacteria per 100 mL of water. Although most strains of coliform bacteria do not cause disease, the fecal coliform test is a reliable way to indicate the likely presence of pathogens, or disease-causing agents, in water.

When dangerous levels of fecal coliform bacteria are discovered in a stream or other body of water, it is important to determine the source of contamination. Finding the source is not always easy because coliform bacteria reside in the intestinal tracts of many animals. The contamination could be coming from human wastes, such as from septic systems that are not operating effectively (discussed later in the chapter); from animal feedlots; or even from the droppings of raccoons, birds, and other wildlife. A new field of science, **bacterial source tracking (BST),** attempts to make the proper identification. BST uses some of the latest techniques in molecular biology to determine subtle differences in strains of *E. coli* on the basis of their animal host. Although this science is still in its infancy, it has successfully identified the source of coliform bacteria in several cases, such as Virginia's Four Mile Run near Washington, D.C.

Sediment Pollution

Clay, silt, sand, and gravel are sediments suspended and carried in water. When a river flows into a lake or ocean, its flow velocity decreases, and the sediments often settle out. Over time, as sediments accumulate, new land is formed. A river delta is a flat, low-lying plain created from these sediments. River deltas, with their abundant wildlife and waterways for trade routes, have always been important settlement sites for people. Today, river deltas are among the most densely populated areas in the world. Sediments are also deposited on land when a river overflows its banks during a flood.

Sediment pollution consists of excessive amounts of suspended soil particles that eventually settle out and accumulate on the bottom of a body of water. Sediment pollution comes from erosion of agricultural lands, forest soils exposed by logging, degraded stream banks, overgrazed rangelands, strip mines, and construction. Control of soil erosion reduces sediment pollution in waterways.

Sediment pollution reduces light penetration, covers aquatic organisms, brings insoluble toxic pollutants into the water, and fills in waterways. When sediment particles are suspended in the water, they make the water turbid (cloudy), which in turn decreases the distance that light penetrates. Because the base of the food web in an aquatic ecosystem consists of photosynthetic algae and plants that require light for photosynthesis, turbid water lessens the ability of producers to photosynthesize. Extreme turbidity reduces the number of photosynthesizing organisms, which in turn causes a decrease in the number of aquatic organisms that feed on the primary producers (**Figure 22.4**). Sediment that settles out of the water and forms a layer over coral reefs or shellfish beds can clog the gills and feeding structures of many aquatic animals.

Sediments adversely affect water quality by carrying toxic chemicals, both inorganic and organic, into the water. The sediment particles provide surface area to which some insoluble, toxic compounds adhere, so that when sediments get into wa-

Figure 22.4   **Sediment pollution.**

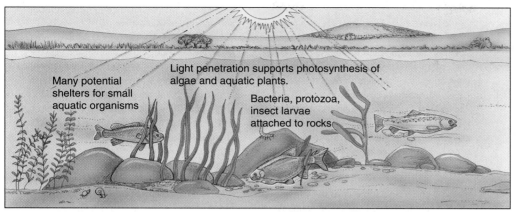

(a) Stream ecosystem with low level of sediment.

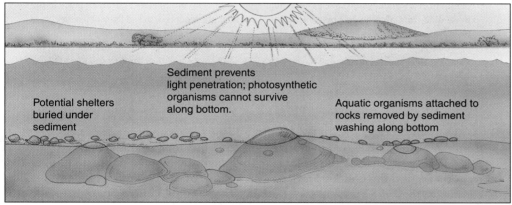

(b) Same stream with high level of sediment.

ter, the toxic chemicals get in as well. Disease-causing agents are also transported into water via sediments.

On the basis of the growing recognition that most of the toxic pollutants in water are stored in and released from sediments, the U.S. Environmental Protection Agency (EPA) evaluated data from more than 21,000 sampling stations in 1363 U.S. watersheds. The resulting survey, which was released in 1998, determined that sediments in 7% of watersheds are so seriously contaminated with toxic pollutants that eating fishes from those waterways would potentially threaten human health.

When sediments settle out of solution, they fill in waterways. This problem is particularly serious in lakes and channels through which ships must pass. Thus, sediment pollution may adversely affect the shipping industry.

Inorganic Plant and Algal Nutrients

Inorganic plant and algal nutrients are chemicals such as nitrogen and phosphorus that stimulate the growth of plants and algae. Inorganic plant and algal nutrients are essential for the normal functioning of healthy ecosystems but are harmful in larger concentrations. Nitrates and phosphates come from such sources as human and animal wastes, plant residues, atmospheric deposition, and fertilizer runoff from agricultural and residential land. Inorganic plant and algal nutrients encourage excessive growth of algae and aquatic plants. Although algae and aquatic plants are the base of the food web in aquatic ecosystems, their excessive growth disrupts the natural balance between producers and consumers and causes other problems, including enrichment, bad odors, and a high BOD. The high BOD occurs when the excessive numbers of algae die and are decomposed by bacteria.

ENVIRONEWS

Harmful Algal Blooms

When some pigmented marine algae experience population explosions or blooms, their great abundance frequently colors the water orange, red, or brown. One familiar type of bloom, known as **red tide,** may cause serious environmental harm and threaten the health of humans and animals (see Figure 7.13). Some of the algal species that form red tides produce toxins that attack the nervous systems of fishes, leading to massive fish kills. Water birds such as cormorants suffer and sometimes die when they eat the contaminated fishes. The toxins work their way up the food web to marine mammals and people. In 1997 more than 100 monk seals, one-third of the endangered species' total numbers, died from algal toxin poisoning off the West African coast. Humans may suffer if they consume algal toxins in shellfish or fishes. Even nontoxic algal species may wreak havoc when they bloom, as they shade aquatic vegetation and upset food web dynamics.

No one knows what triggers red tides, which are becoming more common and more severe, but many experts think the blame lies with coastal pollution. Wastewater and agricultural runoff to coastal areas contains increasingly larger quantities of nitrogen and phosphorus, two nutrients that stimulate algal growth. Changes in ocean temperatures, such as those attributed to global warming, may trigger algal blooms. In addition, a possible connection exists between red tide outbreaks in Florida's coastal waters and the arrival of dust clouds from Africa. These dust clouds, which sometimes blow across the Atlantic Ocean, enrich the water with iron and seem to trigger algal blooms.

Few control measures are in place to prevent the blooms or to end them when they occur. Newer technologies such as satellite monitoring and weather-tracking systems allow better prediction of conditions likely to stimulate blooms.

The Dead Zone in the Gulf of Mexico Every spring and summer, fertilizer runoff from midwestern fields and manure runoff from livestock operations in such states as Iowa, Wisconsin, and Illinois eventually find their way into the Mississippi River and, from there, into the Gulf of Mexico. The amount of such runoff is considerable. According to the 1998 Senate Agriculture Hearings, in the United States livestock produce approximately 20 times the feces and urine that humans do, yet even a decade later, many of these wastes are still not covered by water quality laws and do not go to sewage treatment plants.

These nutrients (nitrogen and phosphorus) are largely responsible for a huge dead zone in the Gulf of Mexico (**Figure 22.5**). The dead zone extends from the sea floor

Figure 22.5 **The dead zone in the Gulf of Mexico.** Pollution from the Mississippi River's water basin (tan) is responsible for the dead zone (red).

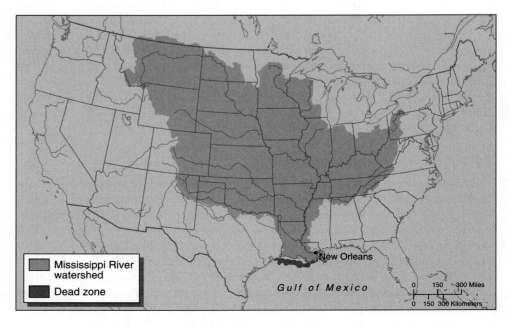

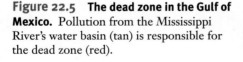
Mississippi River watershed

Dead zone

New Orleans

Gulf of Mexico

0   150   300 Miles

0 150 300 Kilometers

up into the water column, sometimes to within a few meters of the surface. Floods and droughts change its size and shape. It generally persists from March or April, as snowmelt and spring rains flow from the Mississippi River into the Gulf, to September. It is most severe in June, July, and August. Although the size of the dead zone varies with weather conditions, overall it seems to be growing. In 2002 it was more than 22,000 km^2 (8500 mi^2) (about the size of Massachusetts), which is the largest dead zone in the Gulf of Mexico since measurements began in the 1980s.

Other than bacteria that thrive in oxygen-free environments, no life exists in the dead zone. The water does not contain enough dissolved oxygen to support fishes or other aquatic organisms. Fishes, shrimp, and other active swimmers avoid the area, but bottom dwellers such as sea stars, brittle stars, worms, and clams suffocate and die.

This oxygen-free condition, known as **hypoxia,** occurs when algae grow rapidly because of the presence of nutrients in the water. When these algae die, they sink to the bottom and are decomposed by bacteria, which deplete the water of dissolved oxygen, leaving too little for other sea life. Although hypoxia has been reported in more than 146 coastal areas around the world, the dead zone in the Gulf of Mexico is one of the largest. (Dead zones in the Black Sea and the Baltic Sea are reportedly larger.)

In 2001 the EPA released recommendations to reduce nitrogen runoff coming down the Mississippi River by 30% by 2015, because of the threat of enriched nutrients to commercial fisheries and to the marine environment in general. The challenge is to modify farming methods so that less fertilizer is needed, because then most of the fertilizer would remain on the farmers' fields. (The Mississippi River drains all or part of 31 states and two Canadian provinces, and its watershed contains more than half of all U.S. farms.) In 2004 the EPA added phosphorus pollution as a cause of the dead zone and recommended a focus on cleaning up phosphorus as well as nitrogen.

Other potential inorganic nutrient sources, such as sewage treatment plants and airborne nitrogen oxides from automobile emissions, also must be addressed. Restoring former wetlands in the Mississippi River watershed would reduce the nitrate and phosphate load from fertilizers entering the Gulf of Mexico. Recall from the chapter introduction that wetlands retain nutrients such as nitrogen and phosphorus. The EPA recognizes that the dead zone problem is immense in scope and will take billions of dollars and decades of effort to fix (see **Figure 22.6**).

Figure 22.6   Cartoon decrying dead oceans. What does the last panel of this cartoon suggest about influences on public attitudes toward the environment? Do you agree? (Rustle the Leaf Comics, properties © 2005 and TM of GO NATUR L STUDIOS, LLC. All Rights Reserved.)

Organic Compounds

Organic compounds are chemicals that contain carbon atoms; a few examples of natural organic compounds are sugars, amino acids, and oils. Most of the thousands of organic compounds found in water are human-produced chemicals; these synthetic chemicals include pesticides, solvents, industrial chemicals, and plastics. (Several examples of toxic organic compounds sometimes found in polluted water are given in **Table 22.2**). Some organic compounds seep from landfills into surface water and groundwater. Others, such as pesticides, leach downward through the soil into groundwater or get into surface water by runoff from farms and residences. Some industries dump organic compounds directly into waterways.

A comprehensive study of synthetic organic pollutants in U.S. waterways was completed in 2002. Researchers from the U.S. Geological Survey (USGS) collected samples of water from 139 streams in 30 states. Most of the streams were downstream from cities and agricultural areas such as dairies and pig farms. The scientists tested the water samples for 95 different organic compounds, such as antibiotics, ibuprofen, acetaminophen, insect repellents, antimicrobial substances, fragrances, caffeine, and steroids such as hormones from birth control pills and hormone therapy. Low concentrations—in the range of **parts per billion**—of 82 of these organic compounds were detected. At least one organic chemical was present in 80% of the streams, whereas more than one-third of the water samples contained traces of 10 or more organic compounds. The effects on human health of ingesting drinking water containing traces of these chemicals are generally unknown. However, the hormone contaminants cause problems in many aquatic organisms (see Chapter 7). Of the organic chemicals that were tested, 33 are suspected to be **endocrine disrupters** and cause hormonal effects. The USGS, which has monitored U.S. streams, rivers, and groundwater since the 1970s, has recently expanded its water testing to include more than 600 chemical contaminants.

There are several ways to control the presence of organic compounds in our water. Everyone, from individual homeowners to large factories, should take care to prevent organic compounds from ever finding their way into water. Alternative organic compounds, which are less toxic and degrade more readily so that they are not as persistent in the environment, can be developed and used. Tertiary water treatment, considered later in this chapter, effectively eliminates many synthetic organic compounds in water.

Inorganic Chemicals

Inorganic chemicals are contaminants that contain elements other than carbon; examples include acids, salts, and heavy metals. Inorganic chemicals do not easily degrade, or break down. When they are introduced into a body of water, they remain there for a long time. Many inorganic chemicals find their way into both surface water and groundwater from sources such as industries, mines, irrigation runoff, oil drilling, and urban runoff from storm sewers. Some of these inorganic pollutants are toxic to aquatic organisms. Their presence may make water unsuitable for drinking or other purposes.

Table 22.2 Some Synthetic Organic Compounds Found in Polluted Water

Compound	Some Reported Health Effects
Aldicarb (pesticide)	Attacks nervous system
Benzene (solvent)	Associated with blood disorders (bone marrow suppression); leukemia
Carbon tetrachloride (solvent)	Possibly causes cancer; liver damage; may also attack kidneys and vision
Chloroform (solvent)	Possibly causes cancer
Dioxins (TCDD) (chemical contaminants)	Some cause cancer; may harm reproductive, immune, and nervous systems
Ethylene dibromide (EDB) (fumigant)	Probably causes cancer; attacks liver and kidneys
Polychlorinated biphenyls (PCBs) (industrial chemicals)	Attack liver and kidneys; possibly cause cancer
Trichloroethylene (TCE) (solvent)	Probably causes cancer; induces liver cancer in mice
Vinyl chloride (plastics industry)	Causes cancer

Here we consider the heavy metals lead and mercury, two inorganic chemicals that sometimes contaminate water and accumulate in the tissues of humans and other organisms (see the discussion of bioaccumulation and biological magnification in Chapter 7). Arsenic, another heavy metal, is discussed later in this chapter.

Lead People used to think of lead poisoning as affecting only inner-city children who ate paint chips that contained lead. Lead-based paint was banned in the United States in 1978, but the EPA estimates that more than three-fourths of U.S. homes still contain some lead-based paint. Although lead-based paint remains an important source of lead poisoning in children, lead lurks in many other places in the environment as well.

Lead-containing anti-knock agents in gasoline were outlawed in the United States in 1986; prior to 1986 lead dust was released into the atmosphere when the fuel was burned, and that lead still contaminates the soil, particularly in inner cities near major highways. Children living in the inner city may be at risk when they play outdoors in their schoolyards and backyards. Lead contaminates the soil, surface water, and ground-water when incinerator ash is dumped into ordinary sanitary landfills. It may be spewed into the atmosphere from old factories that lack air pollution control devices. We ingest additional amounts of lead from pesticide and fertilizer residues on produce, from food cans soldered with lead, and even from certain types of dinnerware on which our food is served. Low amounts of lead also originate from natural sources such as volcanoes and wind-blown dust.

Millions of U.S. residents, many of them children, have damaging levels of lead in their bodies. At least 2% of U.S. children age one to five have blood lead levels that exceed 10 micrograms (μg) per deciliter (dL) of blood. A blood lead level above 10 μg per dL is considered dangerous, and recent data suggest that even lower concentrations of lead are harmful to the brain.

The three groups of people at greatest risk from lead poisoning are middle-aged men, pregnant women, and young children. Middle-aged men with high levels of lead are more likely to develop **hypertension,** or high blood pressure. High lead levels in pregnant women increase the risk of miscarriages, premature deliveries, and stillbirths. Even children with low levels of lead in their blood may suffer from a variety of mental and physical impairments, including partial hearing loss, hyperactivity, attention deficit, lowered IQ, and learning disabilities. A 1996 study by the University of Pittsburgh School of Medicine reported a link between male juvenile delinquency and high bone lead concentrations. (Because lead accumulates in bones, bone lead levels more accurately reflect long-term exposure than do blood lead levels.) In 2001 the American Medical Association reported a link between murder rates and lead levels in the air.

According to the EPA, in the mid-1990s more than 10% of all large and medium-sized municipal water supplies contained lead levels that exceeded the maximum permitted by the Safe Drinking Water Act (discussed later in this chapter). In addition, tap water often contains higher levels of lead than are in municipal water supplies; the extra lead comes from the corrosion of old lead water pipes or of lead solder in newer pipes.

Mercury Mercury is a metal that vaporizes at room temperatures; this characteristic poses special environmental challenges. Small amounts of mercury occur naturally in the environment, but most mercury pollution comes from human activities. According to the EPA, coal-fired power plants release the largest amount (40%) of mercury into the environment. Coal contains traces of mercury that vaporize and are released into the atmosphere with the flue gases when the coal is burned. This mercury then moves from the atmosphere to the water via precipitation. The technology exists to control mercury emissions from coal-burning power plants, but it is expensive, and the trapped mercury would have to be properly disposed in a hazardous waste landfill or it could recontaminate the environment. In 2005, the EPA issued the nation's first proposal for regulating mercury emissions from power plants. Environmentalists think that the EPA proposal does not adequately address the problem. They recommend that Congress adopt a more aggressive plan to reduce mercury emissions.

Municipal waste and medical waste incinerators also release mercury (when incinerators burn materials containing mercury.) Fluorescent lights and thermostats are examples of municipal wastes that contain mercury, whereas thermometers and blood-pressure cuffs are examples of medical waste. The EPA now regulates mercury emissions from municipal and medical incinerators. Many hospitals, recognizing the risks posed by mercury, are switching away from mercury-containing equipment (including thermometers).

Significant amounts of mercury are released into the environment during the smelting of metals such as lead, copper, and zinc. Mercury is used in a variety of industrial processes, such as chemical plants that manufacture chlorine and caustic soda. Some of this mercury vaporizes, thereby entering the atmosphere. In addition, when industries release their wastewater, some metallic mercury may enter natural bodies of water along with the wastewater. Mercury sometimes enters water by precipitation after household trash containing batteries, paints, and plastics is burned in incinerators.

Once mercury enters a body of water, it settles into the sediments where bacteria convert it to methyl mercury compounds, a more toxic form that readily enters the food web. Mercury bioaccumulates in the muscles of albacore tuna, swordfish, sharks, king mackerel, and marine mammals—the top predators of the open ocean. Human exposure to mercury is primarily from eating fishes and marine mammals containing high levels of mercury. At least 48 states in the United States have released health advisories on the human consumption of mercury-tainted seafood from lakes and reservoirs.

Methyl mercury compounds remain in the environment for a long time and are highly toxic to organisms, including humans. Prolonged exposure to methyl mercury compounds causes kidney disorders and severely damages the nervous and cardiovascular systems. The exposure of developing human fetuses to mercury is linked to a variety of conditions, such as diminished cognitive function, cerebral palsy, and developmental delays. Methyl mercury compounds are unusual in that they can cross the body's blood-brain barrier (many materials do not pass from the blood to the cerebrospinal fluid and brain). Low levels of mercury in the brain cause neurological problems such as headache, depression, and quarrelsome behavior.

Radioactive Substances   **Radioactive substances** contain atoms of unstable isotopes that spontaneously emit radiation. Radioactive substances get into water from several sources, including the mining and processing of radioactive minerals such as uranium and thorium. Many industries use radioactive substances; although nuclear power plants and the nuclear weapons industry use the largest amounts, medical and scientific research facilities also employ them. It is possible for radiation to inadvertently escape from any of these facilities, polluting the air, water, and soil. Accidents at nuclear power plants may release into the atmosphere large quantities of radiation, which eventually contaminate soil and water. Radiation from natural sources can pollute groundwater.

Since the mid-1980s, low levels of radioactive substances have been measured in the wastewater of several sewage treatment plants in the United States. The EPA reports that radioactive materials may concentrate in sludge (a slimy solid mixture formed during the treatment of sewage). Guidelines from the EPA help municipal sewage treatment plants identify radioactive materials in sewage sludge and, when present, reduce or eliminate the contamination.

Thermal Pollution

Thermal pollution occurs when heated water produced during certain industrial processes is released into waterways. Many industries, such as steam-generated electric power plants, use water to remove excess heat from their operations. Afterward, the heated water is allowed to cool a little before it is returned to waterways, but its temperature is still warmer than it was originally. The result is that the waterway is warmed slightly.

A body of water's rise in temperature has several chemical, physical, and biological effects. Chemical reactions, including decomposition of wastes, occur faster, depleting the water of oxygen. Moreover, less oxygen dissolves in warm water than in cool water (**Figure 22.7**), and the amount of oxygen dissolved in water has important effects on aquatic life. When the level of dissolved oxygen is lowered due to thermal pollution, a fish ventilates its gills more frequently to obtain enough oxygen. Gill ventilation, however, requires an increased consumption of oxygen. This situation puts a great deal of stress on the fish as it tries to obtain a greater supply of oxygen from a smaller supply dissolved in the water.

Other subtle changes may take place in the activities and behavior of aquatic organisms in thermally polluted water because temperature affects reproductive cycles, digestion rates, and respiration rates. At warmer temperatures, fishes require more food to maintain body weight. They typically have shorter life spans and smaller populations. In cases of extreme thermal pollution, fishes and other aquatic organisms die.

REVIEW

1. What is water pollution?

2. What are the eight main groups of water pollutants? Give an example of each main type.

3. What is biochemical oxygen demand (BOD)? How is BOD related to sewage?

4. What is artificial eutrophication? What causes it?

 ## Water Quality Today

LEARNING OBJECTIVE

• Contrast point source pollution and nonpoint source pollution.

Water pollutants come from both natural sources and human activities. For example, some of the mercury that contaminates the biosphere is from natural sources in Earth's crust; the remainder comes from human activities. Nitrate pollution has both natural and human sources—the nitrate that occurs in soil and the inorganic fertilizers added to it, respectively. Although natural sources of pollution are sometimes of local concern, human-generated pollution is generally more widespread.

The sources of water pollution are classified into two types: point source pollution and nonpoint source pollution. **Point source pollution** is discharged into the environment through pipes, sewers, or ditches from specific sites such as factories or sewage treatment plants. Point source pollution is relatively easy to control legislatively, but accidents still occur. A cyanide spill contaminated the Tisza and Danube rivers in Europe in 2000, killing millions of fishes and shutting off downstream water supplies. The cyanide was in a holding basin at a gold mine in Romania, but heavy snow and rain caused the basin to overflow into the river. Although cyanide disperses to nonlethal levels rapidly, the heavy metals present in the holding basin will contaminate the rivers and Black Sea, which they empty into, for many years.

Nonpoint source pollution, also called *polluted runoff*, is caused by land pollutants that enter bodies of water over large areas rather than at a single point. Nonpoint source pollution occurs when precipitation moves over and through the soil, picking up and carrying away pollutants that eventually are deposited in lakes, rivers, wetlands, groundwater, estuaries, and the ocean. Although nonpoint sources are diffuse, their cumulative effect is often huge. Nonpoint source pollution includes agricultural runoff (such as fertilizers, pesticides, livestock wastes, and salt from irrigation), mining wastes (such as acid mine drainage), municipal wastes (such as inorganic plant and algal nutrients), and construction sediments. Soil erosion from fields, logging operations, eroding stream banks, and construction sites is a major cause of nonpoint source pollution.

Three major sources of human-induced water pollution that we now examine in greater detail are agriculture, municipalities (that is, domestic activities), and industries. We then consider groundwater pollution and water pollution in other countries.

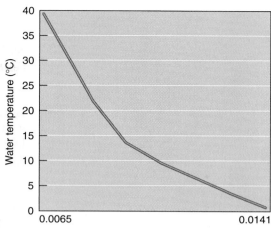

Figure 22.7   Dissolved oxygen in water at various temperatures. As water temperature rises, its capacity to contain dissolved oxygen goes down. These data are for water in contact with air at 760 mm mercury pressure.

point source pollution: Water pollution that can be traced to a specific origin.

nonpoint source pollution: Pollutants that enter bodies of water over large areas rather than being concentrated at a single point of entry.

Water Pollution from Agriculture

According to the EPA, agriculture is the leading source of water quality impairment of surface waters nationwide: 72% of the water pollution in rivers is attributed to agriculture. Agricultural practices produce several types of pollutants that contribute to nonpoint source pollution. Fertilizer runoff causes water enrichment. Animal wastes and plant residues in waterways produce high BODs and high levels of suspended solids as well as water enrichment. Amid growing concern about polluted runoff from animal wastes, the EPA unveiled a largely voluntary program in 1998 that asked the 450,000 livestock operations in the United States to develop Comprehensive Nutrient Management Plans by 2008. These plans are to consider safe ways to handle and store manure so that it does not become polluted runoff.

Chemical pesticides used in agriculture may leach into the soil and from there into water. These chemicals are highly toxic and adversely affect human health as well as the health of aquatic organisms. The National Water Quality Assessment Program, an ongoing study of pesticides and their degradation products, indicates that pesticides are widespread in U.S. rivers, streams, and groundwater. More than 95% of the river and stream samples and almost 50% of groundwater samples they examined contained at least one pesticide. Many water samples contained a mixture of pesticides. More than 50% of all stream samples contained five or more pesticides, and about 10% of all streams contained 10 or more pesticides.

Soil erosion from fields and rangelands causes sediment pollution in waterways. In addition, some agricultural chemicals that are not very soluble in water, such as certain pesticides, find their way into waterways by adhering to sediment particles. Thus, soil conservation methods both conserve soil and reduce water pollution.

Municipal Water Pollution

Although sewage is the main pollutant produced by cities and towns, municipal water pollution also has a nonpoint source: urban runoff from storm sewers (**Figure 22.8**; see

Figure 22.8 **Urban runoff.** The largest single pollutant in urban runoff is organic waste, which removes dissolved oxygen from water as it decays. Fertilizers cause excessive algal growth, which further depletes the water of oxygen. Other everyday pollutants include used motor oil, which is often illegally poured into storm drains, and heavy metals. These pollutants may be carried from storm drains on streets to streams and rivers.

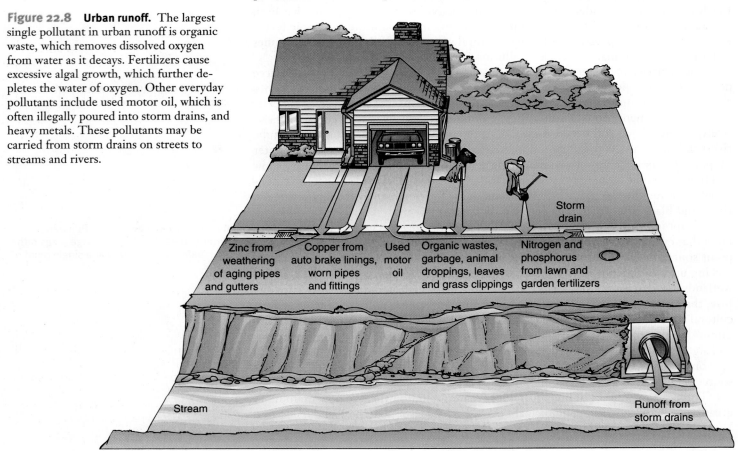

Zinc from weathering of aging pipes and gutters

Copper from auto brake linings, worn pipes and fittings

Used motor oil

Organic wastes, garbage, animal droppings, leaves and grass clippings

Nitrogen and phosphorus from lawn and garden fertilizers

Storm drain

Stream

Runoff from storm drains

"You Can Make a Difference: Preventing Water Pollution"). The water quality of urban runoff from city streets is often worse than that of sewage. Urban runoff carries salt from roadways, untreated garbage, animal wastes (especially from dogs), construction sediments, and traffic emissions (via rain that washes pollutants out of the air). It often may contain such contaminants as asbestos, chlorides, copper, cyanides, grease, hydrocarbons, lead, motor oil, organic wastes, phosphates, sulfuric acid, and zinc.

Some 1100 cities across the United States, such as New York, San Francisco, Pittsburgh, and Boston, have **combined sewer systems** in which human and industrial wastes are mixed with urban runoff from storm sewers before flowing into the sewage treatment plant. A problem arises when there is heavy rainfall or a large snowmelt because even the largest sewage treatment plant can process only a given amount of wastewater each day. When too much water enters the system, the excess, known as **combined sewer overflow,** flows into nearby waterways without being treated. Combined sewer overflow, which contains raw sewage, has been illegal since passage of the Clean Water Act of 1972 (discussed shortly), but cities have only recently begun to address the problem. According to the EPA, 1.2 trillion gallons of combined sewer overflow are discharged into U.S. waterways every year.

Some cities, such as St. Paul, Minnesota, have installed two separate sewers, one for sewage and industrial wastes and one for urban runoff. However, such an installation is expensive and requires that every street be dug up. Other cities, such as Birmingham, Michigan, have kept their combined sewer systems but installed huge retention basins to hold the overflow until it is treated. The basin in Birmingham, which was installed in 1998, holds 20.8 million L (5.5 million gal). Such tanks are less

YOU CAN MAKE A DIFFERENCE

Preventing Water Pollution

Although individuals produce little water pollution, the collective effect of municipal water pollution, even in a small neighborhood, can be quite large. There are many things you can do to protect surface waters and groundwater from water pollution. Here are some specific dos and don'ts that you should adopt; many municipalities have specific regulations or requirements that cover these.

1. Many household chemicals, such as oven cleaners, mothballs, drain cleaners, and paint thinners, are quite toxic. Use such products sparingly, and try to substitute less hazardous chemicals wherever possible. When disposing of unwanted hazardous household chemicals, contact the solid waste management office in your county for information about hazardous waste collection centers in your area. Never put these chemicals down a drain or toilet, because they may disrupt your septic system or contaminate sewage sludge produced at municipal sewage treatment facilities. Never pour these chemicals on the ground because they may contaminate runoff when it rains. Here are some safer alternatives (note that these chemicals are not nontoxic; they are simply less toxic than many commercial products):

   a. Ammonia, to clean appliances and windows.
   b. Bleach, to disinfect. Never mix ammonia and bleach because the mixture releases toxic chlorine gas.
   c. Borax, to remove stains and mildew.
   d. Baking soda, to remove stains, deodorize, and clean household utensils.
   e. Mineral oil, to polish furniture and wax floors.
   f. Vinegar, to clean surfaces, polish metals, and remove stains and mildew.

2. Never throw unwanted medicines down the toilet. Several studies have shown that traces of a variety of drugs are showing up in tap water.

3. Never pour used motor oil or antifreeze down storm drains or on the ground. Recycle these chemicals by dropping them off at a service station or local hazardous waste collection center.

4. Pick up pet waste and dispose of it in the garbage or toilet. If left on the ground, it eventually washes into waterways where it can increase BOD and fecal coliform levels.

5. Drive less. The air pollution emissions from automobiles eventually get into groundwater and surface water. Toxic metals and oil by-products deposited on the road by automobiles—the average automobile annually leaks more than one quart of petroleum products onto roads and parking lots—are washed into surface waters by precipitation.

6. If you are a homeowner, replace some of your grass lawn with trees, shrubs, and ground covers, which absorb up to 14 times more precipitation and require little or no fertilizer. To reduce erosion, use mulch to cover bare ground (see Chapter 15).

7. Use fertilizer sparingly because excess fertilizer leaches into groundwater or waterways. If you hire a professional lawn care service to apply fertilizer and pesticides, use one that monitors and applies chemicals when they are needed. Many of these companies do calendar spraying, applying chemicals every few months regardless of need.

8. Never apply fertilizer near a body of water. Always allow a buffer zone of at least 20 to 40 feet.

9. Make sure that gutters and downspouts drain onto water-absorbing grass or graveled areas instead of paved surfaces.

10. Clean up spilled oil, brake fluid, and antifreeze and sweep sidewalks and driveways instead of hosing them off. Dispose of the dirt properly; do not sweep the dirt into gutters or storm drains.

11. Likewise, do not let grass clippings or leaves wash into gutters or storm drains.

12. Use pesticides sparingly, both indoors and outdoors. Dispose of unwanted pesticides at hazardous waste collection centers.

13. Replace paved driveways and sidewalks with porous surfaces, such as interlocking bricks or stones, and build wood decks instead of concrete patios. These features allow precipitation to seep into the ground, thereby decreasing runoff.

expensive to install than separate sewer systems, but there are concerns that after several days of heavy rain or snow, the basin itself could overflow. So far, the Birmingham retention basin has overflowed less than six times a year, as compared to an average of 60 times a year that Birmingham experienced combined sewer overflow before the retention basin was installed.

Industrial Wastes in Water

Different industries generate different types of pollutants. Food processing industries produce organic wastes that are readily decomposed but have a high BOD. In addition to a high BOD, pulp and paper mills produce toxic compounds and sludge. The paper industry, however, has begun to adopt new manufacturing methods, such as the production of paper without the use of chlorine as a bleaching agent, that produce significantly less toxic effluents.

Many industries in the United States treat their wastewater with advanced treatment methods. The electronics industry produces wastewater containing high levels of heavy metals such as copper, lead, and manganese but uses special techniques such as ion exchange and electrolytic recovery to reclaim those heavy metals. Plates with commercial value are produced from the recovered metals that would otherwise have become a component of hazardous sludge. Although U.S. industries do not usually dump highly toxic wastes into water, disposal is still sometimes a problem.

CASE IN POINT

Green Chemistry

For many industrial, agricultural, and domestic processes that generate water pollution (**Figure 22.9**), the traditional solution has been to try to remove pollutants from the waste stream. This can be a difficult, expensive, and energy-intensive process, particularly for complex, biologically potent chemicals that occur, often in small amounts, in many settings. Pesticides, pharmaceuticals, cosmetics, dyes, and other chemicals enter wastewater streams from a variety of sources. Caffeine, for example, is found in fish in Puget Sound (near Seattle Washington), and musks—the molecules used to provide scent in deodorants, aftershaves, and perfumes—in catfish taken from Lake Mead (near Las Vegas, Nevada). Unused medicines, both over-the-counter and prescription, are dumped into toilets and pass through wastewater treatment systems without being altered or removed. Eventually they are found in rivers and lakes—and in the plants and animals that live there.

These chemicals—there are thousands of them—occur in relatively small amounts compared to other pollutants we often worry about, such as oil, organic matter, and nitrogen. So why should we be concerned? First, they are very difficult to remove. Most organic matter eventually breaks down into carbon dioxide; synthetic molecules, on the other hand, can be extremely hard to break down. Indeed, as in the case of dyes, longevity is something that manufacturers *want* in their products. Second, the environmental impacts of these chemicals can be substantial, even at very low concentrations. Synthetic chemicals can disrupt many species' growth, especially reproductive development. Even at low concentrations, some chemicals can affect the ratio of male to female fish and cause developmental damage, including deformities and cancer.

From a systems perspective, a variety of ways exist to deal with this problem. Several fall under the general term **green chemistry**, or chemistry designed with minimizing environmental impacts in mind. One approach of green chemistry is to find new ways to remove the chemicals from the waste stream. Researchers at the Institute for Green Oxidation Chemistry at Carnegie Mellon University have been working on this problem for years. One promising solution is a class of chemicals called TAMLs (tetra-amido macrocyclic ligands) that serve as catalysts to speed the breakdown of synthetic chemicals.

Another approach is to change chemical processes and use fewer synthetic chemicals. This branch of green chemistry is explored at laboratories around the world. For example,

Pulp Mill
Colored lignin fragments,
organochlorines

Farm
Herbicides,
insecticides
animal waste
and medicines

Sewer
Dyes,
cosmetics,
drugs

Figure 22.9 **Sources of synthetic pollutants in water.** Water pollution containing small amounts of synthetic chemicals can come from a variety of sources: industry (for example, pulp mills), agriculture, and domestic (for example, sewers). (Scientific American 2006)

the production of plastic containers can involve the release of many complex chemicals. Alternatively, biodegradable containers can be produced from sugars, with little or no chemical by-products. Such bottles have the added advantage that they can last long enough to be useful, but not for the decades that a plastic bottle would remain in the environment.

Perchloroethylene (or "perc") is a hydrocarbon responsible for the highly recognizable odor of dry cleaners. It is also a common water and air contaminant associated with a range of health problems. In parts of southern California, it is no longer legal to use perc in new or upgraded dry cleaning facilities. Several companies now provide alternatives, including less damaging hydrocarbons, silicon-based cleaners and even a new "wet cleaning" approach that uses water and alternative soaps to clean delicate fabrics without harming them.

Green chemistry has the potential to change or replace many highly toxic industrial or household chemicals. The challenge is to find alternatives that work as well and cost the same as or less than traditional synthetic chemicals. ■

Groundwater Pollution

Roughly half of the people in the United States obtain their drinking water from groundwater, which is also withdrawn for irrigation and industry. In recent years the quality of the nation's groundwater has become a concern. The most common pollutants, such as pesticides, fertilizers, and organic compounds, seep into groundwater from municipal sanitary landfills, underground storage tanks, backyards, golf courses, and

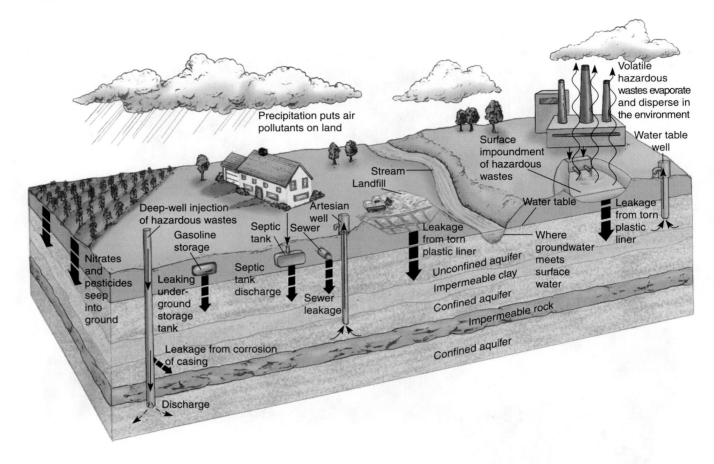

Precipitation puts air pollutants on land

Volatile hazardous wastes evaporate and disperse in the environment

Water table well

Surface impoundment of hazardous wastes

Stream
Landfill

Deep-well injection of hazardous wastes

Gasoline storage

Septic tank

Artesian well

Sewer

Water table

Leakage from torn plastic liner

Nitrates and pesticides seep into ground

Leaking underground storage tank

Septic tank discharge

Sewer leakage

Leakage from torn plastic liner

Where groundwater meets surface water

Unconfined aquifer

Impermeable clay

Confined aquifer

Impermeable rock

Leakage from corrosion of casing

Confined aquifer

Discharge

Figure 22.10 **Sources of groundwater contamination.** Agricultural practices, sewage (both treated and untreated), landfills, industrial activities, and septic systems are some of the sources of groundwater pollution. Once groundwater is contaminated, it does not readily cleanse itself by natural processes. We do not know the degree of groundwater contamination because access is difficult and because contaminants do not disperse quickly (groundwater moves very slowly). (Figure is not drawn to scale.)

intensively cultivated agricultural lands (**Figure 22.10**). More than 250,000 underground petroleum storage tanks may be leaking at service stations in the United States. Cleanup of leaking gas tanks is expensive—$500,000 or more per tank—but state gasoline taxes help pay for the cleanup in some states (such as California).

Nitrates sometimes contaminate shallow groundwater—30.5 m (100 ft) or less from the surface—with fertilizer being the most common source. High nitrate levels are a concern in some rural areas, where 80 to 90% of the residents use shallow groundwater for drinking water. When nitrates get into the human body, they are converted to nitrites, which reduce the blood's ability to transport oxygen. This condition is one of the causes of cyanosis (the "blue baby" syndrome), a serious disorder in young children. The level of nitrate in drinking water is monitored from municipal systems, so it is generally not of concern. If you drink well water, you should probably have its nitrate level checked periodically.

Contamination of groundwater is a relatively recent environmental concern. People used to think that the underlying soil and rock through which surface water must seep to become groundwater filtered out any contaminants, thereby ensuring the purity of groundwater. This assumption proved false when groups began to monitor the quality of groundwater and discovered contaminants at certain sites. It appears that the natural capacity of soil and rock to remove pollutants from groundwater varies widely from one area to another.

Currently, most of the groundwater supplies in the United States are of good quality and do not violate standards established to protect human health. There are some local problems, however, that have led to well closures and raised public health concerns. For example, in 1996 Santa Monica, California, closed 7 of its 11 municipal wells when methyl tertiary butyl ether (MTBE), a gasoline additive that reduces tailpipe emissions, was detected in the groundwater. MTBE, added to gasoline in the United States since about 1979, may cause cancer. The groundwater became polluted primarily from leaking underground gas-storage tanks. MTBE persists in groundwater for years and is expensive to remove. The EPA under the Clinton administration tried to ban MTBE as a gasoline additive, but that effort was canceled under President Bush's presidency. However, some states now ban the addition of MTBE to gasoline.

Cleanup of polluted groundwater is costly, takes years, and in some cases is not technically feasible. Compounding the cleanup problem is the challenge of safely disposing of the toxic materials removed from groundwater, which, if not handled properly, could contaminate groundwater once again.

Water Pollution in Other Countries

According to the World Health Organization, an estimated 1.4 billion people lack access to safe drinking water, and about 2.9 billion people do not have access to adequate sanitation systems; most of these people live in rural areas of developing countries. Worldwide, at least 250 million cases of water-related illnesses occur each year, with 5 million or more of these resulting in death—1.8 million from diarrheal illnesses alone.

Municipal water pollution from sewage is a greater problem in developing countries, many of which lack water treatment facilities, than in highly developed nations. Sewage from many densely populated cities in Asia, Latin America, and Africa is dumped directly into rivers or coastal harbors.

Almost every nation in the world faces problems of water pollution. For an international perspective on water pollution, let us examine some specific issues in South America, Europe, Asia, and Africa.

Lake Maracaibo, Venezuela   Lake Maracaibo in Venezuela is the largest lake in South America (**Figure 22.11**). It receives fresh water from several rivers, and water flows from it into the Caribbean Sea. Larger than the state of Connecticut, Lake Maracaibo suffers from the effects of oil pollution and human wastes as well as contamination from farms and factories. About 10,000 oil wells tap the oil and natural gas reserves under the shallow lake. An underwater network of old oil pipes about 15,400 km (9600 mi) long leaks oil into the lake. Fertilizers and other agricultural chemicals drain into the lake from nearby farms, providing the nutrients for an overgrowth of algae. Until recently, raw sewage from the city of Maracaibo's 1.6 million people and many smaller communities was discharged directly into the water, contributing to the nutrient overload. Modern sewage-treatment facilities were installed during the 1990s to take care of human wastes, but Lake Maracaibo's other pollution problems must still be addressed.

Figure 22.11   Lake Maracaibo, Venezuela. Drilling platforms are located throughout the lake, and pipelines join the oil wells to refineries on land. Shown are storage and export facilities in the foreground and the town of Cabimas in the background. An oil tanker is docked at the leftmost pier. (© Yann-Arthus-Bertrand/Corbis Images)

Po River, Italy   The Po River, which flows across northern Italy, empties into the Adriatic Sea. The Po is Italy's equivalent of the Mississippi River, and it is heavily polluted. Many cities, including Milan with 1.3 million residents, dump their treated and untreated sewage into the Po. Industry is responsible for half the pollutants that enter the Po. Italian agriculture, including large poplar plantations, relies heavily on chemicals and is responsible for massive amounts of nonpoint source pollution. Soil erosion has resulted in so much sediment deposition at the mouth of the river that the Po River Delta is advancing about 81 hectares (200 acres) into the Adriatic Sea each year.

More than 17 million people—almost one-third of all Italians—live in the Po River Basin. The health of many Italians is potentially threatened because the Po is the source of their drinking water. In addition, pollution from the Po has jeopardized tourism and fishing in the Adriatic Sea. Pollution has temporarily closed swimming at some beaches. Although Italy recognizes the problems of the Po and would like to

Figure 22.12 **Ganges River.** Bathing and washing clothes in the Ganges River are common practices in India. The river is contaminated by raw sewage discharged directly into the river at many different locations. (Mike Barlow/Dembinsky Photo Associates)

do something about them, the improvement of water quality will be difficult to implement because the river is under the jurisdiction of dozens of local and regional governments. The cleanup of the Po will require the implementation of a national plan over a period of several decades.

Ganges River, India   The Ganges River is a holy river that symbolizes the spirituality and culture of the Indian people. The river is widely used for bathing and washing clothes (**Figure 22.12**). It is highly polluted. Little of the sewage and industrial waste produced by the 350 million people who live in the Ganges River Basin is treated. Another major source of contamination is the 35,000 human bodies that are cremated annually in the open air in Varanasi, the holy city of the Hindus. (The Hindus cremate the body to free the soul; dumping the ashes into the Ganges increases the chances of the soul getting into heaven.) Incompletely burned bodies are dumped into the Ganges River, where their decomposition adds to the BOD of the river. In addition, people who cannot afford cremation costs for their dead dump human remains into the river.

The Indian government has initiated the Ganga Action Plan, an ambitious cleanup project that includes construction of water treatment plants in 29 large cities in the river basin. In addition, 32 electric crematoriums are being set up along the banks. Although the government has spent about $100 million constructing sewage treatment plants in major cities along the Ganges, most are not completed or are not working effectively. Most cities in the river basin still discharge raw sewage into the river. Costs have escalated, and many delays in the Ganga Action Plan have occurred. According to critics, the government plan has not worked in part because it has not tried to get people involved at the community level. People along the Ganges are not well informed about what the government is trying to accomplish, why it is important, and how they can help.

Kwale, Kenya   Many Africans have serious health problems from drinking surface water contaminated with disease-causing organisms. The World Bank and the U.N. Development Program sponsored a hand pump project to address the water safety problem for many Kenyans. For example, Kwale, Kenya, has had cholera and serious diarrhea outbreaks in the past. With the installation of village wells with hand pumps, it was hoped that clean groundwater would be available to the inhabitants of Kwale. Unfortunately, an acute drought lowered water levels in the early 2000s, drying up many of the wells. Pit latrines contaminated the water in many of the wells, leading to an outbreak of cholera in 2002. Flooding has led to cholera deaths in Kwale as recently as 2007.

Arsenic Poisoning in Bangladesh   The same kind of water development program that was implemented in Kenya was put into practice in Bangladesh. During the 1980s, world health organizations, concerned about illnesses caused by drinking contaminated surface water in Bangladesh, funded the installation of more than 2.5 million wells with hand pumps. Tragically, the groundwater in many of these wells is contaminated with high levels of naturally occurring arsenic. Tens of thousands of people now suffer from chronic arsenic poisoning, which initially causes skin lesions, particularly on the hands and feet. Eventually, when the cumulative dose is large enough, arsenic poisoning leads to death from cancer. Health authorities do not know how many people are at risk. Estimates range from hundreds of thousands to as many as 70 million.

International agencies are currently providing funding to test all the wells. Meanwhile, the Bangladeshi people continue to drink the groundwater because they have no other option. Scientists have developed an inexpensive bucket filtration system to remove arsenic from well water using inexpensive iron shavings. However, the arsenic-rich sediments produced by the filtration system would have to be safely disposed of if this procedure were to be widely adopted.

Bangladesh is not the only place in the world with naturally occurring arsenic contamination of groundwater. Arsenic is also found in shallow aquifers in parts of Argentina, Chile, China, Hungary, India, Mexico, Mongolia, Romania, Taiwan, Thailand, the western United States, and Vietnam.

1. What is point source pollution?
2. How do point source and nonpoint source pollution differ?

 Improving Water Quality

LEARNING OBJECTIVES

- Describe how most drinking water is purified in the United States and discuss the chlorine dilemma.
- Distinguish among primary, secondary, and tertiary treatments for wastewater.
- Define *primary* and *secondary sludge*.

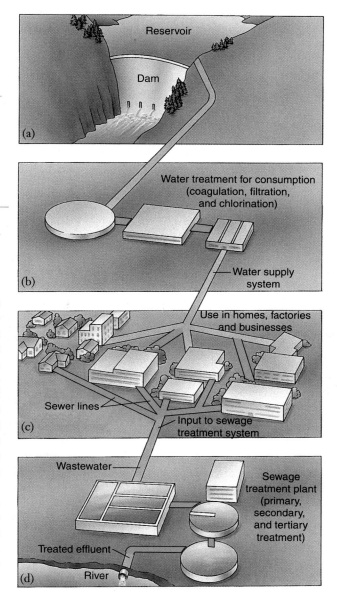

Figure 22.13 **Water treatment for municipal use.** (a) The water supply for a town may be stored in a reservoir, as shown, or obtained from groundwater. (b) The water is treated before use so that it is safe to drink. (c) After use, municipal sewer lines collect the wastewater. (d) The quality of the wastewater is fully or partially restored by sewage treatment before the treated effluent is dispersed into a nearby body of water.

Water quality is improved by removing contaminants from the water supply before and after it is used. Technology assists in both processes.

Purification of Drinking Water

The United States has nearly 60,000 municipal water facilities that serve 232 million people. Surface-water sources of municipal water supplies include streams, rivers, and lakes. Often a dam is built across a river or stream to form an artificial lake, or reservoir. Reservoirs accumulate water when there is an adequate supply and store it for use during periods of drought.

In the United States, most municipal water supplies are treated before being used so that the water is safe to drink (**Figure 22.13**). Turbid water is treated with a chemical coagulant (aluminum sulfate) that causes the suspended particles to clump together and settle out. The water is then filtered through sand to remove remaining suspended materials as well as many microorganisms. A few cities, such as Cincinnati, pump the water through activated carbon granules to remove much of the organic compounds dissolved in the water.

In the final purification step before distribution in the water system, the water is disinfected to kill any remaining disease-causing agents. The most common way to disinfect water is to add chlorine. A small amount of chlorine is left in the water to provide protection during its distribution through many kilometers of pipes. Other disinfection systems use ozone or ultraviolet (UV) radiation in place of chlorine.

The Chlorine Dilemma During the 19th century waterborne, disease-causing organisms often contaminated drinking water supplies in the United States. The discovery that chlorine kills these organisms allowed 20th-century Americans to drink water with little fear of contracting typhoid, cholera, or dysentery. The addition of chlorine to our drinking water supply has undoubtedly saved millions of lives.

At the same time, chlorine byproducts, formed when chlorine reacts with organic matter in treated wastewater, are tentatively linked to several kinds of cancer (rectal, pancreatic, and bladder), an increased risk of miscarriages, and possibly rare birth defects. As a result, use of chlorine to disinfect drinking water has triggered a debate over the costs and benefits of chlorinating water. The concern is whether there is a long-term hazard from low levels of chlorine in drinking water.

Because there are few viable alternatives to chlorination, the EPA was initially reluctant to reduce the level of chlorine permissible in drinking water, despite the evidence of potential risks. The EPA did not want what happened in Peru to occur in the United States. In 1991 a terrible cholera epidemic swept much of Peru, infecting more than 300,000 people and killing at least 3500. This outbreak occurred when Peruvian officials decided to stop chlorinating much of the country's drinking water after they learned of the slightly increased cancer risk due to chlorination. Peru has since resumed chlorinating its drinking water.

After a detailed review of current evidence linking chlorine to cancer, the EPA proposed in 1994 that water treatment facilities reduce the maximum permissible level of chlorine in drinking water. One alternative to chlorination is to use *chloramine*, a disinfectant produced by combining chlorine with ammonia. Chloramine does not form potentially harmful byproducts, although preliminary studies suggest that its use may cause an increase in lead levels in drinking-water systems. (Chloramine may make lead atoms adhering to water pipes more soluble in the water.) Another alternative is to filter water through activated carbon granules, as is done in Cincinnati; one-third less chlorine is then needed in the final step. Much of Europe has adopted another alternative to chlorination, the use of UV disinfection.

Fluoridation   Small amounts of fluoride have been added to most municipal drinking water since the mid-1940s to reduce tooth decay. Fluoride is also added to many toothpastes for the same reason. This practice is somewhat controversial, with opponents questioning the safety and effectiveness of fluoride and supporters saying it is completely safe and effective in preventing decay. More than 40 years of research have failed to link fluoridation at the levels practiced in the United States to cancer, kidney disease, birth defects, or any other serious medical condition. Most dental health officials think fluoride is the main reason for the 50 to 60% decrease in tooth decay observed in children during the past several decades. This observation is based on comparisons of cavity rates in schoolchildren between cities with fluoridation and without.

As of 2002, 66% of U.S. public water supplies were fluoridated. Currently, fluoridation is more common in the eastern half of the country than in the western half, although California mandated fluoridation in 1995, as did Nevada in 2002.

Municipal Sewage Treatment

Wastewater, including sewage, usually undergoes several treatments at a sewage treatment plant to prevent environmental and public health problems. The treated wastewater is then discharged into rivers, lakes, or the ocean.

primary treatment: Treating wastewater by removing suspended and floating particles by mechanical processes.

Primary treatment removes suspended and floating particles, such as sand and silt, by mechanical processes such as screening and gravitational settling (**Figure 22.14**,

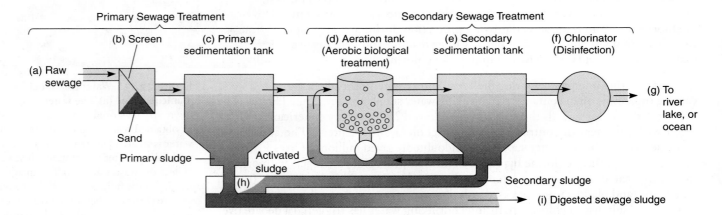

Figure 22.14   Primary and secondary sewage treatment. (a) Raw sewage enters the water treatment plant from the municipal sewage system. (b) Large debris, such as wood, metals, and plastics are removed, and sand settles to the bottom. (c) In the primary sedimentation tank, suspended solids and floating particles sink to the bottom. (d) Aeration tanks mix the partially treated wastewater with air (oxygen) to support bacteria that consume the suspended organic wastes. (e) The cleanest water is taken from the surface and placed in a secondary sedimentation tank, where any remaining suspended particles settle to the bottom. (f) The cleanest water is taken from the surface and disinfected by chlorination or ultraviolet light to kill any disease-causing bacteria. (g) The treated water is discharged to a river or other natural water source. (h) Sludge is drawn from the bottom of the primary and secondary sedimentation tanks and pumped to a digester, where bacteria consume the organic wastes. (i) The digested sewage sludge is disposed of in a sanitary landfill, incinerated, or converted into fertilizer.

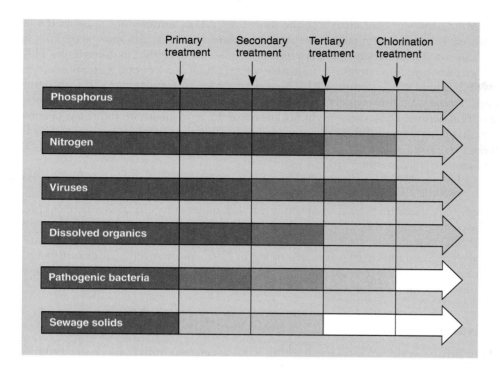

Figure 22.15 Effectiveness of primary, secondary, and tertiary sewage treatment. This figure shows the relative concentrations of various water pollutants after each treatment. The most intense color represents the greatest concentration of a given pollutant, whereas white means all of that specific pollutant is removed. Note how ineffective secondary treatment is in removing certain contaminants, such as phosphorus and nitrogen. Also note that even after tertiary treatment, some pollutants are still detectable, though at greatly reduced levels.

left side). The solid material that settles out at this stage is **primary sludge.** Primary treatment does little to eliminate the inorganic and organic compounds that remain suspended in the wastewater. The wastewater treatment facilities for about 11% of the U.S. population have primary treatment only.

Secondary treatment uses microorganisms (aerobic bacteria) to decompose the suspended organic material in wastewater **(Figure 22.14, right side).** One of the several types of secondary treatment is *trickling filters*, in which wastewater trickles through aerated rock beds that contain bacteria and other microorganisms, which degrade the organic material in the water. In another type of secondary treatment, the *activated sludge process*, wastewater is aerated and circulated through bacteria-rich particles; the bacteria degrade suspended organic material. After several hours, the particles and microorganisms are allowed to settle out, forming secondary sludge, a slimy mixture of bacteria-laden solids. Water that has undergone primary and secondary treatment is clear and free of organic wastes such as sewage. The wastewater treatment facilities for about 62% of the U.S. population have both primary and secondary treatments.

Even after primary and secondary treatments, wastewater still contains pollutants, such as dissolved minerals, heavy metals, viruses, and organic compounds **(Figure 22.15).** Advanced wastewater treatment methods, or tertiary treatment, include a variety of biological, chemical, and physical processes. Tertiary treatment reduces phosphorus and nitrogen, the nutrients most commonly associated with enrichment. Tertiary treatment purifies wastewater for reuse in communities where water is scarce. The wastewater treatment facilities for about 27% of the U.S. population have primary, secondary, and tertiary treatments.

Disposal of Sludge A major problem associated with wastewater treatment is disposal of the primary and secondary sludge formed during primary and secondary treatments. Five possible ways to handle sludge are anaerobic digestion, application to soil as a fertilizer, incineration, ocean dumping, and disposal in a sanitary landfill. In anaerobic digestion, the sludge is placed in large circular digesters and kept warm (about 35°C, or 95°F), which allows anaerobic bacteria to break down the organic material into gases such as methane and CO_2. The methane can be trapped and burned to heat the digesters.

secondary treatment: Treating wastewater biologically to decompose suspended organic material; secondary treatment reduces the water's biochemical oxygen demand.

primary and secondary sludge: The solids remaining after sewage treatment has been completed.

tertiary treatment: Advanced wastewater treatment methods that are sometimes employed after primary and secondary treatments.

After a few weeks of digestion, the sludge resembles humus and can be used as a fertilizer. It has the advantage of being rich in plant nutrients, although sometimes it contains too many heavy metals from industrial effluents to be used commercially. This happens when sewer systems mix industrial waste, which may contain toxic substances, with household waste. Farmers have long used sludge to fertilize hay and feed-grain crops. However, many farmers are reluctant to use it on crops for direct human consumption because consumers might not purchase the food grown in sludge out of concern that it may pose a threat to human health. In 1996 the National Research Council of the National Academy of Sciences announced that properly treated sludge could be used safely to fertilize food crops.

Although sludge can be used to condition soil, it is generally treated as a solid waste. Dried sludge is often incinerated, which may contribute to air pollution, although sometimes the heat is used constructively, to generate electricity, for example. In the past, coastal cities such as New York dumped their sludge into the ocean. In 1988 the U.S. Congress passed the **Ocean Dumping Ban Act,** which barred ocean dumping of sludge and industrial waste, beginning in 1991. Alternatively, sludge is disposed of in sanitary landfills (see Chapter 24). As landfill space becomes more costly, many cities are looking for other ways to handle sludge.

Individual Septic Systems

Many private residences, particularly in rural areas, use individual septic systems for sewage disposal instead of using municipal sewage treatment. Household sewage is piped to the septic tank, where particles settle to the bottom (**Figure 22.16a**). Grease and oils form a scummy layer at the top, where bacteria decompose much of it. Wastewater containing suspended organic and inorganic material then flows into the drain field through a network of small, perforated pipes set in trenches of gravel or crushed stone (**Figure 22.16b**). The drain field is located just below the soil's surface, and bacteria decompose the remaining organic material in the well-aerated soil. The purified wastewater then percolates into the groundwater or evaporates from the soil.

Septic tank systems require care to operate properly. Household chemicals such as bleach and drain cleaners must be used sparingly because they could kill the bacteria that break down the organic wastes. A kitchen garbage disposal should not be used because it could overload the system. Every two to five years, depending on use, the sludge that collects at the bottom of the septic tank is removed and taken to a municipal sewage treatment plant for proper disposal. If a septic system is not maintained properly, it can malfunction or overflow, releasing bacteria and nutrients into groundwater or waterways.

ENVIRONEWS

Water Pollution and Hurricane Katrina

When Hurricane Katrina breeched the levees around New Orleans in 2005, anything the water reached became a potential source of contamination. In the midst of an excess of water, remaining New Orleans residents faced severe shortages of safe drinking water.

Included in the flood's range were thousands of homes containing pesticides, cleaners, paints, foods (BOD), motor oil, and engine coolant. Flooded businesses included dry cleaners (perchloroethylene and other solvents), gas stations (batteries, oil, grease, coolant, and engine cleaners), printers (inks and solvents), and medical centers (medicines, disinfectant cleaners). Contaminants were picked up from outdoors as well: oil spilled on the roads, soil, construction and yard debris, animal droppings, pesticides, and fertilizers.

Just weeks after the flooding, a team of researchers from Texas Tech University found that several soil contaminants, including arsenic and lead, had been mobilized by the Katrina flooding. This was not new contamination, since the arsenic and lead were already present. However, the flood waters disturbed the soils, and allowed the chemicals to become easily airborne and inhaled or ingested, as when children eat with dirty hands.

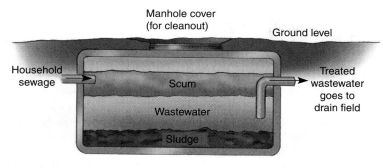

(a) The septic tank works much like primary treatment in municipal sewage treatment. Sewage from the house is piped to the septic tank, where particles settle to the bottom.

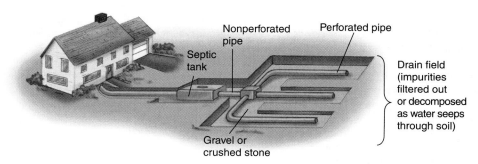

(b) Wastewater containing suspended organic and inorganic material flows into the drain field and gradually seeps into the soil.

Figure 22.16 Septic tank systems. Many private homes in rural areas use a septic tank and drainage field. Both are located underground.

REVIEW

1. How is most drinking water purified in the United States?

2. Why is chlorine added to drinking water? What is the potential problem with adding chlorine to drinking water?

3. Distinguish among primary sewage treatment, secondary sewage treatment, tertiary sewage treatment, and septic tanks.

4. What is sludge?

Laws Controlling Water Pollution

LEARNING OBJECTIVES

- Compare the goals of the Safe Drinking Water Act and the Clean Water Act.
- Define *maximum contaminant level* and specify the water legislation it relates to.
- Define *national emission limitation* and specify the water legislation it relates to.

Many governments have passed legislation to control water pollution. Point source pollutants lend themselves to effective control more readily than nonpoint source pollutants. Governments generally control point source pollution in one of two ways—by imposing penalties on polluters (a common approach in the United States) or by taxing polluters to pay for the cleanup (common in Japan).

Although most countries have passed laws to control water pollution, monitoring and enforcement are difficult, even in highly developed countries. Typically, too few resources are allotted for enforcement. For example, in 1996 the 70 enforcement agents in the Office of Drinking Water of the EPA were expected to handle more than 80,000 complaints about drinking water safety. (See "Meeting the Challenge: Using Citizen Watchdogs to Monitor Water Pollution.")

The United States has attempted to control water pollution through legislation since the passage of the **Refuse Act** of 1899, which was intended to reduce the release

MEETING THE CHALLENGE

Using Citizen Watchdogs to Monitor Water Pollution

Insufficient staff and lack of funds prevent well-intentioned government agencies from effectively monitoring and enforcing environmental laws such as the Safe Drinking Water Act. For example, the San Francisco Bay Conservation and Development Commission is California's federally designated management agency charged with patrolling 1600 km (1000 mi) of shoreline and 1500 km² (600 mi²) of water to ascertain that no one is illegally polluting San Francisco Bay. In addition, it handles hundreds of cases arising from its monitoring activities. The agency is severely understaffed and cannot adequately protect the bay. San Francisco Bay, like most other aquatic ecosystems, is endangered by the combined impact of many different pollution sources rather than from a single disaster, such as occurred when the Exxon Valdez spilled oil in Alaska (see Chapter 11). Thus, continual monitoring is necessary to ensure

that many small polluters do not collectively do irreparable harm to the bay.

A growing number of private citizens have become actively involved in monitoring and enforcing environmental laws to protect waterways in their communities. Provisions in the Clean Water Act, the Safe Drinking Water Act, and other key environmental laws allow citizens to file suit when the government does not enforce the laws. Citizen action groups also pressure firms to clean up.

In San Francisco Bay, some 300 citizen watchdogs called Bay Keepers monitor the bay from boats, airplanes, and helicopters. Law students at local universities advise the keepers on issues of litigation. The San Francisco Bay Keeper program is modeled after the Hudson River Keeper Program, which first organized in 1966 as a coalition of commercial and recreational fisherman who wanted to reclaim the Hudson River from its polluters. In 1983, using money obtained from successful lawsuits against polluters, the Hudson River Keeper program hired its first full-time River

Keeper. A network of community fishermen and environmentalists inform the River Keeper of any suspicious activity on the river. The River Keeper monitors water quality in a boat, attends board meetings, educates the public, and employs litigation as a last resort. The River Keeper position is described as part investigator, scientist, lawyer, lobbyist, and public relations agent.

Dozens of River, Sound, Bay, Inlet, Channel, and Coast Keeper groups have organized across the United States, from the Cook Inlet Keeper in Alaska to the Pensacola Gulf Coast Keeper in Florida. The umbrella organization for keeper groups is the Water Keeper Alliance, which organized in 1992. Its philosophy is based on the idea that daily vigilance by citizens is required to protect a community's natural resources. The Water Keeper Alliance helps new keeper programs organize, both in the United States and other countries. Keeper programs have been established in Australia, Bolivia, Canada, Colombia, the Czech Republic, Mexico, and the United Kingdom.

of pollutants into navigable rivers. The two federal laws that have the most impact on water quality today are the Safe Drinking Water Act and the Clean Water Act.

Safe Drinking Water Act

Prior to 1974, individual states set their own standards for drinking water, which, of course, varied a great deal from state to state. In 1974 the **Safe Drinking Water Act** was passed, which set uniform federal standards for drinking water to guarantee safe public water supplies throughout the United States. This law required the EPA to determine the **maximum contaminant level**, which is the maximum permissible amount of any water pollutant that might adversely affect human health. The EPA oversees the states to ensure that they adhere to the maximum contaminant levels for specific water pollutants. The EPA noted in 1998 that 40,000 water systems reported violations of public safe drinking water laws for that year; this number, though large, represents less than 25% of U.S. water systems. Of these, 9600 water systems (about 6%) had significant violations.

Most water suppliers take few or no steps to prevent the contamination of the watershed or groundwater from which they draw. The vast majority of water utilities do not use modern water treatment technologies such as activated carbon granules or UV disinfection to reduce chemical contamination by pesticides, arsenic, and chlorine disinfection by-products. Also, the average water pipe in the United States is 100 or more years old before it is replaced. Many aging pipes are cracked, which permits contaminated water to seep into them and increases the risk of waterborne diseases.

The Safe Drinking Water Act was amended in 1986 and again in 1996. The 1996 version requires municipal water suppliers to tell consumers what contaminants are

maximum contaminant level: The upper limit for the concentration of a particular water pollutant in water intended for human consumption.

present in their city's water and whether these contaminants pose a health risk. The law requires the EPA to review risks from radon and arsenic in drinking water and to revise its drinking water standards for each contaminant accordingly.

Clean Water Act

The **Clean Water Act** affects the quality of rivers, lakes, aquifers, estuaries, and coastal waters in the United States. Originally passed as the Water Pollution Control Act of 1972, it was amended and renamed the Clean Water Act of 1977; additional amendments were made in 1981 and 1987. Congress will likely reauthorize the Clean Water Act sometime in the next few years. The Clean Water Act has two basic goals: to eliminate the discharge of pollutants in U.S. waterways and to attain water quality levels that make these waterways safe for fishing and swimming. Under the provisions of this act, the EPA is required to set up and monitor **national emission limitations**.

Overall, the Clean Water Act effectively improved the quality of water from point sources, despite the relatively low fines it imposes on polluters. It is not hard to identify point sources, which must obtain permits from the **National Pollutant Discharge Elimination System (NPDES)** to discharge untreated wastewater.

According to the EPA, nonpoint source pollution is a major cause of water pollution. However, nonpoint source pollution is much more difficult and expensive to control than point source pollution. The 1987 amendments to the Clean Water Act expanded the NPDES to include nonpoint sources, such as sediment erosion from construction sites.

To date, U.S. environmental policies have failed to effectively address nonpoint source pollution, which requires regulating land use, agricultural practices, and many other activities. Such regulation necessitates the interaction and cooperation of many government agencies, environmental organizations, and private citizens. Such coordination is enormously challenging but necessary to reduce nonpoint source pollution.

The United States has improved its water quality in the past several decades, thereby demonstrating that the environment recovers once pollutants are eliminated. Much remains to be done, however. The EPA's 2002 National Water Quality Inventory indicated that water pollution has increased in U.S. rivers, lakes, estuaries, and coastal areas in recent years. According to the report, based on 2000 data from the states, 39% of the nation's rivers, 45% of its lakes, and 51% of its estuaries were too polluted for swimming, fishing, or drinking.

national emission limitation: The maximum permissible amount of a water pollutant that can be discharged from a sewage treatment plant, factory, or other point source.

Laws That Protect Groundwater

Several federal laws attempt to control groundwater pollution. The Safe Drinking Water Act contains provisions to protect underground aquifers that are important sources of drinking water. In addition, the Safe Drinking Water Act regulates underground injection of wastes in an effort to prevent groundwater contamination. The **Resource, Conservation, and Recovery Act** deals with the storage and disposal of hazardous wastes and helps prevent groundwater contamination (see Chapter 24). Several miscellaneous laws related to pesticides, strip mining, and cleanup of abandoned hazardous waste sites indirectly protect groundwater. The many laws that directly or indirectly affect groundwater quality were passed at different times and for different reasons. These laws provide a disjointed and, at times, inconsistent protection of groundwater. The EPA makes an effort to coordinate all these laws, but groundwater contamination still occurs.

REVIEW

1. What are the main goals of the Safe Drinking Water Act? the Clean Water Act?

2. How do maximum contaminant levels and national emission limitations differ?

REVIEW OF LEARNING OBJECTIVES WITH KEY TERMS

- **Define *water pollution*.**

Water pollution consists of any physical or chemical change in water that adversely affects the health of humans and other organisms.

- **List and briefly describe eight categories of water pollutants.**

Sewage is the release of wastewater from drains or sewers (from toilets, washing machines, and showers); it includes human wastes, soaps, and detergents. **Disease-causing agents**, such as bacteria, viruses, protozoa, and parasitic worms, are transmitted in sewage. **Sediment pollution**, primarily from soil erosion, increases water turbidity, thereby reducing photosynthetic productivity in the water. **Inorganic plant and algal nutrients**, such as nitrogen and phosphorus, contribute to **enrichment**, the fertilization of a body of water. Many **organic compounds**, such as pesticides, solvents, and industrial chemicals, are quite toxic to organisms. **Inorganic chemicals** include toxins such as lead and mercury. **Radioactive substances** include the wastes from mining, refining, and using radioactive metals. Radioactive substances may concentrate in sewage sludge. **Thermal pollution** occurs when heated water, produced during many industrial processes, is released into waterways.

- **Discuss how sewage is related to eutrophication, biochemical oxygen demand (BOD), and dissolved oxygen.**

Sewage supplies nutrients that contribute to eutrophication and a high biochemical oxygen demand. **Eutrophication**, the nutrient enrichment of lakes, estuaries, or slow-moving streams, results in high photosynthetic productivity, supporting an overpopulation of algae. Eutrophication kills fishes and causes a decline in water quality as these algae die and decompose. **Biochemical oxygen demand (BOD)** is the amount of oxygen needed by microorganisms to decompose biological wastes into carbon dioxide, water, and minerals. A large amount of sewage generates a high BOD, which lowers the level of dissolved oxygen in the water.

- **Distinguish between oligotrophic and eutrophic lakes and explain how humans induce artificial eutrophication.**

An **oligotrophic** lake has a low level of nutrients and is therefore unenriched. A **eutrophic** lake has a high level of nutrients and is enriched. Eutrophic lakes tend to fill in rapidly as dead organisms settle to the bottom. **Artificial eutrophication**, overnourishment of an aquatic ecosystem by nutrients such as nitrates and phosphates, is due to human activities such as agriculture and discharge from sewage treatment plants.

- **State the purpose of the fecal coliform test.**

The **fecal coliform test** is a water quality test for the presence of fecal bacteria, which indicates a chance that pathogenic organisms may be present as well. Thus, the common intestinal bacterium *Escherichia coli* provides an indirect measure of disease-causing organisms.

- **Contrast point source pollution and nonpoint source pollution.**

Point source pollution is water pollution that can be traced to a specific spot. **Nonpoint source pollution** consists of pollutants that enter bodies of water over large areas rather than being concentrated at a single point of entry.

- **Describe how most drinking water is purified in the United States and discuss the chlorine dilemma.**

Most municipal water supplies are treated before being used so that the water is safe to drink. Water is usually treated with aluminum sulfate to cause suspended particles to clump and settle out, filtered through sand, and disinfected by adding chlorine. Because there is concern over whether low levels of chlorine in drinking water pose a health hazard, the EPA has proposed that water treatment facilities reduce the maximum permissible level of chlorine in drinking water.

- **Distinguish among primary, secondary, and tertiary treatments for wastewater.**

Primary treatment is treating wastewater by removing suspended and floating particles by mechanical processes. **Secondary treatment** is treating wastewater biologically to decompose suspended organic material; secondary treatment reduces the water's biochemical oxygen demand. **Tertiary treatment** is advanced wastewater treatment methods that are sometimes employed after primary and secondary treatments.

- **Define *primary* and *secondary sludge*.**

Primary and **secondary sludge** consist of the solids remaining after sewage treatment has been completed. Primary sludge is formed during primary treatment, and secondary sludge is formed during secondary treatment. One of the most pressing problems of wastewater treatment is sludge disposal.

- **Compare the goals of the Safe Drinking Water Act and the Clean Water Act.**

The **Safe Drinking Water Act** sets uniform federal standards for drinking water to guarantee safe public water supplies throughout the United States. The **Clean Water Act** has two basic goals: to eliminate the discharge of pollutants in U.S. waterways and to attain water quality levels that make these waterways safe to fish and swim in.

- **Define *maximum contaminant* level and specify the water legislation it relates to.**

Maximum contaminant level is the upper limit for the concentration of a particular water pollutant in water intended for human consumption. The Safe Drinking Water Act determines maximum contaminant levels for water pollutants that might affect human health.

- **Define *national emission limitation* and specify the water legislation it relates to.**

National emission limitation is the maximum permissible amount of a water pollutant that can be discharged from a sewage treatment plant, factory, or other point source. The Clean Water Act instructs the EPA to set up and monitor national emission limitations.

Thinking About the Environment

1. What is sustainable water use?

2. What is water pollution? Why is wastewater treatment an important part of sustainable water use?

3. Explain why untreated sewage may kill fishes when it is added directly to a body of water.

4. How do midwestern farmers threaten the livelihood of fishermen in the Gulf of Mexico?

5. How can coastal hypoxia be reversed?

6. Distinguish between oligotrophic and eutrophic lakes.

7. Contrast organic compounds and inorganic chemicals as types of water pollution.

8. Tell whether each of the following represents point source pollution or nonpoint source pollution: fertilizer runoff from farms, thermal pollution from a power plant, urban runoff, sewage from a ship, erosion sediments from deforestation.

9. What is the source of arsenic contamination of groundwater in Bangladesh?

10. Compare the potential pollution problems of groundwater and surface water used as sources of drinking water.

11. Why is chlorine added to drinking water? Why does the Environmental Protection Agency recommend that public water treatment facilities find alternatives to chlorine?

12. What is sludge? How is it disposed of?

13. Is the Clean Water Act related in any way to the quality of public drinking water in the United States? Explain your answer.

14. Taking a systems perspective, suggest how to prevent BOD from a dairy farm from reaching an adjacent river.

Quantitative questions for this chapter are found on our Web site.

Take a Stand

Visit our Web site at http://www.wiley.com/college/raven (select Chapter 22 from the Table of Contents) for links to more information about the water chlorination issue. Consider the opposing views of those who support and those who oppose the chlorination of water, and debate the issue with your classmates. You will find tools to help you organize your research, analyze the data, think critically about the issues, and construct a well-considered argument. *Take a Stand* activities can be done individually or as a team, as oral presentations, written exercises, or Web-based (e-mail) assignments.

Additional online materials relating to this chapter, including a Student Testing Section with study aids and self-tests, Environmental News, Activity Links, Environmental Investigations, and more, are also on our Web site.

23

The Pesticide Dilemma

Salad Vac. This machine controls certain insect pests without the use of chemical pesticides. Photographed in Salinas, California. (Richard Steven Street/Streetshots)

Picture a giant vacuum cleaner slowly moving over rows of strawberry or vegetable crops and sucking insects off the plants. **Edgar Shaw,** an entomologist (a biologist who studies insects) with the Research Division of Driscoll Strawberry Associates of California, invented such a machine as a substitute for the chemical poisons we call pesticides (see photograph). Increasing public concern about pesticide residues on food and in groundwater supplies has caused farmers to look seriously at other ways to control pests—even by vacuuming insects.

Each vacuuming by the "salad-vac," or "bug-vac," eliminates the need for one application of pesticide. The salad-vac sucks insects through whirling fan blades, killing them instantly. The dead insects are sprayed out the back of the salad-vac to fertilize the soil. Dozens of California growers use the farm-sized vacuum cleaners to remove and kill lygus bugs, leafhoppers, Colorado potato beetles, and other insect pests from their strawberries and other crops. The salad-vac removes and kills some beneficial insects as well but fewer than would be killed if chemical insecticides were sprayed.

In few areas of environmental sciences is taking a systems perspective more compelling than in the case of pesticides. Humans' interactions with other plants and animals are in some ways no different than other species–species relationships. We are consumers—eating producers (plants) and other consumers (animals). We compete with other consumers (which we often call pests). We have a mutualistic relationship with bees and apple trees: We eat apples and honey, and we provide homes for the bees near the apples. We plant, water, and fertilize apples . . . and protect them from other consumers (worms, birds) and competing plants (weeds).

However, while our relationship with bees and apples is similar to those found in natural systems, there are some profound differences in the details. We often dedicate large areas to growing single crop—in this case, apple orchards. In a system without humans, we would find a combination of apples and other plants, and a range of animals that eat apples. We must rely on energy inputs and chemicals to maintain the "unnatural" agricultural system. The "bug-vac" above uses mechanical energy to remove insects from crops. More commonly, chemical pesticides are used to kill weeds and insects.

Agriculture is the sector that uses the most pesticides worldwide—approximately 85% of the estimated 2.6 million metric tons (2.9 million tons) used each year. Highly developed countries use about three-fourths of all

pesticides, but pesticide use is increasing most rapidly in developing countries.

In this chapter you will examine the types and uses of pesticides, their benefits, and their disadvantages. Pesticides have saved millions of lives by killing insects that carry disease and by increasing the amount of food we grow. Modern agriculture depends on pesticides to produce blemish-free fruits and vegetables at a reasonable cost to farmers (and therefore to consumers).

Pesticides cause environmental and health problems, and it appears that in many cases their harmful effects outweigh their benefits. Pesticides rarely affect only the targeted pest species, and the balance of a natural system, such as predator–prey relationships, is upset. Certain pesticides concentrate at higher levels of the food chain. Humans who apply and work with pesticides may be at risk for pesticide poisoning (short-term) and cancer (long-term), and people who eat traces of pesticide on food are concerned about the long-term effects. In this chapter you will also consider some alternatives to pesticides and discuss the pesticide laws that protect our health and the environment.

World View and Closer to You...
Real news footage relating to pesticides around the world and in your own geographic region is available at www.wiley.com/college/raven by clicking on "World View and Closer to You."

What Is a Pesticide?

LEARNING OBJECTIVE

- Distinguish between narrow-spectrum and broad-spectrum pesticides, and describe the various types of pesticides, such as insecticides and herbicides.

Any organism that interferes in some way with human welfare or activities is a **pest**. Some weeds, insects, rodents, bacteria, fungi, nematodes (microscopic worms), and other pest organisms compete with humans for food; other pests cause or spread disease. The definition of pest is subjective. A mosquito may be a pest to you, but it is not a pest to the bat or bird that eats it. People try to control pests, usually by reducing the size of the pest population. **Pesticides** are the most common way of doing this, particularly in agriculture. Pesticides can be grouped by their target organisms—that is, by the pests they are supposed to eliminate. **Insecticides** kill insects, **herbicides** kill plants, **fungicides** kill fungi, and **rodenticides** kill rodents, such as rats and mice.

The ideal pesticide would be a **narrow-spectrum pesticide** that would kill only the organism for which it was intended and not harm any other species. The perfect pesticide would readily break down, either by natural chemical decomposition or by biological organisms, into safe materials such as water, carbon dioxide, and oxygen. The ideal pesticide would stay exactly where it was put and would not move around in the environment.

Unfortunately, there is no such thing as an ideal pesticide. Most pesticides are **broad-spectrum pesticides.** Some pesticides do not degrade readily or else break down into compounds as dangerous as, if not more dangerous than, the original pesticide. And most pesticides move around a great deal throughout the environment.

broad-spectrum pesticide: A pesticide that kills a variety of organisms, including beneficial organisms, in addition to the target pest.

First-Generation and Second-Generation Pesticides

Before the 1940s, pesticides were of two main types, inorganic compounds (also called minerals) and organic compounds. Inorganic compounds that contain lead, mercury, and arsenic are extremely toxic to pests but are not used much today, in part because of their chemical stability in the environment. Natural processes do not degrade inorganic compounds, which persist and accumulate in the soil and water. This accumulation poses a threat to humans and other organisms, which, like the target pests, are susceptible to poisoning by inorganic compounds.

Plants, which have been fighting pests longer than humans, have evolved several natural organic compounds that are poisonous, particularly to insects. Such plant-derived pesticides are called **botanicals.** Examples of botanicals include nicotine from tobacco, pyrethrin from chrysanthemum flowers **(Figure 23.1)**, and rotenone from roots of the derris plant, all of which are used to kill insects. Botanicals are easily degraded by microorganisms and do not persist for long in the environment. However, they are highly toxic to aquatic organisms and to bees (beneficial insects that pollinate crops).

Synthetic botanicals are human-made insecticides produced by chemically modifying the structure of natural botanicals. An important group of synthetic botanicals is the pyrethroids, which are chemically similar to pyrethrin. Pyrethroids do not persist in the environment; they are slightly toxic to mammals and bees but toxic to fishes. Allethrin is an example of a pyrethroid.

In the 1940s many synthetic organic pesticides began to be produced. Earlier pesticides, both inorganic compounds and botanicals, are called **first-generation pesticides** to distinguish them from the vast array of synthetic poisons in use today, called **second-generation pesticides**. The insect-killing ability of dichlorodiphenyltrichloroethane (DDT), the first of the second-generation pesticides, was recognized in 1939 **(Figure 23.2).** There are currently about 20,000 registered commercial pesticide products, consisting of combinations of about 675 active chemical ingredients.

Figure 23.1 Pesticide derived from plants. Chrysanthemum flowers, shown here as they are harvested in Rwanda, are the source of the insecticide pyrethrin. Botanicals are plant chemicals used as pesticides. (Robert E. Ford/Terraphotographics/BPS)

Figure 23.2   Applying DDT in 1945. Pesticides such as DDT, shown as it was sprayed to control mosquitoes at New York's Jones Beach State Park in 1945, were used in ways that would be unacceptable now. The sign on the truck reads, in part, "D.D.T. Powerful insecticide. Harmless to humans." The harmful environmental effects of DDT were not known until many years later. (© UPI/Bettmann/Corbis Images)

The Major Groups of Insecticides

Insecticides, the largest category of pesticides, are usually classified into groups based on chemical structure. Three of the most important groups of second-generation insecticides are the chlorinated hydrocarbons, organophosphates, and carbamates.

DDT is a **chlorinated hydrocarbon,** an organic compound containing chlorine. After DDT's insecticidal properties were recognized, many more chlorinated hydrocarbons were synthesized as pesticides. Generally speaking, chlorinated hydrocarbons are broad-spectrum insecticides. Most are slow to degrade and persist in the environment (even inside organisms) for many months or even years. They were widely used from the 1940s until the 1960s, but since then many have been banned, or their use has largely been restricted, mainly because of problems associated with their persistence in the environment and impacts on humans and wildlife. Three chlorinated hydrocarbons still in use in the United States are endosulfan, lindane, and methoxychlor. Many people first became aware of the problems with pesticides in 1963, when Rachel Carson published her book, *Silent Spring* (see Figure 2.4).

Organophosphates, organic compounds that contain phosphorus, were developed during World War II as an outgrowth of German research on nerve gas. Organophosphates are more poisonous than other types of insecticides, and many are highly toxic to birds, bees, and aquatic organisms. The toxicity of many organophosphates in mammals, including humans, is comparable to that of some of our most dangerous poisons—arsenic, strychnine, and cyanide. Organophosphates do not persist in the environment as long as chlorinated hydrocarbons do. As a result, organophosphates have generally replaced the chlorinated hydrocarbons in large-scale uses such as agriculture, although many are not widely available to consumers because of their high level of toxicity. Methamidophos, dimethoate, and malathion are three examples of organophosphates.

Carbamates, the third group of insecticides, are broad-spectrum insecticides derived from carbamic acid. Carbamates are generally not as toxic to mammals as the organophosphates, although they still show broad, nontarget toxicity. Two common carbamates are carbaryl and aldicarb.

The Major Kinds of Herbicides

Herbicides are chemicals that kill or inhibit the growth of unwanted vegetation such as weeds in crops or lawns. Like insecticides, herbicides can be classified into groups on the basis of chemical structure, but this method is cumbersome because there are at least 12 different chemical groups used as herbicides. It is easier to group herbicides according to how they act and what they kill. **Selective herbicides** kill only certain types of plants, whereas nonselective herbicides kill all vegetation. Selective herbicides can be further classified according to the types of plants they affect. **Broad-leaf herbicides** kill plants with broad leaves but do not kill grasses; **grass herbicides** kill grasses but are safe for most other plants.

Two common herbicides with similar structures are *2,4-dichlorophenoxyacetic acid (2,4-D)* and *2,4,5-trichlorophenoxyacetic acid (2,4,5-T)*. Both were developed in the United States in the 1940s. These broad-leaf herbicides, which are similar in structure to a natural growth hormone in plants, disrupt the plants' natural growth processes; they kill plants such as dandelions but do not harm grasses. You may recall from Chapter 19 that many of the world's important crops, such as wheat, corn, and rice, are cereal grains, which are grasses. Both 2,4-D and 2,4,5-T kill weeds that compete with these crops, although 2,4,5-T is no longer used in the United States. The Environmental

Protection Agency (EPA) banned most uses of 2,4,5-T in 1979 because of possible harmful side effects to humans that became apparent after its use in the Vietnam War.

REVIEW

1. What is a pest? What is a pesticide?

2. Explain the difference between narrow-spectrum and broad-spectrum pesticides.

Benefits and Problems with Pesticides

LEARNING OBJECTIVES

- Relate the benefits of pesticides in disease control and crop protection.
- Explain why monocultures are more susceptible to pest problems.
- Summarize the problems associated with pesticide use, including development of genetic resistance; creation of imbalances in the ecosystem; persistence, bioaccumulation, and biological magnification; and mobility in the environment.
- Describe the pesticide treadmill and explain how resistance management may help with this problem.

Each day a war is waged as farmers, struggling to produce bountiful crops, battle insects and weeds. Similarly, health officials fight their own war against the ravages of human diseases transmitted by insects. One of the most effective weapons in the arsenals of farmers and health officials is the pesticide.

Although pesticides have their benefits, they have several problems. First, many pest species evolve a resistance to pesticides after repeated exposure to them. Second, pesticides affect numerous species in addition to the target pests, generating imbalances in the ecosystem (including agricultural fields) and posing a threat to human health. Finally, the ability of some pesticides to resist degradation and to readily move around in the environment causes even more problems for humans and other organisms.

Benefit: Disease Control

Insects transmit several devastating human diseases. Fleas and lice carry the microorganism that causes typhus in humans. Malaria, also caused by a microorganism, is transmitted to millions of humans each year by female *Anopheles* mosquitoes **(Figure 23.3)**. According to the World Health Organization (WHO), approximately 300 million to 500 million people currently suffer from malaria, and as many as 2.7 million people, mostly

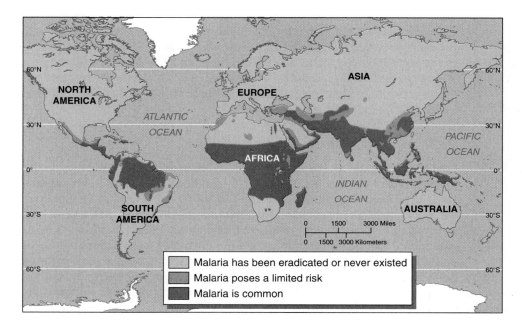

Figure 23.3   Location of malaria. Insecticides sprayed to control mosquitoes in these locations have saved millions of lives.

children in developing countries, die from the disease each year. Because only a few antimalarial drugs are available to treat malaria, the focus of controlling this disease is on eliminating the mosquitos that carry it. Important steps in mosquito control include limiting stagnant pools of water where mosquitos reproduce and applying pesticides.

Pesticides, particularly DDT, have helped control the population of mosquitoes, thereby reducing the incidence of malaria. Consider Sri Lanka. In the early 1950s, more than two million cases of malaria were reported in Sri Lanka each year. When spraying of DDT was initiated to control mosquitoes, malaria cases dropped to almost zero. When DDT spraying was discontinued in 1964, malaria reappeared almost immediately. By 1968, its annual incidence had increased to greater than one million cases per year. Since then, DDT use has resumed in many places, but not in the huge amounts and broad applications that caused so much damage in the 1960s. Today, DDT-impregnated mosquito nets are used to keep the pests off people, and indoor walls are sprayed with DDT.

Benefit: Crop Protection

Although it is difficult to make exact assessments, it is widely estimated that pests eat or destroy more than one-third of the world's crops. Given our expanding population and world hunger, it is easy to see why control of agricultural pests is desirable.

Pesticides reduce the amount of a crop lost through competition with weeds, consumption by insects, and diseases caused by plant **pathogens** (microorganisms, such as fungi and bacteria, that cause disease). Although many insect species are beneficial from a human viewpoint (two examples are honeybees, which pollinate crops, and ladybugs, which prey on crop-eating insects), many are considered pests. Of these, about 200 species have the potential to cause large economic losses in agriculture. For example, the Colorado potato beetle is one of many insects that voraciously consume the leaves of the potato plant, reducing the plant's ability to produce large tubers for harvest (see Figure 19.11).

Serious agricultural losses are minimized in the United States and other highly developed nations primarily by the heavy application of pesticides. Pesticide use is usually justified economically, in that farmers save an estimated $3 to $5 in crops for every $1 that they invest in pesticides. In developing countries where pesticides are not used in appreciable amounts, the losses due to agricultural pests can be considerable.

Why are agricultural pests found in such great numbers in our fields? Part of the reason is that agriculture is usually a monoculture, in which only one variety of one crop species is grown on large tracts of land. The cultivated field represents a simple ecosystem, from which much of the "system" has been removed by humans. In contrast, forests, wetlands, and other natural ecosystems are complex and contain many different species, including predators and parasites that control pest populations and plant species that pests do not use for food. A monoculture reduces the dangers and accidents that might befall a pest as it searches for food. A Colorado potato beetle in a forest would have a hard time finding anything to eat, but a 500-acre potato field is like a big banquet table set just for the pest. It eats, prospers, and reproduces. In the absence of many natural predators and in the presence of plenty of food, the population thrives and grows, and more of the crop becomes damaged.

Problem: Evolution of Genetic Resistance

The prolonged use of a particular pesticide can cause a pest population to develop genetic resistance to the pesticide. In the 50 years during which pesticides have been widely used, at least 520 species of insects and mites have evolved genetic resistance to certain pesticides **(Figure 23.4)**. Many pests now have multiple resistance to several pesticides, and at least 17 species, such as diamondback moths and palm thrips, are resistant to all major classes of insecticides that farmers are legally allowed to use on them. Insects are not the only pests to evolve genetic resistance; at least 84 weed species are currently resistant to certain herbicides. Some weeds, such as annual ryegrass and canary grass, are resistant to all available herbicides.

monoculture: The cultivation of only one type of plant over a large area.

genetic resistance: Any inherited characteristic that decreases the effect of a pesticide on a pest.

Figure 23.4 **Genetic resistance.** The number of species exhibiting genetic resistance to pesticides has increased dramatically. More than 520 insect and mite species have evolved resistance to insecticides.

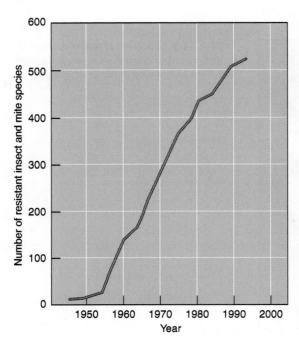

How does genetic resistance to pesticides occur? Every time a pesticide is applied to control a pest, some survive. The survivors, because of certain genes they already possess, are genetically resistant to the pesticide, and they pass on this trait to future generations. Thus, *evolution*—any cumulative genetic change in a population of organisms—occurs, and subsequent pest populations contain larger percentages of pesticide-resistant pests than before. Insects and other pests are constantly evolving. The short generation times (the period between the birth of one generation and that of another) and large populations characteristic of most pests favor rapid evolution, which allows the pest population to quickly adapt to the pesticides used against it. As a result, an insecticide that kills most of an insect population becomes less effective after prolonged use because the survivors and their offspring are genetically resistant.

Manufacturers of chemical pesticides have often responded to genetic resistance by recommending that the pesticide be applied more frequently or in larger doses. Alternatively, they recommend switching to a new, often more expensive, pesticide. These responses result in a predicament known as the **pesticide treadmill**. Over time, the pesticide treadmill results in increased pesticide use, higher production costs, and declines in crop yields.

pesticide treadmill: A predicament faced by pesticide users, in which the cost of applying pesticides increases (because they have to be applied more frequently or in larger doses) while their effectiveness decreases (as a result of increasing genetic resistance in the target pests).

Resistance Management   **Resistance management** is a relatively new approach to dealing with genetic resistance. It involves efforts to delay the evolution of genetic resistance in insect pests or weeds. Strategies of resistance management vary depending on the pest species involved.

resistance management: Strategies for managing genetic resistance in order to maximize the period in which a pesticide is useful.

One strategy of resistance management for insect pests is to maintain a nearby "refuge" of untreated plants where the insect pest can avoid being exposed to the insecticide. Those insects that live and grow in the refuge remain susceptible to the insecticide. When susceptible insects migrate into the area being treated with insecticide, they mate with members of the genetically resistant population. This interbreeding delays the development of genetic resistance in the population as a whole.

Avoiding repeated use of the same herbicide on the same field is a strategy of resistance management that slows the development of weed resistance to herbicides. After herbicides are applied, the field is scouted to see if any weed plants survived the herbicide application. These weeds are resistant to the herbicide, and they should be removed from the field before they flower. Cultural methods that prolong the usefulness of herbicides include mechanically pulling weeds and planting seed that is certified free of weed seeds.

Problem: Imbalances in the Ecosystem

When a pesticide is applied, a new and powerful limiting factor enters an often delicately balanced system. Beneficial insects are killed as effectively as pest insects. In a study of the effects of spraying the insecticide dieldrin to kill Japanese beetles, scientists found a large number of dead animals in the treated area, such as birds, rabbits, ground squirrels, cats, and beneficial insects. (Use of dieldrin in the United States has since been banned.) Pesticides do not have to kill organisms to harm them. Quite often the stress of carrying pesticides in its tissues makes an organism more vulnerable to predators, diseases, or other stressors in its environment.

Because the natural enemies of pests often starve or migrate in search of food after pesticide is sprayed in an area, pesticides are indirectly responsible for a large reduction in the populations of these natural enemies. Pesticides also kill natural enemies directly because predators consume a lot of the pesticide when consuming the pests. After a brief period, the pest population rebounds and gets larger than ever, partly because no natural predators are left to keep its numbers in check.

Despite a 33-fold increase in pesticide use in the United States since the 1940s, crop losses due to insects, diseases, and weeds have not changed appreciably (**Table 23.1**). Increasing genetic resistance to pesticides and the destruction of the natural enemies of pests

Table 23.1 Percentage of Crops Lost Annually to Pests in the United States

Period	Insects	Diseases	Weeds
2006	13.0	12.5	12.0
1989-1999	13.0	12.0	12.0
1974	13.0	12.0	8.0
1951-1960	12.9	12.2	8.5
1942-1951	7.1	10.5	13.8

Source: USDA Agricultural Research Service.

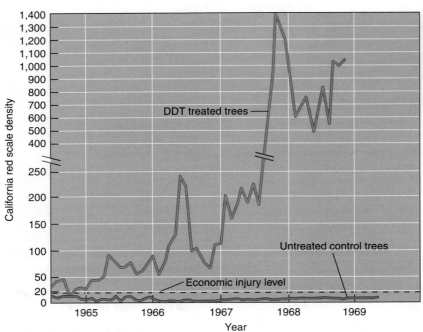

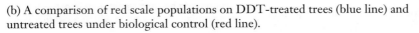

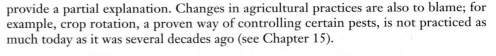

(b) A comparison of red scale populations on DDT-treated trees (blue line) and untreated trees under biological control (red line).

(a) Red scale on oranges.
(Courtesy Max Badgley)

Figure 23.5 **Pesticide use and new pest species.** An infestation of red scale insects on lemons occurred after DDT was sprayed to control a different pest. Prior to DDT treatment, red scale did not cause significant economic injury to citrus crops.

bioaccumulation: The buildup of a persistent pesticide or other toxic substance in an organism's body.

biological magnification: The increased concentration of toxic chemicals such as certain pesticides in the tissues of organisms at higher trophic levels in food webs.

Figure 23.6 **Peregrine falcon feeding chicks.** Biological magnification of DDT in the tissues of falcons and other birds of prey caused a decline in their reproductive success in the United States during the 1960s and early 1970s. DDT caused the birds to lay eggs with extremely thin, fragile shells, causing the chicks' deaths. (DDT use was banned in the United States in 1972.) (Thomas D. Mangelsen/Still Pictures/Peter Arnold, Inc.)

provide a partial explanation. Changes in agricultural practices are also to blame; for example, crop rotation, a proven way of controlling certain pests, is not practiced as much today as it was several decades ago (see Chapter 15).

Creation of New Pests   In some instances, use of a pesticide has resulted in a pest problem that did not exist before. Creation of new pests—that is, turning minor pest organisms into major pests—is possible because the pesticide kills most of a certain pest's natural predators, parasites, and competitors, allowing the pest's population to rebound. The use of DDT to control certain insect pests on lemon trees was documented as causing an outbreak of a scale insect (a sucking insect that attacks plants) that was not a problem before spraying **(Figure 23.5).** In a similar manner, the European red mite became an important pest on apple trees in the northeastern United States, and beet armyworms became an important pest on cotton, both after the introduction of pesticides.

Problems: Persistence, Bioaccumulation, and Biological Magnification

As discussed in Chapter 7, some pesticides, particularly chlorinated hydrocarbons, are extremely **persistent** in the environment and may take many years to break down into less toxic forms. When a persistent pesticide is ingested, it is stored, usually in fatty tissues. Over time, the organism may accumulate high concentrations of the pesticide, a phenomenon known as **bioaccumulation**.

Organisms at higher levels on food webs tend to have greater concentrations of bioaccumulated pesticide stored in their bodies than those lower on food webs. The increase in pesticide concentrations as the pesticide passes through successive levels of the food web is known as **biological magnification** **(Figure 23.6;** also see Figure 7.6).

Problem: Mobility in the Environment

Another problem associated with pesticides is that they do not stay where they are applied but tend to move through the soil, water, and air, sometimes long distances **(Figure 23.7).** Pesticides that are applied to agricultural lands and then wash into rivers and streams when it rains can harm fishes. If the pesticide level in the aquatic ecosystem is high enough, the fishes may be killed. If the level is sublethal (that is, not enough to kill the fishes), the fishes may still suffer from undesirable effects such as bone degeneration. These effects may decrease their competitiveness and increase their chances of being preyed on.

Pesticide mobility is also a problem for humans. In 1994 the Environmental Working Group (EWG), a private environmental organization, analyzed herbicides in drinking water by evaluating 20,000 water tests performed by state and federal government inspectors. Their study revealed that 14.1 million U.S. residents drink water containing traces of five widely used herbicides. Because these herbicides are often used

Figure 23.7   Intended versus actual pathways of pesticides.

(a) A helicopter sprays pesticides on a crop—and everything else in its pathway—in California. (Laurence Migdale/Photo Researchers, Inc.)

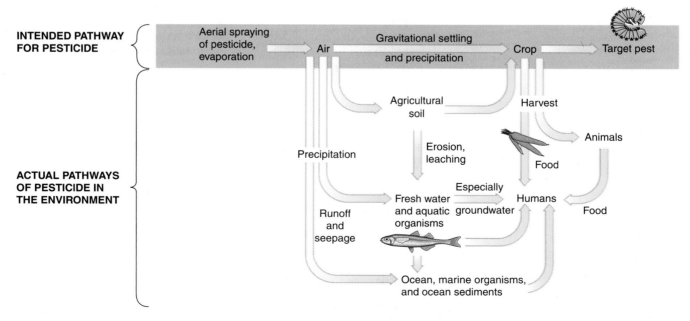

(b) The intended pathway of pesticides in the environment (shown at top of figure) is quite different from the actual pathways (shown at bottom).

on corn and soybeans, the study focused on the midwestern states where these crops are commonly grown. The study concluded that 3.5 million people living in the Midwest face a slightly elevated cancer risk because of their exposure to the herbicides. Since the study, the EPA has reduced the use of these herbicides (alachlor, metachlor, atrazine, cyanazine, and simazine).

From 1996 to 1998 the EWG conducted a different study in California that revealed pesticide mobility in the atmosphere. They collected nearly 100 air samples in several California counties and submitted them to a certified laboratory for analysis. Almost two-thirds of the samples contained small amounts of pesticides that had drifted from farm fields.

CASE IN POINT

Economic Development and Pesticides in Central America

In the 1980s, the USAID (U.S. Agency for International Development), in an effort to build and stabilize economies in Central America, promoted the production of cash-crops for export to the United States. Crops included snow peas, broccoli, lettuce, and berries—crops in great demand in the United States, but that do not grow well in the colder climates found in the northern United States.

Several years later, the plan seemed to backfire as the U.S. Department of Agriculture (USDA) began to find unacceptable levels of pesticides on imports—including pesticides that were banned or even inappropriate for the crops they had been used on. When the USDA began turning away shipments of produce from Central America, it became clear that different parts of the U.S. government were working at cross purposes.

The problem was that the USAID efforts did not account for the decentralized agricultural system in which crops were being raised. An individual farmer would grow snow peas, for example, on a small plot of land, using whatever pesticides she had available. The pesticide chlorothalonil, used to reduce blemishes, was of particular concern. A middleman (or "coyote") would come by in a pickup truck and buy a few crates of snow peas from each of a number of farmers, then drive to another location where many such loads of snow peas were combined—and it became impossible to tell which batch came from which truckload, let alone which farmer! A batch of snow peas might be rejected by the actions of one or a few farmers who were unaware that they were using pesticides incorrectly.

As a remedy, the EPA partnered with the USAID to develop an information campaign for Central American farmers, coupled to a program to manage exports of pesticides to Central America. The program has been largely successful, as the United States imports increasing amounts of fruits and vegetables from Central America every year. ∎

REVIEW

1. What are two important benefits of pesticide use?
2. What are persistence, bioaccumulation, and biological magnification?
3. How do pesticides move around in the environment?
4. What is the pesticide treadmill?

Risks of Pesticides to Human Health

LEARNING OBJECTIVE

- Discuss pesticide risks to human health, including short-term effects and long-term effects.
- Explain endocrine disruption.

Exposure to pesticides, which is greater than most people realize and often occurs without their knowledge, can damage human health **(Figure 23.8).** Pesticide poisoning caused by short-term exposure to high levels of pesticides can result in harm to organs and even death, whereas long-term exposure to lower levels of pesticides can cause certain types of cancer. There is concern that exposure to trace amounts of certain pesticides could disrupt the human endocrine (hormone) system. Chapter 7 discusses the elevated risk of pesticides to children (see Figure 7.10).

Short-Term Effects of Pesticides

Pesticides poison approximately 67,000 people in the United States each year. Most of these are farm workers or others whose occupations involve daily contact with large quantities of pesticides **(Figure 23.9)**. A person with a mild case of pesticide poisoning may exhibit symptoms such as nausea, vomiting, and headaches. More serious cases, particularly organophosphate poisonings, may result in permanent damage to the nervous system and other body organs. Although the number is low in the United States, people do die from overexposure to pesticides. Almost any pesticide can kill a human if the dose is large enough.

The WHO estimates that, globally, pesticides poison more than four million people each year; of these, about 300,000 die. The incidence of pesticide poisoning is highest in developing countries, in part because they often use dangerous pesticides that are banned or greatly restricted by highly developed nations. Pesticide users in developing nations often are not trained in the safe handling and storage of pesticides, and safety regulations are generally more lax there.

Long-Term Effects of Pesticides

Many studies of farm workers and workers in pesticide factories, many of whom are exposed to low levels of pesticides over many years, show an association between cancer and long-term exposure to pesticides. A type of lymphoma (a cancer of the lymph system) is associated with the herbicide 2,4-D. Other pesticides are linked to a variety of cancers, such as leukemia and cancers of the brain, lungs, and testicles. Although the issue of whether certain pesticides cause breast cancer remains unresolved, researchers have noted a correlation between breast cancer and a high level of one or more pesticides in the breast's fatty tissue.

Long-term exposure to at least one pesticide may have resulted in sterility in thousands of farm workers on banana and pineapple plantations. More than 26,000 workers in 12 countries sued their employers or pesticide manufacturers, and most of the defendants have paid settlements but have not admitted liability.

A 2001 study that compared records of pesticide applications in 1984 in California's Central Valley with state health records for the same area showed that pregnant women who live near pesticide applications have higher rates of miscarriage. Other studies have indicated that the children of agricultural workers are at greater risk for birth defects, particularly stunted limbs. Evidence also suggests that exposure to pesticides may compromise the body's ability to fight infections.

Long-term exposure to pesticides may increase the risk of Parkinson's disease, which afflicts about one million people in the United States. When researchers injected rats with repeated doses of rotenone, a plant-derived pesticide, the rats developed the symptoms of Parkinson's disease, including difficulty in walking, tremors, and abnormal protein deposits in their brains. Many pesticides have chemical structures similar to rotenone, which suggests that other pesticides may increase the risk of Parkinson's disease as well. It is important to remember that this connection between pesticide exposure and Parkinson's disease is tentative and will require additional research to prove or disprove. To date, no study has connected Parkinson's disease in humans with exposure to a specific pesticide.

Figure 23.8   Risk of pesticide exposure to human health. What does this cartoon suggest about the relative costs and benefits of broad- and narrow-spectrum pesticides? Is this a fair representation? Why/why not? (© Ziggy and Friends, Inc. Reprinted with permission/Universal Press Syndicate)

Figure 23.9   Farm workers harvesting bell peppers. Most people who suffer from pesticide-related illnesses are farm workers or others whose occupations involve daily contact with pesticides. Photographed in California. (Inga Spence/Visuals Unlimited)

Table 23.2 Some Pesticides That Are Known Endocrine Disrupters*

Pesticide	General Information
Atrazine	Herbicide; still used
Chlordane	Insecticide; banned in United States in 1988
DDT (dichlorodiphenyl-trichloroethane)	Insecticide; banned in United States in 1972
Endosulfan	Insecticide; still used
Kepone	Insecticide; banned in United States in 1977
Methoxychlor	Insecticide; still used

* Based on experimental research with laboratory animals.

Pesticides as Endocrine Disrupters

In the 1990s, many articles were published that linked certain pesticides and other persistent toxic chemicals with reproductive problems in animals (**Table 23.2**; see Chapter 7 discussion on endocrine disrupters). River otters exposed to synthetic chemical pollutants were found to have abnormally small penises. Female sea gulls in Southern California exhibited behavioral aberrations by pairing with one another rather than with males during the mating season. In many cases scientists have reproduced the same abnormal symptoms in the laboratory, providing support that certain persistent chemicals cause the defects.

The suggestion in the 1996 book, *Our Stolen Future*, by Theo Colborn, Dianne Dumanoski, and John Myers, that persistent toxic chemicals in the environment are disrupting human hormone systems ignited a barrage of media attention. **Theo Colborn,** a senior scientist at the World Wildlife Fund, hypothesized that ubiquitous chemicals in the environment are linked to disturbing trends in human health. These include increases in breast and testicular cancer, increases in male birth defects, and decreases in sperm counts.

Although a few studies suggest that certain chemicals in the environment may affect human health, a direct cause-and-effect relationship between these chemicals and adverse effects on the human population has not been established. Some scientists think the potential danger is so great that persistent chemicals such as DDT should be internationally banned immediately. Other scientists are more cautious in their assessment of the danger, but still think the problem should not be disregarded.

Our Stolen Future ignited public concern and triggered universities, governments, and industries to investigate how synthetic chemicals interfere with the actions of human hormones. It will take years before these studies tell us if these chemicals are acting as endocrine disrupters in humans. Meanwhile, an international effort is underway to ban the production and use of nine pesticides suspected of being endocrine disrupters (see section on the global ban of persistent organic pollutants later in this chapter).

REVIEW

1. What are some of the long-term effects of pesticides on human health?
2. Why are endocrine disruptors of particular concern?

Alternatives to Pesticides

LEARNING OBJECTIVE

- Describe alternative ways to control pests, including cultivation methods, biological controls, pheromones and hormones, reproductive controls, genetic controls, quarantine, integrated pest management, and food irradiation.

Given the many problems associated with pesticides, it is clear that they are not the final solution to pest control. Fortunately, pesticides are not the only weapons in our arsenal. Alternative ways to control pests include cultivation methods, biological controls, pheromones and hormones, reproductive controls, genetic controls, quarantine, and irradiation. A combination of these methods in agriculture, often including a limited use of pesticides as a last resort, is known as *integrated pest management (IPM)*. IPM is the most effective way to control pests. (Also see Chapter 19 for a discussion of the growing popularity of organic foods, which are grown in the absence of pesticides and commercial fertilizers.)

Using Cultivation Methods to Control Pests

Sometimes agricultural practices are altered in such a way that a pest is adversely affected or discouraged from causing damage. Although some practices, such as the insect vacuum mentioned at the beginning of the chapter, are relatively new, other cultivation methods that discourage pests have been practiced for centuries.

One way to reduce damage from crop pests is to *intercrop* mixtures of plants, for example, by alternating rows of different plants. When corn was interplanted with molasses grass in an experiment in Kenya, only about 5% of the corn crop was damaged, as compared with about 39% damage in the control field of corn. Molasses grass repels some insects, and attracts wasps that lay their eggs inside corn borers, insects that destroy the stems of corn plants.

A technique used with success in alfalfa crops is *strip cutting*, in which only one segment of the crop is harvested at a time. The unharvested portion of the crop provides an undisturbed habitat for natural predators and parasites of the pest species. The same type of benefit is derived from keeping strips of unplowed plants (that is, wild plants, including weeds) along the margins of fields. A German study found that pest mortality was about 50% higher in oilseed rape fields with margin strips of other plants, as compared to fields without margins. The margins provided a refuge for three parasites of the pollen beetle, which is a significant pest on oilseed rape plants.

The proper timing of planting, fertilizing, and irrigating promotes healthy, vigorous plants that are more resistant to pests, because they are not stressed by other environmental factors. The rotation of crops controls pests: When corn is not planted in the same field for two years in a row, the corn rootworm is effectively controlled.

Biological Controls

As an example of a **biological control**, suppose that an insect species is accidentally introduced into a country where it was not found previously and becomes a pest. It might be possible to control this pest by going to its place of origin, and looking at how the pest fit into its native ecosystem. Usually, some other organism—often a wasp or disease—is an exclusive predator or parasite of the pest species. That predator or parasite, if successfully introduced, may lower the population of the pest species so that it is no longer a problem.

The pest species typically does not evolve genetic resistance to the biological control agent in the same way it does to a pesticides, since both pest and predator are living organisms that are responsive to *natural selection*. As the pest evolves a way to resist the biological control agent, the agent in turn may evolve some sort of countermeasure against the pest. (Recall the discussion of the predator–prey "arms race" in Chapter 4; a similar arms race occurs here between the pest and its biological control agent.)

Cottony-cushion scale provides an example of successfully using one organism to control another. The cottony-cushion scale is a small insect that sucks the sap from the branches and bark of many fruit trees, including citrus trees. It is native to Australia but was accidentally introduced to the United States in the 1880s. A U.S. entomologist went to Australia and returned with several possible biological control agents. One, the vedalia beetle, was effective in controlling scales, which it eats voraciously and exclusively. Within two years of its introduction, the vedalia beetle had significantly reduced the cottony-cushion scale in citrus orchards. Today both the cottony-cushion scale and the vedalia beetle are present in low numbers, and the scale is not considered an economically important pest.

More than 300 species have been introduced as biological control agents to North America. The Agricultural Research Service of the USDA is constantly investigating possible biological controls for insect and weed pests. Although some examples of biological control are quite spectacular, finding an effective parasite or predator is usually difficult. And just because a parasite or predator is identified does not mean it will become successfully established in a new environment. Slight variations in environmental conditions such as temperature and moisture can alter the effectiveness of the biological control organism in its new habitat.

Care must be taken to ensure that the introduced control agent does not attack unintended hosts and become a pest itself. To guard against this possibility when the control agent is an insect, scientists put the insects in cages with samples of important crops, ornamentals, and native plants to determine if the control agent will eat the

biological control: A method of pest control that involves the use of naturally occurring disease organisms, parasites or predators to control pests.

plants when they are starving. Despite such tests, organisms introduced as biological controls sometimes cause unintended problems in their new environment, and once they are introduced, they cannot be recalled. For example, a weevil was introduced in 1968 to control the Eurasian musk thistle, a noxious weed that arrived in North America in the mid-19th century. Since then, the weevil has expanded its host range to include a North American thistle listed as a threatened species. The weevils significantly reduce seed production in the native thistles.

Biological Control Agents   Insects are not the only biological control agents (indeed, cats have long been kept to control rodents!). When DDT was banned in the United States, alternative solutions to mosquito problems were explored. One that has been successful in several locations has been to spray nematodes (microscopic worms) that attack mosquito larvae. Other nematodes are often sprayed on corn to kill corn borers, which eat through and destroy maturing corn stalks. The nematodes can be sprayed much like a chemical pesticide, but with the advantage that the nematode is usually specific to a single pest and harmless to other animals. Nematodes can also be effective against a range of pests, including weevils, grasshoppers, and locusts.

Another noninsect biological control agent targets desert locusts: a fungal spore known as the "Green Muscle." Locusts periodically increase in number and swarm across the African Sahel, threatening crops across 12 million hectares (30 million acres) (see Figure 5.21 for a map showing the Sahel). During a major swarm in 1988, African farmers spent $300 million to spray massive quantities of pesticides into the environment to bring the locust population under control. Many people were concerned about the adverse ecological and health effects of using such large quantities of pesticides. The most recent locust outbreak occurred during the 2004 to 2005 growing season, just before the Green Muscle became available. If it is used widely in the future, the Green Muscle might prevent locusts from swarming again.

Bacteria and viruses that harm insect pests have also been used successfully as biological controls. *Bacillus popilliae*, which causes milky spore disease in insects, is applied as a dust on the ground to control the larval (immature) stage of Japanese beetles. The common soil bacterium *Bacillus thuringiensis*, or *Bt*, produces a natural pesticide toxic to some insects, such as the cabbage looper, a green caterpillar that damages many vegetable crops, and the corn earworm. When eaten by insect larvae, *Bt* toxin damages the intestinal tract, killing the young insect. The toxin does not persist in the environment and is not known to harm mammals, birds, or other noninsect species.

Pheromones and Hormones

pheromone: A natural substance produced by animals to stimulate a response in other members of the same species.

Pheromones are commonly called sexual attractants because they are often produced to attract members of the opposite sex for mating. Each insect species produces its own specific pheromone, so once the chemical structure is known, it is possible to make use of pheromones to control individual pest species. Pheromones lure insects such as Japanese beetles to traps, where they are killed. Alternatively, pheromones are released into the atmosphere to confuse insects so that they cannot locate mates.

Insect hormones are natural chemicals produced by insects to regulate their own growth and metamorphosis, which is the process in which an insect's body changes through several stages to an adult. Specific hormones must be present at certain times in the life cycle of the insect; if they are present at the wrong time, the insect develops abnormally and dies. Many such insect hormones are known, and synthetic hormones with similar structures have been made. Entomologists are actively pursuing the possibility of using these substances to control insect pests.

A synthetic version of the insect hormone ecdysone, which causes molting, was the first hormone approved for use. Known as MIMIC, the hormone triggers abnormal molting in insect larvae (caterpillars) of moths and butterflies. Since MIMIC affects some beneficial insects, its use has risks.

Reproductive Controls

Like biological controls, reproductive controls of pests involve the use of organisms. Instead of using another species to reduce the pest population, reproductive control strategies suppress pests by sterilizing some of its members. In the **sterile male technique,** large numbers of males are sterilized in a laboratory, usually with radiation or chemicals. Males are sterilized rather than females because male insects of species selected for this type of control mate several to many times, but females of that species mate only once. Thus, releasing a single sterile male may prevent successful reproduction by several females, whereas releasing a single sterile female would prevent successful reproduction by only that female.

The sterilized males are released into the wild, where they reduce the reproductive potential of the pest population by mating with normal females, which then lay eggs that never hatch. As a result, the population of the next generation is much smaller.

One disadvantage of the sterile male technique is that to be effective it must be carried out continually. If sterilization is discontinued, the pest population rebounds to a high level in a few generations (which, you will recall, are short). The procedure is expensive, for it requires the rearing and sterilization of large numbers of insects in a laboratory or production facility. During the 1990 Mediterranean fruit fly (medfly) outbreak in California, as many as 400 million sterile male medflies were released each week. Such extraordinary measures are taken because the medfly is so destructive. The adult medfly lays its eggs on 250 different fruits and vegetables. When the eggs hatch, the maggots (larvae) feed on the fruits and vegetables and turn them into a disgusting mush.

Genetic Controls

Traditional selective breeding has been used to develop many varieties of crops that are genetically resistant to disease organisms or insects. Traditional breeding of crop plants typically involves identifying individual plants that are in an area where the pest is common but that are not damaged by the pest. These individuals are then crossed with standard crop varieties in an effort to produce a pest-resistant version. It may take as long as 10 to 20 years to develop a resistant crop variety, but the benefits are usually worth the time and expense.

Although traditional selective breeding has resulted in many disease-resistant crops and has enabled decreased use of pesticides, there are potential problems. Fungi, bacteria, and other plant pathogens evolve rapidly and quickly adapt to the disease-resistant host plant—meaning that the new pathogen strains can cause disease in the formerly disease-resistant plant variety. Plant breeders are in a continual race to keep one step ahead of plant pathogens.

Genetic engineering offers promise in breeding pest-resistant plants more quickly (see Chapter 19). For example, a gene from the soil bacterium *Bt* (already discussed in the section on biological controls) has been introduced into several plants, such as corn and cotton. Caterpillars that eat cotton leaves from these **genetically modified (GM)** plants die or exhibit stunted growth.

 CASE IN POINT

Bt, Its Potential and Problems

Bt has been marketed since the 1950s, but it was not sold on a large scale until recently, mainly because there are many different varieties of the *Bt* bacterium, and each variety produces a slightly different protein toxin. Each is toxic to only a small group of insects. The *Bt* variety that works against corn borers would not be effective against potato beetles. As a result, *Bt* is not economically competitive compared to chemical pesticides, each of which could kill many different kinds of pests on many different crops.

Genetic engineers have greatly increased the potential of *Bt*'s toxin as a natural pesticide by modifying the gene coding for the toxin so that it affects a wider range of insect pests. They then inserted the *Bt* gene that codes for the toxin into at least 18 crop species, including corn, potato, and cotton. *Bt* corn, one of the first GM crops, was engineered to produce a continuous supply of toxin, which provides a natural defense against insects such as the European corn borer. Similarly, *Bt* tomato and *Bt* cotton are more resistant to pests such as the tomato pinworm and the cotton bollworm. Early ecological risk assessment studies, performed before EPA approved their use, indicated that genetically modified crops are essentially like their more conventional counterparts that are produced by selective breeding. Genetically modified crops do not become invasive pests or persist in the environment longer than crops that are not genetically modified. Significantly fewer pesticide applications are needed for GM crops with the *Bt* gene.

The future of the *Bt* toxin as an effective substitute for chemical pesticides is not completely secure, however. Beginning in the late 1980s, farmers began to notice that chemically applied *Bt* was not working as well against the diamondback moth as it had in the past. All farmers who reported this reduction in effectiveness had used *Bt* frequently and in large amounts. In 1996 many farmers growing *Bt* cotton reported that their crop succumbed to the cotton bollworm, which *Bt* is supposed to kill. It appears that certain insects, such as the diamondback moth and possibly the cotton bollworm, may have evolved resistance to this natural toxin. If *Bt* is used in greater and greater amounts, it is likely that more insect pests will develop genetic resistance to it, greatly reducing *Bt*'s potential as a natural pesticide. Scientists are developing resistance management strategies to slow the evolution of pest resistance to genetically engineered crops that make *Bt*. ■

Quarantine

Governments attempt to prevent the importation of foreign pests and diseases by practicing **quarantine,** or restriction of the importation of exotic plant and animal material that might harbor pests. If a foreign pest is accidentally introduced, quarantine of the area where it is detected helps prevent its spread. If a foreign pest is detected on a farm, the farmer may be required to destroy the entire crop.

Quarantine is an effective, though not foolproof, means of control. The USDA has blocked the accidental importation of medflies on more than 100 separate occasions. On the few occasions when quarantine failed and these insects successfully passed into the United States, millions of dollars of crop damage were incurred in addition to the millions spent to eradicate the pests. Eradication efforts include the use of helicopters to spray the insecticide malathion over hundreds of square kilometers and the rearing and releasing of millions of sterile males to breed the medfly out of existence.

Many experts think the repeated finds of medflies in California indicate that, rather than being accidentally introduced each year, the medfly has become established in the state. If so, there are potentially disastrous consequences for California's $18 billion agricultural economy. Other countries could stop importing California produce or require expensive inspections and treatments of every shipment to prevent the importation of the medfly into their countries.

The Systems Approach: Integrated Pest Management

Many pests are not controlled effectively with a single technique; rather, a combination of control methods is often more effective. **Integrated pest management (IPM)** combines a variety of biological, cultivation, and pesticide controls tailored to the conditions and crops of an individual farm, campus, city, or greenhouse. **Figure 23.10** provides an illustrated example of some options for IPM to control corn pests.

Biological and genetic controls, including pest-resistant GM crops, are used as much as possible, and conventional pesticides are used sparingly and only when other

integrated pest management: A combination of pest control methods that, if used in the proper order and at the proper times, keep the size of a pest population low enough that it does not cause substantial economic loss.

methods fail (see Meeting the Challenge: Reducing Agricultural Pesticide Use by 50% in the United States on page 551). When pesticides are required, the least toxic pesticides are applied in the lowest possible quantities and at carefully scheduled times. Thus, IPM allows the farmer to control pests with a minimum of environmental disturbance and often at a minimal cost.

To be effective, IPM requires a thorough knowledge of the system, including life cycles, feeding habits, travel, and nesting habits of the pests as well as all their interactions with their hosts and other organisms. The timing of planting, cultivation, and treatments with biological controls is critical and is determined by carefully monitoring the concentration of pests. Integrated pest management optimizes natural controls by using agricultural techniques that discourage pests. Integrated pest management is an important part of sustainable agriculture (see Chapter 19).

Figure 23.10   Integrated pest management (IPM)

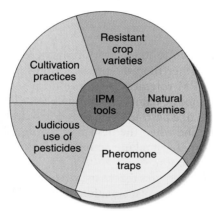

(a) IPM takes a systems approach to management of pests, providing interventions that impact various parts of pests' lifecycles.

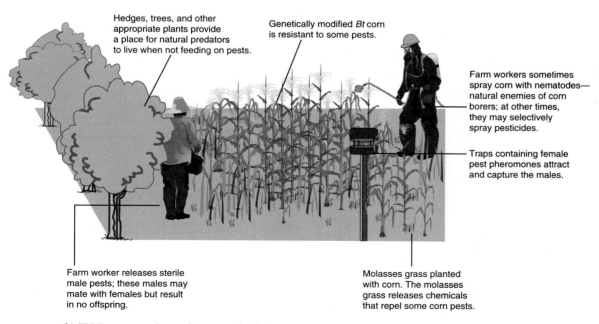

(b) IPM) to control corn borers and other corn pests.

There are two fundamental premises associated with IPM. First, IPM is the management rather than the eradication of pests. Farmers who have adopted the principles of IPM tolerate a low level of pests in their fields and accept a certain amount of economic damage from the pests. These farmers do not spray pesticides at the first sign of a pest. Instead, they periodically sample the pest population in the field to determine when the pest population reaches an economic injury threshold at which the benefit of taking action (such as the judicious use of pesticides) exceeds the cost of that action.

Second, IPM requires that farmers be educated so that they know what strategies will work best in their particular situations **(Figure 23.11)**. Managing pests is more complex than trying to eradicate them. The farmer must know what pests to expect on each crop and what to do to minimize their effects. The farmer must also know what beneficial species will assist in controlling the pests and how to encourage these beneficial species.

Cotton, which is attacked by many insect pests, has responded well to IPM. Cotton has the heaviest insecticide application of any crop: Although only about 1% of agricultural land in the United States is used for this crop, cotton accounts for almost 50% of all the insecticides used in agriculture! Applying simple techniques, such as planting a strip of alfalfa adjacent to the cotton field, lessens the need for chemical pesticides. Lygus bugs, a significant pest of cotton, move from the cotton field to the strip of alfalfa, which they prefer as a food. Thus, less damage is done to the cotton plants.

U.S. farmers have increasingly adopted IPM since the 1960s, but the overall proportion of farmers using IPM is still small. IPM is not more widespread, in part because the knowledge needed to use IPM is more sophisticated than that required to use pesticides.

Integrated pest management has been most successful in controlling insect pests. Scientists are now trying to develop IPM techniques to control weeds with a minimal use of herbicides. There is also a widely recognized need to develop IPM techniques to reduce pesticide use in urban and suburban environments.

Irradiating Foods

It is possible to prevent insects and other pests from damaging harvested food without using pesticides. The food is harvested and then exposed to ionizing radiation, which kills many microorganisms, such as salmonella, a bacterium that causes food poisoning. This is called **food irradiation** or **cold pasteuriza-**

Figure 23.11   Rice production and pesticide use in Indonesia, 1972–1990. The decline in pesticide use in the late 1980s and early 1990s, which was sponsored by the Indonesian government, did not cause a decrease in rice yields. Instead, rice production increased during the four years following the new policy. Indonesia was the first Asian country to widely embrace integrated pest management (IPM). It phased out pesticide subsidies, banned the use of dozens of pesticides on rice, and trained more than 200,000 farmers in IPM techniques.

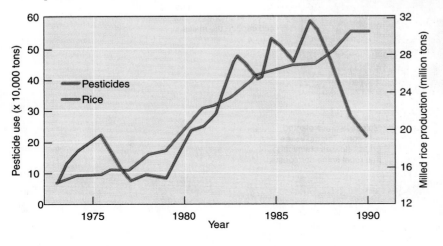

MEETING THE CHALLENGE

Reducing Agricultural Pesticide Use by 50% in the United States

Pesticides have benefited farmers (by increasing agricultural productivity) and consumers (by lowering food prices). But pesticide use has had its price—not necessarily in economic terms, for it is difficult to assign monetary values to many of its effects—but in terms of health problems and damage to agricultural and natural ecosystems.

Is it feasible to ban all pesticides known to cause cancer in laboratory animals? Probably not, at least not now. In many cases, substitute pesticides either do not exist, are less effective, or are considerably more expensive. A pesticide ban would increase food prices, although estimates on the magnitude of that increase vary considerably from one study to another. A pesticide ban would cause considerable economic hardship for certain growers, although other farmers might benefit. For example, some insects are more troublesome in certain areas than in others. A farmer growing a crop in an area where an insect was harmful might not be able to afford the crop losses that would occur without the use of a banned pesticide. Growers in areas where the insect was less of a problem could then increase their production of that crop, benefiting financially from the first farmer's loss.

Since it is impractical to ban large numbers of pesticides right now, is there another way to provide greater protection to the environment and human health without reducing crop yields? Governments in Sweden, Denmark, the Netherlands, and the Canadian Province of Ontario think so. Sweden achieved a 50% reduction in pesticide use in 1992 and is now on a second program to reduce pesticide use another 50%. Denmark, the Netherlands, and the Province of Ontario implemented similar programs to reduce pesticide use 50% during the 1990s. Strategies to reduce pesticide use include removing subsidies that encourage pesticide use, applying pesticide only when needed, using improved application equipment (to reduce the amount of pesticide applied), and adopting integrated pest management (IPM) practices.

Too often pesticides are applied unnecessarily to prevent a possible buildup of pests. **Calendar spraying** is the regular use of pesticides regardless of whether pests are a problem. Instead of calendar spraying, pesticide use can be decreased by continually monitoring pests so that pesticides are applied only when pests become a problem; this technique is known as **scout-and-spray**. A 1991 study at Cornell University determined that a monitoring program might reduce the use of insecticides on cotton by 20%, for example; scout-and-spray is now recognized as important to maintaining the nation's cotton crop.

The use of aircraft to apply pesticide is extremely wasteful because 50 to 75% of the pesticide does not reach the target area, but instead drifts in air currents until it settles on soil or water (see Figure 23.7). Pesticides applied on land by traditional methods drift in air currents. Advances in the design of equipment for applying pesticides could reduce pesticide use considerably. A rope-wick applicator reduces herbicide use on soybean fields approximately 90%.

Pesticide use is reduced considerably through alternative pest control strategies. The widespread adoption of IPM would make it feasible, based on current available technologies, for the United States to reduce pesticide use by 50% within a 5- to 10-year period, assuming that Congress passed the necessary legislation. The economic results of such a reduction would benefit both farmers and consumers.

tion. Numerous countries, such as Canada, much of Western Europe, Japan, Russia, and Israel, extend the shelf lives of foods with irradiation. The U.S. Food and Drug Administration (FDA) approved this process for fruits and vegetables as well as fresh poultry in 1986, and the first irradiated food was sold in the United States in 1992. In 2000 the USDA approved the irradiation of additional raw meats, such as ground beef, steaks, and pork chops, to eliminate bacterial contamination.

The irradiation of foods is somewhat controversial. Some consumers are concerned because they fear that irradiated food is radioactive; it is not. You are not exposed to radiation when you eat irradiated foods.

Critics of irradiation are concerned because irradiation forms traces of certain chemicals called *free radicals*, some of which are carcinogenic in laboratory animals. Critics also point out that we do not know the long-term effects of eating irradiated foods. Others are concerned about the potential security risk from misuse of the radiation source, usually cobalt-60 or cesium–137. Proponents of irradiation argue that free radicals normally occur in food and are produced by cooking methods such as frying and broiling. They assert that more than 1000 investigations of irradiated foods, conducted worldwide for more than three decades, have demonstrated it is safe. Furthermore, irradiation lessens the need for pesticides and food additives.

REVIEW

1. What is an example of using cultivation methods to control pests? of using pheromones and hormones?

2. What is integrated pest management? Why is IPM considered a systems approach?

Laws Controlling Pesticide Use

LEARNING OBJECTIVE

- Briefly summarize the three U.S. laws that regulate pesticides: the Food, Drug, and Cosmetics Act; the Federal Insecticide, Fungicide, and Rodenticide Act; and the Food Quality Protection Act.

The federal government has passed several laws to regulate pesticides in the interest of protecting human health and the environment. These include the Food, Drug, and Cosmetics Act; the Federal Insecticide, Fungicide, and Rodenticide Act; and the Food Quality Protection Act. The EPA and, to a lesser extent, the FDA and the USDA oversee the implementation of these laws. In addition, the EPA has some responsibility to regulate pesticides if they lead to violations of the Endangered Species Act.

Food, Drug, and Cosmetics Act

The **Food, Drug, and Cosmetics Act (FDCA),** passed in 1938, recognized the need to regulate pesticides found in food but did not provide a means of regulation. The FDCA was made more effective in 1954 with passage of the **Pesticide Chemicals Amendment.** This amendment, also called the *Miller Amendment*, required the establishment of acceptable and unacceptable levels of pesticides in food.

An amended FDCA, passed in 1958, contained an important section, the **Delaney Clause,** which stated that no substance capable of causing cancer in test animals or in humans would be permitted in processed food. Processed foods are prepared in some way, such as frozen, canned, dehydrated, or preserved, before being sold. The Delaney Clause recognized that pesticides tend to concentrate in condensed processed foods, such as tomato paste and applesauce.

The Delaney Clause, though commendable for its intent, contained two inconsistencies. First, it did not cover pesticides on raw foods such as fresh fruits and vegetables, milk, meats, fish, and poultry. As an example of this double standard, residues of a particular pesticide might be permitted on fresh tomatoes but not in tomato ketchup. Second, because the EPA lacked sufficient data on the cancer-causing risks of pesticides used for a long time, the Delaney Clause applied only to pesticides registered after strict tests were put into effect in 1978. This situation gave rise to one of the paradoxes of the Delaney Clause. There were cases in which a newer pesticide that posed minimal risk was banned because of the Delaney Clause, but an older pesticide, which the newer one was to have replaced, was still used even though it was many times more dangerous.

When the Delaney Clause was passed, the technologies for detecting pesticide residues could reveal only high levels of contamination. Modern scientific techniques are so sensitive that it is almost impossible for any processed food to meet the Delaney standard. As a result, the EPA found it difficult to enforce the strict standard required by the Delaney Clause. In 1988 the EPA began granting exceptions that permitted a "negligible risk" of one case of cancer in 70 years for every one million people. However, the EPA was taken to court because of its failure to follow the Delaney Clause as currently written, and the U.S. courts decided in 1994 that no exceptions could be granted unless Congress modified the Delaney Clause. (One of the key provisions of the 1996 Food Quality Protection Act, discussed shortly, did just that.)

Federal Insecticide, Fungicide, and Rodenticide Act

The **Federal Insecticide, Fungicide, and Rodenticide Act (FIFRA)** was originally passed in 1947 to regulate the effectiveness of pesticides—that is, to prevent people from buying pesticides that did not work. FIFRA has been amended over the years to require testing and registration of the active ingredients of pesticides. Any pesticide that does not meet the tolerance standards established by the FDCA must be denied registration by FIFRA.

Table 23.3 Worst-Case Estimates of Risk of Cancer from Pesticide Residues on Food

Food	Cancer Risk*
Tomatoes	$8.8 \times 10^{-4\dagger}$
Beef	6.5×10^{-4}
Potatoes	5.2×10^{-4}
Oranges	3.8×10^{-4}
Lettuce	3.4×10^{-4}
Apples	3.2×10^{-4}
Peaches	3.2×10^{-4}
Pork	2.7×10^{-4}
Wheat	1.9×10^{-4}
Soybeans	1.3×10^{-4}

* Note that these values are highly uncertain— see chapter 7. Four assumptions were made in arriving at these values: (1) the entire U.S. crop (of tomatoes, for example) is treated (2) with *all* pesticides that are registered for use on that crop; (3) the pesticides are applied the maximum number of times (4) at the maximum rate, or amount, each time.

†As an example of how to interpret these figures, tomatoes are estimated to cause an average of 8.8 deaths from cancer for every 10,000 people.

In 1972 the EPA was given the authority to regulate pesticide use under the terms of the FDCA and FIFRA. Since that time, the EPA has banned or restricted the use of many chlorinated hydrocarbons. In 1972 the EPA banned DDT for almost all uses. Aldrin and dieldrin were outlawed in 1974 after more than 80% of all dairy products, fish, meat, poultry, and fruits were found to contain residues of these insecticides. The banning of kepone occurred in 1977 and of chlordane and heptachlor in 1988.

A two-year study by the National Research Council concluded in 1987 that U.S. laws regarding pesticide residues in food were not adequate to protect the public from cancer-causing pesticides (**Table 23.3**). It included several recommendations that were made into law—an amended FIFRA—in 1988. The 1988 law required reregistration of older pesticides, which subjected them to the same toxicity tests that new pesticides face.

The 1988 law had its critics. Although it was stricter than previous legislation, it represented a compromise between agricultural interests, including pesticide manufacturers, and those opposed to all uses of pesticides. The new law did not address an important issue, the contamination of groundwater by pesticides. Nor did the law address the establishment of standards for pesticide residues on foods and the safety of farm workers who are exposed to high levels of pesticides.

FIFRA also did not require pesticide companies to disclose the inert ingredients in their formulations. Many pesticide products contain as much as 99% inert ingredients, which are not supposed to have active properties against the target organism. The National Coalition for Alternatives to Pesticides has examined the EPA's chemical ingredient database and determined that 394 of the more than 2500 chemicals listed as "inert" were once listed as active ingredients. Many inert ingredients are generally recognized as safe (examples include pine oil, ethanol, silicone, and water). Others are known toxins (examples include asbestos, benzene, formaldehyde, lead, and cadmium). In fact, more than 200 inert ingredients are classified as hazardous air and water pollutants, and 21 are known or suspected to cause cancer.

Food Quality Protection Act

The **Food Quality Protection Act** of 1996 amended both the FDCA and FIFRA. It revised the Delaney Clause by establishing identical pesticide residue limits—those that pose a negligible risk—for both raw produce and processed foods. The law requires that the increased susceptibility of infants and children to pesticides be considered when establishing pesticide residue limits for some 9700 pesticide uses on specific crops. The pesticide limits are established for all health risks, not just cancer. For example, the EPA must develop a program to test pesticides for endocrine-disrupting properties. Another key provision of the Food Quality Protection Act is that it reduces the time it takes to ban a pesticide considered dangerous, from 10 years to 14 months.

REVIEW

1. What three laws regulate pesticides in the United States? What are the goals of each law?

The Manufacture and Use of Banned Pesticides

LEARNING OBJECTIVE

- Define *persistent organic pollutant* and describe the purpose of the Stockholm Convention on Persistent Organic Pollutants.

Some U.S. companies manufacture pesticides that are banned or heavily restricted in the United States and export them to developing countries, particularly in Asia, Africa, and Latin America. International trade of banned or restricted pesticides is notoriously difficult to monitor. The Foundation for Advancements in Science and Education began documenting the extent of the trade in hazardous pesticides in 1991 by examining customs records from U.S. ports. The latest available data indicate that from 1997 to 2000, U.S. companies exported over 30,500 metric tons (32,500 tons) of pesticides forbidden from use in the United States. This amount includes pesticides considered too dangerous to use in the United States as well as pesticides that the EPA has not evaluated. Other highly developed nations also export banned pesticides.

The U.N. Food and Agriculture Organization (FAO) is attempting to help developing nations become more aware of dangerous pesticides. It established a "red alert" list of more than 50 pesticides banned in five or more countries. The FAO further requires that the manufacturers of these pesticides inform importing countries about why such pesticides are banned. The United States supports these international guidelines and exports banned pesticides only with the informed consent of the importing country. However, this information often does not trickle down to the local level, and many foreign farmers never receive any guidelines or training on the safe storage and application of pesticides.

Another concern relating to banned pesticides is that unwanted stockpiles of leftover, deteriorated pesticides are accumulating, particularly in developing countries. The United Nations estimates that more than 100,000 tons of these obsolete pesticides are stockpiled in developing countries. They are often stored in drums at waste sites in the countryside because developing countries have few or no hazardous waste disposal facilities. Over time, chemicals leach from such waste sites into the soil, and from there they get into waterways and groundwater (see Chapter 24).

The Importation of Food Tainted with Banned Pesticides

The fact that many dangerous pesticides are no longer being used in the United States is no guarantee that traces of those pesticides are not in our food. Although many pesticides are restricted or banned in the United States, they are widely used in other parts of the world. Much of our food—some 1.2 million shipments annually—is imported from other countries, particularly those in Latin America. Some produce contains traces of banned pesticides such as DDT, dieldrin, chlordane, and heptachlor. The 2007 discovery of contaminated pet food and toothpaste from China highlighted the risk of imported foods.

It is not known how much of the food coming into the United States is tainted with pesticides. The FDA monitors toxic residues on incoming fruits and vegetables, but it inspects only about 1% of the food shipments that enter the United States each year. In addition, the General Accounting Office reports that some food importers illegally sell food after the FDA has found it is tainted with pesticides. When caught, these companies face fines that are not severe enough to discourage such practice.

persistent organic pollutants: A group of persistent, toxic chemicals that bioaccumulate in organisms and can travel thousands of kilometers through air and water to contaminate sites far removed from their source.

The Global Ban of Persistent Organic Pollutants

The **Stockholm Convention on Persistent Organic Pollutants,** which went into effect in 2004, is an important international treaty. It seeks to protect human health and the environment from the 12 most toxic chemicals, classified as **persistent organic pollutants (POPs)** (**Table 23.4**; also see the discussion of long-distance transport of air pollution in Chapter 20). Nine of these POPs are pesticides. Some POPs disrupt the endocrine system, others cause cancer, and still others adversely affect the developmental processes of organisms.

The Stockholm Convention requires that countries develop plans to eliminate the production and use of intentionally produced POPs. A notable exception to this requirement is that DDT can still be produced and used to control mosquitoes that carry the malaria pathogen in countries where no affordable alternatives exist. (DDT is inexpensive, and many of these countries cannot afford safer alternatives.)

Although the Stockholm Convention goes a long way toward the goal of a global ban on these chemicals, it applies only to countries that have ratified the treaty.

Table 23.4 Persistent Organic Pollutants: The "Dirty Dozen"

Persistent Organic Pollutant	Use
Aldrin	Insecticide
Chlordane	Insecticide
DDT (dichlorodiphenyltrichloroethane)	Insecticide
Dieldrin	Insecticide
Endrin	Rodenticide and insecticide
Heptachlor	Fungicide
Hexachlorobenzene	Insecticide; fire retardant
Mirex(TM)	Insecticide
Toxaphene(TM)	Insecticide
PCBs (polychlorinated biphenyls)	Industrial chemical
Dioxins	By-product of certain manufacturing processes
Furans (dibenzofurans)	By-product of certain manufacturing processes

To be effective, more countries, including the United States, will have to ratify it and adhere to its rules.

REVIEW

1. What is the Stockholm Convention on Persistent Organic Pollutants?

REVIEW OF LEARNING OBJECTIVES WITH KEY TERMS

• **Distinguish between narrow-spectrum and broad-spectrum pesticides, and describe the various types of pesticides, such as insecticides and herbicides.**

Pesticides are toxic chemicals used to kill pests such as insects (**insecticides**), weeds (**herbicides**), fungi (**fungicides**), and rodents (**rodenticides**). The ideal pesticide would be a **narrow-spectrum pesticide** that kills only the target organism. Most pesticides are **broad-spectrum pesticides**, which kill a variety of organisms, including beneficial organisms, in addition to the target pest.

• **Relate the benefits of pesticides in disease control and crop protection.**

Pesticides help prevent diseases transmitted by insects. Pesticides reduce crop losses from pests, thereby increasing agricultural productivity. Pesticides reduce competition with weeds, consumption by insects, and diseases caused by plant **pathogens** such as certain fungi and bacteria.

• **Explain why monocultures are more susceptible to pest problems.**

A **monoculture** is the cultivation of only one type of plant over a large area. Thus it represents a system that is out of balance compared to what would be found without human intervention. Agricultural fields are monocultures that provide abundant food for pest organisms. Many natural predators are absent from monocultures.

• **Summarize the problems associated with pesticide use, including development of genetic resistance; creation of imbalances in the ecosystem; persistence, bioaccumulation, and biological magnification; and mobility in the environment.**

Genetic resistance is any inherited characteristic that decreases the effect of a pesticide on a pest. Pesticides affect species other than those for which they are intended, causing imbalances in ecosystems; in some instances the use of a pesticide has resulted in a pest problem that did not exist before. A pesticide that demonstrates **persistence** takes a long time to break down into less toxic forms. **Bioaccumulation** is the buildup of a persistent pesticide or other toxic substance in an organism's body. **Biological magnification** is the increased concentration of toxic chemicals such as certain pesticides in the tissues of organisms at higher trophic levels in food webs. Many pesticides move through the soil, water, and air, sometimes for long distances.

• **Describe the pesticide treadmill and explain how resistance management may help with this problem.**

The **pesticide treadmill** is a predicament faced by pesticide users, in which the cost of applying pesticides increases (because they have to be applied more frequently or in larger doses) while their effectiveness decreases (as a result of increasing genetic resistance in the target pest). **Resistance management** consists of strategies for managing genetic resistance to maximize the period in which a pesticide is useful.

• **Discuss pesticide risks to human health, including short-term effects and long-term effects.**

Some serious health problems are associated with pesticide use. Humans may be poisoned by short-term exposure to a large amount of pesticide. Lower levels of many pesticides may pose a long-term threat of cancer. Certain persistent pesticides may interfere with the actions of natural hormones.

• **Explain endocrine disruption.**

Endocrine disruption occurs when a chemical interferes with or mimics a hormone associated with growth and development in humans or other animals. Endocrine disrupters include atrazine and DDT, and problems include abnormal physical features—such as deformed reproductive organs—and aberrant behavior.

• **Describe alternative ways to control pests, including cultivation methods, biological controls, pheromones and hormones, reproductive controls, genetic controls, quarantine, integrated pest management, and food irradiation.**

Cultivation techniques such as strip cutting, interplanting, and crop rotation are effective in controlling pests. **Biological control** is a method of pest control that involves the use of naturally occurring disease organisms, parasites, or predators to control pests. A **pheromone** is a natural substance produced by animals to stimulate a response in other members of the same species; pheromones can be used to lure insects to traps or to confuse insects so that they cannot locate mates. Insect hormones are natural substances produced by insects to regulate their own growth and metamorphosis; a hormone present at the wrong time in an insect's life cycle disrupts its normal development. Reproductive controls include reducing the pest population by sterilizing some of its members (the **sterile male technique**). Genetic control of pests involves producing varieties of crops and livestock animals that are genetically resistant to pests; some **genetically modified (GM)** crops contain the *Bacillus thuringiensis* (*Bt*) gene that codes for a toxin used against insect pests. **Quarantine** involves restricting the importation of exotic plant and animal material that might harbor pests; if a foreign pest is accidentally introduced, quarantine of the area where it is detected helps prevent its spread. **Integrated pest management (IPM)** is a systems approach that combines several pest control methods that, if used in the proper order and at the proper times, keep the size of a pest population low enough that it does not cause substantial economic loss. **Irradiation** of food controls pests after food is harvested.

• **Briefly summarize the three U.S. laws that regulate pesticides: the Food, Drug, and Cosmetics Act; the Federal Insecticide, Fungicide, and Rodenticide Act; and the Food Quality Protection Act.**

The **Food, Drug, and Cosmetics Act (FDCA)**, as originally passed, recognized the need to regulate pesticides in food but did not provide a means of regulation. The *Miller Amendment* required the establishment of acceptable and unacceptable levels of pesticides in food, and the **Delaney Clause** stated that no substance capable of causing cancer in laboratory animals or in humans would be permitted in processed food. The **Federal Insecticide, Fungicide, and Rodenticide Act (FIFRA)** was amended over the years to require testing and registration of the active ingredients of pesticides; the 1988 version required the reregistration of older pesticides, which subjected them to the same toxicity tests that new pesticides face. The **Food Quality Protection Act** amended both the FDCA and FIFRA and revised the Delaney Clause by establishing identical pesticide residue limits—those that pose a negligible risk—for both raw and processed foods.

• **Define *persistent organic pollutant* and describe the purpose of the Stockholm Convention on Persistent Organic Pollutants**

Persistent organic pollutants (POPs) are a group of persistent, toxic chemicals that bioaccumulate in organisms and can travel thousands of kilometers through air and water to contaminate sites far removed from their source. The **Stockholm Convention on Persistent Organic Pollutants** seeks to protect human health and the environment from the 12 most toxic chemicals on Earth.

Thinking About the Environment

1. Distinguish among insecticides, herbicides, fungicides, and rodenticides.

2. Describe the general characteristics of each of the following groups of insecticides: chlorinated hydrocarbons, organophosphates, and carbamates.

3. What is the dilemma referred to in the title of this chapter?

4. Overall, do you think the benefits of pesticide use outweigh its disadvantages? Give at least two reasons for your answer.

5. Sometimes pesticide use increases the damage done by pests. Explain.

6. The widely used herbicide Roundup is starting to lose its effectiveness in killing certain weeds. Explain why.

7. How is the buildup of insect resistance to insecticides similar to the increase in bacterial resistance to antibiotics?

8. How does genetic change in response to biological control agents differ from genetic resistance to pesticides?

9. Biological control is often much more successful on a small island than on a continent. Offer at least one reason why this might be the case.

10. It is more effective to use the sterile male technique when an insect population is small than when it is large. Explain.

11. Define integrated pest management (IPM). List five tools of IPM, and give an example of each.

12. How is IPM related to ecological concepts such as food webs and energy flow?

13. Which of the following uses of pesticides do you think are most important? Which are least important? Explain your views.

    a. Keeping roadsides free of weeds

    b. Controlling malaria

    c. Controlling crop damage

    d. Producing blemish-free fruits and vegetables

14. Why is pesticide misuse increasingly viewed as a global environmental problem?

15. Propose a multi-part, systems-based approach to controlling mice in a home. Your approach may include a rodenticide.

Quantitative questions about this chapter are found on our Web site.

Take a Stand

Visit our Web site at http://www.wiley.com/college/raven (select Chapter 23 from the Table of Contents) for links to more information about issues surrounding the genetic engineering of the *Bt* gene into crops. Consider the opposing views of those who support and those who oppose the use of genetic engineering for this purpose and debate the issue with your classmates. You will find tools to help you organize your research, analyze the data, think critically about the issues, and construct a well-considered argument. *Take a Stand* activities can be done individually or as a team, as oral presentations, written exercises, or Web-based (e-mail) assignments.

Additional online materials relating to this chapter, including a Student Testing Section with study aids and self-tests, Environmental News, Activity Links, Environmental Investigations, and more, are also on our Web site.

24

Solid and Hazardous Wastes

Obsolete computer equipment. (© Don Mason/Corbis)

I n the United States, Canada, and other highly developed countries the average computer is replaced every 18 to 24 months, not because it is broken but because rapid technological developments and new generations of software make it obsolete. Old computers may still be in working order, but they have no resale value and are even difficult to give away. As a result, they sit in warehouses, garages, and basements—or are frequently thrown away with the trash and end up in landfills or incinerators. According to the Grass Roots Recycling Network, there were 300 million to 600 million obsolete personal computers in the United States by 2004 (latest data available).

This disposal represents a huge waste of the high-quality plastics and metals (aluminum, copper, tin, nickel, palladium, silver, and gold) that make up computers. Computers also contain the toxic heavy metals lead, cadmium, mercury, and chromium, which could potentially leach from landfills into soil and groundwater. Most computers contain three to eight pounds of lead, for example. Several

states have passed legislation requiring businesses and residents to eCycle consumer electronics—that is, recycle PCs, color monitors, cell phones, and color televisions.

Electronic waste—or e-waste—has provided a business opportunity for certain companies, such as Viatek Solutions in Tampa, Florida. Viatek disassembles old computers, monitors, fax machines, printers, and other electronic equipment that they obtain from businesses and individuals. The company recycles the glass from monitors, plastics from casings, and metal from wires and circuit boards. They then ship the components to glass, plastic, and metal recyclers in the United States and abroad. For example, the plastics in broken components are recycled to make such items as park benches and shelving.

Currently, about 10% of discarded computer and electronic components are eCycled in the United States. Although companies such as Viatek Solutions handle obsolete computers in the United States, many U.S. computers to be recycled are shipped overseas to developing countries such as India, Pakistan, and China. There the computers are disassembled, often by methods potentially dangerous to the workers taking them apart. For example, circuit boards are often burned to obtain the small amount of gold in them, and burning releases hazardous fumes into the air.

Some highly developed countries have been more progressive than the United States in

dealing with their computer waste. The European Union implemented a Waste Electrical and Electronic Equipment plan in 2005 to recover, recycle, and dispose of electronic waste and remove some of the most hazardous chemicals. Similarly, Japan and other industrialized nations have put such policies into place.

Like it or not, humans produce solid waste, and it is our responsibility to deal with our waste in safe, cost-effective, and environmentally sensitive ways. Using a system perspective, for example, we could design computers to allow more reuse, particularly of the cabinets that house the computer components. There is no single solution to the challenge of solid waste disposal today. A combination of source reduction, reuse, recycling, composting, burning, and burying in sanitary landfills is currently the optimal way to manage solid waste. In this chapter you will examine the problems and the opportunities associated with the management of solid waste. You will then consider hazardous waste, which contains toxic materials and requires special disposal.

World View and Closer to You...
Real news footage relating to solid and hazardous wastes around the world and in your own geographic region is available at www.wiley.com/college/raven by clicking on "World View and Closer to You."

Solid Waste

LEARNING OBJECTIVES

- Distinguish between municipal solid waste and nonmunicipal solid waste.
- Describe the features of a modern sanitary landfill and relate some of the problems associated with sanitary landfills.
- Describe the features of a mass burn incinerator and relate some of the problems associated with incinerators.

The United States generates more solid waste, per capita, than any other country. (Canada is a close second.) Each person in the United States produces an average of 2.1 kg (4.6 lb) of solid waste per day. This amount corresponded to a total of 215 million metric tons (238 million tons) in 2005. And the problem worsens each year as the U.S. population increases.

The solid waste problem was made abundantly clear by several highly publicized instances of garbage barges wandering from port to port and from country to country, trying to find someone willing to accept their cargo. In 1987 the tugboat *Break of Dawn* towed a garbage barge from New York to North Carolina. When North Carolina refused to accept the solid waste, the *Break of Dawn* set off on a journey of many months. In total, six states and three countries rejected the waste, which was eventually returned to New York to be incinerated.

Another example of our solid waste problem is the story of the *Khian Sea*, a Bahamian ship hired by the city of Philadelphia in 1986. It was to transport 14,000 tons of incinerator ash to the Bahamas, but before it got to port, it was denied entry because the Bahamians worried that the ash might contain toxic chemicals. For the next few years, the barge tried unsuccessfully to dump the ash in Puerto Rico, the Dominican Republic, the Netherlands Antilles, Honduras, and Guinea-Bissau on the western coast of Africa. In 1988 about 4000 tons were dumped on a beach in Haiti, which demanded that it be reloaded onto the barge. The barge took off without reloading the ash and then disappeared for several months, only to reappear in Singapore under a new name and minus its cargo, which was dumped somewhere in the Indian Ocean. In 2002 the Pennsylvania Department of Environmental Protection transported the ash remaining in Haiti back to Pennsylvania for permanent disposal.

Waste generation is an unavoidable consequence of prosperous, high-technology, industrial economies. It is a problem not only in the United States but also in Canada and other highly developed nations. Many products that could be repaired, reused, or recycled are simply thrown away. Others, including paper napkins and disposable diapers, are supposed to be used once and then discarded. Packaging, which not only makes a product more attractive and more likely to sell, but protects it and keeps it sanitary, also contributes to waste. Nobody likes to think about solid waste, but the fact is that it is a concern of modern society—we keep producing it, and places to dispose of it safely are limited.

municipal solid waste: Solid materials discarded by homes, office buildings, retail stores, restaurants, schools, hospitals, prisons, libraries, and other commercial and institutional facilities.

nonmunicipal solid waste: Solid waste generated by industry, agriculture, and mining.

Figure 24.1 Composition of municipal solid waste, 2005. (EPA)

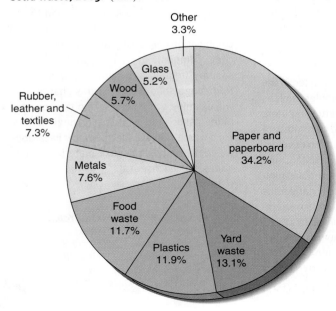

Types of Solid Waste

Municipal solid waste is a heterogeneous mixture composed primarily of paper and paperboard; yard waste; plastics; food waste; metals; materials such as rubber, leather, and textiles; wood; and glass (**Figure 24.1**). The proportions of the major types of solid waste in this mixture change over time. Today's solid waste contains more paper and plastics than in the past, whereas the amounts of glass and steel have declined.

Municipal solid waste is a relatively small portion of all the solid waste produced. **Nonmunicipal solid waste**, which includes wastes from mining (mostly waste rock, about 75% of the total solid waste production), agriculture (about 13%), and industry (about 10%), is produced in substantially larger amounts than municipal solid waste (about 2%). Thus, most solid waste generated in the United States is from nonmunicipal sources.

Disposal of Solid Waste

Solid waste has been traditionally regarded as material that is no longer useful and that should be disposed of. There are four ways to get rid of solid waste: dump it, bury it, burn it, or compost it.

The old method of solid waste disposal was dumping. Open dumps were unsanitary, malodorous places in which disease-carrying vermin such as rats and flies proliferated. Methane gas was released into the surrounding air as microorganisms decomposed the solid waste, and fires polluted the air with acrid smoke. Liquid that oozed and seeped through the solid waste heap ultimately found its way into the soil, surface water, and groundwater. Hazardous materials that were dissolved in this liquid often contaminated soil and water.

Sanitary Landfills   Open dumps have been replaced by **sanitary landfills**, which receive about 54% of the solid waste generated in the United States today (**Figure 24.2**). Sanitary landfills differ from open dumps in that the solid waste is placed in a hole, compacted, and covered with a thin layer of soil every day (**Figure 24.3**). This process reduces the number of rats and other vermin usually associated with solid waste, lessens the danger of fires, and decreases the amount of odor. If a sanitary landfill is operated in accordance with solid waste management-approved guidelines, it does not pollute local surface and groundwater. Safety is ensured by layers of compacted clay and plastic sheets at the bottom of the landfill, which prevent liquid waste from seeping into groundwater. Newer landfills possess a double liner system (plastic, clay, plastic, clay) and use sophisticated systems to collect **leachate** (liquid that seeps through the solid waste) and gases that form during decomposition.

Sanitary landfills charge "tipping fees" to accept solid waste. This money helps offset the landfill's operating costs and lets the jurisdiction charge lower property taxes for homes and businesses located in close proximity to the sanitary landfill. Tipping fees vary widely from one state to another. Sanitary landfills in Nevada, for example, currently charge tipping fees of $11 per ton of solid waste, whereas Massachusetts landfills charge $69 per ton. Some jurisdictions cannot handle all their waste, and so they export their solid waste to nearby states with lower tipping fees. For example, sanitary landfills in Ohio and Indiana accept solid waste from New York and New Jersey.

The location of an "ideal" sanitary landfill is based on a variety of factors, including the geology of the area, soil drainage properties, and the proximity of nearby bodies of water and wetlands. The landfill should be far enough away from centers of dense population that it is inoffensive but close enough that high transportation costs are not required. Landfill designs should take into account an area's climate, such as rainfall, snowmelt, and the likelihood of flooding.

Problems Associated With Sanitary Landfills.   Although the operation of sanitary landfills has improved over the years with passage of stricter and stricter guidelines, few landfills are ideal. Most sanitary landfills in operation today do not meet current legal standards for new landfills.

One problem associated with sanitary landfills is the production of methane gas by microorganisms that decompose organic material anaerobically (in the absence of oxygen). This methane may seep through the solid waste and accumulate in underground pockets, creating the possibility of an explosion. It is even possible for methane to seep into basements of nearby homes, which is an extremely dangerous situation. Conventionally, landfills collected the methane and burned it off in flare systems. A growing number of landfills have begun to use the methane for gas-to-energy projects. About 425 landfills in the United States currently use methane gas to generate electricity.

sanitary landfill: The most common method of disposal of solid waste, by compacting it and burying it under a shallow layer of soil.

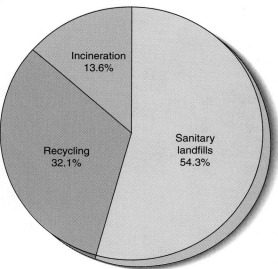

Figure 24.2   U.S. disposal of municipal solid waste in 2005. (EPA)

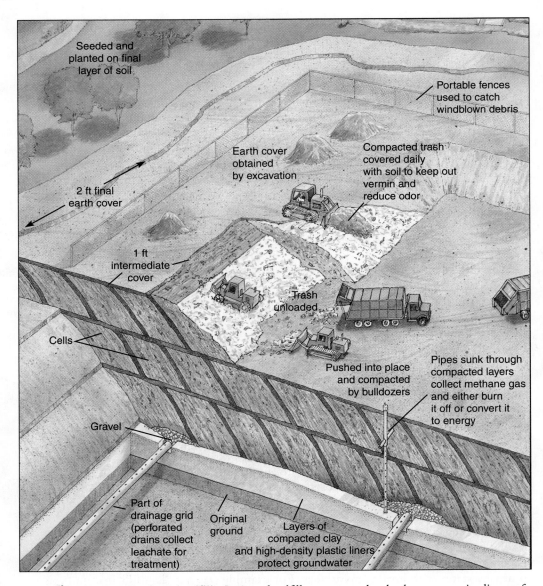

Figure 24.3   Sanitary landfill. Sanitary landfills constructed today have protective liners of compacted clay and high-density plastic and sophisticated leachate collection systems that minimize environmental problems such as contamination of groundwater. Solid waste is spread in a thin layer, compacted into small sections, called cells, and covered with soil.

Another problem associated with sanitary landfills is the potential contamination of surface water and groundwater by leachate that seeps from unlined landfills or through cracks in the lining of lined landfills. New York's Fresh Kills Landfill, which was the largest in the United States before it closed in 2002, continues to produce an estimated one million gallons of leachate each day. Because household trash contains hazardous chemicals such as heavy metals, pesticides, and organic compounds that can seep into groundwater and surface water, the leachate must be collected and treated, even though the sanitary landfill has closed.

Landfills are not an indefinite remedy for waste disposal because they are filling up. From 1988 to 2005 the number of U.S. landfills in operation decreased from 8000 to 1654. Many of the landfills closed because they had reached their capacity. Other landfills closed because they did not meet state or federal environmental standards.

Fewer new sanitary landfills are being opened to replace the closed ones, although new landfills are generally much larger than in the past. The reasons for fewer new landfills are many and complex. Many desirable sites are already taken. Also, people living near a proposed site are usually adamantly opposed to a landfill near their homes.

Recall from Chapter 12 that the opposition of people to the location of hazardous facilities near their homes is known as the not-in-my-backyard, or *NIMBY*, response. This attitude is partly the result of past problems with landfills, ranging from offensive odors to dangerous contamination of drinking water. It is also caused by the fear that a nearby facility will lower property values.

Once a sanitary landfill is full, closing it involves considerable expense. Because groundwater pollution and gas explosions remain possible for a long time, the Environmental Protection Agency (EPA) currently requires owners to continuously monitor a landfill for 30 years after the landfill is closed. In addition, no homes or other buildings can be built on a closed sanitary landfill for many years.

The Special Problem of Plastic.   The amount of plastic in our solid waste is growing faster than any other component of municipal solid waste. More than half of this plastic is from packaging. Plastics are chemical **polymers** composed of chains of repeating carbon compounds. The properties of the many types of plastics—polypropylene, polyethylene, and polystyrene, to name a few—differ on the basis of their chemical compositions.

Most plastics are chemically stable and do not readily break down, or decompose. This characteristic, which is essential in the packaging of certain products, such as food, causes long-term problems. Indeed, most plastic debris disposed of in sanitary landfills will probably last for centuries.

In response to concerns about the volume of plastic waste, some areas have actually banned the use of certain types of plastic, such as the polyvinyl chloride employed in packaging. Special plastics that have the ability to degrade or disintegrate have been developed. Some of these are **photodegradable;** that is, they break down only after being exposed to sunlight, which means they will not break down in a sanitary landfill. Other plastics are **biodegradable**—that is, they are decomposed by microorganisms such as bacteria. Whether biodegradable plastics actually break down under the conditions found in a sanitary landfill is not yet clear, although several studies indicate that they probably do not. Many factors, such as temperature, amount of oxygen present, and composition of the microbial community, affect biodegradation of plastics and other organic materials. (Other waste management options for plastic are discussed later in this chapter.)

The Special Problem of Tires.   One of the most difficult materials to manage is rubber. Discarded tires—about 290 million each year in the United States—are made of vulcanized rubber, which cannot be melted and reused for tires. At least 265 million old tires have accumulated in tire dumps, as well as along roadsides and in vacant lots. Disposal of tires in sanitary landfills is a real problem because tires, being relatively large and light, have a tendency to move upward through the accumulated solid waste. After a period, they work their way to the surface of the landfill. These tires are a fire hazard, creating fires difficult to extinguish. Old tires also collect rainwater, providing a good breeding place for mosquitoes. Accordingly, most states either ban tires from sanitary landfills or require that they be shredded to save space and prevent water from pooling in them. (Other waste management options for tires are discussed later in the chapter.)

Incineration   When solid waste is incinerated, two positive things are accomplished. First, the volume of solid waste is reduced up to 90%. The ash that remains is, of course, much more compact than solid waste that has not been burned. Second, incineration produces heat that can make steam to warm buildings or generate electricity. In 2005 the United States had 88 waste-to-energy incinerators, which burned about 14% of the nation's solid waste. In comparison, less than 1% of U.S. solid waste was incinerated in 1970. Waste-to-energy incinerators produce substantially less carbon dioxide emissions than equivalent power plants that burn fossil fuels (**Figure 24.4**). (Recall from Chapter 21 that carbon dioxide is a potent greenhouse gas.)

Figure 24.4   Carbon dioxide emissions per kilowatt-hour of electricity generated. Waste-to-energy incinerators release less carbon dioxide into the atmosphere than do equivalent power plants that burn fossil fuels. (EcoBalance)

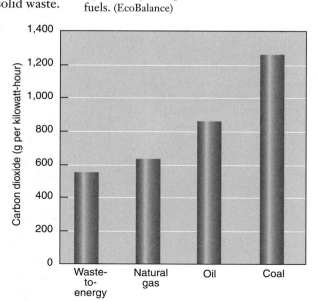

Figure 24.5   Tires that will be burned to generate electricity. This mountain in Westley, California, contains four million to six million old tires. The power plant that burns them supplies electricity to 3500 homes. (The person wearing red gives a sense of scale.) (Jose Azel/Aurora Photos)

Some materials are best removed from solid waste before incineration occurs. Glass does not burn, and when it melts, it is difficult to remove from the incinerator. Although food waste burns, its high moisture content often decreases the efficiency of incineration, so it is better to remove it before incineration. Removal of batteries, thermostats, and fluorescent lights is desirable because it eliminates most mercury emissions produced during combustion. The best materials for incineration are paper, plastics, and rubber.

Paper is a good candidate for incineration because it burns readily and produces a great amount of heat. Several studies have examined the economic and environmental costs and benefits of various waste paper management options. Many of these studies have concluded that waste-to-energy incineration is better than recycling, which in turn is better than disposal in a sanitary landfill. (The studies do not reach unanimous conclusions because economists do not agree about the cash value that should be applied for environmental benefits and costs. For example, estimates of the environmental cost of emitting 1 kg of carbon dioxide range from $1 to more than $50.) One potential environmental complication associated with burning paper is the presence of hazardous compounds in the ink and paper that might be emitted during incineration. Some types of paper release *dioxins* into the atmosphere when burned; dioxins are discussed later in the chapter.

Plastic produces a lot of heat when it is incinerated. In fact, one kilogram of plastic waste yields almost as much heat as a kilogram of fuel oil. As with paper, the pollutants that might be emitted during the incineration of plastic are of some concern. Polyvinyl chloride, a common component of many plastics, may release dioxins and other hazardous compounds when incinerated.

One of the best uses for old tires is incineration because burning rubber produces much heat. Some electric utilities in the United States and Canada burn tires instead of or in addition to coal (**Figure 24.5**). Tires produce as much heat as coal and often generate less pollution. In 2003, 45% of all tires discarded annually were incinerated.

Types of Incinerators. The three types of incinerators are mass burn, modular, and refuse-derived fuel. Most **mass burn incinerators** are large and are designed to recover the energy produced from combustion (**Figure 24.6**). **Modular incinerators** are smaller incinerators that burn all solid waste. They are assembled at factories and so are less expensive to build. In **refuse-derived fuel incinerators,** only the combustible portion of solid waste is burned. First, noncombustible materials such as glass and metals are removed by machine or by hand. The remaining solid waste, including plastic and paper, is shredded or shaped into pellets and burned.

Problems Associated with Incineration. The combustion of any fuel, whether it is coal or municipal solid waste, yields some air pollution. The possible production of hazardous air pollutants is the main reason people oppose incineration. Incinerators can pollute the air with carbon monoxide, particulates, heavy metals such as mercury, and other hazardous materials unless air pollution-control devices are used. Such devices include **lime scrubbers,** towers in which a chemical spray neutralizes acidic gases,

mass burn incinerator: A large furnace that burns all solid waste except for unburnable items such as refrigerators.

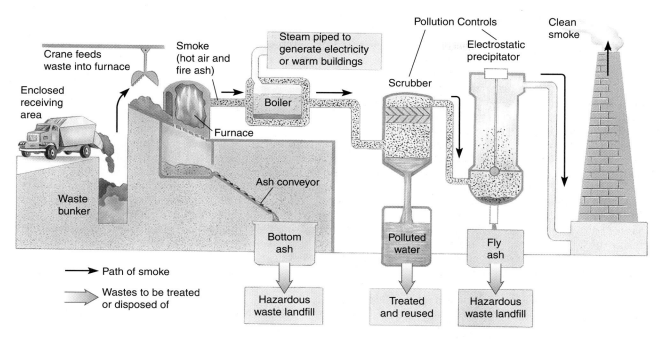

Figure 24.6 Mass burn, waste-to-energy incinerator. Modern incinerators have pollution-control devices such as lime scrubbers and electrostatic precipitators to trap dangerous and dirty emissions.

and **electrostatic precipitators,** which give ash a positive electrical charge so that it adheres to negatively charged plates rather than going out the chimney (see Figure 20.11 for diagrams of a scrubber and an electrostatic precipitator).

Incinerators produce large quantities of ash that must be disposed of properly. Two kinds of ash are produced, bottom ash and fly ash. **Bottom ash,** or slag, is the residual ash left at the bottom of the incinerator when combustion is completed. **Fly ash** is the ash from the flue (chimney) that is trapped by air pollution-control devices. Fly ash usually contains more hazardous materials, including heavy metals and possibly dioxins, than bottom ash.

Currently, both types of incinerator ash are best disposed of in specially licensed hazardous waste landfills (discussed later in this chapter). What happens to the hazardous materials in incinerator ash when it is placed in an average sanitary landfill is unknown, but there are concerns that the hazardous materials could contaminate groundwater.

As in the case of sanitary landfills, site selection for incinerators is controversial. People may recognize the need for an incinerator, but they do not want it near their homes. Another drawback of incinerators is their high cost. Prices have escalated because costly pollution-control devices are now required.

Composting Yard waste, such as grass clippings, branches, and leaves, is a substantial component of municipal solid waste (see Figure 24.1). As space in sanitary landfills becomes more limited, other ways to dispose of yard waste are being implemented. One of the best ways is to convert organic waste into soil conditioners such as compost or mulch (see "You Can Make a Difference: Practicing Environmental Principles," in Chapter 15). Food scraps, sewage sludge, and agricultural manure are other forms of solid waste that can be used to make compost. Compost and mulch are used for landscaping in public parks and playgrounds or as part of the daily soil cover at sanitary landfills. Compost and mulch are also sold to gardeners.

Composting as a way to manage solid waste first became popular in Europe. Many municipalities in the United States have composting facilities as part of their

MEETING THE CHALLENGE

Municipal Solid Waste Composting

The sanitary landfill has received most of the nation's solid waste for the past several decades, but as you have seen, sanitary landfills are not a long-term solution to waste disposal. Much of the bulky waste in sanitary landfills—paper, yard refuse, food wastes, and such—is organic and, given the opportunity, could decompose into compost. In sanitary landfills, little of this material breaks down. Rapid and complete decomposition requires the presence of oxygen, and in a sanitary landfill, garbage is compacted under a layer of soil, so little oxygen is available.

Municipal solid waste composting is the large-scale composting of the entire organic portion of a community's garbage. Because approximately two-thirds of all household garbage is organic (paper, yard wastes, food wastes, and wood), municipal solid waste composting substantially reduces demand for sanitary landfills. Numerous city and county governments are currently composting leaves and yard wastes in an effort to reduce the amount of solid waste sent to landfills. Although this endeavor is undeniably beneficial, municipal solid waste compost-

ing encompasses much more than yard wastes. It involves composting food wastes, paper, and anything else in the solid waste stream that is organic.

Initial composting occurs quickly—in three to four days—because conditions such as moisture and the carbon-nitrogen ratio are continually monitored and adjusted (by adding water or fertilizer, for example) for maximum decomposition. The decay process is carried out by billions of bacteria and fungi, which convert the organic matter into carbon dioxide, water, and humus. So many decomposers eat, reproduce, and die in the compost heap that the drum heats up, killing off potentially dangerous organisms such as disease-causing bacteria. When the material emerges, it is placed outside for several months to cure, during which time additional decomposition occurs. Finally, it is sold as compost.

The potential market for compost is huge. Professional nurseries, landscapers, greenhouses, and golf courses use compost. Also, tons of compost could be used to reclaim the 167 million hectares (413 million acres) of badly eroded farmland in the United States. Compost could improve the fertility of badly eroded rangeland, forestland, and strip mines. There is

no shortage of markets for compost, and should certain technical problems be resolved, composting on a large scale could become economically feasible.

Technical problems include concerns over the presence of pesticide residues and heavy metals in the compost. Pesticides sprayed on urban and suburban landscapes would naturally find their way into compost material on leaves, grass clippings, and other yard wastes. However, several studies indicate that most pesticides are either decomposed by bacteria and fungi during composting or broken down by the high temperatures in the compost heap.

More troubling is the concern over heavy metals, such as lead and cadmium. Heavy metals can enter compost from sewage sludge, which may contain industrial wastewater, or consumer products such as batteries. (Sewage sludge is often added to compost because it is a rich source of nitrogen for the decomposing microorganisms.) Two ways to reduce heavy metal contamination in municipal compost are sorting out heavy metal sources before everything is dumped into the composting drum and requiring industries to pretreat their industrial wastewater before it gets to the sewage treatment facility.

comprehensive solid waste management plans, and many states have banned yard waste from sanitary landfills. This trend is likely to continue, making composting even more desirable (see Meeting the Challenge: Municipal Solid Waste Composting).

REVIEW

1. What is the difference between municipal solid waste and nonmunicipal solid waste?
2. What are three features of a sanitary landfill?
3. What are the main features of a mass burn incinerator?

Waste Prevention

LEARNING OBJECTIVES

- Summarize how source reduction, reuse, and recycling help reduce the volume of solid waste.
- Define *integrated waste management*.

Given the problems associated with sanitary landfills and incinerators, it makes sense to do whatever we can to lessen the need for these waste disposal methods. The three goals of waste prevention, in order of priority, are (1) reduce the amount of waste as much as possible, (2) reuse products as much as possible, and (3) recycle materials as much as possible.

Reducing the amount of waste includes purchasing products that have less packaging and that last longer or are repairable (**Figure 24.7**). Consumers can also decrease their consumption of products to reduce waste. Before deciding to purchase a product, a consumer should ask, "Do I really *need* this product, or do I merely *want* it?" Many U.S. consumers have participated actively for more than a decade in efforts to convert their throwaway economy into a waste prevention economy. Individual efforts have focused mainly on recycling, however, and much remains to be done in the areas of waste reduction and reuse (see Meeting the Challenge: Reusing and Recycling Old Automobiles).

We already discussed dematerialization, reuse, and recycling in Chapter 16, in the context of resource conservation. Now let us examine the impact of these practices on solid waste.

Reducing the Amount of Waste: Source Reduction

The most underutilized aspect of waste management is **source reduction**. Source reduction is accomplished in a variety of ways, such as substituting raw materials that introduce less waste during the manufacturing process and reusing and recycling wastes at the plants where they are generated. Innovations and product modifications can reduce the waste produced after a consumer has used a product. Dry-cell batteries, for example, contain much less mercury today than they did in the early 1980s. The 35% weight reduction in aluminum cans since the 1970s is another example of source reduction.

The **Pollution Prevention Act** of 1990 was the first U.S. environmental law to focus on the reduced generation of pollutants at their point of origin rather than the reduction of pollutants or repair of damage caused by such substances. The act was written to increase the adoption of cost-effective source reduction measures. It requires the EPA to develop source reduction models, and it requires manufacturing facilities to report to the EPA annually on their source reduction and recycling activities.

Dematerialization, the progressive decrease in the size and weight of a product as a result of technological improvements, is an example of source reduction only if the new product is as durable as the one it replaced. If smaller, lighter products have shorter life spans and must be replaced more often, source reduction is not accomplished.

Reusing Products

One example of reuse is refillable glass beverage bottles. Years ago, refillable beverage bottles were used a great deal in the United States. Today they are rarely used. For a glass bottle to be reused, it must be considerably thicker (and heavier) than one-use bottles. Because of the increased weight, transportation costs are higher. In the past, reuse of glass bottles made economic sense because there were many small bottlers scattered across the United States, minimizing transportation costs. Today there are approximately one-tenth as many bottlers. Because of the centralization of bottling beverages, it is economically difficult to go back to the days of refillable bottles. The price of beverages might have to increase to absorb increased transportation costs and the energy used to sterilize dirty bottles.

Although the quantity of reusable glass bottles in the United States has declined, certain countries still reuse glass to a large extent. In Japan, almost all beer and sake bottles are reused as many as 20 times. Bottles in Ecuador may remain in use for 10 years or longer. European countries such as Denmark, Finland, Germany, the Netherlands, Norway, Sweden, and Switzerland have passed legislation that promotes the refilling of beverage containers. Parts of Canada and about 11 states in the United States also have deposit laws.

Figure 24.7 **The six steps of wasteful packaging.** What harmful environmental effects of wasteful packaging are depicted here? (© Steve Greenberg)

source reduction: An aspect of waste management in which products are designed and manufactured in ways that decrease the volume of solid waste and the amount of hazardous waste in the solid waste stream.

MEETING THE CHALLENGE

Reusing and Recycling Old Automobiles

In the United States, about 35 million motor vehicles leave service each year. Most are exported to developing countries, but about 11 million cars and trucks are discarded. Although by weight about 75% of a discarded car is easily reused as secondhand parts or recycled as scrap metal, the remaining 25% is not easy to recycle and usually ends up in sanitary landfills. Because automobiles typically contain about 600 materials—glass, metals, plastics, fabrics, rubber, foam, leather, and so on—identifying ways to reuse or recycle old parts is complex. Economics is an important aspect of the problem, for reuse and recycling companies must make money in their recycling endeavors.

How does a car disassembly factory work? Workers typically begin disassembling a used car by draining all fluids—such as antifreeze, gasoline, transmission fluid, oil, and brake fluid—and recycling the fluids or processing them for disposal. Reusable parts and components, such as the engine, tires, and battery, are then removed, cleaned, tested, and inventoried before being sold as used parts. Body shops, new and used car dealers, repair shops, and auto and truck fleets are the main buyers of used parts. In addition to being reused, some parts are disassembled for their materials. For example, catalytic converters are disassembled because they contain valuable amounts of platinum and rhodium.

An automotive recycling facility then sends the remaining vehicle "shell" to a scrap processor for "scrapping." At the scrap processing facility, a giant machine shreds the entire automobile into small pieces. Magnets and other machines sort the pieces into piles of steel, iron, copper, aluminum, and "fluff," which consists of the remaining materials, such as plastic, rubber, upholstery, and glass.

About 37% of the iron and steel scrap reprocessed in the United States comes from old automobiles. Recycling iron and steel saves energy and reduces pollution. According to the Environmental Protection Agency, recycling scrap iron and steel produces 86% less air pollution and 76% less water pollution than mining and refining an equivalent amount of iron ore.

Recycling plastic is one of the biggest challenges in auto recycling. Plastic is lightweight, and, as a result, automakers use a lot of plastic to improve fuel efficiency. Because no industry standards for plastic parts currently exist, the kinds and amounts of plastic from which cars are made vary. As many as 15 plastics comprise some dashboards, and because many of these plastics are chemically incompatible, they cannot be melted together for recycling.

Auto manufacturers around the world have begun to address the challenge of reusing and recycling old cars. Japan and the European Union have mandated that by 2015, 95% of each discarded car must be recoverable. Toyota has developed a way to recover urethane foam and other shredded materials to make silencer padding. Daimler Chrysler is developing a Composite Concept Vehicle that will have completely recyclable body sections. Honda, Mercedes-Benz, Peugeot, Toyota, Volkswagen, Volvo, and other auto manufacturers have started to design cars so that each part of an old automobile can be reused or recycled.

Recycling Materials

It is possible to collect and reprocess many materials found in solid waste into new products of the same or a different type. Recycling is preferred over landfill disposal because it conserves our natural resources and is more environmentally benign. Every ton of recycled paper saves 17 trees, 7000 gallons of water, 4100 kilowatt-hours of energy, and three cubic yards of landfill space. Recycling also has a positive effect on the economy by generating jobs and revenues (from selling the recycled materials). However, recycling does have environmental costs. It uses energy (as does any human activity), and it generates pollution (as does any human activity). For example, the de-inking process in paper recycling requires energy and produces a toxic sludge that contains heavy metals.

The many materials in municipal solid waste must be separated from one another before recycling. It is easy to separate materials such as glass bottles and newspapers, but the separation of materials in items with complex compositions is difficult. Some food containers are composed of thin layers of metal foil, plastic, and paper, and trying to separate these layers is a daunting prospect, to say the least.

The number of U.S. communities with recycling programs increased remarkably during the 1990s but leveled off somewhat in the early 2000s. Recycling programs include curbside collection, drop-off centers, buy-back programs, and deposit systems. The annual recycling rate in 2005 of aluminum and steel cans, plastic bottles, glass containers, newspapers, and cardboard was 87 kg (39.4 lb) per person (**Figure 24.8**).

Recyclables are usually sent to a **materials recovery facility**, where they are either hand-sorted or separated using a variety of technologies, including magnets, screens, and conveyor belts, and prepared for re-manufacturing. Currently, the United States recycles about 32% of its municipal solid waste, including composting of yard trimmings; this value is higher than in other highly developed nations. (Recall, however, that the United States also generates more municipal solid waste than any other country.)

Figure 24.8   The solid waste produced by an average U.S. family of four in one year. The cans, bottles, and newspapers on the left are what the average U.S. household recycles. The trashcans and bags are filled with the solid waste they discard after recycling. (Jose Azel/Aurora Photos)

Most people think recycling involves merely separating certain materials from the solid waste stream, but that is only the first step. For recycling to work, there must be a market for the recycled goods, and the recycled products must be used in preference to virgin products. Prices paid by processors for old newspapers, used aluminum cans, used glass bottles, and the like vary significantly from one year to the next, depending largely on the demand for recycled products. In some places, recycling—particularly curbside collection—is not economically feasible.

Recycling Paper   The United States currently recycles about 50% of its paper and paperboard. Many highly developed countries have greater recycling rates. Denmark, for example, recycles 97% of its paper. Part of the reason paper is not recycled more in the United States is that many older paper mills are not equipped to process waste paper. The number of mills that can process waste paper has increased in recent years, in part because of consumer demand. Most new mills in the United States are located near cities to take advantage of a local supply of scrap paper.

E N V I R O N E W S

The U.S.-China Recycling Connection

Most of the materials that Americans recycle—from scrap metal to old cardboard boxes to used soda bottles—are redeveloped into products in the United States, but a growing amount is exported abroad. During the early 2000's, China became the biggest importer of America's recyclable materials, collectively called scrap. When the scrap arrives in China, it becomes the raw materials for Chinese factories, paper mills, and steel mills. According to the Institute of Scrap Recycling Industries, scrap is now the third largest product exported from the United States to China, after airplanes and semiconductors.

China is experiencing an economic boom but does not have the natural resource base that countries such as the United States are lucky to possess. To fuel its economic growth, China relies on scrap—used paper to replace its dearth of wood pulp and steel scrap to replace its dearth of iron ore. Some of the scrap shipped from the United States to China makes a round-trip, returning to the United States as auto parts, polyester shirts, and toys. Because Chinese workers are paid much less than American workers, products made in China are generally less expensive for American consumers than the equivalent domestic products. The economic downside of the U.S. scrap-Chinese product cycle is that Chinese imports reduce the number of jobs available in the United States.

In addition to a slow increase in paper recycling in the United States, there is a growing demand for U.S. waste paper in other countries. China, Mexico, Taiwan, and Korea import large quantities of waste paper and cardboard from the United States.

Recycling Glass Glass is another component of solid waste appropriate for recycling. The United States currently recycles about 25% of its glass containers. Recycled glass costs less than glass made from virgin materials. Glass food and beverage containers are crushed to form **cullet,** which glass manufacturers can melt and use to make new products. Although cullet is much more valuable when glass containers of different colors are separated before being crushed, cullet made from a mixture of colors has some good uses; for example, it is used to make glassphalt, a composite of glass and asphalt that makes an attractive roadway (**Figure 24.9**).

Recycling Aluminum The recycling of aluminum is one of the best success stories in U.S. recycling, largely because of economic factors (**Figure 24.10**). Making a new aluminum can from a recycled one requires a fraction of the energy it would take to make a new can from raw metal. Because energy costs for new cans are high, there is a strong economic incentive to recycle aluminum. According to the EPA, in 2005 about 45% of discarded aluminum beverage cans were recycled, saving about 15 million barrels of oil.

Recycling Metals Other Than Aluminum Other recyclable metals include lead, gold, iron and steel, silver, and zinc. One obstacle to recycling metal products discarded in municipal solid waste is that their metallic compositions are often unknown. It is also difficult to extract metal from products, such as stoves, that contain other materials besides metal (plastic, rubber, or glass, for instance). In contrast, any waste metal produced at factories is recycled easily because its composition is known.

The economy has a large influence on whether metal is recycled or discarded. Greater recycling generally occurs when the price of metallic ores are more expensive than the price of recycled metals. Thus, although the supply of metal waste is fairly constant, the amount of recycling varies from year to year.

Figure 24.9 Glassphalt. Shown is a close-up of a Baltimore street paved with glassphalt, which contains a mixture of broken glass from different-colored containers. At night the road sparkles, as light from automobile headlights reflects off the pieces of glass. (Courtesy Baltimore Department of Public Works)

One exception to this generalization is steel. Before the 1970s, almost all steel was produced in large mills that processed raw ores. Starting in the 1970s and continuing to the present, "mini-mills" that produce steel products from up to 100% scrap became increasingly important. These mills are located near many cities in the United States so that they can process local scrap more profitably (because they do not have to pay to transport the scrap long distances). Mini-mills usually have electric arc furnaces that are energy-efficient and less polluting than the furnaces in old steel mills. According to the Institute of Scrap Recycling Industries, new steel products contain an average of 56% recycled scrap steel.

Recycling Plastic Less than 20% of plastic is recycled, in part because, depending on the economic situation, it is sometimes less expensive to make it from raw materials (petroleum and natural gas) than to recycle it. In other words, plastic recycling—indeed, all recycling—is affected by economics. Some local and state governments support or require the recycling of plastic.

Polyethylene terphthalate (PET), the plastic used in soda bottles, is recycled more than any other plastic. According to the EPA, 34% of the 9.5 billion PET bottles sold annually are recycled to make such diverse products as carpeting, automobile parts, tennis ball felt, and polyester cloth (**Figure 24.11**); it takes about 25 plastic bottles to make one polyester sweater.

Polystyrene (one form of which is Styrofoam) is an example of a plastic that has great recycling potential but is currently not recycled

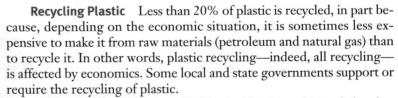

Figure 24.10 Recycling aluminum. Bales of crushed aluminum beverage cans are ready for processing in Columbus, Ohio. (Shari Lewis/© AP/Wide World Photos)

appreciably. Cups, tableware, and packaging materials made of polystyrene can be recycled into a variety of products, such as coat hangers, flower pots, foam insulation, and toys. Because approximately 2.3 billion kg (5 billion lb) of polystyrene are produced each year, large-scale recycling would make a major dent in the amount of polystyrene that ends up in landfills.

The fact that there are so many kinds of plastic presents a challenge in recycling them. Forty-six plastics are common in consumer products, and many products contain multiple kinds of plastic. A plastic ketchup bottle, for example, may have up to six layers of different plastics bonded together. To allow for effective recycling of high-quality plastic, the different types must be meticulously sorted or separated. If two or more resins are recycled together, the resultant plastic is of lower quality.

Low-quality plastic mixtures are used to make a construction material similar to wood. Because of its durability, this "plastic lumber" is particularly useful for outside products, such as fence posts, planters, highway retaining walls, picnic tables, and park benches.

Recycling Tires Although about 290 million tires are discarded in the United States each year—a huge quantity potentially available for recycling—relatively few kinds of products are made from old tires. Uses for old tires include retread tires; playground equipment; trashcans, garden hoses, and other consumer products; and rubberized asphalt for pavement. More recently, rubber from old tires has been used to make carpets, roofing materials, and molded products. Research in product development continues, and almost all states now have tire-recycling programs. According to the EPA, 36% of tires are currently recycled to make other products.

Integrated Waste Management

The most effective way to deal with solid waste is through a combination of techniques. In **integrated waste management**, a variety of options that minimize waste, including the three R's of waste prevention (reduce, reuse, and recycle), are incorporated into an overall waste management plan (**Figure 24.12**). Even on a large scale, recycling and source reduction will not entirely eliminate the need for disposal facilities such as incinerators and landfills. However, recycling and source reduction will substantially reduce the amount of solid waste requiring disposal in incinerators and landfills.

REVIEW

1. What is source reduction?

2. How do source reduction, reuse, and recycling reduce the volume of solid waste?

3. What is integrated waste management?

 ## Hazardous Waste

LEARNING OBJECTIVES

- Define *hazardous waste* and briefly characterize representative hazardous wastes (dioxins, PCBs, and radioactive wastes).
- Contrast the Resource Conservation and Recovery Act and the Comprehensive Environmental Response, Compensation, and Liability Act (the Superfund Act).
- Explain how green chemistry is related to source reduction.

Hazardous waste accounts for about 1% of the solid waste stream in the United States. Hazardous waste includes dangerously reactive, corrosive, explosive, or toxic chemicals. The chemicals may be solids, liquids, or gases.

Hazardous waste has held national attention since 1977, when it was discovered that hazardous waste from an abandoned chemical dump had contaminated homes and possibly people in **Love Canal**, a small neighborhood on the edge of Niagara Falls, New York. Lois Gibbs, a housewife in Love Canal, led a successful crusade to evacuate the area after she discovered what seemed a high number of serious illnesses, particularly among

Figure 24.11 Recycled plastic. A wide variety of products are made from recycled polyethylene terphthalate (PET), shown, and other plastics. (Courtesy National Association for PET Container Resources [NAPCOR])

integrated waste management: A combination of the best waste management techniques into a consolidated, systems-based program to deal effectively with solid waste.

hazardous waste: Any discarded material that threatens human health or the environment.

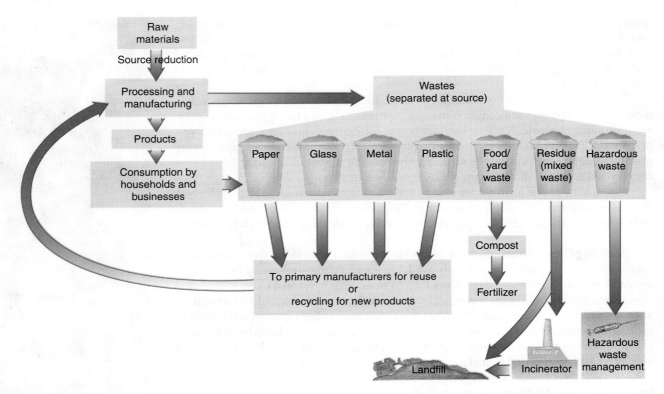

Figure 24.12 **Integrated waste management.** Source reduction, reuse, recycling, and composting are part of integrated, systems-based waste management, in addition to incineration and disposal in landfills.

children, in the neighborhood. (Today, Ms. Gibbs is executive director for the Center for Health, Environment, and Justice.) As a result of the publicity, Love Canal became synonymous with chemical pollution caused by negligent hazardous waste management. In 1978 it became the first location ever declared a national emergency disaster area because of hazardous waste; some 700 families were evacuated.

How did the Love Canal disaster come about? From 1942 to 1953, a local industry, Hooker Chemical Company, disposed of about 20,000 metric tons (22,000 tons) of toxic chemical waste in the 914-m-long (3000-ft-long) Love Canal. When the site was filled, Hooker added topsoil and donated the land to the local board of education. A school and houses were built on the site, which began oozing hazardous waste several years later. Over 300 chemicals, many of them carcinogenic, have been identified in Love Canal's hazardous waste.

Tons of contaminated soil were removed during the cleanup that followed, but because the canal was so huge, the federal government decided to contain the waste and construct drainage trenches to prevent hazardous wastes from leaking from the site. In 1990, after almost 10 years of cleanup, the EPA and the New York Department of Health declared the area safe for resettlement. The nearby housing area was renamed Black Creek Village. Today, the canal is a 40-acre mound covered by clay and surrounded by a chain-link fence and warning signs.

Whether the site caused adverse health effects in residents remains unanswered because of a lack of reliable scientific data. On average, residents of Love Canal seem to have had more health problems, from miscarriages and birth defects to psychological disorders.

The Love Canal episode resulted in passage of the federal Superfund law, which held polluters accountable for the cost of cleanups (discussed shortly). Love Canal also generated an immediate concern about hazardous waste that has been with us ever since. For example, a 1997 California study reported that women living within a quarter mile of untreated hazardous waste sites (Superfund sites) were at greater risk of having babies with serious birth defects, such as malformed neural tubes and defective hearts.

Other countries have the same problems with hazardous waste management. How should we deal with the bewildering array of hazardous waste continually generated and released in ever increasing amounts into the environment by mining, industrial processes, incinerators, military activities, and thousands of small businesses? How do we clean up the hazardous materials that have already contaminated our world?

Table 24.1 Examples of Hazardous Waste

Hazardous Material	Some Possible Sources
Acids	Ash from power plants and incinerators; petroleum products
CFCs (chlorofluorocarbons)	Coolant in air conditioners and refrigerators
Cyanide	Metal refining; fumigants in ships, railway cars, and warehouses
Dioxins	Emissions from incinerators and pulp and paper plants
Explosives	Old military installations
Heavy metals	Paints, pigments, batteries, ash from incinerators, sewage sludge with industrial waste, improper disposal in landfills
Arsenic	Industrial processes, pesticides, additives to glass, paints
Cadmium	Rechargeable batteries, incineration, paints, plastics
Lead	Lead–acid storage batteries, stains and paints, TV picture tubes and electronics discarded in landfills
Mercury	Coal-burning power plants; paints, household cleaners (disinfectants), industrial processes, medicines, seed fungicides
Infectious waste	Hospitals, research labs
Nerve gas	Old military installations
Organic solvents	Industrial processes; household cleaners, leather, plastics, pet maintenance (soaps), adhesives, cosmetics
PCBs (polychlorinated biphenyls)	Older appliances (built before 1980); electrical transformers and capacitors
Pesticides	Household products
Radioactive waste	Nuclear power plants, nuclear medicine facilities, weapons factories

Types of Hazardous Waste

More than 700,000 chemicals are known to exist. How many are hazardous is unknown because most have never been tested for toxicity, but without a doubt, hazardous substances number in the thousands. According to the EPA report, *Chemical Hazard Data Availability Study*, only 7% of the 3000 chemicals used in large quantities (more than 500 tons annually) in U.S. commerce have undergone comprehensive studies of potential health and environmental effects. Hazardous chemicals include a variety of acids, dioxins, abandoned explosives, heavy metals, infectious waste, nerve gas, organic solvents, polychlorinated biphenyls (PCBs), pesticides, and radioactive substances (**Table 24.1**). Many of these chemicals have already been discussed, particularly in Chapters 7, 12, 20, 22, and 23, which examine endocrine disrupters, radioactive waste, air pollution, water pollution, and pesticides. Here we discuss dioxins, PCBs, and radioactive wastes produced during the Cold War.

Dioxins **Dioxins** are a group of 75 similar chemical compounds formed as un-wanted by-products during the combustion of chlorine compounds. Some of the known sources of dioxins are medical waste and municipal waste incinerators, iron ore mills, copper smelters, cement kilns, metal recycling, coal combustion, pulp and paper plants that use chlorine for bleaching, and chemical accidents. Incineration of medical and municipal wastes accounts for 70 to 95% of known human emissions of dioxins. In Japan, where nearly 75% of its solid waste is burned in incinerators, the air contains nearly 10 times the amount of dioxins found in other highly developed countries. Hospital waste incinerators are probably the largest dioxin polluters because they are so numerous (there are more than 6000 of them in the United States), and they generally have unsophisticated pollution controls.

Dioxins are emitted in smoke and then settle on plants, the soil, and bodies of water; from there they are incorporated into the food web. When humans and other animals ingest dioxins, they are stored and accumulate in their fatty tissues (see discussion of bioaccumulation and biological magnification in Chapter 7). Humans are primarily exposed when they eat contaminated meat, dairy products, and fish. Because dioxins are so widely distributed in the environment, virtually everyone has dioxins in his or her body fat.

Just how dangerous dioxins are to humans is somewhat controversial. Dioxins are known to cause several kinds of cancer in laboratory animals, but the data are conflicting

ENVIRONEWS

Handling Nanotechnology Safely

Nanotechnology is in the news a lot these days. **Nanomaterials,** which are unique materials and devices designed on the ultra-small scale of atoms or molecules, have numerous possible applications. For example, nanoparticles of cadmium selenide might be injected into cancerous tissue, where it would accumulate inside cancer cells; when exposed to ultraviolet radiation, these nanoparticles glow, which would maked it easier for surgeons to excise the cancerous tissues and leave the healthy tissues intact. Nanocrystals have the potential to be used in thin-film solar panels to convert solar energy to electricity. Silica nanoparticles embedded in glass could make a heat-resistant glass able to withstand temperatures of up to 1800°F for several hours.

Despite the potential of nanotechnology, particles on the nanometer scale (a nanometer is one-billionth of a meter) might pose significant health, safety, and environmental risks. No one knows for sure. The EPA has adopted a precautionary approach (see Chapter 7) and decided to regulate nanomaterials that might adversely affect the environment. This means that the burden of proof about product safety will fall on companies that sell nanotechnology. Similarly, the Food and Drug Administration will have to oversee regulation of nanotechnology with potential health and safety risks.

on their cancer-causing ability in humans. A 1997 study of residents of Seveso, Italy, who were exposed to high levels of dioxin after a chemical accident in 1976, revealed a statistically significant increase in cancer deaths. Another study published in 2002 linked high levels of dioxin exposure to an increased incidence of breast cancer in Italian women living near the accident site. According to the EPA, dioxins probably cause several kinds of cancer in humans.

Other concerns center on the effects of dioxins on the human reproductive, immune, and nervous systems. Dioxins may delay fetal development and cause cognitive damage, lead to endometriosis in women (the growth of uterine tissue in abnormal locations in the body), and decrease sperm production in men. Dioxins are also linked to an increased risk of heart disease. Because human milk contains dioxins, nursing infants, who feed almost exclusively on milk, are considered particularly at risk.

PCBs **Polychlorinated biphenyls (PCBs)** are a group of 209 industrial chemicals composed of carbon, hydrogen, and chlorine. These clear or light yellow, oily liquids or waxy solids were manufactured in the United States between 1929 and 1979. PCBs were used as cooling fluids in electrical transformers, electrical capacitors, vacuum pumps, and gas-transmission turbines. They were also in hydraulic fluids, fire retardants, adhesives, lubricants, pesticide extenders, inks, and other materials.

The first evidence that PCBs were dangerous occurred in 1968, when Japanese who ate rice bran oil contaminated with PCBs experienced liver and kidney damage. A similar mass poisoning, also attributed to PCB-contaminated rice oil, occurred in Taiwan in 1979. Since then, toxicity tests conducted on animals indicate that PCBs harm the skin, eyes, reproductive organs, and gastrointestinal system. PCBs are endocrine disrupters because they interfere with hormones released by the thyroid gland. Several studies suggest that children exposed to PCBs before birth have certain intellectual impairments, such as poor reading comprehension, memory problems, and difficulty paying attention. Several studies suggest that PCBs may be carcinogenic.

PCBs are chemically stable and resist chemical and biological degradation. Like dioxins, PCBs accumulate in fatty tissues and are subject to biological magnification in food webs. The general human population is mainly exposed to PCBs by eating food that became contaminated through biological magnification. One way that PCBs enter aquatic food webs is by benthic invertebrates that live in contaminated sediments. (PCBs tend to bind to organic particles in aquatic sediments.) Small fish eat these invertebrates, and as larger fish eat the smaller fish, the PCBs bioaccumulate. Human populations whose diets consist primarily of fish and marine mammals, such as the Inuit of northern Canada, are exposed to large amounts of PCBs.

Prior to the EPA ban in the 1970s, PCBs were dumped in large quantities into landfills, sewers, and fields. Such improper disposal is one of the reasons PCBs are still a threat today. Also, when sealed electrical transformers and capacitors leak or catch fire, PCB contamination of the environment occurs.

High-temperature incineration is one of the most effective ways to destroy PCBs. However, incineration is not practical for the removal of PCBs that have leached into the soil and water because, among other difficulties, the cost of incinerating large quantities of soil is prohibitively high.

Several bacteria can degrade PCBs. However, when PCB-eating bacteria are sprayed on the surface of the soil, they cannot decompose the PCBs that have already leached into the soil or groundwater systems. These microorganisms show promise in removing PCBs from the environment, but additional research is needed to make the biological degradation of PCBs practical. (Additional discussion about using bacteria to break down hazardous waste is found later in this chapter.)

CASE IN POINT

Hanford Nuclear Reservation

U.S. nuclear weapons facilities are no longer actively manufacturing nuclear weapons, but they present us with a greater challenge—reducing and managing radioactive and toxic wastes that have accumulated at numerous sites around the United States since the 1940s. Every step in the production of nuclear warheads generated radioactive and chemical wastes. We focus our discussion on the Hanford Nuclear Reservation, a 1400 km² (560 mi²) area on the Columbia River in south-central Washington State (**Figure 24.13a**). Hanford, the main production site for plutonium used in nuclear weapons, is the largest, most seriously contaminated site in the U.S. nuclear weapons infrastructure.

The immensity of the cleanup task at Hanford is daunting. Tons of highly radioactive solid and liquid wastes were stored or dumped into trenches, pits, tanks, ponds, and underground cribs—a total of 1700 waste sites. (These methods of disposal were standard practice at the time.) Two concrete pools of water store more than 100,000 spent fuel rods. As they corrode, the rods release highly radioactive uranium, plutonium, cesium, and strontium into the water. Because these pools are leaking, soil and groundwater have been contaminated, and the Columbia River is in danger. Ultimately, the fuel rods will be placed in Yucca Mountain, an underground geologic repository that has not opened yet (see Chapter 12); until then, they will remain at Hanford.

Figure 24.13 Hanford Nuclear Reservation.

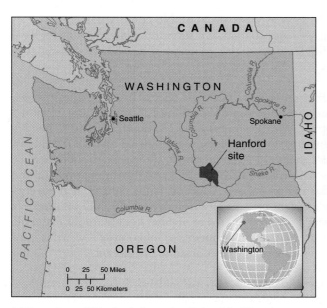

(a) Location of Hanford along the Columbia River in Washington State.

(b) A worker at the Waste Receiving and Processing Facility crushes drums of low-level nuclear waste. The crushed drums will then be packed into larger drums for permanent burial in the hazardous waste landfill at Hanford. (Peter Essick/Aurora Photos)

The Columbia River is also threatened by millions of gallons of toxic chemical and radioactive liquid wastes stored in 174 large underground tanks. Liquids in some of these tanks are so reactive that they boiled for years from the heat of their own radioactivity or chemical activity, but most tanks are now covered by semisolid crusts that formed from chemical reactions within the mixtures. Some of the tanks are potentially explosive: Chemical reactions in some of the tanks produce hydrogen and other hazardous gases. Vents allow the gas to escape and reduce the danger of ignition. Many of these tanks were designed to last 10 to 20 years and are now leaking their poisons into the ground.

Cleanup of this hazardous waste is complicated because the extent of radioactive pollution and the kinds of hazardous mixtures present are not well known. (Although huge amounts of radioactive and chemical wastes were produced over many years, environmental records of hazardous waste production generally were not kept until the 1970s.) As a result, scientists and engineers are assessing the damage, prioritizing the cleanup process, and determining how best to proceed for each type of contamination (**Figure 24.13b**). Most experts believe that the cleanup, which is under the direction of the U.S. Department of Energy (DOE), will take at least three decades to complete and cost hundreds of billions of tax dollars. Currently, the cleanup is requiring 7,000 workers and costing $1.7 billion a year.

The cleanup has sparked legal battles over environmental concerns and workers' health. In fact, cleanup may be a more dangerous occupation than working at Hanford when it was at full operation producing the nation's nuclear arms stockpile. Some of the cleanup workers have developed serious chronic illnesses because they come into close contact with toxic materials, such as beryllium, that were used to make bombs. Inhalation of beryllium particles can cause an incurable lung disease.

After the cleanup is finished, Hanford will remain hazardous for hundreds or even thousands of years, in part because we do not have the technologies to address the widespread soil contamination. Also, much of the buried radioactive waste may be left in place because there is no other place to move it. The DOE will have to establish and maintain a long-term monitoring program at the site to limit human exposure to remaining hazards. ■

Management of Hazardous Waste

Humans have the technology to manage hazardous waste in an environmentally responsible way, but it is extremely expensive. Although great strides have been made in educating the public about the problems of hazardous waste, we have only begun to address many of the issues of hazardous waste disposal. No country currently has an effective hazardous waste management program, but several European countries have led the way by producing smaller amounts of hazardous waste and by using fewer hazardous substances.

Chemical Accidents When a chemical accident occurs in the United States, whether at a factory or during the transport of hazardous chemicals, the National Response Center (NRC) is notified. Most chemical accidents reported to NRC involve oil, gasoline, or other petroleum spills. The remaining accidents involve more than 1000 other hazardous chemicals, such as PCBs, ammonia, sulfuric acid, and chlorine.

Chemical safety programs have traditionally stressed accident mitigation and the addition of safety systems to existing procedures. More recently, industry and government agencies have focused on accident prevention through the **principle of inherent safety,** in which industrial processes are redesigned to involve less hazardous materials so that dangerous accidents are prevented. The principle of inherent safety, which is an example of systems thinking, is an important aspect of source reduction.

Current Management Policies Currently, two federal laws dictate how hazardous waste should be managed: (1) the Resource Conservation and Recovery Act, which is concerned with managing hazardous waste being produced now, and (2) the Superfund Act, which provides for the cleanup of abandoned and inactive hazardous waste sites.

The **Resource Conservation and Recovery Act (RCRA)** was passed in 1976 and amended in 1984. Among other things, RCRA instructs the EPA to identify which waste is hazardous and to provide guidelines and standards to states for hazardous waste management programs. RCRA bans hazardous waste from land disposal unless it is treated to meet EPA's standards of reduced toxicity. In 1992 the EPA initiated a major reform of RCRA to expedite cleanups and streamline the permit system to encourage hazardous waste recycling.

In 1980 the **Comprehensive Environmental Response, Compensation, and Liability Act (CERCLA),** commonly known as the Superfund Act, established a program to clean up abandoned and illegal hazardous waste sites across the United States. At many of these sites, hazardous chemicals have migrated deep into the soil and have polluted groundwater. The greatest threat to human health from hazardous waste sites comes from drinking water laced with such contaminants.

Cleaning Up Existing Hazardous Waste: The Superfund Program   The federal government estimates that the United States has more than 400,000 hazardous waste sites with leaking chemical storage tanks and drums (both above and below ground), pesticide dumps, and piles of mining waste (**Figure 24.14a**). This estimate does not include the hundreds or thousands of hazardous waste sites at military bases and nuclear weapons facilities.

By 2007, the CERCLA inventory listed 10,753 sites. This means that the EPA has identified them as qualifying for cleanup (**Figure 24.14b**). (These sites do not include the 1,010 sites that have been cleaned up and removed from the CERCLA inventory since 1980.) CERCLA sites are not identified according to any particular criteria. Some are dumps that local or state officials have known about for years, whereas concerned citizens identify others. The sites in the inventory are evaluated and ranked to identify those with extremely serious hazards. The ranking system uses data from preliminary assessments, site inspections, and expanded site inspections, which include contamination tests of soil and groundwater and sampling of hazardous waste.

The sites that pose the greatest threat to public health and the environment are placed on the **Superfund National Priorities List,** which means that the federal government will assist in their cleanup. As of 2006, a total of 1558 sites were on the National Priorities List. The five states with the greatest number of sites on the national priorities list as of 2006 are New Jersey (115 sites), California (93 sites), Pennsylvania (93 sites), New York (86 sites), and Michigan (65 sites).

As of 2006, a total of 966 hazardous waste sites had been cleaned up enough to be deleted from the National Priorities List, and 422 other sites were partially corrected. These sites, designated "construction completes," have had extensive cleanups but remain on the Superfund list because of continued problems with groundwater contamination. The average cost of cleaning up a site is $20 million.

One reason for the urgency about cleaning up the sites on the National Priorities List is their locations. Most were originally in rural areas on the outskirts of cities. With the growth of cities and their suburbs, residential developments now surround many of the dumps. One in three Americans lives within 5 km (3 mi) of one or more Superfund sites.

Because the federal government cannot assume major responsibility for cleaning up every old dumping ground in the United States, the current landowner, prior owners, and anyone who has dumped waste or has transported waste to a particular site share cleanup costs. For some sites, many parties are considered liable for cleanup costs. The cleanup process has been mired in litigation, mostly by companies, charged with polluting, who are suing each other. Despite the urgency of cleaning up sites on the National Priorities List, it will take many years to complete the job.

Although critics decry the slow pace and high cost of cleaning up Superfund sites, the very existence of CERCLA is a deterrent to further polluting. Companies that produce hazardous waste are now fully aware of the costs of liability and cleanup and are more likely to take steps to dispose of their hazardous wastes in proper fashion.

Figure 24.14   **Cleaning up hazardous waste.**

(a) Hazardous waste in deteriorating drums at a site near Washington, D.C. The metal drums in which much of the waste is stored have corroded and started to leak. Old hazardous waste dumps are commonplace around the United States. (Courtesy U.S. Department of Agriculture)

(b) Cleanup of a hazardous waste site near Minneapolis, Minnesota. Removal and destruction of the wastes are complicated by the fact that usually nobody knows what chemicals are present. (Gary Milburn/Tom Stack & Associates)

The Biological Treatment of Hazardous Contaminants A variety of methods are employed to clean up soil contaminated with hazardous waste. Because most of these processes are prohibitively expensive, innovative approaches such as bioremediation and phytoremediation are being developed to deal with hazardous waste. **Bioremediation** is the use of bacteria and other microorganisms to break down hazardous waste into relatively harmless components. **Phytoremediation** is the use of plants to absorb and accumulate hazardous materials from the soil. (The prefix *phyto-* comes from the Greek word meaning "plant.")

More than 1000 species of bacteria and fungi have been demonstrated to clean up various forms of organic pollution. Bioremediation takes a little longer to work than traditional hazardous waste disposal methods, but it accomplishes the cleanup at a fraction of the cost. In bioremediation, the contaminated site is exposed to an army of microorganisms, which gobble up the poisons such as petroleum and other hydrocarbons and leave behind harmless substances such as carbon dioxide, water, and chlorides. Bioremediation encourages the natural processes in which bacteria consume organic molecules such as hydrocarbons. During bioremediation, conditions at the hazardous waste site are modified so that the desired bacteria will thrive in large enough numbers to be effective. Environmental engineers might pump air through the soil (to increase its oxygen level) and add a few soil nutrients such as phosphorus or nitrogen. They might install a drainage system to pipe any contaminated water that leaches through the soil back to the surface for another exposure to the bacteria.

In phytoremediation, plant species known to remove specific hazardous materials from the soil are grown at a contaminated site. As the roots penetrate the soil, they selectively absorb the toxins, which accumulate in both root and shoot tissues. Later, the plants are harvested and disposed of in a hazardous waste landfill. Alternatively, some plants break down a hazardous chemical into more benign chemicals.

The field of phytoremediation is in its infancy, but already, specific plants are known to remove such hazardous materials as trinitrotoluene (TNT), radioactive strontium and uranium, selenium, lead, and other heavy metals. Researchers in England have identified three species of marsh plants that degrade herbicides and other pesticides; in this case, the actual organisms that attack the pesticide compounds are bacteria that live in symbiotic association with the plants' roots. Some researchers are working to genetically engineer certain plants to do an even better job of accumulating toxins.

Phytoremediation is much cheaper than conventional methods to clean up hazardous waste sites, but it does have limitations. Plants cannot remove contaminants present in the soil deeper than their roots normally grow. There is also concern that insects and other animals might eat the plants, thereby introducing the toxins into the food web.

Managing the Hazardous Waste We Are Producing Now Many people think, incorrectly, that establishment of the Superfund has eliminated the problem of hazardous waste. The Superfund deals only with hazardous waste produced in the past; it does nothing to eliminate the large amount of hazardous waste produced today. There are three ways to manage hazardous waste: (1) source reduction, (2) conversion to less hazardous materials, and (3) long-term storage.

As with municipal solid waste, the most effective of the three methods is *source reduction*—that is, reducing the amount of hazardous materials used in industrial processes and substituting less hazardous or nonhazardous materials for hazardous ones. Source reduction relies on an increasingly important subdiscipline of chemistry known as **green chemistry** (see Case in Point: Green Chemistry in Chapter 22).

For example, chlorinated solvents are widely used in electronics, dry cleaning, foam insulation, and industrial cleaning. It is sometimes possible to accomplish source reduction by substituting a less hazardous water-based solvent for a chlorinated solvent. Reducing solvent emissions also results in substantial source reduction of chlorinated solvents. Most chlorinated solvent pollution gets into the environment by evaporation during industrial processes. Installing solvent-saving devices benefits the environment and also provides economic gains, because smaller amounts of chlorinated solvents must be purchased. No matter how efficient source reduction becomes, however, it will never entirely eliminate hazardous waste.

green chemistry: A subdiscipline of chemistry in which commercially important chemical processes are redesigned to significantly reduce environmental harm.

The second best way to deal with hazardous waste is to reduce its toxicity by chemical, physical, or biological means, depending on the nature of the hazardous waste. One method to detoxify organic compounds is high-temperature incineration. The high heat of combustion reduces these dangerous compounds, such as pesticides, PCBs, and organic solvents, into safe products such as water and carbon dioxide. The incineration ash is hazardous and must be disposed of in a landfill designed specifically for hazardous materials. One method to reduce the toxicity of hazardous waste is incineration using a *plasma torch*, which produces such high temperatures (up to 10,000°C) that hazardous waste is almost completely converted to harmless gases. In comparison, conventional incinerators produce temperatures no higher than 2000°C.

Hazardous waste that is produced in spite of source reduction and that is not completely detoxified must be placed in long-term storage. Hazardous waste landfills are subject to strict environmental criteria and design features. They are located as far as possible from aquifers, streams, wetlands, and residences. These landfills have many special features, including several layers of compacted clay and high-density plastic liners at the bottom of the landfill to prevent leaching of hazardous substances into surface water and groundwater (**Figure 24.15**). *Leachate* is collected and treated to remove contaminants. The entire facility and nearby groundwater deposits are carefully monitored to make sure there is no leaking. Only solid chemicals (not liquids) that have been treated to detoxify them as much as possible are accepted at hazardous waste landfills. These chemicals are placed in sealed barrels before being stored in the hazardous waste landfill.

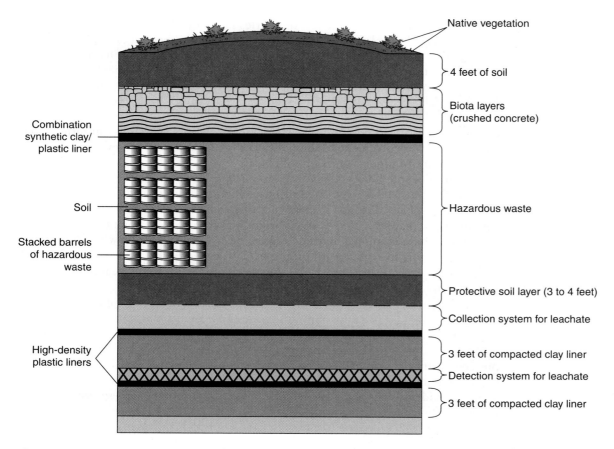

Figure 24.15 Cutaway view through a hazardous waste landfill. The bottom of this hazardous waste landfill has two layers of compacted clay, each covered by a high-density plastic liner. (Some hazardous waste landfills have three layers of compacted clay.) A drain system allows liquid leachate to collect in a basin where it can be treated, and a leak detection system is installed between the clay liners. Barrels of hazardous waste are placed above the liners and covered with soil. (Adapted from Rocky Mountain Arsenal Remediation Venture Office)

Few landfills are certified to handle hazardous waste. Currently, there are only 23 commercial hazardous waste landfills in the United States, although many larger companies are licensed to treat their hazardous waste on-site. As a result, much of our hazardous waste is still placed in sanitary landfills, burned in incinerators that lack the required pollution control devices, or discharged into sewers.

Some liquid hazardous wastes, such as certain organic compounds, fuels, explosives, and pesticides, are stored in the Earth's crust by **deep well injection.** These wells extend into impermeable rock layers several thousand feet below the surface. The Safe Drinking Water Act and the Underground Injection Control Program regulate the placement and number of such wells, which can be located only where there is no danger of groundwater contamination.

REVIEW

1. What is hazardous waste?
2. How are the Resource Conservation and Recovery Act and the Comprehensive Environmental Response, Compensation, and Liability Act alike? What is the focus of each?

 ## Environmental Justice

LEARNING OBJECTIVE

- Define *environmental justice* and discuss the issue of international waste management as it relates to environmental justice.

Opposition to the environmental inequities faced by low-income minority communities in both rural and urban areas has increased in recent years. Many studies indicate that poor minority neighborhoods are more likely to have hazardous waste facilities, sanitary landfills, sewage treatment plants, and incinerators in their neighborhoods. Beginning in the late 1970s, environmental sociologist Robert Bullard at Clark Atlanta University identified patterns of injustice in Houston. For example, Bullard found that six of Houston's eight incinerators were located on less expensive land in predominantly black neighborhoods. Such communities often have a limited involvement in the political process and may not even be aware of their exposure to higher levels of pollutants.

Because people in low-income communities frequently lack access to sufficient health care, they may not be treated adequately for exposure to environmental hazards. The high incidence of asthma in many minority communities is an example of a health condition that may be caused or exacerbated by exposure to environmental pollutants. Few studies have examined how environmental pollutants interact with other socioeconomic factors to cause health problems. Those studies that have been done often fail to conclusively tie exposure to environmental pollutants to the health problems of poor and minority communities.

A 1997 health study of residents in San Francisco's Bayview-Hunters Point area found that hospitalizations for chronic illnesses were almost four times higher than the state average. Bayview-Hunters Point is heavily polluted. It has 700 hazardous waste facilities, 325 underground oil storage tanks, and two Superfund sites. Yet the study did not demonstrate that an increased exposure to toxic pollutants was the primary cause of illnesses in the residents. Although anecdotal evidence abounds, currently little scientific evidence shows to what extent a polluted environment is responsible for the disproportionate health problems of poor and minority communities. Lead contamination is a notable exception (see the discussion of lead in Chapter 22).

In addition to concerns about pollution in their neighborhoods, low-income communities may not receive equal benefits from federal cleanup programs. A 1992 paper published in the *National Law Journal* reported that hazardous waste sites in white communities were cleaned up faster and better than were those in communities of blacks, Hispanics, or other minorities.

Environmental Justice and Ethical Issues

There is an increasing awareness that environmental decisions such as where to locate a hazardous waste landfill have important ethical dimensions. The most basic ethical dilemma centers on the rights of the poor and disenfranchised versus the rights of the rich and powerful. Whose rights should have priority in these decisions? The challenge of **environmental justice** is to find and adopt equitable solutions that respect all groups of people, including those not yet born. Viewed ethically, environmental justice is a fundamental human right. Although we may never completely eliminate environmental injustices of the past, we have a moral imperative to prevent them today so that the negative effects of pollution do not disproportionately affect any segment of society.

> **environmental justice:** The right of every citizen, regardless of age, race, gender, social class, or other factor, to adequate protection from environmental hazards.

In response to these concerns, a growing environmental justice movement has emerged at the grassroots level as a strong motivator for change. Advocates of the environmental justice movement are calling for special efforts to clean up hazardous sites in low-income neighborhoods, from inner-city streets to Indian reservations. Many environmental justice groups base their demands on the inherent "rightness" of their position. Other groups want science to give their demands legitimacy. Many advocates cite the need for more research on human diseases that environmental pollutants may influence.

Mandating Environmental Justice at the Federal Level

In 1994 President Clinton signed an executive order requiring all federal agencies to develop strategies and policies to ensure that their programs do not discriminate against poor and minority communities when decisions are made about where future hazardous facilities are located. The first response to President Clinton's initiative came in 1997 when the Nuclear Regulatory Commission (NRC) rejected a request to build a uranium processing plant near two minority neighborhoods in northern Louisiana. The commission decided that racial considerations were a factor in site selection because the applicant had ruled out all potential sites near predominantly white neighborhoods. This NRC decision sent a message that the U.S. government will protect the rights of vulnerable members of society. Since then, however, there have been several failed attempts to pass state and federal legislation pertaining to environmental justice.

Environmental Justice and International Waste Management

Environmental justice applies to countries as well as to individuals. Although there are ways to reduce and dispose of waste in an environmentally sound manner, industrialized countries have sometimes chosen to send their waste to other countries. (As industrialized nations develop more stringent environmental standards, disposing of hazardous waste at home becomes much more expensive than sending it to a developing nation, where property values and labor costs are lower.) Some waste is exported for legitimate recycling, but other waste is exported strictly for disposal.

The export of both solid and hazardous wastes by the United States, Canada, Japan, and the European Union is one of the most controversial aspects of waste management today. The recipients of this waste are usually developing nations in Africa, Central and South America, and the Pacific Rim of Asia, although Eastern and Central Europe and countries of the former Soviet Union are also common sites. The governments, industries, and citizens of these countries are often inexperienced and ill equipped to contain and monitor such materials. As a result, the waste often causes the same types of environmentally hazardous sites in developing nations that industrialized nations are trying to clean up at home.

In 1989 the U.N. Environment Program developed a treaty, the **Basel Convention,** to restrict the international transport of hazardous waste. The treaty allows countries to export hazardous waste only with the prior informed consent of the importing country as well as of any countries that the waste passes through in transit. The Basel Convention went into effect in 1992 after it was ratified by the required number of countries. (The United States signed the treaty in 1989 but has not yet ratified it because Congress has not passed the necessary legislation.) In 1995 the Basel Convention was amended to ban the export of *any* hazardous waste from industrialized to developing countries.

REVIEW

1. What is environmental justice?

2. Is environmental justice a local issue, an international issue, or both? Explain.

 REVIEW OF LEARNING OBJECTIVES WITH KEY TERMS

• **Distinguish between municipal solid waste and nonmunicipal solid waste.**

Municipal solid waste consists of solid materials discarded by homes, office buildings, retail stores, restaurants, schools, hospitals, prisons, libraries, and other commercial and institutional facilities. **Nonmunicipal solid waste** consists of solid waste generated by industry, agriculture, and mining.

• **Describe the features of a modern sanitary landfill and relate some of the problems associated with sanitary landfills.**

A **sanitary landfill** is the most common method of disposal of solid waste, by compacting it and burying it under a shallow layer of soil. The location of a sanitary landfill must take into account the geology of the area, soil drainage properties, the proximity of nearby surface waters and wetlands, and distance from population centers. Despite design features such as high-density plastic liners and **leachate** collection systems, most sanitary landfills have the potential to contaminate soil, surface water, and groundwater.

• **Describe the features of a mass burn incinerator and relate some of the problems associated with incinerators.**

A **mass burn incinerator** is a large furnace that burns all solid waste except for unburnable items such as refrigerators. Most mass burn incinerators recover the energy produced from combustion. One drawback of incineration is the great expense of installing pollution control devices on the incinerators. **Lime scrubbers** produce a chemical spray that neutralizes acidic gases. **Electrostatic precipitators** give ash a positive electrical charge so that it adheres to negatively charged plates. These controls reduce the toxicity of the gaseous emissions from incinerators but make the ash that remains behind more hazardous. Incinerators produce two kinds of ash, **bottom ash** (the residual ash left at the bottom of the incinerator) and **fly ash** (ash from the flue, or chimney), which must be disposed of safely. Fly ash is more hazardous than bottom ash.

• **Summarize how source reduction, reuse, and recycling help reduce the volume of solid waste.**

The three goals of waste prevention are to reduce the amount of waste, reuse products, and recycle materials as much as possible. **Source reduction** is an aspect of waste management in which products are designed and manufactured in ways that decrease the volume of solid waste and the amount of hazardous waste in the solid waste stream. One example of reuse is refillable glass beverage bottles. Recycling involves collecting and reprocessing materials into new products. Many communities recycle paper, glass, metals, and plastic.

• **Define *integrated waste management*.**

Integrated waste management is a combination of the best waste management techniques into a consolidated, systems-based program to deal effectively with solid waste.

• **Define *hazardous waste* and briefly characterize representative hazardous wastes (dioxins, PCBs, and radioactive wastes).**

Hazardous waste consists of any discarded chemical that threatens human health or the environment. **Dioxins** are hazardous chemicals formed as unwanted by-products during the combustion of many chlorine compounds. **Polychlorinated biphenyls (PCBs)** are hazardous, oily, industrial chemicals composed of carbon, hydrogen, and chlorine. Radioactive and chemical wastes have accumulated at numerous U.S. nuclear weapons facilities around the United States; the largest, most seriously contaminated site is Hanford Nuclear Reservation in Washington State.

• **Contrast the Resource Conservation and Recovery Act and the Comprehensive Environmental Response, Compensation, and Liability Act (the Superfund Act).**

The **Resource Conservation and Recovery Act (RCRA)** instructs the EPA to identify which waste is hazardous and to provide guidelines and standards to states for hazardous waste management programs. The **Comprehensive Environmental Response, Compensation, and Liability Act**, also known as the **Superfund Act**, addresses the challenge of cleaning up abandoned and illegal hazardous waste sites in the United States. The sites that pose the greatest threat to public health and the environment are placed on the **Superfund National Priorities List**.

• **Explain how green chemistry is related to source reduction.**

Green chemistry is a subdiscipline of chemistry in which commercially important chemical processes are redesigned to significantly reduce environmental harm. Source reduction is the best way to reduce hazardous waste, and green chemists redesign processes with source reduction in mind.

• **Define *environmental justice* and discuss the issue of international waste management as it relates to environmental justice.**

Environmental justice is the right of every citizen, regardless of age, race, gender, social class, or other factor, to adequate protection from environmental hazards. The export of solid and hazardous wastes by highly developed countries to developing countries is controversial. Developing countries are often inexperienced and ill equipped to handle such materials; the waste often causes the same environmentally hazardous sites in developing nations that industrialized nations are trying to clean up at home. The **Basel Convention** restricts the international transport of hazardous waste; it allows countries to export hazardous waste only with prior consent of the importing country as well as that of any countries that the waste passes through in transit.

Thinking About the Environment

1. What is solid waste?

2. Compare the advantages and disadvantages of disposing of waste in sanitary landfills and by incineration.

3. List what you think are the best ways to treat each of the following types of solid waste, and explain the benefits of the processes you recommend: paper, plastic, glass, metals, food waste, yard waste.

4. How do industries such as Goodwill, which accepts donations of clothing, appliances, and furniture for resale, affect the volume of solid waste?

5. It could be argued that a business that collects and sells its waste paper is not really recycling unless it buys products made from recycled paper. Explain.

6. Why is creating a demand for recycled materials sometimes referred to as "closing the loop"?

7. How does recycling link the world's largest economy (the United States) to the world's fastest growing economy (China)?

8. What are dioxins, and how are they produced? What harm do they cause?

9. What are PCBs, and what harm do they cause?

10. Suppose hazardous chemicals were suspected to be leaking from an old dump near your home. Outline the steps you would take to (1) have the site evaluated to determine if there is a danger and (2) mobilize the local community to get the site cleaned up.

11. What are the goals, strengths, and weaknesses of the Superfund program?

12. The Organization for African Unity has vigorously opposed the export of hazardous waste from industrialized countries to developing nations. They call this practice "toxic terrorism." Explain.

13. What is integrated waste management? Why must a sanitary landfill always be included in any integrated waste management plan?

14. Compare integrated pest management, discussed in Chapter 23, to integrated waste management. How does each reduce potential damage to the environment?

15. How does the system of integrated waste management depicted in Figure 24.12 compare to a natural ecosystem?

16. Does the cartoon below refer to highly developed or developing countries? How do you think a sanitary landfill would differ in the two types of countries?

Quantitative questions relating to this chapter are on our Web site.

(*For Better or Worse* © 2006 Lynn Johnston Productions. Distributed by Universal Press Syndicate. Reprinted with permission. All rights reserved.)

Take a Stand

Visit our Web site at http://www.wiley.com/college/raven (select Chapter 24 from the Table of Contents) for links to more information about whether recycling programs should be required components of municipal waste management plans. You will find tools to help you organize your research, analyze the data, think critically about the issues, and construct a well-considered argument. *Take a Stand* activities can be done individually or as a team, as oral presentations, written exercises, or Web-based (e-mail) assignments.

Additional online materials relating to this chapter, including a Student Testing Section with study aids and self-tests, Environmental News, Activity Links, Environmental Investigations, and more, are also on our Web site.

25

Tomorrow's World

Child planting a tree. Such activities teach tomorrow's generation about environmental issues, such as the importance of forests. (P. Cenini/18405/Food and Agriculture Organization)

I n this text you were presented with a broad overview of today's environmental issues, including a perspective on how ecosystems and other environmental systems operate. Now we wish to speak directly to you, not as students but as citizens. On a global level, nearly one in four people lives in extreme poverty, and one in eight cannot get enough or the right kind of foods. Each year, millions of children die of malnutrition and disease; half of these deaths are attributable to environmental factors. To compound the problem of poverty, the human population continues to grow.

Humans are not managing the world's resources sustainably; instead, we are depleting them rapidly. We have cut about a third of the forests since 1950 without replacing them. We are driving the world's species of plants, animals, fungi, and microorganisms to extinction at a rate thousands of times faster than in the past 65 million years.

During the 20th century, we lost as much as one-fifth of the world's topsoil. Because

of soil salinization, desertification, urban sprawl, erosion, and other factors, we are feeding well over twice as many people as in 1950 on only 80% of the agricultural lands that were being cultivated then. The challenge of feeding almost seven billion people increases with each year.

Global climate change is underway, with about 20 percent more carbon dioxide (CO_2), the most important greenhouse gas, having been added to the total amount in Earth's atmosphere since 1950. As CO_2 increases, the climate warms, with potentially disastrous outcomes. Much of the CO_2 entering our atmosphere comes from our addiction to fossil fuels. Human society requires energy to function but we must obtain that energy in a way that reduces the combustion of fossil fuels.

Worldwide, more and more people are living in urban areas. This influx into cities has resulted in unsafe, unhealthy environments for many residents, particularly in developing countries. Urbanization has also put a

severe strain on surrounding rural areas, which have been degraded to support the high density of people. The challenge is to create sustainable cities.

This chapter contains a five-point strategy for what must be done in the near future to confront these critical global issues. Read, think, discuss, and come to your own conclusions. And then, if our world is to have a peaceful, prosperous, and sustainable future, act.

World View and Closer to You...
Real News footage relating to the future of the environment around the world and in your own geographic region is available at www.wiley.com/college/raven by clicking on "World View and Closer to You."

Living Sustainably

LEARNING OBJECTIVES

- Define *environmental sustainability*.
- Discuss how the natural environment is linked to sustainable development.
- Briefly describe the consumption habits of people in highly developed countries and relate consumption to carrying capacity.

If the world's resources were considered a bank account, then we are living off the principal and not off the interest. Living off the interest alone, without touching the principal, would be sustainable. Our way of living, however, is clearly unsustainable. We exist as we do by rapidly exhausting the quantity and quality of the natural resources that will be available to people in the future **(Figure 25.1)**.

Environmental sustainability is a concept that people have discussed for many years. *Our Common Future*, the Brundtland Report of the World Commission on Environment and Development (1987), presented the closely related concept of **sustainable development**. The goals of sustainable development are improved living conditions for all people while maintaining a healthy environmental system in which natural resources are not overused and excessive pollution is not generated. Sustainable development balances economic growth with environmental conservation. The authors of the report pointed out that the concept of *needs* in the definition of sustainable development includes meeting the needs of the world's poor, because unless their needs are met, there can be no overall sustainability.

Our Common Future also observed that the environment's ability to meet present and future needs is directly related to the state of technology and social organization existing at a given time and in a given place. The number of people existing, their degree of affluence (that is, their level of **consumption**), and their choices of technology all interact to produce the total effect of a given society, or of society at large, on the sustainability of the environment.

According to *Our Common Future*:

> *When the [20th] century began, neither human numbers nor technology had the power to radically alter planetary systems. As the century closes, not only do vastly increased human numbers and their activities have that power, but major unintended changes are occurring in the atmosphere, in soils, in waters, among plants and animals, and in the relationships among all of these. The rate of change is outstripping the ability of scientific disciplines and our current capabilities to assess and advise. It is frustrating the attempts of political and economic institutions, which evolved in a different, more fragmented world, to adapt and cope.*

Even with the use of the best technologies we could imagine, the productivity of Earth still has its limits, and the extent of our use of Earth's productivity cannot be expanded indefinitely. To live within these limits, population size must be held at a sustainable level, and the wealthy must first stabilize their use of natural resources and then reduce this use to a level that can be maintained. The world does not contain enough resources to sustain everyone at the level of consumption enjoyed in the United States, Canada, Europe, and Japan, although countries such as China are rapidly catching up **(Figure 25.2)**. However, suitable strategies exist to reduce these levels of consumption without concurrently reducing the real quality of life.

Although we can not accurately predict the consequences of particular kinds of economic development, all economic development inevitably must take place within the **carrying capacity** of the ecosystems that support it. The carrying capacity of a given ecosystem is determined ultimately by its ability to absorb wastes and renew itself. In understanding Earth's carrying capacity for the human population and in designing appropriate economic development strategies, we should bear in mind the great disparities between living standards and expectations in different areas.

environmental sustainability: The ability to meet humanity's current needs without compromising the ability of future generations to meet their needs.

sustainable development: Economic development that meets the needs of the present without compromising the ability of future generations to meet their own needs.

consumption: The human use of materials and energy; generally speaking, people in highly developed countries are extravagant consumers.

carrying capacity (K): The maximum number of individuals of a given species that a particular environment can support for an indefinite period, assuming there are no changes in the environment.

Figure 25.1 **Earth under attack.** How are Earth's natural resources—and our use or overuse of them—critically linked to environmental sustainability? (Bizarro © Dan Piraro/Reprinted with special permission of King Features Syndicate)

Figure 25.2 Consumption in China.
China is now the world's largest consumer of basic products such as coal, steel, and grain. The world's supply of natural resources is finite—and their use generates a huge amount of waste. Shown is a man in Haikou, China, transporting plastic containers for recycling. About half of the plastic in China is currently recycled; the rest is discarded. (© China Photo/Reuters/Corbis Images)

Kai N. Lee is a professor of environmental studies at Williams College in Massachusetts. In his book *Compass and Gyroscope: Integrating Science and Politics for the Environment*, Lee discusses sustainability this way:

Against this background it is possible to see that sustainable development is not a goal, not a condition likely to be attained on Earth as we know it. Rather, it is more like freedom or justice, a direction in which we strive, along which we search for a life good enough to warrant our comforts. Freedom and justice are easily taken for granted, although many have died in their pursuit and defense. A more materialist goal such as sustainability is harder to imbue with romance and ideology. But the enormous changes our species has wrought leave us a difficult choice: either to accept our humanity in the company of the whole human race and the natural world we jointly share, or to concede that being human is too difficult for the richest, most advanced beings in history.

REVIEW

1. What is environmental sustainability?
2. How is the natural environment—Earth's living organisms and ecosystems—an essential part of sustainable development?

Sustainable Living: A Plan of Action

LEARNING OBJECTIVES

- Define *poverty* and briefly describe this global problem.
- Discuss problems relating to loss of forests and declining biological diversity, including the important ecosystem services that these resources provide.
- Describe the extent of food insecurity, and relate at least two ways to increase food production sustainably.
- Define the *enhanced greenhouse effect,* and explain how stabilizing climate is related to energy use.
- Describe at least two problems in megacities in the developing world, and relate how these problems could be addressed.

There is no shortage of suggestions for ways to address the world's many environmental problems (see the Millennium Development Goals discussed in Chapter 9). We have organized this section around the five recommendations for sustainable living presented in *Plan B 2.0: Rescuing a Planet under Stress and a Civilization in Trouble*, published in 2006 by **Lester R. Brown.** We think that if people, both individuals and collectively as governments, focus their efforts and financial support on Brown's plan, the quality of human life will be much improved. Brown's five recommendations for sustainable living are

1. eliminating poverty and stabilizing the human population,
2. protecting and restoring Earth's resources,
3. providing adequate food for all people,
4. mitigating climate change, and
5. designing sustainable cities.

Seriously addressing these recommendations now offers us hope for the kind of future that we all want for our children and grandchildren.

Recommendation 1: Eliminating Poverty and Stabilizing the Human Population

The ultimate goal of economic development is to improve the quality of human life, making it possible for humans throughout the world to enjoy long, healthy, and fulfilling lives. A serious complication lies in the fact that the distribution of the world's resources is unequal. Those who live in highly developed countries, a rapidly shrinking 19% of the global population, control about 79% of the world's finances, as measured by summing gross domestic products. The *per capita GNI PPP* (gross national income in purchasing power parity) in highly developed countries in 2006 (latest data available) was about $27,790; the per-capita GNI PPP of the other 81% of the people in the world was about $4,950. At present, nearly 3.5 billion people, more than half of the world's population, live on less than $2 per day.

For many of the world's women and children, life is an endless struggle for survival, centering on the daily requirements for firewood, clean water, and food. Such **poverty**, along with the enormous pressures that human population growth, consumption rates, and application of technologies are putting on the world's productive capacity, are global problems. These problems cannot be solved without modifying the standard of living enjoyed by most people in highly developed countries.

We who live in the United States, Canada, and other highly developed countries enjoy an abundance of what the world has to offer. We are collectively the wealthiest people who have ever existed, with the highest standard of living (shared with a few other rich countries). Because the United States, with less than 5% of the world's people, controls approximately 25% of the world's economy, it is obvious that we depend on many other nations for our prosperity. In our actions we often seem to miss this relationship and to underestimate our effects on the environment that supports us.

Failing to confront the problem of poverty around the world continues to make it impossible to attain global sustainability and is morally indefensible. For example, most people would find unacceptable that about 29,000 infants and children under the age of five die each day (2006 data from United Nations Children's Fund [UNICEF]). Most of these deaths could have been prevented by access to adequate supplies of food or basic medical techniques and supplies. For us to allow so many to starve, to go hungry, and to live in absolute poverty is to threaten the future of the global ecosystem that sustains us all. Everyone must have a reasonable share of Earth's productivity, or our civilization will eventually come unraveled. As U.S. President **Franklin Delano Roosevelt** said so well in his second inaugural address in 1937, "The test of our progress is not whether we add more to the abundance of those who have much; it is whether we provide enough for those who have too little."

Improving the quality of life in lower-income countries will require increasing their economic growth so that issues of health, nutrition, and education can be addressed adequately. The role of women requires special attention, since women are often disproportionately disadvantaged in poor countries, and the improvement of their status can make a significant contribution to the stability and prosperity of those communities. As **Nafis Sadik,** head of the U.N. Population Fund from 1987 to 2000, has pointed out, women hold a paradoxical place in many societies. As part of their traditional duties as mothers and wives, they are expected to bear the whole responsibility for childcare; at the same time, they are often expected to contribute significantly to the family income by direct labor **(Figure 25.3).** In many developing countries women have few rights and little legal ability to protect their property, their rights to their children, their income, or anything else. Improving the status of women is a crucial aspect of sustainable development.

Raising the standard of living for poor countries will require that special attention be given to the poorest segments of all populations—that is, to those without much hope. In this context, the universal education of children and the reduction of illiteracy are of critical importance in raising and maintaining appropriate standards of living in every country.

One of the greatest barriers to equalizing the gap between highly developed and developing nations is the lack of trained professionals in the latter group. Only about 10% of the world's scientists and engineers live in less developed countries. Considering that a majority

poverty: A condition in which people are unable to meet their basic needs for adequate food, clothing, shelter, education, or health.

Figure 25.3 A woman harvesting tomatoes in Mali. In addition to caring for their children, women are the farmers in much of Africa. (Larry C. Price)

of those scientists and engineers live in only four nations (China, Brazil, Mexico, and India), it is apparent that for most developing countries, trained technical personnel are virtually nonexistent. How are these countries to decide, on the basis of their limited knowledge, whether to join international agreements concerning the environment or how best to manage their own natural resources? The training and employment of professional scientists and engineers in developing countries is a matter of high priority.

Goods, services, and money flow throughout the world and greatly affect the nations among which they move. We have entered a new era of global trade, within which we must establish new guidelines for national, corporate, and individual behavior even though our understanding of the dynamics of global systems is imperfect. For example, the flow of money from developing countries to highly developed countries has exceeded the flow in the other direction for many years. Former West German Chancellor **Willy Brant** termed this phenomenon "a blood transfusion from the sick to the healthy." A world that values environmental sustainability and social justice for everyone must find ways to reverse this flow for the common good. Debts from the poorest countries should be forgiven more readily than is the case now, and international development assistance should be enhanced.

If each nation accepts the duty to live sustainably, to help other nations do so, and to support international laws and treaties that promote sustainability, the world as a whole will function increasingly well. At the same time, ways and means must be found to increase the income of poor countries because an improving economy is generally one of the most important factors that leads away from poverty. The strengthening of the United Nations as an effective force for global sustainability, by focusing on the whole array of problems that will determine the future of the world, would contribute greatly to the creation of a sustainable, healthy, peaceful, and prosperous world.

Stabilizing the Human Population

You have seen that the poorer people living in developing countries have far less than their share of the world's resources, and most cannot live sustainably, given their current circumstances. In addition, population growth rates are generally highest where poverty is most intense. This overall situation is clearly unstable and must be corrected by the determination and adoption of acceptable levels of population and consumption for all regions.

The world population grew from 2.5 billion in 1950 to 6.6 billion in 2006; it is continuing to increase at a rate of nearly 80 million additional people each year. Recently (in 2004), the United Nations calculated that world population will reach approximately 9.1 billion people in 2050. To restrict the global population, there must be sustained worldwide attention to family planning. If we pay consistent attention to overpopulation and devote the resources necessary to make family planning available for everyone, the human population will stabilize (**Figure 25.4**). If we do not continue to emphasize family planning measures, we simply will not achieve population stability. All governments of the world must pay sustained attention to the need for family planning; it is important in both developing countries and in highly developed countries, which consume most of the world's resources even though they constitute a minority of the world's population.

To stay within Earth's carrying capacity, it will be necessary to reach a stable population and to reduce excessive consumption and waste. These factors must be managed in an integrated fashion and coupled with educational programs everywhere, so that people will have the opportunity to understand that Earth's carrying capacity is not unlimited. Although we do not know what Earth's carrying capacity for humans is, if the population continues to grow, we will exceed that carrying capacity at some point (recall the discussion of carrying capacity in Chapter 8). There is no hope for a peaceful world

Figure 25.4   Family planning in the Ivory Coast. Women at the Adjame Market in Abidjan learn about family planning and birth control. (Karen Kasmauski/National Geographic Society)

without overall population stability, and no hope for regional economic sustainability without regional population stability.

The Links among Immigration, Poverty, and Population   Another important way in which we are linked to the developing world is massive immigration of poor people from the tropics and subtropics into the industrialized nations of the temperate zone. The U.S. Customs and Border Protection estimates that in 2004 (latest data available), more than 1.1 million unauthorized migrants were apprehended at or near U.S. borders, suggesting that many others may have entered successfully (see Chapter 8). The U.S. government and other organizations such as the Pew Hispanic Center estimate that immigration now accounts for at least 50 percent of population growth in the United States.

The same pattern of migration to flee poverty can be seen worldwide. According to the Natural Resources and Environmental Institute in Cairo, in 2005 an estimated 30 million people were environmental refugees who left their homes in search of food, often crossing national borders. The number of such people seeking to enter the United States, Canada, and other highly developed countries is likely to increase greatly. This pattern is the direct result of mounting populations, economic pressures, and environmental degradation in the developing countries. If we continue to ignore the driving forces that underlie immigration, we will never succeed in decreasing the numbers.

Recommendation 2: Protecting and Restoring Earth's Resources

To build and maintain a sustainable society, it is necessary to preserve the productive natural systems that support us. Renewable resources such as forests, biodiversity, soils, fresh water, and fisheries must be treated in ways that ensure their long-term productivity. Their capacity for renewal must be understood and respected.

The conservation of nonrenewable resources, such as oil, natural gas, and minerals, is obvious, although discoveries of new supplies of nonrenewable resources have often given us the illusion that they are inexhaustible. However, it is the potentially renewable resources—productive living systems such as forests, shrublands, grazing lands, and wetlands—that have been badly damaged over the past 200 years.

The World's Forests   The world's forests, which are being cut, burned, or seriously altered at a frightening rate, are being lost for two principal reasons. First, they are converted to cash. Like natural resources all over the world, forests are being exploited and sold. The cutting of old-growth forests in Oregon, Washington, Alaska, Canada, and Siberia is subsidized by the central government in each case. Until concepts of natural productivity and the central role of environmental sustainability become important aspects of economic calculations, susceptible natural resources will continue to be consumed unsustainably, driven by short-term economics.

The second reason that the world's forests are being lost is the pressure created by rapid population growth and widespread poverty. In many developing countries, the forests have traditionally served as a "safety valve" for the poor, who by consuming small tracts of forest on a one-time basis and moving on, find a source of food, shelter, and clothing for themselves and their families. But now the numbers of people in developing countries are too great for their forests to support.

Tropical rain forests—biologically the world's richest terrestrial areas—have been reduced to less than half their original area. We know very little about replacing most tropical forests with productive agriculture and forestry. Short-term exploitation by logging or clearing (often by burning) often results in irreversible destruction of the tropical soil's potential productivity. Methods of forest clearing that were suitable when population levels were lower and forests had time to recover from temporary disturbances simply do not work any longer: They convert a potentially sustainable resource (forests growing in infertile soil) into an unsustainable one.

Clearing the forests and prairies of Eurasia and North America traditionally led to the establishment of productive farms, for the soil was rich. In contrast, clearing the forests of tropical Africa or Latin America often produces wastelands. The relative infertility of many tropical soils, their thin and easily disturbed surface layer of organic matter, and the high temperature and precipitation levels of tropical regions often combine to make the attainment of sustainable agriculture or forestry systems difficult.

Tropical forests are rapidly being cleared and destroyed, not only because of the needs of the people who live in or near them, but also because of the demands of the global economy. Many products—foods such as beef, bananas, coffee, and tea; medicines; and hardwoods—come to the industrialized world from the tropics. As timber is harvested or trees are cleared for other reasons, however, only a very limited amount of replanting is taking place. It is estimated that only one tree is planted for every ten that are cut in the Latin American tropics. Few nations have forestry plans, and there is almost no coordination of forestry policies.

Loss of Biodiversity Over the next few decades, we can expect the rate of extinction to climb from dozens of species a day to hundreds of species a day, the great majority unknown to science. How big a loss is this? Unfortunately, we still have a limited knowledge about the world's **biological diversity**. An estimated five-sixths of all kinds of organisms have not yet been recognized and described scientifically. Earth has but one living library; people have read few of its books and don't even have a complete catalogue of the volumes it contains—but the library is being burned unread.

biological diversity: The number and variety of Earth's organisms.

At the most basic level, we are part of Earth's web of life, having evolved within it and as part of it. We are entirely dependent on that web, with all its interactions, for our survival and for anything that we wish to attain individually or collectively in the future. In view of our utter dependence on the web of life, many people have questioned whether it is morally acceptable for us to continue destroying it so rapidly **(Figure 25.5).** Can we justify, in ethical terms, the fact that we are driving species to extinction that are, as far as we know, our only living companions in the universe?

Pragmatically, we have a clear interest in protecting Earth's biological diversity and managing it sustainably. We obtain from living organisms all our food, most medicines, many building and clothing materials, biomass for energy, and numerous other products. In addition, communities of organisms and ecosystems provide an enormous array of **ecosystem services** without which we would not survive. These services include the protection of watersheds and soils; the development of fertile agricultural lands; the determination of local climate and, through processes such as *carbon sequestration*, of global climate; and the maintenance of habitats for beneficial animals and plants. In view of these ecosystem services, the ultimate reason for caring for the community of life on Earth is a selfish one.

Economic development will succeed only if it is carried out in such a way as to maintain the sustainable productivity of the biosphere. The human population of 6.6 billion people uses an estimated 32% of annual land-based net primary productivity (see Chapter 3). We also use approximately 55% of accessible, renewable supplies of fresh water. This puts enormous and unprecedented pressures on the life-support systems that nature provides and on the sustainability of Earth's renewable resources: forests, soils, fresh water, fisheries, and biological diversity. Preserving Earth's biological diversity makes the whole Earth system stronger, and the more variety that is preserved, the more interesting and beautiful the world is.

Some 80% of the species of plants, animals, fungi, and microorganisms on which we all depend are found in developing countries. How will these relatively poor countries sustainably manage and conserve these precious resources? Biological diversity is an intrinsically local problem, and each

Figure 25.5 Endangered green sea turtle in the Atlantic Ocean. After mating in shallow water, the females lay their eggs on beaches from Cape Canaveral to Palm Beach. Coastal development is one of the ways humans affect sea turtles adversely. (Dave B. Fleetham/Tom Stack & Associates)

nation must address it for the sake of its own people's future, as well as for the world at large. Biological diversity, like most challenges of sustainable development, can be addressed adequately only if we provide international assistance where needed, including help in training scientists and engineers from developing countries.

Biological diversity and human cultural diversity are intertwined: They are in fact two sides of the same coin. **Cultural diversity** is Earth's variety of human communities, each with its individual languages, traditions, and identities **(Figure 25.6)**. Cultural diversity enriches the collective human experience. Unfortunately, cultural diversity sometimes promotes distrust—a feeling of "us" against "them." For that reason, the United Nations Educational, Scientific, and Cultural Organization (UNESCO) supports the protection of minorities, peace, and equity in the context of cultural diversity.

Sustainable Development and Protecting and Restoring Earth's Resources Sustainable development must provide real improvements in the quality of human life. At the same time it must maintain the life-support systems on which our lives, and the lives of all other species, are based. It is not possible to ignore the functions of these biological and physical systems in achieving economic development; if we do ignore them, we ultimately degrade the quality of human life as well. The economic development strategies most suitable for a given locality depend on the biological, physical, and human factors in operation there, and we must strive to understand these if our decisions are to be effective.

The basic message is that whatever actions we take, they must be taken within the carrying capacity of individual ecosystems, and ultimately that of Earth itself. The transition to a sustainable world depends on changing individual attitudes and practices in intelligent ways based on scientific information. Throughout the world, communities are finding that sustainable activities preserve their local environments and improve their standards of living. Getting the best available information to rural communities is essential to help the people who live there make the most appropriate choices for themselves. Once attitudes have shifted at the individual and community levels, then national strategies can be devised that safeguard the natural resources of each country for the future. And finally, because many environmental problems extend across national boundaries, alliances among countries are necessary to preserve the organisms, lands, and waters on which our livelihoods depend.

Recommendation 3: Providing Adequate Food for All People

A humanitarian crisis exists that rarely makes the evening news: Globally, more than 852 million people lack access to the food needed for healthy, productive lives. This estimate, according to the U.N. Food and Agriculture Organization, includes a high percentage of children. Children are particularly susceptible to food deficiencies because their brains and bodies cannot develop properly without adequate nutrition. The World Health Organization (WHO) estimates that 182 million children under the age of five are seriously underweight for their age. Most malnourished people live in rural areas of the poorest developing nations. The link between poverty and **food insecurity** is inescapable.

Improving agriculture is one of the highest priorities involved in achieving future global sustainability. In a rich country like the United States, where food is inexpensive, it is difficult to appreciate this goal. We in highly developed nations fail to acknowledge the inadequate food productivity in much of the world and the serious need for additional food for rapidly growing populations in developing countries.

Many improvements in agriculture will be needed to feed the world's people adequately in the future. In general, per capita grain production has kept pace with human population growth over the past 45 years. However, per capita grain production and other examples of expanded agricultural productivity have taken place at high environmental costs that often are not sustainable. Moreover, the global population continues to expand, putting additional pressure on food production.

Figure 25.6 Biological diversity and human cultural diversity in Brazil. Humans in increasing numbers are encroaching on the traditional environments of organisms such as this scarlet macaw and of native peoples such as this Amazonian boy. (Loren McIntyre)

food insecurity: The condition in which people live with chronic hunger and malnutrition.

Figure 25.7   Damage to soil resources in Kenya. Extensive withdrawal of water contributed to erosion in this once-fertile area. Careful stewardship of the land prevents such damage. (Fred Hoogervorst/Panos Pictures)

enhanced greenhouse effect: The additional warming produced by increased levels of gases that absorb infrared radiation.

Farmlands and grazing lands must be managed efficiently but sustainably so that they produce as much as possible and thus reduce the pressure on natural lands everywhere. This is especially true because worldwide we are growing our food and other agricultural products (such as cotton and biomass) on an area about the size of South America. Little additional land that is not currently under cultivation is suitable for agriculture. One way to increase the productivity of agricultural land is *multi-cropping*, or growing more than one crop per year. For example, winter wheat and summer soybean crops are grown in some areas of the United States. However, multi-cropping can be accomplished only in regions where water supplies are adequate for irrigation. Also, care must be taken to prevent a decline in soil fertility from such intensive use.

The negative environmental effects of agriculture, including loss of soil fertility, soil erosion, aquifer depletion, soil and water pollution, and air pollution, must be brought under control by using an array of methods **(Figure 25.7)**. Many strategies exist to retard the loss of topsoil and degradation of agricultural lands, conserve water, conserve energy, and reduce reliance on agricultural chemicals.

For example, *conservation tillage*, in which residues from previous crops are left in the soil, partially covering it and helping to hold it in place, is one important element of sustainable agriculture. The practice of conservation tillage, which protects the topsoil, is growing rapidly throughout the world.

Replacing furrow irrigation, where farmers flow water down small trenches between crop rows, with *drip irrigation* can reduce water use by about 50 percent. Improving water efficiency is crucial, because in many places, lack of water is the overriding factor reducing agricultural productivity.

Supporting local agriculture so that crops are grown near where they are consumed conserves energy. By one estimate, most food eaten in the United States travels about 2400 kilometers (1500 miles) from farm to plate, using massive quantities of fossil fuels in the process. In the United States, there is a growing local food movement; several colleges and universities, for example, are making a concerted effort to purchase food grown locally.

Precision farming, in which global positioning systems (GPSs) and satellite images are used to determine the most appropriate levels of fertilizer application for individual small areas, provides an important key to agriculture of the future. Precision farming allows a farmer to apply fertilizer to those areas of the field where it is needed and to skip fertilizing those areas that don't need it. *Integrated pest management*, involving the use of beneficial insects and improved cultural practices, with the application of pesticides only as a last resort, must be applied more widely. *Genetic engineering* should be used to develop nutritious crops with the precise combination of characteristics suitable for growing in a given location.

We must develop sustainable agricultural systems that provide improved dietary standards, such as the inclusion of high-quality protein in diets in developing countries. Several methods of animal protein production are particularly promising. India has dramatically increased its dairy production by feeding cows cornstalks, wheat straw, and even grass collected along roads. Using these plant materials, which would normally "go to waste," is far more efficient than feeding the cows high-quality grain. China's expanding use of *aquaculture* is another example of efficient protein production. The carp that are raised in Chinese aquaculture are very efficient at converting food into high-quality protein. In China, fish production by aquaculture now exceeds poultry production!

Recommendation 4: Mitigating Climate Change

One of the most widely discussed consequences of the pressure we are exerting on the global environment is climate change caused by the **enhanced greenhouse effect**. Both highly developed and developing countries contribute to major increases in CO_2 in the atmosphere, as well as to the increasing amounts of methane, nitrous oxide, tropospheric ozone, and CFCs. The most important greenhouse gas, CO_2, is mainly produced when we burns fossil fuels—coal, oil, and natural gas.

Although Earth's climate has been relatively stable during the present interglacial period (the past 10,000 years), evidence has mounted that human activities are causing it to change. According to the Intergovernmental Panel on Climate Change, the average global temperature increased by almost one degree Celsius during the 20th century; more than half of that warming occurred during the past 30 years. The scientific literature overwhelmingly indicates that Earth's climate will continue to change rapidly during the 21st century.

The outcomes of these likely changes, which were discussed in Chapter 21, are serious, because modern society has evolved and successfully adapted to conditions as they are. Keeping in mind that the change from the last ice age to the present was accompanied by an increase in global temperatures of five degrees Celsius puts the consequences of the present change, the most rapid of the last 10,000 years, into perspective.

We must address climate change in an aggressive and coordinated fashion, but how do we modify the actions of billions of people? How do we get all nations of the world to adopt the many approaches necessary to effect change? Many policymakers say that we should wait until scientific knowledge of climate change is complete. This reasoning is flawed because the Earth system is extremely complex, and there are limits to our ability to predict the consequences of climate change. We will never have a complete scientific understanding of Earth as a system; we should act on our current understanding of the system.

For example, we often say that an increase in atmospheric CO_2 leads to climate warming. However, the increase in CO_2, like all human impacts, is not a simple cause-and-effect relationship but instead a cascade of interacting responses that ripple through the Earth system **(Figure 25.8)**. Increasing atmospheric CO_2 also affects how plants grow, but different species react to increasing CO_2 in different ways; we can expect changes in plant community composition as the plants that are more competitive in the new conditions thrive and replace less competitive plants. We cannot begin to predict how these vegetation changes will affect humans or the rest of the biosphere.

As another example of the cascading effects of a single change (increasing levels of atmospheric CO_2), the higher levels of atmospheric CO_2 are now resulting in increased levels of dissolved CO_2 in the ocean. This change profoundly alters ocean chemistry, which in turn affects marine organisms, possibly leading to the extinction of some or many species. Again, our incomplete scientific knowledge prevents us from saying precisely what will happen in the ocean system. Can we afford the risk?

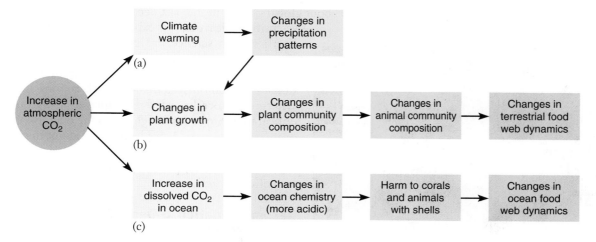

Figure 25.8 **Cascading responses through the Earth system.** (a) Most people know that, an increase in atmospheric CO_2 leads to global warming, but this phenomenon is far from a simple cause-and-effect relationship. (b), (c) Two highly simplified examples of how increasing CO_2 may cause a cascade of interacting responses throughout the Earth system. (b) Effects of increased atmospheric CO_2 on land plants. (c) Effects of increased atmospheric CO_2 on the ocean. Both (b) and (c) lead to increased extinctions.

Studying global changes in the past—such as the Vostok ice core that measured atmospheric CO_2 for the past 420,000 years—reveals that the Earth system reached critical thresholds that triggered abrupt changes. In other words, increasing CO_2 did not cause gradual climate changes in the past. Many climate scientists are concerned that our effect on climate has the potential to cause abrupt change, and therefore, CO_2 reductions must be urgently addressed.

Stabilizing the Climate Requires a Comprehensive Energy Plan   A comprehensive energy plan must be implemented for the entire world. This plan would involve phasing out fossil fuels in favor of renewable energy (such as solar and wind power), increasing energy conservation, and improving energy efficiency; there would also be limited roles for nuclear power and carbon sequestration. Many national and local governments as well as corporations and environmentally aware individuals are setting goals to cut carbon emissions from fossil fuels. Other nations, however, have not recognized the urgency of the global climate problem. We need a global consensus to address climate change, even though each country will have to implement a national energy policy that is custom tailored to include its specific renewable energy resources.

At present, the 81 percent of the world's people who live in developing countries use about 40 percent of the world's commercial energy, the bulk of which is the fossil fuels—coal, oil, and natural gas. The number of people in developing countries is growing rapidly, and even without any increase in their standard of living, the people in these countries will use much larger quantities of energy in the future. Of special concern in this connection is China, a rapidly industrializing country of more than 1.3 billion people, with the world's second largest economy and huge deposits of coal **(Figure 25.9).** If we reach a global consensus on limiting CO_2 in the atmosphere, those of us who live in highly developed nations will find ourselves insisting that developing countries not burn coal as they industrialize. Such a strategy seems unlikely to prevail unless we all share in paying for cutting carbon emissions. Highly developed nations must realize that implementing a comprehensive energy plan for the developing world is a necessary ingredient for their own future security, as well as for global climate stability.

The global energy plan must also involve reducing the profligate use of fossil fuels in the United States, Canada, and other highly developed countries. This energy use cannot be sustained, much less used as a model for the rest of the world. Wind power may emerge as a critical part of energy policies in the United States and Canada. According to *Plan B 2.0*, a single advanced-design wind turbine in the Midwest can produce up to $100,000 worth of electricity that can be used to heat and cool buildings, cook food, and power electric automobiles.

We have learned much about global change in the past decade or so, but as in all scientific endeavors, our expanding knowledge of Earth system science has generated new questions. The challenges of mitigating climate change require immediate investment in additional research on systems-level interactions and synergies. More knowledge of global change is critical to improve our insights into the operations of the Earth system. As we learn more about these environmental systems, we hope our greater understanding will spur people around the world into effective action.

Figure 25.9 Increasing energy consumption in China. Trucks in China deliver coal straight from mines to homes. (Peter Essick/Aurora Photos)

Recommendation 5: Designing Sustainable Cities

At the beginning of the Industrial Revolution, in approximately 1800, only 3 percent of the world's people lived in cities, and 97 percent were rural, living on farms or in small towns. In the two centuries since then, population distribution has changed radically—

toward the cities. More people live in Mexico City today than were living in all the cities of the world 200 years ago. This is a staggering difference in the way people live.

Almost 50 percent of the world's population now lives in cities, and the percentage continues to grow. For example, although the population of Brazil has grown rapidly to its present 187 million people, the population in rural areas of Brazil is actually declining. In industrialized countries like the United States and Canada, almost 80 percent of the people live in cities.

CASE IN POINT

Jakarta, Indonesia

There is an urgent need to improve the environment and quality of life in cities, particularly in rapidly growing **megacities** of developing countries. Consider Jakarta, Indonesia, with a population of 13.2 million in 2006. Jakarta is plagued with many of the problems found in other rapidly growing megacities in the developing world.

The air in Jakarta is badly polluted with the exhaust from cars, buses, and motorbikes that transport about two million commuters into the city each day (see Chapter 20 introduction). Water pollution is also a critical problem. At least 95 percent of human wastes produced in the city are not cleaned up at sewage treatment plants. Instead, human sewage and garbage are dumped directly into nearby rivers **(Figure 25.10)**. Jakarta's municipal water supply is so polluted that piped water must be boiled before drinking.

As groundwater has been depleted to meet the city's needs, parts of Jakarta have subsided so that many areas are more flood-prone, particularly during the rainy season. Illegal squatter settlements proliferate; here the poorest inhabitants build dwellings on vacant land using whatever materials they can scavenge. As in other megacities around the world, squatter settlements in Jakarta have the worst water, sewage, and solid waste problems.

Although we have painted a grim picture of Jakarta, it should be noted that improvements are slowly beginning to occur. As part of an ambitious long-term transportation upgrade, Jakarta developed a busway with dedicated bus lines that reduce commuter times and encourage people to commute by bus rather than by automobile. A monorail and subway system are also planned. ■

How to Design a Sustainable City City planners around the world are trying a variety of approaches to make cities more livable for their inhabitants. Like Jakarta, many cities are developing urban transportation systems to reduce the use of cars and the problems associated with them, such as congested roads, large areas devoted to parking, and air pollution. Urban transportation can be as elaborate and expensive as mass transit subways and light rails or as inexpensive as bicycle and pedestrian pathways. Investing in urban transportation other than building more highways encourages commuters to use forms of transportation other than automobiles. To encourage mass transit, some cities also tax people using highways into and out of cities during business hours.

When a city is built around people instead of cars, such as establishing parks and open spaces instead of highways and parking lots, urban residents have an improved quality of life. Also, commuters spend less time sitting in traffic congestion. Air pollution, including climate-warming CO_2 emissions, is substantially reduced. Bicycle transportation also provides healthy exercise. To encourage bicycle use in the Netherlands, city streets have dedicated bike lanes **(Figure 25.11)**.

Some cities encourage their residents to tend small garden plots on vacant lots in their neighborhoods. Other residents grow small plots of vegetables in their yards or on rooftops. Small-scale farming in cities provides multiple benefits. Urban farms provide food for a city's inhabitants. Urban farmers in Singapore, for example, grow about one-fourth of the vegetables consumed by its residents. Gardening also provides healthy exercise and, when multiple people garden on a vacant lot, promotes a sense of community with neighbors who might otherwise remain strangers. Also, there is something innately satisfying to the human spirit in tending plants and watching them grow and flourish.

Figure 25.10   Water pollution in Jakarta, Indonesia. (© Jose Tuste Raga/Corbis)

megacity: A city with more than 10 million inhabitants.

Figure 25.11   Bicycles parked at a train station in Amsterdam, the Netherlands. Each resident in the Netherlands rides a bicycle an average of 573 mi (917 km) per year. Photographed in 2007. (Koen van Weel/Reuters/Landov LLC)

Water scarcity is a major issue for many cities of the world. Some city planners think that innovative approaches must be adopted where water resources are scarce. These approaches would replace the traditional one-time water use that involves water purification before use, treatment of sewage and industrial wastes after use, and then discharge of the treated water. Some cities, such as Singapore, recycle some of their wastewater after it has been treated.

In cities where growth has outstripped the ability to treat sewage, the *composting toilet* is a promising alternative to sewage treatment. The composting toilet converts human wastes and table scraps into compost without using water. It is relatively inexpensive and has several advantages: Composting toilets reduce water use, reduce energy use (required to pump water and treat sewage), reduce the amount of sewage needing treatment (or reduce water pollution in cities without adequate water treatment), and produce compost, which can be added to soil as a conditioner.

Effectively dealing with the problems in squatter settlements is an urgent need. Evicting squatters does not address the underlying problem of abject poverty. Instead, cities should incorporate some sort of plan for the eventual improvement of squatter settlements. Providing basic services, such as clean water to drink, transportation (so people can leave their shelters to find gainful employment), and garbage pickup would help improve the quality of life for the poorest of the poor.

REVIEW

1. What are the five recommendations for sustainable living discussed in this chapter?
2. What is the global extent of poverty?
3. What are two ecosystem services that natural resources such as forests and biological diversity provide?
4. What is the global extent of food insecurity?
5. How is stabilizing climate related to energy use?
6. What are two serious problems in megacities in the developing world?

Changing Personal Attitudes and Practices

LEARNING OBJECTIVES

- Explain how consumption overpopulation threatens environmental sustainability.
- Relate how sustainable consumption addresses the problem of consumption overpopulation.

Any long-term improvement in the condition of the world must start with individuals—our values, attitudes, and practices. Each of us is part of many interrelated systems, and our individual choices and actions impact the rest of those systems. In other words, each of us makes a difference, and it is ultimately our collective activities that make the world what it is.

In the richer parts of the world, apathy, ignorance, or incentives to wasteful consumption have negative effects. **Consumption overpopulation** reflects the growing idea that many of the world's environmental and resource problems stem from the lifestyles of people living in highly developed nations **(Figure 25.12)**. We interpret lifestyle broadly, to include goods and services (including energy) for food, clothing, housing, travel, recreation, and entertainment. In evaluating consumption overpopulation, all aspects of the production, use, and disposal of these goods and services must be taken into account, including the environmental costs of resource extraction, production, transport, trade, and waste management. Such an analysis provides a sense of what it means to consume sustainably versus unsustainably.

Like sustainable development, **sustainable consumption** is a concept that forces us to address whether our present actions undermine the environment's long-term ability to meet the needs of future generations. Factors that affect sustainable consumption include population, economic activities, technology choices, social values, and government policies.

consumption overpopulation: A situation that occurs when each individual in a population consumes too large a share of resources.

sustainable consumption: The use of goods and services that satisfy basic human needs and improve the quality of life but that minimize the use of resources so they are available for future generations.

Figure 25.12   Consumption overpopulation. Advertising promotes consumption of products that are often not needed. Highly developed nations like the United States consume more than 50 percent of the world's resources, produce 75 percent of its pollution and waste, and represent only 19 percent of its total population. Photographed in Times Square, New York City. (Luis Veiga/The Image Bank/Getty Images, Inc)

In the less developed countries, survival may be such an overwhelming concern that sustainable consumption may seem irrelevant. Sustainable consumption requires the eradication of poverty, which means that poor people in developing countries must increase their consumption of certain essential resources. This increased consumption is not sustainable unless the consumption patterns of people in highly developed countries change.

Widespread adoption of sustainable consumption will not be easy. It will require major changes in the consumption patterns and lifestyles of most people in highly developed countries. Some examples of promoting sustainable consumption include switching from motor vehicles to public transport and bicycles and developing durable, repairable, recyclable products.

Making individual changes may seem disruptive and daunting. However, although sustainable consumption contains an inherent threat to "business as usual," it also offers new and exciting opportunities. Many scientists and population experts advocate that we adopt sustainable consumption before it is forced on us by an environmentally degraded, resource-depleted world.

The Role of Education

As people adopt new lifestyles, they must be educated so that they understand the reasons for changing practices that may be highly ingrained or traditional. People are generally concerned about the environment, but their concerns do not naturally translate into action. Furthermore, most people believe that individual actions do not really make a difference, whereas, in fact, they are the only factors that do so, both individually and collectively. If people understand the way natural systems function, they can appreciate their own place in them and value sustainable actions. Formal education and informal education are important both in bringing about change and in contributing to the sustainable management of resources. Accurate information must be made available widely; the media have an important part to play in such efforts. No national plan can be successful without a clear statement of its strategic goals, accompanied by an educational program that lays out the fundamental reasons why these goals are important and what individuals can do to help realize them.

In the United States, all citizens will be called on throughout their lives to take actions based on their understanding of the environment. Many of these actions reflect personal choices, such as the amount of water or the kind of transportation to use. In a democracy, many choices that require environmental knowledge are also made at a local, national, and global level. Should we ratify the Convention on Biological Diversity or remain one of a handful of countries that have not done so? Should we support the Kyoto Protocol on greenhouse gases? Should we promote world trade—and what kind of trade? How important is it to reduce sulfur emissions from coal-fired plants? To answer these and similar questions, we can read and study. To help our fellow citizens understand these issues, we can:

1. Set up environmental curricula in primary and secondary schools and in colleges.
2. Encourage the activities of environmental organizations.
3. Support institutions such as natural history museums, zoos, aquaria, and botanical gardens, all of which promote conservation and sustainability.
4. Encourage the inclusion of relevant material in the programs of churches, social groups, and other institutions that should logically emphasize them.

Most people are interested in the environment—their own local environment—in their own way. We must work to create a democratic society in which everyone will have the opportunity to learn and to contribute.

REVIEW

1. How does sustainable consumption address the problem of overpopulation?

What Kind of World Do We Want?

LEARNING OBJECTIVE

• Write a one-page essay describing what kind of world you want to leave for your children and your children's children.

Perhaps the most important single lesson to have learned from this book is that those of us who live in highly developed countries are at the core of the problems facing the global environment today. Highly developed countries consume a disproportionate share of resources and must act forcefully to reduce their levels of consumption if we are all going to achieve sustainability. We in highly developed countries continue to strive to increase our high standard of living, from a level considered utopian by most people on Earth. We drain resources from the entire globe and thus contribute to a future in which neither our children nor our grandchildren will be able to live in anything like the affluence that we experience now.

The citizens of highly developed nations often seem to assume implicitly that overall prosperity is a result of science and technology, and that the environment will take care of itself. If we are to improve the world situation, we must radically change our view of the world and adopt new ways of thinking, or we will suffer together **(Figure 25.13).** We must learn to understand, respect, and work with one another, regardless of the differences that exist among us. The most heartening aspect of the situation that we confront is that people, given the motivation, do have the ability to make substantial changes.

At the deepest level, the most critical environmental problems, from which all others arise, are our own attitudes and values. We are out of touch and out of balance with the natural world, and until we reconnect and readjust in some significant way, all solutions will be stopgaps. As a society we do not feel we are a part of the global ecosystem; we feel separate, above it, and therefore in a position to consume and to abuse it without thought of consequences.

This book has been about consequences. In the last 40 years or so, we have come to recognize the nature of the impact of human activities on the biosphere. With that knowledge, we have found solutions to many environmental problems. We now understand this impact enough to know that we cannot continue to act as we have been acting and expect any sort of viable future for our species. If all we do, as a result of this new knowledge, is to make some shifts in consumer choices and write a few letters, it will not be nearly enough.

Your generation must become the next pioneers. A pioneer is one who ventures into unexplored territory, a process simultaneously terrifying and profoundly exciting. The unexplored territory in this case is the development of a different way for humans to exist in the world. No models exist for this kind of change. You must forge a new revolution—akin in scope and effect to the Agricultural Revolution or the Industrial Revolution, yet totally different because it must be deliberate. You must help create the political will for it with your numbers and your commitment. You must create the economic power with your thoughtful decisions as both consumers and leaders. You must create social change with your acceptance and respect of the differences among peoples.

This change will require reconnecting with the natural environment. That means, at a personal level, taking opportunities to go outside (even if it is a city park or a backyard garden), to listen to the wind, and to look at the exquisite variety of plants, insects, and other life forms with which we coexist. Humans evolved in nature. Our immensely multidimensional brains evolved as we interacted with growing things, weather patterns, and other animals. The world we have created now screens us from all that. The sophisticated devices we imagined and manufactured—things such as plasma televisions, laptop computers, and highly engineered automobiles—have come to define our world. One of your challenges will be to use technology as a tool but not to let it define your interaction with the world.

The new environmental revolution will require that we revalue ourselves according to a different set of ideals. Wealth and material possessions have come to mean success, at a tremendous cost to the planet. Such ideas are deeply embedded and extremely compelling and will be difficult to change. You may have to throw off some of the myths of Western culture, such as the belief that "faster" and "more" inevitably mean "better." You will at least need to examine those myths thoughtfully and to decide what they mean to you.

It will require reinventing economic constructs, such as building the cost of environmental impact and damage into our accounting systems, which will cause market forces to work in favor of environmental protection. Business activities must involve developing cooperative partnerships with people all over the world and making decisions based on long-term benefits to the environment.

This responsibility may seem daunting, but it is not overwhelming, and the rewards will be great. The choices we make now will have a greater impact on the future than those that any generation has had before. Even choosing to do nothing will have profound consequences for the future. At the same time, it is an incredible opportunity. **Margaret Mead** (1901–1978), the noted American anthropologist, once said, "Never doubt that a small group of thoughtful, committed people can change the world; indeed, it's the only thing that ever has!" This is a time in history when the best of human qualities—vision, courage, imagination, and concern—will play a critical role in establishing the nature of tomorrow's world.

Figure 25.13 **Humans cause and exacerbate almost all environmental problems.** What warning does this comic make? (Bizarro © Dan Piraro/Reprinted with special permission of King Features Syndicate)

REVIEW

1. How would you describe the world you want your children to live in?

2. How does the future world you envision differ from the realities of today?

REVIEW OF LEARNING OBJECTIVES WITH KEY TERMS

• **Define *environmental sustainability*.**

Environmental sustainability is the ability to meet humanity's current needs without compromising the ability of future generations to meet their needs. Currently, we face enormous environmental problems. Nearly one in four people lives in poverty, and nearly one in eight people doesn't have enough food or the right kinds of food to grow and function properly. Furthermore, we are not managing the world's resources sustainably.

• **Discuss how the natural environment is linked to sustainable development.**

Sustainable development is economic development that meets the needs of the present without compromising the ability of future generations to meet their own needs. Earth's productivity is limited, and our use of it at the current level cannot be extended indefinitely. Sustainable development must allow for the maintenance of the life-support systems on which our lives and the lives of all other species are based.

• **Briefly describe the consumption habits of people in highly developed countries and relate consumption to carrying capacity.**

Consumption is the human use of materials and energy; generally speaking, people in highly developed countries are extravagant consumers. **Carrying capacity** is the maximum number of individuals of a given species that a particular environment can support for an indefinite period, assuming there are no changes in the environment. The world does not contain enough resources to sustain more than six billion people at the level of consumption enjoyed in the United States.

• **Define *poverty* and briefly describe this global problem.**

Poverty is a condition in which people are unable to meet their basic needs for adequate food, clothing, shelter, education, or health. Currently, almost 3.5 billion people, more than half of the world's population, live on less than $2 per day. For many of the world's women and children, life is an endless struggle for survival, centering on the daily requirements for firewood, clean water, and food.

• **Discuss problems relating to loss of forests and declining biological diversity, including the important ecosystem services that these resources provide.**

The world's forests are being cut, burned, or seriously altered for timber and other products that the global economy requires. Also, rapid population growth and widespread poverty in many developing countries is putting pressure on their forests, as poor people try to eke out a living in them. **Biological diversity**, the number and variety of Earth's organisms, is declining at an alarming rate. Humans are part of Earth's web of life and are entirely dependent on that web, with all its interactions, for our survival. Organisms and their ecosystems provide important **ecosystem services**.

• **Describe the extent of food insecurity, and relate at least two ways to increase food production sustainably.**

Food insecurity is the condition in which people live with chronic hunger and malnutrition. Globally, more than 800 million people lack access to the food needed for healthy, productive lives. Conservation tillage, in which residues from previous crops are left in the soil, partially covering it and helping to hold it in place, is an important aspect of sustainable agriculture. Replacing furrow irrigation, where farmers flow water down small trenches between crop rows, with drip irrigation can reduce water use; in many places, lack of water is the overriding factor reducing agricultural productivity.

• **Define the *enhanced greenhouse effect*, and explain how stabilizing climate is related to energy use.**

The **enhanced greenhouse effect** is the additional warming produced by increased levels of gases that absorb infrared radiation. Both highly developed and developing countries contribute to major increases in CO_2 in the atmosphere, as well as to the increasing amounts of the greenhouse gases methane, nitrous oxide, tropospheric ozone, and CFCs. An increase in atmospheric CO_2, mostly produced when fossil fuels are burned, leads to climate warming and associated climate changes. To stabilize climate, we must phase out fossil fuels in favor of renewable energy, increased energy conservation, and improved energy efficiency, along with nuclear power and carbon sequestration.

• **Describe at least two problems in megacities in the developing world, and relate how these problems could be addressed.**

A **megacity** is a city with more than 10 million inhabitants. The air in megacities in the developing world is badly polluted with the exhaust from motor vehicles, which also congest roads and require large areas for parking. Many cities are developing urban transportation systems to reduce the use of cars and the problems associated with them. Illegal squatter settlements proliferate in megacities; here the poorest inhabitants build dwellings on vacant land using whatever materials they can scavenge. Squatter settlements have the worst water, sewage, and solid waste problems. Cities should develop plans to providing basic services, such as clean water to drink, transportation, and garbage pickup to improve the quality of life in squatter settlements.

• **Explain how consumption overpopulation threatens environmental sustainability.**

Consumption overpopulation is a situation that occurs when each individual in a population consumes too large a share of resources. Many of the world's environmental and resource problems that threaten environmental sustainability stem from consumption overpopulation—the extravagant lifestyles of people living in highly developed nations.

• **Relate how sustainable consumption addresses the problem of consumption overpopulation.**

Sustainable consumption is the use of goods and services that satisfy basic human needs and improve the quality of life but that minimize the use of resources so that they are available for future generations. If people in highly developed countries adopt sustainable consumption, then the problems associated with consumption overpopulation are lessened.

Thinking About the Environment

1. What is environmental sustainability? How are people in highly developed countries not living sustainably? How are people in developing countries not living sustainably?

2. How are the natural environment and sustainable development linked?

3. What are the consumption habits of people in highly developed countries? How is consumption related to human carrying capacity?

4. How pervasive is poverty around the world?

5. What important ecosystem services does biological diversity provide?

6. How pervasive is food insecurity around the world?

7. How is stabilizing climate related to fossil fuel use?

8. At home, place an uncracked egg in a small bowl and cover it with vinegar. Allow it to sit in the vinegar for 24 hours. Take the egg out and examine the shell. How does this simple experiment demonstrate a possible effect of increased atmospheric CO_2 on marine organisms with calcium carbonate shells?

9. Describe three serious problems associated with megacities in the developing world.

10. Explain how consumption overpopulation threatens environmental sustainability.

11. What is sustainable consumption? How can it address the problem of consumption overpopulation?

12. Discuss four specific environmental goals that you hope are achieved in your lifetime.

13. Explain why the global increase in CO_2 is not a simple cause-and-effect relationship with climate warming but instead a cascade of interacting responses that ripple through the Earth system.

Quantitative questions relating to this chapter are on our Web site.

Take a Stand

Visit our Web site at http://www.wiley.com/college/raven (select Chapter 25 from the Table of Contents) for links to more information about actions that contribute to community sustainability. Examine three examples of such actions, and decide with your classmates which one you could implement in your college community. You will find tools to help you organize your research, analyze the data, think critically about the issues, and construct a well-considered plan. *Take a Stand* activities can be done individually or as a team, as oral presentations, written exercises, or Web-based (e-mail) assignments.

Additional online materials relating to this chapter, including a Student Testing section with study aids and self-tests, Environmental News, Activity Links, Environmental Investigations, and more, are also on our Web site.

Review of Basic Chemistry

I

Elements

All matter, living and nonliving, is composed of chemical **elements,** substances that cannot be broken down into simpler substances by chemical reactions. There are 92 naturally occurring elements, ranging from hydrogen (the lightest) to uranium (the heaviest). In addition to the naturally occurring elements, about 20 elements heavier than uranium have been made in laboratories by bombarding elements with subatomic particles.

Instead of writing out the name of each element, chemists use a system of abbreviations called **chemical symbols**—usually the first one or two letters of the English or Latin name of the element. The symbol stands for one atom of the element. For example, O is the symbol for one atom of oxygen, C for one atom of carbon, Cl for one atom of chlorine, and Na for one atom of sodium (its Latin name is *natrium*). Note that the first letter of a chemical symbol is capitalized, and the second letter (if there is one) is not (Table A.1).

Table A.1 Symbols for Some Elements That Are Important in Environmental Science

Element	Symbol	Element	Symbol
Aluminum	Al	Lithium	Li
Bromine	Br	Magnesium	Mg
Calcium	Ca	Mercury	Hg
Carbon	C	Nitrogen	N
Chlorine	Cl	Oxygen	O
Copper	Cu	Phosphorus	P
Fluorine	F	Plutonium	Pu
Helium	He	Potassium	K
Hydrogen	H	Sodium	Na
Iodine	I	Sulfur	S
Iron	Fe	Uranium	U
Lead	Pb	Zinc	Zn

Atoms The **atom** is the smallest subdivision of an element that retains the characteristic chemical properties of that element. Atoms are unimaginably small, much smaller than the tiniest particle visible under a light microscope.

An atom is composed of smaller components called **subatomic particles**—protons, neutrons, and electrons (Table A.2). **Protons** have a positive electrical charge; **neutrons** are uncharged particles with about the same mass as protons. Protons and neutrons make up almost all the mass of an atom and are concentrated in the atomic nucleus. **Electrons** have a negative electrical charge and an extremely small mass (only about 1/1800 of the mass of a proton). The electrons, which behave like waves as well as particles, spin in the space surrounding the atomic nucleus.

Table A.2 Some Properties of Subatomic Particles

Subatomic Particle	Electric Charge	Location
Proton	Positive, +1	Nucleus
Neutron	Neutral	Nucleus
Electron	Negative, −1	Outside nucleus

Each kind of element has a fixed number of protons in the atomic nucleus. This number, called the **atomic number**, determines the chemical identity of the atom. The total number of protons plus neutrons in the atomic nucleus is termed the **atomic mass**. For example, the element oxygen has eight protons and eight neutrons in the nucleus; it therefore has an atomic number of 8 and an atomic mass of 16: $^{16}_{8}O$.

When an atom is uncombined, it generally contains the same number of electrons as protons. Some kinds of chemical combinations and certain other circumstances change the number of electrons but chemical reactions do not affect anything in the atomic nucleus. Because electrons and protons have equal but opposite charges, an uncombined atom is electrically neutral.

The Periodic Table Elements are organized in a **periodic table** (Figure A.1). The periodic table provides information about each element's chemical behavior as well as its atomic mass and the structure of its atoms. The box for each element shows the name, atomic number, symbol, and atomic mass.

Element Name
Atomic number
Symbol
Atomic mass

Examine the element magnesium, with an atomic number of 12. The atomic number indicates that magnesium has 12 protons in each atomic nucleus and 12 electrons outside the nucleus. The atomic mass of magnesium, 24.305 atomic mass units (amu), is the average of the masses of its naturally occurring isotopes (discussed shortly). The number of neutrons in the typical magnesium atom is obtained by subtracting the atomic number from the atomic mass: $24 - 12 = 12$ neutrons.

Notice that the elements are arranged on the periodic table in order of increasing atomic number. The chemical properties of the elements are a regularly repeated function of their atomic numbers. For example, the elements in the far right column are called **noble gases**. These elements—helium, neon, argon, krypton, xenon, and radon—have similar chemical properties. All are colorless, odorless gases that form very few compounds with other elements.

Periodic Table of the Elements

Atomic masses are based on carbon-12. Elements marked with † have no stable isotopes. The atomic mass given is that of the isotope with the longest known half-life.

Atomic number → 11
Symbol → **Na**
Name → Sodium
Atomic mass → 22.99

Transition Elements

Noble Gases

Group

Period

Group 1 1A	2 2A		3 3B	4 4B	5 5B	6 6B	7 7B	8 8B	9 8B	10	11 1B	12 2B	13 3A	14 4A	15 5A	16 6A	17 7A	18 8A

Period 1:
1 **H** Hydrogen 1.008 ; 2 **He** Helium 4.003

Period 2:
3 **Li** Lithium 6.941 ; 4 **Be** Beryllium 9.012 ; 5 **B** Boron 10.81 ; 6 **C** Carbon 12.01 ; 7 **N** Nitrogen 14.01 ; 8 **O** Oxygen 16.00 ; 9 **F** Fluorine 19.00 ; 10 **Ne** Neon 20.18

Period 3:
11 **Na** Sodium 22.99 ; 12 **Mg** Magnesium 24.31 ; 13 **Al** Aluminum 26.98 ; 14 **Si** Silicon 28.09 ; 15 **P** Phosphorus 30.97 ; 16 **S** Sulfur 32.07 ; 17 **Cl** Chlorine 35.45 ; 18 **Ar** Argon 39.95

Period 4:
19 **K** Potassium 39.10 ; 20 **Ca** Calcium 40.08 ; 21 **Sc** Scandium 44.96 ; 22 **Ti** Titanium 47.87 ; 23 **V** Vanadium 50.94 ; 24 **Cr** Chromium 52.00 ; 25 **Mn** Manganese 54.94 ; 26 **Fe** Iron 55.85 ; 27 **Co** Cobalt 58.93 ; 28 **Ni** Nickel 58.69 ; 29 **Cu** Copper 63.55 ; 30 **Zn** Zinc 65.39 ; 31 **Ga** Gallium 69.72 ; 32 **Ge** Germanium 72.61 ; 33 **As** Arsenic 74.92 ; 34 **Se** Selenium 78.96 ; 35 **Br** Bromine 79.90 ; 36 **Kr** Krypton 83.80

Period 5:
37 **Rb** Rubidium 85.47 ; 38 **Sr** Strontium 87.62 ; 39 **Y** Yttrium 88.91 ; 40 **Zr** Zirconium 91.22 ; 41 **Nb** Niobium 92.91 ; 42 **Mo** Molybdenum 95.94 ; 43 **Tc** Technetium 98.00† ; 44 **Ru** Ruthenium 101.1 ; 45 **Rh** Rhodium 102.9 ; 46 **Pd** Palladium 106.4 ; 47 **Ag** Silver 107.9 ; 48 **Cd** Cadmium 112.4 ; 49 **In** Indium 114.8 ; 50 **Sn** Tin 118.7 ; 51 **Sb** Antimony 121.8 ; 52 **Te** Tellurium 127.6 ; 53 **I** Iodine 126.9 ; 54 **Xe** Xenon 131.3

Period 6:
55 **Cs** Cesium 132.9 ; 56 **Ba** Barium 137.3 ; 57 **La** Lanthanum 138.9 * ; 72 **Hf** Hafnium 178.5 ; 73 **Ta** Tantalum 180.9 ; 74 **W** Tungsten 183.8 ; 75 **Re** Rhenium 186.2 ; 76 **Os** Osmium 190.2 ; 77 **Ir** Iridium 192.2 ; 78 **Pt** Platinum 195.1 ; 79 **Au** Gold 197.0 ; 80 **Hg** Mercury 200.6 ; 81 **Tl** Thallium 204.4 ; 82 **Pb** Lead 207.2 ; 83 **Bi** Bismuth 209.0 ; 84 **Po** Polonium 209† ; 85 **At** Astatine 210† ; 86 **Rn** Radon 222†

Period 7:
87 **Fr** Francium 223† ; 88 **Ra** Radium 226† ; 89 **Ac** Actinium 227† ** ; 104 **Rf** Rutherfordium 261† ; 105 **Db** Dubnium 262† ; 106 **Sg** Seaborgium 266† ; 107 **Bh** Bohrium 264† ; 108 **Hs** Hassium 277† ; 109 **Mt** Meitnerium 268† ; 110 **Ds** Darmstadtium 271† ; 111 **Rg** Roentgenium 272†

Inner Transition Elements

Lanthanide Series * 6

58 **Ce** Cerium 140.1 ; 59 **Pr** Praseodymium 140.9 ; 60 **Nd** Neodymium 144.2 ; 61 **Pm** Promethium 145† ; 62 **Sm** Samarium 150.4 ; 63 **Eu** Europium 152.0 ; 64 **Gd** Gadolinium 157.3 ; 65 **Tb** Terbium 158.9 ; 66 **Dy** Dysprosium 162.5 ; 67 **Ho** Holmium 164.9 ; 68 **Er** Erbium 167.3 ; 69 **Tm** Thulium 168.9 ; 70 **Yb** Ytterbium 173.0 ; 71 **Lu** Lutetium 175.0

Actinide Series ** 7

90 **Th** Thorium 232.0 ; 91 **Pa** Protactinium 231.0 ; 92 **U** Uranium 238.0 ; 93 **Np** Neptunium 237 ; 94 **Pu** Plutonium 244† ; 95 **Am** Americium 243† ; 96 **Cm** Curium 247† ; 97 **Bk** Berkelium 247† ; 98 **Cf** Californium 251† ; 99 **Es** Einsteinium 252† ; 100 **Fm** Fermium 257† ; 101 **Md** Mendelevium 258† ; 102 **No** Nobelium 259† ; 103 **Lr** Lawrencium 262†

Figure A.1   The periodic table of the elements.

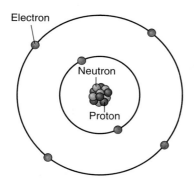

Figure A.2 **A carbon atom.** In this drawing, known as a Bohr model, electrons move in orbitals that correspond to energy levels. Although Bohr models do not depict electronic configurations accurately (electrons do not actually circle the nucleus in fixed concentric pathways), we use Bohr models because of their simplicity.

The Electronic Configuration of an Atom   The way electrons are arranged around an atomic nucleus is referred to as the atom's **electronic configuration**. Knowing the approximate locations of electrons enables chemists to predict how atoms can combine to form different types of chemical compounds.

An atom may have several **energy levels**, or **electron shells**, where electrons are located. The lowest energy level is the one closest to the nucleus (Figure A.2). Only two electrons can occupy this energy level. The second energy level can accommodate a maximum of eight electrons. Although the third and outer shells can each contain more than eight electrons, they are most stable when only eight are present. We may consider the first shell complete when it contains two electrons and every other shell complete when it contains eight electrons.

The energy levels correspond roughly to physical locations of electrons, called **orbitals**. There may be several orbitals within a given energy level. Electrons are thought to whirl around the nucleus in an unpredictable manner, now close to it, now farther away. Orbitals represent the places where electrons are most probably found.

Isotopes   **Isotopes** are atoms of the same element that contain the same number of protons but different numbers of neutrons. Isotopes, therefore, have the same atomic number but different atomic mass numbers. The three isotopes of hydrogen contain zero, one, and two neutrons, respectively. Elements usually occur in nature as mixtures of isotopes (Table A.3).

All isotopes of a given element have essentially the same chemical characteristics. Some isotopes with excess neutrons are unstable and tend to break down, or decay, to a more stable iso-

Table A.3 **Mixture of Isotopes in Naturally Occurring Oxygen**

Isotope	Percentage of Atoms
$^{16}_{8}O$	99.759
$^{17}_{8}O$	0.037
$^{18}_{8}O$	0.204

Figure A.3   **The water molecule.** Each molecule of water consists of two hydrogen atoms bonded to one oxygen atom.

tope (usually of a different element). Such isotopes are termed **radioisotopes** because they emit high-energy radiation when they decay (see Chapter 12).

Molecules   Two or more atoms may combine chemically to form a **molecule**. When two atoms of oxygen combine, for example, a molecule of oxygen is formed. Different kinds of atoms can combine to form **chemical compounds**. A chemical compound is a substance that consists of two or more different elements combined in a fixed ratio. Water is a chemical compound in which each molecule consists of two atoms of hydrogen combined with one atom of oxygen (Figure A.3).

Chemical Bonds   The number and arrangement of electrons in the *outermost* energy level (electron shell) primarily determine the chemical properties of an element. In a few elements (the noble gases), the outermost shell is filled. These elements are chemically inert, meaning that they will not readily combine with other elements. The electrons in the outermost energy level of an atom are referred to as **valence electrons**. The valence electrons are chiefly responsible for the chemical activity of an atom. When the valence (outer) shell of an atom contains fewer than eight electrons (the stable number for the valence shell of most atoms), the atom tends to lose, gain, or share electrons to achieve an outer shell of eight. (The valence shells of the lightest elements, hydrogen and helium, are full when they contain two electrons.)

The elements in a compound are always present in a certain proportion. This reflects the fact that atoms are attached to each other by chemical bonds in a precise way to form a compound. A **chemical bond** is the attractive force that holds two atoms together. Each bond represents a certain amount of potential chemical energy. The atoms of each element form a specific number of bonds with the atoms of other elements—a number dictated by the number of valence electrons (Figure A.4).

Ions   Some atoms have the ability to gain or lose electrons. Because the number of protons in the nucleus remains unchanged, the loss or gain of electrons produces an atom with a net positive or negative charge. Such electrically charged atoms are termed **ions** (Figure A.5).

Chemical Formulas

A **chemical formula** is a shorthand method for describing the chemical composition of a molecule. Chemical symbols are used to indicate the types of atoms in the molecule, and subscript numbers are used to indicate the number of each atom present.

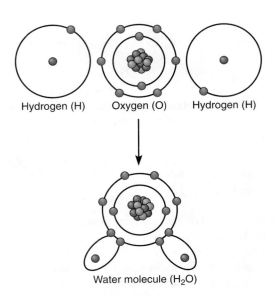

Figure A.4 **Electronic configuration of a water molecule.** A molecule of water is formed when two hydrogen atoms share their valence electrons with the valence electrons of oxygen.

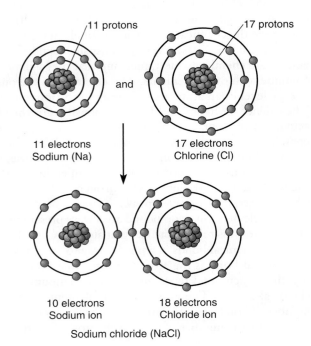

Figure A.5 **Formation of sodium and chloride ions.** Ions gain or lose one or more valence electrons. A sodium atom becomes a positive ion when it donates its single valence electron to chlorine, which has seven valence electrons. With this additional electron, the chlorine atom becomes a negative chloride ion. The attraction between a positively charged sodium ion and a negatively charged chloride ion produces the chemical compound sodium chloride.

The chemical formula for molecular oxygen, O_2, tells us that each molecule consists of two atoms of oxygen. This formula distinguishes it from another form of oxygen, ozone, which has three oxygen atoms and is written O_3. The chemical formula

for water, H_2O, indicates that each molecule consists of two atoms of hydrogen and one atom of oxygen. Note that when a single atom of one type is present it is not necessary to write 1; it is not necessary to write H_2O_1.

Chemical Equations

The chemical reactions that occur between atoms and molecules—for example, between methane and oxygen—can be described on paper by means of **chemical equations**.

Reactant A		Reactant B		Product C		Product D
CH_4 Methane	+	$2O_2$ Oxygen	→	CO_2 Carbon Dioxide	+	$2H_2O$ Water

Methane is broken down in this reaction.

In a chemical reaction, the **reactants** (the substances that participate in the reaction) are written on the left side of the equation and the **products** (the substances formed by the reaction) are written on the right side. The arrow means *yields* and indicates the direction in which the reaction tends to proceed. The number preceding a chemical symbol or formula indicates the number of atoms or molecules reacting. Thus, $2O_2$ means two molecules of oxygen. The absence of a number indicates that only one atom or molecule is present. Thus, the above equation can be translated into ordinary language as "One molecule of methane reacts with two molecules of oxygen to yield one molecule of carbon dioxide and two molecules of water."

Acids and Bases

An **acid** is a compound that ionizes in solution to yield hydrogen ions (H^+)—that is, protons—and negatively charged ions. Acids turn blue litmus paper red and have a sour taste. Hydrochloric acid (HCl) and sulfuric acid (H_2SO_4) are examples of acids.

HCl Hydrochloric acid	$\xrightarrow{\text{in water}}$	H^+ Hydrogen ion	+	Cl^- Chloride ion

The strength of an acid depends on the degree to which it ionizes in water, releasing hydrogen ions. Thus, HCl is a very strong acid because most of its molecules dissociate, producing hydrogen and chloride ions.

Most bases are substances that yield hydroxide ions (OH^-) and positively charged ions when dissolved in water. Bases turn red litmus paper blue. Sodium hydroxide (NaOH) and aqueous ammonia (NH_4OH) are examples of bases.

NaOH Sodium hydroxide	$\xrightarrow{\text{in water}}$	Na^+ Sodium ion	+	OH^- Hydroxide ion

Bases react with hydrogen ions and remove them from solution.

pH

Because the concentration of hydrogen or hydroxide ions is usually small, it is convenient to express the degree of acidity or basicity in a solution in terms of **pH**, formally defined as the negative logarithm of the hydrogen ion concentration. The pH scale is logarithmic, extending from 0, which is the pH of a very strong acid, to 14, which is the pH of a very strong base. The pH of pure water is 7, neither acidic nor basic, but neutral. Even though water does ionize slightly, the concentrations of H^+ ions and OH^- ions are exactly equal; each of them has a concentration of 10^{-7}, which is why we say that water has a pH of 7. Solutions with pH of *less* than 7 are acidic and contain more H^+ ions than OH^- ions. Solutions with a pH *greater* than 7 are basic and contain more OH^- ions than H^+ ions.

Because the scale is logarithmic to base 10, a solution with a pH of 6 has a hydrogen ion concentration that is 10 times greater than that of a solution with a pH of 7, and is much more acidic. A pH of 5 represents another tenfold increase. Therefore, a solution with a pH of 4 is 10×10, or 100 times more acidic than a solution with a pH of 6.

The contents of most animal and plant cells are neither strongly acidic nor basic but are an essentially neutral mixture of acidic and basic substances. Most life cannot exist if the pH of the cell changes very much.

Graphing

II

Much of the study of environmental science involves learning about relationships between variables. For instance, there is a definite relationship between the amount of pollution discharged and the cost of the environmental damage caused by the pollution. Often such relationships can be expressed and understood through graphs.

A graph is a diagram that expresses a relationship between two or more quantities. In some cases there is a definite cause-and-effect relationship, whereas in others the association is not as direct. Graphic presentation of data may not explain the reason for the relationship (as, for example, the worldwide increase in fertilizer use since 1960), but the shape of it can provide clues. A graph puts abstract ideas or experimental data into visual form so that their relationships become more apparent.

Variables

The related quantities displayed on a graph are called **variables**. The simplest sort of graph uses a system of coordinates or axes to represent the values of the variables. Usually the relative size of a variable is represented by its position along the axis. Numbers along the axis allow the reader to estimate the values.

If the relationship being plotted is one of cause and effect, the variable that expresses the cause is called the **independent variable**. Usually this is represented by the horizontal axis, which is called the x-axis. The variable that changes as a result of changes in the independent variables is the **dependent variable**. It is usually represented on the vertical axis, which is called the y-axis. The two axes are arranged at right angles to each other and cross at a point called the origin (**Figure A.6**).

To show the relationship between two variables that are directly related at some specific value, such as point A in Figure A.6, the value on the x-axis (x_1) is extended vertically, and the corresponding value on the y-axis (y_1) is extended horizontally. The relationship between the two variables determines the point A at which these lines cross.

If another pair of points (x_2 and y_2) is chosen, their point of intersection on the graph can also be plotted; this is point B. A line drawn between points A and B can then give information about how all other x- and y-values on this graph should relate to each other.

Types of Relationships

As you may have guessed, this explanation represents a very simple case in which some important assumptions were made. We first assumed that for every x-value there was only one y-value. We further assumed that all of the y-variables were

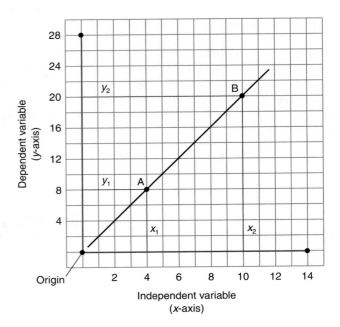

Figure A.6 Typical line graph. Two numbers specify the position of A; these are its x and y coordinates. The x coordinate, 4, is the foot of the perpendicular from A to the x-axis (see x_1). The y coordinate, 8, is the foot of the perpendicular from A to the y-axis (see y_1). Point B is obtained by extending the corresponding values on the x- and y-axes until they intersect, at $x = 10$, $y = 20$.

directly related to all of the x-variables. This is the simplest kind of relationship that a graph can represent. It is called a direct relationship: The y-values get larger as the x-values get larger. An example of this type of graph is shown in **Figure A.7**.

Inverse relationships are also common. In inverse relationships, the y-values get smaller as the x-values get larger (**Figure A.8**). Most relationships found in environmental science are not simple. Over some ranges, a relationship may be direct or inverse, and then it may change as a wider range of variables is considered.

Take a few moments to flip through the pages of this book. You will see many graphs. Some express simple relationships over their entire range of data, whereas others are more complicated, expressing several relationships at once. In some cases there are several lines on the graph, each describing some aspect of the idea being presented. Some data are presented as bar graphs or pie charts instead of lines to illustrate relationships (**Figures A.9 and A.10**). Whatever their form, all these graphs are designed to present important relationships in the clearest possible way. When you learn to interpret information presented graphically, you are well on your way to understanding environmental science.

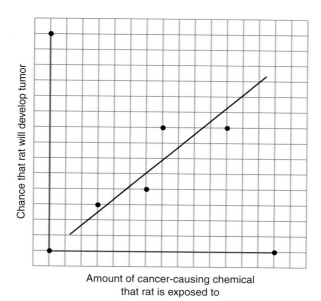

Figure A.7 Direct relationship. In a direct relationship, *y* values increase as *x* values increase, producing an upward-sloping straight line. In science, data points often do not fall exactly on a straight line but are scattered about an ideal line, which is determined mathematically and is drawn to show the general relationship.

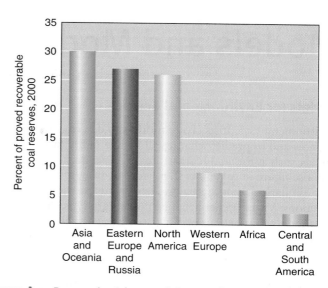

Figure A.9 Bar graph. A bar graph has parallel bars to represent the values in a given set of data. Bar graphs are appropriate for comparing discrete values, as, for example, when comparing the percentage of proved recoverable coal reserves in different parts of the world in 2000.

It is important to keep in mind that how one designs a graph can influence how it is interpreted. For example, in Figure A.7, neither axis is given a scale. One possibility is that it runs from 0 to 100%, in which case there is a very big change with increasing amount. Another possibility is that it runs from 40% to 41%. The graphs would still look the same, but the change is much smaller (indeed, probably not even statistically significant).

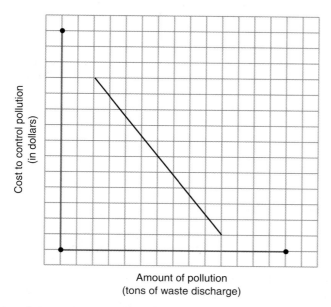

Figure A.8 Inverse relationship. In an inverse relationship, *y* values decrease as *x* values increase, producing a downward-sloping line or curve.

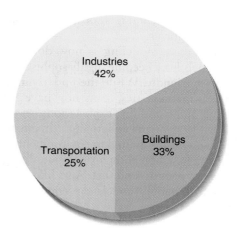

Figure A.10 Pie chart. A pie chart, which is always shown as a circle, is a graphical way to show percentages as a whole, such as major energy consumers in the United States. The size of each section of the pie chart is proportional to its share of the whole.

Models and Modeling

III

Models for Environmental Science

Environmental scientists create **models** to describe the world as it is now and to predict what it will be like in the future. Visual models, such as the atomic structures in Appendix I and the dose-response figure in Appendix II, rely on pictures, graphs, or symbols. Similarly, in conceptual models, words provide qualitative depictions. Visual and conceptual models are typically **descriptive models** that help us organize our thoughts about complex systems, but are not particularly useful for making predictions.

Physical models expand two-dimensional visual models to three dimensions. A globe is a familiar physical descriptive model that can include topographical features (mountains, the ocean floor). However, physical models can also serve as **predictive models**, which we use to run experiments at a small scale to evaluate what will happen on a larger scale. For example, when designing a wetland for wastewater control (see Chapter 22 Introduction), contractors might build a small scale model to experiment with different water flow options.

Mathematical models are particularly useful for environmental scientists. Equations or sets of equations are used to describe the relationships between two or more parameters (model parts). One simple mathematical model is the linear relationship (see Figure A.7). This takes the form Cancer = C × Amount, where C is a constant value that relates cancer rates and amount of chemical (also known as a cancer potency factor; see Chapter 7). In the linear case, doubling the amount doubles the rate of cancer. A common mathematical relationship found in nature is exponential growth (or decay). Compound interest and uncontrolled population increase are examples of exponential growth (see Chapter 8). In **systems**, mathematical relationships are more complex, since they typically involve feedbacks that reduce or amplify change, and relationships can take linear, exponential, and other forms.

Much predictive modeling is now done on computers. Computer models have become highly sophisticated and, at the same time, user friendly. While the underlying basis of computer models is mathematics, many now have graphical interfaces as well as audio and text explanations, so that users need not know or understand the underlying mathematics. Computer models are by no means limited to environmental issues—most video games are based on complex computer programs hidden behind graphics, sounds, and text. Also like computer games, inputs to environmental models can come from a keyboard, mouse, or even joystick.

Geographic Information Systems and System Dynamics Models

Two types of models commonly used by environmental scientists are Geographic Information Systems and System Dynamics models. **Geographic Information Systems (GIS)** are computer-based maps that contain site- or area-specific information. Early GIS maps were physical. For example, a sheet of clear plastic depicting population density in different areas could be superimposed on another sheet depicting the number of endangered species in those areas. Computer-based GIS allows modelers to combine many such virtual sheets and understand the relationships between important **variables**.

Anyone who has played the Sims™ has run a **System Dynamics model**. In the Sims™, a player varies the parameters of a model (for example, the amount of time a character spends at work). The underlying program then combines that parameter with other parameters using mathematical equations that describe relationships. For example, a character who spends less time at work earns less money but may spend more time with friends. Users don't see the mathematical equations; they see words and pictures depicting the results of those equations.

Systems models in environmental science follow the same rules. Equations describe the "dynamic" aspects of the system, or how different parts of the model interact. The user determines initial conditions, and then runs the model. For example, a simple model of a predator–prey relationship may be based on the idea that one coyote needs to eat a certain number of rabbits each week to stay alive. Coupled to birth rates and death rates of rabbits and coyotes, the model can predict possible stable population sizes for coyotes and rabbits. However, coyotes and rabbits live in an environment that includes other organisms as well as physical features. A more complex and realistic model that accounts for dynamic interactions within the system would incorporate other **limiting factors** and **feedbacks**. Once an adequate model is built, users can experiment with the effects of a change in some parameter or parameters. For example, rabbits could be excluded from their historical feeding sites, or coyote birth rates could decline due to some environmental contaminant. A picture can provide a static, descriptive model of this system, while a computer model is needed to understand the dynamic relationships among the parts of the model.

Modeling Challenges

Globes are models that lack the distorting effect of flat maps of the world, and thus are more accurate. In general, we prefer more accurate models to less accurate models, but there are significant tradeoffs. Globes are more expensive to produce than are paper maps. This leads to an important characteristic of models: They are necessarily simplified versions of the world, and as such they are never "right."

Adding complexity to models can improve accuracy, leading to more meaningful predictions. However, more complex models require more time, money, and expertise to develop. More complex models can also be more sensitive to errors and misspecification. Consequently, when building models, environmental scientists are driven by a need for accuracy and completeness, but constrained by usability, technology, cost, and time.

Global circulation models (see Chapter 21) are among the most complex models ever developed. They predict the effects of **greenhouse gases** on the climatic system over decades and even centuries. They include energy and carbon flows, air and ocean movements, and many positive and negative feedback loops. Unfortunately, this means that there are many opportunities for errors in programming or data input, and each run of the model takes an enormous amount of computing time and power.

Units of Measure: Some Useful Conversions

SOME COMMON PREFIXES

Prefix and Symbol	Meaning	Example
Giga- (G)	Billion	1 gigaton = 1,000,000,000 tons
Mega- (M)	Million	1 megawatt = 1,000,000 watts
Kilo- (K)	Thousand	1 kilojule = 1000 joules
Centi- (c)	Hundredth	1 centimeter = 0.01 meter
Milli- (m)	Thousandth	1 milliliter = 0.001 liter
Micro- (μ)	Millionth	1 micrometer = 0.000001 meter
Nano- (n)	Billionth	1 nanometer = 0.000000001 meter
Pico- (p)	Trillionth	1 picocurie = 0.000000000001 curie

LENGTH   Standard Unit = Meter

1 meter (m) = 39.37 in. = 3.28 ft
1 in. = 2.54 cm
1 km = 0.621 mi
1 mi = 1.609 km = 1609 m
1 nautical mile = 1.15 mi = 1.85 km

VOLUME   Standard Unit = Liter

1 liter (L) = 1,000 cm^3 = 1.057 qt (U.S.)
1 gallon (U.S.) = 3.785 L
1 mi^3 = 4.166 km^3
1 acre-foot = 1,233.5 m^3

ENERGY   Standard Unit = Joule

1 joule (J) = 0.24 cal
1 calorie = 4.184 J
1 Calorie = 1000 calories = 1 kcal
1 kilocalorie (kcal) = 4.184 kJ
1 British thermal unit (BTU) = 252 cal
1 kilowatt-hour (kWh) = 3,600,000 J

PRESSURE   Standard Unit = Pascal

1 bar = 10^5 Pa
1 atm = 1.01 bar = 1.01×10^5 Pa
1 millibar = 0.0145 lb/in^2

AREA   Standard Unit = Square Meter (m^2)

1 hectare = 10,000 m^2 = 0.01 km^2 = 2.471 acres
1 acre = 0.405 hectare
1 km^2 = 100 hectares = 0.386 mi^2
1 mi^2 = 640 acres = 259 hectares = 2.59 km^2

MASS   Standard Unit = Kilogram

1 kilogram (kg) = 2.205 lb = 35.3 oz
1 ton = 2000 lb
1 metric ton = 1000 kg = 1.103 ton = 2204.6 lb
1 short ton = 907 kg
1 lb = 453.6 g

ELECTRICAL POWER   Standard Unit = Watt

1 watt (W) = 1 J/second

TEMPERATURE   Standard Unit = Celsius

$°C = (°F - 32) \times \frac{5}{9}$
$°F = °C \times \frac{9}{5} + 32$
$1°C = 1.8°F$

Abbreviations, Formulas, and Acronyms Used in This Text

AID U.S.	Agency for International Development
AIDS	acquired immune deficiency syndrome
ALSC	Adirondack Lakes Survey Corporation
ANWR	Arctic National Wildlife Refuge
ARS USDA	Agricultural Research Service
ATP	adenosine triphosphate
B.C.E.	Before Common Era
BLM	Bureau of Land Management
BMI	Body Mass Index
BOD	biochemical/biological oxygen demand
BRD	Biological Resources Discipline
BST	bacterial source tracking
Bt	*Bacillus thuringiensis*
BTU	British thermal unit
C_6H_6	benzene
CARE	Cooperative for American Relief Everywhere
CDC	Centers for Disease Control and Prevention
CERCLA	Comprehensive Environmental Response, Compensation, and Liability Act
CFCs	chlorofluorocarbons
CH_4	methane
CHP	combined heat and power (cogeneration)
CITES	Convention on International Trade in Endangered Species of Wild Flora and Fauna
CO	carbon monoxide
CO_2	carbon dioxide
CO_3^{2-}	carbonate
CRP	Conservation Reserve Program
DAD	decide, announce, defend
DDT	dichlorodiphenyltrichloroethane
DMS	dimethyl sulfide
DNA	deoxyribonucleic acid
DOD U.S.	Department of Defense
DOE U.S.	Department of Energy
ED_{50}	effective dose-50%
EDB	ethylene dibromide
EFI	European Forest Institute
EIS	environmental impact statement
ENSO	El Niño–Southern Oscillation
EPA U.S.	Environmental Protection Agency
EPI	Environmental Performance Index
ERNS EPA	Emergency Response Notification System
ESA	Endangered Species Act
EU	European Union
FAO	U.N. Food and Agriculture Organization
FDA	U.S. Food and Drug Administration
FDCA	Food, Drug, and Cosmetics Act
FIFRA	Federal Insecticide, Fungicide, and Rodenticide Act
FWS U.S.	Fish and Wildlife Service
GDP	gross domestic product
GFDL	U.S. Geophysical Fluid Dynamics Laboratory
GHP	geothermal heat pump
GIS	Geographic Information Systems
GM	genetically modified
GNP	gross national product
GPP	gross primary productivity
GWe	gigawatt (electric)
HAP	hazardous air pollutants
HCFCs	hydrochlorofluorocarbons
HCPs	habitat conservation plans
HCO_3^-	bicarbonate
HDL	high-density lipoproteins
HFCs	hydrofluorocarbons
HIV	human immune deficiency virus
HNO_2	nitrous acid
HNO_3	nitric acid
H_2S	hydrogen sulfide
H_2SO_4	sulfuric acid
IAEA	International Atomic Energy Agency
ICPR	International Commission for Protection of the Rhine
IIASA	International Institute for Applied Systems Analysis
IMF	International Monetary Fund
IPCC U.N.	Intergovernmental Panel on Climate Change
IPM	integrated pest management
IRCA	Immigration Reform and Control Act
IRRI	International Rice Research Institute
IUCN	World Conservation Union (formerly International Union for the Conservation of Nature and Natural Resources)
K	carrying capacity
kcal	kilocalorie

kJ	kilojoule
kWh	kilowatt-hour
LD_{50}	lethal dose-50%
LDC	less developed country
LTER	long-term ecological research
MDG	Millennium Development Goal
medfly	Mediterranean fruit fly
MIC	methyl isocyanate
MOX	mixed oxide fuel
mpg	miles per gallon
mph	miles per hour
MTBE	methyl tertiary butyl ether
MVP	Millennium Villages Project
MWe	Megawatt (electric)
NASA	National Aeronautics and Space Administration
NDP	net domestic product
NGO	Nongovernmental Organization
NEPA	National Environmental Policy Act
NH_3	ammonia
NIMBY	not in my backyard
NIMTOO	not in my term of office
NMFS	National Marine Fisheries Service
NNP	net national product
NO	nitric oxide
NO_x	collective name for nitrogen oxides
N_2O	nitrous oxide
NO_2	nitrogen dioxide
NO_2^-	nitrite
NO_3^-	nitrate
NOAA	National Oceanic and Atmospheric Administration
NPP	net primary productivity
NPS	National Park Service
NRC	National Research Council
NRC	National Response Center
NRC	Nuclear Regulatory Commission
NRCS	National Resources Conservation Service (formerly Soil Conservation Service)
NWS	National Weather Service
O_3	ozone
OH^-	hydroxide ion
OPEC	Organization of Petroleum Exporting Countries
OTEC	ocean thermal energy conversion
PANs	peroxyacyl nitrates
PBMR	pebble-bed modular reactor
PCBs	polychlorinated biphenyls
perc	perchloroethylene
PET	polyethylene terphthalate
PM 2.5	particulate matter less than 2.5 micrometers in diameter
PM 10	particulate matter less than 10 micrometers in diameter
PO_4^{3-}	phosphate
POET	four variables in urban centers (population, organization, environment, and technology)
POPs	persistent organic pollutants
ppb	parts per billion
PPP	purchasing power parity
ppm	parts per million
PV	photovoltaic
r	growth rate
RCRA	Resource Conservation and Recovery Act
SARS	severe acute respiratory syndrome
SCAQMD	South Coast Air Quality Management District
SMCRA	Surface Mining Control and Reclamation Act
SO_2	sulfur dioxide
SO_3	sulfur trioxide
SO_4^{2-}	sulfate
SO_x	collective name for sulfur oxides
STDs	sexually transmitted diseases
SUV	sports utility vehicle
TGD	Three Gorges Dam
TCDD	tetrachlorodibenzodioxin (a dioxin)
TCE	trichloroethylene
TNT	trinitrotoluene
2,4-D	2,4-dichlorophenoxyacetic acid
2,4,5-T	2,4,5-trichlorophenoxyacetic acid
U.N.	United Nations
UNAIDS	U.N. Program on AIDS
UNCLOS	U.N. Convention on the Law of the Sea
UNFCCC	U.N. Framework Convention on Climate Change
USAID	U.S. Agency for International Development
USDA	U.S. Department of Agriculture
USFS	U.S. Forest Service
UV	ultraviolet
VOC	volatile organic compound (or chemical)
W/m^2	Watts per square meter
WHO	World Health Organization
WRP	Wetlands Reserve Program
WWF	World Wide Fund for Nature (also known as World Wildlife Fund in the United States, Canada, and Australia)

Glossary

abiotic   Nonliving. Compare *biotic*.

abyssal benthic zone   The ocean's benthic environment that extends from a depth of 4000 to 6000 meters. Compare *hadal benthic zone*.

acid   A substance that releases hydrogen ions (protons) in water. Acids have a sour taste and turn blue litmus paper red. Compare *base*.

acid deposition   Sulfur dioxide and nitrogen dioxide emissions that react with water vapor in the atmosphere to form acids that return to the surface as either dry or wet deposition. See *acid precipitation*.

acid mine drainage   Pollution caused when sulfuric acid and dangerous dissolved materials such as lead, arsenic, and cadmium wash from coal and metal mines into nearby lakes and streams.

acid precipitation   Precipitation that is acidic as a result of both sulfur and nitrogen oxides forming acids when they react with water in the atmosphere; partially due to the combustion of coal; includes acid rain, acid snow, and acid fog.

acre-foot   The amount of water needed to cover an acre of land one foot deep; equal to 1233 m^3 (326,000 gallons), which is enough to supply eight people for one year.

active solar heating   A system in which a series of collection devices mounted on a roof or in a field are used to absorb solar energy. Pumps or fans distribute the collected heat. Compare *passive solar heating*.

acute toxicity   Adverse effects that occur within a short period after exposure to a toxicant. Compare *chronic toxicity*.

adaptation   An evolutionary modification that improves the chances of survival and reproductive success of a population in a given environment.

adaptive radiation   The evolution of a large number of related species from an unspecialized ancestral organism.

additivity   A phenomenon in which two or more pollutants interact in such a way that their combined effects are what one would expect given the individual effects of each component of the mixture. Compare *antagonism* and *synergism*.

aerobic cellular respiration   The process by which cells use oxygen to break down organic molecules with the release of energy that can be used for biological work. Compare *anaerobic cellular respiration*.

aerosol   Tiny particles of natural and human-produced air pollution that are so small they remain suspended in the atmosphere for days or even weeks.

aerosol effect   Atmospheric cooling that occurs where and when aerosol pollution is the greatest.

age structure   The number and proportion of people at each age in a population.

agroforestry   Concurrent use of forestry and agricultural techniques to improve degraded soil and offer economic benefits.

A-horizon   The topsoil; located just beneath the O-horizon of the soil. The A-horizon is rich in various kinds of decomposing organic matter.

air pollution   Various chemicals (gases, liquids, or solids) present in high enough levels in the atmosphere to harm humans, other animals, plants, or materials. Excess noise and heat are also considered air pollution.

air toxic   See *hazardous air pollutant*.

albedo   The proportional reflectance of Earth's surface; glaciers and ice sheets have high albedos and reflect most of the sunlight hitting their surfaces, whereas the ocean and forests have low albedos.

alcohol fuel   A liquid fuel such as methanol or ethanol that may be used in internal combustion engines; mixing gasoline with 10% ethanol produces a cleaner-burning mixture known as gasohol.

algae   Unicellular or simple multicellular photosynthetic organisms; important producers in aquatic ecosystems.

alpine tundra   A distinctive ecosystem located in the higher elevations of mountains, above the tree line; characteristic vegetation includes grasses, sedges, and small tufted plants. Compare *tundra*.

alternative agriculture   See *sustainable agriculture*.

altitude   The height of a thing above sea level.

amino acids   Organic compounds that are linked together to form proteins. See *essential amino acid*.

ammonification   The conversion of nitrogen-containing organic compounds to ammonia (NH_3) and ammonium ions (NH_4^+) by certain bacteria (ammonifying bacteria) in the soil; part of the *nitrogen cycle*.

anaerobic cellular respiration   The process by which cells break down organic molecules in the absence of oxygen to release energy that can be used for biological work. Compare *aerobic cellular respiration*.

antagonism   A phenomenon in which two or more pollutants interact in such a way that their combined effects are less severe than the sum of their individual effects. Compare *additivity* and *synergism*.

anthracite   The highest grade of coal; has the highest heat content and burns the cleanest of any grade of coal. Also known as *hard coal*. Compare *bituminous coal*, *subbituminous coal*, and *lignite*.

anticline   An upward folding of rock layers, or strata.

aquaculture   The rearing of aquatic organisms (fishes, seaweeds, and shellfish), either freshwater or marine, for human consumption. See *mariculture*.

aquatic   Pertaining to the water. Compare *terrestrial*.

aqueduct   A large pipe or conduit constructed to carry water from a distant source.

aquifer   The underground caverns and porous layers of underground rock in which groundwater is stored. See *confined aquifer* and *unconfined aquifer*.

aquifer depletion   The removal by humans of more groundwater than can be recharged by precipitation or melting snow. Also called *groundwater depletion*.

arctic tundra   See *tundra*.

arid land   A fragile ecosystem in which lack of precipitation limits plant growth. Arid lands are found in both temperate and tropical regions. Also called *desert*. Compare *semi-arid land*.

artesian aquifer   See *confined aquifer*.

artificial eutrophication   Overnourishment of an aquatic ecosystem by nutrients such as nitrates and phosphates. In artificial eutrophication, the pace of eutrophication is rapidly accelerated due to human activities such as agriculture and discharge from sewage treatment plants. Also called *cultural eutrophication*. See *eutrophication*.

artificial insemination   A technique in which sperm collected from a suitable male of a rare species is used to artificially impregnate a female (perhaps located in another zoo).

assimilation   The incorporation of a substance into the cells of an organism.

asthenosphere   The region of the mantle where rocks become hot and soft.

atmosphere   The gaseous envelope surrounding Earth.

atom   The smallest quantity of an element that retains the chemical properties of that element; composed of protons, neutrons, and electrons.

atomic mass   A number that represents the sum of the number of protons and neutrons in the nucleus of an atom. The atomic mass represents the relative mass of an atom. Compare *atomic number*.

atomic number   A number that represents the number of protons in the nucleus of an atom. Each element has its own characteristic atomic number. Compare *atomic mass*.

autotroph   See *producer*.

Baby Boom   The large wave of births that followed World War II, from 1945 to 1962.

background extinction   The continuous, low-level extinction of species that has occurred throughout much of the history of life. Compare *mass extinction*.

bacteria   Unicellular, prokaryotic microorganisms. Most bacteria are decomposers, but some are autotrophs, and some are parasites.

bacterial source tracking   Using molecular biological techniques to identify the source of dangerous bacteria in a stream or other body of water.

base   A compound that releases hydroxide ions (OH^-) when dissolved in water. A base turns red litmus paper blue. Compare *acid*.

benthic environment   The ocean floor, which extends from the intertidal zone to the deep ocean trenches.

benthos   Bottom-dwelling marine organisms that fix themselves to one spot, burrow into the sand, or simply walk about on the ocean floor. Compare *plankton* and *nekton*.

B-horizon   The light-colored, partially weathered soil layer underneath the A-horizon; subsoil. The B-horizon contains much less organic material than the A-horizon.

bioaccumulation   The buildup of a persistent toxic substance, such as certain pesticides, in an organism's body, often in fatty tissues. Also called *bioconcentration*.

biocentric preservationist   A person who believes in protecting nature because all forms of life deserve respect and consideration.

biochemical oxygen demand (BOD)   The amount of oxygen needed by microorganisms to decompose (by aerobic respiration) the organic material in a given volume of water. Also called *biological oxygen demand*.

bioconcentration   See *bioaccumulation*.

biodegradable   Referring to a chemical pollutant capable of being decomposed (broken down) by organisms or by other natural processes. Compare *nondegradable*.

biodiversity   See *biological diversity*.

biodiversity hotspots   Relatively small areas of land that contain an exceptional number of endemic species and are at high risk from human activities.

biogas   A clean fuel, usually composed of a mixture of gases, whose combustion produces fewer pollutants than either coal or biomass. Biogas is produced from the anaerobic digestion of organic materials.

biogeochemical cycle   Process by which matter cycles from the living world to the nonliving physical environment and back again. Examples of biogeochemical cycles include the *carbon cycle, nitrogen cycle, phosphorus cycle,* and *sulfur cycle*.

biological amplification   See *biological magnification*.

biological control   A method of pest control that involves the use of naturally occurring disease organisms, parasites, or predators to control pests. Also called *biological pest control*.

biological diversity   The number, variety, and variability of Earth's organisms; consists of three components: genetic diversity, species richness, and ecosystem diversity. Also called *biodiversity*.

biological magnification   The increased concentration of toxic chemicals such as PCBs, heavy metals, and certain pesticides in the tissues of organisms at higher trophic levels in food webs. Also called *biological amplification*.

biological oxygen demand (BOD)   See *biochemical oxygen demand*.

biological pest control   See *biological control*.

biomass   (1) A quantitative estimate of the total mass, or amount, of living material. Often expressed as the dry weight of all the organic material that comprises organisms in a particular ecosystem. (2) Plant and animal materials used as fuel.

biome   A large, relatively distinct terrestrial region characterized by a similar climate, soil, plants, and animals, regardless of where it occurs on Earth; because it is so large in area, a biome encompasses many interacting ecosystems.

bio-prospecting   Investigation of organisms in a forest, coral reef, or other natural area as potential sources of chemical products, including drugs, flavorings, fragrances, and natural pesticides.

bioremediation   A method employed to clean up a hazardous waste site that uses microorganisms to break down the toxic pollutants. Compare *phytoremediation*.

biosphere   The parts of Earth's atmosphere, ocean, land surface, and soil that contain all living organisms.

biotic   Living. Compare *abiotic*.

biotic potential   See *intrinsic rate of increase*.

birth rate   The number of births per 1000 people per year. Also called *crude birth rate* and *natality*.

bitumen   See *tar sand*.

bituminous coal   The most common form of coal; produces a high amount of heat and is used extensively by electric power plants. Also called *soft coal*. Compare *anthracite, subbituminous coal,* and *lignite*.

bloom   Large algal population caused by the sudden presence of large amounts of essential nutrients (such as nitrates and phosphates) in surface waters.

boreal forest   A region of coniferous forest (such as pine, spruce, and fir) in the Northern Hemisphere; located just south of the tundra; also called *taiga*.

botanical   A plant-derived chemical used as a pesticide. See *synthetic botanical*.

bottom ash   The residual ash left at the bottom of an incinerator when combustion is completed. Also called *slag*. Compare *fly ash*.

breeder nuclear fission   A type of nuclear fission in which nonfissionable U-238 is converted to fissionable Pu-239.

broad leaf herbicide   An herbicide that kills plants with broad leaves but does not kill grasses (such as corn, wheat, and rice).

broad-spectrum pesticide   A pesticide that kills a variety of organisms in addition to the pest against which it is used. Most pesticides are broad-spectrum pesticides. Also called *wide-spectrum pesticide*. Compare *narrow-spectrum pesticide*.

brownfield   An urban area of abandoned, vacant factories, warehouses, and residential sites that may be contaminated from past uses.

bycatch   Unwanted fishes, dolphins, and sea turtles that are caught along with commercially valuable fishes and then dumped, dead or dying, back into the ocean.

calendar spraying   The regular use of pesticides regardless of whether pests are a problem or not. Compare *scout-and-spray*.

calorie   A unit of heat (thermal energy); the amount of heat required to raise 1 g of water 1°C.

cancer   A malignant tumor anywhere in the body. Cancers tend to spread throughout the body.

cancer potency   An estimate of the expected increase in cancer associated with a unit increase in exposure to a chemical.

carbamates   A class of broad-spectrum pesticides that are derived from carbamic acid.

carbohydrate   An organic compound containing carbon, hydrogen, and oxygen in the ratio of 1C:2H:1O. Carbohydrates include sugars and starches, molecules metabolized readily by the body as a source of energy.

carbon cycle   The worldwide circulation of carbon from the abiotic environment into organisms and back into the abiotic environment.

carbon management   Ways to separate and capture the carbon dioxide produced during the combustion of fossil fuels and then sequester it away from the atmosphere.

carbon sequestration   Placing carbon that has been produced when generating usable energy from fossil fuels into some sort of permanent storage.

carcinogen   Any substance that causes cancer or accelerates its development.

carnivore   An animal that feeds on other animals; flesh-eater. See *secondary consumer*. Compare *herbivore* and *omnivore*.

carrying capacity (K)   The maximum number of individuals of a given species that a particular environment can support sustainably (long term), assuming there are no changes in that environment.

cell   The basic structural and functional unit of life. Simple organisms are composed of single cells, whereas complex organisms are composed of many cells.

cellular respiration   A process in which the energy of organic molecules is released within cells. Compare *aerobic cellular respiration* and *anaerobic cellular respiration*.

CFCs   See *chlorofluorocarbons*.

chain reaction   A reaction maintained because it forms as products the very materials that are used as reactants in the reaction. For example, a chain reaction occurs during nuclear fission when neutrons collide with other U-235 atoms and those atoms are split, releasing more neutrons, which then collide with additional U-235 atoms.

chaparral   A biome with a Mediterranean climate (mild, moist winters and hot, dry summers). Chaparral vegetation is characterized by small-leaved evergreen shrubs and small trees.

chemosynthesis   The biological process by which certain bacteria take inorganic compounds from their environment and use them to obtain energy and make carbohydrate molecules; light is not required for this process. Compare *photosynthesis*.

chlorinated hydrocarbon   A synthetic organic compound that contains chlorine and is used as a pesticide (for example, DDT) or an industrial compound (for example, PCBs).

chlorofluorocarbons Human-made organic compounds composed of carbon, chlorine, and fluorine that had many industrial and commercial applications but were banned because they attack the stratospheric ozone layer. Also called *CFCs*.

chlorophyll A green pigment that absorbs radiant energy for photosynthesis.

C-horizon The partly weathered layer in the soil located beneath the B-horizon. The C-horizon borders solid parent material.

chronic bronchitis A disease in which the air passages (bronchi) of the lungs become permanently inflamed, causing breathlessness and chronic coughing.

chronic toxicity Adverse effects that occur after a long period of exposure to a toxicant. Compare *acute toxicity*.

circumpolar vortex A mass of cold air that circulates around the southern polar region, in effect isolating it from the warmer air in the rest of the world.

clay The smallest inorganic soil particles. Compare *silt* and *sand*.

clean-coal technologies New methods for burning coal that do not contaminate the atmosphere with sulfur oxides and that produce significantly fewer nitrogen oxides.

clearcutting A forest management technique that involves the removal of all trees from an area at a single time. Compare *seed-tree cutting*, *selective cutting*, and *shelterwood cutting*.

climate The average weather conditions that occur in a place over a period of years. Includes temperature and precipitation. Compare *weather*.

closed system A system that does not exchange energy with its surroundings. Compare *open system*.

coal A black, combustible solid composed mainly of carbon, water, and trace elements found in Earth's crust. Coal was formed from the remains of ancient plants that lived millions of years ago. See *fossil fuel*.

coal gasification The technique of producing a synthetic gaseous fuel (such as methane) from solid coal.

coal liquefaction The process by which solid coal is used to produce a synthetic liquid fuel similar to oil.

coastal wetlands Marshes, bays, tidal flats, and swamps that are found along a coastline. See *mangrove forest*, *salt marsh*, and *wetlands*.

coevolution The interdependent evolution of two or more species that occurs as a result of their interactions over a long period. Flowering plants and their animal pollinators are an example of coevolution because each has profoundly affected the other's characteristics.

cogeneration An energy technology that involves recycling "waste" heat so that two useful forms of energy (electricity and either steam or hot water) are produced from the same fuel. Also called *combined heat and power* (*CHP*).

combined heat and power See *cogeneration*.

combined sewer overflow A problem that arises in a combined sewer system when too much water (from heavy rainfall or snowmelt) enters the system. The excess flows into nearby waterways without being treated.

combined sewer system A municipal sewage system in which human and industrial wastes are mixed with urban runoff from storm sewers before flowing into the sewage treatment plant.

combustion The process of burning by which organic molecules are rapidly oxidized, converting them into carbon dioxide and water with an accompanying release of heat and light.

command and control regulation Pollution control laws that require specific technologies or limits. Compare *incentive-based regulation*.

commensalism A type of symbiosis in which one organism benefits, and the other one is neither harmed nor helped. See *symbiosis*. Compare *mutualism* and *parasitism*.

commercial extinction Depletion of the population of a commercially important species to the point that it is unprofitable to harvest.

commercial harvest The collection of commercially important organisms from the wild. Examples include the commercial harvest of parrots (for the pet trade) and cacti (for houseplants).

commercial inorganic fertilizer See *fertilizer*.

community An association of different species living together at the same time in a defined habitat with some degree of mutual interdependence. Compare *ecosystem*.

community stability The ability of a community to withstand environmental disturbances.

compact development The design of cities so that tall, multiple-unit residential buildings are close to shopping and jobs, all of which are connected by public transportation.

comparative risk analysis Analysis of any of a range of federal, state, regional, and local projects designed to prioritize environmental regulation decisions by comparing or ranking risks along several dimensions including impacts to human health and ecosystems.

competition The interaction among organisms that vie for the same resources in an ecosystem (such as food, living space, or other resources). See *interspecific competition* and *intraspecific competition*.

competitive exclusion The concept that no two species with identical living requirements can occupy the same ecological niche indefinitely. Eventually, one species will be excluded by the other as a result of interspecific competition for a resource in limited supply.

compost A natural soil and humus mixture that improves soil fertility and soil structure.

confined aquifer A groundwater storage area trapped between two impermeable layers of rock. Also called *artesian aquifer*. Compare *unconfined aquifer*.

conifer Any of a group of woody trees or shrubs (gymnosperms) that bear needle-like leaves and seeds in cones.

conservation The sensible and careful management of natural resources. Compare *preservation*.

conservation-based pricing Water supply pricing structures that reward consumers for using less water; often comes in the form of low prices for water use up to some level, and stepped-up prices as use increases.

conservation biology A multidisciplinary science that focuses on the study of how humans impact organisms and on the development of ways to protect biological diversity; includes *in situ* and *ex situ conservation*.

conservation easement A legal agreement that protects privately owned forest or other property from development for a specified number of years.

conservation tillage A method of cultivation in which residues from previous crops are left in the soil, partially covering it and helping to hold it in place until the newly planted seeds are established. See *no-tillage*. Compare *conventional tillage*.

conservationist A person who supports the conservation of natural resources.

consumer An organism that cannot synthesize its own food from inorganic materials and therefore must use the bodies of other organisms as sources of energy and body-building materials. Also called *heterotroph*. Compare *producer*.

consumption The human use of materials and energy; generally speaking, people in highly developed countries are extravagant consumers, and their use of resources is greatly out of proportion to their numbers.

consumption overpopulation A situation in which each individual in a population consumes too large a share of resources, that is, much more than is needed to survive; results in pollution, environmental degradation, and resource depletion. See *overpopulation*. Compare *people overpopulation*.

containment building A safety feature of nuclear power plants that provides an additional line of defense against any accidental leak of radiation.

continental shelf The submerged, relatively flat ocean bottom that surrounds continents. The continental shelf extends out into the ocean to the point where the ocean floor begins a steep descent.

contour plowing Plowing that matches the natural contour of the land, that is, the furrows run around rather than up and down a hill; lessens erosion, as on a hillside. See *strip cropping*.

contraceptive A device or drug used to intentionally prevent pregnancy.

control An essential part of every scientific experiment in which the experimental variable remains constant. The control provides a standard of comparison to verify the results of an experiment. See *variable*.

conventional tillage The traditional method of cultivation in which the soil is broken up by plowing before seeds are planted. Compare *conservation tillage*.

cooling tower Part of an electric power generating plant within which heated water is cooled.

coral reef Structure built from accumulated layers of calcium carbonate ($CaCO_3$); found in warm, shallow seawater. The living portion is composed principally of red coralline algae or of colonies of millions of tiny coral animals.

Coriolis effect   The tendency of moving air or water to be deflected from its path to the right in the Northern Hemisphere and to the left in the Southern Hemisphere. Caused by the direction of Earth's rotation.

corridor   See *wildlife corridor.*

cost–benefit analysis   A mechanism that helps policy-makers make decisions about environmental issues. Compares estimated costs of a particular action with potential benefits that would occur if that action were implemented.

crop rotation   The planting of different crops in the same field over a period of years. Crop rotation reduces mineral depletion of the soil because the mineral requirements of each crop vary.

crude birth rate   See *birth rate.*

crude death rate   See *death rate.*

crude oil   See *petroleum.*

cultural eutrophication   See *artificial eutrophication.*

culture   The ideas and customs of a group of people at a given period; culture, which is passed from generation to generation, evolves over time.

datum (pl. *data*)   The information, or facts, with which science works and from which conclusions are inferred.

DDT   Dichlorodiphenyltrichloroethane; a chlorine-containing organic compound that has insecticidal properties; because it is slow to degrade and therefore persists in the environment and inside organisms, DDT has been banned in the United States and many other countries.

death rate   The number of deaths per 1000 people per year. Also called *crude death rate* and *mortality.*

debt-for-nature swap   The cancellation of part of a country's foreign debt in exchange for their agreement to protect certain lands (or other resources) from detrimental development.

decommission   To dismantle an old nuclear power plant after it closes. Compare *entombment.*

decomposer   A heterotroph that breaks down organic material and uses the decomposition products to supply it with energy. Decomposers are organisms of decay. Also called *saprotroph.* Compare *detritivore.*

deductive reasoning   Reasoning that operates from generalities to specifics and can make a relationship among data more apparent. Compare *inductive reasoning.*

deep ecology worldview   An understanding of our place in the world based on harmony with nature, a spiritual respect for life, and the belief that humans and all other species have an equal worth.

deep well injection   Disposal of liquid hazardous wastes in deep repositories located several thousand feet below Earth's surface.

deforestation   The temporary or permanent clearance of large expanses of forests for agriculture or other uses.

degradation   The natural or human-induced process that decreases the future ability of the land (or soil) to support crops or livestock.

delta   A deposit of sand or soil at the mouth of a river.

demand-side management   A way that electric utilities can meet future power needs by helping consumers conserve energy and increase energy efficiency.

dematerialization   The decrease in the size and weight of a product as a result of technological improvements that occur over time.

demographics   The applied branch of sociology that deals with population statistics; provides information on the populations of various countries or groups of people.

demographic transition   The process whereby a country moves from relatively high birth and death rates to relatively low birth and death rates.

demography   The applied branch of sociology that deals with population statistics and provides information on the populations of various countries or groups of people.

denitrification   The conversion of nitrate (NO_3^-) to nitrogen gas (N_2) by certain bacteria (denitrifying bacteria) in the soil; part of the *nitrogen cycle.*

density-dependent factor   An environmental factor whose effects on a population change as population density changes; density-dependent factors tend to retard population growth as population density increases and enhance population growth as population density decreases.

density-independent factor   An environmental factor that affects the size of a population but is not influenced by changes in population density.

derelict land   Land area that was degraded by mining.

desalination   See *desalinization.*

desalinization   The removal of salt from ocean or brackish (somewhat salty) water. Also called *desalination.*

desert   See *arid land.*

desertification   Degradation of once-fertile rangeland, agricultural land, or tropical dry forest into nonproductive desert. Caused partly by soil erosion, deforestation, and overgrazing.

detritivore   An organism (such as an earthworm or crab) that consumes fragments of dead organisms. Also called *detritus feeder.* Compare *decomposer.*

detritus   Organic matter that includes dead organisms (such as animal carcasses and leaf litter) and wastes (such as feces).

detritus feeder   See *detritivore.*

deuterium   An isotope of hydrogen that contains one proton and one neutron per atom. Compare *tritium.*

developed country   See *highly developed country.*

developing country   A country not highly industrialized and characterized by a high fertility rate, high infant mortality rate, and low per-capita income. Developing countries fall into two subcategories: *moderately developed* and *less developed.* Compare *highly developed country.*

development, economic   See *economic development.*

dilution   A technique of soil remediation that involves running large quantities of water through contaminated soil to leach out pollutants such as excess salt.

dioxin   Any of a family of mildly to extremely toxic chlorinated hydrocarbon compounds that are formed as by-products in certain industrial processes.

disease   A departure from the body's normal healthy state as a result of infectious organisms, environmental stressors, or some inherent weakness.

dispersal   The movement of individuals among populations, from one region or country to another. See *immigration* and *emigration.*

distillation   A process in which water is evaporated and then recondensed to purify or separate it from other components of a complex mixture. Saltwater or brackish water may be distilled to remove the salt from the water. Compare *reverse osmosis.*

DNA   Deoxyribonucleic acid. Present in a cell's chromosomes, DNA contains all of an organism's genetic information.

domestication   The adaptation of wild plants and animals for cultivation and use by humans; due to their association with humans, domesticated plants and animals are so altered from their original ancestors that it is doubtful they could survive and compete successfully in the wild.

dose   In toxicology, the amount of a toxicant that enters the body of an exposed organism.

dose-response curve   In toxicology, a graph that shows the effect of different doses on a population of test organisms.

doubling time   The time it takes for a population to double in size, assuming that its current rate of increase doesn't change.

dragline   In coal mining, a huge shovel that takes enormous chunks out of a mountain to reach underground coal seams.

drainage basin   A land area that delivers water into a stream or river system. Also called *watershed.*

drip irrigation   See *microirrigation.*

dry deposition   A form of *acid deposition* in which dry, sulfate-containing particles settle out of the air. Compare *wet deposition.*

Dust Bowl, American   A semiarid region of the Great Plains that became desert-like as a result of human mismanagement, extended drought, and severe dust storms during the 1930s.

dust dome   A dome of heated air that surrounds an urban area and contains a lot of air pollution. Compare *urban heat island.*

dynamic equilibrium   The condition in which the rate of change in one direction is the same as the rate of change in the opposite direction.

ecological footprint   The amount of land and ocean needed to supply an individual with food, energy, water, housing, transportation, and waste disposal.

ecological niche   See *niche.*

ecological pyramid   A graphic representation of the relative energy value at each trophic level. See *pyramid of biomass, pyramid of energy,* and *pyramid of numbers.*

ecological restoration   See *restoration ecology.*

ecological risk assessment The process by which the ecological consequences of human activities are estimated. See *risk assessment*.

ecological succession See *succession*.

ecologically sustainable forest management Forest management that seeks not only to conserve forests for the commercial harvest of timber and nontimber forest products, but also to sustain biological diversity, prevent soil erosion, protect the soil, and preserve watersheds that produce clean water. Also called *sustainable forestry*.

ecology The study of systems that includes interrelationships among organisms and between organisms and their environment.

economic development An expansion in a government's economy, viewed by many as the best way to raise the standard of living.

economics The study of how people (individuals, businesses, or countries) use their limited economic resources to fulfill their needs and wants. Economics encompasses the production, consumption, and distribution of goods.

ecosystem The interacting system that encompasses a community and its nonliving, physical environment. In an ecosystem, all of the biological, physical, and chemical components of an area form a complex interacting network of energy flow and materials cycling. Compare *community*.

ecosystem diversity Biological diversity that encompasses the variety among ecosystems, such as forests, grasslands, deserts, lakes, estuaries, and oceans. Compare *genetic diversity* and *species richness*.

ecosystem management A conservation focus that emphasizes restoring and sustaining ecosystem quality rather than the conservation of individual species.

ecosystem services Important environmental benefits that ecosystems provide to people, including clean air to breathe, clean water to drink, and fertile soil in which to grow crops.

ecotone The transitional zone where two ecosystems or biomes intergrade.

ecotourism A type of tourism in which tourists pay to observe wildlife in natural settings.

ecotoxicology The study of contaminants in the biosphere, including their harmful effects on ecosystems.

edge effect The ecological phenomenon in which ecotones between adjacent communities often have more kinds of species or greater population densities of certain species than either adjoining community.

effective dose-50% (ED$_{50}$) In toxicology, the dose that causes 50% of a population to exhibit whatever biological response is under study.

efficiency An economics term used to describe getting the greatest amount of goods or services from a limited set of resources. Resources are allocated efficiently when two or more individuals or firms would be unwilling to further trade resources with each other.

E-horizon A heavily leached soil area that sometimes develops between the A- and B-horizons.

electromagnetic spectrum The continuous range of wavelengths of electromagnetic energy; includes radio waves, microwaves, infrared waves, visible light, ultraviolet radiation, x rays, and gamma rays.

electrostatic precipitator An air pollution control device that gives ash a positive electrical charge so that it adheres to negatively charged plates.

El Niño–Southern Oscillation (ENSO) A cycling of alternating warming and cooling of surface waters of the tropical eastern Pacific Ocean that affects both ocean and atmospheric circulation patterns and results in unusual weather in areas far from the tropical Pacific. See *La Niña*.

embryo transfer A technique in which a female of a rare species is treated with fertility drugs, which cause her to produce multiple eggs; some of these eggs are collected, fertilized with sperm, and surgically implanted into a female of a related but less rare species, which later gives birth to offspring of the rare species.

emigration A type of dispersal in which individuals leave a population and thus decrease its size. Compare *immigration*.

emission charge A government policy that controls pollution by charging the polluter for each given unit of emissions, that is, by establishing a tax on pollution.

emission reduction credit (ERC) A waste-discharge permit that can be bought and sold by companies that produce emissions. See *marketable waste-discharge permit*.

emissions scenarios In climate modeling, assumptions about the amounts, rates, and mix of greenhouse gases in the future.

emphysema A disease in which the air sacs (alveoli) in the lungs become irreversibly distended, decreasing the efficiency of respiration and causing breathlessness and wheezy breathing.

endangered species A species whose numbers are so severely reduced that it is in imminent danger of becoming extinct in all or a significant part of its range. Compare *threatened species*.

endemic species Localized, native species that are not found anywhere else in the world.

endocrine disrupter A chemical that interferes with the actions of the endocrine system (the body's *hormones*). Includes certain plastics such as polycarbonate; chlorine compounds such as PCBs and dioxin; the heavy metals lead and mercury; and some pesticides such as DDT, kepone, chlordane, and endosulfan.

energy The capacity or ability to do work.

energy conservation Saving energy by reducing energy use and waste. Carpooling to work or school is an example of energy conservation. Compare *energy efficiency*.

energy density The amount of energy contained within a fixed mass of an energy source. Gasoline has a higher energy density than does dry wood, which in turn has a higher energy density than wet wood.

energy efficiency A measure of the fraction of energy used relative to the total energy available in a given source. Efficiency ranges from 0 to 100%; use of natural gas for heating has an efficiency of close to 100%, while the efficiency of burning natural gas to generate electricity has a maximum efficiency of about 60%, and is usually lower. Compare *energy conservation*.

energy flow The passage of energy in a one-way direction through an ecosystem.

energy intensity A statistical estimate of energy efficiency, as, for example, a country's or region's total energy consumption divided by its gross national product.

enhanced greenhouse effect See *greenhouse effect*.

enrichment (1) The process by which uranium ore is refined after mining to increase the concentration of fissionable U-235. (2) The fertilization of a body of water, caused by the presence of high levels of plant and algal nutrients such as nitrogen and phosphorus; see *eutrophication*, which is a type of enrichment.

entombment An option after the closing of an old nuclear power plant in which the entire power plant is permanently encased in concrete. Compare *decommission*.

entropy A measure of the randomness or disorder of a system.

environment All the external conditions, both abiotic and biotic, that affect an organism or group of organisms.

environmental ethics A field of applied ethics that considers the moral basis of environmental responsibility and how far this responsibility extends; environmental ethicists try to determine how we humans should relate to nature.

environmental impact statement (EIS) A document that summarizes the potential and expected adverse impacts on the environment associated with a project, as well as alternatives to the proposed project. Typically mandated by law for public and/or private projects.

environmentalist A person who works to solve environmental problems such as overpopulation; pollution of Earth's air, water, and soil; and depletion of natural resources. Environmentalists are collectively known as the environmental movement.

environmental justice Deals with concerns that populations at high risk due to social or economic factors also face elevated impacts from environmental hazards.

environmental movement See *environmentalist*.

environmental resistance Limits set by the environment that prevent organisms from reproducing indefinitely at their intrinsic rate of increase; includes the limited availability of food, water, shelter, and other essential resources, as well as limits imposed by disease and predation.

environmental science The interdisciplinary study of how humanity interacts with other organisms and the nonliving physical environment.

environmental stressor An environmental factor, whether natural or human-induced, that taxes an organism's ability to thrive.

environmental sustainability See *sustainability*.

environmental worldview A worldview that helps us make sense of how the environment works, our place in the environment, and right and wrong environmental behaviors.

epicenter The site at Earth's surface located directly above an earthquake's focus.

epidemiology The study of the effects of toxic chemicals and diseases on human populations.

epiphyte A small organism that grows on another organism but is not parasitic on it. Small plants that live attached to the bark of a tree's branches are epiphytes.

essential amino acid Any of the ten amino acids that must be obtained in the diet because humans cannot synthesize them from simpler materials.

estuary A coastal body of water that connects to the ocean, in which fresh water from a river mixes with saltwater from the ocean.

ethanol A colorless, flammable liquid, C_2H_5OH. Also called *ethyl alcohol*.

ethics The branch of philosophy that deals with human values. See *environmental ethics*.

ethyl alcohol See *ethanol*.

euphotic zone The upper reaches of the pelagic environment, from the surface to a maximum depth of 150 m in the clearest open ocean water; sufficient light penetrates the euphotic zone to support photosynthesis.

eutrophic lake A lake enriched with nutrients such as nitrates and phosphates and consequently overgrown with plants or algae; water in a eutrophic lake contains little dissolved oxygen. Compare *oligotrophic lake*.

eutrophication The enrichment of a lake, estuary, or slow-flowing stream by nutrients that cause increased photosynthetic productivity. Eutrophication that occurs naturally is a slow process in which the body of water gradually fills in and converts to a marsh, eventually disappearing. See *enrichment* and *artificial eutrophication*.

evaporation The conversion of water from a liquid to a vapor. Also called *vaporization*.

evolution The cumulative genetic changes in populations that occur during successive generations. Evolution explains the origin of all the organisms that exist today or have ever existed.

ex situ conservation Conservation efforts that involve conserving biological diversity in human-controlled settings. See *conservation biology*. Compare *in situ conservation*.

exosphere The outermost layer of the atmosphere, bordered by the thermosphere and interplanetary space.

exponential population growth The accelerating population growth that occurs when optimal conditions allow a constant rate of increase over time. When the increase in population number versus time is plotted on a graph, exponential population growth produces a characteristic J-shaped curve.

externality In economics, the effect (usually negative) of a firm that does not have to pay all the costs associated with its production.

extinction The elimination of a species from Earth; occurs when the last individual member of a species dies.

extrapolation In toxicology or epidemiology, estimating the expected effects at some dose of interest from the effects at known doses.

facultative parasite A normally saprotrophic organism that, given the opportunity, becomes parasitic. Compare *obligate parasite*.

fall turnover A mixing of the lake waters in temperate lakes, caused by falling temperatures in autumn. Compare *spring turnover*.

family planning Providing the services, including information about birth control methods, to help people have the number of children they want.

famine Widespread starvation caused by a drastic shortage of food. Famine is caused by crop failures that are brought on by drought, war, flood, or some other catastrophic event.

fault A fracture in the crust along which rock moves forward and backward, up and down, or from side to side. Fault zones are often found at *plate boundaries*.

fecal coliform test A water quality test for the presence of *E. coli* (fecal bacteria common in the intestinal tracts of people and animals). The presence of fecal bacteria in a water supply indicates a chance that pathogenic organisms may be present as well.

fertilizer A material containing plant nutrients that is put on the soil to enhance plant growth. Organic fertilizers include such natural materials as animal manure, crop residues, bone meal, and compost, whereas commercial inorganic fertilizers are manufactured from inorganic chemical compounds.

first law of thermodynamics Energy cannot be created or destroyed, although it can be transformed from one form into another. See *thermodynamics*. Compare *second law of thermodynamics*.

fission A nuclear reaction in which large atoms of certain elements are each split into two smaller atoms with the release of a large amount of energy. Also called *nuclear fission*. Compare *fusion*.

flood plain The area bordering a river that is subject to flooding.

flowing-water ecosystem A freshwater ecosystem such as a river or stream in which the water flows. Compare *standing-water ecosystem*.

fluidized-bed combustion A clean-coal technology in which crushed coal is mixed with particles of limestone in a strong air current during combustion; the limestone neutralizes the acidic sulfur compounds produced during combustion.

fly ash The ash from the flue (chimney) that is trapped by electrostatic precipitators. Compare *bottom ash*.

flyway An established route that ducks, geese, and shorebirds follow during their annual migrations.

focus The site, often far below Earth's surface, where an earthquake begins.

food chain The successive series of organisms through which energy flows in an ecosystem. Each organism in the series eats or decomposes the preceding organism in the chain. Compare *food web*.

food insecurity The condition in which people live with chronic hunger and malnutrition.

food security The condition in which people do not live in hunger or fear of starvation.

food web A complex interconnection of all the food chains in an ecosystem. Compare *food chain*.

forest decline A gradual deterioration (and often death) of many trees in a forest; may be the result of a combination of environmental stressors, such as acid precipitation, toxic heavy metals, and surface-level ozone.

forest edge The often sharp boundary between the forest and surrounding farmlands or residential neighborhoods.

forest management See *ecologically sustainable forest management*.

fossil fuel Combustible deposits in Earth's crust, composed of the remnants (fossils) of prehistoric organisms that existed millions of years ago. Coal, oil (petroleum), and natural gas are the three types of fossil fuel.

fragmentation See *habitat fragmentation*.

freshwater wetlands Lands that are usually covered by shallow fresh water for at least part of the year and that have a characteristic soil and water-tolerant vegetation; include *marshes* and *swamps*.

frontier attitude The attitude of most Americans during the 1700s and early 1800s that, because the natural resources of North America were seemingly inexhaustible, there was no reason not to conquer and exploit nature as much and as quickly as possible.

fuel cell A device that directly converts chemical energy into electricity without the intermediate step of needing to produce steam and use a turbine and generator; the fuel cell requires hydrogen from a tank or other source and oxygen from the air.

full-cost accounting The process of evaluating and presenting to decision makers the relative benefits and costs of various alternatives.

fundamental niche The potential ecological niche that an organism could have if there were no competition from other species. See *niche*. Compare *realized niche*.

fungicide A toxic chemical that kills fungi.

fusion A nuclear reaction in which two smaller atoms are combined to make one larger atom with the release of a large amount of energy. Also called *nuclear fusion*. Compare *fission*.

game farming See *wildlife ranching*.

gas hydrates See *methane hydrates*.

gasohol See *alcohol fuel*.

gender inequality The social construct that results in women not having the same rights, opportunities, or privileges as men.

gene A segment of DNA that serves as a unit of hereditary information.

genetically modified (GM) organism An organism that has had its genes intentionally manipulated.

genetic diversity   Biological diversity that encompasses the genetic variety among individuals within a single species. Compare *ecosystem diversity* and *species richness*.

genetic engineering   The process of taking a specific gene from one cell and placing it into another cell where it is expressed.

genetic resistance   An inherited characteristic that decreases the effect of a pesticide on a pest. Over time, the repeated exposure of a pest population to a pesticide causes an increase in the number of individuals that can tolerate the pesticide.

geothermal energy   The natural heat within Earth that arises from ancient heat within Earth's core, from friction where continental plates slide over one another, and from decay of radioactive elements; geothermal energy can be used for space heating and to generate electricity.

geothermal heat pump (GHP)   A device that heats and cools buildings by taking advantage of the difference in temperature between Earth's surface and subsurface at depths from about 1 to 100 m.

germplasm   Any plant or animal material that may be used in breeding; includes seeds, plants, and plant tissues of traditional crop varieties and the sperm and eggs of traditional livestock breeds.

global commons   Those resources of our environment that are available to everyone but for which no single individual has responsibility.

global distillation effect   The process whereby volatile chemicals evaporate from land as far away as the tropics and are carried by air currents to higher latitudes, where they condense and fall to the ground.

grain stockpiles   See *world grain carryover stocks*.

gray water   Water that has already been used for a relatively non-polluting purpose, such as showers, dishwashing, and laundry; gray water is not potable, but can be reused for toilets, plants, or car washing.

green architecture   The practice of designing and building homes with environmental considerations such as energy efficiency, recycling, and conservation of natural resources in mind.

green chemistry   A subdiscipline of chemistry in which commercially important chemical processes are redesigned to significantly reduce environmental harm.

green power   Electricity produced from renewable sources such as solar power, wind, biomass, geothermal, and small hydroelectric plants.

green revolution   The period during the 20th century when plant scientists developed genetically uniform, high-yielding varieties of important food crops such as rice and wheat.

greenhouse effect   The increase of heat in a system where energy enters (often as light), is absorbed as heat, and released some time later; because the heat has a residence time within the system, the overall temperature of the system will be higher than its surroundings.

greenhouse gas   A gas that absorbs infrared radiation; includes carbon dioxide, methane, nitrous oxide, chlorofluorocarbons, and tropospheric ozone, all of which are accumulating in the atmosphere as a result of human activities, thereby increasing Earth's temperature.

gross primary productivity   The rate at which energy accumulates in an ecosystem (as biomass) during photosynthesis. Compare *net primary productivity*.

groundwater   The supply of fresh water under Earth's surface. Groundwater is stored in underground caverns and porous layers of underground rock called aquifers. Compare *surface water*.

groundwater depletion   See *aquifer depletion*.

growth rate (*r*)   The rate of change of a population's size, expressed in percent per year. In populations with little or no dispersal, it is calculated by subtracting the death rate from the birth rate. Also called *natural increase* in human populations.

gyre   A circular, prevailing wind that generates circular ocean currents.

habitat   The local environment in which an organism, population, or species lives.

habitat fragmentation   The division of habitats that formerly occupied large, unbroken areas into smaller areas by roads, fields, cities, and other land-transforming activities.

hadal benthic zone   The ocean's benthic environment that extends from 6000 meters to the bottom of the deepest ocean trenches. Compare *abyssal benthic zone*.

half-life   See *radioactive half-life*.

hard coal   See *anthracite*.

hazard   A condition that has the potential to cause harm; differs from *risk* in not having an associated probability.

hazardous air pollutant   A potentially harmful air pollutant that may pose long-term health risks to people who live and work around chemical factories, incinerators, or other facilities that produce or use it; also called *air toxic*.

hazardous waste   Any discarded material (solid, liquid, or gas) that threatens human health or the environment; may be flammable, chemically reactive, corrosive, and/or toxic.

herbicide   A toxic chemical that kills plants.

herbivore   An animal that feeds on plants or algae. See *primary consumer*. Compare *carnivore* and *omnivore*.

heterocysts   Oxygen-excluding cells of nitrogen-fixing cyanobacteria.

heterotroph   See *consumer*.

high-grade ore   An ore that contains relatively large amounts of a particular mineral. Compare *low-grade ore*.

high-input agriculture   See *industrialized agriculture*.

high-level radioactive waste   Any radioactive solid, liquid, or gas that initially gives off large amounts of ionizing radiation. Compare *low-level radioactive waste*.

highly developed country   An industrialized country that is characterized by a low fertility rate, low infant mortality rate, and high per-capita income. Also called *developed country*. Compare *developing country*.

horizons, soil   See *soil horizons*.

hormone   A chemical messenger produced by a living organism in minute quantities to regulate its growth, reproduction, and other important biological functions. Also see *endocrine disrupter*.

hot spot   (1) A rising plume of magma that flows from deep within Earth's rocky mantle through an opening in the crust. (2) An area of great species diversity that is at risk of destruction by human activities.

Hubbert's peak   See *peak oil*.

humus   Black or dark brown decomposed organic material.

hydrocarbons   A diverse group of organic compounds that contain only hydrogen and carbon.

hydrogen bond   A bond between water molecules, formed when the negative (oxygen) end of one water molecule is attracted to the positive (hydrogen) end of another water molecule. Hydrogen bonding is the basis for many of water's physical properties.

hydrologic cycle   The water cycle, which includes evaporation, precipitation, and flow to the seas. The hydrologic cycle supplies terrestrial organisms with a continual supply of fresh water.

hydrology   The science that deals with Earth's waters, including the availability and distribution of fresh water.

hydropower   A form of renewable energy that relies on flowing or falling water to generate mechanical energy or electricity.

hydrosphere   Earth's supply of water (both liquid and frozen, fresh and salty).

hydrothermal reservoir   Large underground reservoir of hot water and possibly also steam; some of the hot water or steam may escape to the surface, creating hot springs or geysers.

hydrothermal vent   A hot spring on the seafloor where a solution of hot, mineral-rich water rises to the surface. Many hydrothermal vents support thriving communities.

hypertension   High blood pressure.

hypothesis   An educated guess that might be true and is testable by observation and experimentation. Compare *theory*.

hypoxia   Low dissolved oxygen concentrations that occur in many bodies of water when nutrients stimulate the growth of algae that subsequently die and are decomposed by oxygen-using bacteria. Hypoxia often causes too little oxygen for other aquatic life.

illuviation   The deposition of material in the lower layers of soil from the upper layers; caused by leaching.

immigration   A type of dispersal in which individuals enter a population and thus increase its size. Compare *emigration*.

incentive-based regulation   Pollution control laws that work by establishing emission targets and providing industries with incentives to reduce emissions. Compare *command and control regulation*.

indicator species   An organism that provides an early warning of environmental damage. Examples include lichens, which are sensitive to air pollution, and amphibians, which are sensitive to pesticides and other environmental contaminants.

inductive reasoning   Reasoning that uses specific examples to draw a general conclusion or discover a general principle. Compare *deductive reasoning.*

industrial ecosystem   A complex web of interactions among various industries, in which "wastes" produced by one industrial process are sold to other companies as raw materials. Finding profitable uses for a company's wastes is an extension of sustainable manufacturing. See *sustainable manufacturing.*

industrial smog   The traditional, London-type smoke pollution, which consists principally of sulfur oxides and particulate matter. Compare *photochemical smog.*

industrialized agriculture   Modern agricultural methods that use large inputs of energy (from fossil fuels), mechanization, water, and agrochemicals (fertilizers and pesticides) to produce large crop and livestock yields. Also called *high-input agriculture.* Compare *sustainable agriculture* and *subsistence agriculture.*

infant mortality rate   The number of infant deaths per 1000 live births. (An infant is a child in its first year of life.)

infectious disease   A disease caused by a microorganism (such as a bacterium or fungus) or infectious agent (such as a virus). Infectious diseases can be transmitted from one individual to another.

infrared radiation   Electromagnetic radiation with a wavelength longer than that of visible light, but shorter than that of radio waves. Most of the energy absorbed by Earth is reradiated as infrared radiation, which is absorbed by greenhouse gases. Humans perceive infrared radiation as invisible waves of heat.

inherent safety   See *principle of inherent safety.*

inorganic chemical   A chemical that does not contain carbon and is not associated with life. Inorganic chemicals that are pollutants include mercury compounds, road salt, and acid drainage from mines.

inorganic fertilizer   See *fertilizer.*

inorganic plant nutrient   A nutrient such as phosphate or nitrate that stimulates plant or algal growth. Excessive amounts of inorganic plant nutrients, which may come from animal wastes and plant residues as well as fertilizer runoff, can cause both soil and water pollution.

insecticide   A toxic chemical that kills insects.

in situ conservation   Conservation efforts that concentrate on preserving biological diversity in the wild. See *conservation biology.* Compare *ex situ conservation.*

integrated pest management (IPM)   A combination of pest control methods (biological, chemical, and cultivation) that, if used in the proper order and at the proper times, keep the size of a pest population low enough that it does not cause substantial economic loss.

integrated waste management   A combination of the best waste management techniques into a consolidated, systems-based program to deal effectively with solid waste.

intercropping   A form of intensive subsistence agriculture that involves growing several crops simultaneously on the same field. See *polyculture.*

interspecific competition   Competition between members of different species. See *competition.* Compare *intraspecific competition.*

intertidal zone   The shoreline area between the low tide mark and the high tide mark.

intraspecific competition   Competition among members of the same species. See *competition.* Compare *interspecific competition.*

intrinsic rate of increase   The exponential growth of a population that occurs under constant conditions.

invasive species   Foreign species whose introduction causes economic or environmental harm.

ionizing radiation   Radiation that contains enough energy to eject electrons from atoms, forming positively charged ions. Ionizing radiation can damage living tissue.

IPAT equation   A model that shows the mathematical relationship between environmental impacts and the forces that drive them (number of people, affluence per person, and environmental effects of technologies used to obtain and consume resources).

isotope   An alternate form of the same element that has a different atomic mass; an isotope has a different number of neutrons but the same number of protons and electrons.

K selection   A reproductive strategy in which a species typically has a large body size, slow development, long life span, and does not devote a large proportion of its metabolic energy to the production of offspring. Compare *r selection.*

kelps   Large brown algae that are common in relatively shallow, cooler temperate marine water along rocky coastlines.

kerogen   See *oil shales.*

keystone species   A species that is crucial in determining the nature and structure of the entire ecosystem in which it lives; other species of a community depend on or are greatly affected by the keystone species, whose influence is much greater than would be expected by its relative abundance.

kilocalorie   A unit of heat equivalent to the energy required to raise the temperature of 1 kilogram of water by $1°C$.

kilojoule   A unit of energy; one kilocalorie equals 4.184 kilojoules.

kinetic energy   The energy of a body that results from its motion. Compare *potential energy.*

krill   Tiny shrimplike animals that are important in the Antarctic food web.

Kyoto Protocol   An international treaty that stipulates that highly developed countries must cut their emissions of CO_2 and other gases that cause climate change by an average of 5.2% by 2012.

land degradation   See *degradation.*

landscape   A region that includes several interacting ecosystems.

landscape ecology   A subdiscipline in ecology that focuses on connections among ecosystems in a particular area.

land-use planning   The process of deciding the best uses for undeveloped land in a given area.

La Niña   A periodic occurrence in the eastern Pacific Ocean in which the surface water temperature becomes unusually cool, and westbound trade winds become unusually strong; often occurs after an El Niño event. Part of *El Niño–Southern Oscillation.*

latitude   The distance, measured in degrees north or south, from the equator.

lava   Magma (molten rock) that reaches Earth's surface.

leachate   The liquid that seeps through solid waste at a sanitary landfill or other waste disposal site.

leaching   The process by which dissolved materials (nutrients or contaminants) are washed away or filtered down through the various layers of the soil.

less developed country   A developing country with a low level of industrialization, a high fertility rate, a high infant mortality rate, and a low per-capita income (relative to highly developed countries). Compare *moderately developed country* and *highly developed country.*

lethal dose-50% (LD_{50})   In toxicology, the dose lethal to 50% of a population of test animals.

lignite   A low-grade brown or brown-black coal that has a soft, woody texture (softer than subbituminous coal). Compare *anthracite, bituminous coal,* and *subbituminous coal.*

lime scrubber   An air pollution control device in which a chemical spray neutralizes acidic gases. See *scrubbers.*

limiting resource   Any environmental resource that, because it is scarce or at unfavorable levels, restricts the ecological niche of an organism.

limnetic zone   The open-water area away from the shore of a lake or pond that extends down as far as sunlight penetrates (and therefore photosynthesis occurs). Compare *littoral zone* and *profundal zone.*

lipid   A diverse group of organic molecules that are metabolized by cellular respiration to provide the body with a high level of energy; commonly called fats and oils.

liquefied petroleum gas   A mixture of liquefied propane and butane. Liquefied petroleum gas is stored in pressurized tanks.

lithosphere   Earth's outermost rigid rock layer that is composed of seven large plates, plus a few smaller ones.

littoral zone   The shallow-water area along the shore of a lake or pond. Compare *limnetic zone* and *profundal zone.*

loam   An ideal agricultural soil that has an optimum combination of soil particle sizes: approximately 40% each of sand and silt, and about 20% of clay.

low-grade ore   An ore that contains relatively small amounts of a particular mineral. Compare *high-grade ore*.

low-input agriculture   See *sustainable agriculture*.

low-level radioactive waste   Any radioactive solid, liquid, or gas that gives off small amounts of ionizing radiation. Compare *high-level radioactive waste*.

magma   Molten rock formed within Earth. Compare *lava*.

malnutrition   Poor nutritional status; results from dietary intake either below or above required needs. Compare *overnutrition* and *undernutrition*.

manganese nodule   A small rock that contains manganese and other minerals; common on parts of the ocean floor.

mangrove forest   Swamps of mangrove trees that grow along many tropical coasts.

marginal cost of pollution   The cost in environmental damage of a unit of pollution emitted into the environment.

marginal cost of abatement   The cost to reduce of a unit of pollution.

mariculture   The rearing of marine organisms (fishes, seaweeds, and shellfish) for human consumption; a subset of aquaculture.

marine snow   Organic debris that drifts into the darkened regions of the oceanic province from the upper, lighted regions.

marketable waste-discharge permit   A permit issued under a government policy that controls pollution by allowing the holder to pollute a given amount. Holders are not allowed to produce more emissions than the permit allows. See *emission reduction credit*.

marsh   A treeless wetland dominated by grasses; freshwater marshes are found inland along lakes and rivers, and salt marshes are found in bays and rivers near the ocean or in protected coastal areas. See *salt marsh*.

mass burn incinerator   A large furnace that burns all solid waste except for unburnable items such as refrigerators.

mass extinction   The extinction of numerous species during a relatively short period of geologic time. Compare *background extinction*.

maximum contaminant level   The maximum permissible amount (by law) of a water pollutant that might adversely affect human health.

megacity   A city with more than 10 million inhabitants.

meltdown   The melting of a nuclear reactor vessel. A meltdown would cause the release of a substantial amount of radiation into the environment. See *reactor vessel*.

mesosphere   The layer of the atmosphere between the stratosphere and the thermosphere. It is characterized by the lowest atmospheric temperatures.

metal   A malleable, lustrous element that is a good conductor of heat and electricity. See *mineral*. Compare *nonmetal*.

methane   The simplest hydrocarbon, CH_4, which is an odorless, colorless, flammable gas.

methane hydrates   Reserves of ice-encrusted natural gas located in porous rock in the arctic tundra (under the permafrost) and in the deep ocean sediments of the continental slope and ocean floor. Also called *gas hydrates*.

methanol   A colorless, flammable liquid, CH_3OH. Also called *methyl alcohol*.

methyl alcohol   See *methanol*.

microclimate   Local variation in climate produced by differences in elevation, in the steepness and direction of slopes, and in exposure to prevailing winds.

microirrigation   A type of irrigation that conserves water. In microirrigation, pipes with tiny holes bored into them convey water directly to individual plants. Also called *drip* or *trickle irrigation*.

mineral   (1) An element, inorganic compound, or mixture that occurs naturally in Earth's crust. See *metal* and *nonmetal*. (2) An inorganic nutrient ingested (by animals) or absorbed (by plants) in the form of salt. Compare *vitamin*.

mineral reserve   A mineral deposit that has been identified and is currently profitable to extract. See *total resources*. Compare *mineral resource*.

mineral resource   Any undiscovered mineral deposits or known deposits of low-grade ore that are currently unprofitable to extract. Mineral resources are potential resources that may be profitable to extract in the future. See *total resources*. Compare *mineral reserve*.

mixed oxide fuel   A nuclear reactor fuel that contains a combination of uranium oxide and plutonium oxide; the plutonium can come from reprocessed spent fuel or from other plutonium stockpiles, including dismantled weapons.

model   (1) A formal statement that describes a situation and can be used to understand the present or predict the future course of events. (2) A simulation, using powerful computers, that represents the overall effect of competing factors to describe an environmental situation in numerical terms.

moderately developed country   A developing country with a medium level of industrialization, a high fertility rate, a high infant mortality rate, and a low per-capita income (all relative to highly developed countries). Compare *less developed country* and *highly developed country*.

monoculture   The cultivation of only one type of plant over a large area. Compare *polyculture*.

Montreal Protocol   International negotiations that resulted in a timetable to phase out CFC production.

mortality   See *death rate*.

mulch   Material placed on the surface of soil around the bases of plants; helps maintain soil moisture and reduce soil erosion. Organic mulches decompose over time, thereby enriching the soil.

municipal solid waste   Solid waste generated in homes, office buildings, retail stores, restaurants, schools, hospitals, prisons, libraries, and other commercial and institutional facilities. Compare *nonmunicipal solid waste*.

municipal solid waste composting   The large-scale composting of the organic portion of a community's solid waste.

mutation   A change in the DNA (that is, a gene) of an organism. A mutation in reproductive cells may be passed on to the next generation, where it may result in birth defects or genetic disease.

mutualism   A symbiotic relationship in which both partners benefit from the association. See *symbiosis*. Compare *parasitism* and *commensalism*.

mycorrhiza   A mutualistic association between a fungus and the roots of a plant. Most plants form mycorrhizal associations with fungi, which enables plants to absorb adequate amounts of essential minerals from the soil.

narrow-spectrum pesticide   An "ideal" pesticide that kills only the organism for which it is intended and does not harm any other species. Compare *broad-spectrum pesticide*.

natality   See *birth rate*.

national emission limitations   The maximum permissible amount (by law) of a particular pollutant that can be discharged into the nation's rivers, lakes, and oceans from point sources.

National Environmental Policy Act (NEPA)   A U.S. federal law that is the cornerstone of U.S. environmental policy. It requires that the federal government consider the environmental impact of any construction project funded by the federal government.

national marine sanctuary   A marine ecosystem set aside to minimize human impacts and protect unique natural resources and historical sites.

natural capital   All of Earth's resources and processes that sustain living organisms, including humans; natural capital includes minerals, forests, soils, groundwater, clean air, wildlife, and fisheries.

natural gas   A mixture of gaseous, energy-rich hydrocarbons (primarily *methane*) that occurs, often with oil deposits, in Earth's crust. See *fossil fuel*.

natural increase   See *growth rate*.

natural regulation   A controversial park management policy that involves letting nature take its course most of the time, with corrective actions undertaken as needed to adjust for changes caused by pervasive human activities.

natural resources   See *resources*.

natural selection   The tendency of organisms that possess adaptations favorable to their environment to survive and become the parents of the next generation; evolution occurs when natural selection results in genetic changes in a population. Natural selection is the mechanism of evolution first proposed by Charles Darwin.

negative externality   See *externality*.

negative feedback   A situation in which a change in some condition triggers a response that counteracts, or reverses, the changed condition. Compare *positive feedback*.

nekton   Relatively strong-swimming aquatic organisms such as fish and turtles. Compare *plankton* and *benthos*.

neo-Malthusians   Economists who hold that developmental efforts are hampered by a rapidly expanding population. Compare *pronatalists*.

neritic province   Open ocean from the shoreline to a depth of 200 meters. Compare *oceanic province*.

net primary productivity (NPP) Productivity after respiration losses are subtracted; that is, the amount of biomass found in excess of that broken down by a plant's cellular respiration. Compare *gross primary productivity*.

niche The totality of an organism's adaptations, its use of resources, and the lifestyle to which it is fitted. The niche describes how an organism uses materials in its environment as well as how it interacts with other organisms. Also called *ecological niche*. See *fundamental niche* and *realized niche*.

nitrification The conversion of ammonia (NH_3) and ammonium ions (NH_4^+) to nitrate (NO_3^-) by certain bacteria (nitrifying bacteria) in the soil; part of the *nitrogen cycle*.

nitrogen cycle The worldwide circulation of nitrogen from the abiotic environment into organisms and back into the abiotic environment.

nitrogen fixation The conversion of atmospheric nitrogen (N_2) to ammonia (NH_3) by nitrogen-fixing bacteria and cyanobacteria; part of the *nitrogen cycle*.

noise pollution A loud or disagreeable sound, particularly when it results in physiological or psychological harm.

nomadic herding A traditional grazing method in which nomadic herdsmen wander freely over rangelands in search of good grazing for their livestock.

nondegradable Referring to a chemical pollutant (such as the toxic elements mercury and lead) that cannot be decomposed (broken down) by organisms or by other natural processes. Compare *biodegradable*.

nonmetal A nonmalleable, nonlustrous mineral that is a poor conductor of heat and electricity. See *mineral*. Compare *metal*.

nonmunicipal solid waste Solid waste generated by industry, agriculture, and mining. Compare *municipal solid waste*.

nonpoint source pollution Pollutants that enter bodies of water over large areas rather than being concentrated at a single point of entry. Examples include agricultural fertilizer runoff and sediments from construction. Also called *polluted runoff*. Compare *point source pollution*.

nonrenewable resources Natural resources that are present in limited supplies and are depleted by use; include minerals such as copper and tin and fossil fuels such as oil and natural gas. Compare *renewable resources*.

nonurban land See *rural land*.

no-tillage A method of conservation tillage that leaves both the surface and subsurface soil undisturbed. Special machines punch holes in the soil for seeds. See *conservation tillage*. Compare *conventional tillage*.

nuclear energy The energy released from the nucleus of an atom in a nuclear reaction (fission or fusion) or during radioactive decay.

nuclear fission See *fission*.

nuclear fuel cycle The processes involved in producing the fuel used in *nuclear reactors* and in disposing of radioactive wastes (also called nuclear wastes).

nuclear fusion See *fusion*.

nuclear reactor A device that initiates and maintains a controlled nuclear fission chain reaction to produce energy, usually used as electricity.

nutrient cycling The pathways of various nutrient minerals or elements from the environment through organisms and back to the environment; *biogeochemical cycles* are examples of nutrient cycling.

obligate parasite An organism that can exist only as a parasite. Compare *facultative parasite*.

oceanic province That part of the open ocean that is deeper than 200 meters and comprises most of the ocean. Compare *neritic province*.

ocean temperature gradient The differences in temperature at various ocean depths.

Ogallala Aquifer A massive groundwater deposit under eight midwestern states.

O-horizon The uppermost layer of certain soils, composed of dead leaves and other organic matter.

oil See *petroleum*.

oil sand See *tar sand*.

oil shales Sedimentary "oily rocks" that contain a mixture of hydrocarbons known as kerogen; to yield oil, oil shales must be crushed, heated to high temperatures, and refined after they are mined.

oligotrophic lake A deep, clear lake that has minimal nutrients. Water in an oligotrophic lake contains a high level of dissolved oxygen. Compare *eutrophic lake*.

omnivore An animal that eats a variety of plant and animal material. Compare *herbivore* and *carnivore*.

open management A policy in which all fishing boats of a particular country are given unrestricted access to fish in their national waters.

open-pit mining A type of surface mining in which a giant hole (quarry) is dug to extract iron, copper, stone, or gravel. See *surface mining*. Compare *strip mining*.

open system A system that exchanges energy with its surroundings. Compare *closed system*.

optimal amount of pollution In economics, the amount of pollution that is most economically efficient. It is determined by plotting two curves, the marginal cost of pollution and the marginal cost of pollution abatement; the point where the two curves meet is the optimum amount of pollution.

ore Rock that contains a large enough concentration of a particular mineral that it can be profitably mined and the mineral extracted.

organic agriculture Growing crops and livestock without the use of synthetic pesticides or commercial inorganic fertilizers. Organic agriculture makes use of natural organic fertilizers (such as manure and compost) and chemical-free methods of pest control.

organic chemical A compound that contains the element carbon and is either naturally occurring (in organisms) or synthetic (manufactured by humans). Many synthetic organic compounds persist in the environment for an extended period, and some are toxic to organisms.

organic fertilizer See *fertilizer*.

organophosphate A synthetic organic compound that contains phosphorus and is toxic; used as an insecticide.

overburden Overlying layers of soil and rock over mineral deposits. The overburden is removed during surface mining.

overgrazing The destruction of an area's vegetation that occurs when too many animals graze on the vegetation, consuming so much of it that it does not recover.

overnutrition Malnutrition caused by eating food in excess of that required to maintain a healthy body. Compare *undernutrition*.

overpopulation A situation in which a country or geographic area has more people than its resource base can support without damaging the environment. See *people overpopulation* and *consumption overpopulation*.

oxide A compound in which oxygen is chemically combined with some other element.

ozone A pale blue gas (O_3) that is both a pollutant in the lower atmosphere (troposphere) and an essential component that screens out UV radiation in the upper atmosphere (stratosphere).

pandemic A disease that reaches nearly every part of the world, and has the potential to infect almost every person.

paradigm A generally accepted understanding of how some aspect of the world works; for example, evolution is the underlying paradigm of modern biology.

parasitism A symbiotic relationship in which one member (the parasite) benefits and the other (the host) is adversely affected. See *symbiosis*. Compare *mutualism* and *commensalism*.

particulate matter Solid particles and liquid droplets suspended in the atmosphere.

parts per billion (ppb) The number of molecules of a particular substance found in one billion molecules of air, water, or some other material.

parts per million (ppm) Similar to percent (parts per hundred). (1) The number of molecules of a particular substance found in one million molecules of air, water, or some other material. (2) In climate science, the fraction of the atmosphere made up of CO_2. The atmosphere is about 21% O_2, or about 210,000 ppm.

passive solar heating A system that uses the sun's energy without requiring mechanical devices (pumps or fans) to distribute the collected heat. Compare *active solar heating*.

pathogen An agent (usually a microorganism) that causes disease.

PCBs Polychlorinated biphenyls; chlorine-containing organic compounds that enjoyed a wide variety of industrial uses until their dangerous properties, including the fact that they are slow to degrade and therefore persist in the environment, were recognized.

peak oil   The point at which global oil production has reached a maximum rate; by some estimates, peak oil is already past, but best estimates have it occurring around 2020 or later. Also known as "Hubbert's peak," after the U.S. geologist who came up with the idea.

people overpopulation   A situation in which there are too many people in a given geographic area. Even if those people use few resources per person (the minimum amount they need to survive), people overpopulation results in pollution, environmental degradation, and resource depletion. See *overpopulation*. Compare *consumption overpopulation*.

permafrost   Permanently frozen subsoil characteristic of frigid areas such as the tundra.

persistence   A characteristic of certain chemicals that are extremely stable and may take many years to be broken down into simpler forms by natural processes.

persistent organic pollutants (POPs)   A group of persistent, toxic chemicals that bioaccumulate in organisms and can travel long distances through air and water to contaminate sites far removed from their source; some disrupt the endocrine system, cause cancer, or adversely affect the developmental processes of organisms.

pest   Any organism that interferes in some way with human welfare or activities.

pesticide   Any toxic chemical used to kill pests. See *fungicide*, *herbicide*, *insecticide*, and *rodenticide*.

pesticide treadmill   A predicament faced by pesticide users, in which the cost of applying pesticides increases because they have to be applied more frequently or in larger doses, while their effectiveness decreases as a result of increasing genetic resistance in the target pest.

petrochemicals   Chemicals, obtained from crude oil, that are used in the production of such diverse products as fertilizers, plastics, paints, pesticides, medicines, and synthetic fibers.

petroleum   A flammable liquid hydrocarbon mixture found in Earth's crust, formed from the remains of ancient microscopic organisms. When petroleum is refined, the mixture is separated into various hydrocarbon compounds, including gasoline, kerosene, fuel oil, lubricating oils, paraffin, and asphalt. Also called *crude oil*. See *fossil fuel*.

pH   A number from 0 to 14 that indicates the degree of acidity or basicity of a substance.

pH scale   A measure of acidity or basicity of an aqueous solution.

pheromone   A substance secreted into the environment by one organism; influences the development or behavior of other members of the same species.

phosphorus cycle   The worldwide circulation of phosphorus from the abiotic environment into organisms and back into the abiotic environment.

photochemical smog   A brownish orange haze formed by complex chemical reactions involving sunlight, nitrogen oxides, and hydrocarbons. Some of the pollutants in photochemical smog include peroxyacetyl nitrates (PANs), ozone, and aldehydes. Compare *industrial smog*.

photodegradable   Breaking down upon exposure to sunlight.

photosynthesis   The biological process that captures light energy and transforms it into the chemical energy of organic molecules (such as glucose), which are manufactured from carbon dioxide and water. Photosynthesis is performed by plants, algae, and several kinds of bacteria. Compare *chemosynthesis*.

photovoltaic solar cell   A wafer or thin film of a solid state material, such as silicon or gallium arsenide, that is treated with certain metals so that it generates electricity (a flow of electrons) when it absorbs solar energy.

phytoplankton   Microscopic floating algae that are the base of most aquatic food chains. See *plankton*. Compare *zooplankton*.

phytoremediation   A method employed to clean up a hazardous waste site that uses plants to absorb and accumulate toxic materials. Compare *bioremediation*.

pioneer community   The first organisms (such as lichens or mosses) to colonize an area and begin the first stage of ecological succession. See *succession*.

plankton   Small or microscopic aquatic organisms that are relatively feeble swimmers and thus, for the most part, are carried about by currents and waves. Composed of *phytoplankton* and *zooplankton*. Compare *nekton* and *benthos*.

plasma   An ionized gas formed at high temperatures when electrons are stripped from the gas atoms; plasma is formed during fusion reactions.

plate boundary   Any area where two tectonic plates meet. Plate boundaries are often sites of intense geologic activity. See *fault*.

plate tectonics   The study of the processes by which the lithospheric plates move over the asthenosphere.

point source pollution   Water pollution that can be traced to a specific spot (such as a factory or sewage treatment plant) because it is discharged into the environment through pipes, sewers, or ditches. Compare *nonpoint source pollution*.

polar easterly   A prevailing wind that blows from the northeast near the North Pole or from the southeast near the South Pole.

pollutant   A physical, chemical, or biological phenomenon, generated by human activity and emitted into the environment, that has the potential to have harmful effects on the health of organisms including humans.

polluted runoff   See *nonpoint source pollution*.

pollution   Any undesirable alteration of air, water, or soil that harms the health, survival, or activities of humans and other living organisms.

polychlorinated biphenyls   See *PCBs*.

polyculture   A type of intercropping in which several kinds of plants that mature at different times are planted together. See *intercropping*. Compare *monoculture*.

population   A group of organisms of the same species that live in the same geographic area at the same time.

population density   The number of individuals of a species per unit of area or volume at a given time.

population ecology   That branch of biology that deals with the numbers of a particular species that are found in an area and how and why those numbers change (or remain fixed) over time.

population growth momentum   The continued growth of a population after fertility rates have declined, as a result of a population's young age structure; population growth momentum can be either positive or negative but is usually discussed in a positive context.

population momentum   See *population growth momentum*.

positive feedback   A situation in which a change in some condition triggers a response that intensifies the changing condition. Compare *negative feedback*.

potential energy   Stored energy that is the result of the relative position of matter instead of its motion. Compare *kinetic energy*.

poverty   A condition in which people are unable to meet their basic needs for adequate food, clothing, or shelter.

precautionary principle   The idea that no action should be taken or product introduced when the science is inconclusive but unknown risks may exist.

predation   The consumption of one species (the prey) by another (the predator); includes both animals eating other animals and animals eating plants.

preindustrial CO$_2$ level   The level of carbon dioxide in the atmosphere before large-scale use of fossil fuels began in the late 1880s.

preservation   Setting aside undisturbed areas, maintaining them in a pristine state, and protecting them from human activities that might alter the natural state. Compare *conservation*.

prevailing wind   A major surface wind that blows more or less continually.

primary air pollutant   A harmful chemical that enters directly into the atmosphere from either human activities or natural processes (such as volcanic eruptions). Compare *secondary air pollutant*.

primary consumer   An organism that consumes producers. Also called *herbivore*. Compare *secondary consumer*.

primary sludge   A slimy mixture of bacteria-laden solids that settles out from sewage wastewater during primary treatment.

primary succession   An ecological succession that occurs on land that has not previously been inhabited by plants; no soil is present initially. Compare *secondary succession*.

primary treatment   Treatment of wastewater by removing suspended and floating particles (such as sand and silt) by mechanical processes (such as screens and physical settling). Compare *secondary treatment* and *tertiary treatment*.

principle of inherent safety   Chemical safety programs that stress accident prevention by redesigning industrial processes to involve less toxic materials so that dangerous accidents are prevented.

producer An organism (such as a chlorophyll-containing plant) that manufactures complex organic molecules from simple inorganic substances. In most ecosystems, producers are photosynthetic organisms. Also called *autotroph*. Compare *consumer*.

profundal zone The deepest zone of a large lake. Compare *limnetic zone* and *littoral zone*.

pronatalists Those who are in favor of population growth. Compare *neo-Malthusians*.

protein A large, complex organic molecule composed of amino acid subunits; proteins are the principal structural components of cells.

pyramid of biomass An ecological pyramid that illustrates the total biomass (for example, the total dry weight of all organisms) at each successive trophic level in an ecosystem. See *ecological pyramid*. Compare *pyramid of energy* and *pyramid of numbers*.

pyramid of energy An ecological pyramid that shows the energy flow through each trophic level in an ecosystem. See *ecological pyramid*. Compare *pyramid of biomass* and *pyramid of numbers*.

pyramid of numbers An ecological pyramid that shows the number of organisms at each successive trophic level in a given ecosystem. See *ecological pyramid*. Compare *pyramid of biomass* and *pyramid of energy*.

quarantine Practice in which the importation of exotic plant and animal material that might be harboring pests is restricted.

quarry See *open-pit mining*.

r selection A reproductive strategy in which a species typically has a small body size, rapid development, and a short life span, and devotes a large proportion of its metabolic energy to the production of offspring. Compare *K selection*.

radiation The emission of fast-moving particles or rays of energy from the nuclei of radioactive atoms.

radiative forcing The capacity of a gas to affect the balance of energy that enters and leaves Earth's atmosphere; measured in units of power per unit area, usually watts per square meter (W/m^2).

radioactive atoms Atoms of unstable isotopes that spontaneously emit radiation.

radioactive decay The emission of energetic particles or rays from unstable atomic nuclei; includes positively charged alpha particles, negatively charged beta particles, and high-energy, electromagnetic gamma rays.

radioactive half-life The time required for one-half of a radioactive substance to change into a different material.

radioisotope An unstable isotope that spontaneously emits radiation.

radon A colorless, tasteless, radioactive gas produced during the radioactive decay of uranium in Earth's crust.

rain shadow An area on the downwind side of a mountain range with little precipitation. Deserts often occur in rain shadows.

range The area of Earth in which a particular species occurs.

rangeland Land that is not intensively managed and is used for grazing livestock.

rational actor model In economics, the assumption that all individuals try to spend their limited resources in a fashion that maximizes their individual *utilities*.

reactor vessel A huge steel potlike structure encasing the uranium fuel in a nuclear reactor. The reactor vessel is a safety feature designed to prevent the accidental release of radiation into the environment.

realized niche The lifestyle that an organism actually pursues, including the resources that it actually uses. An organism's realized niche is narrower than its fundamental niche because of competition from other species. See *niche*. Compare *fundamental niche*.

reclaimed water Treated wastewater that is reused in some way, such as for irrigation, manufacturing processes that require water for cooling, wetlands restoration, or groundwater recharge.

recycling Conservation of the resources in used items by converting them into new products. For example, used aluminum cans are recycled by collecting, remelting, and reprocessing them into new cans. Compare *reuse*.

red tide A red, orange, or brown coloration of water caused by a bloom, or population explosion, of algae; many red tides cause serious environmental harm and threaten the health of humans and animals.

renewable resources Resources that are replaced by natural processes and can be used forever, provided they are not overexploited in the short term. Examples include fresh water in lakes and rivers, fertile soil, and trees in forests. Compare *nonrenewable resources*.

replacement-level fertility The average number of children a couple must produce to "replace" themselves; the number is greater than two because some children die before reaching reproductive age.

reprocessing Reusing highly radioactive spent nuclear fuel by chemically separating the unused uranium and plutonium from other radioactive waste products.

reservoir An artificial lake produced by building a dam across a river or stream; allows water to be stored for use.

resistance management A relatively new approach to dealing with genetic resistance so that the period in which a pesticide is useful is maximized; involves efforts to delay the development of genetic resistance in pests.

resource partitioning The reduction in competition for environmental resources, such as food, that occurs among coexisting species as a result of each species' niche differing from the others in one or more ways.

resource recovery The process of removing any material—sulfur or metals, for example—from polluted emissions or solid waste and selling it as a marketable product.

resources (1) Any parts of the natural environment that are used to promote the welfare of people or other species; examples include clean air, fresh water, soil, forests, minerals, and organisms. Also called *natural resources*. (2) Anything from the environment that meets a particular species' needs.

response In toxicology, the type and amount of damage caused by exposure to a particular dose.

restoration ecology The field of science in which the principles of ecology are used to help return a degraded environment as close as possible to its former, undisturbed state. Also called *ecological restoration*.

reuse Conservation of the resources in used items by using them over and over again. For example, glass bottles can be collected, washed, and refilled again. Compare *recycling*.

reverse osmosis A desalinization process that involves forcing saltwater through a membrane permeable to water but not to salt. Compare *distillation*.

riparian area The thin patch of vegetation along the bank of a stream or river that interfaces between terrestrial and aquatic habitats; protects the habitat of salmon, trout, and other aquatic species from sedimentation caused by soil erosion. Also called *riparian buffer*.

riparian buffer See *riparian area*.

risk The probability that a particular adverse effect will result from some exposure or condition.

risk assessment The use of statistical methods to quantify the harmful effects on human health or the environment of exposure to a particular danger. Risk assessments can provide a systems perspective when risks are compared with one another. See *ecological risk assessment*.

risk management Determining whether there is a need to reduce or eliminate a particular risk and, if so, what should be done. Based on data from risk assessment as well as political, economic, and social considerations.

rock A naturally formed aggregate, or mixture, of minerals; rocks have varied chemical compositions.

rodenticide A toxic chemical that kills rodents.

runoff The movement of fresh water from precipitation and snowmelt to rivers, lakes, wetlands, and, ultimately, the ocean.

rural land Sparsely populated areas, such as forests, grasslands, deserts, and wetlands. Also called *nonurban land*.

salination See *salinization*.

salinity The concentration of dissolved salts (such as sodium chloride) in a body of water.

salinity gradient The difference in salt concentrations that occurs at different depths in the ocean and at different locations in estuaries.

salinization The gradual accumulation of salt in a soil, often as a result of improper irrigation methods. Most plants cannot grow in salinized soil. Also called *salination*.

salt marsh An estuarine wetland dominated by salt-tolerant grasses.

saltwater intrusion The movement of seawater into a freshwater aquifer located near the coast; caused by *aquifer depletion*. Saltwater intrusion also occurs in low-lying parts of the world due to sea level rise.

salvage logging   A controversial logging method in which trees that are weakened by insects, disease, or fire are harvested. Forestry scientists disprove of salvage logging, particularly in fire-damaged areas, and say it delays or prevents the forest from recovering.

sand   Inorganic soil particles that are larger than clay or silt. Compare *clay* and *silt*.

sanitary landfill   The most common method of disposal of solid waste, by compacting it and burying it under a shallow layer of soil.

saprotroph   See *decomposer*.

savanna   A tropical grassland with widely scattered trees or clumps of trees; found in areas of low rainfall with prolonged dry periods.

science   A human endeavor that seeks to reduce the apparent complexity of the natural world to general principles that can be used to make predictions, solve problems, or provide new insights.

scientific method   The way a scientist approaches a problem (by formulating a hypothesis and then testing it by means of an experiment).

scout-and-spray   The use of pesticides that are applied only when pests become a problem; requires continual monitoring. Compare *calendar spraying*.

scrubbers   Desulfurization systems that are used in smokestacks to decrease the amount of sulfur released in the air by 90% or more. One type is a *lime scrubber*.

sea grasses   Flowering plants that grow in quiet, shallow ocean water in temperate, subtropical, and tropical waters.

secondary air pollutant   A harmful substance formed in the atmosphere when a primary air pollutant reacts with substances normally found in the atmosphere or with other air pollutants. Compare *primary air pollutant*.

secondary consumer   An organism that consumes primary consumers. Also called *carnivore*. Compare *primary consumer*.

secondary sludge   A slimy mixture of bacteria-laden solids that settles out from sewage wastewater during secondary treatment.

secondary succession   An ecological succession that takes place after some disturbance destroys the existing vegetation; soil is already present. Compare *primary succession*.

secondary treatment   Treatment of wastewater biologically to decompose suspended organic material; secondary treatment reduces the water's biochemical oxygen demand; occurs after primary treatment. Compare *primary treatment* and *tertiary treatment*.

second law of thermodynamics   When energy is converted from one form to another, some of it is degraded into a lower-quality, less useful form. Thus, with each successive energy transformation, less energy is available to do work. See *thermodynamics*. Compare *first law of thermodynamics*.

sediment pollution   Excessive amounts of soil particles that enter the water as a result of erosion.

sedimentation   (1) Letting solids settle out of wastewater by gravity during primary treatment. (2) The process in which eroded particles are transported by water and deposited as sediment on river deltas and the sea floor. If exposed to sufficient heat and pressure, sediments can solidify into sedimentary rock.

seed-tree cutting   A forest management technique in which almost all trees are harvested from an area in a single cutting, but a few desirable trees are left behind to provide seeds for the regeneration of the forest. Compare *clearcutting*, *selective cutting*, and *shelterwood cutting*.

seismic waves   Vibrations that travel through rock as a result of an earthquake.

selective cutting   A forest management technique in which mature trees are cut individually or in small clusters while the rest of the forest remains intact so that the forest can regenerate quickly (and naturally). Compare *clearcutting*, *seed-tree cutting*, and *shelterwood cutting*.

semi-arid land   Land that receives more precipitation than a desert but is subject to frequent and prolonged droughts. Compare *arid land*.

sewage   The wastewater released from drains or sewers (from toilets, washing machines, and showers); includes human wastes, soaps, and detergents.

sewage sludge   See *primary sludge* and *secondary sludge*.

shelterbelt   A row of trees planted as a windbreak to reduce soil erosion of agricultural land.

shelterwood cutting   A forest management technique in which all mature trees in an area are harvested in a series of partial cuttings over time; typically two or three harvests occur during a decade. Compare *clearcutting*, *seed-tree cutting*, and *selective cutting*.

shifting cultivation   A traditional form of subsistence agriculature in which short periods of cultivation are followed by longer periods of fallow (land left uncultivated), during which times the natural ecosystems may become reestablished. See *slash-and-burn agriculture*.

sick building syndrome   The condition inside a building in which eye irritations, nausea, headaches, respiratory infections, depression, and fatigue are caused by the presence of air pollution.

silt   Medium-sized inorganic soil particles. Compare *clay* and *sand*.

sink   In environmental science, the part of the natural environment that receives an input of materials; economies depend on sinks for waste products. Compare *source*.

sinkhole   A large surface cavity or depression when an underground cave roof has collapsed; occurs most frequently when a drought or excessive pumping causes a lowering of the water table.

slag   See *bottom ash*.

slash-and-burn agriculture   A type of shifting cultivation in tropical forests in which a patch of vegetation is burned, leaving nutrient minerals in the ash. Crops are planted for a few years until the soil is depleted of nutrient minerals, after which the land must lie fallow for many years to recover. See *shifting cultivation*.

smelting   A process in which ore is melted at high temperatures to help separate impurities from the molten metal.

smog   Air pollution caused by a variety of pollutants. See *industrial smog* and *photochemical smog*.

soft coal   See *bituminous coal*.

soil   The uppermost layer of Earth's crust, which supports terrestrial plants, animals, and microorganisms. Soil is a complex mixture of inorganic minerals (from the parent material), organic material, water, air, and organisms.

soil degradation   See *degradation*.

soil erosion   The wearing away or removal of soil from the land; caused by wind and flowing water. Although soil erosion occurs naturally from precipitation and runoff, human activities such as clearing land accelerate it.

soil horizons   The horizontal layers into which many soils are organized. May include the O-horizon (surface litter), A-horizon (topsoil), E-horizon, B-horizon (subsoil), and C-horizon (partly weathered parent material).

soil pollution   Any physical or chemical change in soil that adversely affects the health of plants and other organisms living in and on it.

soil profile   A vertical section through the soil, from the surface to the parent material, that reveals the soil horizons.

soil remediation   Use of one or more techniques to remove contaminants from the soil.

soil salinization   See *salinization*.

solar energy   Energy from the sun. Solar energy includes both direct solar radiation and indirect solar energy (such as wind, hydropower, and biomass).

solar thermal electric generation   A means of producing electricity in which the sun's energy is concentrated by mirrors or lenses to either heat a fluid-filled pipe or drive a Stirling engine.

solid waste   Any unwanted materials that are discarded as trash. See *municipal solid waste* and *nonmunicipal solid waste*.

source   In environmental science, the part of the environment from which materials move; economies depend on sources for raw materials. Compare *sink*.

source reduction   An aspect of waste management in which products are designed and manufactured in ways that decrease not only the volume of solid waste but also the amount of hazardous materials in the solid waste that remains.

species   A group of similar organisms whose members freely interbreed with one another in the wild to produce fertile offspring; members of one species generally do not interbreed with other species.

species richness   Biological diversity that encompasses the number of species in an area (or community). Compare *genetic diversity* and *ecosystem diversity*.

spent fuel   The used fuel elements that were irradiated in a nuclear reactor.

spoil bank   A hill of loose rock created when the overburden from a new trench is put into the old (already excavated) trench during strip mining.

sprawl, suburban   See *suburban sprawl*.

spring turnover   A mixing of the lake waters in temperate lakes that occurs in spring as ice melts and the surface water reaches 4°C, its temperature of greatest density. Compare *fall turnover*.

stable runoff   The share of runoff from precipitation available throughout the year. Most geographic areas have a heavy runoff during a few months (the spring months, for example) when precipitation and snowmelt are highest. Stable runoff can be depended on every month.

standing-water ecosystem   A body of fresh water surrounded by land and whose water does not flow; a lake or pond. Compare *flowing-water ecosystem*.

sterile male technique   A method of insect control that involves rearing, sterilizing, and releasing large numbers of males of the pest species.

stewardship   The concept that humans share responsibility for the sustainable care and management of our planet.

strata   Layers of rock.

Strategic Petroleum Reserve   An emergency supply of up to one billion barrels of oil stored in underground salt caverns along the coast of the Gulf of Mexico; mandated by the U.S. Energy Policy and Conservation Act.

stratosphere   The layer of the atmosphere between the troposphere and the mesosphere. It contains a thin ozone layer that protects life by filtering out much of the sun's ultraviolet radiation.

stratospheric ozone thinning   The accelerated destruction of ozone in the stratosphere by human-produced chlorine- and bromine-containing chemicals.

streamlining   Shortening the time and/or effort required to perform routine EISs by limiting the scope and schedule of the environmental impact statement (EIS) process.

stressor   See *environmental stressor*.

strip cropping   A type of contour plowing that produces alternating strips of different crops that are planted along the natural contours of the land; lessens erosion, as on a hillside. See *contour plowing*.

strip mining   A type of surface mining in which a trench is dug to extract the minerals, then a new trench is dug parallel to the old one; the overburden from the new trench is put into the old trench, creating a hill of loose rock known as a *spoil bank*. See *surface mining*. Compare *open-pit mining*.

structural trap   An underground geological structure that tends to trap oil or natural gas if it is present.

subbituminous coal   A grade of coal, intermediate between lignite and bituminous, that has a relatively low heat value and sulfur content. Compare *anthracite*, *bituminous coal*, and *lignite*.

subduction   The process in which one tectonic plate descends under an adjacent plate.

sublimation   The property of water in which it changes from a solid to a vapor without going through the liquid phase; freeze drying of various products takes advantage of the fact that water sublimates.

subsidence   The sinking or settling of land caused by aquifer depletion (as groundwater supplies are removed).

subsidy   A form of government support (such as monetary payments, public financing, tax benefits, or tax exemptions) given to a business or institution to promote the activity performed by that group.

subsistence agriculture   Traditional agricultural methods that are dependent on labor and a large amount of land to produce enough food to feed oneself and one's family, with little left over to sell or reserve for hard times. Subsistence agriculture uses humans and draft animals as its main source of energy. Compare *industrialized agriculture*.

subsurface mining   The extraction of mineral and energy resources from deep underground deposits. Compare *surface mining*.

suburban sprawl   A patchwork of vacant and developed tracts around the edges of cities that contains a low population density.

succession   The sequence of changes in a plant community over time. Also called *ecological succession*.

sulfide   A compound in which an element is combined chemically with sulfur.

sulfur cycle   The worldwide circulation of sulfur from the abiotic environment into organisms and back into the abiotic environment.

surface mining   The extraction of mineral and energy resources near Earth's surface by first removing the soil, subsoil, and overlying rock strata (i.e., the overburden). See *strip mining* and *open-pit mining*. Compare *subsurface mining*.

surface water   Fresh water found on Earth's surface in streams and rivers, lakes, ponds, reservoirs, and wetlands. Compare *groundwater*.

survivorship   The probability that a given individual in a population will survive to a particular age; usually presented as a survivorship curve.

sustainability   The ability to meet the current human need for natural resources without compromising the ability of future generations to meet their needs; assumes the environment can function indefinitely without going into a decline from the stresses imposed by human society on natural systems such as fertile soil, water, and air. Also called *environmental sustainability*.

sustainable agriculture   Agricultural methods that rely on beneficial biological processes and environmentally friendly chemicals rather than conventional agricultural techniques. Also called *alternative* or *low-input agriculture*. Compare *industrialized agriculture*.

sustainable city   A city with a livable environment, a strong economy, and a social and cultural sense of community; sustainable cities enhance the well-being of current and future generations of urban dwellers.

sustainable consumption   The use of goods and services that satisfy basic human needs and improve the quality of life but that also minimize the use of nonrenewable and renewable resources so they are available for future generations.

sustainable development   Economic development that meets the needs of the present generation without compromising the ability of future generations to meet their own needs. Also called *sustainable economic development*.

sustainable economic development   See *sustainable development*.

sustainable forest management   See *ecologically sustainable forest management*.

sustainable forestry   See *ecologically sustainable forest management*.

sustainable manufacturing   A manufacturing system based on minimizing waste by industry. Sustainable manufacturing involves such practices as reuse, recycling, and source reduction. See *industrial ecosystems*.

sustainable soil use   The wise use of soil resources, without a reduction in the amount or fertility of soil, so that it is productive for future generations.

sustainable water use   The use of water resources in a fashion that does not harm the essential functions of the hydrologic cycle or the ecosystems on which present and future humans depend.

swamp   A wetland dominated by trees; freshwater swamps are found inland, and saltwater swamps occur along coastal areas. See *mangrove forest*.

symbionts   The partners of a symbiotic relationship.

symbiosis   An intimate relationship between two or more organisms of different species. See *commensalism*, *mutualism*, and *parasitism*.

synergism   A phenomenon in which two or more pollutants interact in such a way that their combined effects are more severe than the sum of their individual effects. Compare *additivity* and *antagonism*.

synfuel   A liquid or gaseous fuel synthesized from coal or other naturally occurring sources and used in place of oil or natural gas. Also called *synthetic fuel*.

synthetic botanical   Any of a group of human-made insecticides that are produced by chemically modifying the structure of natural botanicals. See *botanical*.

synthetic fuel   See *synfuel*.

system   A set of components that interact and function as a whole.

taiga   See *boreal forest*.

tailings   Piles of loose rock produced when a mineral such as uranium is mined and processed (extracted and purified from the ore).

tar sand   An underground sand deposit permeated with a thick, asphalt-like oil known as bitumen. The bitumen can be separated from the sand by heating. Also called *oil sand*.

temperate deciduous forest   A forest biome that occurs in temperate areas where annual precipitation ranges from about 75 cm to 125 cm.

temperate grassland   A grassland characterized by hot summers, cold winters, and less rainfall than is found in a temperate deciduous forest biome.

temperate rain forest   A coniferous biome characterized by cool weather, dense fog, and high precipitation.

temperature inversion   A deviation from the normal temperature distribution in the atmosphere, resulting in a layer of cold air temporarily trapped near the ground by a warmer, upper layer. Also called *thermal inversion*.

terracing   A soil conservation method that involves building dikes on hilly terrain to produce level, terraced areas for agriculture.

terrestrial   Pertaining to the land. Compare *aquatic*.

tertiary treatment   An advanced wastewater treatment method that occurs after primary and secondary treatments and includes a variety of biological, chemical, and physical processes. Compare *primary treatment* and *secondary treatment*.

theory   An integrated explanation of numerous hypotheses, each of which is supported by a large body of observations and experiments. Compare *hypothesis*.

thermal inversion   See *temperature inversion*.

thermal pollution   Water pollution that occurs when heated water produced during many industrial processes is released into waterways.

thermal stratification   The marked layering (separation into warm and cold layers) of temperate lakes during the summer. See *thermocline*.

thermocline   A marked and abrupt temperature transition in temperate lakes between warm surface water and cold deeper water. See *thermal stratification*.

thermodynamics   The branch of physics that deals with energy and its various forms and transformations. See *first law of thermodynamics* and *second law of thermodynamics*.

thermosphere   The layer of the atmosphere between the mesosphere and the exosphere. Temperatures are high due to the absorption of x-rays and shortwave ultraviolet radiation.

threatened species   A species in which the population is low enough for it to be at risk of becoming endangered in the foreseeable future throughout all or a significant portion of its range. Compare *endangered species*.

threshold   In toxicology, the maximum dose that has no measurable effect (or the minimum dose that produces a measurable effect).

tidal energy   A form of renewable energy that relies on the ebb and flow of the tides to generate electricity.

topography   A region's surface features such as the presence or absence of mountains and valleys.

total fertility rate   The average number of children born per woman during her lifetime.

total resources   The combination of a mineral's reserves and resources. See *mineral reserve* and *mineral resource*. Also called *world reserve base*.

toxicant   A chemical with adverse effects on health.

toxicology   The study of harmful chemicals (toxicants) that have adverse effects on health, as well as ways to counteract their toxicity.

toxic waste   A type of hazardous waste harmful to the health of humans or other organisms.

trade wind   A prevailing tropical wind that blows from the northeast (in the Northern Hemisphere) or from the southeast (in the Southern Hemisphere).

transpiration   The loss of water vapor from the aerial surfaces of plants.

trickle irrigation   See *microirrigation*.

tritium   An isotope of hydrogen that contains one proton and two neutrons per atom. Compare *deuterium*.

trophic level   Each level in a food chain. All producers belong to the first trophic level, all herbivores belong to the second trophic level, and so on.

tropical cyclone   Giant, rotating tropical storms with winds of at least 119 kilometers per hour (74 mph); the most powerful tropical cyclones have wind velocities greater than 250 kilometers per hour (155 mph). Called hurricanes in the Atlantic, typhoons in the Pacific, and cyclones in the Indian Ocean.

tropical dry forest   A tropical forest where enough precipitation falls to support trees, but not enough to support the lush vegetation of a tropical rain forest. Many tropical dry forests occur in areas with pronounced rainy and dry seasons.

tropical rain forest   A lush, species-rich forest biome that occurs in tropical areas where the climate is moist throughout the year. Tropical rain forests tend to be characterized by old, infertile soils.

troposphere   The atmosphere from Earth's surface to the stratosphere. It is characterized by the presence of clouds, turbulent winds, and decreasing temperature with increasing altitude.

tundra   The treeless biome in the far north that consists of boggy plains covered by lichens and small plants such as mosses. The tundra is characterized by harsh, cold winters and extremely short summers. Also called *arctic tundra*. Compare *alpine tundra*.

ultraviolet radiation   That part of the electromagnetic spectrum with wavelengths just shorter than visible light; a high-energy form of radiation that can be lethal to organisms at higher levels of exposure. Also called *UV radiation*.

unconfined aquifer   A groundwater storage area located above a layer of impermeable rock. Water in an unconfined aquifer is replaced by surface water that drains from directly above it. Compare *confined aquifer*.

undernutrition   Malnutrition caused when a person receives fewer calories in the diet than are needed to maintain a healthy body. Compare *overnutrition*.

unfunded mandate   A federal requirement imposed on state and local governments without providing a way to pay for the cost of compliance.

upwelling   A rising ocean current that transports colder, nutrient-laden water to the surface.

urban agglomeration   An urbanized core region that consists of several adjacent cities or megacities and their surrounding developed suburbs; an example is the Tokyo–Yokohama–Osaka–Kobe agglomeration in Japan.

urban growth   The rate at which a city's population grows.

urban heat island   Local heat buildup in an area of high population density. Compare *dust dome*.

urbanization   The process in which people increasingly move from rural areas to densely populated cities; also involves the transformation of rural areas into urban areas.

utilitarian conservationist   A person who values natural resources because of their usefulness for practical purposes but uses them sensibly and carefully.

utility   An economic term referring to the benefit that an individual gets from some good or service. Rational actors try to maximize utility.

UV radiation   See *ultraviolet radiation*.

values   The principles that an individual or society considers important or worthwhile.

vapor extraction   A technique of soil remediation that involves injecting or pumping air into soil to remove organic compounds that are volatile (evaporate quickly).

vapor recovery   The removal of unburned gasoline vapors from gasoline containers, including underground tanks at gas stations and automobile gas tanks; recovered vapors are either burned or condensed and recovered.

vaporization   See *evaporation*.

variable   A factor that influences a process. In scientific experiments, all variables are kept constant except for one. See *control*.

vector   (1) An organism that transmits a parasite from one host to another. (2) In genetic engineering, an agent that transfers genetic information from one cell to another.

vitamin   A complex organic molecule required in small quantities for the normal metabolic functioning of living cells. Compare *mineral*.

vitrification   A method of safely storing high-level radioactive liquid wastes in solid form, as enormous glass logs.

voluntary simplicity   A way of life that involves wanting and spending less.

water cycle   See *hydrologic cycle*.

water pollution   Any physical or chemical change in water that adversely affects the health of humans and other organisms.

water table   The uppermost level of an unconfined aquifer, below which the ground is saturated with water.

watershed   See *drainage basin*.

weather   The general condition of the atmosphere (temperature, moisture, cloudiness) at a particular time and place. Compare *climate*.

weathering process   A biological, chemical, or physical process that helps form soil from rock; during weathering, the rock is gradually broken down into smaller and smaller particles.

westerly   A prevailing wind that blows in the mid-latitudes from the southwest (in the Northern Hemisphere) or from the northwest (in the Southern Hemisphere).

Western diseases   A group of noninfectious diseases that are generally more commonplace in industrialized countries, including obesity and heart disease.

Western worldview   An understanding of our place in the world based on human superiority and dominance over nature, the unrestricted use of natural resources, and increased economic growth to manage an expanding industrial base.

wet deposition   A form of *acid deposition* in which acid falls to Earth as precipitation. Compare *dry deposition*.

wetlands   Lands that are transitional between aquatic and terrestrial ecosystems and are covered with water for at least part of the year.

wide-spectrum pesticide   See *broad-spectrum pesticide*.

wilderness   Any area that has not been greatly disturbed by human activities and that humans may visit but do not permanently inhabit.

wildlife corridor   Protected zones that connect unlogged or undeveloped areas; wildlife corridors are thought to provide escape routes and allow animals to disperse so they can interbreed.

wildlife management   An applied field of conservation biology that focuses on the continued productivity of plants and animals; includes the regulation of hunting and fishing and the management of food, water, and habitat for wildlife populations.

wildlife ranching   An alternative use of land in which private landowners maintain herds of wild animals and earn money from tourists, photographers, and sport hunters as well as from animal hides, leather, and meat. Also called *game farming*.

wind   Surface air currents that are caused by the solar warming of air.

wind energy   Electric or mechanical energy obtained from surface air currents caused by solar warming of air. See *wind farm*.

wind farm   An array of wind turbines for utilizing wind energy by capturing it and converting it to electricity.

wise-use movement   A coalition of grassroots organizations that think the government has too many environmental regulations and that federal lands should be used primarily to enhance economic growth.

world grain carryover stocks   The amounts of rice, wheat, corn, and other grains remaining from previous harvests, as estimated at the start of a new harvest; these provide a measure of world food security.

world reserve base   See *total resources*.

worldview   One of many perspectives based on a collection of our basic values; worldviews help us make sense of the world, understand our place in it, and determine right and wrong behaviors. See *environmental worldview*.

yield   In agriculture, the amount of a food crop produced per unit of land.

zero population growth   When the birth rate equals the death rate. A population with zero population growth remains the same size.

zooplankton   The nonphotosynthetic organisms—tiny shrimp, larvae, and other drifting animals—that are part of the plankton. See *plankton*. Compare *phytoplankton*.

zooxanthellae   Algae that live inside coral animals and have a mutualistic relationship with them.

Text, Table & Line Art Credits

CHAPTER 1. **Figure 1-4:** Population Reference Bureau; **Figure 1-9:** U.S. Department of Energy; **Figure 1-19:** From Edmondson, W. T. *The Uses of Ecology: Lake Washington and Beyond.* Seattle: University of Washington Press (1991). Copyright ©1991 by the University of Washington Press. Adapted and redrawn by permission.

CHAPTER 2. **Page 27, exerpt from the "Wilderness Essay,"** a letter written by Wallace Stegner to David Pesonen of the U of California's Wildland Research Center; **page 38, Section on natural resources, the environment, and the national income accounts,** Main source: Levin, J. "The Economy and the Environment: Revising the National Income Accounts," *IMF Survey* (June 4, 1990). Reprinted by permission of the International Monetary Fund, Washington, D.C.; **page 43, list of 8 principles of deep ecology:** Naess, A., and D., Rothenberg. *Ecology, Community and Lifestyle.* Cambridge, UK: Cambridge University Press (2001). **page 43, quote in section on environmental worldviews:** Seed, J., Macy, J., Fleming P., and Naess, A. *Thinking like a Mountain: Towards a Council of All Beings.* Philadelphia, New Society Publishers (1988). **Table 2-1:** Compiled by the author. **Table 2-2:** select EPI scores accessed from Yale University Environmental Performance Index Study.

CHAPTER 3. **Figure 3.16:** After Whittaker, R. H. *Communities and Ecosystems*, 2nd edition. New York: Macmillan (1975).

CHAPTER 4. **Figure 4-15:** Data for graphs adapted from Gause, G. F. *The Struggle for Existence.* Baltimore: Williams & Wilkins (1934). **Figure 4-16:** Adapted from MacArthur, R. H. "Population Ecology of Some Warblers of Northeastern Coniferous Forests." *Ecology*, vol. 39 (1958). **Figure 4-17:** After M. L. Cody and J. M. Diamond, eds., *Ecology and Evolution of Communities.* Harvard University, Cambridge (1975). **Table 4-1:** Adapted from p. 527 of *Climate Change Impacts on the United States*, A report of the National Assessment Synthesis Team, U.S. Global Change Research Program, Cambridge University Press (2001).

CHAPTER 5. **Figures 5-2, 5-3, 5-4, 5-5, and 5-6:** Values are from Schlesinger, W. H. *Biogeochemistry: An Analysis of Global Change*, 2nd edition. Academic Press, San Diego (1997) and based on several sources. **Figure 5-7:** Adapted from Figure 4.15 in Strahler, A., and A. Strahler. *Physical Geography*, 2nd edition. Hoboken, NJ: John Wiley and Sons, Inc. (2002) and based on data from J. T. Kiehl and K. E. Trenberth, "Earth's Annual Global Mean Energy Budget." *Bull. Am. Met. Soc.*, vol. 78 (1997). **Figures 5-10a and 5-11a:** from A. F. Arbogast. *Discovering Physical Geography.* Hoboken, NJ: John Wiley & Sons, Inc. (2007). **Figure 5-15:** After Broecker, W. S. "The Great Ocean Conveyor." Oceanography, vol. 4 (1991). **Figure 5-16:** Data adapted from the

Climate Analysis Center, National Weather Service, and NOAA, Camp Springs, MD. **Figure 5-18:** from Figure 1-8 in deBlij, H. J. and P. O. Muller. *Geography: Realms, Regions and Concepts*, 10th edition. Hoboken, NJ: John Wiley and Sons, Inc. (2002). After Köppen and Geiger, and updated and modified by several geographers.

CHAPTER 6. **Data for all climate graphs** from www.worldclimate.com. **Figure 6-1:** Based on data from the World Wildlife Fund. **Figure 6.2:** Based on Holdridge, L. *Life Zone Ecology.* Tropical Science Center, San Jose, Costa Rica (1967). **Figure 6.25:** Based on pages 15–16 of Palumbi, S. R. *Marine Reserves: A Tool for Ecosystem Management and Conservation*, Pew Oceans Commission (2003).

CHAPTER 7. **Table 7.1:** Josten, M. D. and J. L. Wood. *World of Chemistry*, 2nd Edition. Philadelphia: Saunders College Publishing, (1996). **Table 7.4:** Data compiled from a variety of sources, including Ropeik, D., and G. Gray. *Risk.* Poston: Houghton Mifflin Company (2002). Probabilities calculated by L. Berg. **Figure 7.4:** Adapted from Exhibit 4–6 on page 4–11 in EPA's *Draft on the Environment 2003.* **Figure 7.5(b):** Reprinted with permission from Grier, J. W. "Ban of DDT and Subsequent Recovery of Reproduction in Bald Eagles." Copyright ©1982, American Association for the Advancement of Science. **Figure 7.6:** Based on data from Woodwell, G. M., C. F. Worster, Jr. and P. A. Isaacson. "DDT Residues in an East Coast Estuary: A Case of Biological Concentration of a persistent Insecticide. *Science*, vol. 156 (May 12, 1967). **Figure 7.7(b):** Data from A. R. Woodward, Florida Fish and Wildlife Conservation Commission. **Figure 7.9:** Guilette, E. A., M. M. Meza, M. G. Aguilar, A. D. Soto, and I. E. Garcia. "An Anthropological Approach to the Evaluation of Preschool Children Exposed to Pesticides in Mexico." *Environmental Health Perspectives* (May 1998). **Figure 7.12:** Adapted from *Science and Judgment in Risk Assessment.* Washington, D. C.: National Academy Press (1994). Data in graph from "CDC' s Second National Report on Human Exposure to Environmental Chemicals," published in 2003. **EnviroNews on Pandemic Flu:** compiled from information from the Centers for Disease Control and World Health Organization.

CHAPTER 8. **Figure 8.4b:** After V. C. Scheffer. "The Rise and Fall of a Reindeer Herd." *Sci. Month.*, vol. 73 (1951). **Figure 8-6:** Adapted and redrawn from data by Rolf O. Peterson, Michigan Technological University. **Figure 8-8b:** After R. A. Paynter, Jr. "A New Attempt to Construct Life Tables for Kent Island Herring Gulls. *Bull. Mus. Comp. Zool*, vol. 133, no. 11 (1966). Data collected from Kent Island, Maine, 1934 to 1939. Baby gulls were banded to establish identity. **Figures 8-9, 8-11, 8-15, 8-17, and 8-18 and Tables 8-1 and 8-2:** Data from Population Reference Bureau. **Figure 8-12:** Data from *World Population Prospects, The 2004 Revision*, United Nations Population Division. **Figure 8-19:** Data

from U.S. Census Bureau, as cited in *Population Bulletin*, vol. 67, no. 4 (2006).

CHAPTER 9. **Figure 9.3:** Based on data from U.N. Food and Agricultural Organization and UNICEF. **Figure 9.5:** Adapted from Data Table 12 in *World Resources 2002–2004: Decisions for the Earth: Balance, Voice, and Power*, Washington, D.C.: World Resources Institute (2003). **Figure 9-6a:** U.S. Census Bureau. **Figure 9-7b:** Data from Population Reference Bureau; quote by Kofi Annan, page 210.

CHAPTER 10. **Table 10-1:** "Urban Agglomerations 2005," U.N. Department of Economic and Social Affairs, Population Division. **Figure 10.1:** *World Urbanization Prospects: The 2003 Revision*, U.N. Department of Economic and Social Affairs, Population Division. **Figure 10.3:** Adapted from U.S. Census Bureau. **Figure 10.10:** Adapted from Kaplan, Wheeler, and Holloway. *Urban Geography.* Hoboken, NJ: John Wiley & Sons, Inc. (2004). **Figure 10.14a:** Courtesy of Karl Gude.

CHAPTER 11. **Table 11-1:** compiled by author from EIA, U.S. Department of Energy and USGS. **Table 11-2:** EIA, U.S. Department of Energy. **Table 11.3:** U.S. Department of Energy. **Table 11-4:** National Energy Policy, 2001, U.S. Department of Energy. **Figures 11-1, 11-2, 11-5, 11-10, 11-14 and 11-15:** Compiled from U.S. Department of Energy. **Figure 11-7:** Adapted from Reece, E. "Death of a Mountain." *Harper's Magazine*, April 2005. **Figure 11-19:** Adapted from Hinrichs, R. A., *Energy.* Philadelphia: Saunders College Publishing (1992). **Figures 11-12 (a and b):** after illustrations from Center for Liquified Natural Gas.

CHAPTER 12. **Table 12-2:** adapted from two sources: Hinrichs, R. A. and M. Kleinbach. *Energy: Its Use in the Environment*, 3rd edition, Philadelphia, Harcourt College Publishers (2002) and *Science for Democratic Action*, vol. 8, no. 3 (May 2000). **Table 12-3:** International Atomic Energy Agency and EIA, U.S. Department of Energy.

CHAPTER 13. **Table 13-1:** Data from Kammen, D. M. "The Rise of Renewable Energy." *Scientific American* (September 2006), used with permission. **Table 13-3:** U.S. Department of Energy. **Table 13-4:** data from the Rocky Mountain Institute, 1739 Snowmass Creek, Snowmass, CO 81654, (970) 927-3851, reprinted by permission. **Figure 13-1:** U.S. Department of Energy. **Figure 13-7a:** adapted from Hinrichs, R. A. Energy: Its Use and the Environment, 2nd edition. Philadelphia: Saunders College Publishing (1996).

CHAPTER 14. **Table 14-1:** World Resources Institute and *World Resources: 1998–99.* New York: Oxford University Press (1999). **Figure 14-6:** *Control of Water Pollution from Urban Runoff.* Paris: Organization for Economic Development and Cooperation (1986). **Figure 14-11:** Environmental Systems Research

Institute Incorporated (1999). **Figure 14-14:** Englebert, E. A. and A. Foley, Eds. *Water Scarcity: Impacts on Western Agriculture*. Berkeley, University of California Press (1984). Reprinted by permission. **Figure 14-15:** USGS. **Figure 14.20:** Based on information from USGS Grand Canyon Monitor and Research Center. **Figure in You Can Make a Difference:** Data from Kindler, J. and C. S. Russell. *Modeling Water Demands*. Toronto: Academic Press, Inc. (1984). As reported in Gleick, P. H. "Basic Water Requirements for Human Activities: Meeting Basic Needs." *Water International*, vol. 21 (1996).

CHAPTER 15.   No credits.

CHAPTER 16.   **Table 16-2:** Adapted from Table 6.4 on page 117 in Gardner, G. et al. *State of the World 2003*. New York: W. W. Norton & Company (2003) and based on data from U.S. Geological Survey and Worldwatch. **Figure 16-3:** Adapted from Joesten, M. D., and J. L. Wood. *World of Chemistry*, Second Edition, Philadelphia: Saunders College Publishing (1996). **Figure 16-7:** CRU International (2001) as reported in MMSD. **Figure 16.8b:** Adapted from Rona, P. A., "Resources of the Sea Floor." *Science*, vol. 299 (January 31, 2003).

CHAPTER 17.   **Quote in end-of-chapter questions:** Aldo Leopold, *A Sand County Almanac*. Oxford University Press, Inc. (1991). **Figure 17-7:** Map by Conservation International.

CHAPTER 18.   **Table 18-1:** U.S. Dept. of Interior, U.S. Dept. of Agriculture, and U.S. Dept. of Defense. **Table 18-2:** From Box 1.1 in Noss, R. F., M. A. O'Connell, and D.D. Murphy. *The Science of Conservation Planning: Habitat Conservation Under the Endangered Species Act*. Island Press: World Wildlife Fund (1997). Reprinted with permission. **Figure 18-1:** World Resources Institute, U.N. Food and Agricultural Organization, and The Earth Institute at Columbia University. **Figure 18-2:** Adapted from the map, "Federal Lands in the Fifty States," produced by the Cartographic Division of the National Geographic Society (October 1996). **Figure 18-8:** Data obtained from Marsh, W. M. and J. M. Grossa, Jr., *Environmental Geography*, 2nd edition, John Wiley & Sons, Inc. (2002). **Quote in EnviroNews on Reforestation in Kenya:** By Ole Danbolt Mjoes, the Nobel Committee Chair.

CHAPTER 19.   **Table 19-1:** World production data from U.N. Food and Agricultural Organization. **Table 19-2:** FAO, USDA, and PRB, as reported in Brown, L. R. *Tough Choices: Facing the Challenge of Food Scarcity*. New York: W.W. Norton & Company (1996). **Figure 19-1:** Adapted from FAO data, as reported in Brown, L. R. et al. *Vital Signs 2006–2007*. New York:

W. W. Norton & Company (2006). **Figure 19-2:** Adapted from USDA data. **Figure 19-4:** Adapted from G. H. Heichel, "Agricultural Production and Energy Resources." *American Scientist*, Vol. 64 (January/February 1976). **Figure 19-6:** USDA. **Figure 19-8:** Adapted from D. Livermore, Health Protection Agency's Resistance Monitoring and Reference Laboratory, United Kingdom. **Figure 19-13a:** Data from O'Brien, L., N. Shepherd, and L. Col. *Assessment of the Georges Bank Atlantic Cod Stocks for 2005*. Northeast Fisheries Science Center Ref. Document 06-10 (June 2006).

CHAPTER 20.   **Quote by Ulf Merbold:** From Ulf Merbold of the German Aerospace Liaison. Reprinted by permission. **Tables 20.1 and 20.2:** Compiled by authors. **Table 20-3:** Ozone Policy and Strategies Group, Office of Air and Radiation, Environmental Protection Agency. **Figures 20-4, 20-5:** Environmental Protection Agency; **Figures 20.6 and 20.7:** South Coast AQMD. **Figure 20.11:** a and d, adapted from Joesten, M. D., and J. L. Wood. *World of Chemistry*, Second Edition. Philadelphia: Saunders College Publishing (1996). **Figure 20.13:** Air Quality Planning and Standards, Office of Air and Radiation, EPA. **Figure 20.20:** Adapted from Wania, F. and D. Mackay. "Tracking the Distribution of Persistent Organic Pollutants." *Environmental Science and Technology*, vol. 30 (1996).

CHAPTER 21.   **Figure 21-1:** Global Land-Ocean Temperature Index, Goddard Institute of Space Studies, NASA. **Figure 21-2:** Dave Keeling and Tim Whorf, Scripps Institution of Oceanography, La Jolla, CA. **Figure 21-5 and 21-8:** U.S. Global Change Research Program. **Figure 21.10:** Adapted from Davis, M. B., C. Zabinski, R. L. Peters, and T. E. Lovejoy, eds. "Changes in Geographical Range Resulting from Greenhouse Warming: Effects on Biodiversity in Forests." *Global Warming and Biological Diversity*. R. L. Peters, and T. E. Lovejoy. eds. New Haven, Connecticut: Yale University Press (1992). **Table 21-1:** Carbon Dioxide Information Analysis Center, Environmental Sciences Division, Oak Ridge National Laboratory. **Table 21-2:** From Table 4.21 in *Health and Environment in Sustainable Development: Five Years After the Summit*, World Health Organization (1997) and based on an unpublished assessment prepared by A. J. McMichael et al.

CHAPTER 22.   **Figure 22-1:** Adapted from Joesten, M. D., and J. L. Wood. *World of Chemistry*, 2nd edition. Philadelphia: Saunders College Publishing (1996). **Figure 22.5:** Adapted from Goolsby, D. A. "Mississippi Basin Nitrogen Flux Believed to Cause Gulf Hypoxia." *Eos, Transactions of the American Geophysical Union*, vol. 81, pages 325 to 327 (2000). **Figure 22-9:** Adapted from

Collins, T. J. and Walter, C. "Little Green Molecules," *Scientific American* (2006). **Table 22.1:** Compiled by author. **Table 22-2:** Cancer data based on IARC (WHO International Agency for Research on Cancer) list of carcinogens.

CHAPTER 23.   **Figure 23.3:** From "International Travel and Health. Vaccination Requirements and Health Advice: Situation as of January 1, 1992." Geneva: World Health Organization (1992). Reproduced by permission of the World Health Organization. **Figure 23.4:** Data from Mark Whalon, Michigan State University, as reported in Brown, L. R., Kane, H. and Roodman, D. M., *Vital Signs 1994*. New York: W. W. Norton & Co. (1994). **Figure 23.5b:** Adapted from Debach, P. *Biological Control by Natural Enemies*. New York: Cambridge University Press (1974). **Figure 23.13:** Gardner, G. "IPM and the War on Pests." *World Watch*, vol. 9, no. 2 (March/April 1996). Redrawn by permission of Worldwatch Institute. **Table 23.1:** Adapted from Pimental, D. "Environmental and Economic Effects of Reducing Pesticide Use." *Bioscience*, vol. 41, no. 6 (June 1991). Copyright © by the American Institute of Biological Sciences; 1999 personal communication with D. Pimental. **Table 23.2:** Compiled by author from various sources, including "Chemically-Induced Alterations in Sexual and Functional Development: The Wildlife-Human Connection," vol. XXI, *Advances in Modern Environmental Toxicology*, Colburn, T. and C. Clement. Princeton: Princeton Scientific Publishing Company (1992); and Colburn, T., D. Dumanoski, and J. P. Myers. *Our Stolen Future*. New York: Dutton (1996); **Table 23.3:** Excerpted from "Worst Case Estimates of Risk of Cancer from Pesticide Residues on Food." From *Regulating Pesticides in Food: The Delaney Paradox*. Washington, D.C.: National Academy Press (1987). Copyright © by the National Academy of Sciences, Courtesy of the National Academy Press, Washington, D.C.

CHAPTER 24.   **Figures 24-1 and 24-2:** EPA. **Figure 24-4:** Ecobalance, Inc., and Integrated Waste Services Association. **Figure 24.15:** Adapted from Rocky Mountain Arsenal Remediation Venture Office. **Table 24.1:** Compiled by author from various sources.

CHAPTER 25.   **Quote from Bruntland Report:** *Our Common Future*. World Commission on Environment and Development. Cary, NC: Oxford University Press (1987), page 22. Reprinted by permission of Oxford University. **Quote from Kai N. Lee:** *Compass and Gyroscope: Integrating Science and Politics for the Environment*. Washington, D.C.: Island Press (1993), p. 200. **5 recommendations around which the chapter is organized: Brown, L.** *Plan B 2.0: Rescuing a Planet under Stress and a Civilization in Trouble*. New York: W. W. Norton (2006).

BOX PHOTOS.   Case In Point and Meeting the Challenge: © Royalty Free/Corbis Digital Stock/Corbis; You Can Make A Difference: © Royalty-Free/Corbis.

Index

References to illustrations, either photographs or drawings, are followed by *f*. References to tables are followed by *t*; references to footnotes are denoted by <u>n</u>.